THE FAMILIES
AND GENERA
OF VASCULAR PLANTS

Edited by K. Kubitzki

Volumes published in this series

Volume I	Pteridophytes and Gymnosperms Edited by K.U. Kramer and P.S. Green (1990) Date of publication: 28.9.1990
Volume II	Flowering Plants. Dicotyledons. Magnoliid, Hamamelid and Caryophyllid Families Edited by K. Kubitzki, J.G. Rohwer, and V. Bittrich (1993) Date of publication: 28.7.1993
Volume III	Flowering Plants. Monocotyledons: Lilianae (except Orchidaceae) Edited by K. Kubitzki (1998) Date of publication: 27.8.1998
Volume IV	Flowering Plants. Monocotyledons: Alismatanae and Commelinanae (except Gramineae) Edited by K. Kubitzki (1998) Date of publication: 27.8.1998
Volume V	Flowering Plants. Dicotyledons: Malvales, Capparales and Non-betalain Caryophyllales Edited by K. Kubitzki and C. Bayer (2003) Date of publication: 12.9.2002
Volume VI	Flowering Plants. Dicotyledons: Celastrales, Oxalidales, Rosales, Cornales, Ericales Edited by K. Kubitzki (2004) Date of publication: 21.1.2004
Volume VII	Flowering Plants. Dicotyledons: Lamiales (except Acanthaceae including Avicenniaceae) Edited by J.W. Kadereit (2004) Date of publication: 13.4.2004
Volume VIII	Flowering Plants. Eudicots: Asterales Edited by J.W. Kadereit and C. Jeffrey (2007) Date of publication: 6.12.2006
Volume IX	Flowering Plants. Eudicots: Berberidopsidales, Buxales, Crossosomatales, Fabales p.p., Geraniales, Gunnerales, Myrtales p.p., Proteales, Saxifragales, Vitales, Zygophyllales, Clusiaceae Alliance, Passifloraceae Alliance, Dilleniaceae, Huaceae, Picramniaceae, Sabiaceae Edited by K. Kubitzki (2007) Date of publication: 6.12.2006
Volume X	Flowering Plants. Eudicots: Sapindales, Cucurbitales, Myrtaceae Edited by K. Kubitzki (2011)

The Families and Genera of Vascular Plants

Edited by K. Kubitzki

X

Flowering Plants · Eudicots

Sapindales, Cucurbitales, Myrtaceae

Volume Editor:
K. Kubitzki

With 93 Figures

Springer

Editor
Professor Dr. Klaus Kubitzki
Universität Hamburg
Biozentrum Klein-Flottbek und Botanischer Garten
Ohnhorststraße 18
22609 Hamburg
Germany

ISBN 978-3-642-42351-2 ISBN 978-3-642-14397-7 (eBook)
DOI 10.1007/978-3-642-14397-7
© Springer Heidelberg Dordrecht London New York

Softcover re-print of the Hardcover 1st edition 2011

Cover design: deblik Berlin, Germany

Printed on acid-free paper

Springer is part of Springer Science+Business Media (www.springer.com)

Preface

The present volume includes treatments of the families of the orders Sapindales and Cucurbitales and of the family Myrtaceae as an appendage to the Myrtales, which were dealt with in the previous volume. The contributions once more reflect the enormous progress plant systematics has witnessed since the publication of the first volumes of this series now two decades ago. This can be seen in the greatly improved understanding of the demarcations between and of the relationships among and within the families treated in this volume. The increase in our understanding of the age of the lineages of the flowering plants in connection with the interest of contemporary practitioners in the use of molecular clocks has led to the inclusion, in several contributions, of hypotheses on past dispersal events, often resulting in claims of unexpected long-distance-dispersal events.

Altogether, the volume contains an enormous wealth of interesting information, and I am deeply indebted to all authors for their scholarly contributions. I am also very grateful to all copyright holders who so kindly gave permission to reproduce illustrations published under their responsibility, including the Director and Board of Trustees, Royal Botanic Gardens, Kew; Publications Scientifiques du Muséum national d'Histoire naturelle, Paris; and the editors of Blumea (Leiden, the Netherlands) and of Nuytsia (Perth, Western Australia). The artist Bobbi Angell, New York, deserves my special thanks for the generosity with which she authorized the use of the illustration published under her authorship.

Finally, it is a pleasure to thank the copy editor of the present volume, Monique Delafontaine, for her dedicated editorial work, which greatly improved the manuscript of this volume. I would also like to gratefully acknowledge the enjoyable collaboration in the production of the present volume with the staff of Springer-Verlag, particularly Andrea Schlitzberger, and with SPi Technologies India PvT Ltd.

Hamburg, August 2010 Klaus Kubitzki

Preface

Contents

List of Contributors

Acevedo-Rodríguez, Pedro	Smithsonian Institution, Dept. of Botany, NMNH MRC-166, Washington, DC 20560-0166, USA, acevedop@si.edu
Adema, F.A.C.B.	Nationaal Herbarium Nederland, Universiteit Leiden branch, P.O. Box 9514, 2300RA Leiden, The Netherlands, adema@nhn.leidenuniv.nl
Clayton, Joshua	3 Shawclough Road, Rochdale Lancashire OL12 6LG, UK, moshclayton@hotmail.com
Daly, Douglas C.	Institute of Systematic Botany, The New York Botanical Garden, 200th Street and Kazimiroff Blvd., Bronx, NY 10451–5126, USA, ddaly@nybg.org
de Wilde, J.J.F.E	Nationaal Herbarium Nederland, Universiteit Wageningen branch, Generaal Foulkesweg 37, 6703BL Wageningen, The Netherlands, de.wilde.bakhuizen@hccnet.de
Duretto, Marco	Herbarium, Tasmanian Museum & Art Gallery, Private Bag 4, Hobart, Tasmania 7001, Australia, marco.duretto@tmag.tas.gov.au
Kallunki, Jacquelyn A.	The New York Botanical Garden, 200th Street and Southern Blvd., Bronx, NY 10451–5126, USA, jkallunki@nybg.org
Kubitzki, Klaus	Biozentrum Klein-Flottbek, Ohnhorststr. 18, 22609 Hamburg, Germany, K.Kubitzki@web.de
Lobova, Tatyana A.	Department of Biological Sciences, Old Dominion University, 110 Mills Godwin Building/45th St, Norfolk, VA 23529-0266, USA, tlobova@odu.edu
Mabberley, David J.	Royal Botanic Gardens Kew, Richmond, Surrey TW9 3AB, UK, d.mabberley@kew.org
Miller, Allison J.	Biology Department, Saint Louis University, 3507 LaClede Avenue, St. Louis, MO 63103, USA, amille75@slu.edu
Mitchell, John D.	Institute of Systematic Botany, The New York Botanical Garden, 200th Street and Kazimiroff Blvd., Bronx, NY 10451, USA
Muellner, Alexandra N.	Biodiversität & Klima Forschungszentrum (BiK-F) & Goethe-Universität, Senckenberganlage 25, 60325 Frankfurt am Main, Germany, alexandra.muellner@senckenberg.de
Pell, Susan K.	Brooklyn Botanic Garden, 1000 Washington Ave., Brooklyn, NY 11225, USA, SusanPell@bbg.org
Renner, Susanne S.	Department of Biology, Ludwig Maximilians-Universität München, Menzinger Str. 67, 80638 München, Germany, renner@lrz.uni-muenchen.de

Schaefer, Hanno	Imperial College London, Silwood Park Campus, Ecology and Evolutionary Biology, Buckhorst Road, Ascot, Berkshire SL5 7PY, UK, hanno.schaefer@imperial.ac.uk
Schwarzbach, Andrea E.	Dept. of Biological Sciences, University of Texas, 80 Fort Brown St., Brownsville, TX 78520, USA, andrea.schwarzbach@utb.edu
Sheahan, Mary-Clare	Jodrell Laboratory, Royal Botanic Gardens Kew, Richmond, Surrey TW9 3DS, UK, M.Sheahan@kew.org
Swensen, Susan M.	Department of Biology, Ithaca College, 177 New Science Building, Ithaca, NY 14850-7278, USA, sswensen@ithaca.edu
Tomlinson, Philip B.	Harvard Forest, Harvard University, Petersham, MA 01366, USA, pbtomlin@fas.harvard.edu
van der Ham, R.W.J.M.	Nationaal Herbarium Nederland, Universiteit Leiden branch, P.O. Box 9514, 2300RA Leiden, The Netherlands, ham@nhn.leidenuniv.nl
van Welzen, Peter C.	Nationaal Herbarium Nederland, Universiteit Leiden branch, P.O. Box 9514, 2300RA Leiden, The Netherlands, welzen@nhn.leidenuniv.nl
Wilson, Paul G.	Western Australian Herbarium, Lockede Bay 104, Bentley Delivery Centre, Western Australia 6983, Australia
Wilson, Peter G.	Botanic Gardens Trust, Mrs. Macquaries Road, Sydney NSW 2000, Australia, Peter.Wilson@rbgsyd.nsw.gov.au

Introduction to Sapindales

K. Kubitzki

Conspectus of Families

1. Herbs or low-growing shrubs 2
- Erect shrubs or trees (some Anacardiaceae herbaceous) 4
2. Perennial herbs; nectary glands 5, at base of antesepalous stamens; carpels with distinct stylodia arising from base of ovarioles; ovules solitary, pendulous, epitropous; embryo sac tetrasporic, 16-celled; $n = 5$. 1/4 or 5. E Mediterranean to C Asia **Biebersteiniaceae**
- Low shrubs of saline habitats, rarely (*Tetradiclis*) annual herbs; intrastaminal nectary disk annular or angular; ovary with simple style; ovules 1 or several per carpel, epitropous or apotropous; embryo sac, as far as known, of Polygonum type 3
3. Ovule 1 per carpel, apotropous; fruit drupaceous; $n = 12, 30$. 1/5–8. Old World, Australia **Nitrariaceae**
- Ovules several to many per carpel, epitropous; fruit a loculicidal capsule or a berry; $n = 7, 12, 13$. 3/7–8. E and S Europe to Middle Asia, Mexico **Tetradiclidaceae**
4. Plants usually strongly resinous, with vertical resin canals in the bark and also with resin ducts in the phloem of the larger veins of the leaves and sometimes in wood rays; producing biflavonyls 5
- Plants resinous or not, but without resin ducts in the bark, rays, and leaf veins; biflavonyls 0 6
5. Ovules 2 in each locule, epitropous, collateral or (*Beiselia*) superposed; nodes mostly 5-lacunar 5-trace; flowers actinomorphic and obdiplostemonous, or with the antesepalous stamen whorl reduced; gynoecium of (2)3–5(9–13 in *Beiselia*) connate carpels; style simple with 2–3-lobed or capitate stigma; fruits drupes with 1–5 one-seeded pyrenes or pseudocapsules releasing pyrenes; endotesta lignified; seeds exalbuminous, with hemicellulosic reserves; embryo minute, with folded, usually palmately lobed cotyledons. $n = 11, 13, 23$. 19/640. Pantropical **Burseraceae**
- Ovule solitary in each locule, apotropous, more rarely epitropous; nodes mostly 3-lacunar 3-trace; flowers often monosymmetric, obdiplostemonous or with (1)5–10+ stamens; gynoecium of 4–12 distinct carpels of which usually only one is fertile, or of (2)3(–5) connate carpels; stylodia distinct or more or less connate into a simple style; fruit usually drupaceous with resinous mesocarp; seeds with oily and starchy endosperm; endotegmen lignified, usually thickened; embryo curved, with fleshy cotyledons. $n = 7–12, 14–16, 21$. 81/c. 800. Pantropical, also temperate **Anacardiaceae**
6. Fruit dehiscing with 4 or 8 one-seeded mericarps from a central column; flowers isomerous, 4-merous; testa thin; endosperm 0; trees with alternate, imparipinnate leaves; ellagic acid present. 1/6. Africa, Madagascar **Kirkiaceae**
- Fruit not dehiscing from central column 7
7. Pericycle containing a cylinder of sclerenchyma (*Xanthoceras*, *Guindilia*, and some Acereae excepted); plants containing saponins in idioblasts but no bitter nortriterpenoids; leaves alternate or less often (Acereae, Hippocastaneae) opposite; flowers actinomorphic or obliquely zygomorphic; disk extrastaminal or less often intrastaminal, annular (in *Xanthoceras*, with orange horn-like appendages) or unilateral; petals sometimes (Hippocastaneae, Sapindoideae) with basal scale-like appendage concealing nectary; ovules 1 or 2 per carpel or rarely more, usually apotropous. $n = 10–16, 20$. 141/c. 1,900. Pantropical, with some temperate genera **Sapindaceae** s.l.
- Pericycle without a cylinder of sclerenchyma; producing bitter nortriterpenoids (limonoids or quassinoids) 8
8. Leaves pellucid-punctate and secretory schizogenous cavities scattered through the parenchymatous tissue (not in all Cneoroideae); flowers mostly actinomorphic and obdiplostemonous, sometimes stamens in one cycle and antesepalous; nectary disk intrastaminal; carpels (2)4–5+, more or less connate proximally and usually held together by the joined stylodia, less often completely connate; ovules (1)2–many in each locule, usually epitropous; fruits follicles, drupes, berries, or samaras; producing limonoids, canthin-6-ones, and alkaloids of different types. $n = 7–11, 18+$. 154/c. 1,800. Pantropical and temperate **Rutaceae**
- Leaves not pellucid-punctate 9
9. Stamen filaments not appendaged, usually connate into a staminal tube with anthers in one or two whorls, less often filaments distinct; nodes mostly 5-lacunar 5-trace; ovary (1)2–6(–20)-carpellate, syncarpous; style simple; ovules 1–2 or more per carpel, usually epitropous; seeds often sarcotestal or arillate; seed coat exotegmic with fibres or pachychalazal;

producing limonoids. $n = 8(-180)$. 50/c. 575. Pantropical, some temperate **Meliaceae**

– Stamen filaments distinct, usually with scaly appendage; nodes 3-lacunar; carpels (1)2–5, distinct or basally or ventrally connate; stylodia distinct, conglutinate or connate into a common style; ovule 1 per carpel, epitropous; seeds not fleshy; seed coat usually nondescript, pachychalazal in *Quassia* and *Picrasma*; producing bitter quassinoids, limonoids, and canthin-6-ones. $n = 10–13$. 22/100. Pantropical, some temperate **Simaroubaceae**

Nineteenth century botanists, such as Bentham (in Bentham and Hooker 1862) and Engler (e.g., 1931), tended to treat Sapindales and Rutales (the latter sometimes as Geraniales) as distinct orders, a concept followed by Takhtajan (2009) to the present day; however, a wider ordinal concept with Rutales included in Sapindales, as Terebinthales (Wettstein 1901) or Sapindales (Cronquist 1968), is now broadly supported and accepted. Gene sequence studies (Sheahan and Chase 1996; Gadek et al. 1996; Muellner et al. 2007, among others) have contributed to shaping the present concept of the order and provided support for its monophyly, with increasing indications for Malvales and Brassicales and the little known Huerteales as close relatives of Sapindales (Worberg et al. 2009). The multigene analysis of Wang et al. (2009) has recovered the strongly supported relationship Crossosomatales [Picramniaceae [Sapindales [Huerteales [Brassicales + Malvales]]]]. Insights from morphology and molecular work, particularly a two-gene analysis with a broader sampling of Sapindales (Muellner et al. 2007), suggest the topology presented here (Fig. 1), in which, however, the precise relationship between Simaroubaceae and Meliaceae remains weakly supported.

The androecium is often (basically?) obdiplostemonous (with the carpels in antepetalous position), and the two stamen whorls sometimes (in Burseraceae, Rutaceae, and Sapindaceae) appear in a single cycle (meta-obdiplostemony, Lam 1931), or one cycle is missing. The herbaceous and shrubby, early diverging Nitrariaceae, Tetradiclidaceae, and Biebersteiniaceae are still little known but exhibit variation in ovule curvation and in seed and fruit structure, obviously in adaptation to the challenges of their saline or semiarid habitats. *Kirkia*, formerly included in Simaroubaceae, is now recognised as sister to

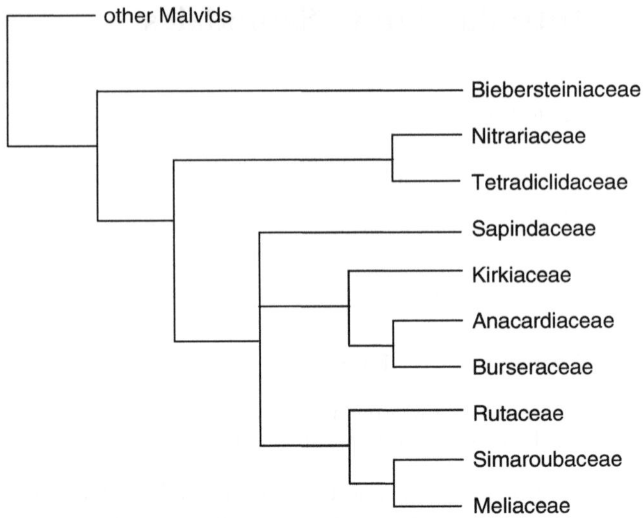

Fig. 1. Phylogenetic relationships of Sapindales families, based on *rbc*L sequence data from Muellner et al. (2007) and Sheahan and Chase (1996)

the Burseraceae/Anacardiaceae clade, with which it shares important similarities in floral structure (Bachelier and Endress 2008). Burseraceae are traditionally distinguished from Anacardiaceae by having two collateral ovules (except for *Beiselia* in which the two ovules are superposed) that are epitropous, in contrast to all other Sapindales. Bachelier and Endress (2009) report, however, that in the earliest developmental stages the ovules in Burseraceae appear apotropous. Thus, the rationale for the use of ovule curvature as a criterion for ordinal distinction becomes questionable.

The close relationship between Burseraceae and Anacardiaceae is well established both by anatomical (Takhtajan 2009), floral morphological (Bachelier and Endress 2009), and molecular evidence. Sapindaceae are treated here to include Aceraceae and Hippocastanaceae, in a return to the practice of several nineteenth century authors (for historical aspects, see the family treatment) and in conformity with the results of recent molecular studies (e.g., Harrington et al. 2005; Buerki et al. 2009), which have also brought to light the peculiar position of *Xanthoceras* as a basal branch of Sapindaceae. Rutaceae, Meliaceae, and Simaroubaceae share the possession of unusual bitter compounds, the limonoids and quassinoids, which are based on degraded triterpenes, the nortriterpenoids. The simplest limonoids are found in Rutaceae, and occur in increasing complexity in Meliaceae and in Rutaceae/Cneoroideae. Cneoroideae comprise genera that until recently had been treated as belonging to either Rutaceae or Simaroubaceae, or had been separated into small satellite families, but the presence of triterpenoid bitter compounds and particularly the results of gene sequence studies have yielded strong arguments for combining them with the Rutaceae. The peculiar apocarpy of Rutaceae and Simaroubaceae, thought by some to be inherited directly from basal angiosperms or Ranunculales, has been revealed to be a phylogenetically secondary condition, as is evidenced by the peculiar postgenital connation of the stylodia that hold together the carpels in the flowering stage (Ramp 1988).

Sapindales are an ancient lineage with a fossil record dating back to the Cretaceous. At least from the Paleocene onward, Meliaceae, Rutaceae, Sapindaceae, Anacardiaceae, and Burseraceae are represented by reliable fossils in the northern hemisphere, particularly in North America and Europe; Simaroubaceae follow in the early Eocene (for documentation, see family treatments in this volume). It is likely that the early evolution of Sapindales took place in North America, and that in the Eocene they dispersed eastward through the warm-temperate belt north of the Sea of Tethys (often erroneously called "paratropical", see Kubitzki and Krutzsch 1996), and from there invaded and diversified in tropical regions.

References

Bachelier, J.B., Endress, P.K. 2008. Floral structure of *Kirkia* (Kirkiaceae) and its position in Sapindales. Ann. Bot. 102: 539–550.

Bachelier, J.E., Endress, P.K. 2009. Comparative floral morphology and anatomy of Anacardiaceae and Burseraceae (Sapindales), with a special focus on gynoecium structure and evolution. Bot. J. Linn. Soc. 159: 499–571.

Bentham, G., Hooker, J.D. 1862. Genera Plantarum, vol. I (1). London: Reeve.

Buerki, S., Forest, F., Acevedo-Rodríguez, P., Callmander, M.W., Nylander, J.A.A., Harrington, M., Sanmartin, I., Küpfer, P., Alvarez, N. 2009. Plastid and nuclear DNA markers reveal intricate relationships at subfamilial and tribal levels in the soapberry family (Sapindaceae). Mol. Phylogen. Evol. 51: 238–258.

Cronquist, A. 1968. The evolution and classification of flowering plants. London: Nelson.

Engler, A. 1931. Cneoraceae, Rutaceae. In: Engler, A., Prantl, K. (eds.) Die natürlichen Pflanzenfamilien, 2nd edn, vol. 19a, pp. 184–359. Leipzig: W. Engelmann.

Gadek, P., Fernando, E.S., Quinn, C.J., Hoot, S.B., Terrazas, T., Sheahan, M.C., Chase, M.W. 1996. Sapindales: molecular delimitation and infraordinal groups. Am. J. Bot. 83: 802–811.

Harrington, M.G., Edwards, K.J., Johnson, S.A., Chase, M.W., Gadek, P.A. 2005. Phylogenetic inference in Sapindaceae sensu lato using plastid *matK* and *rbcL* DNA sequences. Syst. Bot. 30: 366–382.

Kubitzki, K., Krutzsch, W. 1996. Origins of East and Southeast Asian plant diversity, pp. 56–70. In: Floristic characteristics and and diversity of East Asian plants. Beijing: China Higher Education Press, and Berlin: Springer.

Lam, H.J. 1931. Beiträge zur Morphologie der dreizähligen Burseraceae-Canarieae. Ann. Jard. Bot. Buitenzorg 42: 25–56, t. v–vii.

Lam, H.J. 1932. Beiträge zur Morphologie der Burseraceae, insbesondere der Canarieae. Ann. Jard. Bot. Buitenzorg 42: 97–226, t. xi–xvi.

Muellner, A.N., Vassiliades, D.D., Renner, S.S. 2007. Placing Biebersteiniaceae, a herbaceous clade of Sapindales, in a temporal and geographic context. Pl. Syst. Evol. 266: 233–252.

Ramp, E. 1988. Struktur, Funktion und systematische Bedeutung des Gynoeciums bei den Rutaceae und Simaroubaceae. Inaug.-Diss., Philos. Fak. II, Univ. Zürich: ADAG.

Sheahan, M.C., Chase, M.W. 1996. A phylogenetic analysis of Zygophyllaceae R.Br. based on morphological, anatomical and *rbc*L DNA sequence data. Bot. J. Linn. Soc. 122: 279–300.

Takhtajan, A. 2009. Flowering plants, 2nd edn. Dordrecht: Springer.

Wang, H., Moore, M.J., Soltis, P.S., Bell, C.D., Brockington, S.F., Alexandre, R., Davis, C.D., Latvis, M., Manchester, S.R., Soltis, D.E. 2009. Rosid radiation and the rapid rise of angiosperm-dominated forests. Proc. Natl. Acad. Sci. USA 106: 3853–3858.

Wettstein, R. v. 1901. Handbuch der systematischen Botanik, vol. 1. Leipzig and Wien: F. Deuticke.

Worberg, A., Alford, M.H., Quandt, D., Borsch, T. 2009. Huerteales sister to Brassicales plus Malvales, and newly circumscribed to include *Dipentodon*, *Gerrardina*, *Huertea*, *Perrottetia*, and *Tapiscia*. Taxon 58: 468–478.

Introduction to Cucurbitales

K. Kubitzki

Conspectus of Families

1. Cambium initials not storied; flowers with lobed or crenate, intra- or interstaminal nectary disk on top of ovary (but see *Octomeles*/Datiscaceae); anthers dorsifixed; [flowers obdiplostemonous; fruit a drupe, samara, or capsule; endosperm 0; cotyledons reduced or 0]. 4/34. Pantropical **Anisophylleaceae**
– Cambium initials and secondary xylem and phloem storied; floral nectaries 0 (*Octomeles* excepted); anthers usually basifixed 2
2. Flowers hypogynous; ovule 1 per carpel, pendent; placentation apical; stylodia unbranched; stipules present, caducous; ellagitannins present 3
– Flowers (hemi)epigynous; ovules usually many per carpel; placentation parietal; stylodia sometimes branched; stipules present or not; ellagitannins 0 4
3. Leaves opposite or whorled; stipules lateral, small, caducous; fertile stamens 10; carpels 5 or 10, each with a long, slender stylodium stigmatic over its entire surface; pollen 3-aperturate. 1/15+. Worldwide **Coriariaceae**
– Leaves alternate; stipules intrapetiolar, caducous; fertile stamens 5; carpel 1(2); stylodium (stylodia) with capitate stigma(s); pollen 2-colporate. 1/6. Southwest Pacific region **Corynocarpaceae**
4. Tendril-bearing dioecious or less often monoecious climbers or trailers, rarely tendrils 0; young stems nearly always with 2 rings of bicollateral bundles; stamens 3–5, often 4 of them joined or connate in 2 pairs; gynoecium(1)3(–5)-carpellate, (semi)inferior; stylodia free or connate into a single style; fruit usually a soft- or hard-shelled berry; seeds flat; bitter cucurbitacins widespread. About 97/960, tropical, some reaching temperate regions **Cucurbitaceae**
– Tendrils 0; bundles never bicollateral; fruit capsular or rarely (Begoniaceae) baccate; seeds not flat; seed coat with operculum; cucurbitacins absent, except for roots of *Datisca* 5
5. Leaves simple with mostly large stipules, usually asymmetrical; monoecious, rarely dioecious perennials or rarely annuals or halfshrubs; placentation axile, sometimes parietal; seeds with collar cells arranged in transverse ring around operculum. $2n = 16$–156 (no clear base number recognisable). 2/>1,500. Tropical and subtropical regions of the World and temperate parts of Asia, but not in Australia **Begoniaceae**
– Leaves estipulate, simple, lobed, pinnate or pinnatifid, not asymmetrical; (andro)dioecious trees or perennial herbs; placentation parietal; seeds without collar cells around operculum. $n = 11$, c. 23. 3/4. E Mediterranean to SE Asia and Papuasia, and California, Baja California **Datiscaceae**

Recognition of the close relationship among the core families of Cucurbitales (Datiscaceae incl. Tetramelaceae, Begoniaceae, and Cucurbitaceae) dates back to the 19th century, although in the more recent pre-molecular era these families usually have been included in more comprehensive groupings named Violales or Parietales (for more details, see Matthews and Endress 2004, and Zhang et al. 2006). The addition of Coriariaceae, Corynocarpaceae, and Anisophylleaceae to the core Cucurbitales is an outcome of molecular studies (Chase et al. 1993; Swensen 1996; Setoguchi et al. 1999; Schwarzbach and Ricklefs 2000, among many others). The inclusion of Apodanthaceae, recently favoured by several authorities (e.g., Stevens 2001), is presently not supported (APG III; S.S. Renner, Oct. 2009).

In early molecular studies of the order, using the *rbc*L gene, these families were not fully resolved and topologies were often contradictory. Still, in recent multigene analyses covering also other orders, statistical support for the branches within Cucurbitales is generally lower than in other angiosperm clades (e.g., Wang et al. 2009). Nevertheless, the analysis of nine loci from three genomes of all Cucurbitales families by Zhang et al. (2006) has resolved Cucurbitales as monophyletic and served as a basis for an understanding of morphological and sexual system evolution within the order, but did not resolve the relationships among all families (Fig. 2). Fagales are now generally viewed as the closest relatives of Cucurbitales; both orders share the essentially unisexual and epigynous flowers.

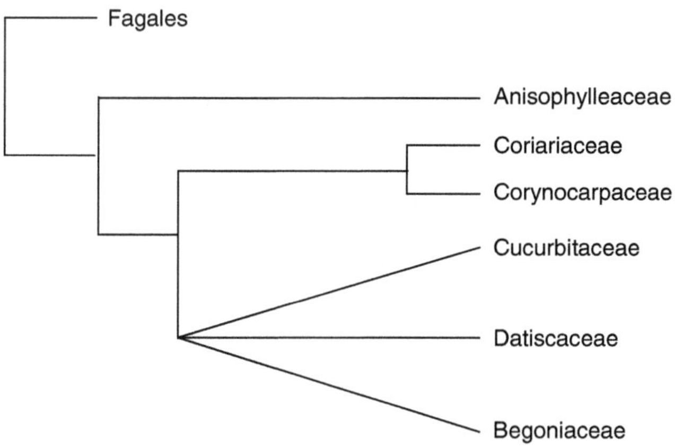

Fig. 2. Phylogenetic relationships of Cucurbitales families, based on the multigene sequence analysis of Zhang et al. (2006)

The strongly supported multigene analysis of Wang et al. (2009) has recovered the relationship Fabales[Rosales[Cucurbitales + Fagales]].

Anisophylleaceae, formerly included in Rhizophoraceae, are firmly established as the sister group to all remaining Cucurbitales, from which they differ significantly in reproductive and vegetative morphology (Schwarzbach and Ricklefs 2000). Matthews et al. (2001) and Matthews and Endress (2004) have pointed to similarities in floral structure that exist between Anisophylleaceae and Cunoniaceae, but at the same time have also revealed morphological traits in common between Anisophylleaceae and core Cucurbitales, such as unisexual flowers and inferior ovaries. Anisophylleaceae share with other Cucurbitales some anatomical characters of the wood, such as nonbordered or minimally bordered perforation plates and wide rays not accompanied by uniseriate rays, traits that are conservative and less likely affected by ecology. Anisophylleaceae have retained, however, characters that are more conservative than those in the other families of the order, such as absence of storying, presence of tracheids, and heterogeneous rays (Carlquist and Miller 2001). Thus, it appears that this family is correctly placed in Cucurbitales, and that its similarities with Cunoniaceae are due to convergence.

Among the remaining families, Coriariaceae and Corynocarpaceae stand out with 1-ovulate carpels, apical placentation, and superior ovaries, the latter trait, in view of the topology of Zhang et al. (2006), certainly derived. Cucurbitaceae, Datiscaceae s.l. (i.e., including Tetramelaceae), and Begoniaceae have epigynous flowers (as do Anisophylleaceae), essentially basifixed introrse (or latrorse) anthers, trimerous gynoecia, bifurcate carpels, and a peculiar extended neck over the roof of the ovaries or instead (in *Begonia* and many Cucurbitaceae) a narrow neck at this site (Matthews and Endress 2004). It is notable that Cucurbitaceae share with *Coriaria* and *Corynocarpus* a rare combination of wood anatomical traits (vertical parenchyma scanty vasicentric, banded, and ray adjacent, and rays with upright cells strongly predominant; Carlquist and Miller 2001). In the molecular topology, Cucurbitaceae place as sister to Datiscaceae and Begoniaceae, but the precise relationship between the latter remains unresolved.

In view of the amount of morphological differentiation both in Cucurbitaceae and in Begoniaceae, the difference in the numbers of genera recognised in the two families is surprising, if not paradoxical. By the middle of the nineteenth century, the development of taxonomic concepts in both families had reached a comparable level. Further development in Cucurbitaceae led to a steady consolidation of taxonomic concepts, and until the present, the family has remained a field of dynamic systematics activities. Begoniaceae, in contrast, never recovered from A. de Candolle's degradation of Klotzsch's 41 genera to sections, in which he has been followed by all students of the family to the present day. Although in principle Klotzsch's concept survives in the sectional classification of the family, this never has attracted much interest by botanists (for a notable exception, see Doorenbos et al. 1998); instead, they sometimes resorted to an alphabetic sequence of the 1,400 species of *Begonia*, and the family became a field mainly of floristic, rather than systematics activity. It is true that the decisive differences among begonias are difficult to observe and put into words, many of them being included in the unpopular area of inflorescence morphology. Nevertheless, I am convinced that Klotzsch's generic concepts would have been further developed had his genera not disappeared out of the focus of botanists through their degradation to sections.

References

Carlquist, S., Miller, R.B. 2001. Wood anatomy of *Corynocarpus* is consistent with cucurbitalean placement. Syst. Bot. 26: 54–65.

Chase, M.W., Soltis, D.E., Olmstead, R.G., Morgan, D., Les, D.H. and 37 further authors. 1993. Phylogenetics of seed plants: an analysis of nucleotide sequences from the plastid gene *rbc*L. Ann. Missouri Bot. Gard. 80: 528–580.

Doorenbos, J., Sosef, M.S.M., de Wilde, J.J.F.E. 1998. The sections of Begonia including descriptions, keys and species lists. Wageningen Agric. Univ. Papers 98-2, 266 pp.

Matthews, M.L., Endress, P.K. 2004. Comparative floral structure and systematics in Cucurbitales (Corynocarpaceae, Coriariaceae, Tetramelaceae, Datiscaceae, Begoniaceae, Cucurbitaceae, Anisophylleaceae). Bot. J. Linn. Soc. 145: 129–185.

Matthews, M.L., Endress, P.K., Schönenberger, J., Friis, E.M. 2001. A comparison of floral structures of Anisophylleaceae and Cunoniaceae and the problem of their systematic position. Ann. Bot. 88: 439–455.

Schwarzbach, A.E., Ricklefs, R.E. 2000. Systematic affinities of Rhizophoraceae and Anisophylleaceae, and intergeneric relationships within Rhizophoraceae, based on chloroplast DNA, nuclear ribosomal DNA, and morphology. Am. J. Bot. 87: 547–564.

Setogushi, H., Kosuge, H., Tobe, H. 1999. Molecular phylogeny of Rhizophoraceae based on *rbc*L gene sequences. J. Plant Res. 112: 443–455.

Stevens, P.F. 2001 onward. Angiosperm phylogeny website, version 03.04.09. http://www.mobot.org/MOBOT/research/APweb/

Swensen, S.M. 1996. The evolution of actinorhizal symbioses: evidence for multiple origins of the symbiotic association. Am. J. Bot. 83: 1503–1512.

Wang, H., Moore, M.J., Soltis, P.S., Bell, C.D., Brockington, S.F., Alexandre, R., Davis, C.D., Latvis, M., Manchester, S.R., Soltis, D.E. 2009. Rosid radiation and the rapid rise of angiosperm-dominated forests. Proc. Natl. Acad. Sci. USA 106: 3853–3858.

Zhang, Li-Bing, Simmons, M.P., Kocyan, A., Renner, S.S. 2006. Phylogeny of the Cucurbitales based on DNA sequences of nine loci from three genomes: implications for morphological and sexual systems evolution. Mol. Phylogen. Evol. 39: 305–322.

Anacardiaceae

Anacardiaceae R. Br. (1818), nom. cons.

S.K. Pell, J.D. Mitchell, A.J. Miller, and T.A. Lobova

Trees, shrubs, rarely subshrubs, lianas, frequently with contact dermatitis-causing exudate; vertical resin canals present in bark and in phloem of petioles and large veins of leaves, also widely present in fruits, flowers, and other tissues. Leaves alternate, rarely opposite or whorled, simple or pinnately compound, very rarely palmate or bipinnately compound, sessile or petiolate; leaflets opposite, subopposite, or alternate, entire, serrate, dentate, or crenate; stipules absent. Inflorescences terminal and/or axillary, thyrsoid, paniculate, racemose, or spicate, rarely cauliflorous, rarely flowers solitary; bracts and prophylls caducous or persistent. Flowers actinomorphic, unisexual or bisexual (plants dioecious, monoecious, andromonoecious, polygamous, or hermaphrodite); pedicels often articulate; hypanthium sometimes present; perianth usually 2-whorled, rarely 1-whorled or absent, imbricate or valvate; sepals (3–)4–5, usually basally fused, rarely bracteate or calyptriform, caducous to persistent, sometimes accrescent in fruit; petals (3)4–5(–8), rarely 0, caducous to persistent, rarely accrescent in fruit; androecium usually actinomorphic, rarely zygomorphic; stamens (1–)5–10(–>100), in 1 or 2 whorls, rarely more whorls, in some genera only 1 or 2 stamens fertile; filaments distinct, rarely basally connate; anthers tetrasporangiate, dorsi- or basifixed, usually longitudinally dehiscent, introrse, rarely extrorse; disk intrastaminal, rarely extrastaminal or 0; gynoecium 1-carpellate or syncarpous and 2–12-carpellate; rarely, the carpels distally distinct and the gynoecium appearing apocarpous; ovary usually superior, rarely inferior, 1–5(–12)-locular; ovule 1 per locule, apotropous, attached basally, apically, or laterally; stylodia 2–5 (–12) or style simple, apical or lateral, erect or recurved, rarely sigmoid; stigmas capitate, discoid, lobate, or spathulate, rarely punctiform. Fruits drupes or samaras (rarely syncarps, utricles, or baccates), fleshy or dry, occasionally subtended by a fleshy hypocarp or an accrescent, chartaceous or fleshy calyx or corolla; mesocarp sometimes with prominent black resin canals. Seeds 1–5(–12); endosperm scant or absent; embryo curved or straight (rarely horseshoe-shaped or pyramidal); cotyledons usually planoconvex or flat and distinct, usually equal in size, rarely fused or ruminate, sometimes bilobed.

Approximately 81 genera and 800 species in dry to moist, mostly lowland habitats in the tropics and subtropics worldwide, but also extending into the temperate zone.

Vegetative Morphology. The family consists primarily of trees and shrubs, with a few subshrubs, scandent trees, and lianas, and rarely herbaceous suffrutexes. Succulent stems occur in dry habitats (e.g., *Cyrtocarpa*, *Spondias purpurea*). Some arid- or cold-adapted genera have thorns (e.g., *Schinopsis*, *Schinus*, *Searsia*). A geoxylic suffrutex habit (massive woody underground trunk usually with annual or short-lived aerial shoots) is found particularly in the Zambezian region of Africa (e.g., *Lannea edulis*, *L. gossweileri*, *L. katangensis*, *L. virgata*, *Ozoroa nitida*, *Searsia kirkii*) (White 1976) and the Cerrado region of central South America (e.g., *Anacardium corymbosum*, *A. humile*, *A. nanum*) (Lopez-Naranjo 1977; Mitchell and Mori 1987). Water storage roots have also been reported for the family (e.g., *Spondias tuberosa*). The nodes are usually trilacunar or occasionally unilacunar. Many representatives of Anacardiaceae have a turpentine-smelling exudate that may turn black with exposure to air. The exudate may be milky, red, orange, yellow, or clear.

The leaves are deciduous or evergreen, estipulate and usually alternate (opposite in *Bouea*,

Blepharocarya). Most taxa have imparipinnate leaves (rarely paripinnate, bipinnate in *Spondias bipinnata*), usually with opposite leaflets (rarely alternate in, e.g., *Pseudospondias, Sorindeia, Thyrsodium*), while others have trifoliolate leaves (e.g., *Rhus, Searsia, Smodingium, Toxicodendron*) or simple or unifoliolate leaves (e.g., *Anacardium, Cotinus, Heeria, Lithrea, Malosma, Rhus*); very rarely the simple leaves are palmate (*Campylopetalum*). Leaf margins can be entire, dentate, serrate, or crenate, prominently revolute (e.g., *Abrahamia, Anacardium*), or rarely spinose (e.g., *Comocladia*). Various forms of domatia are sometimes present in the secondary vein axils abaxially. Both hairy tuft domatia (e.g., *Choerospondias, Dracontomelon, Mauria, Rhodosphaera, Toxicodendron*) and marsupiform domatia (e.g., *Pleiogynium*) are found in the family. See Wilkinson (1979) and O'Dowd and Willson (1991) for reviews of leaf domatia. Cataphylls occur in a few genera (e.g., *Astronium, Buchanania, Harpephyllum, Mangifera, Pistacia*).

Leaf architecture within Anacardiaceae is extremely diverse. Primary leaf venation is pinnate, rarely palmate (e.g., *Campylopetalum*). Secondary venation is most commonly eucamptodromous, brochidodromous (usually festooned), craspedodromous, semi-craspedodromous, or cladodromous (which is usually diagnostic of Anacardiaceae when present) and rarely exmedially reticulodromous (e.g., *Rhus thouarsii*). An intramarginal vein is rarely present (e.g., *Spondias, Solenocarpus*). Some genera have mixed secondary venation patterns either throughout (e.g., in *Comocladia glabra* lamina, craspedodromous alternates with brochidodromous veins) or directionally (e.g., *Gluta* and *Campnosperma* laminas are apically brochidodromous and basally eucamptodromous). Intercostal tertiary fabric is frequently random reticulate, polygonal-reticulate, mixed alternate-opposite, or opposite-percurrent. Intersecondary veins are frequently present, but the consistency varies in many taxa. Epimedial tertiaries are frequently present: they may be perpendicular to the primary vein, or varying from parallel to variously angled relative to the secondary veins. In several genera (e.g., *Abrahamia, Spondias*) the tertiary veins are admedially branched. A diagonally oriented, admedially branched, trunked tertiary is characteristic of several species of *Sorindeia* and *Buchanania*. In *Comocladia*, the tertiary veins are perpendicular to the secondary veins in the intercostal region. In some Anacardioideae (e.g., *Comocladia, Rhus, Toxicodendron*), the apparently blindly ramifying tertiary veins are interconnected by quaternary veins. Rarely (e.g., *Abrahamia, Rhus perrieri* (=*Baronia* or possibly segregate), *Melanococca*), the tertiaries are truly freely ramified (i.e., areoles absent). Marginal veins are rarely of secondary gauge (e.g., *Drimycarpus, Lithrea*). A fimbrial vein is typically present, and occasionally the marginal ultimate tertiary venation is looped (e.g., *Spondias bivenomarginalis*). Areoles vary from being clearly defined (e.g., *Anacardium, Tapirira*) to being highly variable in shape and pattern (e.g., *Spondias*). Freely ending veinlets (FEVs) are commonly highly branched (either dichotomously or dendritically) or rarely one- to two-branched. Sometimes the FEVs are terminated by highly branched sclereids (e.g., *Sorindeia, Spondias radlkoferi*). Some taxa are characterized by having FEVs terminated by prominent tracheoid idioblasts (e.g., *Comocladia, Harpephyllum, Melanococca, Pleiogynium, Spondias*). Terminology for leaf architecture is based on the *Manual of Leaf Architecture* (Leaf Architecture Working Group 1999) and subsequent revisions by the Leaf Architecture Working Group (Ellis et al. 2009).

Trichomes are common throughout the family, usually simple, unicellular or multicellular, sessile or stalked, glandular or non-glandular. Two types of trichomes were described in detail for *Rhus* subgenus *Rhus*: acicular and bulbous gland type (Hardin and Phillips 1985). Stellate trichomes are characteristic of *Lannea* and occur rarely in some other taxa (e.g., *Campnosperma, Pseudosmodingium, Semecarpus, Trichoscypha*) (Aguilar-Ortigoza and Sosa 2004a). Lepidote scales are rarely present in the family, but are characteristic of *Campnosperma*.

Vegetative Anatomy. Wood and bark anatomy of Anacardiaceae has been extensively studied by many authors, such as Dadswell and Ingle (1948), Kryn (1952), Roth (1969, 1981), Young (1974), Wannan (1986), Yunus et al. (1990), Gregory (1994), Terrazas (1994, 1995), Giménez and Moglia (1995), and Baas et al. (2000). Resin canals are common in the wood of numerous genera. They develop schizogenously, lysigenously, or schizolysigenously in the cortex, pith of the stem, phloem, and wood rays (see Venning 1948 for an ontogeny review; therein, resin canals are referred to as laticiferous canals). Resin canals are also found in Anacardiaceae leaves and reproductive structures. Resin canals run parallel to the phloem in leaf petioles and in major lamina veins, and are absent only in the most minor veins. They are also associated with the phloem of every vascular bundle in the reproductive structures of most genera studied (Copeland 1959). Important resin duct and gum duct anatomy studies include examples from *Anacardium* (Nair et al. 1983), *Holigarna* (Nair et al. 1952a), *Lannea* (Venkaiah and Shah 1984; Venkaiah 1992), *Mangifera* (Joel and Fahn 1980), *Rhus* (Fahn 1979), *Semecarpus* (Nair et al. 1952b; Bhatt and Ram 1992), *Toxicodendron* (Vassilyev 2000), *Trichoscypha* (Den Outer and Van Veenedaal 1986), and *Amphipterygium* and *Orthopterygium* (Figueroa 2001).

Leaf anatomy of Anacardiaceae was concisely covered by Metcalfe and Chalk (1950), and selected genera have been investigated by Goris (1910), Silva (1973), Paula and Alves (1973), Arrillaga-Maffei et al. (1973), Gibson (1981), and Muñoz (1990). A detailed survey of the leaf surface anatomy with special emphasis on epidermal features was undertaken by Wilkinson (1971), who later presented a detailed study of *Gluta* (Wilkinson 1983).

Inflorescences. Inflorescence structure is quite diverse in Anacardiaceae, but basically appears axillary and much branched. Inflorescences are often pseudoterminal (e.g., *Spondias*, *Tapirira*), but can be truly terminal (e.g., *Apterokarpos*, *Cotinus*, *Dobinea*, *Heeria*, *Rhus*), or rarely cauliflorous (e.g., *Semecarpus*, *Sorindeia*, *Trichoscypha*). Most are thyrso-paniculate, but racemes and spikes have also been described. Rarely female inflorescences are reduced to solitary flowers (e.g., *Choerospondias*, *Operculicarya*, *Sclerocarya*). Inflorescences of *Schinus* species from the high Andes and Patagonia are often reduced to just a few fascicles.

Members of Anacardioideae tend to have more condensed inflorescences as compared to Spondioideae. These are usually in the form of condensed panicles, thyrses, or thyrsoids, some being extremely condensed (e.g., *Blepharocarya*, *Laurophyllus*). The bracts subtending inflorescences may be very large (e.g., *Trichoscypha*), prominently foliose (e.g., *Anacardium*, *Dobinea*), condensed into a cupule in female flowers (*Blepharocarya*, *Laurophyllus*), or brightly colored (e.g. white to pink foliose bracts in *Anacardium spruceanum*). In some species there is a gradual transition from leaves to bracts subtending the terminal inflorescence (e.g., *Anacardium*, *Dobinea*, *Mauria*).

A detailed review of some South American genera was completed by Barfod (1988), but his claim that thyrsoids evolved from panicles in the family is not supported by the predominant thyrsoid inflorescence type found in the sister family, Burseraceae (Wannan and Quinn 1992). According to Wannan (Wannan et al. 1987; Wannan and Quinn 1992), the cupule of the female inflorescence of *Laurophyllus* is derived from a panicle, whereas the cupule of *Blepharocarya* is derived from a thyrsoid. It is important to note that inflorescence morphology of Anacardiaceae has not been thoroughly studied for a majority of genera, and conflicting reports of inflorescence structure are common. Much research remains to be conducted on whether Anacardiaceae inflorescences are wholly monopodial or are partially sympodial. For a review of Anacardiaceae inflorescence morphology, see Wannan et al. (1987), Barfod (1988), Endress and Stumpf (1991), and Wannan and Quinn (1991, 1992).

Floral Structure. Flowers are sessile or pedicellate, the pedicel frequently articulated and glabrous, or variously pubescent. Hypogyny is the normal state in the family, with some taxa being rarely perigynous (e.g., *Melanochyla*, *Thyrsodium*, Fig. 5E) or epigynous with (*Holigarna*) or without (*Drimycarpus*) a well-developed hypanthium. Very rarely the receptacle partially surrounds the gynoecium, or is apparently hemi-inferior due to the ovary being partially immersed in the receptacle (*Semecarpus*). Most genera have a biseriate perianth; occasionally, the corolla is absent and very rarely the entire perianth is absent. The calyx is usually green, occasionally the same color as the corolla, or becoming colorful in the genera that have an accrescent calyx that facilitates wind dispersal of the fruit (e.g., *Astronium*, *Loxostylis*, *Parishia*). The corolla is usually imbricate or valvate, rarely apert or contorted in bud. The corolla can be greenish, yellow, cream-colored, pink, red, purple, or rarely brownish. The petals are generally reflexed or patent, sometimes erect at anthesis. Their venation is often inconspicuous, frequently parallel, or sometimes prominently dichotomously branching (e.g., *Pseudosmodingium*). The corolla is usually campanulate, rarely trumpet-shaped (e.g., *Anacardium*); the petal tips are sometimes apiculate. A hypanthium is present in three genera (*Amphipterygium*, *Melanochyla*, *Thyrsodium*, Fig. 5E).

A nectariferous disk is usually present and intrastaminal, rarely extrastaminal (*Mangifera*, *Swintonia*) or lacking (e.g., *Anacardium*, *Gluta*, *Pistacia* male flowers). The disk is variously colored (green, orange, yellow), shaped (e.g., campanulate, saucer-shaped), and lobed (often corresponding to the location of the stamens). It is usually glabrous, but can be variously pubescent, occasionally papillose. In *Mangifera* the disk is often discontinuous (Fig. 3F). In *Thyrsodium* it is adnate to the hypanthium.

Most genera are diplo- or haplostemonous; very rarely, there is a proliferation of stamens (*Gluta*, Fig. 3E), or extreme staminal reduction as seen in *Anacardium* (Fig. 3G) and *Mangifera*, where there are one or more fertile stamens and a series of staminodes, and *Fegimanra* (Fig. 3B), which has a single stamen. Ronse Decraene and Smets (1995) stated that Anacardiaceae are strictly diplostemonous as compared to Burseraceae, which are all obdiplostemonous, but obdiplostemony was recently reported in the cashew family by Bachelier and Endress (2007, 2009). In diplostemonous species the filaments are often prominently unequal. They are usually subulate and can be basally connate, forming a staminal tube of variable height (*Anacardium*). Their insertion is below or outside the disk, except in *Mangifera* and *Swintonia* where they are inside the disk. The anthers are dithecate, usually dorsifixed, less commonly basifixed or versatile, and are usually introrse, rarely extrorse. They are elliptical, sagittate, or orbicular in shape. The anther connective is usually inconspicuous, occasionally prominently apiculate, sometimes glandular, rarely prominently bilobed (*Androtium*). Staminodes are usually present but reduced in female flowers or sometimes absent; rudimentary anthers are frequently sagittate, rarely absent, or without thecae.

The gynoecium is usually superior (rarely, flowers are perigynous as in *Melanochyla* and *Thyrsodium*), or inferior (*Drimycarpus*, *Holigarna*, Fig. 4E), or apparently hemi-inferior due to the ovary being partially immersed in the receptacle (*Semecarpus*). Carpels are fused but nearly apocarpous in *Androtium* and *Buchanania*. The ovary is usually sessile or rarely subtended by a gynophore. The style or the stylodia are usually apical, often subapical or lateral, rarely approaching gynobasic (e.g., *Anacardium*, *Mangifera*), usually erect, rarely patent (e.g., *Searsia*, *Trichoscypha*), sigmoid (e.g., *Anacardium*), apically decurved (e.g., *Fegimanra*), or recurved. The stylodia are basally connate (e.g., *Baronia*, *Heeria*, *Lithrea*) or distinct, rarely apically connate forming a stigmatic head (*Allospondias lakonensis*, *Dracontomelon*), which is more characteristic of Burseraceae (Bachelier and Endress 2009). The stigmas are usually capitate, rarely punctiform (*Anacardium*, *Gluta*), discoid (*Campnosperma*), variously lobed, rarely sessile, often papillose. In male flowers a pistillode is usually present and reduced, sometimes minute or absent (e.g., *Amphipterygium*, *Astronium*, *Campylopetalum*, *Pistacia*, *Sclerocarya*, *Searsia*). See Wannan and Quinn (1991) for a detailed review of Anacardiaceae flower structure, Endress and Stumpf (1991) for androecium structure in Sapindales, Bachelier and Endress (2007) for a detailed study of *Amphipterygium* and *Pistacia* inflorescence and floral structure, and Bachelier and Endress (2009) for a comparative discussion of Anacardiaceae and Burseraceae flowers with emphasis on the gynoecium.

Embryology. Anthers are tetrasporangiate with longitudinal dehiscence. During the maturation of the anther, 1–3 middle layers become obliterated by the end of meiosis (*Pistacia vera*). The tapetum is secretory, the cells becoming binucleate and subsequently being absorbed. The mature anther wall is represented by the epidermis and the endothecium with fibrous bands. Simultaneous cytokinesis follows meiotic divisions in the microspore mother cells. The mature pollen grains are binucleate (Aleksandrovski and Naumova 1985). High levels of pollen sterility are found in cultivated *Mangifera* and *Spondias* (Juliano 1937; Maheshwari et al. 1955).

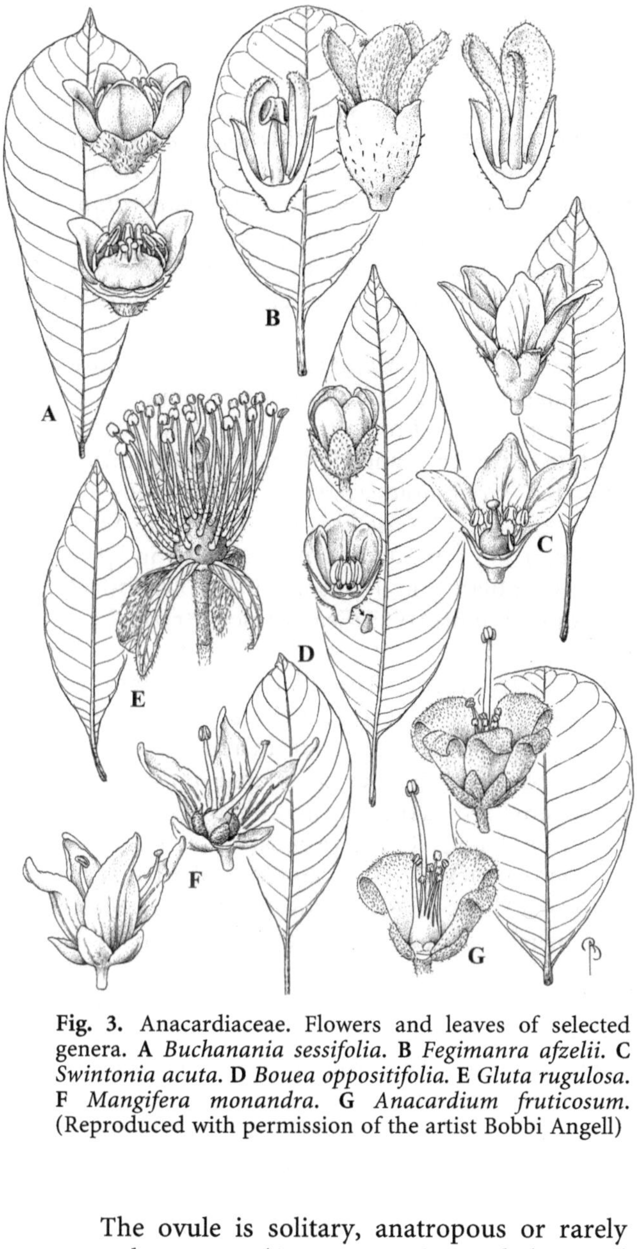

Fig. 3. Anacardiaceae. Flowers and leaves of selected genera. **A** *Buchanania sessifolia*. **B** *Fegimanra afzelii*. **C** *Swintonia acuta*. **D** *Bouea oppositifolia*. **E** *Gluta rugulosa*. **F** *Mangifera monandra*. **G** *Anacardium fruticosum*. (Reproduced with permission of the artist Bobbi Angell)

The ovule is solitary, anatropous or rarely campylotropous (*Semecarpus* in Bachelier and Endress 2009), apotropous ("syntropous" of Bachelier and Endress 2009), apical, lateral or basal, with dorsal or ventral raphe, crassinucellate, with one (e.g., *Anacardium*) or two either entirely or partially developed integuments (e.g., *Lannea*, *Pistacia*, *Rhus*, *Schinus*, *Toxicodendron*) (Geesink et al. 1981). The micropyle is formed by the inner integument, or by both, as in *Schinus molle*. The funicle is long, massive, and bent. A funicular obturator and a cup-like hypostase are present. In addition, a ponticulus (a small bridge between the pollen tube track and the dorsal surface of the ovule) is present in all Anacardiaceae studied by Bachelier and Endress (2009). A nucellar cap of 2–4 cell layers is developed in *Anacardium*, *Lannea*, *Rhus*, *Schinus*, and *Toxicodendron* (Kelkar 1958a, 1958b; Copeland 1961; Grundwag and Fahn 1969; Aleksandrovski and Naumova 1985). The variations and details of ovule structure are summarized by Plisko (1996).

Linear and T-shaped megaspore tetrads are usually formed. Degeneration of megaspore mother cell, megaspore, and embryo sac, and anomalies during meiosis have been observed in *Anacardium*, *Mangifera*, and *Pistacia* (Maheshwari et al. 1955; Copeland 1961; Grundwag and Fahn 1969). An 8-nucleate embryo sac of the Polygonum type develops from a chalazal megaspore (Johri 1963). Cases of development of two embryo sacs occur in *Lannea coromandelica*, *Pistacia chinensis*, *P. vera*, *Searsia mysorensis*, and *Toxicodendron diversilobum*. In *Anacardium*, *Pistacia*, *Rhus*, and *Toxicodendron*, the occurrence of chalazogamy has been observed (Copeland 1961; Grundwag and Fahn 1969; Aleksandrovski and Naumova 1985).

Embryo development is Onagrad-type, with Euphorbia- (e.g., *Lannea*, *Rhus*, and *Semecarpus*), Asterad-, and Penaea-type (*Anacardium*) variations represented in the family. Exact embryogeny is hard to determine in *Pistacia* because of high variations in the orientation of walls during cell divisions in the proembryo (Aleksandrovski and Naumova 1985). During embryo development in *Anacardium* and *Pistacia*, the hypocotyl is turned at a right angle to the cotyledons (Copeland 1961). Chlorophyllous embryos have been reported to occur in *Buchanania latifolia*, *Cotinus coggygria*, *Heeria argentea*, *Pistacia lentiscus*, *P. mutica*, *P. vera*, and *Protorhus longifolia*. Non-chlorophyllous embryos have been reported for *Anacardium occidentale*, *Mangifera indica*, and species of *Ozoroa* (Yakovlev and Zhukova 1973; von Teichman and van Wyk 1996). Endosperm is nuclear, usually absorbed by the growing embryo, or present as a thin layer in *Pistacia* and *Schinus*.

Nucellar polyembryony is well known in *Mangifera* (up to 50 embryos in a seed of *M. indica*; Sachar and Chopra 1957); synergid polyembryony is found in *P. vera* and *Lannea coromandelica* (Grundwag and Fahn 1969). Low seed production frequently occurs in *Anacardium*,

Mangifera, and *Pistacia*. Parthenocarpy is common in *Pistacia* and *Rhus*.

Pollination. Anacardiaceae are primarily entomophilous, but some exceptions are found. *Anacardium* species are pollinated by butterflies and moths (Free and Williams 1976; Mitchell and Mori 1987), and secondarily by bats (Gardner 1977; Dobat and Peikert-Holle 1985). Some *Mangifera* are pollinated by flies, perhaps thrips (Kostermans and Bompard 1993), and secondarily by flying foxes (these bats are feeding on the floral nectar and pollen) (Dobat and Peikert-Holle 1985). A few genera are wind-pollinated (*Amphipterygium, Campylopetalum, Dobinea, Orthopterygium, Pistacia*). There is a possibility of heteranthery in *Anacardium* and *Mangifera*, which both have species with emergent large stamens and a set of smaller stamens, both with pollen (Vogel 1978; Mitchell and Mori 1987). Overall, much more research is needed to complete an understanding of pollination biology in the family.

Pollen Morphology. The pollen grains of Anacardiaceae are generally tricolporate, spheroidal; the colpi are long and narrow. The pollen grains vary in size from 17×15 μm in *Schinus polygama* to 46–56×33 μm in *Spondias mombin* and 49×33 μm in *Semecarpus anacardium*. The reticulations are coarse, with high, upstanding ridges enclosing large polygonal lacunae. The sexine can be finely grano-rugulate, striate-reticulate, striate-perforate or reticulate. The germ pore shape varies from spherical to oblong, and its surface can be smooth, ragged, or psilate (for further information, see Heimsch 1940; Erdtman 1952; Marticorena 1968; Anzótegui 1971; Huang 1972; Baksi 1976; Ibe and Leis 1979; Olivera et al. 1998).

The unusual pollen grains of *Campylopetalum* and *Dobinea* were used as a primary argument for the segregation of Podoaceae (Erdtman 1952; Forman 1954). *Pistacia* and the sister genera *Amphipterygium* and *Orthopterygium* (see Jiménez-Reyes and Figueroa 2001) were similarly segregated as distinct families (Pistaciaceae and Julianiaceae, respectively), in part on the basis of their aberrant pollen morphologies. These five genera have pollen with a higher number of smaller, shallower colpi that appear appropriately structured for wind pollination, which may explain their deviating morphology from the rest of the family.

Karyology. Chromosome numbers for the family range from $n = 7$ (*Campylopetalum, Dobinea*) to $n = 30$ (*Semecarpus*). Some counts indicate intraspecific variation (e.g., *Anacardium, Pistacia, Rhus, Searsia*), and polyploidy is common in the family (e.g., *Lannea, Mangifera, Rhus, Searsia*). Chromosome numbers reported for the cashew (*Anacardium occidentale*) range from $n = 12$ to $n = 29$. Members of Anacardioideae are usually $n = 15$, but exceptions include *Mangifera* ($n = 20$), *Pistacia* ($n = 12$ or 14), *Searsia dentata* ($n = 16$), and *Schinopsis* and *Schinus* ($n = 14$). Spondioideae vary in chromosome number: $n = 11$ in *Buchanania*, $n = 13$ in *Sclerocarya*, $n = 14$ or 20 in *Lannea*, $n = 16$ in *Spondias* (Goldblatt and Johnson 1979–2008; http://mobot.mobot.org/W3T/Search/ipcn.html).

Fruit and Seed. The fruits are drupes or samaras (rarely syncarps, utricles, or baccates), and all appear to be derived from a fundamentally drupaceous fruit type as elucidated in the family phylogeny (see Taxonomy and Phylogeny section for more detail). They are most often 1-locular, but incompletely 2-, 3-, 4-, and 5-locular fruits are not rare. Those of *Pleiogynium* are 5–12-locular and usually contain 1–5, very rarely up to 12 seeds. The pericarp is multilayered and well differentiated within the family. The exocarp varies in thickness, and can have a lignified outer epidermis (some Anacardioideae) or subepidermal sclereids (*Dracontomelon, Pentaspadon*). It is pubescent or glabrous, and is variably colored. In some taxa (e.g., *Lithrea, Toxicodendron*), the exocarp is brittle and chartaceous, and separates from the mesocarp at maturity. The mesocarp is usually fleshy and resinous, can be waxy or oily, and is often pulpy and edible (e.g., *Bouea, Mangifera, Spondias*), dry (e.g., *Amphipterygium, Loxopterygium, Pachycormus, Schinopsis*), or thin (e.g., *Pistacia, Solenocarpus*), and sometimes contains dermatitis-causing liquid in usually black resin canals of varying thickness (e.g., *Anacardium, Gluta, Lithrea, Mangifera, Toxicodendron*). In *Melanochyla* both the mesocarp and endocarp contain black resin. Anatomically, the mesocarp is rather uniform in structure, and includes scattered vascular bundles and resin canals, and often fiber bundles or sclerenchyma zones.

The endocarp can be bony, fibrous-woody, or chartaceous. Wannan and Quinn (1990) describe two structurally distinct endocarp types in the family, the Spondias type with a mass of lignified and irregularly oriented sclerenchyma, and the Anacardium type, which is discretely layered and has palisade-like sclereids. The Spondias type is characteristic of the Spondioideae and Anacardioideae members *Campnosperma* and *Pentaspadon*, while the Anacardium type endocarp characterizes the rest of the family (except possibly genera that have not yet been investigated). The presence of Spondias type endocarp in *Canarium* (Burseraceae) suggests that this type may be plesiomorphic and the Anacardium type apomorphic (Wannan and Quinn 1990).

Endocarps open by various mechanisms at germination: irregular splitting in the stone wall, regular splitting from one or two sides, or ingenious opening devices such as shutters, stoppers, plugs, caps, or lids (Hill 1933, 1937). The latter opening mechanisms, usually called opercula, are located in the endocarp, and sometimes involve participation of the mesocarp. They are found only in Spondioideae. Opercula are usually visible on the surface of the endocarp, except in fruits of *Spondias* and *Harpephyllum* where they are covered by projections of the endocarp and mesocarp. During germination, the whole operculum becomes detached and is pushed off by the growing radicle (e.g., *Antrocaryon*, *Dracontomelon*, *Pleiogynium*, *Sclerocarya*), or the operculum splits into two equal halves that are pushed apart by the radicle like a pair of shutters (*Haematostaphis*, *Pseudospondias*) (Hill 1933, 1937). *Choerospondias* fruits, although not considered to be operculate, have pits in the endocarp but lack the sealing caps; fibrous coverings occur over the pits instead.

Seeds vary in size from 2 mm to more than 10 cm. They are generally ellipsoid, ovoid, falcate, lenticular, or reniform. Species of *Mangifera* have labyrinth seeds—the extreme form of rumination in which the seed coat deeply encroaches on the endosperm and embryo (Boesewinkel and Bouman 1984). The seed coat s.s. (derived from the integuments) is usually undifferentiated and/or reduced, while the chalaza is well developed and forms the greater part of the seed coat s.l. Pachychalazal seeds with undifferentiated seed coat have been found in *Mangifera* and *Heeria*; partially pachychalazal seeds with some traces of lignification in the seed coat have been found in subfamily Spondioideae, *Campnosperma*, *Ozoroa*, *Pistacia*, and *Rhus*, and both types of seeds, some endotegmic, in subfamily Anacardioideae (von Teichman 1991). von Teichman (1991) suggested that the evolutionary trend of seeds within Anacardiaceae is from pachychalazal, to partially pachychalazal, and ultimately to seeds without extensive chalaza. The embryo of Anacardiaceae is oily, curved or straight, with two expanded cotyledons. Endosperm is scant.

Both epigeal and hypogeal germination are found in the family (sometimes within one genus), with great diversity of seedling features. Epigeal and phanerocotylar seedlings are described in *Anacardium*, *Buchanania*, *Dracontomelon*, *Lannea*, *Parishia*, *Pistacia*, *Rhus*, *Spondias*, and *Toxicodendron*. Hypogeal and cryptocotylar seedlings are described in *Gluta*, *Lannea*, *Mangifera*, *Melanochyla*, *Pistacia*, *Rhus*, and *Semecarpus* (de Vogel 1980; Kamilya and Paria 1995). Epigeal and cryptocotylar seedlings are found in *Astronium graveolens*, *Koordersiodendron*, and *Swintonia* (de Vogel 1980; Carmello-Guerreiro and Paoli 1999). For further information, see von Teichman (1998) and Carmello-Guerreiro and Paoli (1999).

Dispersal. The majority of Anacardiaceae have fleshy drupaceous fruits that are animal-dispersed, but there are also additional fruit modifications adapted for different mechanisms of dispersal. Four genera, *Anacardium*, *Fegimanra*, *Holigarna*, and *Semecarpus*, have an enlarged edible hypocarp subtending or partially enveloping the drupe. One species of *Anacardium*, *A. microsepalum*, which grows in the flooded forests of the Amazon, lacks the hypocarp and may be fish-dispersed (Mitchell and Mori 1987). Water dispersal has been reported or purported for species of three genera, *Mangifera*, *Poupartiopsis*, and *Spondias*. The variety of structural adaptations for wind dispersal seen throughout Anacardioideae include subtending enlarged sepals (*Astronium*, *Loxostylis*, *Myracrodruon*, *Parishia*), subtending enlarged petals (*Gluta*, *Swintonia*), trichome-covered margins on a globose fruit (*Actinocheita*), trichome-covered margins on a flattened fruit (*Blepharocarya*, *Ochoterenaea*), *Ulmus*-like samaras encircled with a marginal wing (*Campylopetalum*, *Cardenasiodendron*, *Dobinea*, *Laurophyllus*, *Pseudosmodingium*, *Smodingium*), samaras with a single wing (*Faguetia*, *Loxopterygium*, *Schinopsis*), dry samaroid syncarps (multiple fruit, *Amphipterygium*, *Orthopterygium*), dry achene-like drupe without a wing (*Apterokarpos*), and elongated ciliate pedicels of sterile flowers on broken segments of the infructescence that function much like a tumbleweed (*Cotinus*). The dry utricle fruits of *Pachycormus* are most likely wind-dispersed, but there is no report of this in the literature. The major seed dispersers of fleshy-fruited species are birds (e.g., *Metopium*, *Rhus*, *Schinus*, *Searsia*, *Toxicodendron*), bats (e.g., *Anacardium*, *Antrocaryon*, *Campnosperma*, *Mangifera*, *Spondias*, *Thyrsodium*), and primates (e.g., *Anacardium*, *Antrocaryon*, *Mangifera*, *Pseudospondias*, *Sclerocarya*, *Sorindeia*, *Spondias*, *Trichoscypha*). In addition to these, there are also reports in the literature of the following animal dispersers: elephants and ruminants (e.g., *Antrocaryon*, *Pseudospondias*), deer (e.g., *Anacardium*, *Rhus*, *Spondias*), and coyotes, coatis, foxes, peccaries, reptiles, and tapirs (e.g., *Spondias*) (Gautier-Hion et al. 1985; Mitani et al. 1994; Fragoso 1997; Altrichter et al. 1999; Li et al. 1999; Birkinshaw 2001; Poulsen et al. 2001). Squirrels and rodents mostly serve as seed predators, rather than dispersers (Gautier-Hion et al. 1985). A study of local mammal extinction due to the bush meat trade in Cameroon revealed a sharp decline in seed dispersal of *Antrocaryon* (Wang et al. 2007).

Phytochemistry and Toxicity. Toxic compounds and other chemicals within members of Anacardiaceae have been widely investigated (see review in Aguilar-Ortigoza and Sosa 2004b). Several of these studies focused on the medicinal activity of phenolics (Corthout et al. 1994), esters (Corthout et al. 1992, Galvez et al. 1992), and tannins (Corthout et al. 1991; Galvez et al. 1991; Viana et al. 1997). Others have dealt with toxic components such as contact dermatitis-causing compounds (see Mitchell 1990 for a review; Rivero-Cruz et al. 1997; Drewes et al. 1998), and those responsible for causing nut allergies (Jansen et al. 1992; Fernandez et al. 1995). Some of the compounds in Anacardiaceae have been shown to be defensive in function. These include antimicrobials (Saxena et al. 1994), and antifungal and/or insect- and herbivore-repelling compounds (Chen and Wiemer 1984; Cojocaru et al. 1986).

Contact dermatitis-causing compounds are present in approximately 32 genera of Anacardiaceae (Mitchell 1990; Aguilar-Ortigoza and Sosa 2004b). Most of the poisonous substances are phenols, primarily catechols and resorcinols that accumulate in the resin canals. Pentadecylcatechols (often referred to as urushiols) are the dermatitis-inducing component of poison ivy, poison oak, poison sumac, and the Asian lacquer tree (*Toxicodendron* spp.); they are also found in *Gluta*, *Holigarna*, *Lithrea*, *Metopium*, *Semecarpus*, and *Smodingium*. Heptadecylcatechols are found in *Gluta*, *Lithrea*, *Metopium*, *Semecarpus*, and *Toxicodendron*. Salicylic acid derivatives (anacardic acid, etc.) have been identified in *Anacardium occidentale* and *Pentaspadon*. Other poisonous compounds found in the family include bhilawanols (*Semecarpus*), glutarenghol, laccol (*Toxicodendron*), moreakol (*Gluta usitata*) (Behl and Captain 1979), thitsiol (*Gluta*), renghol (*Gluta*, *Semecarpus*), and semecarpol (a monohydroxy phenol found in *Semecarpus*) (Behl and Captain 1979).

The compounds mentioned above are variously structured oleoresins that may cause an immune system reaction upon binding with skin proteins (Mitchell 1990). Humans and other animals allergic to these compounds can have anywhere from a very mild to a deadly reaction depending upon the location of contact, species encountered, and severity of their allergy. The chemistry of the offending compounds has been researched for many taxa (see above and, e.g., Hill et al. 1934; Backer and Haack 1938; Loev 1952; Tyman and Morris 1967; Johnson et al. 1972; Gross et al. 1975; Halim et al. 1980; Stahl et al. 1983; Gambaro et al. 1986), but the cause of the toxicicity in others is unknown.

Taxonomy and Phylogeny. The family Anacardiaceae was first proposed by Lindley in 1830, but its members have been variously placed in other families including Blepharocaryaceae, Comocladiaceae, Julianiaceae, Pistaciaceae,

Podoaceae, Rhoaceae, Schinaceae, Spondiadaceae, and Terebinthaceae. Engler (1892) divided the family into five tribes, Dobineeae (=Dobineae), Mangifereae (=Anacardieae), Rhoideae (=Rhoeae), Semecarpeae, and Spondieae (=Spondiadeae). He circumscribed them using one vegetative and several floral and fruit characters, including number of carpels, insertion of the ovule on the placenta, number of staminal whorls, leaf complexity, number of locules in the ovary and fruit, embryo morphology, and stylar insertion on the ovary. The tribal circumscription was revised by Mitchell and Mori (1987) who updated Ding Hou's (1978) modification of Engler's classification. However, this classification has more recently been questioned. Wannan and Quinn (1990, 1991) used floral and pericarp structure, wood anatomy, and biflavonoid data to investigate the classification of Anacardiaceae. They tentatively identified two groups, A and B, which were each divided into two subgroups, 1 and 2. Engler's tribes Anacardieae, Dobineae, Rhoeae, and Semecarpeae, with the exception of *Androtium*, *Buchanania*, *Campnosperma*, and *Pentaspadon*, were placed in group A, while group B contains all of Spondiadeae plus the four genera named above (two genera each from Anacardieae and Rhoeae). Wannan and Quinn (1991) designated two genera, *Faguetia* and *Pseudoprotorhus* (=Sapindaceae, *Filicium*), as not assignable to any group.

In the first molecular investigation of Anacardiaceae, Terrazas (1994) used sequences of the chloroplast gene *rbc*L, together with data on morphology and wood anatomy to interpret the phylogeny of the family. Her combined *rbc*L-morphology phylogeny elucidated a monophyletic Anacardiaceae comprised of two groups. The first group, clade A2, contains Spondiadeae plus *Pentaspadon*, and is united by the morphological synapomorphy multicellular stalked glands on the leaves. The second group, clade A1, contains the remaining genera in the four other tribes, and is supported by the morphological and wood anatomical synapomorphies, unicellular stalked leaf glands and the presence of both septate and nonseptate fibers. Based on the combined phylogeny, Terrazas (1994) informally proposed splitting the family into two subfamilies, Anacardioideae and Spondioideae. Pell found a similar division of the family based on analysis of three plastid genes, and described and circumscribed the two subfamilies (Pell 2004; Mitchell et al. 2006). This classification is followed herein, with the exception of the subfamilial placement of *Buchanania*: recent molecular data have helped resolve ambiguous morphological data, and suggest that this genus is a member of subfamily Spondioideae. It should be noted that recent, more extensively sampled analyses of cpDNA (*trn*LF and *rps*16) and nrDNA (ETS) data (Pell, unpublished data) indicate that subfamily Spondioideae (formerly recognized as tribe Spondiadeae) is polyphyletic. However, because the two subfamilies as circumscribed by Mitchell et al. (2006) are structurally recognizable, and formal subfamilial rankings have not yet been assigned to the newly resolved clades, the two-subfamily system is used here to divide the genera.

The phylogenies elucidated by Pell (2004, and unpublished data) added support for several relationships within the family. The position within Anacardiaceae of several formerly segregated families including Podoaceae (*Campylopetalum* and *Dobinea*), Julianiaceae (*Amphipterygium* and *Orthopterygium*), and Pistaciaceae (*Pistacia*) has been solidified by molecular data. The core members of former tribe Anacardieae (*Anacardium*, *Bouea*, *Fegimanra*, *Gluta*, *Mangifera*) form a clade. *Rhus* s.l. is polyphyletic, with up to five different origins, and is in urgent need of taxonomic revision (Pell 2004; Pell et al. 2008; see also Miller et al. 2001; Yi et al. 2004, 2007). Moffett (2007) published the new combinations in *Searsia* for the mostly African former *Rhus* species, but much work remains to be done.

Anacardiaceae are most closely allied with Burseraceae. Both families share the synapomorphies vertical intercellular secretory canals in the primary and secondary phloem, and often horizontal ones in the wood rays, and the ability to synthesize biflavonyls (Wannan et al. 1985; Wannan 1986; Wannan and Quinn 1990, 1991; Terrazas 1994). Additionally, in these families the ovules are often attached to a short, broad placental obturator, and the plants are strongly resinous. The close relationship of Anacardiaceae and Burseraceae has been suggested by numerous authors based upon morphological, anatomical, and biochemical data (Gundersen 1950; Cronquist 1981; Wannan 1986; Takhtajan 1987; Thorne 1992), further supported by DNA sequence data (Gadek et al. 1996; APG 1998, 2003, 2009; Savolainen et al. 2000a, b; Pell 2004). Nevertheless, Anacardiaceae are distinguished from Burseraceae by having a single apotropous ovule per locule versus two epitropous ovules per locule, respectively. Additional features that are not universal, but often effectively separate the two families, include chemical and fruit characteristics. Burseraceae lack chemical compounds that are present in many Anacardiaceae, including 5-deoxyflavonoids and contact dermatitis-causing compounds. All Anacardiaceae fruits are indehiscent, while many Burseraceae fruits are dehiscent; some Burseraceae have stipules or pseudostipules that in Anacardiaceae are lacking; many Burseraceae have a terminal pulvinulus subtending the lamina of the leaflets, which is absent in Anacardiaceae.

Historically the family Anacardiaceae has been placed within the higher taxonomic rankings of Burserales, Rutales, Sapindales, or Terebinthinae. Most modern authors consider it a member of Sapindales, and recent molecular studies at the ordinal level (Gadek et al. 1996) and above (Chase et al. 1993; Bremer et al. 1999; Savolainen et al. 2000a, 2000b; APG 2003, 2009) have supported this classification.

Distribution. The family is native to the Western Hemisphere (from southern Canada to Patagonia); Africa; southern Europe; temperate, subtropical and tropical Asia; tropical and subtropical Australia; and most of the Pacific Islands. Anacardiaceae are noticeably absent from the floras of northern Eurasia, southwestern Australia, New Zealand, the Galapagos Islands, northern North America, and extreme desert and high-elevation habitats. The primary centers of diversity are in Mexico, South America, southern and equatorial Africa, Madagascar, Indochina, and Malesia. The Paleotropics are richer in species number than are the Neotropics.

Paleobotany. Anacardiaceae have rich fossil records because of their woody growth form, and past and current wide distribution. Anacardiaceae pollen and wood first appear in the Paleocene, 65 to 55 million years ago (Hsu 1983; Muller 1984), and are found throughout the world. Fossils occur mostly in the western Unites States south to Panama from the Eocene and Oligocene (Taylor 1990; Ramírez and Cevallos-Ferriz 2002). Leaf fossil material of four species of *Rhus* and one putative species of *Cotinus* with its distinctive cladodromous secondary venation have been identified from the Eocene in the Florissant fossil beds of Colorado (Meyer 2003). Silicified wood fragments from the lower Miocene formations of central west Sardinia have been allied with *Sclerocarya birrea* (Biondi 1981). Many purported Anacardiaceae leaf and wood fossils have been determined to belong to taxa outside of the family, and many others are of dubious affinity.

Fossil fruits assigned to *Antrocaryon* have been found in the 3 million year-old deposits of the Lower Omo Valley (Bonnefille and Letouzey 1976), and the early-middle Miocene (>16 Ma) Bakate Formation (Tiffney et al. 1994) in Ethiopia. Fruits with possible affinities to *Pistacia* (see Taylor 1990) and *Dracontomelon* have also been found (Collinson 1983). Fossils with affinities to *Tapirira* include fruits (Reid 1933), flowers preserved in amber, and fossil wood (see Taylor 1990 for a review). Six genera from the London Clay flora have been provisionally assigned to Anacardiaceae, and appear to be related to the Spondioideae due to the presence of opercula-like structures on the fossilized endocarps (Collinson 1983). Miocene fruit fossils of *Loxopterygium* have recently been identified in Andean Ecuador (Burnham and Carranco 2004). Numerous *Anacardium* fossils were reported by Berry (1924, 1929) from North and South America, and recently, Manchester et al. (2007) reported *Anacardium* fossils from the Eocene of Europe, greatly expanding the range of this genus. Some endocarp fossils originally assigned to *Dracontomelon* have subsequently been transferred to extinct genera (Manchester 1994).

Economic Importance. The major agricultural food products of Anacardiaceae are cashews (*Anacardium*), mangos (*Mangifera*), pink peppercorns (*Schinus*), and pistachios (*Pistacia*). However, numerous taxa have other edible parts of high regional value (e.g., *Antrocaryon*, *Buchanania*, *Choerospondias*, *Cyrtocarpa*, *Harpephyllum*, *Lannea*, *Ozoroa*, *Rhus*, *Searsia*, *Spondias*, *Tapirira*, and *Trichoscypha*). *Sclerocarya birrea* has recently become economically important outside of its native range of sub-Saharan Africa and Madagascar, due to the export of a liquor, Amarula Cream, made from its fruit, marula (Hall et al. 2002). The cashew (*Anacardium occidentale*) yields three major economic products: seed ("cashew"), hypocarp ("cashew apple"), and mesocarp resin ("cashew nutshell liquid"). The cashew seed is eaten raw and roasted, powdered to make a beverage, and is used in confections. Cashew apples are more important locally than globally, and are eaten fresh, juiced, and fermented to make alcoholic beverages. The cashew nutshell liquid is used industrially in the manufacturing of various polymers such as plastics, adhesives, lubricants, and resins. Examples of specific products are wallboard and break linings. Of these products, the seed is by far the most economically important. Major countries of production are Brazil, India, Indonesia, Guinea, Mozambique, Nigeria, Tanzania, and Vietnam (Behrens 1996). Cashew is native to lowland South America, and was brought to India by the Portuguese; other early colonialists introduced the cashew of commerce into cultivation throughout the lowland tropics of the New and Old Worlds.

No species of Anacardiaceae ranks as a major, internationally important timber tree but many have an important role in smaller timber markets, and are valued for their quality wood and rot resistance (Record 1939; Hess 1949). One of the most prized rot-resistant Anacardiaceous timber trees comes from species of the South American genus *Schinopsis* (quebracho), which has been used extensively in Argentina for railroad ties (Kerr 1935). *Astronium* (gonçalo alvez) and *Myracrodruon* are also of significant importance for exported lumber. Other timber genera are locally important and used for making matchsticks, cabinetry, bows, charcoal, housing, axe-handles, furniture, firewood, and kitchenware.

Many Anacardiaceae species are also valued for their horticultural appeal. Specimens of *Cotinus*, *Harpephyllum*, *Lannea*, *Pistacia*, *Rhodosphaera*, *Rhus*, *Schinus*, *Searsia*, *Smodingium*, and *Toxicodendron* are planted for their beautiful inflorescences, infructescences, evergreen foliage, and/or fall foliage. A few agricultural and horticultural species have escaped cultivation and become invasive in their non-native areas. Japanese wax tree, *Toxicodendron succedaneum*, is an Asian species that was originally cultivated in Brazil but escaped after introduction, and is now invasive. Brazilian pepper tree or pink peppercorn, *Schinus terebinthifolia*, is another notoriously problematic species where it occurs in the Everglades of central and southern Florida, in the Hawaiian Islands, and various other parts of the subtropics and tropics (Gilman 1999; Mitchell 2004). More recently, *Pistacia chinensis* has become naturalized and invasive in Texas (McWilliams 1991).

Anacardiaceae have long been known for their medicinal properties. *Spondias* and *Rhus* are used extensively by native populations for everything from healing broken bones to treating colds. Other taxa are used to treat fever (e.g., *Buchanania*, *Comocladia*), hepatitis (*Haematostaphis*), gastrointestinal illness (e.g., *Anacardium*, *Antrocaryon*, *Heeria*, *Lannea*, *Ozoroa*, *Pseudospondias*, *Schinus*, *Sorindeia*), respiratory disease (e.g., *Astronium*), skin disease and/or wounds (e.g., *Buchanania*, *Lannea*, *Metopium*, *Ozoroa*, *Schinus*, *Sclerocarya*, *Searsia*, *Sorindeia*, *Trichoscypha*), venereal disease (e.g., *Buchanania*, *Lannea*), various pregnancy-related conditions (e.g., *Metopium*, *Ozoroa*, *Trichoscypha*), and as an astringent (e.g., *Anacardium*, *Astronium*, *Mangifera*) (Morton 1981; Burkill 1985; Mitchell 2004).

Modern medicine has investigated many of these ethnobotanical uses and isolated several active compounds from various plant structures. Cardol, a compound extracted from the pericarp of the cashew, has been shown to exhibit antifilarial activity useful against filariasis, a major tropical disease caused by filarial parasites that has affected more than 400 million people worldwide (Suresh and Raj 1990). Three anacardic acids isolated from the juice of the "apple" (swollen hypocarp) of the cashew have been shown to have significant in vitro cytotoxicity against BT-20 breast carcinoma cells (Kubo et al. 1993). Three anacardic acids isolated from the cashew nut shell oil provide potent antibacterial activity against *Streptococcus mutans* (a component of plaque), the bacterium that adheres to the smooth surface of the tooth and facilitates the accumulation of other oral microorganisms that produce organic acids that destroy enamel, leading to the formation of cavities (Muroi and Kubo 1993).

Anacardiaceae contact dermatitis is responsible for a great deal of lost work worldwide. In the US state of California, *Toxicodendron diversilobum* (poison oak) dermatitis costs the state ca. 1% of its workers' compensation budget, and nearly one third of US forest fire fighters are disabled by this dermatitis when responding to fires on the US west coast (Epstein 1994). Several of the Asian contact dermatitis-causing taxa are used for their tannins and in the lacquerware industry. The resins of *Toxicodendron vernicifluum* and *Gluta* species are used in Burma, China, Japan, Thailand, and Vietnam to create decorative, long-lasting wooden art pieces such as trays, jewelry boxes, vases, picture frames, and furniture. Resin collected from the trees is refined and applied to fine wood, increasing the woods' chemical, heat, and humidity resistance. Unfortunately, the oleoresins' activity is not completely suppressed upon drying, and lacquerware can continue to cause much discomfort in unsuspecting admirers for years (Kullavanijaya and Ophaswongse 1997; Prendergast et al. 2001; Rodriguez et al. 2003). Other industrial uses of Anacardiaceae species include the production of dyes for marking laundry (e.g., *Lannea*, *Semecarpus*) and automobile break linings (*Anacardium*) (Mitchell and Mori 1987).

Key to the Subfamilies

1. Trees, shrubs, rarely vines or perennial herbs; exudate often present and sometimes causing contact dermatitis; leaves simple, unifoliolate, or multifoliolate; 1 or more whorls of perianth sometimes absent; androecium haplo-, diplo-, or greater than diplostemonous; carpels 1–3 (5 in *Androtium*) and fused; ovary 1-locular (often by abortion; 2-locular in *Campnosperma*); ovule apical, basal, or lateral; exocarp usually thin; opercula 0; fruit sometimes wind-dispersed **I. Subfam. Anacardioideae**
– Trees or shrubs; exudate often present and not causing contact dermatitis (very rarely causing dermatitis in *Spondias*); leaves multifoliolate (rarely simple/unifoliolate in *Buchanania*, *Haplospondias*, *Lannea*, and *Sclerocarya*); perianth always present; androecium diplostemonous; carpels 4–5 (1 in *Solenocarpus*, more than 5 in *Buchanania* and *Pleiogynium*); ovary (1)4–5(+)-locular; ovule apical to subapical (basal or sublateral in *Buchanania*); exocarp thick; opercula often present; fruit never wind-dispersed **II. Subfam. Spondioideae** (Key p. 21)

Key to the Genera of Anacardioideae

1. Leaves simple or unifoliolate 2
– Leaves compound 41
2. Leaves opposite 3
– Leaves alternate or subopposite 8
3. Leaves always opposite; style 1, stigma 1 4
– Leaves opposite, subopposite, alternate, or verticillate, more than one position usually present on plant; stylodia 1–3; stigmas 3 5
4. Trees; evergreen; bud scales present; leaves coriaceous, margins always entire; flowers bisexual, perianth always biseriate **11. *Bouea***
– Subshrubs to shrubs; deciduous; bud scales absent; leaves not coriaceous (chartaceous to membranaceous), margins serrate (at least at base); flowers unisexual, perianth biseriate in male flowers and 0 in female flowers 7
5. Style 1; drupe ellipsoidal and symmetrical; cotyledons usually fused and ruminate **1. *Abrahamia*** p. p.
– Stylodia 3; drupe reniform and oblique; cotyledons not fused or ruminate 6
6. Leaves often pubescent abaxially, young leaves green; pedicel articulate; ovule lateral or basal; widespread sub-Saharan Africa and Arabian Peninsula **42. *Ozoroa*** p. p.
– Leaves glabrous abaxially, young leaves orange-red; pedicel non-articulate; ovule subapical; endemic to southeast South Africa **47. *Protorhus***
7. Leaves palmately lobed and veined; pistillode absent; disk present in male flowers **13. *Campylopetalum***
– Leaves not palmately lobed and veined; pistillode present; disk absent in male flowers **17. *Dobinea*** p. p.
8. Ovary inferior 9
– Ovary superior 10
9. Prominent marginal secondary vein absent; petiole with spur-like, caducous or persistent appendages; corolla valvate; ovule apical **25. *Holigarna***
– Prominent marginal secondary vein present; petiole appendages 0; corolla imbricate; ovule basal **18. *Drimycarpus***
10. Stilt roots frequently present; hypanthium present **33. *Melanochyla***
– Stilt roots usually absent (very rarely present in *Gluta* and *Campnosperma*); hypanthium absent 11
11. Unicarpellate; style 1; stigma 1, undivided 12
– Tricarpellate or >3 carpels; stylodia 1 or more; stigma >1 16
12. Stamen one; staminodes absent; perianth always 4-merous **21. *Fegimanra***
– Stamens one or more, if only one, then stamen fertile; staminodes present; perianth usually more than 4-merous 13
13. Domatia often present in secondary vein axils abaxially; staminal tube present; drupe usually subtended by fleshy hypocarp; mesocarp woody **4. *Anacardium***
– Domatia absent; staminal tube absent; hypocarp 0; mesocarp fleshy or resinous 14
14. Plants hermaphrodite; calyx calyptriform, circumscissile or bursting irregularly at anthesis, caducous; disk absent; gynophore present **22. *Gluta***
– Plants andromonoecious; calyx imbricate, persistent; disk present, extrastaminal (very rarely intrastaminal in *Mangifera*); gynophore absent 15
15. Glandular ridges on petals; androecium usually zygomorphic, stamens 5(–10), only 1–2 (3–5 or all) fertile,

the staminodes much reduced; enlarged petals absent in fruit **31. *Mangifera***
- Glandular ridges absent from petals; androecium always actinomorphic, stamens 5, staminodes 0; enlarged petals subtending fruit **57. *Swintonia***

16. Connective apically bilobed, prolonged and dilated; carpels 5 **5. *Androtium***
- Connective not apically bilobed, prolonged or dilated; carpels 3 17

17. Perianth absent in female flowers; disk absent in male flowers, present in female flowers; drupe peltate on an accrescent, obovate, or rounded bract that aids in wind dispersal **17. *Dobinea*** p. p.
- Perianth present in female flowers; disk present in all flowers; drupe not as above 18

18. Fleshy hypocarp subtending drupe **54. *Semecarpus***
- Fleshy hypocarp absent 19

19. Fruits with fleshy mesocarp 20
- Fruits with waxy, dry, or resinous (not fleshy) mesocarp 25

20. Leaves linear to lanceolate; perianth of tepals in male flowers, biseriate in female flowers **23. *Haplorhus***
- Leaves various; perianth biseriate in all flowers 21

21. Leaves with peltate or lobed scales; drupes often incompletely bilocular with only one locule fertile **12. *Campnosperma***
- Leaves without peltate or lobed scales; drupes 1-locular 22

22. Perianth usually 4-merous 23
- Perianth usually 5-merous 24

23. Style 1; stigma 1, capitate to trilobed; India and Sri Lanka **39. *Nothopegia***
- Stylodia 3–4 (very rarely 6); stigmas 3, capitate to shallowly bilobed; sub-Saharan Africa **60. *Trichoscypha*** p. p.

24. Androecium haplostemonous or less 31
- Androecium diplostemonous or greater 39

25. Fruit with elongated wing-like structure 26
- Fruits without elongated wing-like structure 27

26. Leaves entire; fruit a laterally winged samara; South America **51. *Schinopsis*** p. p.
- Leaves serrate; fruit a syncarp subtended by a winged peduncle; Mexico to Costa Rica **3. *Amphipterygium*** p. p.

27. Some flowers in inflorescence aborting, and their pedicels elongating and becoming villous; infructescence wind-dispersed like a tumbleweed **16. *Cotinus***
- Flowers, inflorescence, and infructescence not as above 28

28. Inflorescence highly condensed, characterized by fusion of higher-order branches into flattened broad segments; style 1; ovule apical **26. *Laurophyllus***
- Inflorescence branches not fused; stylodia 3 or style branches 3; ovule basal 29

29. Leaves simple; exocarp readily separating from mesocarp; mesocarp waxy and white with black resin canals 30
- Leaves dissected and simple, both on same plant; exocarp and mesocarp not as above **10. *Bonetiella***

30. Marginal secondary vein absent from leaflets; hairy tuft domatia sometimes in secondary vein axils abaxially; androecium haplostemonous; eastern Asia **59. *Toxicodendron*** p. p.
- Leaflets with marginal secondary vein; hairy tuft domatia 0; androecium diplostemonous; Brazil, Paraguay, Argentina, Uruguay, and Chile **27. *Lithrea*** p. p.

31. New World 32
- Old World 33

32. Leaves evergreen or deciduous; exocarp red with glandular trichomes **50. *Rhus* subg. *Lobadium*** p. p. and ***R. chiangii***
- Leaves always evergreen; exocarp white, glandular trichomes 0 **30. *Malosma***

33. Madagascar 34
- Africa, Asia, Europe 36

34. Fruit ellipsoidal, symmetrical; cotyledons usually fused and ruminate; endocarp bony **1. *Abrahamia*** p. p.
- Fruit sigmoid or reniform, asymmetrical; cotyledons not as above; endocarp thin (chartaceous or cartilaginous) 35

35. Inflorescence pendent; corolla pink to red; style 3-branched **36. *Micronychia***
- Inflorescence erect; corolla white, yellow, or green; stylodia 3 **8. *Baronia***

36. Leaves trisect, inflorescence cauliflorous, thorns present **53. *Searsia*** p. p. (=*S. problematodes*)
- Leaves entire, inflorescence terminal and/or axillary, thorns absent 37

37. Exudate watery; leaves strongly discolorous with fine silvery pubescence abaxially; fruits 25–30 mm at longest axis **24. *Heeria***
- Exudate milky or brown; leaves not as above; fruits 4–14 mm at longest axis 38

38. Leaves often pubescent abaxially, young leaves green; pedicel articulate; ovule lateral or basal; widespread sub-Saharan Africa and Arabian Peninsula **42. *Ozoroa*** p. p.
- Leaves glabrous abaxially, young leaves orange-red; pedicel non-articulate; ovule subapical; endemic to southeast South Africa **47. *Protorhus***

39. Calyx deeply lobed; stylodia usually 3; exocarp generally separating from mesocarp at maturity **52. *Schinus*** p. p.
- Calyx shallowly lobed; style always 1; exocarp not separating from mesocarp at maturity 40

40. Plants dioecious (very rarely monoecious); stamens 10–20; Africa **56. *Sorindeia*** p. p.
- Plants hermaphrodite; stamens 10; Central to South America **32. *Mauria*** p. p.

41. Leaves opposite **9. *Blepharocarya***
- Leaves alternate 42

42. Leaves trifoliolate 43
- Leaves greater than trifoliolate (multifoliolate) 48

43. Fruit winged 44
- Fruit not winged 45

44. Inflorescence terminal and/or axillary thyrses; perianth 5-parted, biseriate; fruit a samara with a marginal wing; South Africa **55. *Smodingium***
- Inflorescence axillary, male flowers arranged in pendent thyrses, female flowers tightly arranged and subtended by an involucre; calyx 6–8-parted in male flowers, 0 in female flowers, corolla 0; fruit a syncarp subtended by a winged peduncle; Mexico to Costa Rica **3. *Amphipterygium*** p. p.
45. Exudate turning black with exposure to air; exocarp readily separating from mesocarp; mesocarp waxy and white or gray with black resin canals 46
- Exudate not turning black with exposure to air; exocarp and mesocarp not as above 47
46. Marginal secondary vein present; hairy tuft domatia 0; androecium diplostemonous; Brazil, Bolivia, Paraguay, Argentina, Uruguay, and Chile **27. *Lithrea*** p. p.
- Marginal secondary vein absent; hairy tuft domatia sometimes in secondary vein axils abaxially; androecium haplostemonous; southern Canada to Guatemala, and eastern Asia **59. *Toxicodendron*** p. p.
47. Terminal inflorescence branches spicate; exocarp reddish with glandular trichomes; southern Canada to Panama **50. *Rhus*** subg. ***Lobadium*** p. p.
- Terminal inflorescence branches not spicate; exocarp not as above, if reddish, not with glandular trichomes; Africa, Sicily, Middle East to China **53. *Searsia*** p. p.
48. Fruit likely wind-dispersed (samaras, dry drupes with enlarged trichomes, dry achene-like drupes, utricles) 49
- Fruit likely animal-dispersed (fleshy drupes or dry but not winged) 63
49. Plants caudiciform; fruit a utricle **43. *Pachycormus***
- Plants not caudiciform; fruit not a utricle (various) 50
50. Fruit subtended by stiffened enlarged sepals 51
- Fruit not as above 54
51. Perianth 4-merous **44. *Parishia***
- Perianth 5-merous 52
52. Pistillodes present and reduced; fruit globose to subglobose **38. *Myracrodruon***
- Pistillodes absent; fruit not as above 53
53. Rachis alate; flowers large with showy petals; stylodia lateral; South Africa **29. *Loxostylis***
- Rachis not alate; flowers inconspicuous; stylodia apical; Mexico to South America **7. *Astronium***
54. Perianth absent in female flowers; flowers wind-pollinated; fruit a syncarp subtended by a winged peduncle 55
- Perianth present in female flowers; flowers insect-pollinated; fruit not as above 56
55. Male inflorescence pendant and large; sepals 5–8, longer than stamens; Mexico to Costa Rica **3. *Amphipterygium*** p. p.
- Male inflorescence erect and small; sepals 3–8, shorter than stamens; Peru **41. *Orthopterygium***
56. Gynoecium and fruits with long trichomes on surface and/or margins 57
- Gynoecium and fruits glabrous, or trichomes inconspicuous if present 58
57. Leaflets entire; fruit a samara, laterally compressed with trichomes occurring prominently on the margins; inflorescence corymbose; Venezuela, Panama, and Colombia south to Bolivia **40. *Ochoterenaea***
- Leaflets broadly pinnately lobed (leaf fern-like); fruit a globose drupe covered with trichomes; inflorescence not corymbose; Mexico **2. *Actinocheita***
58. Fruit not winged **6. *Apterokarpos***
- Fruit winged 59
59. Fruit with an elongated wing 60
- Fruit compressed and encircled by a winged margin 62
60. Samara wing terminal and ending in style remnants; Madagascar **20. *Faguetia***
- Samara wing lateral and not ending in style remnants; South America 61
61. Plants polygamodioecious; leaves usually deciduous; wing chartaceous with conspicuous venation, consisting of exocarp tissue **28. *Loxopterygium***
- Plants dioecious or monoecious; leaves usually evergreen; wing stiffened and thick, lacking conspicuous venation, consisting of exocarp and mesocarp **51. *Schinopsis*** p. p.
62. Flowers sessile; stylodia 3; ovule basal; drupe with two unequal, narrow lateral wings; Bolivia **14. *Cardenasiodendron***
- Flowers pedicellate; style 3-branched; ovule apical; drupe with two equal, broad lateral wings; Mexico **48. *Pseudosmodingium***
63. Flowers apetalous, wind-pollinated; stylodia plumose **46. *Pistacia***
- Flowers with petals, insect-pollinated; stylodia not plumose 64
64. Perianth 3–4-merous 65
- Perianth 5-merous 66
65. Leaflets usually spinose; perianth 3(–4)-merous; West Indies, Mexico, Belize, and Guatemala **15. *Comocladia***
- Leaflets entire; perianth always 4-merous; sub-Saharan Africa **60. *Trichoscypha*** p. p.
66. Hypanthium present **58. *Thyrsodium***
- Hypanthium absent 67
67. Domatia in axils of secondary veins abaxially; androecium usually consisting of one whorl of fertile stamens and one whorl of staminodes **45. *Pentaspadon***
- Domatia usually 0 (sometimes present in *Toxicodendron* and *Mauria*); androecium of uniform stamens 68
68. Androecium haplostemonous 69
- Androecium diplostemonous or greater 75
69. Exudate usually turning black with exposure to air; leaves often with black spots where injured; exudate toxic 70
- Exudate not turning black with exposure to air; leaves without conspicuous black spots; exudate not toxic (possibly toxic in *Apterokarpos*) 71
70. Plants polygamodioecious; leaves usually deciduous (rarely evergreen); stylodia 3; exocarp white to gray or yellow, easily separating from mesocarp; mesocarp white or gray with black resin canals **59. *Toxicodendron*** p. p.

- Plants dioecious; leaves evergreen; style 1; exocarp orange to brown, not easily separating from mesocarp; mesocarp resin canals not conspicuous **35. *Metopium***
71. Drupe dry, with longitudinal grooves; northeastern Brazil **6. *Apterokarpos***
- Drupe fleshy, lacking longitudinal grooves; not in Brazil 72
72. Leaves palmately compound (usually pentafoliolate) **53. *Searsia*** p. p.
- Leaves pinnately compound 73
73. Leaflet venation lacking areolation; exocarp black **34. *Melanococca***
- Leaflet venation with areolation; exocarp red to reddish brown 74
74. Erect or scandent shrubs or small trees (up to 7 m); fruit covered with glandular trichomes; seed fills most of the locule **50. *Rhus*** p. p.
- Medium to large trees (up to 20 m); fruit glabrous; seed fills only a small portion of the locule **37. *Mosquitoxylum***
75. Neotropics 76
- Paleotropics 78
76. Leaflets with a prominent marginal secondary vein; fruits pale gray to whitish **27. *Lithrea*** p. p.
- Leaflets lacking marginal secondary vein; fruit color variable but not as above 77
77. Leaflets petiolulate; corolla valvate or subvalvate; exocarp not easily separating from mesocarp at maturity **32. *Mauria*** p. p.
- Leaflets sessile to subsessile; corolla imbricate; exocarp easily separating from mesocarp at maturity **52. *Schinus*** p. p.
78. Prominent diagonally oriented admedial tertiary veins in leaflets often present; androecium often greater than diplostemonous; Africa and Madagascar **56. *Sorindeia*** p. p.
- Admedial tertiary veins in leaflets, if present, not prominent and not diagonally oriented; androecium diplostemonous; Australia, New Caledonia, and New Guinea 79
79. Leaves paripinnate or imparipinnate often with very reduced terminal leaflet, leaflets usually alternate; anther connective glandular; style 1; mesocarp fleshy **19. *Euroschinus***
- Leaves usually imparipinnate, terminal leaflet not reduced, leaflets opposite or subopposite; anther connective not glandular; stylodia 3; mesocarp woody **49. *Rhodosphaera***

Key to the Genera of Spondioideae

1. Leaves simple or unifoliolate 2
- Leaves compound 5
2. Stellate trichomes present; perianth 4-merous, stylodia 4 **71. *Lannea*** p. p.
- Stellate trichomes absent; perianth 4–5-merous; style 1 or stylodia 3 3
3. Leaves always simple; outside of Africa 4
- Leaves 1(–9)-foliolate; E Africa **78. *Sclerocarya*** p. p. (=*S. gillettii*)
4. Carpels 1, possibly pseudomonomerous; restricted to Yunnan, China **68. *Haplospondias***
- Carpels 4–6, distinct or basally distinct and syncarpous at mid ovary; widespread in tropical Asia, Malesia, Australia, Micronesia, Melanesia, and Samoa **63. *Buchanania***
5. Leaves bipinnately compound **80. *Spondias*** p. p. (=*S. bipinnata*)
- Leaves once compound 6
6. Stellate trichomes present **71. *Lannea*** p. p.
- Stellate trichomes absent 7
7. Hairy tuft domatia in secondary vein axils abaxially 8
- Hairy tuft domatia 0 14
8. Domatia hairy and marsupiform; stylodia and locules 5–12 9
- Domatia hairy and not marsupiform; stylodia and locules 1–5 10
9. Leaves evergreen; calyx valvate; stylodia and locules 5–12 **74. *Pleiogynium***
- Leaves deciduous; calyx slightly imbricate; stylodia and locules 5 **80. *Spondias*** p. p. (=*S. petelotii*)
10. Endocarp with opercula; fruits maturing dark red, (blue-)black, or dark brown 11
- Endocarp sometimes with 5 pores but lacking external opercula; fruits maturing yellowish or red to purplish 13
11. Leaflet apex usually emarginate or rounded, often with a pubescent mucro; fruits maturing dark red; endocarp not lobed, one apical operculum **67. *Haematostaphis***
- Leaflet apex usually acuminate, not mucronate; fruits maturing dark brown or (blue-)black; endocarp slightly 5-lobed, 3–5 subapical opercula 12
12. Plants hermaphrodite, carpels connate at apex and base only; perianth 5-merous; fruit depressed-globose or depressed-obovoid, 1–5 locules fertile; endocarp with 5 simple opercula or lids **66. *Dracontomelon*** p. p.
- Plants dioecious; carpels connate at base only; perianth 3–4-merous; fruit oblong-obovoid to ellipsoid, 1(–2) locule fertile; endocarp with 3–4 (–5) bipartite opercula (split into 2 equal halves by the radicle emergence) **77. *Pseudospondias*** p. p.
13. Scandent shrubs or climbers; leaflets always toothed; stylodia connivent, stigma discoid; fruits maturing red to purplish, endocarp crustaceous, 1-locular, lacking pores **73. *Pegia***
- Trees; leaflets usually entire but sometimes toothed; stylodia separate; stigmas capitate; fruits maturing green to yellow, the endocarp bony, 5-locular, with 4–5 pores toward the apex **64. *Choerospondias***
14. Leaf rachis alate 15
- Leaf rachis not alate 17
15. Leaf rachis broadly alate, leaflets microphyllous; fruits solitary, globose, endocarp relatively smooth, interrupted only by the opercula **72. *Operculicarya*** p. p.

- Leaf rachis narrowly alate, leaflets medium sized to large; fruits in racemes or panicles, (oblique-)oblong, endocarp surrounded by a fibrous matrix or interrupted by several irregular cavities 16
16. Leaflets falcate; stylodia 4–5; endocarp interrupted by several irregular cavities, opercula internal; South Africa **69. *Harpephyllum***
- Leaflets not falcate; stylodia 3; endocarp surrounded by a thick fibrous matrix, opercula not apparent; Madagascar **76. *Poupartiopsis***
17. Leaflets with an intramarginal vein (rarely a marginal secondary) 18
- Leaflets without an intramarginal or marginal secondary vein 19
18. Trees; stylodia (4–)5; drupe 4–5-locular, endocarp surrounded by a fibrous matrix **80. *Spondias*** p. p.
- Scandent trees; style 1; drupe 1-locular, endocarp woody and smooth (fibrous matrix lacking) **79. *Solenocarpus*** p. p. (=*S. indica*)
19. Drupe depressed-globose 20
- Drupe oblong, (ob)ovoid, or globose but not depressed 21
20. Plants polygamodioecious; leaves deciduous; drupe radially symmetric (pedicel scar on stone usually centered); fruit maturing yellow or light orange, endocarp surface pitted **62. *Antrocaryon***
- Plants hermaphrodite; leaves evergreen; drupe oblique (pedicel scar on stone markedly excentric); fruit maturing black or brown, endocarp surface relatively smooth **66. *Dracontomelon*** p. p.
21. Female flowers usually solitary; stamens 10–16(23); pistillode absent **78. *Sclerocarya*** p. p.
- Female flowers, if present, not solitary; stamens usually 10; pistillode present, or absent and flowers hermaphroditic 22
22. Scandent trees or hemi-epiphytes; style 1; fruit with single persistent style **79. *Solenocarpus*** p. p. (=*S. philippinensis*)
- Trees; stylodia 3–5; fruit with more than one persistent stylodia or style 0 23
23. Perianth usually 3–4-merous; continental Africa **77. *Pseudospondias*** p. p.
- Perianth usually 5-merous; outside of continental Africa 24
24. Pistil(ode) sparsely to densely pubescent; endocarp cartilaginous to crustaceous (rarely bony, e.g., *Tapirira mexicana*) 25
- Pistil(ode) usually glabrous; endocarp bony 26
25. Flowers usually unisexual; ovary not lobed; fruits maturing dark purple to black; cotyledons with purple striations; Neotropics **81. *Tapirira***
- Flowers bisexual; ovary 5-lobed; fruits maturing yellow; cotyledons without purple striations; Malesia **70. *Koordersiodendron***
26. Corolla valvate; external opercula absent; tropical Asia **62. *Allospondias***
- Corolla imbricate; external opercula present; Neotropics and Indian Ocean islands 27
27. Anther connective usually glandular; Neotropics **65. *Cyrtocarpa***
- Anther connective eglandular; Indian Ocean islands 28
28. Opercula 1; embryo horseshoe-shaped **72. *Operculicarya*** p. p.
- Opercula 2–5; embryo curved **75. *Poupartia***

I. Subfam. Anacardioideae Link (1831)

1. *Abrahamia* Randrianasolo & P.P. Lowry II

Abrahamia Randrianasolo & P.P. Lowry II (ined.).
Protorhus Engl. (1881), p. p.

Dioecious shrubs or trees with clear to milky white to red resinous exudate. Leaves alternate, opposite, or subopposite, simple, petiolate, entire; secondary veins parallel, usually evenly spaced. Inflorescences terminal and/or axillary panicles. Flowers pedicellate, articulate; perianth (4–)5(–6)-parted; sepals minute, connate; petals imbricate; androecium haplostemonous; filaments subulate; anthers dorsifixed; pistillode present or 0 in male flowers; staminodes reduced; disk glabrous, annular to cup-shaped with crenulate margin; carpels 3; style short, 3-branched; stigmas capitate; ovules 3, subapical, only 1 fertile. Drupe ellipsoid to oblong, 3-locular or unilocular; exocarp often beige, longitudinally striated; mesocarp fleshy; endocarp chartaceous to cartilaginous. Seed cotyledons ruminate, usually connate.

Nineteen species endemic to Madagascar.

2. *Actinocheita* F.A. Barkley

Actinocheita F.A. Barkley, Ann. Missouri Bot. Gard. 24: 2 (1937).
Rhus L. (1753), p. p.

Polygamodioecious shrubs or trees. Leaves deciduous, alternate, imparipinnate, petiolate; leaflets sessile, densely pubescent. Inflorescences axillary panicles. Perianth 5-parted, densely pubescent; petals imbricate; androecium haplostemonous; anthers ovoid; filaments thickened and longer than anthers, pilose; pistillode reduced; staminodes reduced; disk modified into a gynophore; carpels 3, densely pubescent, trichomes often surpassing stylodia; stylodia 3, short; stigmas capitate; ovule basal. Drupe probably wind-dispersed, globose, 1-locular; exocarp reddish brown, covered with very long-villous, violet-reddish trichomes.

A single species, *A. filicina*, endemic to south central Mexico.

3. ***Amphipterygium*** Schiede ex Standl.

Amphipterygium Schiede ex Standl., Contr. U.S. Natl. Herb. 23: 672 (1923); X.M.C. Figueroa, Ibugana Bol. IBUG 13(1): 27–47 (2005), rev.
Hypopterygium Schltdl. (1843).
Juliania Schltdl. (1843).

Dioecious shrubs or trees with milky exudate. Leaves deciduous, alternate, imparipinnate, trifoliolate, or rarely unifoliolate; leaflets opposite. Inflorescences axillary; male flowers arranged in pendent thyrses, female flowers tightly arranged and subtended by an involucre. Flowers perigynous; calyx 6–8-parted in male flowers, 0 in female flowers; corolla 0; androecium haplostemonous; filaments short; pistillode and staminodes 0; disk 0; carpels 3; stylodia 3; stigmatic lobes 3–4 per stylodium, revolute; ovule basal. Fruit a samaroid syncarp, 1-locular; exocarp brown; endocarp bony; peduncle much-dilated with tapering margins.

Four to five species in western Mexico south to northwestern Costa Rica.

Together with *Orthopterygium*, this genus is often segregated into the family Julianiaceae, but morphological and molecular data place it well within Anacardiaceae.

4. ***Anacardium*** L. Fig. 3G

Anacardium L., Sp. Pl. 383 (1753); Mitchell, Brittonia 44: 331–338 (1992), rev.; Mitchell & Mori, Mem. N. Y. Bot. Gard. 42: 1–76 (1987), rev.
Acajou (Tournefort) Adans. (1763).
Rhinocarpus Bert. & Balb. ex Humb., Bonpl. & Kunth (1824).

Andromonoecious subshrubs or trees with contact dermatitis-causing exudate. Leaves generally evergreen, alternate, simple, sessile to petiolate, entire, chartaceous to coriaceous; domatia usually present in secondary vein axils abaxially. Inflorescences terminal and/or axillary thyrsoids. Flowers pedicellate, articulate; perianth 5-parted; calyx imbricate; corolla of reflexed petals; stamens (6–)8–10(–12), fertile stamens 1(–4), much larger and exserted; filaments basally connate into a staminal tube; pistillode very reduced; staminodes 0 or in bisexual and male flowers reduced; disk 0; gynoecium pseudomonomerous; style straight to sigmoid (sometimes approaching gynobasic), apical or lateral; stigma punctiform, obscure; ovule basal. Drupe reniform, 1-locular, subtended by hypocarp, sigmoid to pyriform (rarely 0), white, green, yellow, or red; pericarp woody with large rectangular cavities containing caustic oils; exocarp greenish to brown to black. Seed with reniform embryo. $n = 12, 29$.

Eleven or more species, Honduras south to Paraguay, Brazil and Bolivia. *Anacardium occidentale* is cultivated pantropically.

5. ***Androtium*** Stapf

Androtium Stapf in Hook., Icon. Pl., t. 2763 (1903).

Hermaphrodite trees. Leaves evergreen, alternate, spiral, simple, petiolate, entire and recurved; primary and secondary veins prominent on ad- and abaxial surfaces. Inflorescences axillary panicles, pubescent. Flowers pedicellate, articulate, perianth (4–)5-parted, imbricate; petals pink with white tips; androecium diplostemonous; filaments subulate; anthers basifixed; connective apically bilobed, prolonged and dilated; disk glabrous, cupular and crenulate; carpels 5, connate at the base, pilose, only 1 (larger) fertile; style obscure; stigma oblique, slightly lateral; ovules 5, basal, only 1 fertile. Drupe lentiform, apex depressed, 5-locular.

A single species, *A. astylum* in peninsular Malaysia and Borneo.

Androtium is very similar in its gynoecium morphology to *Buchanania*, which has recently been shown to be a member of Spondioideae. Future studies may reveal that *Androtium* should also be recognized in this subfamily.

6. ***Apterokarpos*** C.T. Rizzini

Apterokarpos C.T. Rizzini, Leandra 5(6): 40 (1975).
Loxopterygium Hook.f. (1862), p. p.

Dioecious shrubs or trees. Leaves deciduous, alternate, imparipinnate, petiolate; leaflets opposite to subopposite. Inflorescences terminal panicles. Flowers subsessile to pedicellate; perianth

5-parted, imbricate; androecium haplostemonous; anthers dorsifixed; pistillode 0; staminodes reduced; disk glabrous; carpels 3; style simple; stigma 3-lobed; ovule 1. Drupe dry, oblique, laterally compressed, obovoid to discoid with persistent calyx; 1-locular.

A single species, *A. gardneri*, endemic to the Caatinga of northeastern Brazil.

7. *Astronium* Jacq.

Astronium Jacq., Enum. Syst. Pl. 10 (1760); F.A. Barkley, Phytologia 16: 107–152 (1969), rev.

Dioecious trees with clear contact dermatitis-causing exudate; bark sometimes with exfoliating patches. Leaves deciduous, alternate, imparipinnate, petiolate; leaflets opposite (alternate), petiolulate; cataphylls sometimes present. Inflorescences terminal and/or axillary thyrsoids. Flowers pedicellate, articulate; perianth 5-parted, imbricate; calyx larger in female flowers; corolla greenish white or yellowish and turning pink with age; androecium haplostemonous; stamens alternating with petals and lobes of disk; pistillode 0; staminodes reduced; disk glabrous, 5-lobed, very thin; carpels 3; stylodia 3, recurved; stigmas capitate; ovule apical or basal. Fruit fusiform, glabrous baccate or nutlet-like, wind-dispersed by accrescent, chartaceous sepals, often with persistent stylodia, 1-locular; mesocarp resinous; endocarp thin, brittle when dry. Seed with straight or pyramidal embryo. $n = 15$.

Seven or more species in Mexico south to Paraguay and northern Argentina.

8. *Baronia* Baker

Baronia Baker, J. Bot. 20: 67 (1882).
Rhus sect. *Baronia* (Baker) H. Perrier (1944).

Dioecious or hermaphrodite trees, rarely shrubs, with clear to milky or red exudate. Leaves evergreen or deciduous, alternate to subopposite, simple, petiolate, entire. Inflorescences terminal and/or axillary panicles. Flowers articulate; perianth 5-parted, imbricate; corolla white, yellow, or green; stamens 5(–4); anthers dorsifixed; pistillode rudimentary; staminodes reduced; disk glabrous, cup-shaped, crenulate; ovary glabrous; carpels 3; stylodia 3, short, basally connate, apical to lateral; stigmas capitate; ovule lateral. Drupe widely depressed, ovoid or sigmoid, with persistent calyx; exocarp with longitudinal striations; endocarp chartaceous.

One to three species endemic to Madagascar.

This genus is often treated as a subgeneric taxon of *Searsia* (*Rhus* s.l.), but molecular phylogenetic data (Pell et al. 2008) show that the three species (*B. taratana*, *Rhus perrieri*, and *R. thouarsii*) are far removed from *Searsia* (and *Rhus* s.s.), and indicate that two of the species (*R. perrieri* and *R. thouarsii*), recognized by some in *Protorhus*, should perhaps be placed in a new segregate genus.

9. *Blepharocarya* F. Muell.

Blepharocarya F. Muell., Fragm. 11: 15 (1878); Jessup, Fl. Australia 25: 170–187 (1985); B.S. Wannan, J.T. Waterhouse & C.J. Quinn, Bot. J. Linn. Soc. 95: 61–72 (1987), rev.

Dioecious trees with contact dermatitis-causing exudate. Leaves opposite, paripinnate or imparipinnate; leaflets opposite or subopposite. Inflorescences terminal and/or axillary highly condensed thyrsoids. Male inflorescences bracteate glomerules usually consisting of three 7-flowered dichasial cymes; female flowers inside a cupule of partly coherent and connate, bracteate, cymose branches; perianth 4-parted, more or less sepaloid; calyx basally connate; corolla imbricate in male flowers; androecium diplostemonous; anthers dorsifixed, introrse; pistillode columnar, apically pilose; staminodes 0; disk pilose, raised-annular in female flowers, flat-annular in male; carpels 1; style lateral, excentric filiform, dilated at stigma; ovule pendulous, lateral. Infructescence a cupule opening by numerous branched valves; drupe dry, indehiscent, laterally compressed, subreniform, ciliate; 1-locular.

Two species, Australia, one in monsoon forests of the Northern Territory, and one in dry and rain forests on the east coast of northern Queensland.

10. *Bonetiella* Rzed.

Bonetiella Rzed., Ciencia (Mexico) 16: 139 (1957).

Polygamodioecious shrubs with contact dermatitis-causing exudate. Leaves deciduous, alternate, simple, petiolate, entire, linear to trifid or pinnatifid. Inflorescences axillary panicles. Flowers subsessile to shortly pedicellate; perianth 5-parted, imbricate; corolla greenish white; androecium haplostemonous; pistillode reduced; staminodes reduced; disk glabrous, 5-lobed; carpels 3; stylodia 3, short, subapical and unequal; stigmas 3; ovule basal. Drupe laterally compressed and subreniform, 1-locular; exocarp with numerous glands near remnant stylodia, yellowish; mesocarp thin, resinous; endocarp fibrous.

A single species, *B. anomala*, endemic to northern to central Mexico.

Morphological and molecular evidence suggests that *Bonetiella* is closely allied with *Pseudosmodingium*.

11. *Bouea* Meisn. Fig. 3D

Bouea Meisn., Pl. Vasc. Gen. 1: 75; 2: 55 (1837).

Andromonoecious trees with prominent terminal and axillary buds with bud scales. Leaves evergreen, opposite-decussate, simple, petiolate. Inflorescences axillary (rarely also terminal) panicles. Flowers pedicellate, non-articulate; perianth 3–5-parted; calyx valvate; corolla imbricate, white, greenish, or yellow, petals keeled along midrib; androecium haplostemonous; filaments subulate, glabrous; anthers basifixed with an apiculate connective; pistillode very reduced; staminodes 0; disk glabrous, small, flat or concave; gynoecium pseudomonomerous; style short; stigma rounded and flat, sometimes 2–3-grooved; ovule basal. Drupe subglobose to ellipsoid, 1-locular; exocarp yellow, orange, or red; mesocarp fleshy, edible; endocarp fibrous. Seed with straight embryo.

Three or more species in SE Asia south to western Malesia.

12. *Campnosperma* Thwaites

Campnosperma Thwaites in Hook., J. Bot. Kew Gard. Misc. 6: 65, t. 1 (1854), nom. cons.

Polygamodioecious trees with *Terminalia*-branching, often trunk buttressed or with stilt roots, and with contact dermatitis-causing exudate. Leaves evergreen, alternate, simple, sessile to petiolate (petioles sometimes with auriculate appendages), entire, coriaceous, peltate or lobed scales present ad- and abaxially; stellate trichomes sometimes present ad- and abaxially. Inflorescences axillary panicles. Perianth (3)4(5)-parted; corolla imbricate, white, greenish, yellow; stamens in two whorls of unequal length; anthers dorso-basifixed; pistillode very reduced; staminodes reduced; disk glabrous; round and flat in male flowers, cupular in female flowers; carpels 2; style short or obscure; stigma flattened, discoid, irregularly lobed; ovule pendulous, apical. Drupe subglobose or ovoid, incompletely 2-locular; exocarp generally red to black; endocarp woody. Seed cotyledons faintly plano-convex or flat; embryo curved.

Thirteen or more species: two from Honduras to northwest Ecuador and Amazonia; eleven or more in Madagascar (4+), the Seychelles (1), Sri Lanka (1), southeast Asia, Malesia, Micronesia, and Melanesia.

13. *Campylopetalum* Forman

Campylopetalum Forman, Kew Bull. 4: 555 (1954).

Dioecious herbaceous suffrutex. Leaves deciduous, opposite, decussate, simple, petiolate, palmately trilobed; domatia in secondary vein axils abaxially. Inflorescences terminal panicles, subtended by a pair of pinnately veined entire margined bracts in male plants and by palmately lobed bracts in female plants. Flowers pedicellate; perianth present in male flowers, 0 in female flowers, but female flowers subtended by a single bract; calyx apert, cup-shaped, with teeth-like lobes; corolla valvate, 4-parted; androecium diplostemonous; anthers with extrorse dehiscence, slightly pubescent; pistillode and staminodes 0; disk glabrous in male flowers, extremely reduced to 0 in female flowers; carpels 1; style filiform; stigma not differentiated; ovule pendulous, basal. Drupe subtended by an accrescent bract that aids in wind dispersal, 1-locular; exocarp brown; $n = 7$.

A single species, *C. siamense*, is endemic to northern Thailand.

Together with *Dobinea*, this genus is often placed in a separate family, Podoaceae, but molecular and morphological data suggest that it belongs in Anacardiaceae.

14. *Cardenasiodendron* F.A. Barkley

Cardenasiodendron F.A. Barkley, Lloydia 17: 242 (1954).
Loxopterygium Hook.f. (1862), p. p.

Dioecious trees with contact dermatitis-causing exudate. Leaves deciduous, alternate, imparipinnate, petiolate; leaflets opposite or subopposite, petiolulate; venation craspedodromous. Inflorescences terminal and/or axillary, compound panicles with ultimate branches spicate. Flowers sessile, subtended by three triangular bracts; perianth 5-parted, imbricate; androecium haplostemonous; stamens alternating with lobes of disk; pistillode minute; disk glabrous, 5-lobed; carpels 3; stylodia 3; stigmas 3; ovule basal. Samara obovate to subreniform with two unequal wings and persistent calyx, 1-locular. Seed obliquely curved.

A single species, *C. brachypterum*, endemic to Bolivia.

15. *Comocladia* P. Br.

Comocladia P. Br., Civ. Nat. Hist. Jamaica 124 (1756).

Polygamodioecious shrubs or trees, usually not branching, with contact dermatitis-causing exudate turning black with exposure to air. Leaves alternate, imparipinnate, petiolate; leaflets opposite, petiolulate, entire or toothed to spinose. Inflorescences axillary panicles; perianth 3–4-parted, imbricate; calyx light red; corolla red to purple; androecium haplostemonous; filaments subulate to filiform, inserted at notches in disk; anthers basifixed; pistillode very reduced; staminodes very reduced; disk glabrous, cup-shaped, slightly lobed; carpels 3; stylodia 0 or 3, short; stigmas 3; ovule pendulous, basal. Drupe oblong-ellipsoidal with persistent calyx, 1-locular; exocarp yellow, red, or black; mesocarp fleshy. Seed oblong; cotyledons fleshy.

Sixteen or more species widespread in central Mexico south to Guatemala, Belize, and the Greater and Lesser Antilles.

16. *Cotinus* Miller

Cotinus Miller, Gard. Dict. Abr. Ed., 4 (1754).
Rhus L. (1753), p. p.

Dioecious or gynodioecious (polygamodioecious or monoecious), shrubs or trees with contact dermatitis-causing exudate. Leaves deciduous, alternate, simple, petiolate, entire, oblong, secondary venation cladodromous. Inflorescences terminal panicles. Flowers pedicellate; perianth 5-parted, imbricate; corolla greenish or yellowish white; androecium haplostemonous (diplostemonous); filaments subulate; pistillode present; staminodes present or 0; disk glabrous; carpels 3; stylodia 3, lateral; stigmas 3; ovule pendulous, basal. Drupe obliquely ovoid, 1-locular, with persistent calyx; fruiting panicles wind-dispersed like a tumbleweed, aided by elongated plumose pedicels of numerous aborted flowers. Seed reniform. $n = 15$.

Four or more species: one in temperate, southern United States and northern Mexico; one in central to southern Europe, east to China; and two species in south-central China.

Two Mexican species, *C. chiangii* and *C. carranzae*, are problematic and need further taxonomic investigation. We recognize the former species in *Rhus* here, but it may represent a distinct evolutionary lineage apart from other former tribe Rhoeae members. An additional species, *C. kanaka*, was transferred from *Rhus* in *Flora of India*, but this affinity is somewhat uncertain.

17. *Dobinea* Buch.-Ham. ex D. Don

Dobinea Buch.-Ham. ex D. Don, Prod. Fl. Nep. 249 (1825).
Podoön Baill. (1887).

Dioecious shrubs. Leaves opposite or alternate, simple, petiolate, serrate. Inflorescences terminal, pyramidal panicles. Flowers pedicellate (female pedicels very short), non-articulate; perianth 4-parted in male flowers, 0 in female flowers but these flowers each subtended by a leafy bract; calyx cupular with tiny dentate lobes; corolla yellowish; androecium diplostemonous, glabrous; pistillode very reduced, style pilose; staminodes 0; disk 0 in male flowers, glabrous and annular in female flowers; carpels 1; style filiform, curved; stigma not differentiated; ovule pendulous, basal or sublateral. Drupe orbicular and compressed, peltate on an accrescent, obovate or rounded bract that aids in wind dispersal, 1-locular; bract and fruit with prominent venation; style persistent and curled; exocarp whitish to brown. $n=7$.

Two species in northern India, Nepal, Bhutan, and China.

Together with *Campylopetalum siamense*, this genus is often placed in a separate family, Podoaceae, but molecular and morphological data suggest that it belongs in Anacardiaceae.

18. *Drimycarpus* Hook. f.

Drimycarpus Hook.f. in Benth. & Hook.f., Gen. Pl. 1: 424 (1862).
Semecarpus L. f. (1781), p. p.
Holigarna Buch.-Ham. ex Roxb. (1820), p. p.

Polygamodioecious trees. Leaves evergreen, alternate, simple, petiolate, entire. Inflorescences terminal and/or axillary panicles or racemes. Flowers articulate, epigynous; perianth (4–) 5-parted, imbricate; corolla white, greenish, or yellow; androecium haplostemonous; anthers dorsifixed; pistillode reduced; staminodes reduced; disk glabrous; ovary inferior; carpels 3; style short; stigmas 3, capitate; ovule pendulous, basal to sublateral. Drupe transverse oblong, 1-locular; mesocarp resinous; endocarp coriaceous. Seed with straight embryo.

Three or more species in India to Myanmar, Indo-China to Borneo.

19. *Euroschinus* Hook. f.

Euroschinus Hook.f. in Benth. & Hook., Gen. Pl. 1: 422 (1862).

Polygamodioecious trees. Leaves evergreen, alternate, paripinnate (imparipinnate with terminal leaflet very reduced), petiolate; leaflets alternate, often falcate, with hairy tuft domatia sometimes in vein axils abaxially; cataphylls sometimes present. Inflorescences terminal and/or axillary thyrsiform panicles. Flowers sessile or short pedicellate, articulate; calyx apert, fused at base; corolla imbricate; androecium diplostemonous; connective glandular; pistillode reduced; staminodes reduced; disk glabrous, 5-lobed; carpels 3; style short; stigmas 3; ovule pendulous, subapical. Drupe obliquely ovoid, 1-locular; exocarp green, gray, blue, purple or blue-black; mesocarp fleshy.

Nine or more species: one in New Guinea, one in Australia, seven in New Caledonia.

20. *Faguetia* Marchand

Faguetia Marchand, Rév. Anacard.: 174 (1869).

Dioecious, or possibly monoecious, trees. Leaves evergreen, alternate, imparipinnate, petiolate; leaflets opposite, petiolulate, falcate. Inflorescences axillary, branched cymes. Flowers pedicellate; perianth 4(–5)-parted in male flowers, 4-parted in female flowers; calyx subvalvate; corolla imbricate, venation prominent; androecium haplostemonous; filaments distinct, alternating with petals; pistillode very reduced; staminodes reduced; disk glabrous, annular, 4–6-lobed; carpels (2)3; style 0; stigma 2–3-lobed, capitate; ovule lateral to subbasal. Samara large, dry, flattened with a terminal wing; exocarp (and possibly mesocarp) with copious oleoresins.

A single species, *F. falcata* Marchand, endemic to eastern Madagascar.

21. *Fegimanra* Pierre ex Engl. Fig. 3B

Fegimanra Pierre ex Engl. in Engl. & Prantl, Natürl. Pflanzenfam. III, 5: 458 (1896).

Andromonoecious or monoecious trees with contact dermatitis-causing exudate. Leaves evergreen, alternate, simple, petiolate, entire, subcoriaceous to coriaceous. Inflorescences terminal and/or axillary thyrsoids. Flowers short pedicellate; perianth 4-parted; calyx apert; corolla imbricate, forming a tube basally, apically spreading; stamen 1; filament subulate, villous; anther basi- or dorsifixed; pistillode very reduced; staminode may be present but only slightly reduced (fertility of stamen in female/bisexual flower is in doubt); disk 0: gynoecium pseudomonomerous or possibly truly monomerous; style apically decurved, lateral; stigma minutely bilobed or punctiform; ovule lateral. Drupe asymmetrically reniform and subtended by a small hypocarp.

Three species in Upper Guinean West Africa and Cameroon south to Gabon.

22. *Gluta* L. Fig. 3E

Gluta L., Mant. Pl. 2: 293 (1771).
Melanorrhoea Wall. (1829).

Hermaphrodite trees, rarely large shrubs with contact dermatitis-causing exudate. Leaves evergreen or deciduous, alternate, simple, sessile to petiolate, entire. Inflorescences axillary panicles. Flowers pedicellate, articulate or non-articulate; calyx calyptriform, circumscissile or bursting irregularly at anthesis, caducous; corolla imbricate and/or contorted, rarely valvate; petals (4)5 (8), white, often red or yellow at base and/or changing to pink; stamens 4, 5, 10 to 100+, sometimes inserted on conical torus; filaments glabrous or pubescent; anthers dorsifixed; disk

0; gynoecium sessile or supported by a gynophore atop the torus, obovoid, ellipsoid or subglobose, glabrous or pubescent, pseudomonomerous; style filiform, lateral to sublateral; stigma punctiform; ovule basal. Drupe smooth to wrinkled, sessile or stalked, globose to reniform, 1-locular; some species wind-dispersed by persistent, accrescent, wing-like petals; exocarp brown to purplish red to black. Seed cotyledons sometimes incompletely fused; embryo straight (rarely slightly curved).

Thirty or more species in Madagascar (1), India, Myanmar to Indo-China and Malesia.

23. *Haplorhus* Engl.

Haplorhus Engl., Bot. Jahrb. 1: 419 (1881).

Dioecious trees. Leaves evergreen, alternate, simple, sessile to very short petiolate, linear to lanceolate. Inflorescences axillary panicles. Flowers sessile; perianth 5-parted; epicalyx and red to purple tepals of female flowers imbricate; male flowers subtended by bracts, tepals pink, imbricate; androecium haplostemonous; anthers basifixed; pistillode 0; staminodes 0; disk glabrous; carpels 3; stylodia 3, short; stigmas capitate; ovule pendulous, basal. Drupe obliquely ovoid, 1-locular; exocarp red; mesocarp thin, fleshy; endocarp cartilaginous.

A single species, *H. peruviana*, endemic to dry inter-Andean valleys of central Peru south to northern Chile.

24. *Heeria* Meisn.

Heeria Meisn., Gen. Comm.: 55 (1837).
Anaphrenium E. Mey. ex Endl. (1841), p. p.

Dioecious shrubs or small trees with watery exudate. Leaves evergreen, alternate, simple, petiolate, entire, coriaceous, strongly discolorous with fine silvery trichomes and parallel venation prominent abaxially. Inflorescences terminal or axillary, spiciform panicles, with lignified, unicellular trichomes on subtending bracts, pedicels, and abaxial surface of calyx. Flowers pedicellate, articulate; perianth 5-parted; corolla imbricate, recurved at anthesis, cream-colored; androecium haplostemonous; pistillode reduced; staminodes reduced; disk glabrous, orange; lobed in male flowers, annular in female flowers; carpels 3; stylodia 3, excentric, distinct or basally connate; stigmas capitate; ovule basal to sublateral. Drupe globose, depressed, 1-locular; pericarp coriaceous and slightly shriveled; exocarp yellowish green, longitudinally rugose-striate. Seed large, recalcitrant.

A single species, *H. argentea*, endemic to South Africa.

25. *Holigarna* Buch.-Ham. ex Roxb. Fig. 4

Holigarna Buch.-Ham. ex Roxb., Hort. Beng.: 22 (1814).

Polygamodioecious or dioecious trees with contact dermatitis-causing exudate turning black with exposure to air. Leaves alternate, spiral, simple, petiolate (4 spur-like caducous or persistent appendages along petiole margins), entire. Inflorescences terminal and/or axillary panicles. Flowers pedicellate, epigynous, with a well-developed hypanthium; perianth 5-parted; calyx imbricate, cupular; corolla valvate; androecium

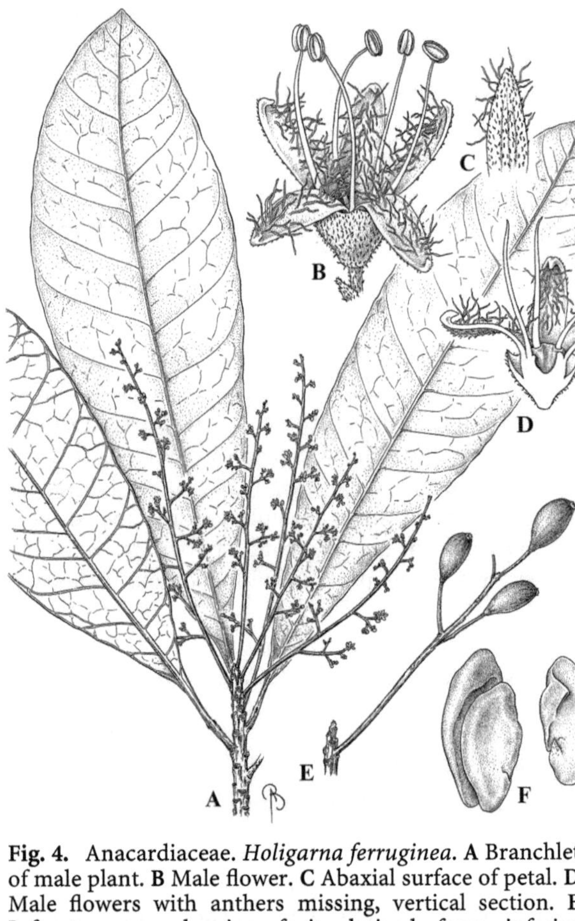

Fig. 4. Anacardiaceae. *Holigarna ferruginea*. **A** Branchlet of male plant. **B** Male flower. **C** Abaxial surface of petal. **D** Male flowers with anthers missing, vertical section. **E** Infructescence showing fruit derived from inferior ovary. **F** Embryo with large cotyledons. (Reproduced with permission of the artist Bobbi Angell)

haplostemonous; disk glabrous, obscure in bisexual flowers; carpels 3–5; ovary inferior; stylodia 3–5, apical; stigmas clavate; ovule pendulous, apical or lateral. Drupe sometimes enclosed within obconic and turbinate hypocarp, 1-locular; mesocarp contains an abundance of black resin.

Seven or more species in India, Bangladesh, Myanmar, and Indochina.

26. *Laurophyllus* Thunb.

Laurophyllus Thunb., Nov. Gen. Pl. 6: 104 (1792).
Botryceras Willd. (1860).

Dioecious shrubs. Leaves evergreen, alternate, simple, petiolate, elliptic-oblong, serrate. Inflorescences terminal and/or axillary, highly condensed panicles characterized by fusion of higher order branches into flattened broad segments. Flowers pedicellate; perianth 4–5-lobed almost to the base, imbricate; petals white, lanceolate in female flowers; androecium haplostemonous; filaments filiform; pistillode and staminodes 0; disk glabrous, fleshy and broad; carpels 1; style lateral or excentric; stigma bilobed or trilobed, capitate; ovule pendulous, apical or subapical. Samara hard, ovoid, laterally compressed with a marginal wing, 1-locular; pericarp with prominent venation.

A single species, *L. capensis*, endemic to wooded hillsides and stream banks of the eastern Cape of South Africa.

27. *Lithrea* Hook.

Lithrea Hook., Bot. Misc. 3: 175 (1833), sphalm., nom. cons.
Lithraea Miers ex Hook. & Arn. (1826), partim.; F.A. Barkley, Phytologia 8(7): 329–365 (1962), rev.
Schinus L. (1753, 1754), p. p.

Dioecious shrubs or trees with contact dermatitis-causing exudate. Leaves evergreen, alternate, imparipinnate or unifoliolate, petiolate; rachis often alate; leaflets sessile, entire; marginal secondary vein prominent. Inflorescences terminal and/or axillary panicles. Flowers pedicellate, articulate; perianth 5-parted; calyx apert to slightly imbricate, lobes minutely deltate; corolla valvate in bud, greenish-white to greenish-yellow; androecium diplostemonous; pistillode reduced; staminodes reduced; disk glabrous, 10-lobed; carpels 3; stylodia 3, basally connate, apical; stigmas capitate; ovule basal. Drupe globose, 1-locular; exocarp pale gray to whitish, smooth, brittle and easily separating from mesocarp at maturity; mesocarp resinous, attached to endocarp; endocarp bony. $n = 15$.

Three species in Brazil, Bolivia, Paraguay, Argentina, Uruguay, and Chile.

28. *Loxopterygium* Hook. f.

Loxopterygium Hook.f. in Benth. & Hook., Gen. Pl. 1: 419 (1862); F.A Barkley, Lloydia 25: 109–122 (1962), rev.

Polygamodioecious trees with contact dermatitis-causing exudate, clear or white and turning black with exposure to air. Leaves usually deciduous (*L. sagotii* may be evergreen), alternate, imparipinnate, petiolate; leaflets opposite or alternate, petiolulate, entire to crenate or serrate. Inflorescences axillary or rarely terminal thyrsoids. Flowers pedicellate; perianth 5-parted, imbricate; corolla yellowish green; androecium haplostemonous; pistillode reduced; staminodes reduced; disk glabrous, annular and 5-lobed; carpels 3; stylodia 3, unequal, lateral; stigmas capitate or discoid; ovule pendulous or short-funiculate, basal to lateral. Samara falcate, 1-locular; lateral wing chartaceous with prominent venation, stigmas persistent in fruit; endocarp bony. Seed with curved embryo. $n = 15$.

Three species with disjunct distributions from Venezuela south to Argentina, absent from Amazonia; *L. sagotii* in Venezuela and the Guianas; *L. huasango* in southwestern Ecuador to northwestern Peru; *L. grisebachii* in Bolivia south to northwestern Argentina.

29. *Loxostylis* Spreng. f. ex Rchb.

Loxostylis Spreng. f. ex Rchb., Ic. Exot. Cent. 3: 3, t. 205 (1827).

Dioecious trees with contact dermatitis-causing exudate. Leaves evergreen, alternate, imparipinnate, petiolate; rachis alate; leaflets opposite, sessile, entire. Inflorescences terminal panicles. Flowers pedicellate, non-articulate; perianth 4–5-parted, imbricate; calyx fused at base, initially green, turning pink to dark red in flower and fruit; corolla white to cream-colored, caducous; androecium haplostemonous; stamens unequal in length; pistillode 0; staminodes reduced; disk glabrous, with five 2-lobed projections;

ovary papillose; carpels 1–4; stylodia 1–4, lateral; stigmas capitate; ovule basal. Drupe ellipsoidal, resinous and flattened, subtended by enlarged, dark red calyx, 1-locular; exocarp brown, brittle and easily separating from mesocarp at maturity, rippled.

A single species, *L. alata* Spreng. f. ex Rchb., endemic to rocky outcroppings of eastern South Africa.

30. *Malosma* Nutt. ex Abrams

Malosma Nutt. ex Abrams, Fl. Los Angeles 3: 220 (1917).
Rhus subg. *Malosma* Nutt. ex Torr. & A. Gray (1838).
Rhus sect. *Venenatae* Engl. (1881), p. p.

Polygamodioecious shrubs or trees. Leaves evergreen, alternate, simple, petiolate, entire, longitudinally plicate. Inflorescences terminal thyrsoids. Flowers pedicellate; perianth 5-parted, imbricate; corolla whitish; androecium haplostemonous; pistillode reduced; staminodes reduced; disk glabrous; carpels 3; stylodia 3, short; stigmas 3; ovule basal. Drupe laterally compressed, glabrous, 1-locular; exocarp white; mesocarp thick, waxy; endocarp bony.

A single species, *M. laurina*, in chaparral of southern California south to central Baja California, Mexico.

Malosma is segregated from *Rhus* on the basis of having a white exocarp and lacking glandular trichomes.

31. *Mangifera* L. Fig. 3F

Mangifera L., Sp. Pl.: 200 (1753); A.J.G.H. Kostermans & J.M. Bompard, The mangoes: their botany, nomenclature, horticulture and utilization. Academic Press, San Diego, CA, (1993), rev.

Andromonoecious trees with contact dermatitis-causing exudate. Leaves evergreen, alternate, simple, petiolate, entire, lanceolate; apex acute to acuminate; cataphylls sometimes present. Inflorescences terminal and/or axillary thyrsoids. Flowers pedicellate, articulate; perianth 5-parted, imbricate; corolla white to pink, often with yellow or pink glandular ridges on adaxial surface; stamens 5(–10), only 1–2 (3–5 or all) fertile, the others much reduced; filaments distinct or connate at base; anthers dorsifixed; pistillode reduced; disk glabrous, extrastaminal (intrastaminal), papillose or glabrous, 4–5-lobed; gynoecium pseudomonomerous; style lateral, sometimes approaching gynobasic; stigma punctiform; ovule basal. Drupe subglobose, sigmoid, subreniform or ovoid, 1-locular; exocarp green, yellow, orange, red to brown or black; mesocarp fleshy; endocarp woody-fibrous. Seed oblong-ovoid, compressed; cotyledons usually plano-convex (sometimes lobed and the radicle ascendant), rarely the cotyledons and inner seed coat form a labyrinth; testa chartaceous; embryo curved. n=20 or 30, polyploidy common.

Sixty-nine species in tropical Asia and India east to Malesia and the Solomon Islands; greatest diversity in western Malesia. *Mangifera indica* is cultivated pantropically.

32. *Mauria* Kunth

Mauria Kunth, Ann. Sci. Nat. I, 2: 338 (1824).

Hermaphrodite, sometimes cleistogamous, or less frequently polygamodioecious shrubs or trees with contact dermatitis-causing exudate. Leaves evergreen or deciduous, alternate, simple, trifoliolate, or imparipinnate, petiolate; leaflets opposite, petiolulate, entire or weakly toothed; hairy tuft domatia sometimes present in secondary vein axils abaxially. Inflorescences terminal and/or axillary panicles or pleiothyrsoids; flowers pedicellate; perianth 5-parted; calyx short-cupulate; corolla valvate or subvalvate; androecium diplostemonous; stamens sometimes of unequal lengths; filaments subulate; anthers dorsifixed, connective extended slightly above anthers; disk glabrous, 10-crenulate; carpels 3; style short; stigma 3-lobed; ovule pendulous, lateral or subapical. Drupe laterally compressed, oblique, crowned by vestigial style, 1-locular; exocarp orange, red, to brown; mesocarp thin, fleshy; endocarp chartaceous. Seed with flattened embryo.

Ten to fifteen Andean and Central American species: El Salvador south to eastern Venezuela and extreme northern Argentina.

33. *Melanochyla* Hook. f.

Melanochyla Hook.f., Fl. Brit. Ind. 2: 38 (1876); Ding Hou, Fl. Males. I, 8: 490–499 (1978).

Dioecious trees with short buttresses or stilt roots and contact dermatitis-causing exudates turning black with exposure to air. Leaves alternate, simple, petiolate, entire, papillose abaxially. Inflorescences terminal and/or axillary panicles, rarely fascicles. Flowers pedicellate, articulate, perigynous; perianth 5-parted; corolla imbricate, campanulate, white, yellow, or dark brown; petals pubescent ad- and abaxially; androecium haplostemonous; filaments distinct or lower part adnate to petals, densely pubescent; anthers dorsifixed; pistillode present or 0 in male flowers, very reduced; staminodes reduced; disk glabrous, rim-like, 4–5-lobed; carpels 3; style short; stigmas 3; ovule pendulous, apical to subapical. Drupe ellipsoid, ovoid, subglobose, depressed globose, oblong; adnate to base of hypanthium, subtended by an accrescent calyx; 1-locular; exocarp yellow, covered with rusty-brownish trichomes; mesocarp and endocarp with black resin. Seed cotyledons distinct; embryo straight.

Thirty species in Thailand, Sumatra, Peninsular Malaysia, and Borneo.

34. *Melanococca* Blume

Melanococca Blume, Mus. Bot. Lugd. Bat. 1: 236 (1850).
Duckera F.A. Barkley (1942).
Rhus sect. *Melanocarpae* Engl. (1881).
Rhus subg. *Melanococca* (Blume) Brizicky (1963).

Dioecious trees with white exudate turning black with exposure to air. Leaves alternate, imparipinnate, petiolate; leaflets petiolulate, entire; tertiary and quaternary veins freely ramified (i.e., no areolation). Inflorescences terminal and/or axillary panicles, branches tomentose. Flowers short pedicellate; perianth 5-parted, imbricate; sepal margins pubescent; corolla cream-white, rarely pink; petals basally hairy adaxially; androecium haplostemonous; anthers dorsifixed; pistillode reduced; staminodes reduced; disk glabrous, 10-lobed; carpels 3; style obscure; stigmas 3, more or less united; ovule basal. Drupe subglobose, 1-locular; exocarp black, not separating from mesocarp when ripe; stigmas persistent in fruit.

A single species, *M. tomentosa* (=*Rhus taitensis*), in Australia (Queensland), Philippines, and Java east to Tahiti.

Melanococca is segregated from *Rhus* on the basis of having a black exocarp that does not separate from the mesocarp at maturity.

35. *Metopium* P. Br.

Metopium P. Br., Civ. Nat. Hist. Jamaica: 177 (1756); F.A. Barkley, Ann. Missouri Bot. Gard. 24: 265–499 (1937), rev.
Rhus L. (1753), p. p.

Dioecious trees or shrubs with prominent resin ducts and contact dermatitis-causing exudate turning black with exposure to air. Leaves evergreen, imparipinnate, petiolate; leaflets petiolulate, entire; mature leaflets often speckled with black spots. Inflorescences axillary panicles, lax. Flowers pedicellate, non-articulate; perianth 5-parted, imbricate; calyx fused; corolla yellow-green with dark veins; androecium haplostemonous; anthers basi- or dorsifixed; pistillode reduced; staminodes reduced; disk glabrous, 5-lobed; carpels 3; style short; stigma 3-lobed; ovule pendulous, basal. Drupe ellipsoidal to obovoid, 1-locular; exocarp orange to brown, glabrous. Seed compressed, somewhat quadrangular; funicle expanded, covering one margin; embryo oriented vertically but with a curved radicle.

Three species in West Indies, southern Florida (US), Mexico, and northern Central America.

36. *Micronychia* Oliv.

Micronychia Oliv. in Hook., Icon. Pl. 14: 27, t. 1337 (1881); Randrianasolo, Adansonia III, 22: 145–155 (2000), rev.; Randrianasolo & Lowry II, Adansonia III, 31: 157–168 (2009), key.

Dioecious trees with clear to milky exudate. Leaves evergreen, alternate to subopposite, simple, sessile or petiolate, entire. Inflorescences terminal and/or axillary panicles, often pendulous. Perianth 5-parted; calyx very small, sepals more or less basally connate; corolla imbricate, white, cream-colored, yellow, pink, or red; androecium haplostemonous; filaments straight to sigmoid; anthers dorsifixed; pistillode reduced; staminodes reduced; disk glabrous, annular to cup-shaped; carpels 3; style 3-branched, lateral; stigmas capitate; ovule apical. Drupe small, sigmoid, 1-locular; exocarp with longitudinal striations visible when dry; mesocarp fleshy.

Ten species endemic to Madagascar.

37. *Mosquitoxylum* Krug & Urb.

Mosquitoxylum Krug & Urb., Notizbl. Königl. Bot. Gart. Berlin 1: 78 (1895); F.A. Barkley & M.J. Reed, Am. Midl. Nat. 24: 666–679 (1940), rev.

Polygamodioecious (androdioecious) trees. Leaves evergreen (more or less), alternate, imparipinnate; leaflets opposite or subopposite, short-petiolulate, entire. Inflorescences terminal and/or axillary panicles with spicate branches. Flowers sessile or short-pedicellate, each subtended by 3 deltoid bracts; perianth 5-parted, imbricate; corolla greenish white, or cream-colored; androecium haplostemonous; pistillode reduced; staminodes in female flower very reduced; disk glabrous, 5-lobed; carpels 3; style short, distally 3-branched, excentric; stigmas 3; ovule sublateral. Drupe obliquely ovoid, compressed, 1-locular; exocarp red, glabrous.

A single species, *M. jamaicense*, southern Mexico south to northwestern Ecuador and Jamaica.

Morphological and molecular evidence suggests that *Mosquitoxylum* is closely related to *Rhus*.

38. *Myracrodruon* Allem.

Myracrodruon Allem. in Trab. Comm. Sc. Expl. Ceará, Secc. Bot. 3, tt. 1, 2 (1862); D.A. Santin & H. de F. Leitão Filho, Revista Brasil. Bot. 14: 133–145 (1991), rev. *Astronium* Jacq. (1760), p. p.

Dioecious trees with contact dermatitis-causing exudate. Leaves deciduous, alternate, imparipinnate, petiolate; leaflets petiolulate, entire or serrate. Inflorescences terminal and/or axillary thyrsoids. Flowers pedicellate, articulate; perianth 5-parted, imbricate; corolla greenish-white or cream-colored to purplish; androecium haplostemonous; pistillode reduced; staminodes reduced; disk glabrous, 5-lobed in male flowers and annular in female flowers; carpels 3; stylodia 3, apical; stigmas capitate; ovule lateral or basal. Drupe subglobose, subtended by persistent stiffened and expanded sepals, 1-locular; exocarp brown.

Two or more species in sub-Amazonian Brazil, Bolivia, Paraguay, and northern Argentina.

This genus is segregated from *Astronium* by its ovule position and fruit morphology.

39. *Nothopegia* Blume

Nothopegia Blume, Mus. Bot. Lugd. Bat. 1: 203 (1850).
Glycycarpus Dalz. (1849).

Dioecious (androdioecious) trees with contact dermatitis-causing, milky exudate turning black with exposure to air. Leaves alternate to subopposite, simple, petiolate, entire, lanceolate to elliptic. Inflorescences axillary or pseudoterminal racemes. Flowers short pedicellate; perianth 4-parted, imbricate; corolla of distinct, twisted white petals; androecium haplostemonous; filaments pubescent to nearly glabrous; anthers pubescent; pistillode very reduced; staminodes reduced; disk glabrous, 4–5-lobed; carpels 3; style apical; stigma 1–3-lobed; ovule pendulous, apical to lateral. Drupe depressed globose or pyriform with persistent calyx and style, 1-locular; exocarp red to dark purple, sometimes longitudinally striate; mesocarp fleshy.

Ten or more species in India and Sri Lanka.

40. *Ochoterenaea* F.A. Barkley

Ochoterenaea F.A. Barkley, Bull. Torrey Bot. Club 69: 442 (1942).

Dioecious trees with milky exudate. Leaves alternate, imparipinnate, petiolate; leaflets opposite, sessile to very short petiolulate, entire, lanceolate, membranaceous, puberulent ab- and adaxially. Inflorescences terminal corymbose thyrsoids. Flowers pedicellate; perianth 5-parted, valvate; androecium haplostemonous; pistillode reduced, stylode simple; staminodes reduced; disk glabrous, annular; carpels 3, 1 fertile; stylodia 3, basally connate, lateral; stigmas capitate; ovule basal. Samara laterally compressed with long, violet trichomes on the margins, 1-locular by abortion.

A single species, *O. colombiana*, in Panama, Andean Venezuela and Colombia south to Bolivia.

Without nomenclatural conservation, the species name may change if *Rhus samo* is shown to be an earlier basionym as expected.

41. *Orthopterygium* Hemsl.

Orthopterygium Hemsl. in Phil. Trans. R. Soc. London B, 199: 190 (1907).

Dioecious shrubs or trees with milky exudate. Leaves deciduous, alternate, imparipinnate, petiolate; leaflets opposite, sessile to very short petiolulate, crenate. Inflorescences terminal; male flowers arranged in pendent or erect panicles, female flowers tightly arranged in 3-flowered clusters subtended by an involucre (2 flowers abort); male flowers pedicellate, females sessile; calyx 3–8-parted in male flowers, 0 in female flowers; corolla 0; androecium haplostemonous; pistillode and staminodes 0; disk 0; carpels 3; stylodia slightly basally connate, apical; stigmas 3; ovule pendulous, basal. Syncarp samaroid, with little-dilated peduncle with parallel margins, 1-locular; exocarp brown.

A single species, *O. huaucui*, endemic to western Peru.

Together with *Amphipterygium*, this genus is often segregated into the family Julianiaceae, but morphological and molecular data place it well within Anacardiaceae.

42. ***Ozoroa*** Delile

Ozoroa Delile in Ann. Sci. Nat. II, 20: 91, t. 1 (1843); R.B. Fernandes, Garcia de Orta (Lisboa) 14: 19–60 (1966), rev.
Heeria Meissn. (1837), p. p.
Anaphrenium E. Mey. ex Endl. (1841), p. p.

Dioecious subshrubs, shrubs, or trees, with milky exudate. Leaves evergreen, alternate, opposite, or whorled, simple, petiolate, entire. Inflorescences terminal and/or axillary panicles. Flowers pedicellate, articulate; perianth 5-parted, imbricate; corolla patent or inflexed at the apex, white, yellow, or reddish, often pubescent; androecium haplostemonous; filaments subulate; anthers dorsifixed; pistillode very reduced; staminodes reduced; disk glabrous, crenulate; carpels 3; stylodia 3, slightly connate at base or distinct; stigmas capitate; ovule pendulous, lateral or basal. Drupe transversely reniform, subglobose or globose, 1-locular; exocarp black shiny or red turning black; mesocarp fleshy; endocarp bony. Seed obovate, compressed.

Forty or more species in sub-Saharan Africa and Yemen.

43. ***Pachycormus*** Coville

Pachycormus Coville in Cent. Dict., rev. ed., 6708 (1911).
Rhus L. (1753), p. p.
Bursera Jacq. ex L. (1762), p. p.

Dioecious trees with short trunk and crooked branches, caudiciform; outer white to yellow bark exfoliating, revealing green inner bark, with milky exudate drying clear. Leaves deciduous, alternate, imparipinnate, petiolate; leaflets opposite to subopposite, sessile to short-petiolulate, entire to irregularly lobed, elliptic. Inflorescences axillary panicles. Flowers pedicellate; perianth 5-parted, white to dark pink; calyx subvalvate; corolla exduplicate-valvate; androecium diplostemonous; pistillode very reduced; staminodes reduced; disk present but not well known, may be alternating with stamens; carpels 3; stylodia 3; stigmas capitate. Utricle pubescent; 1-locular.

A single species, *P. discolor*, endemic to central Baja California, Mexico.

44. ***Parishia*** Hook. f.

Parishia Hook.f., Trans. Linn. Soc. 23: 169, t. 26 (1860); Ding Hou, Flora Males. I, 8: 541–545 (1978).

Dioecious trees with contact dermatitis-causing, white to brown exudate turning black with exposure to air. Leaves deciduous, alternate, imparipinnate, petiolate; leaflets opposite to subopposite, petiolulate, entire. Inflorescences terminal and/or axillary panicles. Flowers pedicellate, articulate; perianth 4-parted, imbricate; androecium haplostemonous; filaments filiform, thin, glabrous; anthers dorsi- to basifixed; pistillode very reduced; staminodes reduced; disk pubescent; carpels 3; stylodia 3(4), very short; stigmas 3(4); ovule 1. Drupe wind-dispersed by persistent, enlarged calyx with 4 wing-like lobes, 1-locular; exocarp yellowish to reddish to brown, covered with dense brown trichomes; endocarp cartilaginous. Seed with straight embryo.

Five species in Myanmar, Thailand, and western Malesia.

45. ***Pentaspadon*** Hook. f.

Pentaspadon Hook.f., Trans. Linn. Soc. 23: 168, t. 24 (1860); Ding Hou, Flora Males. I, 8: 520–524 (1978).
Microstemon Engl. (1881).

Hermaphrodite trees with contact dermatitis-causing exudate turning black with exposure to air. Leaves deciduous, alternate, imparipinnate, petiolate; leaflets opposite to subopposite, sessile or petiolulate, entire, with hairy tuft domatia in secondary vein axils abaxially. Inflorescences axillary panicles. Flowers very short pedicellate, articulate; perianth 5-parted, imbricate; corolla white to pink; androecium diplostemonous or haplostemonous, sometimes (Malesian taxa) in two series with a whorl of stamens alternating with a whorl of staminodes, staminodes with or without antherodia; filaments subulate, papillose; anthers basi- or dorsifixed; disk glabrous,

10-lobed; carpels 1; style short; stigma subglobose or slightly 2-lobed; ovule pendulous, lateral. Drupe elongate ovoid, 1-locular; exocarp tan or purple turning black; mesocarp resinous; endocarp thin, seed oily.

Six species in Southeast Asia, Malesia, and the Solomon Islands.

This genus is of somewhat uncertain subfamilial affiliation, and may actually belong in subfamily Spondioideae.

46. *Pistacia* L.

Pistacia L., Sp. Pl.: 1025 (1753).
Lentiscus (Tourn.) L. (1735).
Terebinthus P. Br. (1735).

Dioecious shrubs or trees with clear exudate. Leaves evergreen or deciduous, alternate, paripinnate and/or imparipinnate, rarely simple or trifoliolate, petiolate; rachis sometimes alate; leaflets opposite or subopposite, petiolulate; cataphylls sometimes present. Inflorescences axillary thyrsoids, panicles, racemes, or spikes; perianth 0, but flowers surrounded by 1–3 small bracts and (1)2–7 tepal-like bracteoles; androecium: 3–5(–8) stamens; filaments short; anthers basifixed; pistillode reduced or 0; staminodes reduced or 0; disk glabrous, often reduced to a patch or 0; carpels (2)3; style short, 3-branched, apical; stigmas (2)3, bilobed or simple, recurved; ovule pendulous from a basal funicle. Drupe globose or ovoid, 1-locular; exocarp chartaceous, red to purplish; mesocarp thin; endocarp woody. $n = 12$, 14, 15.

Twelve species in Texas, USA, south to Nicaragua; Mediterranean Europe, and North and East Africa; Southwest and Central Asia (former Soviet Republics) east to Afghanistan and temperate central and southern China, south to peninsular Malaysia and the Philippines. *Pistacia vera* is cultivated worldwide in dry, warm climates.

47. *Protorhus* Engl.

Protorhus Engl., Bot. Jahrb. 1: 420 (1881).
Anaphrenium E. Mey. ex Endl. (1841), p. p.

Polygamodioecious shrubs or trees with milky, possibly contact dermatitis-causing, exudate. Leaves evergreen, opposite to subopposite, simple, petiolate, linear oblong or narrowly elliptic; young leaves bright orange. Inflorescences terminal and/or axillary panicles. Flowers pedicellate, non-articulate; perianth 5-parted, imbricate; corolla greenish white to yellow; androecium haplostemonous; pistillode reduced; staminodes reduced, pink; disk glabrous, urceolate in males, saucer-shaped and orange in females; carpels 1–3, only 1 fertile; stylodia 3, basally connate or distinct; stigmas capitate; ovules 1–3, pendulous, subapical, only 1 fertile. Drupe reniform, asymmetrical, 1-locular; exocarp purple to black, slightly ridged at maturity, wrinkled when dry; mesocarp fleshy, resinous; endocarp chartaceous to cartilaginous. Seed cotyledons easily separable; embryo oblong.

A single species, *P. longifolia*, endemic to eastern South Africa.

48. *Pseudosmodingium* Engl.

Pseudosmodingium Engl., Bot. Jahrb. 1: 419 (1881); F.A. Barkley & M.J. Reed, Am. Midl. Nat. 24: 666–679 (1940), rev.; C.J. Aguilar-Ortigoza & V. Sosa, Rhodora 106(928): 348–359 (2004), rev.

Dioecious or polygamodioecious trees with contact dermatitis-causing exudate. Leaves deciduous, alternate, imparipinnate, petiolate; leaflets opposite or alternate, sessile or petiolulate, entire to serrate. Inflorescences axillary panicles. Flowers pedicellate, non-articulate; perianth 5-parted, imbricate; corolla white and prominently veined; androecium haplostemonous; filaments filiform; pistillode extremely reduced; staminodes slightly reduced; disk glabrous; carpels 3; style 3-branched; stigmas 3; ovule pendulous, apical. Samara with two broad lateral wings, reniform in outline, 1-locular; exocarp yellow to brown or reddish brown, glabrous. Seed reniform; cotyledons slender.

Four species endemic to central and southern Mexico.

49. *Rhodosphaera* Engl.

Rhodosphaera Engl., Bot. Jahrb. Syst. 1: 423 (1881).

Polygamodioecious trees with bark shedding in scaly patches; exudate white, gummy. Leaves alternate, imparipinnate (paripinnate), petiolate; leaflets opposite to subopposite, petiolulate, entire, oblong or elliptic, base asymmetrical; hairy tuft domatia sometimes present in secondary vein axils abaxially. Inflorescences terminal and/or axillary panicles, pyramidal, dense. Flowers pedicellate; perianth 5-parted, imbricate; corolla dark pink to red; androecium diplostemonous; anthers dorsifixed; pistillode present or 0, reduced; staminodes reduced; disk glabrous, cupular, 10-lobed; carpels 3; stylodia 3, distinct, short, apical; stigmas capitate; ovule pendulous, basal. Drupe 1-locular; exocarp shiny brown; mesocarp woody; endocarp bony. Seed cotyledons flat.

A single species, *R. rhodanthema*, endemic to Queensland and northern New South Wales, Australia.

50. *Rhus* L.

Rhus L., Sp. Pl. 1: 265 (1753); F.A. Barkley, Ann. Missouri Bot. Gard. 24: 265–499 (1937), rev.; D. Young, Systematics of *Rhus* subg. *Lobadium* sect. *Styphonia*. Ph.D. dissertation, Claremont Graduate School, Claremont, CA, (1975); Ding Hou, Flora Males. I, 8: 534–541 (1978).
Rhus subg. *Rhus* L. (1754).
R. sect. *Sumac* DC. (1825), p. p.
R. subg. *Lobadium* (Raf.) Torr. & A. Gray (1838), p. p.
R. sect. *Trichocarpae* Engl. (1881), p. p.
R. subg. *Schmaltzia* (Desvaux ex de Candolle) C.K. Schneider (1907).
Schmaltzia Desv. ex Small emend. F.A. Barkley & Reed (1940).
Excluding: *R.* sect. *Baronia* H. Perrier (1944); *R.* sect. *Gerontogeae* Engl. (1881); *R.* sect. *Melanocarpae* Engl. (1881); *R.* sect. *Venenatae* Engl. (1881); *R.* subg. *Melanococca* (Blume) Brizicky (1963); *R.* subg. *Thezera* (DC.) K. Koch (1853).

Polygamodioecious or rarely hermaphrodite shrubs or trees, rarely sarmentose, sometimes with dichotomous branching. Leaves evergreen or deciduous, alternate, imparipinnate, trifoliolate, or unifoliolate, petiolate; rachis sometimes alate; leaflets opposite or subopposite, subsessile to petiolulate. Inflorescences terminal and/or axillary thyrsoids or compound spikes. Flowers sessile or short pedicellate; perianth 5-parted, imbricate; androecium haplostemonous; filaments subulate; anthers dorsifixed; pistillode reduced; staminodes reduced; disk glabrous; carpels 3, only 1 fertile; stylodia 1–3, united at base; stigmas 3; ovule basal (pendulous, apical or lateral). Drupe globose, 1-locular; exocarp red to brown, usually with both glandular and non-glandular trichomes. $n = 15$ or 16, polyploidy is common.

Thirty-five or more species: twenty-seven or more from southern Canada south to Panama and Cuba; one in North Africa to Mediterranean Europe east to Asia where six or more species are found east to China, Japan and Korea, south to Java and the Philippines; one endemic to the Hawaiian islands.

In much need of taxonomic revision. One species recognized here within *Rhus*, *R. chiangii*, may be a distinct genus from Mexico. Infrageneric classification includes subgenera *R.* subg. *Rhus* with pedicellate flowers and thyrsoid inflorescences, and *R.* subg. *Lobadium* with sessile or subsessile flowers and pseudospicate inflorescences. Much taxonomic work remains to be done within *Rhus*, particularly in Mexico and Asia. See Yi et al. (2004, 2007) for recent molecular and biogeographical assessments of *Rhus*. Useful taxonomic revisions of *Rhus* are cited above, but many treatments include taxa that should be or have since been segregated.

51. *Schinopsis* Engl.

Schinopsis Engl. in Mart., Fl. Brasil. 12(2): 403 (1876); T. Meyer & F.A. Barkley, Lilloa 33(11): 207–257 (1973), rev.
Quebrachia Griseb. (1874).
Loxopterygium Hook.f., p. p.

Dioecious or monoecious trees, sometimes with thorns, and with contact dermatitis-causing exudate turning black with exposure to air. Leaves usually evergreen, alternate, imparipinnate or unifoliolate (rarely both on the same plant) or rarely trifoliolate or paripinnate, petiolate; rachis sometimes alate; leaflets opposite, sessile or petiolulate, entire. Inflorescences terminal and/or axillary panicles. Flowers sessile or pedicellate, articulate or non-articulate; perianth 5-parted, imbricate; corolla greenish to white; petals with a prominent midvein; androecium haplostemonous; anthers dorsifixed; pistillode very reduced; staminodes reduced; disk glabrous, 5-lobed; carpels 3; stylodia 0 or 3, lateral; stigmas 3 or solitary and sessile; ovule pendulous, subapical. Samara 1-locular; exocarp and mesocarp expanded into a flattened lateral wing, green or red to brown; endocarp bony; calyx persistent. $n=14$.

Seven species in dry forests of northern Peru, and sub-Amazonian and eastern Brazil south to central Argentina. Often the dominant canopy tree in Chaco forests of Bolivia, Paraguay, and northern Argentina.

52. *Schinus* L.

Schinus L., Sp. Pl.: 388 (1753); F.A. Barkley, Brittonia 5: 160–198 (1944), and in Lilloa 28: 5–110 (1957), rev.
Duvaua Kunth (1824), p. p.

Dioecious shrubs or trees, rarely subshrubs, rarely with thorns, and with contact dermatitis-causing exudate. Leaves evergreen or deciduous, alternate, unifoliolate or imparipinnate (paripinnate), petiolate; rachis often alate; leaflets opposite or alternate, sessile to subsessile. Inflorescences terminal and/or axillary, spike-like pseudoracemes, panicles, or pleiothyrses, rarely reduced to a few fascicles. Flowers pedicellate, articulate; perianth (4–)5-parted, imbricate; androecium diplostemonous, in two whorls distinctly unequal in length; filaments subulate; pistillode very reduced; staminodes reduced; disk glabrous, 8–10-lobed, patelliform in male flowers, disk-shaped in female flowers; carpels 3; stylodia (1–)3; stigmas capitate; ovule pendulous, lateral to apical. Drupe small, globose, 1-locular; exocarp light purple to dark red, sometimes densely pubescent, thin at maturity, separating from rest of pericarp; mesocarp resinous, fleshy, adhering to the bony endocarp. Seed compressed. n=14, 15.

Thirty or more species from Ecuador south to Patagonia but excluding Amazonia. Two species, *S. molle* and *S. terebinthifolia*, are cultivated throughout the tropical, subtropical, and warm temperate regions of the world.

Barkley (l.c. 1944, see also l.c. 1957) recognized two subgenera: *S.* subg. *Duvaua* with unifoliolate leaves, often thorny, and subg. *Schinus* (as subg. *Euschinus*) with compound leaves and lacking thorns.

53. *Searsia* F.A. Barkley

Searsia F.A. Barkley, Am. Midl. Nat. 28: 472 (1942); R.O. Moffett, Bothalia 37: 165–175 (2007), rev.
Terminthia Bernh. (1838).
Rhus sect. *Sumac* DC. (1825), p. p.
Rhus subg. *Thezera* (DC.) K. Koch (1853); R.O. Moffett, Fl. South Africa 19, 3 (1993), rev.
Rhus sect. *Gerontogeae* Engl. (1881).

Monoecious (polygamous) subshrubs, shrubs, or trees, stems conspicuously lenticellate, thorns sometimes present. Leaves evergreen or deciduous, alternate, imparipinnate, usually trifoliolate; leaflets opposite or subopposite, sessile or petiolulate. Inflorescences terminal and/or axillary (one sp. cauliflorous) panicles or racemes. Flowers subsessile to pedicellate, non-articulate; perianth (4)5(6)-parted, imbricate; calyx greenish to red; corolla greenish yellow to whitish or red; petals generally twice as long as sepals; androecium haplostemonous; anthers dorsifixed; filaments subulate; pistillode usually 0, rarely present in male flowers, reduced; staminodes reduced; disk glabrous, 5(–10)-crenulate, cup-shaped; carpels 3; stylodia 3 (very rarely 4), distinct or connate basally, apical, erect or patent; stigmas capitate; ovule 1(–3), basal, pendulous. Drupe globose, ovoid, or laterally compressed, 1(–3)-locular; exocarp shiny light brown to black or red; mesocarp resinous and adhering to endocarp; endocarp bony; stylodia persistent or caducous. Seed ovoid or reniform, compressed. n=14, 15, 16.

One hundred-twenty or more species, in Sicily, the Middle East, Arabian Peninsula, Socotra, Africa, India, Nepal, Bhutan, Myanmar, and China. *S. lancea* is widely cultivated in warm regions.

This genus was formerly recognized within *Rhus* but was segregated by Barkley (1942). Moffett (2007) published most of the necessary new combinations in *Searsia*. It is separated from *Rhus* by a combination of having ternate leaves and a mesocarp adherent to the endocarp at maturity.

54. *Semecarpus* L. f.

Semecarpus L. f., Suppl. Pl.: 285 (1781).
Oncocarpus A. Gray (1854).

Dioecious (polygamous) shrubs or trees, rarely monopodial, with contact dermatitis-causing, milky exudate turning black with exposure to air. Leaves evergreen or deciduous, alternate, simple, petiolate, entire, coriaceous; leaf shape and size vary widely. Inflorescences terminal and/or axillary (cauliflorous) panicles. Flowers sessile or pedicellate, articulate; perianth (4) 5-parted; calyx imbricate (rarely valvate); corolla imbricate, rarely valvate; androecium haplostemonous; filaments subulate and glabrous; anthers dorsifixed; pistillode present or 0 in male flowers; staminodes reduced; disk glabrous to densely pubescent; ovary sometimes seemingly semi-inferior due to envelopment by hypocarp; carpels

3; stylodia 3, basally adnate, apical; stigmas variously shaped; ovule apical. Drupe subglobose, ovoid to oblong (rarely transversely elongate), immersed in a cupular, obconical, or disk-shaped hypocarp, 1-locular; exocarp yellow, gray, or brown; mesocarp fleshy, resinous; endocarp crustaceous; hypocarp yellow to red. Seed with straight embryo. n=29, 30.

Seventy to seventy-five species in moist or dry forests, along the banks of fast-flowing rivers (rheophytes), or on cliffs, in India and Sri Lanka, Nepal, Bhutan, tropical and subtropical East Asia to Melanesia, Micronesia, and tropical Australia.

55. ***Smodingium*** E. Mey.

Smodingium E. Mey. in Drège, Zwei Pfl. Docum.: 222 (1843).

Dioecious subshrubs to small trees, sometimes semiscandent, with contact dermatitis-causing exudate turning black with exposure to air. Leaves alternate, trifoliolate, petiolate; leaflets petiolulate, lanceolate, sharply toothed. Inflorescences terminal and/or axillary thyrses. Flowers pedicellate, articulate; perianth 5-parted; corolla imbricate, reflexed at maturity, white to cream-colored; androecium haplostemonous; anthers dorsifixed; pistillode reduced; staminodes reduced; disk glabrous, orange, 10-crenulate; carpels 3; stylodia 3, basally adnate, recurved, lateral; stigmas capitate; ovule pendulous, lateral. Samara dry, laterally compressed with a marginal wing, suborbicular in outline, 1-locular; exocarp light brown; calyx and stylodia persistent. Seed reniform.

A single species, *S. argutum*, endemic to eastern South Africa.

56. ***Sorindeia*** Thou.

Sorindeia Thou., Gen. Nov. Madag.: 23 (1806); Breteler, Adansonia III, 25: 93–113 (2003), rev.
Dupuisia A. Rich. (1832).

Dioecious or possibly very rarely monoecious trees, usually small and slender, branched or monopodial, shrubs or lianas, with light-colored contact dermatitis-causing exudate. Leaves alternate, imparipinnate (unifoliolate), petiolate; leaflets opposite to alternate, petiolulate; in most species tertiary veins collect into an oblique vein that is directed admedially toward the secondary vein axil. Inflorescences terminal and/or axillary, or cauliflorous thyrsoids. Flowers pedicellate, articulate or non-articulate; perianth 5-parted; calyx fused in lower half or more; corolla valvate or rarely imbricate; color widely variable: white, yellow, greenish-yellow, pink, or red when open; red to purple or green in bud; androecium of 10–20 stamens; filaments subulate, inserted outside of and on the disk, shorter than anthers; anthers dorsifixed; pistillode 0; staminodes 5–10; disk glabrous, crenulate; carpels 3; style short; stigma capitate or 3-lobed; ovule pendulous, subapical or apical. Drupe ellipsoid or asymmetrically ovoid, 1-locular; exocarp yellow, orange, red to black-purple; mesocarp fleshy, thin; endocarp chartaceous or woody; stigma persistent. Seed ellipsoidal. $n = 16$.

Nine species in tropical Africa and Madagascar. The species number was reduced from 80+ by Breteler (l.c.).

57. ***Swintonia*** Griff. Fig. 3C

Swintonia Griff. in Proc. Linn. Soc. 1: 283 (1846).

Andromonoecious trees with contact dermatitis-causing exudate. Leaves evergreen, alternate, simple, petiolate, entire, papillose abaxially. Inflorescences terminal and/or axillary panicles. Flowers pedicellate, articulate or non-articulate; perianth 5-parted, imbricate; corolla white to light green or yellowish, turning red when enlarged (prior to drying in mature fruit); androecium haplostemonous; filaments filiform or subulate; anthers dorsifixed; pistillode very reduced, pubescent; staminodes 0; disk glabrous, extrastaminal, 5-lobed; gynoecium pseudomonomerous; style cylindrical, distinct; stigma capitate; ovule basal. Drupe wind-dispersed by persistent, enlarged, wing-like petals, 1-locular; endocarp coriaceous. Seed with straight embryo.

Twelve species, Andaman Islands, Myanmar east to Malesia (Sumatra, Malay Peninsula, Borneo, and Philippines).

58. ***Thyrsodium*** Salzm. ex Benth. Fig. 5

Thyrsodium Salzm. ex Benth., Hook., J. Bot. Kew Gard. Misc. 4: 17 (1852); Mitchell & Daly, Brittonia 45: 115–129 (1993), rev.
Garuga Roxb. (1814), p. p.
Kunthia Benth. & Hook. (1862), p. p.

Dioecious trees with milky exudate. Leaves evergreen, alternate to subopposite, imparipinnate, petiolate; leaflets opposite or alternate, petiolulate, entire. Inflorescences terminal and/or axillary thyrsoids. Flowers non-articulate, perigynous; perianth 5-parted; calyx valvate; corolla imbricate; white, greenish, yellow; androecium haplostemonous; stamens opposite the sepals; filaments very short; anthers sometimes pubescent; pistillode reduced; staminodes reduced; disk glabrous, adnate to the hypanthium, or 0; carpels 3; style simple or 2–3-branched, apical; stigmas 1–3 or 2–3-lobed; ovule lateral. Drupe globose, obovoid, oblong or ellipsoid, 1-locular; mesocarp fleshy; endocarp crustaceous. Seed with straight embryo.

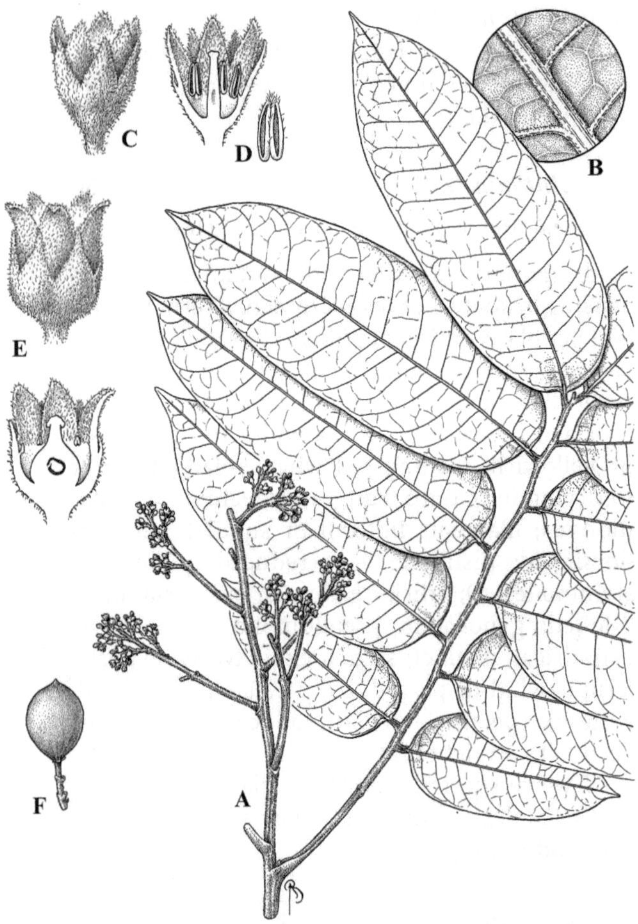

Fig. 5. Anacardiaceae. *Thyrsodium bolivianum*. **A** Flowering branchlet. **B** Abaxial leaflet surface. **C** External view and vertical section of male flower, showing cupular hypanthium. **D** Anther. **E** External view and vertical section of female flower. **F** Fruit. (Reproduced with permission of the artist Bobbi Angell)

Six to seven species east of the Andes in Colombia, Peru, Bolivia, southern and eastern Venezuela, the Guianas, and Amazonian and eastern Brazil.

59. ***Toxicodendron*** Mill.

Toxicodendron Mill., Gard. Dict. Abr. Ed., 4 (1754); Gillis, Rhodora 73: 72–159, 161–237, 370–443, 465–540 (1971), rev.
Rhus sect. *Sumac* DC. (1825), p. p.
Rhus subg. *Toxicodendron* (Mill.) K. Koch (1853); emend. Schneider (1907).
Rhus sect. *Trichocarpae* Engl. (1881), p. p.
Rhus sect. *Venenatae* Engl. (1881).

Polygamodioecious shrubs, trees, or lianas with contact dermatitis-causing, white exudate turning black with exposure to air. Leaves deciduous, rarely evergreen, alternate, imparipinnate, usually multifoliolate, often trifoliolate, very rarely unifoliolate, petiolate; leaflets opposite to subopposite, sessile or petiolulate, entire, serrate or lobed; hairy tuft domatia sometimes present in secondary vein axils abaxially. Inflorescences axillary panicles. Flowers pedicellate, non-articulate; perianth (4)5(6)-parted; calyx fused at base; corolla imbricate; white to greenish; androecium haplostemonous; anthers dorsifixed; pistillode reduced, stylodium 1; staminodes reduced; disk glabrous, annular and lobed; carpels 3; stylodia 3, short; stigmas capitate; ovule basal. Drupe globose, often laterally compressed, 1-locular; exocarp yellowish to white or pale gray, sometimes pubescent, separating from mesocarp at maturity; mesocarp white waxy, striated with resin canals; endocarp bony. $n=15$, polyploidy is common.

Twenty-two species from southern Canada south to Bolivia; India and Nepal; Bhutan and Myanmar; and temperate East Asia to New Guinea.

Several taxa belonging to *Toxicodendron* have not yet been transferred and are currently recognized in other genera including *Rhus*. Three sections are recognized within the genus: *Simplicifolia*, *Toxicodendron*, and *Venenata* (Gillis l.c.).

60. ***Trichoscypha*** Hook. f.

Trichoscypha Hook.f. in Benth. & Hook.f., Gen. Pl. 1: 423 (1862); Breteler, Adansonia III, 23: 247–264 (2001) et ibid. 26: 97–127 (2004), rev.
Emiliomarcelia Th. & H. Dur. (1909).

Dioecious trees, sarmentose trees, shrubs or lianas with white to pink, contact dermatitis-causing exudate turning black with exposure to air. Leaves evergreen, alternate, imparipinnate (unifoliolate), petiolate; leaflets opposite or alternate, petiolulate, lanceolate to oblong, entire. Inflorescences terminal and/or axillary, or cauliflorous thyrsoid panicles; the female inflorescence with a thicker, shorter axis than the male inflorescence. Flowers sessile or pedicellate; non-articulate; perianth 4(5, very rarely 6)-parted; calyx valvate, basally fused; corolla imbricate to valvate, reflexed at maturity; white to red; androecium haplostemonous; filaments filiform; anthers dorsifixed; pistillode present or 0 in male flowers; staminodes reduced; disk glabrous to densely pubescent, cup-shaped; carpels 3–4 (very rarely 6); stylodia (1–)3–4 (very rarely 6), erect or patent, compressed; stigmas capitate to shallowly bilobed (sessile); ovule pendulous, apical. Drupe ellipsoid, slightly oblique, ovoid, subglobose, or turbinate, 1-locular; exocarp red to purple or black, glabrous to densely pubescent; mesocarp fleshy; endocarp thin, coriaceous or crustaceous. n=24.

Thirty-two species in Tropical West and central Africa, and eastern to southern Africa. Breteler (l.c. 2001, 2004) completed a revision of the genus in which he recognized fourteen species in Upper Guinea, and eighteen in Lower Guinea and Congolia.

II. Subfam. Spondioideae Kunth ex Arn. (1832).

61. *Allospondias* (Pierre) Stapf

Allospondias (Pierre) Stapf in Hook., Icon. Pl.: 2667 (1900).
Spondias L. (1753), p. p.
Poupartia Comm. ex Juss. (1789), p. p.

Trees, dioecious (androdioecious) or with hermaphrodite flowers. Leaves deciduous, alternate, imparipinnate, petiolate; leaflets opposite or alternate, sessile to petiolulate, entire. Inflorescences terminal panicles, pubescent. Flowers pedicellate, articulate; perianth 5-parted; calyx imbricate; corolla valvate, white; androecium diplostemonous; pistillode reduced; staminodes reduced; disk glabrous; ovary prominently lobed (carpels only fused on one plane); carpels 5; stylodia 4–5, apically connate or distinct; stigmas capitate or spathulate; ovules 5, apical, 1–4 fertile. Drupe subglobose, 2–5-locular; exocarp red; mesocarp fleshy; endocarp woody with matrix of fibers expanding into mesocarp.

Two species, Myanmar, Thailand, Indochina, southern and tropical China.

A. laxiflora may represent a distinct genus due to differences in the connation of the stylodia (distinct), shape of stigmas (capitate), absence of endocarp lobing, number of locules (2), and the absence of four parenchyma-filled cavities.

62. *Antrocaryon* Pierre

Antrocaryon Pierre in Bull. Mens. Soc. Linn. Paris II, 3: 23 (1898); R.B. Fernandes, Garcia de Orta, Bot., Lisboa, 2: 107–110 (1975), rev.
Poupartia Comm. ex Juss. (1789), p. p.

Polygamodioecious trees. Leaves deciduous, alternate, imparipinnate, petiolate; leaflets opposite or subopposite, sessile or petiolulate, entire. Inflorescences axillary panicles, often emerging with new leaves. Flowers pedicellate; perianth 5-parted; calyx slightly imbricate or apert; corolla imbricate (valvate in *A. klaineanum*), pubescent; androecium diplostemonous; pistillode very reduced; staminodes reduced; disk glabrous, 10-lobed; carpels 5; stylodia 5, recurved, subapical and excentric; stigmas capitate; ovules 5, apical or subapical. Drupe plum-like or apple-shaped and depressed at apex, 5-locular; mesocarp strong smelling and edible; endocarp woody, angled with 5 apical opercula. $n = 12$.

Three species in tropical Africa, 1 in Amazonian Brazil, Colombia, and Peru.

63. *Buchanania* Spreng. Fig. 3A

Buchanania Spreng. in Schrad., J. Bot. 2: 234 (1800).
Launzan Buch.-Ham. (1799).
Cambessedea Kunth (1824).
Coniogeton Blume (1826).

Hermaphrodite trees. Leaves alternate, simple, sessile to petiolate, entire, subcoriaceous to coriaceous; cataphylls sometimes present. Inflorescences terminal and/or axillary panicles. Flowers hermaphrodite, pedicellate, articulate or non-articulate; perianth (4)5(6)-parted, imbricate; corolla greenish to white; androecium diplostemonous; filaments articulated in some species; anthers basifixed and usually sagittate; disk glabrous, cupular and crenulate; carpels 4–6, basally connate, only 1 fertile, the fertile one sometimes pubescent; stylodia 4–6, short; stigmas oblique, truncate, sterile carpels with functional stigmas, fertile carpel without functional stigma (stigma vestigial); ovules 4–6, basal or sublateral, only 1 fertile. Drupe lentiform, 1-locular, with persistent or caducous calyx; exocarp red or brown; endocarp bony, splitting vertically in half at maturity in some species. $n = 11$.

Twenty-five to thirty or more species in tropical Asia, Malesia, Australia, Micronesia, Melanesia, and Samoa, with greatest diversity in Malesia.

64. *Choerospondias* B.L. Burtt & A.W. Hill

Choerospondias B.L. Burtt & A.W. Hill, Ann. Bot. II, 1: 254 (1937).
Spondias L. (1753), p. p.
Poupartia Comm. ex Juss. (1789), p. p.

Polygamodioecious trees. Leaves evergreen or deciduous, alternate, imparipinnate, petiolate; leaflets opposite, petiolulate, entire or sometimes serrate; hairy tuft domatia present in secondary vein axils abaxially. Inflorescences terminal and/or axillary; female flowers in racemes (or solitary), male flowers in panicles. Flowers pedicellate; calyx cupular with tiny lobes; corolla imbricate; petals 5, purplish; androecium diplostemonous; filaments connate at the base and connate with the disk; anthers dorsifixed; pistillode reduced; staminodes reduced; disk glabrous; carpels 5; stylodia 5, short; stigmas capitellate to capitate; ovules 5, apical. Drupe ovoid or oblong, 5-locular; exocarp yellow; mesocarp thin, fleshy; endocarp woody with 5 apical shallow orifices, each covered by a membrane (possibly rudimentary opercula).

A single species, *C. axillaris*, in northeastern India, Nepal east to China, Taiwan, Thailand, Vietnam, and Japan.

65. *Cyrtocarpa* Kunth in Humb., Bonpl. & Kunth

Cyrtocarpa Kunth in Humb., Bonpl. & Kunth, Nov. Gen. Sp., Qu. Ed., 7: 20, t. 609 (1824); Mitchell & Daly, Ann. Missouri Bot. Gard. 78: 184–189 (1991), rev.
Bursera Jacq. ex L. (1762), p. p.

Polygamodioecious trees with somewhat succulent branchlets and exudate. Leaves deciduous, alternate, imparipinnate (paripinnate), petiolate; leaflets opposite, occasionally subopposite, sessile to short-petiolulate, entire. Inflorescences terminal and/or axillary panicles or pseudospikes. Flowers pedicellate; perianth 5-parted, imbricate; petals usually patent at anthesis; androecium diplostemonous; anther sometimes with glandular connective; pistillode reduced to five stylodes; staminodes reduced; disk glabrous, annular, crenulate and fleshy; carpels 5; stylodia (3–)5, short; stigmas capitate; ovule pendulous, subapical or apical. Drupe obliquely obtuse-oblong, 1–3(–5)-locular; exocarp purple or yellow to orange; mesocarp fleshy; endocarp bony with 1–5 opercula. Seed cotyledons reniform.

Five species in dry forests to open arid habitats: 1 endemic to southern Baja California; 2 in western Mexico; 1 in northern Colombia east to Guyana, Venezuela, and northern Brazil; 1 endemic to the Caatinga of northeast Brazil.

66. *Dracontomelon* Blume

Dracontomelon Blume, Mus. Bot. Lugd. Bat. 1: 231, t. 42 (1850).
Comeurya Baill. (1872).
Dracontomelum orthographic variant

Hermaphrodite trees with buttresses and white exudate. Leaves evergreen, alternate, imparipinnate, petiolate; leaflets opposite, subopposite or alternate, petiolulate, entire; hairy tuft domatia sometimes present in secondary vein axils abaxially. Inflorescences terminal and/or axillary panicles. Flowers hermaphrodite, pedicellate, articulate; perianth 5-parted; sepals imbricate, essentially distinct; petals valvate, imbricate at apex, white, greenish, or yellow; androecium diplostemonous; filaments subulate; anthers dorsifixed or medifixed; disk glabrous or pubescent; carpels 5; stylodia 5, basally and apically connate; stigmas capitate, receptive face lateral; ovules 1–5, apical, usually not all fertile. Drupe globose, depressed globose, ovoid, or ellipsoid, 5-locular or appearing fewer due to abortion; exocarp orange, brown to black; mesocarp fleshy; endocarp woody with 5 apical opercula. Seed with straight embryo. $n = 18$.

Eight species, India to Myanmar, Indo-China, tropical China, Malesia, and Fiji.

67. *Haematostaphis* Hook. f.

Haematostaphis Hook. f., Trans. Linn. Soc. 23: 169, t. 25 (1860).

Dioecious trees with clear exudate. Leaves alternate, imparipinnate, prominently multifoliolate, petiolate; leaflets opposite or alternate, petiolulate, emarginate at apex; small hairy tuft domatia present in secondary vein axils abaxially. Inflorescences terminal panicles, large; perianth 3-parted, imbricate; corolla whitish; androecium diplostemonous, in two whorls distinctly unequal in length; anthers dorsifixed; pistillode present or 0 in male flowers; staminodes reduced; disk glabrous; carpels 3; styles 3; stigmas 3; ovules 1–2, apical. Drupe with red exocarp; mesocarp thin, fleshy; endocarp woody with a bipartite, apical operculum.

A single species, *H. barteri*, tropical West Africa south to Nigeria.

68. ***Haplospondias*** Kosterm.

Haplospondias Kosterm., Kedondong, Ambarella, Amra. The Spondiodeae (Anacardiaceae) in Asia and the Pacific area. Published by the author; printed by Bina Karya 78 Printing Works, Bogor, Indonesia (1991).
Spondias L. (1753), p. p.
Bouea Meisn. (1837), p. p.

Hermaphrodite trees. Leaves alternate, simple, petiolate, entire. Inflorescences terminal, pyramidal thyrses. Flowers hermaphrodite, pedicellate; perianth (4)5-parted; calyx cupular with tiny teeth-like lobes; corolla valvate, reflexed at maturity; androecium diplostemonous; filaments glabrous; anthers dorsifixed; disk glabrous, 10-lobed; carpel 1; style very thick and gradually forming from the apex of the ovary; stigma a long, bilobed extension of the style. Fruit unknown.

A single species, *H. brandisiana* (=*H. haplophylla*), endemic to Yunnan, China, and Myanmar. This species is known from a single collection.

69. ***Harpephyllum*** Bernh. ex Krauss

Harpephyllum Bernh. ex Krauss, Flora 27: 349 (1844).

Dioecious trees. Leaves evergreen, alternate, imparipinnate, petiolate; rachis narrowly alate; leaflets sessile, opposite, entire, falcately narrowly ovate; cataphylls present. Inflorescences axillary panicles. Flowers pedicellate, non-articulate; perianth 4–5-parted, imbricate; corolla white to yellow, androecium diplostemonous (7–10 stamens); filament apex subulate, base flattened; anthers oblong-ovate; pistillode reduced; staminodes reduced; disk glabrous, cup-shaped, crenulate; carpels 4–5; stylodia 4–5, short; stigmas capitate; ovules 2, pendulous. Drupe oblong-obovoid, 4-locular (2 fertile); exocarp red; mesocarp subfleshy; endocarp woody or bony with spinose projections and 1–2 internal opercula. Seed compressed.

A single species, *H. caffrum*, in the Cape Province, Transvaal, and Natal, South Africa. Cultivated outside its native range.

70. ***Koordersiodendron*** Engl.

Koordersiodendron Engl., Meded. Lands Plantent. 19: 411 (1898).

Hermaphrodite trees. Leaves evergreen, alternate, imparipinnate, petiolate; leaflets opposite or subopposite, petiolulate, entire. Inflorescences axillary panicles. Flowers hermaphrodite, pedicellate, articulate; perianth 5-parted; calyx fused at base; corolla imbricate, white, greenish, or yellow; androecium diplostemonous; anther connective slightly protruding; disk glabrous; carpels 5, incompletely connate; stylodia 5, short; stigmas small; ovules 1–5, apical, usually only 1 fertile. Drupe broadly ellipsoid, oblique at base, 1(–3)-locular by abortion; exocarp yellow; endocarp cartilaginous. Seed with straight embryo.

A single species, *K. pinnatum*, in Borneo, the Philippines, Sulawesi, Maluku, and New Guinea.

71. ***Lannea*** A. Rich. in Guillem.

Lannea A. Rich. in Guillem., Fl. Seneg. Tent.: 153, t. 42 (1831).
Odina Roxb. (1832).
Scassellatia Chiov. (1932).

Dioecious, androdioecious or polygamodioecious subshrubs, shrubs, or trees with inflorescences and young stems and leaves covered with stellate trichomes. Leaves deciduous, alternate, imparipinnate, trifoliolate, or rarely unifoliolate (rarely all leaf types on same plant), petiolate; leaflets opposite or subopposite, sessile or petiolulate, entire or rarely serrate. Inflorescences terminal and/or axillary; female flowers in racemes or spikes; male flowers in panicles that are reduced to appear spike-like; inflorescence often emerging before the foliage. Flowers pedicellate, articulate; perianth 4(5)-parted, imbricate; corolla recurved at maturity, white, greenish, yellow, or red; androecium diplostemonous; filaments subulate; anthers dorsi- to basifixed; pistillode reduced; staminodes reduced; disk glabrous, patelliform, 8-crenulate; carpels 4, 1(2) fertile; stylodia 3–4, short, subapical; stigmas subglobose or capitate; ovules pendulous, apical. Drupe obovoid or ovoid, laterally compressed or subglobose, 1–4-locular (1–3 sterile); exocarp red to purple, brown, or black; mesocarp thin; endocarp woody, alveolate with ridges, 1–2 apical opercula; style bases persistent. Seed reniform, laterally compressed; embryo straight. $n = (14)\ 15\ (20)$.

Forty or more species from sub-Saharan Africa to Socotra; one of these (*L. coromandelica*) is distributed from Pakistan to Indo-China. Cultivated in Southeast Asia.

72. *Operculicarya* H. Perrier

Operculicarya H. Perrier, Mem. Mus. Hist. Nat. II, 118 (7): 248 (1944); Eggli, Bull. Mus. Natl. Hist. Nat., Paris IV, 17, sect. B, Adansonia: 149–158 (1995), rev.; Randrianasolo & Lowry, Adansonia III, 28: 359–371 (2006), rev.

Dioecious shrubs or small trees with swollen trunks (especially basally) and gummy exudate. Leaves deciduous, alternate, imparipinnate, nearly sessile to short petiolate, usually microphyllous; rhachis usually alate; leaflets opposite to subopposite, sessile to short-petiolulate, entire. Female inflorescence of up to 4 flowers or reduced to a solitary flower, rarely multiflorate, axillary or on tip of short shoot; male inflorescence arising from short shoots and often reduced to one flower or in spicate to paniculate solitary or fascicled inflorescences. Flowers sessile or pedicellate, articulate; perianth (4)5(6)-parted, imbricate or apert; corolla creamy-yellow; androecium diplostemonous; anthers introrse, dorsifixed; pistillode very reduced, 3–5-cleft; staminodes reduced; disk glabrous, crenate; carpels 5; stylodia 5, often 4 are very reduced; stigma 1, shortly glandular; ovule apical. Drupe subglobose, 1(2)-locular; endocarp stony, with one operculum; exocarp red to black. Seed with horseshoe-shaped embryo.

Eight species in Madagascar, the Comoros, and Aldabra.

73. *Pegia* Coleb.

Pegia Coleb., Trans. Linn. Soc. 15: 364 (1827).
Tapirira Aubl. (1775), p. p.
Phlebochiton Wall. (1835).

Polygamous or polygamodioecious scandent shrubs, sarmentose trees, or lianas. Leaves evergreen, alternate, imparipinnate, petiolate; leaflets opposite or subopposite, petiolulate, entire or crenate. Inflorescences terminal and/or axillary panicles. Flowers pedicellate, articulate; perianth (4)-5-parted; calyx valvate; corolla imbricate or subvalvate, whitish; androecium diplostemonous; filaments filiform; pistillode reduced; staminodes reduced; disk glabrous, 5-lobed; carpels (4)5, only 1 fertile; stylodia (4)5, apical; stigmas capitate; ovule 1. Drupe slightly reniform, 1-locular; exocarp red to purple; endocarp crustaceous. Seed with straight embryo.

Two species, India to Myanmar, Indo-China, tropical China, and Malesia.

74. *Pleiogynium* Engl.

Pleiogynium Engl. in DC., Monogr. Phan. 4: 255 (1883).

Dioecious trees. Leaves evergreen, alternate, imparipinnate (very rarely paripinnate), petiolate; leaflets opposite to subopposite, petiolulate, entire, sometimes with hairy marsupiform domatia. Inflorescences axillary panicles (sometimes female inflorescence racemose or spiciform). Flowers pedicellate, articulate; perianth (4)5(6)-parted; calyx valvate; corolla imbricate, green, white, yellow, or brown; androecium diplostemonous; filaments filiform-subulate; anther with glandular connective; pistillode reduced, of 5 carpellodes; staminodes reduced; disk glabrous; carpels (5)8–12; stylodia (5)8–12, short, forming a marginal ring; stigmas spathulate; ovules pendulous. Drupe depressed globose or turbinate, (5) 8–12-locular; exocarp membranous, red to brown or black and depressed at opercula locations; outer mesocarp fleshy, inner mesocarp woody; endocarp bony with 5–12 opercula. Seed with curved embryo.

Two species, from Indo-China to Malesia, South Pacific Islands, and northeastern Australia.

75. *Poupartia* Comm. ex Juss.

Poupartia Comm. ex Juss., Gen.: 372 (1789); H. Perrier, Mém. Mus. Natl. Hist. Nat. 18: 243–269 (1944), rev.; F. Friedmann, Flore des Mascareignes 77: 5–9 (1997).
Spondias L. (1753), p. p.
Shakua Bojer (1837).
Sclerocarya Hochst. (1844), p. p.

Polygamodioecious or dioecious trees with exudate. Leaves evergreen or deciduous, alternate, imparipinnate, petiolate; leaflets opposite, petiolulate, entire. Inflorescences axillary to pseudoterminal; male flowers arranged in spicate racemes; female flowers in few-flowered panicles or racemes; male flowers sessile, female flowers pedicellate; perianth (4)5(6)-parted, imbricate; androecium diplostemonous; filaments distinct; anthers dorsifixed; pistillode minute or 0 in male flowers; staminode reduced; disk glabrous; carpels (1)2–5; stylodia 2–5, short, subapical; stigmas capitate; ovules (1)2–5. Drupe (1)2–5-locular; mesocarp fleshy; endocarp bony with 2–5 apical opercula. Seed with curved embryo. $n = 12$.

Seven species in Madagascar and the Mascarene Islands.

Distinct from *Operculicarya* by its curved versus horseshoe-shaped embryo, and 2–5 versus 1 opercula.

76. ***Poupartiopsis*** Capuron ex J.D. Mitch. & Daly

Poupartiopsis Capuron ex J.D. Mitch. & Daly, Syst. Bot. 31: 338 (2006).

Dioecious trees with milky exudate. Leaves evergreen, alternate, imparipinnate, petiolate; rhachis narrowly alate; leaflets opposite or subopposite, petiolulate. Inflorescences terminal and/or axillary panicles. Flowers (sub)sessile; perianth imbricate, 5-parted; androecium diplostemonous; filaments subulate, inserted at base of disk; anthers dorsi- to basifixed, connective with discolorous, slightly bulbous apex; pistillodes in male flowers; staminodes reduced; disk glabrous; carpels 3; stylodia 3; stigmas 3; ovules pendulous, subapical. Drupe lightweight and buoyant, large oblong to ovoid, 3-locular; mesocarp thin; endocarp bony with long hooked spinose processes parallel to the axis of the fruit enclosed in a fibrous matrix.

A single species, *P. spondiocarpus*, endemic to coastal forests of eastern Madagascar.

77. ***Pseudospondias*** Engl.

Pseudospondias Engl. in DC., Monogr. Phan. 4: 258 (1883).
Haematostaphis Hook.f. (1860), p. p.

Dioecious shrubs or trees with exfoliating bark. Leaves alternate, imparipinnate, petiolate; leaflets opposite or alternate, petiolulate, lateral leaflets asymmetrical at base, with strongly arcuate venation. Inflorescences axillary panicles. Flowers pedicellate, non-articulate; perianth 3–4(5)-parted, imbricate; corolla whitish; androecium diplostemonous, often in two whorls of unequal length; pistillode reduced to four lobes; staminodes reduced; disk glabrous, annular, crenulate; carpels 3–4(5), usually only 1(2) fertile; stylodia 3–4(5), very short, subapical; stigmas nearly sessile; ovules pendulous, apical. Drupe oblong-obovoid, 3–4(5)-locular; exocarp blue-black; mesocarp resinous; endocarp woody with 3–4(5) bipartite, apical opercula. $n = 15$.

Two species in riverine forests and rain forests of sub-Saharan Africa.

78. ***Sclerocarya*** Hochst.

Sclerocarya Hochst. in Flora 27, Bes. Beil. 1 (1844).
Poupartia Comm. ex Juss. (1789), p. p.

Dioecious (very rarely gynodioecious) trees. Leaves deciduous, alternate, imparipinnate, rarely unifoliolate, petiolate; leaflets opposite or occasionally subopposite, petiolulate (sessile). Inflorescences terminal and/or axillary; female inflorescence often reduced to a solitary flower; male flowers arranged in spicate racemes. Flowers pedicellate, non-articulate; perianth 4–5-parted, imbricate; calyx yellow to red; corolla yellow to purplish-pink to red; stamens (8–10–)15–25(–30); filaments subulate; anthers basi-dorsifixed; pistillode barely evident to 0 in male flowers; staminodes many (21–26); disk glabrous, patelliform; carpels (1)2(–4); stylodia 2–3, short, subapical; stigmas capitate; ovules pendulous, apical, funicle very long. Drupe obovoid-subglobose, 2–3(4)-locular; exocarp yellow to orange; mesocarp fleshy, thick; endocarp bony with (1)2–3(4) apical opercula; stylodia often persistent. Seed obclavate, compressed; testa chartaceous, brownish. $n = 13$.

Two to three species in sub-Saharan Africa and Madagascar.

79. ***Solenocarpus*** Wight & Arn.

Solenocarpus Wight & Arn., Prodr.: 171 (1834); Kostermans, Kedondong, Ambarella, Amra, The Spondiodeae (Anacardiaceae) in Asia and the Pacific area. Published by the author; printed by Bina Karya 78 Printing Works, Bogor, Indonesia (1991), rev.
Spondias L. (1753), p. p.
Pegia Coleb. (1827), p. p.

Hermaphrodite hemi-epiphytes, sarmentose trees, or lianas. Leaves deciduous, alternate, imparipinnate, petiolate; leaflets opposite to subopposite, petiolulate, oblique; *S. indica* with intramarginal vein, *S. philippinensis* with eucamptodromous secondary venation. Inflorescences terminal panicles. Flowers pedicellate; perianth 5-parted; calyx cupular with tiny lobes; corolla valvate in bud, distally reflexed at maturity; androecium diplostemonous; anthers dorsifixed; disk glabrous; carpels 1; style partially lateral or apical, clavate; stigma bilobed; ovule pendulous. Drupe 1-locular, mesocarp thin, fleshy; endocarp woody.

Two species, one in southern India and one in Malesia.

80. *Spondias* L. Fig. 6

Spondias L., Sp. Pl.: 200 (1753); Kostermans, Kedondong, Ambarella, Amra. The Spondiodeae (Anacardiaceae) in Asia and the Pacific area. Published by the author; printed by Bina Karya 78 Printing Works, Bogor, Indonesia (1991), rev.
Evia Commerson ex Blume (1850).
Warmingia Engl. (1874).

Polygamodioecious or hermaphrodite (rarely dioecious), strongly protandrous trees (very rarely hemi-epiphytes) with contact dermatitis-causing exudate. Leaves deciduous, alternate, imparipinnate (bipinnate), petiolate; leaflets opposite, subopposite or alternate, sessile to petiolulate. Inflorescences terminal and/or axillary panicles (racemes); often appearing before leaves or with young leaves. Flowers pedicellate, articulate; calyx slightly imbricate or apert; (4)5-lobed; corolla valvate, (4)5(6)-parted, cucullate; white, cream-colored, purple, or red; androecium diplostemonous, sometimes in two whorls of unequal length; filaments filiform or subulate; anthers dorsifixed; pistillodes and staminodes reduced; disk glabrous or papillose, annular and lobed; carpels (3–)5; stylodia (3–)5; stigmas capitate to spathulate; ovules apical. Drupe globose, obovoid, oblong or ellipsoid, (1–)5-locular; exocarp yellow-orange, red-purple, or greenish; mesocarp fleshy; endocarp bony, usually with a fibrous outer layer (very rarely spiny) projecting into the fleshy mesocarp. Seed with curved embryo. $n = 16$.

Sixteen or more species: eight from Mexico south to southeastern Brazil and Bolivia; eight in India and Sri Lanka east to tropical China and South Pacific Islands; Neotropical species naturalized in West Africa, the West Indies, and Southeast Asia. Several species cultivated pantropically.

A new monotypic genus, ***Attilaea abalak*** E. Martinez & Ramos, has been described from the Yucatan Peninsula in Mexico, and was suggested to be closely related to *Spondias* (Martinez and Ramos 2007). This newly described taxon shares the ubiquitous intramarginal vein of *Spondias* and is very similar to *S. purpurea*, but differs from it by having two rather than five carpels, and being scandent rather than erect in habit. This bicarpellate gynoecium differs from all other Anacardiaceae. We have been unable to examine enough material to confirm the separate generic status of this taxon, but do believe it belongs within the family in subfam. Spondioideae.

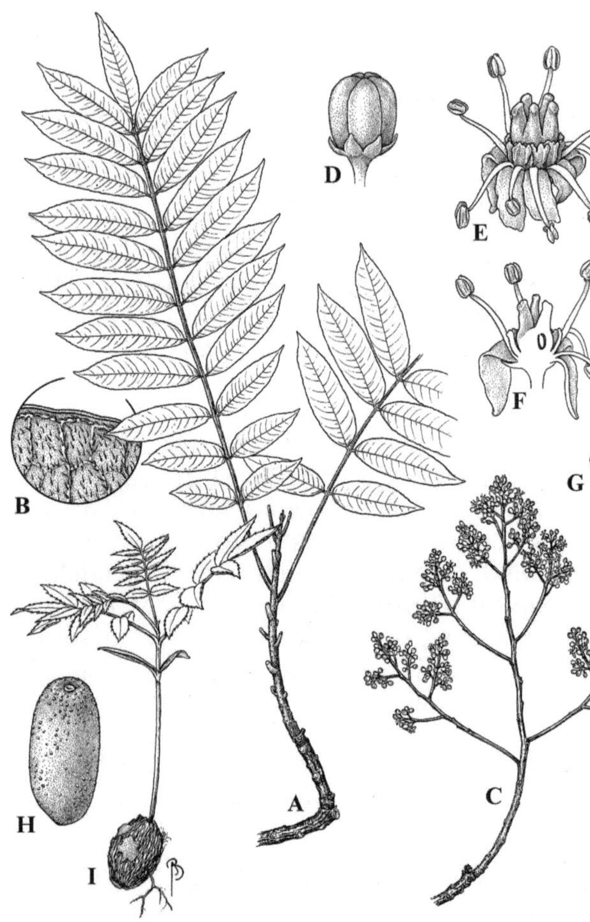

Fig. 6. Anacardiaceae. *Spondias testudinis*. **A** Leafy branchlet. **B** Abaxial leaflet surface. **C** Inflorescence. **D** Flower bud. **E** Flower. **F** Longitudinal section of flower. **G** Transverse section of ovary. **H** Fruit. **I** Seedling. (Reproduced with permission of the artist Bobbi Angell)

81. *Tapirira* Aubl. Fig. 7

Tapirira Aubl., Hist. Pl. Guiane 1: 470, t. 188 (1775).
Mauria Kunth (1824), p. p.

Polygamodioecious trees. Leaves evergreen, alternate, imparipinnate to paripinnate, petiolate; leaflets opposite or subopposite, petiolulate, entire. Inflorescences terminal and/or axillary panicles. Flowers pedicellate, non-articulate; perianth 5-parted, imbricate; corolla greenish yellow or cream-colored; stamens (8–)10, in two whorls of unequal length; pistillode reduced; staminodes reduced; disk glabrous, (8–)10-lobed; carpels (4) 5; stylodia (4)5; stigmas capitate; ovule apical or subapical. Drupe globose, oblong-oblique or ellipsoid, 1-locular, with persistent calyx; exocarp black or purple; endocarp bony or cartilaginous, usually brittle when dry; mesocarp thin, fleshy. Seed with curved embryo; cotyledons with purple striations.

Eight or more species from southern Mexico to southeastern Brazil, Bolivia, and Paraguay.

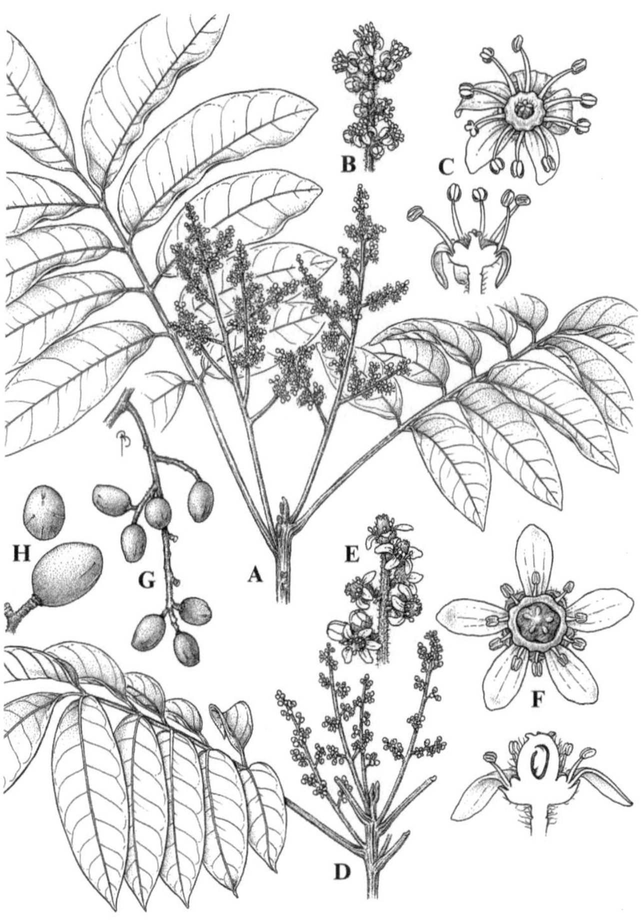

Fig. 7. Anacardiaceae. *Tapirira chimalapana*. **A–C** Male plant. **A** Flowering branchlet. **B** Portion of inflorescence. **C** Flowers, in plan and in longitudinal section. **D–F** Female plant. **D** Flowering branchlet. **E** Portion of inflorescence. **F** Flowers, in plan and in vertical section. **G** Part of infructescence. **H** Fruit, apical and side view. (Reproduced with permission of the artist Bobbi Angell)

Selected Bibliography

Aguilar-Ortigoza, C.J., Sosa, V. 2004a. Taxonomic revision of the genus *Pseudosmodingium* (Anacardiaceae). Rhodora 106(928): 348–359.

Aguilar-Ortigoza, C.J. Sosa, V. 2004b. The evolution of toxic phenolic compounds in a group of Anacardiaceae genera. Taxon 53: 357–364.

Aleksandrovski, E.S, Naumova, T.N. 1985. Family Anacardiaceae. In: Yakovlev, M.S. (ed.) Comparative embryology of flowering plants. Brunelliaceae - Tremandraceae (in Russian). Leningrad: Nauka, pp. 166–173.

Altrichter, M., Sáenz, J., Carrillo, E. 1999. Chanchos cariblancos (*Tayassu pecari*) como depredadores y dispersores de semillas en el Parque Nacional Corcovado, Costa Rica. Brenesia 52: 53–59.

Angiosperm Phylogeny Group (APG). 1998. An ordinal classification for the families of flowering plants. Ann. Missouri Bot. Gard. 85: 531–553.

Angiosperm Phylogeny Group APG II. 2003. An update of the Angiosperm Phylogeny Group classification for the orders and families of flowering plants: APG II. Bot. J. Linn. Soc. 141: 399–436.

Angiosperm Phylogeny Group APG III. 2009. An update of the Angiosperm Phylogeny Group classification for the orders and families of flowering plants: APG III. Bot. J. Linn. Soc. 161: 105–121.

Anzótegui, L.M. 1971. El pollen de las Anacardiaceae del N.E. de la Argentina. Ameghiniana 8: 329–340.

Arrillaga-Maffei, B.R., Ziliani, G., Ren, J. 1973. Anacardiáceas de Uruguay, Bol. 126. Montevideo: Universidad de la Republica.

Baas, P., Wheeler, E., Chase, M. 2000. Dicotyledonous wood anatomy and the APG systems of angiosperm classification. Bot. J. Linn. Soc. 134: 3–17.

Bachelier, J.B., Endress, P.K. 2007. Development of inflorescences, cupules, and flowers in *Amphipterygium*, and comparison with *Pistacia* (Anacardiaceae). Int. J. Plant Sci. 168: 1237–1253.

Backer, H.J., Haack, N.H. 1938. Le principe toxique des fruits de Renghas (*Semecarpus heterophylla* Bl.). Recueil Travaux Chimiques Pays-Bas 57: 225–232.

Baksi, S.K. 1976. Pollen morphology of the genera *Gluta* Linnaeus and *Melanorrhoea* Wallich (Anacardiaceae). In: Ferguson, I.K., Muller, J. (eds.) The evolutionary significance of the exine. London: Academic Press, pp. 379–395.

Barfod, A. 1988. Inflorescence morphology of some South American Anacardiaceae and the possible phylogenetic trends. Nord. J. Bot. 8: 3–11.

Barkley, F.A. 1942. A key to the genera of Anacardiaceae. Am. Midl. Nat. 28(2): 465–474.

Behl, P.N., Captain, R.M. 1979. Skin-irritating and sensitizing plants found in India. Ram Nagar, New Delhi: S. Chand & Co.

Behrens, R. 1996. Cashew as an agroforestry crop: prospects and potentials. Margraf: Weikersheim.

Berry, E.W. 1924. An Oligocene cashew nut from South America. Am. J. Sci. 8: 126–126.

Berry, E.W. 1929. An *Anacardium* from the Eocene of Texas. J. Wash. Acad. Sci. 19: 37–39.

Bhatt, J.R., Ram, H.Y.M. 1992. Development and ultrastructure of primary secretory ducts in the stem of *Semecarpus anacardium* (Anacardiaceae). IAWA Bull. n.s. 13(2): 173–185.

Biondi, E. 1981. *Arganioxylon sardum* N. Gen., N. Sp. et *Sclerocaryoxylon chiarugii* N. Gen., N. Sp,: Bois fossiles du Miocène de la Sardaigne (Italie). Rev. Palaeob. Paly. 34: 301–320.

Birkinshaw, C. 2001. Fruit characteristics of species dispersed by the Black Lemur (*Eulemur macaco*) in the Lokobe forest, Madagascar. Biotropica 33: 478–486.

Boesewinkel, F.D., Bouman, F. 1984. The seed: structure. In: Johri, B.M. (ed.) Embryology of angiosperms. New York: Springer.

Bonnefille, R., Letouzey, R. 1976. Fruits fossiles d'*Antrocaryon* dans la vallée de l'Omo (Ethiopie). Adansonia II, 16: 65–82.

Bremer, K., Bremer, B., Thulin, M. 1999. Introduction to phylogeny and systematics of flowering plants, 5th ed. Uppsala: Uppsala University.

Burkill, H.M. 1985. Useful plants of West Tropical Africa. Vol. 1: Families A–D. Kew, London: Royal Botanic Gardens.

Burnham, R.J., Carranco, N.L. 2004. Miocene winged fruits of *Loxopterygium* (Anacardiaceae) from the Ecuadorian Andes. Am. J. Bot. 91: 1767–1773.

Carmello-Guerreiro, S.M., Paoli, A.A.S. 1999. Morfologia e desenvolvimento pós-seminal de *Schinus terebinthifolius* Raddi, *Lithraea molleoides* (Vell.) Engl., *Myracrodruon urundeuva* Fr. Allem. *Astronium graveolens* Jacq. (Anacardiaceae). Naturalia, São Paulo 24: 127–138.

Chase, M.W., Soltis, D.E., Olmstead, R.G., Morgan, D., Les, D.H., and 37 further authors 1993. Phylogenetics of seed plants: an analysis of nucleotide sequences from the plastid gene *rbc*L. Ann. Missouri Bot. Gard. 80: 528–580.

Chen, T.K., Wiemer, D.F. 1984. A volatile leafcutter ant repellent from *Astronium graveolens*. Naturwissenschaften 71: 97–98.

Cojocaru, M., Droby, S., Glotter, E., Goldman, A., Gottlieb, H.E., Jacoby, B., Prusky, D. 1986. 5-(12-Heptadecenyl)-resorcinol, the major component of the antifungal activity in the peel of mango fruit. Phytochemistry 25: 1093–1095.

Collinson, M.E. 1983. Fossil plants of the London Clay. London: Palaeontology Association.

Copeland, H.F. 1959. The reproductive structures of *Schinus molle* (Anacardiaceae). Madroño 15: 14–25.

Copeland, H.F. 1961. Observations on the reproductive structures of *Anacardium occidentale*. Phytomorphology 11: 315–325.

Corthout, J., Pieters, L.A., Claeys, M., Vanden Berghe, D.A., Vlietinck, A.J. 1991. Antiviral ellagitannins from *Spondias mombin*. Phytochemistry 30: 1129–1130.

Corthout, J., Pieters, L.A., Claeys, M., Vanden Berghe, D.A., Vlietinck, A.J. 1992. Antiviral caffeoyl esters from *Spondias mombin*. Phytochemistry 31: 1979–1981.

Corthout, J., Pieters, L.A., Claeys, M., Geerts, S., Vanden Berghe, D., Vlietninck, A.J. 1994. Antibacterial and molluscicidal phenolic acids from *Spondias mombin*. Planta Medica 60: 460–463.

Cronquist, A. 1981. An integrated system of classification of flowering plants. New York: Colombia University Press.

Dadswell, H.E., Ingle, H.D. 1948. The anatomy of timbers of the southwest Pacific area: I. Anacardiaceae. Aust. J. Sci. Res., ser. B1 4: 391–415.

Den Outer, R.W., Van Veenendaal, W.L.H. 1986. Distribution and development of secretory ducts in *Trichoscypha* (Anacardiaceae). Acta Bot. Neerl. 35: 425–435.

de Vogel, E.F. 1980. Seedlings of dicotyledons. Wageningen: Ctr. Agr. Publ. Doc.

Ding Hou. 1978. Anacardiaceae. Flora Malesiana I, 8(3): 395–548.

Dobat, K., Peikert-Holle, T. 1985. Blüten und Fledermäuse. Frankfurt am Main: Waldemar Kramer.

Drewes, S.E., Horn, M.M., Mabaso, N.J. 1998. *Loxostylis alata* and *Smodingium argutum* – a case of phytochemical bedfellows? S. Afr. J. Bot. 64: 128–129.

Ellis, B., Daly, D.C., Hickey, L.J., Johnson, K.R., Mitchell, D.J., Wing, S.L. 2009. The Manual of Leaf Architecture. New York: Cornell University Press.

Endress, P.K., Stumpf, S. 1991. The diversity of stamen structures in 'lower' Rosidae (Rosales, Fabales, Proteales, Sapindales). Bot. J. Linn. Soc. 107: 217–293.

Engler, A. 1892. Anacardiaceae. In: Engler, A., Prantl, K. (eds.) Die natürlichen Pflanzenfamilien III, 5: 138–178. Leipzig: W. Engelmann.

Epstein, W.L. 1994. Occupational poison ivy and oak dermatitis. Dermatol. Clin. 12(3): 511–516.

Erdtman, G. 1952. Pollen morphology and plant taxonomy. Waltham, MA: Chronica Botanica Company.

Fahn, A. 1979. Secretory tissues in plants. London: Academic Press.

Fernandez, C., Fiandor, A., Martineez-Garate, A., Martinez Quesada, J. 1995. Allergy to pistachio: crossreactivity between pistachio nut and other Anacardiaceae. Clin. Exp. Allergy 25: 1254–1259.

Figueroa, X.M.C. 2001.La familia Julianiaceae en el estado de Jalisco, Mexico. Jalisco, Mexico: Universidad de Guadalajara.

Forman, L.L. 1954. A new genus from Thailand. Kew Bull. 4: 555–564.

Fragoso, J.M.V. 1997. Tapir-generated seed shadows: scale-dependent patchiness in the Amazon rain forest. J. Ecol. 85: 519–529.

Free, J.B., Williams, I.H. 1976. Insect pollination of *Anacardium occidentale* L., *Mangifera indica* L., *Blighia sapida* Koenig and *Persea americana* Mill. Trop. Agric. (Trinidad) 53(2): 125–139.

Gadek, P.A., Fernando, E.S., Quinn, C.J., Hoot, S.B., Terrazas, T., Sheahan, M.C., Chase, M.W. 1996. Sapindales: molecular delimitation and infraordinal groups. Am. J. Bot. 83: 802–811.

Galvez, J., Zarzuelo, A., Crespo, M.E., Utrilla, M.P., Jimenez, J., Spiessens, C., de Witte, P. 1991. Antidiarrhoeic activity of *Sclerocarya birrea* bark extract and

its active tannin constituent in rats. Phytoth. Res. 5: 276–278.
Galvez, J., Zarzuelo, A., Busson, R., Cobbaert, C., de Witte, P. 1992. (-)-Epicatechiin-3-galloyl ester: a secretagogue compound from the bark of *Sclerocarya birrea*. Planta Medica 58: 174–175.
Gambaro, V., Chamy, M.C., von Brand, E., Gambarino, J.A. 1986. 3-(pentadec-10-enyl)-catechol, a new allergenic compound from *Lithraea caustica* (Anacardiaceae). Planta Medica 44: 20–22.
Gardner, A.L. 1977. Feeding habits. In: Baker, R.J., Jones, J.K. Jr., Carter, D.C. (eds.) Biology of bats of the New World, Family Phyllostomidae. Part III. Special Publications. Lubbock, TX: The Museum, Texas Technical University Press.
Gautier-Hion, A., Duplantier, J.-M., Quris, R., Feer, F., Sourd, C., Decoux, J.-P., Dubost, G., Emmons, L., Erard, C., Hecketsweiler, P., Moungazi, A., Roussilhon, C., Thiollay, J.-M. 1985. Fruit characters as a basis of fruit choice and seed dispersal in a tropical forest vertebrate community. Oecologia 65: 324–337.
Geesink, A.J., Leeuwenberg, M., Ridsdale, C.E., Veldkamp, J.F. 1981. Thonner's analytical key to the families of flowering plants. Leiden: University of Leiden Press.
Gibson, A.C. 1981. Vegetative anatomy of *Pachycormus* (Anacardiaceae). Bot. J. Linn. Soc. 83: 273–284.
Gillis, W.T. 1971. The systematics and ecology of poison-ivy and poison-oaks. Rhodora 73: 72–159; 161–237; 370–443; 465–540.
Gilman, E.F. 1999. *Schinus terebinthifolius*. University of Florida Cooperative Extension Service Fact Sheet FPS-542.
Giménez, A.M., Moglia, G. 1995. Estructura cortical de Anacardiaceas Argentinas. Invest. Agr., Sist. Recur. For. 4(2): 189–203.
Goldblatt, P., Johnson, D.E. (eds.). 1979-2008. Index to plant chromosome numbers. St. Louis, MO: Missouri Botanical Garden. http://mobot.mobot.org/W3T/Search/ipcn.html
Goris, M.A. 1910. Contribution à l'étude des Anacardiacées de la tribu Mangiférées. Ann. Sci. Nat. (Bot.) IX, 11: 1–29.
Gregory, M. 1994. Bibliography of systematic wood anatomy of dicotyledons. IAWA J. l, suppl. 1: 1–266.
Gross, M., Baer, H., Fales, H.M. 1975. Urushiols of poisonous Anacardiaceae. Phytochemistry 14:2263–2266.
Grundwag, M., Fahn, A. 1969. The relation of embryology to the low seed set in *Pistacia vera* (Anacardiaceae). Phytomorphology 19: 225–235.
Gundersen, A. 1950. Families of dicotyledons. Waltham, MA: Chronica Botanica Company.
Halim, H., Locksley, H.D., Memon, J.J. 1980. Vesicant principles of poison ivy and related plants: synthesis of the urushiols, 1,2-dihydroxy-3-((Z)-pentadec-8-enyl)benzene and 1,2-dihydroxy-3-pentadecylbenzene. J. Chem. Soc., Perkin Trans. 1: 2331–2337.
Hall, J.B, O'Brien, E.M., Sinclair, F.L. 2002. *Sclerocarya birrea*: a monograph. Bangor: University of Wales.
Hardin, J.W., Phillips, L.L. 1985. Atlas of foliar surface features in woody plants, VII. *Rhus* subg. *Rhus* (Anacardiaceae) of North America. Bull. Torrey Bot. Club 112(1): 1–10.
Heimsch, C.H. Jr. 1940. Wood anatomy and pollen morphology of *Rhus* and allied genera. J. Arnold Arb. 21: 279–291.
Hess, W.R. 1949. Identification of New World timbers. Part II. Anacardiaceae. Trop. Woods 87: 11–34.
Hill, A.W. 1933. The method of germination of seeds enclosed in a stony endocarp. Ann. Bot. 47: 873–887.
Hill, A.W. 1937. The method of germination of seeds enclosed in a stony endocarp II. Ann. Bot. II, 1: 239–256.
Hill, G.A., Mattacotti, V., Graham, W.D. 1934. The toxic principle of the poison ivy. J. Am. Chem. Soc. 56: 2736–2738.
Hsu, J. 1983. Late Cretaceous and Cenozoic vegetation in China, emphasizing their connections with North America. Ann. Missouri Bot. Gard. 70: 490–508.
Huang, T. 1972. Anacardiaceae. In: Pollen flora of Taiwan. Taipei: National Taiwan University, Botany Department Press.
Ibe, R.A., Leis, R.A. 1979. Pollen morphology of Anacardiaceae of northeastern North America. Bull. Torrey Bot. Club 106: 140–144.
Jansen, A., de Raadt, J.L., van Toorenenbergen, A.W., van Wijk, R.G. 1992. Allergy to pistachio nuts. Allergy Proc. 13: 255–258.
Jiménez-Reyes, N., Cuevas Figueroa, X.M. 2001. Morfología del pollen de *Amphipterygium* Schiede ex Standley (Julianiaceae). Bol. IBUG 8(1/2): 65–73.
Joel, D.M., Fahn, A. 1980. Ultrastructure of the resin ducts of *Mangifera indica* L. (Anacardiaceae). 1. Differentiation and senescence of the shoot ducts. Ann. Bot. 46: 225–233.
Johnson, R., Baer, H., Kirkpatrick, C., Dawson, C., Khurana, R. 1972. Comparison of the contact allergenicity of the four pentadecylcatechols derived from poison ivy urushiol in humans. J. Allergy Clin. Immunol. 49: 27–35.
Johri, B.M. 1963. Female gametophyte. In: Maheshwari, P. (ed.) Recent advances in the embryology of angiosperms. Delhi: International Society of Plant Morphology.
Juliano, J.B. 1937. Embryos of carabao mango, *Mangifera indica* L. Philipp. Agr. 25: 749–760.
Kamilya, P., Paria, N. 1995. Seedling morphology in taxonomic study of some Indian members of the Anacardiaceae. J. Ind. Bot. Soc. 74: 193–196.
Kelkar, S.S. 1958a. Embryology of *Rhus mysurensis* Heyne. J. Ind. Bot. Soc. 37: 114–122.
Kelkar, S.S. 1958b. A contribution to the embryology of *Lannea coromandelica* (Houtt.) Merr. J. Univ. Bombay 26: 152–159.
Kerr, G.A. 1935. Quebracho forests of South America. Commodities of Commerce Series, no. 9. Washington, D.C.
Kostermans, A.J.G.H., Bompard, J.M. 1993. The mangoes: their botany, nomenclature, horticulture and utilization. Academic Press: London.
Kryn, J.M. 1952. The anatomy of the wood of the Anacardiaceae and its bearing on the phylogeny and relationships of the family. Ann Arbor, MI: Ph.D. dissertation, University of Michigan.
Kubo, I., Ochi, M., Vieira, P.C., Komatsu, S. 1993. Antitumor agents from the cashew (*Anacardium occidentale*) apple juice. J. Agric. Food Chem. 41:1012–1015.

Kullavanijaya, P., Ophaswongse, S. 1997. A study of dermatitis in the lacquerware industry. Contact Dermatitis 36: 244–246.
Leaf Architecture Working Group. 1999. Manual of leaf architecture - morphological description and categorization of dicotyledonous and net-veined monocotyledonous angiosperms. Washington, D.C.: Smithsonian Institution. http://www.peabody.yale.edu/collections/pb/MLA/
Li, X., Baskin, J.M., Baskin, C.C. 1999. Contrasting dispersal phenologies in two fleshy-fruited congeneric shrubs, *Rhus aromatica* Ait. and *Rhus glabra* L. (Anacardiaceae). Can. J. Bot. 77: 976–988.
Loev, B. 1952. The active constituents of poison ivy and related plants. Structure and synthesis. Ph.D. dissertation, New York: Colombia University.
Lopez-Naranjo, H.J. 1977. Hábito de crecimento y estructura de las yemas de *Anacardium humile* St. Hil. Anacardiaceae. Revista Forest. Venez. 27: 159–173.
Maheshwari, P., Sachar, R.C, Chópra, R.N. 1955. Embryological studies in mango, *Mangifera indica*. Proc. 42nd Indian Sci. Congr., Dehli, 3: 233.
Manchester, S.R. 1994. Fruits and seeds of the middle Eocene Nut Beds Flora, Clarno Formation, Oregon. Palaeontogr. Am. 58: 1-205.
Manchester, S.R., Wilde, V., Collinson, M.E. 2007. Fossil cashew nuts from the Eocene of Europe: biogeographic links between Africa and South America. Int. J. Plant Sci. 168: 1199–1206.
Marticorena, C. 1968. Granas de pollen de plantas chilenas - Anaçardiaceae. Gayana 17: 17–21.
Martínez, E., Álvarez, C.H.R. 2007. Un nuevo género de Anacardiaceae de la Peninsula de Yucatán. Acta Bot. Hung. 49: 353–358.
McWilliams, E. 1991. The impending naturalization of *Pistacia chinensis* (Anacardiaceae) in east Texas. Sida 14: 508–511.
Metcalfe, C.R., Chalk, L. 1950. Anatomy of the dicotyledons, vol. 1. London: Oxford University Press.
Meyer, H.W. 2003. The fossils of Florissant. Washington, D.C.: Smithsonian Books.
Miller, A.J., Young, D.A., Wen, J. 2001. Phylogeny and biogeography of *Rhus* (Anacardiaceae) based on ITS sequences. Int. J. Plant Sci. 162: 1401–1407.
Mitani, M., Kuroda, S., Tutin, C.E.G. 1994. Floral lists from five study sites of apes in the African tropical forests. Tropics 3(3/4): 247–348.
Mitchell, J.D. 1990. The poisonous Anacardiaceae genera of the world. Adv. Econ. Bot. 8: 103–129.
Mitchell, J.D. 2004. Anacardiaceae. In: Smith, N., Mori, S., Henderson, A.A., Stevenson, D.W., Heald, S.V. (eds.) Flowering plants of the Neotropics. Princeton, NJ: Princeton University Press.
Mitchell, J.D., Mori, S.A. 1987. The cashew and its relatives (Anacardium: Anacardiaceae). Mem. N. Y. Bot. Gard. 42: 1–76.
Mitchell, J.D., Daly, D., Pell, S.K., Randrianasolo, A. 2006. *Poupartiopsis* gen. nov. and its context in Anacardiaceae classification. Syst. Bot. 31: 337–348.
Moffett, R.O. 2007. Name changes in the Old World *Rhus* and recognition of *Searsia* (Anacardiaceae). Bothalia 37: 165–175.
Morton, J.F. 1981. Atlas of medicinal plants of Middle America - Bahamas to Yucatan. Springfield, IL: Charles C. Thomas.
Muller, J. 1984. Significance of fossil pollen for angiosperm history. Ann. Missouri Bot. Gard. 71: 419–443.
Muñoz, J.D.D. 1990. Flora del Paraguay. St. Louis, MO: Missouri Botanical Garden.
Muroi, H., Kubo, I. 1993. Bacterial activity of anacardic acids against *Streptococcus mutans* and their potentiation. J. Agric. Food Chem. 41: 1780-1783.
Nair, G.M., Venkaiah, K., Shah, J.J. 1983. Ultrastructure of gum-resin ducts in cashew (*Anacardium occidentale*). Ann. Bot. 51: 297–305.
Nair, G.V., Poti, A.N., Pillay, P.P. 1952a. The constituents of lacquer-bearing trees of Travancore-Cochin: Part I - chemical examination of the constituents of *Holigarna arnottiana* Hook. f. J. Sci. Indus. Res. IIB: 294–297.
Nair, G.V., Poti, A.N., Pillay, P.P. 1952b. The constituents of lacquer-bearing trees of Travancore-Cochin: Part II - chemical examination of the latex of *Semecarpus travancorica* Bed. J. Sci. Indust. Res. IIB: 298–299.
O'Dowd, D.J., Willson, M.F. 1991. Associations between mites and leaf domatia. TREE 6(6): 179–182.
Olivera, L., Ludlow-Wiechers, B., Fonseca, R.M. 1998. Anacardiaceae. In: Ludlow-Wiechers, B., Hooghiemstra, H. (eds.) Flora Palinologica de Guerrero. No. 7. Mexico: Facultad de Ciencias, UNAM.
Paula, J.E.D., Alves, J.L.D.H. 1973. Anatomia de *Anacardium spruceanum* Benth. ex Engl. (Anacardiaceae da Amazonia). Acta Amaz. 3: 39–53.
Pell, S.K. 2004. Molecular systematics of the cashew family (Anacardiaceae). Ph.D. dissertation, Baton Rouge, LA: Louisiana State University.
Pell, S.K., Mitchell, J.D., Lowry, P.P., Randrianasolo, A., Urbatsch, L.E. 2008. Phylogenetic split of Malagasy and African taxa of *Protorhus* and *Rhus* (Anacardiaceae) based on cpDNA *trnL-trnF* and nrDNA *ETS* and *ITS* sequence data. Syst. Bot. 33: 375–383.
Plisko, M.A. 1996. Family Anacardiaceae. In: Takhtajan, A.L. (ed.) Comparative seed anatomy. Vol. 5. Dicotyledons. Rosidae I (in Russian). St. Petersburg: Mir i semiya, pp. 445–469.
Poulsen, J.R., Clark, C.J., Smith, T.B. 2001. Seed dispersal by a diurnal primate community in the Dja Reserve, Cameroon. J. Trop. Ecol. 17: 787–808.
Prendergast, H.D.V., Jaeschke, H.F., Rumball, N. 2001. A lacquer legacy at Kew. Kew, London: Royal Botanic Gardens.
Ramírez, J.L., Cevallos-Ferriz, S.R.S. 2002. A diverse assemblage of Anacardiaceae from Oligocene sediments, Tepexi de Rodriguez, Puebla, Mexico. Am. J. Bot. 89: 535–545.
Record, S.J. 1939. American woods of the family Anacardiaceae. Trop. Woods 60: 11–45.
Reid, E.M. 1933. Note on some fossil fruits of Tertiary age from Colombia, South America. Rev. Geogr. Phys. 6: 209–216.
Rivero-Cruz, J.F., Chávez, D., Hernández, B., Anaya, A.L., Mata, R. 1997. Separation and characterization of *Metopium brownei* urushiol components. Phytochemistry 45: 1003–1008.

Rodriguez, J., Howard, B., Robertson, K.R., Nevling, L.I. 2003. In Memoriam Richard Alden Howard, 1917–2003. In: Robertson, K.R. (ed.) ASPT Newslett. 17(2): 5–7.
Ronse Decraene, L.P., Smets, E.F. 1995. The distribution and systematic relevance of the androecial character oligomery. Bot. J. Linn. Soc. 118: 193–247.
Roth, I. 1969. Estructura cortical de algunas especies venezolanas de Anacardiaceae. Acta Biol. Venez. 6: 146–160.
Roth, I. 1981. Structural patterns of tropical barks. Encyclopedia of plant anatomy. Berlin: Borntraeger.
Sachar, R.C., Chopra, R.N. 1957. A study of the endosperm and embryo in *Mangifera* L. Ind. J. Agric. Sci. 27: 219–228.
Savolainen, V., Chase, M.W., Hoot, S.B., Morton, C.M., Soltis, D.E., Bayer, C., Fay, M.F., de Bruijn, A.Y., Sulllivan, S., Qiu, Y.-L. 2000a. Phylogenetics of flowering plants based on combined analysis of plastid atpB and rbcL sequences. Syst. Biol. 49: 306–362.
Savolainen, V., Fay, M.F., Albach, D.C., Backlund, A., van der Bank, M., Cameron, K.M., Johnson, S.A., Lledó, M.D., Pintaud, J.-C., Powell, M., Sheahan, M.C., Soltis, D.E., Soltis, P.S., Weston, P., Whitten, W.M., Wurdack, K.J., Chase, M.W. 2000b. Phylogeny of the eudicots: a nearly complete familial analysis based on *rbc*L gene sequences. Kew Bull. 55: 257–309.
Saxena, G., McCutcheon, A.R., Farmer, S., Towers, G.H. N., Hancock, R.E.W. 1994. Antimicrobial constituents of *Rhus glabra*. J. Ethnopharmacol. 42: 95–99.
Silva, J.D.E. 1973. Catalogo de nervaçao foliar das Anacardiaceae da caatinga – I. Arq. Jard. Bot. Rio de Janeiro 14: 249–256.
Stahl, E., Keller, K., Blinn, C. 1983. Cardanol, a skin irritant in pink pepper (*Schinus terebinthifolius*). Planta Medica 48: 5–9.
Suresh, M., Raj, R.K. 1990. Cardol: the antifilarial principle from *Anacardium occidentale*. Curr. Sci. 59: 477–479.
Takhtajan, A. 1987. Systema Magnoliophytorum. Leningrad: Nauka.
Taylor, D.W. 1990. Paleobiogeographic relationships of angiosperms from the Cretaceous and early tertiary of the North American area. Bot. Rev. 56: 279–417.
Terrazas, T. 1994. Wood anatomy of the Anacardiaceae: ecological and phylogenetic interpretation. Ph.D. dissertation, Chapel Hill, NC: University of North Carolina.
Terrazas, T. 1995. Anatomia sistematica de la familia Anacardiaceae en Mexico. I. La corteza de *Tapirira*. Aublet. Bot. Soc. Bot. Mexico 57: 103–112.
Thorne, R.F. 1992. Classification and geography of the flowering plants. Bot. Rev. 58: 225–348.
Tiffney, B.H., Fleagle, J.G, Bown, T.M. 1994. Early to Middle Miocene angiosperm fruits and seeds from Fejej, Ethiopia. Tertiary Res. 15: 25–42.
Tyman, J.H., Morris, L.J. 1967. The composition of cashew nut-shell liquid (CNSL) and the detection of a novel phenolic ingredient. J. Chrom. 27: 287–288.
Vassilyev, A.E. 2000. Quantitative ultrastructure data of secretory duct epithelial cells in *Rhus toxicodendeon*. Int. J. Plant Sci. 161: 615–630.
Venkaiah, K. 1992. Development, ultrastructure and secretion of gum ducts in *Lannea coromandelica* (Houtt.) Merrill (Anacardiaceae). Ann. Bot. 69: 449–457.
Venkaiah, K., Shah, J.J. 1984. Distribution, development and structure of gum ducts in *Lannea coromandelica* (Houtt.) Merril. Ann. Bot. 54: 175–186.
Venning, F.D. 1948. The ontogeny of the laticiferous canals in the Anacardiaceae. Am. J. Bot. 35: 637–644.
Viana, G.S.B., Bandeira, M.A.M., Moura, L.C., Souza-Filho, M.V.P., Matos, F.J.A., Ribeiro, R.A. 1997. Analgesic and anti-inflammatory effects of the tannin fraction from *Myracrodruon urundeuva* Fr. All. Phytoth. Res. 11(2): 118–122.
Vogel, St. 1978. Evolutionary shifts from reward to deception in pollen flowers. In: Richards, A.J. (ed.) The pollination of flowers by insects. Linn. Soc. Symp. Ser. 6: 89–96. London: Academic Press.
von Teichman, I. 1991. Ontogeny of the seed-coat of *Rhus lancea* L. fil., and pachychalazy in the Anacardiaceae. Bot. J. Linn. Soc. 107: 35–47.
von Teichman, I. 1998. Micromorphological structure of the fruit and seed of *Smodingium argutum* (Anacardiaceae), as an adaptation to its natural habitat. S. Afr. J. Bot. 64: 121–127.
von Teichman, I., van Wyk, A.E. 1996. Taxonomic significance of pericarp and seed structure in *Heeria argentea* (Thunb.) Meisn. (Anacardiaceae), including reference to pachychalazy and recalcitrance. Bot. J. Linn. Soc. 122: 335–352.
Wang, B.C., Sork, V.L., Leong, M.T., Smith, T.B. 2007. Hunting of mammals reduces seed removal and dispersal of the Afrotropical tree *Antrocaryon klaineanum* (Anacardiaceae). Biotropica 39: 340–347.
Wannan, B.S. 1986. Systematics of the Anacardiaceae and its allies. Ph.D. dissertation, Sydney: University of New South Wales.
Wannan, B.S., Quinn, C. 1990. Pericarp structure and generic affinities in the Anacardiaceae. Bot. J. Linn. Soc. 102: 225–252.
Wannan, B.S., Quinn, C. 1991. Floral structure and evolution in the Anacardiaceae. Bot. J. Linn. Soc. 107: 349–85.
Wannan, B.S., Quinn, C. 1992. Inflorescence structure and affinities of *Laurophyllus* (Anacardiaceae). Bot. J. Linn. Soc. 109: 235–245.
Wannan, B.S., Waterhouse, J.T., Gadek, P.A., Quinn, C.J. 1985. Biflavonyls and the affinities of *Blepharocarya*. Biochem. Syst. Ecol. 13: 105–108.
Wannan, B.S., Waterhouse, J.T., Quinn, C.J. 1987. A taxonomic reassessment of *Blepharocarya* F. Muell. Bot. J. Linn. Soc. 95: 61–72.
White, F. 1976. The underground forests of Africa: A preliminary review. Gard. Bull. Straits Settlem. 29: 57–71.
Wilkinson, H.P. 1971. Leaf anatomy of various Anacardiaceae with special reference to the epidermis. Ph.D. thesis, London: University of London.
Wilkinson, H.P. 1979. The plant surface (mainly leaf): Domatia. In: Metcalfe, C.R., Chalk, L. Anatomy of the dicotyledons. Oxford: Clarendon Press.
Wilkinson, H.P. 1983. Leaf anatomy of *Gluta* (L.) Ding Hou (Anacardiaceae). Bot. J. Linn. Soc. 86: 375–403.
Yakovlev, M.S., Zhukova, G.Y. 1973. Angiosperms with green and colorless embryo. Leningrad: Nauka.

Yi, T., Miller, A.J., Wen, J. 2004. Phylogenetic and biogeographic diversification of *Rhus* (Anacardiaceae) in the Northern Hemisphere. Mol. Phylogenet. Evol. 33: 861–879.

Yi, T., Miller, A.J., Wen, J. 2007. Phylogeny of *Rhus* (Anacardiaceae) based on sequences of nuclear *Nia-i3* intron and chloroplast *trnC-trnD*. Syst. Bot. 32(2): 379–391.

Young, D.A. 1974. Comparative wood anatomy of *Malosma* and related genera (Anacardiaceae). Aliso 8: 133–146.

Yunus, M., Yunus, D., Iqbal, M. 1990. Systematic bark morphology of some tropical trees. Bot. J. Linn. Soc. 103: 367–377.

Anisophylleaceae

Anisophylleaceae Ridl. (1922).
Rhizophoraceae subfam. Anisophylloideae A.F.W. Schimper (1893).

A.E. Schwarzbach and P.B. Tomlinson

Aluminum-accumulating trees and shrubs of wet primary tropical forests; indumentum sparse, of short unicellular trichomes; *Combretocarpus* flowers with multicellular peltate trichomes. Leaves alternate, sometimes dimorphic (anisophyllous), estipulate. Inflorescences axillary, racemose to paniculate. Flowers (3)4(5)-merous, small, mostly unisexual by abortion, with smaller male flowers (plants monoecious) but bisexual in *Combretocarpus*; calyx and petals valvate; petals distinct, lobed or laciniate (entire in *Polygonanthus*); stamens twice the number of petals; anthers dorsifixed, introrse; nectary disk intra- and interstaminal, discontinuous, lobed or crenate, on top of the ovary; gynoecium of (3)4(5) united carpels with distinct stylodia; ovary (hemi) inferior, 3–4-locular, with 1 or 2 ovules per loculus. Fruit usually a 1- or few-seeded drupe but dry and winged in *Combretocarpus*. Endosperm 0, the embryo with reduced or no cotyledons; germination hypogeal. $n = 7, 8$.

A family of 4 genera and about 34 species in SE Asia, South America and Africa.

Vegetative Morphology. The distinctive growth habit of *Anisophyllea disticha* and probably *A. scortechini* as well may characterize the family, but in species known only from herbarium material it is difficult to ascertain because the reduced leaves are very small and caducous (Juncosa and Tomlinson 1988b). In its typical expression, the architecture is that of Massart's model, the orthotropic axis bearing only spirally arranged scale leaves, the plagiotropic axes, borne in pseudowhorls, with distinctly dimorphic leaves, producing flattened and markedly dorsiventral branch complexes. Leaves are then uniquely arranged with pairs of scale leaves alternating on the upper surface of the shoot and a corresponding series of foliage leaves on the lower surface (Fig. 8A, B), this arrangement determined by the shoot apex within the bud, i.e., without secondary reorientation (Vincent and Tomlinson 1983). Taxonomic descriptions of most other species are based on flowering branches, which are lateral, plagiotropic shoots apparently with isomorphic leaves. However, the dimorphy of *A. disticha* may be expressed in the juvenile stage of other species and may undergo metamorphism into a distinct adult stage. This feature needs to be examined in much greater detail and in a phylogenetic context.

The morphology of other genera is less well known and *Combretocarpus* has monomorphic leaves. Another feature of the family is the occurrence of serial buds in the foliage leaf axil, with up to six buds increasing in size distally. The venation of foliage leaves varies from divergent in *Anisophyllea* to pinnately veined in *Combretocarpus*. Stipules are absent, except for glandular structures in a stipular position reported for *Anisophyllea disticha* (Keating and Randrianasolo 1988). Stipules ascribed to *Polygonanthus* (Prance et al. 1975) may represent the small caducous leaves of the anisophyllous series. Tiny aerial roots (1–4 cm) are known from *Combretocarpus*.

Vegetative Anatomy. Vessels elements have simple perforations; intervessel pitting is alternate with coalescent apertures. The fibres have distinctly bordered pits. Axial parenchyma is apotracheal and often banded, and varies to paratracheal and irregular. Rays are of two distinct sizes and are 1–20-seriate, with multiseriate ray tails. Lysigenous secretory canals are present in the parenchymatous tissue of *Poga*. Nodes are unilacunar. Stomata are variable but usually paracytic. Leaf venation has been studied by Keating and Randrianasolo (1988); leaf margins are usually entire, with exception of a single species of *Anisophyllea* (*A. meniandii*), which has very simple, non-vascularized glands. Sieve element plastids are S-type (Behnke 1988).

Inflorescence and Floral Morphology. Inflorescences are paniculate or racemose and usually somewhat open-branched. A single bract subtends each branch and flower. In *Anisophyllea disticha*, male and female flowers occur on separate inflorescences, but in most other species and genera in the family the two kinds of flower are mixed in the inflorescences. Floral structure of Anisophylleaceae has been studied by Tobe and Raven (1988b) and Matthews et al. (2001). Except for *Polygonanthus*, the petals are deeply incised (Fig. 8D). Nectariferous lobes are found at both intra- and interstaminal positions, but do not form a continuous annular disk; in *Combretocarpus* they are only intrastaminal. A compitum seems to be lacking in Anisophylleaceae. For the strong similarities between Anisophylleaceae and Cunoniaceae revealed by Matthews et al. (2001), see under "Affinities".

Embryology and Seed Morphology. Pollen grains are 2-celled when shed.

All genera agree in having persistent nucellar tissue at least until early stages of seed development. Otherwise, the genera are very diverse embryologically. The ovules of *Anisophyllea* and *Combretocarpus* are unitegmic. Those of *Poga* and *Polygonanthus* are bitegmic, and the raphe bundle ramifies in the outer integument, whereas the inner integument is very thin (2-layered) and does not contribute to the formation of the seed coat. *Anisophyllea* and *Polygonanthus* have a Polygonum type embryo sac, while that of *Combretocarpus* is Allium type (Tobe and Raven 1987, 1988a). Endosperm formation is of the Nuclear type. The seeds are exalbuminous (Floret 1979; Tobe and Raven 1987), and the embryo is represented by the swollen extended hypocotyl, having either small cotyledons (*Combretocarpus*) or rudimentary and/or no cotyledons (*Anisophyllea* and *Poga*). The seed coat is testal and formed by the outer epidermis alone (*Combretocarpus*), by both the outer epidermis and the multiple inner layer (*Anisophyllea*), or by both the multiple outer layer and the multiple inner layer (*Poga*). Germination is usually described as hypogeal (Tobe and Raven 1987).

Pollen Morphology. Pollen is mainly tricolporoidate, and syncolpate grains are common. Endoapertures, when present, are circular but poorly defined. *Anisophyllea disticha* is unusual in the striate surface of pollen grains (Vezey et al. 1988). *A. manauensis* has tetrad pollen (Prance et al. 1975).

Karyology. *Combretocarpus* has $n = 8$, the other three genera $n = 7$ (Tobe and Raven 1987).

Phytochemistry. Ellagic acid and derivatives thereof are known from *Anisophyllea* (Hegnauer 1973).

Intrafamilial Relationships. As described by Tobe and Raven based on morphological characters, two major branches within the family may be distinguished, one comprising *Poga* and *Polygonanthus*, another with *Anisophyllea* and *Combretocarpus*, the former pair retaining plesiomorphies, the latter possessing derived traits (Tobe and Raven 1987, 1988a). However, recent molecular studies including all genera and based on a multigene dataset have revealed a very different picture. *Anisophyllea* is sister genus to *Poga*, both are sister to *Polygonanthus*, and *Combretocarpus* is sister to the rest of the family (Zhang et al. 2007).

Affinities. Anisophylleaceae have traditionally been included in Rhizophoraceae, although since the treatment of Bentham and Hooker (1865) always as a distinct tribe or subfamily (Schimper 1893). The group was elevated to family rank by Ridley (1922), but without consideration of the non-Malaysian taxa; Melchior (1964) followed this course; however, *Poga* and *Polygonanthus* were not added until 1975. As a result of careful and extended analysis of both families with the addition of much new data, it has been recognized that the two families are widely separated (Juncosa and Tomlinson 1988a, b). Dahlgren (1988) proposed the separation of Rhizophoraceae from Myrtales, where it had traditionally resided, and suggested it should be included in his Celastrales, while Anisophylleaceae were thought to have a

Rosalean affinity. A series of thorough morphological and anatomical studies supported the exclusion of Anisophylleaceae from Rhizophoraceae (included in Raven and Tomlinson 1988). Takhtajan (2009) established an order Anisophyllales that he placed close to his Cunoniales. Molecular studies, in which only *Anisophyllea* and *Combretocarpus* were included (Setogushi et al. 1999; Schwarzbach and Ricklefs 2000), resolved Anisophylleaceae as a member of Cucurbitales. Later it was shown that they are sister to the rest of Cucurbitales (Zhang et al. 2006). In the light of these findings, the strong similarities in floral structure and histology between Cunoniaceae and Anisophylleaceae as described in Matthews et al. (2001) appear as homoplasies. In a more recent comparative morphological study of families of Cucurbitales, Matthews and Endress (2004) have found the prevalence of unisexual flowers and inferior ovaries in all core Cucurbitales as possible synapomorphies of Anisophylleaceae and other Cucurbitales. However, for many other characters Anisophylleaceae are quite distinct from the rest of the order (Matthews and Endress 2004).

Distribution and Habitats. The wide distribution of *Anisophyllea* (South America to Malesia) is in contrast to the more localized distribution of the remaining genera. Curiously, their ranges overlap that of *Anisophyllea* in three separate continents: *Polygonanthus* in South America, *Poga* in Africa, and *Combretocarpus* in SE Asia/Malesia. Members of the family are typically trees of wet lowland tropical forest. Species of *Anisophyllea* range in size from tall trees to small treelets; *A. disticha* is a common understorey component of the Malaysian rainforest, with the distinctive vegetative morphology that gives the genus its name. *Anisophyllea disticha* is widely distributed throughout the Sunda shelf area (except Java), but other Asian species are more localized within three discrete regions of the Malay Peninsula, Sumatra, and Borneo (Ding Hou 1958). *Combretocarpus* is a dominant tree of Bornean freshwater swamps, apparently extinct in peninsular Malaya.

Paleobotany. Pollen of *Combretocarpus* has been described from middle and late Miocene deposits of northwestern Borneo (Anderson and Muller 1975; Morley 1977).

Uses. Only one species (*Anisophyllea griffithii*) is reported to reach commercially exploitable size (Kochummen 1989).

Key to the Genera

1. Flowers bisexual, trimerous; petals linear or irregularly 3–4-lobed. Fruit 3-winged. Leaves always monomorphic, with pinnate venation **4. *Combretocarpus***
– Flowers unisexual or polygamous, rarely bisexual, tetramerous; petals often appendaged. Fruit not winged, usually a drupe. Leaves mostly with divergent venation, sometimes dimorphic 2
2. Flowers strongly heteromorphic; serial buds not conspicuously developed. Fruit with 3–4 edible seeds. Secretory canals present **2. *Poga***
– Flowers not strongly heteromorphic; serial buds conspicuously developed. Fruits usually 1-seeded, inedible. Secretory canals absent 3
3. Petals lobed. Leaves with divergent venation, often dimorphic either in the juvenile stage or as adults **1. *Anisophyllea***
– Petals unlobed. Leaves without divergent venation, monomorphic **3. *Polygonanthus***

Genera of Anisophylleaceae

1. ***Anisophyllea*** R. Br. ex Sabine Fig. 8

Anisophyllea R. Br. ex Sabine, Trans. Hort. Soc. 5: 446 (1824); Ding Hou, Fl. Males. I, 5: 480 (1958).

Monoecious shrubs, treelets or tall trees to 38 m, shoots often distinctly dimorphic (possibly always in juvenile stage). Leaves alternate, spiral on orthotropic shoots, distichous on plagiotropic branches, often asymmetric, foliage leaves with 3–5 longitudinal veins diverging from the base. Inflorescences (supra)axillary, solitary or serial, simple or branched, ebracteate or at most with small bracts. Flowers usually unisexual, rarely bisexual, usually tetramerous; sepals adnate to ovary; petals entire, lobed or lacerate, sometimes with glandular apical appendages; male flowers with stamens twice as many as the petals, the outer stamens longest; nectariferous tissue at base of stamens; pistillode usually 0; female flowers usually with staminodes; ovary 3–5-locular with 1 apical pendulous ovule per loculus; stylodia (3)4. Fruit ellipsoid to pear-shaped, usually 1-seeded; pericarp 3-layered, rather fibrous;

testa leathery; embryo linear and undifferentiated, or represented largely by the hypocotyl.

About 30 species, mostly distributed from tropical Africa to SE Asia/Malesia (Malaysia, Borneo, Sumatra), *A. guianensis* Sandw. and *A. manausensis* Pires & Rodr. in northern South America.

2. *Poga* Pierre

Poga Pierre, Bull. Soc. Linn. Paris 2: 1254 (1896); Hutchinson & Dalziel, Fl. W. Trop. Africa, ed. 2: 282 (1954).

Large tree. Leaves with numerous lateral nerves. Inflorescences catkin-like on specialized leafless branches. Flowers tetramerous, small, unisexual, strongly heteromorphic; petals 3–7-lobed, the lobes with glandular tips; ovary 4-locular; with 1 ovule per loculus. Fruit a drupe up to 6 cm long with 3–4 edible oily seeds. Seeds 2 cm long; seed coat brown, coriaceous; embryo undifferentiated or represented by the hypocotyl, with abundant oil in cortex and medulla.

A single species, *P. oleosa* Pierre, restricted to Equatorial Africa.

The seeds are edible (Inoi nut), but seem little exploited commercially (Vaughan 1970).

3. *Polygonanthus* Ducke

Polygonanthus Ducke, Notizbl. Bot. Gard. Berlin-Dahlem 11: 345 (1932); Prance et al., Acta Amazonica 5: 17–22 (1975).

Small trees. Leaves distichous. Inflorescences racemose but appearing spicate. Flowers usually unisexual, strongly heteromorphic, tetramerous. Male flowers small, with apically inflexed sepals; stamens 8, distinct, inserted at the base of the sepals, exserted or included at anthesis; filaments inflexed in bud, often flattened; glandular lobes of nectary enclosing base of anthers; female flowers larger, few, usually at the base of the inflorescence, petals small, lanceolate; staminodes 8, with reduced anthers, nectary small; ovary half-inferior, forming a tube with the calyx, 4-locular, with 1 ovule per locule; stylodia 4. Fruit careniform, 4-angled.

Two species along river margins in Amazonian Brazil.

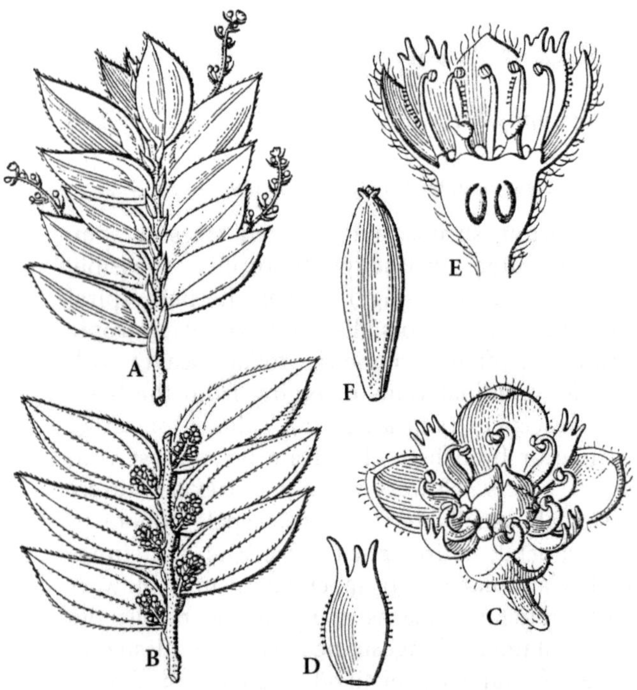

Fig. 8. Anisophylleaceae. *Anisophyllea disticha*. **A** Flowering branch, upper surface. **B** Flowering branch, lower surface. **C** Male flower. **D** Petal. **E** Longitudinal section of female flower with stylodia and staminodes. **F** Dried fruit. (Ding Hou 1958; artwork by R. van Crevel)

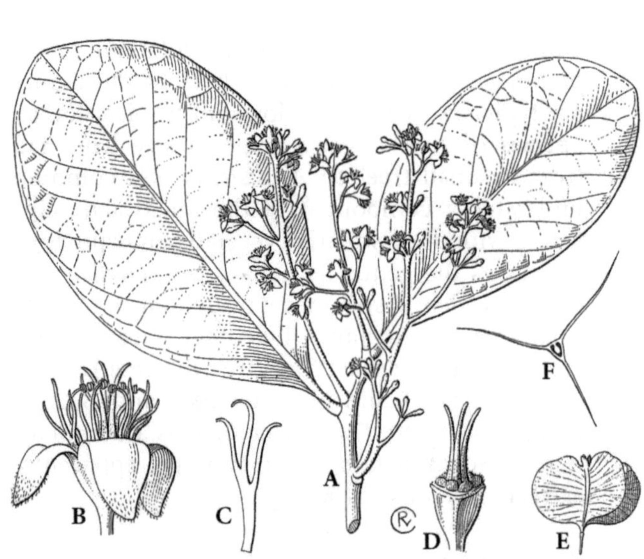

Fig. 9. Anisophylleaceae. *Combretocarpus rotundatus*. **A** Flowering branch. **B** Flower. **C** Petal. **D** Gynoecium. **E** Fruit. **F** Fruit, transverse section. (Ding Hou 1958; artwork by R. van Crevel)

4. *Combretocarpus* Hook. f. Fig. 9

Combretocarpus Hook. f. in Benth. & Hook., Gen. Pl. 1: 683 (1865); Ding Hou, Fl. Males. I, 5: 480 (1958).

Tree to 25 m, easily coppicing or sprouting from fallen branches. Leaves obovate to broadly elliptic with 6–10 pairs of lateral veins diverging from a prominent mid-rib. Inflorescence short racemes; flowers c. 4 mm diameter, yellow, bisexual, usually trimerous, with multicellular peltate trichomes; sepals obtuse, reflexed at anthesis; petals c. 2 mm long, linear, entire or deeply divided; stamens usually 6, obdiplostemonous, inserted between shallow lobes of the nectariferous tissue; ovary inferior, usually 3-ribbed, trilocular, each loculus with 2 ovules; stylodia 3. Fruit 3-winged, 1-seeded by abortion. Seeds elongate, narrow.

One species, *C. rotundatus* (Miq.) Danser, mainly of swamp forests in Sumatra and Borneo, recorded for but possibly extinct in Malaya, often gregarious, possibly by its ability to sucker, and now probably over-exploited because of its useful timber.

Selected Bibliography

Anderson, J.A.R., Muller, J. 1975. Palynological study of a Holocene peat and a Miocene coal deposit from NW Borneo. Rev. Palaeobot. Palynol. 19: 291–351.

Behnke, H.-D. 1988. Sieve element plastids and systematic relationships of Rhizophoraceae, Anisophylleaceae, and allied groups. Ann. Missouri Bot. Gard. 75: 1387–1409.

Dahlgren, R.M.T. 1988. Rhizophoraceae and Anisophylleaceae: summary statement, relationships. Ann. Missouri Bot. Gard. 75: 1259–1277.

Ding Hou 1958. Rhizophoraceae. Flora Malesiana I, 5: 429–493.

Floret, J.-J. 1979. A propos du contenu séminal dans les genres *Anisophyllea* et *Poga* (Rhizophoracées-Anisophylloidées). Adansonia II, 19:109–115.

Hegnauer, R. 1973. Chemotaxonomie der Pflanzen. Vol. 6. Basel: Birkhäuser.

Juncosa, A.M., Tomlinson, P.B. 1988a. A historical and taxonomic synopsis of Rhizophoraceae and Anisophylleaceae. Ann. Missouri Bot. Gard. 75: 1278–1295.

Juncosa, A.M., Tomlinson, P.B. 1988b. Systematic comparison and some biological characteristics of Rhizophoraceae and Anisophylleaceae. Ann. Missouri Bot. Gard. 75: 1296–1318.

Keating, R.C., Randrianasolo, V. 1988. Leaf architecture and relationships in the Rhizophoraceae and Anisophylleaceae. Ann. Missouri Bot. Gard. 75: 1343–1368.

Kochummen, K.M. 1989. Rhizophoraceae. In: Ng, F.S.P. (ed.) Tree Flora of Malaya 4: 302–323.

Matthews, M.L., Endress, P.K. 2004. Comparative floral structure and systematics in Cucurbitales (Corynocarpaceae, Coriariaceae, Tetramelaceae, Datiscaceae, Begoniaceae, Cucurbitaceae, Anisophylleaceae). Bot. J. Linn. Soc. 145: 129–185.

Matthews, M.L., Endress,P.K., Schönenberger, J. Friis, E. M. 2001. A comparison of floral structures of Anisophylleaceae and Cunoniaceae and the problem of their systematic position. Ann. Bot. 88: 439–455.

Morley, R.J. 1977. Palynology of Tertiary and Quaternary sediments in southeast Asia. In: Proc. Petroleum Association 6th Annual Convention, pp. 255–276.

Prance, G.T., Freitas da Silva, M., Albuquerque, B.W., de Jesus da Silva Araújo, I., Medeiros Carreira, L.M., Nogueira Braga, M.M. Macedo, M., da Conceição, P. N., Braga Lisbôa, P.L., Braga, P.I., Lobato Lisbôa, R. C., Queiroz Vilhena, R.C. 1975. Revisão taxonômica das espécies amazônicas de Rhizophoraceae. Acta Amazonica 5: 5–22.

Raven, P.H., Tomlinson, P.B. 1988. Rhizophoraceae-Anisophylleaceae: a symposium. Ann. Missouri Bot. Gard. 75:1258.

Ridley, H.N. 1922. Flora of the Malay Peninsula. Vol. 1. London: L. Reeve.

Schimper, A.F.W. 1893. Rhizophoraceae. In: Engler & Prantl, Die natürlichen Pflanzenfamilien III, 7: 42–56. Leipzig: W. Engelmann.

Schwarzbach, A.E., Ricklefs, R.E. 2000. Systematic affinities of Rhizophoraceae and Anisophylleaceae, and intergeneric relationships within Rhizophoraceae, based on chloroplast DNA, nuclear ribosomal DNA, and morphology. Am. J. Bot. 87: 547–564.

Setogushi, H., Kosuge, H., Tobe, H. 1999. Molecular phylogeny of Rhizophoraceae based on *rbc*L gene sequences. J. Plant Res. 112: 443–455.

Takhtajan, A. 2009. Flowering plants. Dordrecht: Springer.

Tobe, H., Raven, P.H. 1987. Systematic embryology of the Anisophylleaceae. Ann. Missouri Bot. Gard. 74: 1–26.

Tobe, H., Raven, P.H. 1988a. Additional notes on the embryology of *Polygonanthus* (Anisophylleaceae) and relationships of the family. Ann. Missouri Bot. Gard. 75: 1425–1428.

Tobe, H., Raven, P.H. 1988b. Floral morphology and evolution in Anisophylleaceae. Bot. J. Linn. Soc. 98: 1–25.

Vaughan, J.G. 1970. The structure and utilization of oil seeds. London: Chapman & Hall.

Vezey, E.L., Shah, V.P., Skvarla, J.J., Raven, P.H. 1988. Morphology and phenetics of Rhizophoraceae pollen. Ann. Missouri Bot. Gard. 75: 1369–1386.

Vincent, J.R., Tomlinson, P.B. 1983. Architecture and phyllotaxis of *Anisophyllea disticha* (Rhizophoraceae). Garden's Bull. (Singapore) 36: 3–18.

Zhang, L.-B., Simmons, M.P., Kocyan, A., Renner, S.S. 2006. Phylogeny of the Cucurbitales based on DNA sequences of nine loci from three genomes: implications for morphological and sexual system evolution. Mol. Phylogen. Evol. 39: 305–322.

Zhang, L.-B., Simmons, M.P., Renner, S.S. 2007. A phylogeny of Anisophylleaceae based on six nuclear and plastid loci: Ancient disjunctions and recent dispersal between South America, Africa, and Asia. Mol. Phylogen. Evol. 44: 1057–1067.

Begoniaceae

Begoniaceae Bercht. & J. Presl (1820), nom. cons.

J.J.F.E. de Wilde

Perennial or occasionally annual herbs, sometimes suffructescent and even forming poorly branched subshrubs up to 4 m tall; the species monoecious, exceptionally dioecious, terrestrial or epiphytic, evergreen or in some tuberous species with a dormant period devoid of aerial parts or very rarely the plants with deciduous leaves; stems prostrate, erect, reclinate, scrambling or rarely climbing with patent branches or with aerial roots, often rhizomatous and rooting at the nodes, herbaceous, succulent or ligneous; or the plants tuberous and short-stemmed or acaulescent; trichomes simple, stellate or scaly, rarely plants completely glabrous. Leaves petiolate or subsessile, alternate, sometimes distichously arranged, exceptionally opposite or whorled, herbaceous or succulent, occasionally strikingly variegated, peltate or not, simple and entire or variously lobed, or fissured to palmately compound, usually asymmetrical and the halves of the blade generally very unequal; venation palmate or pinnate; occasionally with bulbils in the leaf axils; stipules ovate to triangular, frequently navicular, sometimes large and membranous, persistent or caducous. Inflorescences axillary, cymose, di- or monochasial or combinations of this, exceptionally female flowers paired or solitary, more rarely terminal and then either cymose or raceme-like with cymose branches, bracteate, unisexual or androgynous and then commonly the male flowers reaching anthesis first and early caducous, or in raceme-like bisexual inflorescences the basal branches exclusively with female proterogynous flowers; bracts sometimes persistent, large and enveloping the inflorescence or, more often, early caducous, small; prophylls present or not. Flowers unisexual, often with perianthal dimorphism, actinomorphic or irregular; the pedicel of staminate flowers usually jointed near the base from where eventually the male flower drops after anthesis; perianth segments (tepals) 2(3), 4 or 5, exceptionally up to 11, as a rule distinct though sometimes variably connate, usually white or pink, less often red or yellow, very rarely differentiated in calyx and corolla (*Hillebrandia*); staminate flowers devoid of pistillode; androecium actinomorphic or zygomorphic; stamens 3–>100; filaments distinct or variously clustered; anthers basifixed, 2-celled, basically tetrasporangiate; thecae longidehiscent or opening by an apical slit or pore; connective sometimes extended; pistillate flowers lacking staminodes; ovary inferior or exceptionally semi-inferior (*Hillebrandia*), usually 1–3 (–7) winged or horned, ribbed or angled, more rarely apterous, usually 2–3-carpellate, (1)2–4 (–7)-locular; placentation axile, sometimes parietal, septal or pseudo-axile; placentas entire, bifid or variously branched; stylodia commonly equal in number to the locules, persistent or caducous, often partly connate at the base and once or more times forked towards the apex, the stigmatic tissue covering the apical parts papillate, often in a continuous coiled band or variously shaped; ovules many, anatropous, bitegmic, crassinucellate. Fruit predominantly capsular, with subequal or very unequal wings or horns, occasionally fleshy, berry-like, loculicidally dehiscent (between the stylodia in *Hillebrandia*) or indehiscent. Seeds barrel-shaped, many, minute, testa reticulate composed of a micropylar-hilar part which at germination is lifted off as an operculum (seed lid), collar cells and a chalazal part; embryo relatively large, enveloped by a finally single-layered endosperm.

A pantropical and subtropical family extending into parts of temperate Asia, absent from Australia and the Pacific islands from Fiji to the Galapagos; two genera, one monotypic, the other presently with more than 1,500 named species.

Vegetative Morphology. The majority of the species are perennial, herbaceous and rhizomatous with thickened stems and relatively short internodes. Annuals are rare. The stems are often creeping with the leaves in two ranks, either upright or with the blades almost flattened to the substrate. Sometimes the distal part of the rhizome is upright or the rhizome gives rise to erect stems. The group of begonias which, from underground rhizomes, sends up erect or semi-erect bamboo-like stems with swollen nodes and relatively long internodes is appropriately known as 'cane-like' in horticulture. Upright-stemmed, regularly branched, shrub-like species form another major group of perennial begonias. Some rare species are ligneous, tree-like or lianescent. Tuberous perennials in which the tubers either are of hypocotylar origin or represent thickened portions of creeping rhizomes may be acaulescent or caulescent, periodically shedding their stems and leaves; particular groups of species frequently produce bulbils in the axils of their leaves and, in a rare case, a thickened rhizome apically bears cormlets functioning as diaspores. Finally, there is a group of epiphytic species including what are known as trailing-scandent begonias in horticulture. Among these, a few shed most of their leaves during unfavourable periods and may be called stem succulents. Most species have several leaves, a few have only one or two. The petioled leaves are herbaceous, sometimes thin but usually rather thick and succulent; next to all shades of green, they often show distinctive bright, occasionally metallic colour patterns involving red, brown, silver and black. The leaves are commonly oblique and asymmetric, rarely symmetric, very variable in size and in outline, from 1 cm to more than 50 cm, from orbicular to linear, sometimes peltate. The leaf margins range from entire to leaves in which the lamina is deeply incised between the major veins and sometimes between the secondary veins as well. Plants with deeply dissected leaves are referred to as "fern leaf" begonias (Hallé 1972). Digitately compound leaves also occur. The venation is basically palmate (Figs. 10, 11), but pinnate and palmate-pinnate venation also exists. Stipules are always present, sometimes large and leafy, distinct, equal in dimensions or not, either persistent or early caducous, almost completely encircling the stem, often leaving distinct annular scars.

Vegetative Anatomy. Anatomical observations on the axis are presented by Metcalfe and Chalk (1950), based partly on Hildebrand (1859) and later elaborated upon by Lee (1974). Underneath a 1–4-layered epidermis arises a thin layer of sub-epidermal cork. The cortex is differentiated into a collenchymatous outer zone and an inner cortex composed of thin-walled tissue containing chlorophyllaceous cells, both solitary and clustered calcium oxalate crystals and occasionally anthocyanin. In this ground tissue, a transverse section shows the vascular bundles arranged in a single ring, separated or more or less connate into a closed cylinder. The basic tissue of the secondary xylem consists of septate parenchymatous elements with simple pits. The length walls of the vessels, where in contact, are provided with scalariform bordered pits. The pith is composed of large parenchymatic cells. Pneumatodes, resembling and replacing typical lenticels, are described from the surface of the stem of some *Begonia* species.

The first elaborate study on the anatomy of the leaves was by Fellerer (1892), later summarized and amplified by Solereder (1899). Functional aspects of this anatomy were added by Haberlandt (1916) and elaborated upon by Seybold (1955), Neubauer (1967), Lee and Stone (1979) and Lee (1983). Detailed research on micromorphological and anatomical leaf characters was found useful in more recent taxonomic and phylogenetic studies (Cuerrier et al. 1990, 1991a, b; Arends 1992; Sosef 1994).

The leaves are usually provided with a single-layered epidermis on both surfaces, the cells often large and thin-walled, those on the adaxial surface frequently convex or papillose, those on the abaxial surface often containing anthocyanin. The cuticle may be smooth or with granular, striate or verrucose ornamentation. Usually, a hypoderm is present on one or both surfaces; this may be single- or more-layered and large-celled; in some species it is exceptionally well developed but species groups missing a hypoderm also occur. The stomata are confined to the abaxial surface and are occasionally slightly but distinctly elevated above the surface; they are either solitary and more or less regularly distributed, or from 2–9 in definite groups, usually of the helicocytic type, rarely anisocytic, anomocytic or still otherwise (Payne 1970); each pair of guard cells is surrounded by

3–8 subsidiary cells. Hydathodes are commonly present on the upper surface, most confined to the border and associated with an enlarged vein which terminates in the leaf margin and gives rise to a so-called "begonioid tooth" (Hickey and Wolfe 1975). A study of these complex structures by Brouillet et al. (1987) found them to belong to the epithem type of hydathodes. The lacunar mesophyll consists of one or more layers of parenchyma topped by a single-layered palisade tissue composed of cells which vary considerably in size and shape: up to 36 palisade cells may be found below a single epidermal cell; the chloroplasts present in the palisade cells are various in size and shape and, in certain species and species groups, they are distinctly arranged along the abaxial walls. The silvery sheen characteristic of the leaves of certain species is due to air-filled spaces between the epidermis and palisade layer. Various types of cystotyles (specialized cells with indeterminate contents), sclerites and crystals are fairly frequent in the mesophyll or around the veins. Transverse sections of the petiole show the vascular bundles usually arranged in a ring, and regularly perivascular sclerenchyma is present.

The indumentum is very variable in density and in the types of its constituent trichomes. These include capitate and non-capitate hairs and dentate to stellate scales. The hairs are always multicellular, uniseriate or multiseriate, glandular or non-glandular. The spherical, ellipsoid, club- or hammer-shaped head of capitate hairs may be few- or many-celled, and their stalk may vary in length and structure. Non-capitate hairs include shaggy, shortly spinous, scale-like and 2-armed structures next to whip-like, stellate and tufted hairs. Glandular hairs are comparatively common in the family; they are found both among the capitate and the non-capitate hairs and could be grouped into five types by Cuerrier et al. (1991a). Fimbriate peltate-helicoid membranous scales, provided with a cobweb pattern of thick bands and sitting on a 2-celled stalk (Arends 1992), are confined to the African sections *Tetraphila*, *Baccabegonia* and *Squamibegonia* and a very few species of sect. *Loasibegonia*. Hyaline spherical multicellular sessile trichomes, so-called "Meyen's pearl glands", sometimes present on one or on both leaf surfaces and often associated with the venation, are commonly found in species of sect. *Tetraphila* but also in some species of various American sections.

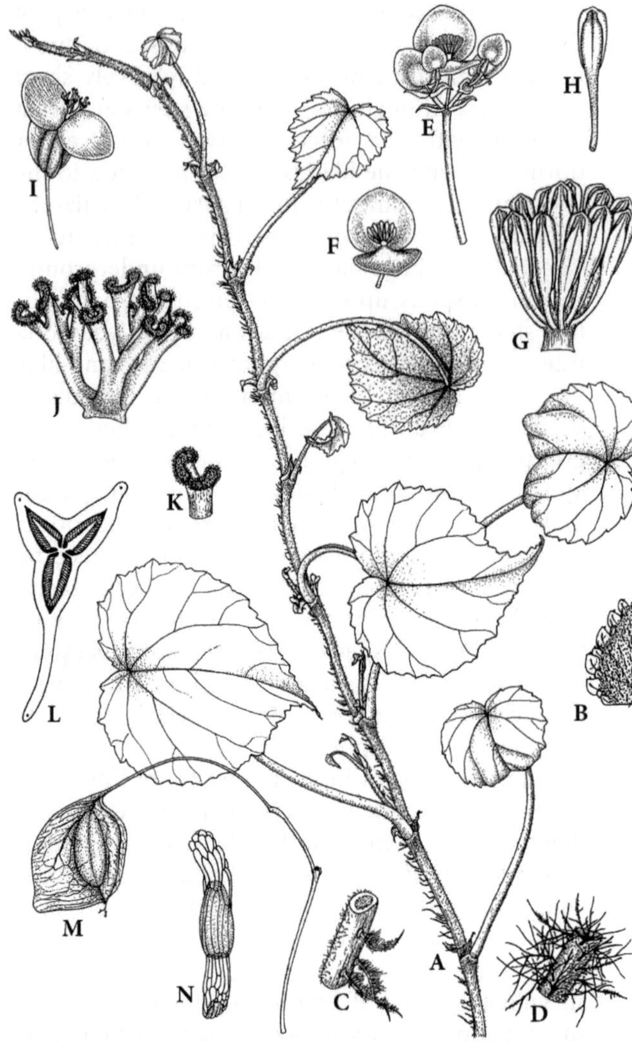

Fig. 10. Begoniaceae. *Begonia thomeana.* **A** Habit. **B** Margin of developing leaf. **C, D** Adventitious roots. **E** Inflorescence. **F** Male flower. **G** Androecium. **H** Stamen. **I** Female flower. **J** Stylodia. **K** Stigma. **L** Ovary, transversally sectioned. **M** Fruit. **N** Seed. (de Wilde 1985; drawn by Ike Zewald)

Inflorescence Structure. The inflorescences have been thoroughly studied by Irmscher (1914, 1925). Next to morphological variation, the way in which the staminate and pistillate flowers are distributed over the inflorescences leads to a great diversity in types. Their structure and architecture was re-examined by Goulet et al. (1994). They considered the inflorescence of Begoniaceae basically thyrsoid, i.e. mixed with a racemose indeterminate main axis but with cymose secondary and ultimate axes. Multiple modifications and reductions lead to the recognition of either terminal or axillary inflorescences. The terminal inflorescences present racemes with cymose branches or they are cymes; axillary inflorescences are without exception cymose. The common type of inflorescence in the family, however, is a dichasial cyme provided with bracts, either a simple 3-flowered dichasium or, after repeated ramification, a highly compound cyme (Figs. 10, 11). Often the inflorescences show a transition from biparous to uniparous cymes and, in that case, the axis represents a series of repeatedly branched lateral axes each terminated by a flower. According to Irmscher (1925), the bracts are homologous with the stipules and hence denoted as pseudo-bracts; they may be small or large, deciduous or persistent; those on the first ramification may form a flattened cup which envelops and conceals the flower-bearing part of the inflorescence, as in section *Squamibegonia* (de Wilde and Arends 1980). The axes of the inflorescences may variously be elongated or contracted and very short or almost nil; sometimes the axes elongate only in the infructescence.

In conjunction with this great wealth of structural variation, there is an often intricate but systematic distribution of the unisexual flowers which, following their differentiation into staminate and pistillate, shows a strong tendency towards spatial and/or temporal segregation. Next to bisexual inflorescences, others are found to be strictly unisexual; in certain species all three types may occur on a same branch but, in the case of unisexuality, usually the male and female inflorescences are isolated in space and/or time. Dioecism occurs but is rare. Cymose bisexual inflorescences predominantly are proterandrous; a staminate flower as a rule terminates the main axis of the inflorescence and this holds also for the usually monochasial ramifications of a higher order; only the ultimate ramification laterally bears one (rarely two) pistillate flower(s). In many-flowered bisexual cymes, this proterandry is often very pronounced; in the phase where staminate flowers reach anthesis, the development of the opposite sex often lags far behind, and it is only after all male flowers have dropped that the pistillate flowers reach anthesis. Sometimes both phases show a certain overlap but, certainly in most cases, geitonogamy and inbreeding become strongly reduced. In bisexual racemose inflorescences, however, the lowermost branches exclusively bear pistillate flowers which reach anthesis first; only when the ovaries and even the seeds therein are well developed do the distally positioned staminate flowers come to anthesis (e.g. sections *Petermannia* and *Symbegonia*). Proterogyny in this case is strict and almost certainly will lead to xenogamy.

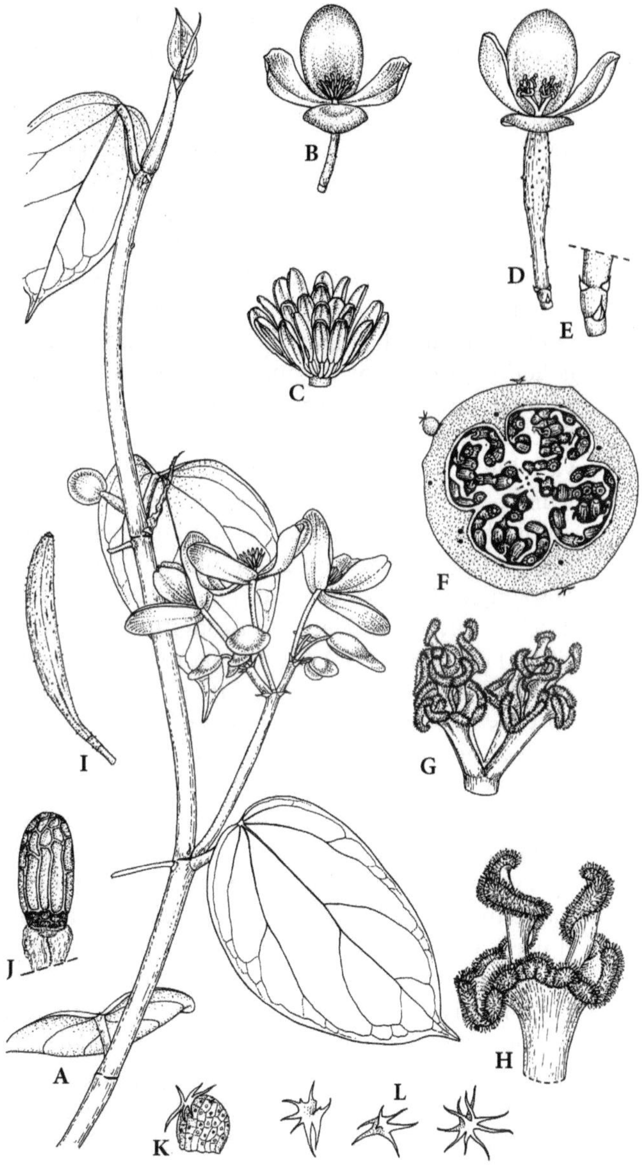

Fig. 11. Begoniaceae. *Begonia molleri.* **A** Flowering shoot with male inflorescence and young 1-flowered female inflorescence. **B** Male flower. **C** Androecium. **D** Female inflorescence. **E** Position of bracts subtending the single flower. **F** Stylodium and stigmas. **G** Ovary, transversally sectioned. **H** Multicellular trichome capped with lepidote-stellate hair. **I** Trichomes of ovary. **J** Fruit. **K** Seed. (de Wilde 2002; drawn by Ike Zewald)

Flower Structure. The flowers of the Begoniaceae are exclusively unisexual and sometimes the perianth shows sexual dimorphism. Staminate flowers contain no rudiments of a gynoecium. Their perianth is petaloid without a clear distinction between calyx and corolla (Figs. 10F, 11B), and consists of 2(3) or 4, rarely 5 (*Hillebrandia*) or exceptionally more segments inserted on top of a perianth cylinder which is considered to be a part of the male flower (de Wilde and Arends 1980), although it is commonly indicated as part of an articulated pedicel; above the cylinder, the segments are distinct or partly connate and arranged in a single bipartite valvate whorl or in two decussate such whorls, and then the larger segments of the outer whorl in bud completely envelop the inner pair; in *Hillebrandia* the perianth is differentiated into two whorls, i.e. 4–5 sepals and 4–5 alternating, much smaller petals. The androecium is actinomorphic or zygomorphic; the 3 to many stamens are distinct or variably connate into a fascicle in which the length of the filaments is variable according to their position in the bundle which, in that way, frequently resembles an amphitheatre; more rarely the filaments are completely connate into a column, e.g. in sections *Heeringia*, *Symbegonia* and *Trachelocarpus*. The 2-celled anthers dehisce with longitudinal lateral slits (Figs. 10G, 11C) but sometimes, and even within a single flower, the slits of a part of the anthers are oriented in an adaxial position, the others abaxial; rarely, opening is by apical pores or short slits; the connective is occasionally cucullate or elongated; the length of the anther in proportion to its filament varies among and within species and this ratio is sometimes of diagnostic value on the sectional level. Invariably, the male flowers after anthesis drop off as a whole from near an articulation at the base of the perianth cylinder.

The pistillate flowers, apart from aberrancies, do not contain stamen rudiments. Similarly to male flowers, the perianth is petaloid. There are 2–6(–10) usually distinct segments, although in sect. *Symbegonia* and in a few species of other sections the segments are variably connate; in many cases, the aestivation of the segments brings about a transverse zygomorphy of the flower. In *Hillebrandia*, the perianth is differentiated into two alternating regular whorls of 5 sepals and 5 tiny petals. The ovary is inferior (Figs. 10I, 11D) and (1)2–3(–6)-locular, or incompletely so. Only in *Hillebrandia* is it apparently semi-inferior, though Matthews and Endress (2004) found it basically to be also inferior. Its shape and the absence or presence of wings or horns and their position, form and dimensions as well as the indumentum are important taxonomic characteristics on the species level and above. On the basis of placentation, two major groups may be distinguished within the family, viz. species with a fundamental parietal placentation in which intrusions from the ovary wall and/or sterile placental tissue partake in the formation of spurious dissepiments, as found in *Hillebrandia* and in *Begonia* sections *Baccabegonia*, *Coelocentrum*, *Mezierea*, *Squamibegonia* and *Tetraphila*, and species with carpellar dissepiments and true axile placentation as in almost all other sections (Reitsma 1984; de Wilde and Arends 1989). In the group first mentioned, the placentation often changes from the bottom towards the top of the ovary in such way that the cavity of the ovary in the lower half is partitioned by partly placental septa and higher up shows a transition towards an open cavity with parietal placentation. In the second group, the often used distinction (even for taxa above the species level) between entire and bilamellate placentae should be handled with care as ovaries are found in which both states and their intermediates occur together. The 2–4(–7) stylodia are persistent or caducous, simple or forked or repeatedly branched (Figs. 10J, 11G), distinct or partly connate; the stigmas are predominantly kidney-shaped or hippocrepiform, more rarely lunate and covered by stigmatic tissue in a straight or often spirally twisted band. A study of the stigmatic papillae showed at least 16 different morphotypes, but this remarkable variation in detail, though diagnostic for some species, was not found suitable for supraspecific delimitation of taxa (Panda and de Wilde 1995).

Floral Anatomy. In conjunction with his own investigations, Arends (1992) summarized research on the vascular supply of the perianth segments in *Begonia*. He concluded that these studies show that usually two or more segments of a flower are supported by a number of vascular bundles and hence can be denoted as sepals, while other segments of the same flower may be supplied by only one, two or three traces and therefore properly may be indicated as petals.

Because the perianth parts are rather often similar in shape, dimensions and colour, they are generally denoted as tepals. The endothecium, a sub-epidermal layer of cells of the loculi of the anther, sometimes shows phylogenetically significant variation (Manning 1996); however, a systematic study of the diversity found in the endothecial thickenings of the anthers of Begoniaceae species showed this to be of moderate taxonomic value within and between sections in *Begonia* (Tebbitt and MacIver 1999).

Elaborate studies of the vascular anatomy and ontogeny of the pistillate flower in Begoniaceae, such as those of Gauthier (1950), Charpentier et al. (1989a, b), Arends (1992) and Sosef (1994), concentrated on the nature of the ovary, viz. on the question whether these inferior ovaries are primarily the product of an appendicular meristem system, or that a major part of the ovary is produced by an axial meristem, or again that both meristem systems act in cooperation. Most observations are in line with the hypothesis that the Begoniaceous ovary consists of carpels and that the margins of each single folded carpel are intimately connate in the centre of the ovary (Jin and Wang 1994). The origin of the ovaries of the baccate-fruited sections and of the unilocular ovaries with parietal placentation found in *Hillebrandia* and in sect. *Coelocentrum* is commonly explained by union of the margins of 2 adjacent carpels and subsequent parietal placentation, whether or not preceded by formation of largely spurious dissepiments of placental origin. Entire axile placentas are supposed to be of bilamellate origin as evidenced by various intermediate stages.

Embryology. The development of the ovules in some *Begonia* species was studied by Boesewinkel and de Lange (1983). For a survey of accumulated data on this topic, these authors refer to Davis (1966) and Corner (1976). The megaspore tetrad is linear and the chalazal cell develops into the embryo sac which becomes surrounded by a somewhat elongated nucellus and by two integuments. Both integuments are two cells thick, except for the micropylar part of the ovule where they are multilayered; the outer integument forms the exostome. After fertilization, most of the cells of the ovule do not divide any longer and further seed development is brought about by cell enlargement. The initially nuclear endosperm around the embryo sac becomes cellular and, after repeated divisions, fills the cavity around the embryo completely; these cells finally become resorbed and the full-grown embryo is surrounded only by a single-layered endosperm. The substantially enlarged outer raphal cells and the cells of the outer layer of the outer integument become tanniferous and finally form the exotestal layer of the seed. The mature ovule is elongate, anatropous, and shows a well developed funicle; it is bitegmic, with a micropyle formed by both two-layered integuments, an orbicular endostome and an exostome of irregular form; it is crassinucellate and free of starch.

Pollen Morphology. van den Berg (1984, 1985) studied the pollen of both *Hillebrandia* and *Begonia* and found 3(4)-colporate monads with compound apertures, ranging in shape from spherical to ± prolate and measuring 12–35×8–14 μm. The compound aperture consists of a very long ectoaperture, the colpus c. 2 μm wide, and an endoaperture which may be a lalongate porus or a ± elliptic colpus. In most *Begonia* species, a margo is present. The exine is very thin, usually 0.4–0.6 μm. TEM study reveals the sexine to be composed of a tectum and irregularly shaped columellae or an alveolar layer. Ornamentation is finely striate in *Hillebrandia*, less clearly so in *Begonia*, and made up of variously shaped and often sharply pointed sculptural elements in species of sect. *Symbegonia*. For fossil pollen, see under Palaeobotany.

Karyology. Legro and Doorenbos (1969, 1971, 1973) and Arends (1992) provided basic knowledge on somatic chromosome numbers in Begoniaceae and summarized the work of previous authors. A critical overview of the accumulated data is not available. In general, *Begonia* chromosomes are small and hard to observe; their counts are hindered by the frequent occurrence of fragments and an often dense texture of the protoplasm and, hence, karyotype morphology is not found very useful for the distinction of taxa. The numbers found range between $2n = 16$ for, e.g., *B. nepalensis* and possibly $2n = 156$ for species of sect. *Begonia*. Legro and Doorenbos counted the somatic numbers of 224 *Begonia* species and found that most sections are characterized by one basic chromosome number, from which the numbers of other species within the section have been derived by polyploidy. Arends (1992), in an exhaustive karyological study of a group of African *Begonia* species within the sect. *Tetraphila*, found a basic number of 18 for this section which, in his opinion, represents a derived and polyploid condition of diploids based on $x = 9$ which do not exist any more. Diploid numbers found so far for species of this section range from $2n = 36$ to 40, while other accessions of the same species show numbers of 71 to 76 which are considered to represent tetraploids. Arends surmised that the tetraploid plants represent interracial autopolyploids. Occasionally within a single homogeneous population, and even within an individual specimen, *Begonia* species show a variable number of somatic chromosomes. This variation is sometimes attributed to either the presence or absence of accessory chromosomes interpreted as B chromosomes. In metaphase chromosomes of species of section *Tetraphila*, their length was found to range between 0.5 and 2.5 μm; in these small chromosomes, the position of the centromere is often indistinct. A clear correlation between ploidy-level and (macro)morphological characters is usually absent; in a few species, however, the possibility was demonstrated to distinguish between 2*x* and 4*x* cytotypes with the use of petiole morphology and the size of pollen grains.

Reproductive System, Pollination and Hybridization. According to East (1940), *Begonia* species are self-compatible. For a number of species, attributed to various sections, this is corroborated by greenhouse experiments and observations in which geitonogamy led to the production of mature fruits and viable seeds. In exceptional cases, e.g. *Begonia rostrata*, *B. wallichiana* and some other species, the flowers in each inflorescence are always arranged in such way that one or more staminate flowers are positioned just above the simultaneously flowering pistillate flower, inducing pollen released from the anthers to drop directly on the stigmas. As a rule, however, *Begonia* species in cultivation rarely spontaneously produce fruits, while hand-mediated geitonogamy often results in seed set. This at least indicates that usually a special vector is needed to transport the pollen. Furthermore, Sosef (1994), experimenting with species belonging to the sections *Loasibegonia* and *Scutobegonia*, concluded that here a self-incompatibility system must be present because mature fruits were rarely obtained from self-pollination. Clearly, there is no general breeding system in *Begonia*. Next to rare cases of dioecism, a limited number of *Begonia* species show a strict differentiation into staminate and pistillate inflorescences, often connected with temporal and/or spatial dioecy in which staminate inflorescences are separated in time or space from pistillate ones. Most commonly, both pistillate and staminate flowers are produced within the same inflorescence; here, dichogamy is almost the rule as the different sexes usually reach anthesis separated in time (but see above for exceptions). The separation of sexes, the distribution of staminate and pistillate flowers over the inflorescence and, in the usual case that a plant bears several inflorescences, the synchronized flowering of these avoiding overlap between pistillate and staminate phases among inflorescence all act to prevent autogamy, to minimize geitonogamy and to favour xenogamy (Burt-Utley 1985).

Within *Begonia*, the transport of pollen leading to pollination is not uniform. Besides the exceptional simple gravitational transport mentioned above for *B. rostrata* and a few other species, many begonias show features which point to insect pollination: e.g. inflorescences situated at the base of the plants underneath the leaves, flowers with distinct sleeping movements, a bright white, red or yellow colouration of the flowers sometimes further supported by coloured bracts, or androecia which in position and conformation mimic the stigmatic part of pistillate flowers. Insect pollination, though scarcely observed and recorded, presumably constitutes the major syndrome leading to fertilization. Only a few *Begonia* species, however, are known in which the flowers are discernibly fragrant, and no nectar production of pistillate flowers has been recorded. It is of interest that, in the absence of a reward to pollinators, several field observations testify to the occurrence of 'mistake' or even deceit pollination induced by the stigmas which mimic the androecium. Pollinating insects, especially bees (Hymenoptera, Apoidea) and hoverflies (Diptera, Syrphidae), attracted by pollen produced in the androecia are 'deceived' and visit the stigmatic part of a pistillate flower, thereby inadvertently acting as a vector for pollination (Seitner 1977; van der Pijl 1978; Ågren and Schemske 1991; Arends 1992; Sosef 1994; Schemske and Ågren 1995). Also of interest is the observation that, both in the field as well as under experimental conditions, shrivelled dried male flowers of a number of species remain attached to the inflorescence for a considerable time and, at this stage, still contain and release plenty of pollen. In some species, e.g. *B. carrieae*, sticky glandular hairs on the inflorescence 'catch' the abscised male flowers. Especially with regard to the epiphytic species in African section *Tetraphila*, this finding leads us to consider wind to be the principal vector for pollen transport in these species (de Wilde 2002).

Both interspecific and intersectional artificial hybrids are frequently produced in the genus *Begonia*, but hybridization certainly is not random. Natural hybridization is much less known.

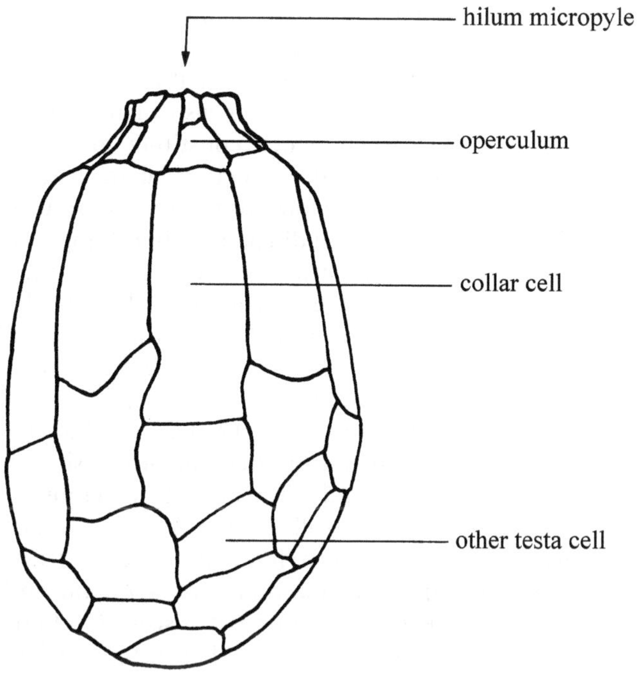

Fig. 13. Begoniaceae. Diagram of a *Begonia* seed with various parts labelled (de Lange and Bouman 1999)

Arends (1992) produced evidence that closely related sympatric species in sect. *Tetraphila* are reproductively isolated. A sterile intersectional hybrid between species of sections *Loasibegonia* and *Filicibegonia* is known from the Crystal Mts in Gabon. Detailed study of species of sect. *Gireoudia* from Central America (Burt-Utley 1985) and of sections *Loasibegonia* and *Scutobegonia* from Africa (Sosef 1994) also revealed taxa suspect of hybrid origin as a number of species involved in natural hybridization. In Asia (Taiwan), *B. buimontana* is a well documented case (Peng and Chen 1991).

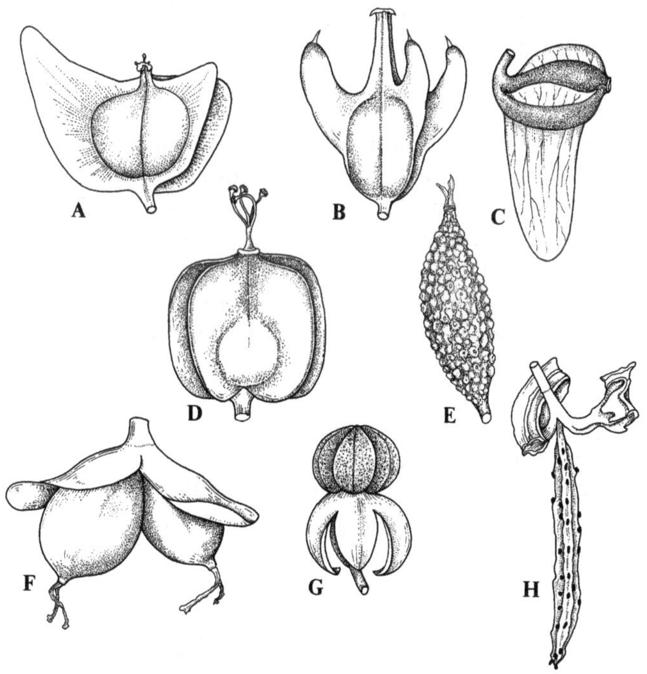

Fig. 12. Begoniaceae. Diversity of fruits in *Begonia*. **A** Winged capsule, membranaceous, dehiscent. **B** Rattleburr, the wings transformed into horns as in sect. *Casparya*. **C** Splash cup as in sect. *Platycentrum*. **D** Indehiscent, somewhat fleshy, characteristic of sect. *Scutobegonia*. **E** Indehiscent, fleshy, as in sect. *Mezierea*. **F** Indehiscent berries supported by persistent pseudobracts characterize sect. *Squamibegonia*. **G** Dehiscent bacciform fruit of sect. *Baccabegonia*. **H** Dehiscent, fusiform, fleshy fruit exhibiting enlarged coloured placentas dotted with seeds as in sect. *Tetraphila*. (Drawn by Ike Zewald)

FRUIT AND SEED. Among the first to be aware of the wealth of variation in fruit types (Fig. 12) was Klotzsch (1855). The basic type in the Begoniaceae, shared by a large majority of the species on all three continents, is a membranaceous or cartilaginous, 2-3-gonous, winged or horned, 2-3-locular capsule with axile placentation (Fig. 12A). The capsules are variously orientated, erect, nodding or pendulous, glabrous or with various indumentum, and frequently crowned by persistent stylodia and stigmas. The (1,2)3(4,6) wings or horns are equal, subequal or unequal in size and shape. In sections *Casparya* and *Sphenanthera*, the wings are transformed into hooks or horns (Fig. 12B). Morphologically, the wings present an outgrowth of the dorsals and sometimes also of the lateral veins of the carpels which constitute the compound ovary. Dehiscence is loculicidal, the locules opening from the base towards the apex by arcuate fissures or by longitudinal slits bordering the wings and, in the process, the pericarp often breaks away from the dorsal wings, leaving the latter intact; more rarely, the seeds are released through basal pores. In Africa, where diversity in fruit type is greatest, the fruits of sections *Loasibegonia* and *Scutobegonia* (Fig. 12D) are indehiscent, 3–4-locular, often remain green or brownish and somewhat fleshy and ultimately disintegrate by rotting of the ovary wall. Also in Africa, apterous fleshy-fruited species are found in which the fruits are either dehiscent (sections *Baccabegonia* and *Tetraphila*, Fig. 12G, H) or indehiscent (sections *Mezierea* and *Squamibegonia*, Fig. 12E, F). These fruits are often attractively coloured scarlet, pink or white and probably destined to be eaten by birds. The monotypic genus *Hillebrandia* has apterous, dry, globose capsules which dehisce between the stylodia.

Supplementary to the light-microscopical study by Seitner (1972), the micromorphology of the seed coat of *Begonia* species in Africa, the Madagascan species included, was thoroughly studied by de Lange and Bouman (1992), followed by a survey of the Neotropical species (de Lange and Bouman 1999). The relatively small seeds of Begoniaceae (Fig. 13) are unique among the seeds of Angiosperms by the presence of a transverse ring of specialized, elongated testa cells, the so-called collar. This collar adjoins the seed lid or operculum, which represents the micropylar-hilar part of the seed. The operculum is lifted off by the embryonic root during germination and the longitudinal walls between the collar cells split, thereby clearing the way for the emerging seedling (de Lange and Bouman 1992). This ground plan shows a great diversity in a number of characters, e.g. size and shape of the seed, the form of the operculum, and the arrangement, shape and cuticular ornamentation of the testa cells. Seed length varies between 195 and 2,240 μm. Diversity in seed size and structure is widest in Africa, where in sect. *Tetraphila* it sometimes is complemented by bright yellow-coloured placentas, arils or swollen funicles. This large variation is in conformity with the finding that, despite their comparatively limited number, the diversity among the African species of *Begonia* is greater than on the other continents. The testa in mature seeds is of simple construction and formed almost exclusively by the outer layer of the outer integument; at its inner face, it is usually lined by a single layer of endosperm (Boesewinkel and de Lange 1983). Cuticular ornamentation of testa cells ranges from smooth in sections such as *Squamibegonia* and *Tetraphila* to strongly pronounced cuticular structures as found in sections *Loasibegonia* and *Scutobegonia*. In particular at the sectional level, and more rarely among species, micromorphological characters of the testa appear useful for taxonomic delimitation.

Dispersal. Differences in type of fruit show a significant correlation with variation in size and micromorphology of the seeds in African *Begonia* species, and the inferred syndromes reflect different adaptations to seed dispersal. This, supplemented with observations on *Begonia* species from the other continents, led de Lange and Bouman (1992) to the recognition of a number of dispersal types within the genus. Most common are species with anemoballistically dispersed seed. Here the pendulous alate capsules, dehiscent by pores or slits, are shaken by wind and gradually release the seeds which sometimes show special adaptations to anemochory, e.g. the extended micropylar and/or chalazal ends of the seed composed of inflated air-filled cells, as in sections *Cristasemen* (Fig. 10N), *Rossmannia*, *Solananthera*, *Trendelenburgia* and *Wageneria*, or the increase of the surface:volume ratio as seen in the scobiform seeds of several American and Asiatic sections. This type of dispersal is commonly, though not exclusively, encountered in species of more dry and exposed habitats. In spite of a complete lack of field observations, zoochory seems another main type of dispersal. This syndrome is characterized by the combination of coloured fleshy fruits, either dehiscent or not, and by a trend towards bigger seeds, loss of cuticular ornamentation and a thicker exotesta, all features which point to endozoochory. It is found particularly in African *Begonia* species and it may be significant that a majority of these species are epiphytic. In our opinion, birds act as the actual dispersers. Interestingly, myrmecochory in species of the epiphytic African section *Tetraphila* was suspected by de Lange and Bouman (1992) and afterwards observed and described by de Wilde (2002). This mode of micro-dispersal on an individual tree may be supplemented by rain wash while, simultaneously, endozoochory presumably by birds acts as vector in macro-dispersal, viz. between trees and over longer distances. A special case is the probably diplochorous dispersal of the seeds of the African sections *Scutobegonia* and *Loasibegonia*. Here the fruits are often hidden under the leaves and in many species characteristically recurved towards the substrate after flowering; these fruits are indehiscent and, when the seeds finally are liberated through disintegration of the pericarp, they are deposited at the base of the parent plant. This adaptation to non-transportation of the seeds, which are amongst the smallest within the genus, is supposedly supplemented by dispersal through rain wash and/or with mud on the legs of passing animals, which may be promoted by the rough ornamentation of the seed coat. The obligate umbrageous habitat alliance of these terrestrial rain forest herbs is clearly linked with a premium on minimization of dispersal of their seeds which, for functional reasons, is supplemented by hydrochorous and epizoochorous dispersal. Other dispersal types include splash cups, rain ballistics in which the two shorter wings are upright and curved to form a cup which is hit by raindrops, thereby bouncing the seeds out of the capsule (Kiew 2005) as known in sect. *Platycentrum* from Asia, and rattle-burrs as in American sect. *Casparya*, in which the fruit wings are replaced by horns which probably are operated by passing animals. Details about the dispersal of the capsular fruited genus *Hillebrandia* are not known but probably do fit into the anemoballistic syndrome.

Phytochemistry. Next to calcium oxalate, *Begonia* species commonly accumulate high quantities of oxalic acid, which probably is responsible for the strong acid reaction of the contents of the cells (Hegnauer 1964). Common phenolics such as proanthocyanidin and several glycosides of cyanidin, but no myricetin or ellagic acid, were recorded by Bopp (1957).

Subdivisions and Relationships. Begoniaceae as a family are well characterized and easy to recognize. Formerly, the family was considered monogeneric, composed of the very natural and homogeneous genus *Begonia*. Klotzsch (1855) subdivided the family into two supposedly phyletic groups, viz. the *Stephanocarpeae* with the stylodia persistent in fruit, and the *Gymnocarpeae* in which the stylodia are deciduous. Within these two tribes he recognized in all 41 genera including 194 species, using primarily floral characters. This dichotomous division of the family was questioned by de Candolle (1859) who refuted the character of the stylodia as reliable but decided to treat most of the genera described by Klotzsch as sections within *Begonia*. de Candolle (1864) maintained 3 genera, viz. *Casparya*, *Begonia* and *Mezierea*, subdivided into 8, 69 and 2 sections respectively, the majority of which had already been coined and circumscribed (as genera) by Klotzsch. Most of these sections show a narrow distribution or are at least restricted to one of the major geographical regions Africa, Asia and Neotropics. Subsequently, Warburg (1894) and Irmscher (1925) largely maintained these sections and their circumscription, though modifications were proposed and new sections added. In the latest monograph of the family, Irmscher (1925) admitted 5 genera, viz. monotypic *Hillebrandia*, confined to the Hawaiian islands, *Begonia* with 60 sections and some 750 species, pantropical, *Semibegoniella* with 2 species from Ecuador, *Begoniella* with 3 species restricted to Colombia and, finally, *Symbegonia* with c. 10 species represented on New Guinea. Doorenbos, Sosef and de Wilde (1998) grouped the 1,403 species of *Begonia* they recognized into 63 sections with, among these, a section *Semibegoniella* comprising the former genus *Begoniella*.

At present, there is consensus on the recognition of only two genera (Forrest and Hollingsworth 2003). The taxonomically and geographically isolated Pacific genus *Hillebrandia* shows by far the most original characters. *Begonia* is very diverse, now includes the American satellite genera *Semibegoniella* and *Begoniella* as well as the former genus *Symbegonia*, and is currently divided into 65 sections comprising an estimated 1,500 species. These sections, also in their modern concepts, are possibly with one single exception limited to one of the continents.

Continental African begonias are currently well studied. With 115 species distributed over 12 sections, the continent is comparatively poor in species but proved to be extremely rich in morphological variation. Keraudren-Aymonin (1983) recognized some 50 *Begonia* species on Madagascar and the Comores but declined to ascertain sections. Only one of these species, viz. *B. oxyloba*, is also widely distributed on the African continent, which endorses the isolated position of the Madagascan species. The Asiatic part of the Palaeotropics is rich in species, though no modern synthesis has been published. Doorenbos et al. (1998) reported c. 580 species accommodated into 18 different sections, still keeping apart *Symbegonia* as a separate genus. Ongoing exploration certainly will yield more new species. From Peninsular Malaysia, only recently 12 new species were described (Kiew 2005). Taxonomic affinities between begonias of Madagascar with these of India and Sri Lanka are not known. Burt-Utley (1985) estimates that *Begonia* is represented in the Neotropics with almost 600 species; presently, these are arranged into 29 sections (Doorenbos et al. 1998).

All current knowledge points to the originality of the characters contained in *Hillebrandia*. Specialization of these led to the overwhelming variation presently encountered in the large genus *Begonia* culminating in, for example, the fusion of the perianth segments as found among others in section *Symbegonia*. The need to group resembling species and to (sub)divide such a large genus is evident. The notion of an infrageneric division of *Begonia* into sections is largely historically determined; a number of the sections certainly deserve generic status. However, as long as many sections are still ill-defined and their phylogenetic relations are insufficiently clear, it is felt that nomenclatural stability is served by preservation of a large single genus *Begonia*.

Affinities. Little is known about the origins of the Begoniaceae which, among themselves, clearly constitute a homogeneous assemblage (Forrest and Hollingsworth 2003). Hutchinson (1959) recognized an order Cucurbitales which contained Cucurbitaceae, Begoniaceae, Datiscaceae and Caricaceae. Melchior (1964) accommodated the Begoniaceae in the vicinity of Caricaceae, Loasaceae and Datiscaceae in an order Violales. Takhtajan (1969) adopted Begoniales to comprise only Datiscaceae and Begoniaceae while stating that it was probably derived from Violales. A study on the embryological and seed characteristics of the Datiscaceae by Boesewinkel (1984) strongly supported a close relationship with the even more specialized Begoniaceae. Cronquist (1988) proposes Violales to contain 9 suborders and, among these, Begoniineae with two families only, viz. Datiscaceae and Begoniaceae, both containing plants without tendrils and not producing cucurbitacins. In the molecular phylogenetic analysis of Zhang et al. (2006), Cucurbitales are resolved as a highly supported clade in which Begoniaceae, Datiscaceae and Cucurbitaceae are sister to Corynocarpaceae and Coriariaceae. A comparative anatomical and morphological study of the floral structure of selected representatives of these families by Matthews and Endress (2004) supports such an order Cucurbitales.

Distribution and Habitats. Begoniaceae are a pantropical family. In sharp contrast to the geographic isolation of the monotypic genus *Hillebrandia*, the genus *Begonia* is widely distributed in almost all tropical and in many subtropical regions. A few species range northwards into the warm temperate zone, among these *B. grandis* which reaches c. 40°N in northern China (Beijing) and in Japan. On New Guinea, *Begonia* is very rich in species. Strikingly, Begoniaceae seem never to have crossed the Torres Street and are absent from tropical Australia. However, a few *Begonia* species occur on the New Hebrides and one species is known from Fiji, extending the area of *Begonia* eastwards well into the Pacific. In Polynesia the family is not recorded. In the Neotropics, *Begonia* ranges from Mexico to northernmost Argentina. It is postulated that *Begonia* migrated into southern North America from South America as early as the Eocene (Raven and Axelrod 1974); in later periods, it probably lost a part of its distribution there as a result of unfavourable abiotic conditions. Begoniaceae are absent from the Galápagos Islands. Irmscher (1925) recognized two centres of diversity for *Begonia*, the one extending from Mexico via Central America into the Andes and into Brazil, the other from the eastern Himalaya via the mountains of Indochina, the Malay Archipelago and the Philippines into New Guinea. Compared to these centres, the West-Indian islands are at best moderately rich in species. Africa is comparatively poorly stocked, probably due to the extinction of species during cooler and dry periods in the Pleistocene. The majority of *Begonia* species show a limited distribution and many species are indeed narrow endemics. In line with this, the distributions of the currently recognized sections are as a rule also restricted and usually confined to only a part of one of the continents, except for the African section *Tetraphila* of which recently a new, still to be named species was discovered in Asia (Thailand and Laos).

Begoniaceae show a wide altitudinal range. Although many species occur just above sea level, the general impression is that a majority of the species favour hills and mountains at altitudes between 200 and 2,000 m. Irmscher (1925) documented *B. gemmipara* as reaching 3,600 m altitude in the eastern Himalaya in Sikkim, India. In Africa, the highest altitude on record is from *B. meyeri-johannis*, collected at 2,900 m in Afromontane rain forest in Rwanda (Klazenga et al. 1994).

Ecologically, many species are confined to wet forests, some preferring deep shade, whereas others are found at forest edges and in clearings. In these habitats, they occur in soil or growing on decaying tree trunks, on shaded damp mossy boulders or on rock faces, and a fair number are obligate epiphytes. Although some species grow on level ground, many more are found along creek beds and on slopes, often preferably in the vicinity of water, and not infrequently on rocks wetted by spray from waterfalls. Few are rheophytes. In western tropical Africa, de Wilde (1988) advocated the use of certain sections as a tool for the reconstruction of the geography of former glacial forest refuges. This idea was intelligently elaborated upon by Sosef (1994) who found many species in these groups to be reliable indicators for former refuges. However, vicariance events, viz. the segregation of an ancestral species into geographically isolated populations leading to speciation, were rarely found. A historical biogeographic analysis of African *Begonia* speciation based on molecular sequence data (Plana et al. 2004) concludes that a considerable proportion of its variation seems to be of pre-Pleistocene origin, but with recent Pleistocene radiations in some lineages. Next to forest dwellers, there exists a considerable group of species which are adapted to more open and often drier habitats. Characteristic for this anemochorous group are the African woodland species which prefer rocky outcrops or comparatively wet rocky places covered by shrub vegetation, or again species which are almost limited to forested termitaries. The Andean section *Eupetalum* comprises tuberous species which seasonally shed their aboveground parts. Thereby, they are adapted to the cool and more dry conditions prevailing at higher altitudes, where they often live in rock crevices. Similar life forms are found in southern Africa where, at altitudes between 800 and 2,000 m, tuberous species which belong in section *Augustia* grow in shaded rock clefts and/or favour the moss cover on (sub)montane rain forest trees and are tuberous epiphytes. On the Yemeni islands of Socotra and nearby Samha, the endemic species *B.* (sect. *Peltaugustia*) *socotrana* and *B. samhaensis* respectively develop around their stem bases a cluster of tunicated bulb-like structures which act as diaspores, and survive dry spells in succulent shrub vegetation on limestone cliffs and valley slopes which probably receive only between 500 and 600 mm rainfall per year.

Palaeobotany. The only known record is from pollen found in the Kamiranzovu Swamp in Rwanda at 1,950 m altitude. Here, in Upper Quaternary peat deposits dated c. 38,000 years B.P., *Begonia* pollen was recognized. At the spot, the lianescent *B. meyeri-johannis* is a notable constituent of the present-day vegetation (Hamilton 1982).

Economic Importance. Wild, domesticated and very often hybridized begonias provide a great many popular flowering pot plants, colourful evergreen houseplants as well as plants used for the outdoor garden. A multitude of publications focus on their cultivation (Tebbitt 2005). Wild *Begonia* species are sometimes grown by hobbyists in contained atmospheres (Thompson and Thompson 1981). The group represents a horticultural value with important economic aspects and, in Europe, Japan, the USA and Australia, nurseries specialize in commercial propagation of a wide range of *Begonia* species, cultivars and hybrids. Among the most important commercially grown plants, Doorenbos (1985) points out four groups of hybrids, viz. tuberous begonias, semperflorens-begonias, winter-flowering begonias and begonias with ornamental foliage. The tuberous begonias are a product of repeated hybridization of *B. boliviensis*, *B. veitchii* and *B. pearcei*, all from South America. Semperflorens-begonias arose from crosses of di- and tetraploid races of the polymorphous *B. cucullata* and *B. schmidtiana*, both from Brazil. Winter-flowering begonias originated through crossing, back-crossing and selection of mutants of *B. socotrana* with the tuberous begonias from South America and with *B. dregei* from South Africa. The begonias with ornamental foliage comprise mainly two hybrid groups, viz. a group 'Rex-cultorum', which originated through crossing *B. rex* with other Asian species, and another group which resulted from hybridization of a number of primarily Mexican species.

Key to the Genera

1. Perianth segments in flowers of both sexes 8 or 10, distinct, in two alternating whorls of 4 or 5 segments each, the segments of the outer whorl (sepals) conspicuously different in shape and dimensions and readily distinguished from the inner whorl (petals); ovary partly inferior **1. *Hillebrandia***
– Perianth segments 2, 3, 4 or 5, exceptionally up to 11, this number usually differing between the ♂ and ♀ flowers of a species, almost distinct or variably connate, in one or in two indistinct whorls but sepals and petals often difficult to distinguish; ovary distinctly inferior **2. *Begonia***

Genera of Begoniaceae

1. *Hillebrandia* Oliv.

Hillebrandia Oliv., Trans. Linn. Soc. London 25: 361, t. 46 (1866).

Herbaceous terrestrial producing shortly branched thick fleshy stems from an irregularly shaped tuberous underground rhizome. Leaves long-petioled, faintly asymmetric, 3-lobed, palmatinerved, hairy. Inflorescences cymose, androgynous. Flowers unisexual, actinomorphic; sepals 4–5, distinct; petals 4–5, alternisepalous, distinct, very small; polyandrous; filaments distinct; anthers dehiscing by lateral slits; ovary semi-inferior, globose, apterous, incompletely 5-locular with parietal/septal placentation, surmounted by 4 or 5 alternipetalous stylodia which are forked and hippocrepiform at the top, the arms covered with a continuous, helically twisted band of stigmatic tissue. Capsules pendulous at maturity, dehiscing with pore-like slits between the bases of the stylodia.

Only *H. sandwicensis* Oliv., endemic to the Hawaiian islands.

2. *Begonia* L. Figs. 10–12

Begonia L., Sp. Pl.: 1056 (1753); A.DC. in DC., Prod. 15: 278–406 (1864); Irmscher in Engl. & Prantl, Nat. Pflanzenfam., ed. 2, 21: 572–588 (1925); Doorenbos, Sosef & de Wilde, Wageningen Agric. Univ. Pap. 98-2: 1-266 (1998).
Begoniella Oliv. (1873).
Symbegonia Warb. (1894).
Semibegoniella C.DC. (1908).
(for exhaustive synonymy, see Smith and Schubert 1958).

Perennial or more rarely annual herbs, sometimes shrubs or occasionally even small trees, terrestrial or epiphytic; stems elongate and herbaceous, often succulent, or woody, frequently rhizomatic, or plants tuberous and either acaulous or short-stemmed, rarely lianoid or climbing with adventitious roots; monoecious, very rarely dioecious. Leaves variously petioled, asymmetric, exceptionally almost symmetric, sometimes peltate, entire, dentate or serrate, lobed, fissured or digitately compound. Inflorescences unisexual or androgynous, usually cymose, sometime racemose or racemose with cymose branches, rarely 1-flowered; cymes dichasial or monochasial. Flowers unisexual; staminate flowers with 2(3)4(5–11) almost distinct to variably connate perianth segments; stamens 3 to many; filaments distinct or variably connate into a column, sometimes the remaining distinct parts uneven in length and arranged into several distinct rows like in an amphitheatre; anthers multiform, opening lengthwise with slits or more rarely with terminal pores; connective frequently enlarged; pistillate flowers with 2–5(6–9) distinct or partially connate, often unequal perianth segments which are sometimes persistent in fruit; ovary inferior, usually winged or horned, occasionally apterous, broadly obovoid, ovoid, globose or fusiform in shape, triangular, square or terete in circumference, 2 or

3(4,5,6)-locular, sometimes the locules incomplete; placentation parietal, septal or axile, occasionally changing from the bottom of the ovary towards the top; stylodia 2–6, persistent or caducous, often connate at the base, one or more times forked towards the apex, more rarely simple; stigmatic tissue very often in a continuous band and helically twisted around the arms, sometimes kidney-shaped. Fruit a usually winged capsule, rarely berry-like and fleshy, loculicidal, more rarely indehiscent.

At least 1,500 spp. in 63 sections, circumtropical with some species into the warm temperate zone.

Selected Bibliography

Ågren, J., Schemske, D.W. 1991. Pollination by deceit in a Neotropical monoecious herb, *Begonia involucrata*. Biotropica 23: 235–241.

Arends, J.C. 1992. The biosystematics of *B. squamulosa* Hook.f. and affiliated species in sect. *Tetraphila* A. DC. Wageningen Agric. Univ. Papers 91-6: 1–223.

Bailey, L.H. 1914. The standard cyclopedia of horticulture I: 469–485.

Berg, R.G. van den 1984. Pollen characteristics of the genera of the Begoniaceae. In: Wilde, J.J.F.E. de (ed.). Studies in Begoniaceae I. Med. Landbouwhogeschool Wageningen 83-9: 55–66.

Berg, R.G. van den 1985. Pollen morphology of the genus *Begonia* in Africa. Agric. Univ. Wageningen Papers 84-3: 5–94.

Boesewinkel, F.D. 1984. Ovule and seed structure in Datiscaceae. Acta Bot. Neerl. 32: 417–425.

Boesewinkel, F.D., Lange, A. de 1983. Development of ovule and seed in *Begonia squamulosa* Hook. f. Acta Bot. Neerl. 32: 417–425.

Bopp, M. 1957. Untersuchungen über die Verteilung und Vererbung von Anthocyanen in den Blättern von Begonien. Planta 48: 631–682.

Bouman, F. 1985. Seed structure of *B. thomeana* C. DC. In: Wilde, J.J.F.E. de (ed.) Studies in Begoniaceae II. Agric. Univ. Wageningen Papers 84-3: 122–124.

Brouillet, L., Bertrand, A., Cuerrier, A., Barabé, D. 1987. Les hydathodes des genres *Begonia* et *Hillebrandia* (Begoniaceae). Can. J. Bot. 65: 34–52.

Burt-Utley, K. 1985. A revision of Central American species of *Begonia* section *Gireoudia*. Tulane Studies Zool. Bot. 25: 1–131.

Candolle, Alph. de 1859. Mémoire sur la famille des Bégoniacées. Ann. Sci. Nat. Bot. IV, 11(2): 93–149.

Candolle, Alph. de 1864. Ordo CLXXII. Begoniaceae. Prodromus systematis naturalis regni vegetabilis 15(1): 266–408. Parisiis: Victoris Masson et filii.

Charpentier, A., Brouillet, L., Barabé, D. 1989a. Organogénèse de la fleur pistillée du *Begonia horticola* (Begoniaceae). Can. J. Bot. 67: 559–572.

Charpentier, A., Brouillet, L., Barabé, D. 1989b. Organogénèse de la fleur pistillée du *Begonia dregei* et de *Hillebrandia sandwicensis* (Begoniaceae). Can. J. Bot. 67: 3625–3639.

Corner, E.J.H. 1976. The seeds of the dicotyledons. Vol. 1. Cambridge: Cambridge University Press.

Cronquist, A. 1988. The evolution and classification of flowering plants, 2nd edn. New York: New York Botanical Garden.

Cuerrier, A., Brouillet, L., Barabé, D. 1990. Numerical taxonomic study of the Begoniaceae using the Mantel test on leaf microcharacters. Taxon 39: 549–560.

Cuerrier, A., Brouillet, L., Barabé, D. 1991a. Micromorphologie foliaire des Begoniaceae. Bull. Mus. Natl. Hist. Nat. Paris IV, 12, sect. B, Adansonia: 297–335.

Cuerrier, A., Brouillet, L., Barabé, D. 1991b. Analyse taxonumérique des charactères foliaires des Begoniaceae. Bull. Mus. Natl. Hist. Nat. Paris IV, 12, sect. B, Adansonia: 337–353.

Davis, G.L. 1966. Systematic embryology of the angiosperms. New York: Wiley.

Doorenbos, J. 1985. Domestication of *Begonia*. Acta Bot. Neerl. 34: 230–231.

Doorenbos, J., Sosef, M.S.M., Wilde, J.J.F.E. de 1998. The sections of *Begonia*. Wageningen Agric. Univ. Papers 98-2:1–266.

East, E.M. 1940. The distribution of self-sterility in flowering plants. Proc. Am. Phil. Soc. 82: 449–518.

Fellerer, C. 1892. Beiträge zur Anatomie und Systematik der Begoniaceae. Thesis, Munich, 239 pp.

Forrest, L.L., Hollingsworth, P.M. 2003. A recircumscription of *Begonia* based on nuclear ribosomal sequences. Plant Syst. Evol. 241: 193–211.

Gauthier, R. 1950. The nature of the inferior ovary in the genus *Begonia*. Contr. Inst. Bot. Univ. Montréal 66: 5–91.

Golding, J., Wasshausen, D.C. 2002. Begoniaceae, 2nd rev. edn. Part I: Annotated species list. Part II: Illustrated key, abridgment and supplement. Smithsonian Inst., Contr. United States Natl. Herb. 43:1–289.

Goulet, I., Barabé, D., Brouillet, L. 1994. Analyse structurale et architecture de l'inflorescence des Begoniaceae. Can. J. Bot. 72: 897–914.

Haberlandt, G. 1916. Blattepidermis und Lichtperzeption. Sitzungsber. Königl. Preuss. Akad. Wiss. 32: 672–687.

Hallé, N. 1972. Les *Begonia* filicifoliés et quatre espèces nouvelles du Gabon (Begoniaceae). Adansonia II, 12: 359–374.

Hamilton, A.C. 1982. Environmental history of East Africa. A study of the Quaternary: 174–181. London: Academic Press.

Hegnauer, R. 1964. Chemotaxonomie der Pflanzen. Vol. 3. Basel: Birkhaeuser.

Hickey, L.J., Wolfe, J.A. 1975. The bases of angiosperm phylogeny: vegetative morphology. Ann. Missouri Bot. Gard. 62: 538–589.

Hildebrand, F.H.G. 1859. Anatomische Untersuchungen über die Stämme der Begoniaceae. Berlin: August Hirschwald.

Hutchinson, J. 1959. The families of flowering plants, 2nd edn. London: Oxford University Press.

Irmscher, E. 1914. Die Verteilung der Geschlechter in den Infloreszenzen der Begoniaceen unter Berücksichtigung der morphologischen Verhältnisse. Bot. Jahrb. Syst. suppl. 50: 556–577.

Irmscher, E. 1925. Begoniaceae. In: Engler, A. (ed.) Die natürlichen Pflanzenfamilien, 2nd edn., 21: 548–588. Leipzig: W. Engelmann.

Jin, X.B., Wang, F.H. 1994. Style and ovary anatomy of Chinese *Begonia* and its taxonomic and evolutionary implications. Cathaya 6: 125–144.

Keraudren-Aymonin, M. 1983. Flore de Madagascar et des Comores. Famille 144 Bégoniacées: 1–108.

Kiew, R. 2005. Begonias of Peninsular Malaysia. Natural History Publications (Borneo) Sdn. Bhd. and Singapore Botanic Gardens.

Klazenga, N., Wilde, J.J.F.E. de, Quené, R.J., 1994. *Begonia* sect. *Mezierea* (Gaud.) Warb., a taxonomic revision. Bull. Jard. Bot. Natl. Belg. 63: 263–312.

Klotzsch, J.F. 1855. Begoniaceen-Gattungen und Arten. Abhandl. Königl. Akad. Wiss. Berlin 1854, 135 pp. and 12 plates.

Lange, A. de, Bouman, F. 1992. Seed micromorphology of the genus *Begonia* in Africa: taxonomic and ecological implications. In: Wilde, J.J.F.E. de (ed.). Studies in Begoniaceae III. Wageningen Agric. Univ. Papers 91-4: 1–82.

Lange, A. de, Bouman, F. 1999. Seed micromorphology of neotropical begonias. Smithsonian Contr. Bot. 90: 1–49.

Lee, Y.S. 1974. A study of stem anatomy in *Begonia* L. Phytologia 27: 464–489.

Lee, D.W. 1983. Unusual strategies of light absorption in rain-forest herbs. In: Givnish, T.J. (ed.) On the economy of plant form and function. Proc. 6th Maria Moors Cabot Symp. Cambridge: Cambridge University Press, pp. 105–131.

Lee, D.W., Stone, B.C. 1979. Abaxial anthocyanin layer in leaves of tropical rainforest plants: enhancer of light capture in deep shade. Biotropica 11: 70–77.

Legro, R.A.H., Doorenbos, J. 1969, 1971, 1973. Chromosome numbers in Begonia 1, 2 and 3. Neth. J. Agric. Sci. 17: 189–202, 19: 176–183, 21: 167–170.

Manning, J.C. 1996. Diversity of endothecial patterns in the angiosperms. In: D'Arcy, W.G., Keating, R.C. (eds.) The anther - form, function and phylogeny. Cambridge: Cambridge University Press, pp. 136–158.

Matthews, M.L., Endress, P.K. 2004. Comparative floral structure and systematics in Cucurbitales (Corynocarpaceae, Coriariaceae, Tetramelaceae, Datiscaceae, Begoniaceae, Cucurbitaceae, Anisophylleaceae). Bot. J. Linn. Soc. 145: 129–185.

Melchior, H. 1964. 34 Reihe Violales. In: Engler, A. Syllabus der Pflanzenfamilien, 12th edn., II. Berlin-Nikolassee: Gebrüder Borntraeger.

Metcalfe, C.R., Chalk, L. 1950. Begoniaceae. In: Anatomy of the dicotyledons, vol. 1. Oxford: Clarendon Press, pp. 691–695.

Muller, J. 1981. Fossil pollen records of extant Angiosperms. Bot. Rev. 47: 1–142.

Neubauer, H.F. 1967. Bemerkungen über den Bau der Begoniaceen. Ber. Deutsch. Bot. Ges. 80: 80–97.

Panda, S., Wilde, J.J.F.E. de 1995. Diversity and taxonomic value of stigmatic surfaces in Begoniaceae: SEM analysis. Acta Bot. Neerl. 44: 139–150.

Payne, W.W. 1970. Helicocytic and allelocytic stomata: unrecognized patterns in the Dicotyledonae. Am. J. Bot. 57: 140–147.

Peng, C.I., Chen, Y.K. 1991. Hybridity and parentage of *Begonia buimontana* Yamamota (Begoniaceae) from Taiwan. Ann. Missouri Bot. Gard. 78: 995–1001.

Pijl, L. van der 1978. Reproductive integration and sexual disharmony in floral functions. In: Richards, A.J. (ed.) The pollination of flowers by insects. London: Academic Press, pp. 79–88.

Plana, V., Gascoigne, A., Forrest, L.L., Harris, D., Pennington, R.T. 2004. Pleistocene and pre-Pleistocene *Begonia* speciation in Africa. Mol. Phylogen. Evol. 31: 449–461.

Raven, P.H., Axelrod, D.I. 1974. Angiosperm biogeography and past continental movements. Ann. Missouri Bot. Gard. 61: 539–673.

Reitsma, J.M. 1984. Placentation in begonias from the African continent. In: Wilde J.J.F.E. de (ed.). Studies in Begoniaceae I. Med. Landbouwhogeschool Wageningen 83-9: 21–53.

Schemske, D.W., Ågren, J. 1995. Deceit pollination and selection on female flower size in *Begonia involucrata*: an experimental approach. Evolution 49: 207–214.

Seitner, P.G. 1972. Some observations on *Begonia* seeds. The Begonian 39: 47–55.

Seitner, P.G. 1977. The role of insects in *Begonia* pollination. Supplement I. The Begonian 44: 213–217.

Seybold, A. 1955. Beiträge zur Optik der Laubblätter. Beitr. Biol. Pfl. 31: 499–513.

Smith, L.B., Schubert, B.G. 1958. Begoniaceae. Flora of Panama VII. Ann. Missouri Bot. Gard. 45: 41–42.

Smith, L.B., Wasshausen, D.C., Golding, J., Karegeannes, C.E. 1986. Begoniaceae. Part I: Illustrated key. Part II: Annotated species list. Smithsonian Contr. Bot. 60: 1–584.

Solereder, H. 1899. Begoniaceae. Systematische Anatomie der Dicotyledonen. Stuttgart: Ferdinand Enke, pp. 449–458.

Sosef, M.S.M. 1994. Refuge begonias. Taxonomy, phylogeny and historical biogeography of *Begonia* sect. *Loasibegonia* and sect. *Scutobegonia* in relation to glacial rain forest refuges in Africa. Wageningen Agric. Univ. Papers 94-1: 1–306.

Takhtajan, A. 1969. Flowering plants – origin and dispersal. Translation from the Russian by C. Jeffrey. Edinburgh: Oliver and Boyd.

Tebbitt, M.C. 2005. Begonias. Cultivation, identification, and natural history. Portland, OR: Timber Press.

Tebbitt, M.C., MacIver, C.M. 1999. The systematic significance of the endothecium in Begoniaceae. Bot. J. Linn. Soc. 131: 203–221.

Thompson, M.L., Thompson, E.J. 1981. Begonias. The complete Reference Guide. New York: Times Books.

Warburg, O. 1894. Begoniaceae. In: Engler, A., Prantl, K. (eds.) Die natürlichen Pflanzenfamilien III, 6a: 121–150. Leipzig: W. Engelmann.

Wilde, J.J.F.E. de 1985. *Begonia* section *Cristasemen* J.J. de Wilde, sect.nov. Studies in Begoniaceae II. Agric. Univ. Wageningen Papers 84-3: 113–129.

Wilde, J.J.F.E. de 1988. Begonia's als indicator voor ijstijdinvloeden in tropisch woud. Bull. Bot. Tuinen Wageningen 21:10–11.

Wilde, J.J.F.E. de 2002. *Begonia* section *Tetraphila* A. DC., a taxonomic revision. Studies in Begoniaceae VII. Wageningen Agric. Univ. Papers 2001-2: 5–258.

Wilde, J.J.F.E. de, Arends, J.C. 1980. *Begonia* section *Squamibegonia* Warb.: a taxonomic revision. Misc. Papers Landbouwhogeschool Wageningen 19: 377–421.

Wilde, J.J.F.E. de, Arends, J.C.1989. *Begonia salaziensis* (Gaud.) Warb., taxonomy and placentation. Acta Bot. Neerl. 38: 31–39.

Zhang, L.-B., Simmons, M.P., Kocyan, A., Renner, S.S. 2006. Phylogeny of the Cucurbitales based on DNA sequences of nine loci from three genomes: implications for morphological and sexual system evolution. Mol. Phylogen. Evol. 39: 305–322.

Biebersteiniaceae

Biebersteiniaceae Endlicher (1841).

A.N. Muellner

Perennial herbs with woody, occasionally tuberous rhizomes. Leaves alternate, pinnate or pinnatisect, with the (pseudo-?)stipules adnate to the petiole. Inflorescences pedunculate panicles or spikes; pedicels with prophylls. Flowers actinomorphic, bisexual, 5-merous; sepals distinct, imbricate; petals distinct, imbricate, sometimes contorted, often denticulate at the apex, sometimes unguiculate, alternate with 5 fleshy extrastaminal nectary glands; stamens 10; filaments shortly connate; anthers tetrasporangiate, dorsifixed, opening longitudinally, introrse; gynoecium 5-carpellate; ovary superior, on short gynophore, deeply 5-lobed; ovules 1 per loculus, pendulous; stylodia arising from the base of the ovarioles, filiform, connected above into a capitate stigma. Fruit a schizocarp, dehiscing into 5 indehiscent 1-seeded, dry, rugose, crustaceous nutlets. Seeds large, rugulose, with a rounded raphe; endosperm scanty; embryo slightly curved; cotyledons foliaceous. $n = 5$.

A unigeneric family with four or five species occurring from the E Mediterranean to W Siberia and Central China.

Vegetative Morphology. The *Biebersteinia* species are perennial, glandular-hairy herbs (Fig. 14) with a sometimes tuberous rhizome (Knuth 1912; Takhtajan 1997). The glandular trichomes have long, multiseriate stalks and multicellular knob-shaped heads (Takhtajan 1997). The rhizome is thickened, tuberous, 0.5–4 cm in diameter, sometimes sparsely to densely furnished with old withered petioles. The stem is erect, sometimes very straight, 11–80 cm long, glandular-hairy, at the base sometimes covered with ovate, brown, membranaceous scales, and sometimes branched. The leaves are lanceolate, bi- or tripinnatisect, sub-tripinnatisect. The (pseudo-?)stipules are lanceolate or semi-ovate and 1–2 cm long.

Vegetative Anatomy. Tutel (1982) investigated the leaf anatomy of the genus. Shen and Huang (1997) conducted morpho-anatomical studies on *Biebersteinia heterostemon*. The stomata are anomyctic.

Inflorescences. The flowers are arranged in paniculate, spicate or pseudocapitate inflorescences. Bracts beneath the inflorescences are foliaceous, ovate, lanceolate-ovate or lanceolate, sometimes acuminate, sometimes strongly pointed, 0.8–2 cm long, pinnately cleft or entire.

Flower Structure. The flowers of *Biebersteinia* have a yellow or red corolla; sometimes they are nodding after anthesis. No comprehensive structural investigation on the flower of *Biebersteinia* has been undertaken yet. It is still unclear whether the extrastaminal glands are modified parts of the androecium. SEM studies of epidermal patterns of the petals were done by Christensen and Hansen (1998).

Embryology. The ovule is anatropous, bitegmic (unitegmic according to Xu Langran and Vassiliades 2008) and crassinucellate. There is a strand of conducting cells between the endosperm and the chalaza. The integuments are about two- to three-layered. The inner layer of the inner integument and the outer layer of the outer integument are tanniniferous (Boesewinkel 1997). The micropyle is formed mainly by the inner integument. The raphal bundle is already differentiated and there is a chalazal tannin layer. There is no obturator. The ovules are slightly twisted (Boesewinkel 1997). The embryo sac of *Biebersteinia* is tetrasporic, 16-nucleate, 13-celled. This development corresponds to the Penaea type (Kamelina and Konnova 1990).

Pollen Morphology. The pollen grains are 3-colporate, sphaeroid-oblate, (30)37–39(48) µm in diameter. The colpi are 18–33 µm long and 0–2 µm wide. The ora are circular and 4–7 µm in diameter. The exine is pertectate and striate, with the striae as broad as the lirae. The tectum is 0.4–0.8 µm thick. The nexine is 0.8–1.2 µm thick. The intine is very thin. Bortenschlager (1967), from whom most of this information is taken, provides further details on the pollen of the different species.

Fossil pollen of *Biebersteinia* is known from the Neomugen Formation of Inner Mongolia (Late Palaeocene, 57.0–54.8 million years; Song et al. 2004). The pollen was assigned to *Biebersteinia heterostemon* on the basis of the structure of its colpi and pores (Song et al. 2004; W.-M. Wang, Chinese Academy of Sciences, pers. comm. 2007).

Karyology. The chromosome number for all four species of *Biebersteinia* has been recorded as $n = 5$ (Aryavand 1975; Constantinidis 1996; Liu et al. 2001).

Fruit and Seed. Boesewinkel (1997), from whom all following information is taken, investigated the fruits and seeds of *Biebersteinia odora*, *B. multifida* and *B. orphanidis* in detail. The fruit is provided with a well-developed endocarp and the carpels are separately released as indehiscent, one-seeded nutlets. The seeds are slightly curved and have a thin endosperm (Fig. 14C, D). The fruit wall has taken over the protective function of the seed coat, of which the testal layer has been resorbed. The inner and probably also the outer layers of the tegmen can be recognised and are provided with reduced wall thickenings. The exotegmic cells are provided with columnar wall thickenings, which are mostly radially oriented and seem to be present at the junction of the cell walls. The endotegmen consists of slightly thickened tannin cells. These tegmic layers have a somewhat tracheidal character. The nucellar remains consist of a layer of crushed cells provided with a cuticle. The seed has a large embryo, and the leaf-like cotyledons are perpendicularly situated to the median plane. The endosperm is thin and granular. In brief, *Biebersteinia* has a crushed seed coat with remains of a thick-walled exo- and endotegmen.

Phytochemistry. In leaf hydrolysates of *Biebersteinia multifida* and *B. odora*, Bate-Smith (1973) found myricetin, procyanidin and prodelphinidin, whereas ellagitannins were lacking. The Mediterranean *Biebersteinia orphanidis* contains six closely related flavone methyl ethers with dihydroxy, dimethoxy A-ring substitution, in which it differs from the Central Asian species (Greenham et al. 2001). Alkaloids of *B. multifida* were studied by Kurbanov and Zharekeev (1974). Poly- and oligosaccarides of *B. multifida* were isolated by Arifkhodzhaev and Rakhimov (1986, 1993, 1994), Arifkhodzhaev et al. (1985) and Meng et al. (1999), who investigated whole-plant extracts of *B. heterostemon* and found a new antibacterial bisabolane-type sesquiterpene glycoside and further bioactive compounds. Tzakou et al. (2001) investigated the fatty acids in leaf tissue of *B. orphanidis*, showing that the species is a C18:3 plant, in which it agrees with Sapindaceae but not Geraniaceae. The search for bioactive compounds in the genus led to the isolation of a bisabalone-type sequiterpene glycoside from *B. heterostemon* (Meng et al. 1999) and extracts with anti-inflammatory and analgesic activities from *B. multifida* roots (Farsam et al. 2000).

Taxonomy and Phylogeny. When describing *Biebersteinia odora*, Stephan (1806) placed his new genus between *Grielum* and *Suriana*. Subsequent authors suggested affinities with Zygophyllaceae, Rosaceae, Geraniaceae and Rutaceae (see Bakker et al. 1998 for the taxonomic history of *Biebersteinia*). Boissier (1867) placed *Biebersteinia* in Geraniaceae and was followed by other workers. *Biebersteinia*, however, never fit well in Geraniaceae in having only one ovule per locule, in its pollen morphology, and in having a gynophore, and Takhtajan (1997) therefore recognized it as a separate order and family (as first suggested by Endlicher 1841). Phylogenetic analyses based on the plastid DNA markers *rbc*L and *atp*B revealed that *Biebersteinia orphanidis* belongs in Sapindales, albeit in an isolated position (Bakker et al. 1998), and the Angiosperm Phylogeny Group ranks *Biebersteinia* as one of the nine families of the Sapindales (APG 1998; APG II 2003). Although the molecular data of Bakker et al. (1998) provided evidence of a sapindalean rather than geranialean affinity of *Biebersteinia*, their analysis included only *B. orphanidis*. An extended genus-level Sapindales sampling and inclusion of all four species of *Biebersteinia* provided strong support for the monophyly of *Biebersteinia* and its inclusion in Sapindales (Muellner et al. 2007). Based on the combined two genes *rbc*L and *atp*B, Bayesian and ML analyses weakly suggested that *Biebersteinia* may be sister to the remainder of Sapindales (Muellner et al. 2007). The placement of *Biebersteinia* in Sapindales, rather than Geraniales, agrees with the phytochemical observations (summarized in Muellner et al. 2007).

Fig. 14. Biebersteiniaceae. *Biebersteinia odora*. **A** Habit. **B** Flower, longitudinal section. **C** Mericarp. **D** Ditto, longitudinal section. (Reiche 1889)

Distribution and Habitats. The four species of *Biebersteinia* occur in temperate mountainous regions from Central Asia to Greece. Three of the species (*Biebersteinia odora*, *B. heterostemon*, *B. multifida*) are adapted to arid or semi-arid environments, while the natural habitat of *B. orphanidis* is open patches in mid-altitude *Abies* and *Cedrus* forests in Greece and Turkey. *Biebersteinia odora* differs from the other three species in being adapted to alpine conditions, frequently growing above 4,500 m near glaciers.

Economic Importance and Uses. All species have medicinal properties and are therefore utilized by local communities (Zhang et al. 1995; Vassiliades and Yannitsaros 2000; Farsam et al. 2000; Miceli et al. 2005). *Biebersteinin heterostemon* is being used in traditional Chinese medical practice to treat a wide range of diseases such as fever, convulsions, encephalitis and dysentery (Jiangsu New Medical College 1977). The ethanol extract of the plant was shown to be hypotensive, analgesic and immunity-regulatory (Zhang 1995).

Only one genus:

Biebersteinia Stephan Fig. 14

Biebersteinia Stephan, Mém. Soc. Imp. Nat. Moscou, ed. 2, 1: 89 (1811).

Description as for the family.

Selected Bibliography

APG (Angiosperm Phylogeny Group). 1998. An ordinal classification for the families of flowering plants. Ann. Missouri Bot. Gard. 85: 531–553.

APG II. 2003. An update of the Angiosperm Phylogeny Group classification for the orders and families of flowering plants: APG II. Bot. J. Linn. Soc. 141: 399–436.

Arifkhodzhaev, A.O., Rakhimov, D.A. 1986. Polysaccarides of saponin containing plants. 3. Polysaccharides of the epigeal organs of *Biebersteinia multifida*. Khimiya Prirodnykh Soedinenii 6: 773–774.

Arifkhodzhaev, A.O., Rakhimov, D.A. 1993. Polysaccarides from saponin containing plants. 4. Structure of glucan-A, glucane-B and glucan-C of *Biebersteinia multifida*. Khimiya Prirodnykh Soedinenii 2: 188–191.

Arifkhodzhaev, A.O., Rakhimov, D.A. 1994. Polysaccharides of saponin-carrying plants. 5. Structural study of glucane-A, glucane-B, glucane-C and their oligosaccharides of *Biebersteinia multifida* plants. Khimiya Prirodnykh Soedinenii 6: 709–714.

Arifkhodzhaev, A.O., Arifkhodzhaev, K.A., Kondratenko, E.S. 1985. Polysaccarides of saponin containing plants. 2. Isolation and characterization of polysaccarides from *Biebersteinia multifida*. Khimia Prirodnykh Soedinenii 6: 755–757.

Aryavand, A. 1975. Contribution à l'étude cytotaxonomique de *Biebersteinia multifida* DC. (Géraniacées). C.R. Hebd. Séances Acad. Sci. (Paris) D 280: 1551–1554.

Bakker, F.R., Vassiliades, D.D., Morton, C., Savolainen, V. 1998. Phylogenetic relationships of *Biebersteinia* Stephan (Geraniaceae) inferred from *rbcL* and *atpB* sequence comparisons. Bot. J. Linn. Soc. 127: 149–158.

Bate-Smith, E.C. 1973. Chemotaxonomy of *Geranium*. Bot. J. Linn. Soc. 67: 347–359.

Boesewinkel, F.D. 1997. Seed structure and phylogenetic relationships of the Geraniales. Bot. Jahrb. 119: 277–291.

Boissier, E. 1867. Biebersteiniae. In: Flora Orientalis. Vol. 1. H. Georg, Basilee, pp. 899–900.

Bortenschlager, S. 1967. Vorläufige Mitteilungen zur Pollenmorphologie der Geraniaceen und ihre systematische Bedeutug. Grana Palynol. 7: 400–468.

Christensen, K.I, Hansen, H.V. 1998. SEM-studies of epidermal patterns of petals in the angiosperms. Opera Bot. 135: 1–91.

Constantinidis, T.A. 1996. *Biebersteinia orphanidis* Boiss. Flora Mediterranea 6: 308–312.

Endlicher, S.L. 1841. Enchiridion Botanicum. Leipzig: W. Engelmann.

Farsam H., Amanlou, M., Reza Dehpour, A., Jahaniani, F. 2000. Anti-inflammatory and analgesic activity of *Biebersteinia multifida* DC. root extract. J. Ethnopharmacol. 71: 443–447.

Greenham, J., Vassiliades, D.D., Harborne, J.B., Williams, C.A., Eagles, J., Grayer, R.J., Veitch, N.C. 2001. A distinctive flavonoid chemistry for the anomalous genus *Biebersteinia*. Phytochemistry 56: 87–91.

Grierson, A.J.C., Long, D.G. 1983-2001. Flora of Bhutan, vol. 1, part 3. Edinburgh: Royal Botanic Garden Edinburgh.

Hegnauer, R. 1989. Chemotaxonomie der Pflanzen, vol. 8. Basel: Birkhäuser, pp. 513–514.

Jiangshu New Medical College. 1977. Dictionary of traditional Chinese medicine. Shanghai: Shanghai People's Publishing House.

Kamelina, O.P., Konnova, V.A. 1990. Embryological characters of the genus *Biebersteinia* Steph. in relation to its systematic position. Doklady Acad. Tajik SSR. 33 (3): 193–195.

Knuth, R. 1912. *Biebersteinia* Steph. In: Engler, A. (ed.) Das Pflanzenreich IV, vol. 129. Leipzig: W. Engelmann, pp. 546–549.

urbanov, D., Zharekeev, B.K. 1974. Alkaloids of *Biebersteinia multifida* and *Peganum harmala* from Karakalpakiya. Khimia Prirodnykh Soedinenii 5: 685–686.

Liu, J.Q., Ho, T.N., Chen, S.L., Lu, A.M. 2001. Karyomorphology of *Biebersteinia* Stephan (Geraniaceae) and its systematic and taxonomic significance. Bot. Bull. Acad. Sin. (Taipei) 42: 61–66.

Meng, J.C., Lu, H., Li, H., Yang, L.,Tan, R.X. 1999. A new antibacterial sesquiterpene glycoside and other bioactive compounds from *Biebersteinia heterostemon*. Spectroscopy Lett. 32: 1005–1012.

Miceli, N., Taviano, M.F., Tzakou, O., Yannitsaros, A., Vassiliades, D., Giuffrida, D., Galati, E.M. 2005. *Biebersteinia orphanidis* Boiss. shows antioxidant and anti-inflammatory activity. Pharmacog. Mag. 1: 54–58.

Muellner, A.N., Vassiliades, D.D., Renner, S.S. 2007. Placing Biebersteiniaceae, a herbaceous clade of Sapindales, in a temporal and geographic context. Plant Syst. Evol. 266: 233–252.

Omurkamzinova, V.B., Maurel, N.D., Bikbulatova, T.N. 1991. Flavonoids from *Biebersteinia multifida*. Khimiya Prirodnykh Soedinenii 5: 720–721.

Reiche, K. 1889. Geraniaceae. In: Engler & Prantl, Die natürlichen Pflanzenfamilien III, 4. Leipzig: W. Engelmann, pp. 1–14.

Shen, S, Huang, R. 1997. Cytological and morpho-anatomical studies of *Biebersteinia heterostemon* Maxim. Acta Biol. Plat. Sin. 13: 5–8.

Song, Z.-C., Wang, W.-M., Fei, H. 2004. Fossil pollen records of extant angiosperms in China. Bot. Rev. 70: 425–458.

Stephan, F. 1806. Déscription de deux nouveaux genres de plantes. In: Mém. Soc. Nat. Mosc. 1: 125–128.

Takhtajan, A. 1997. Diversity and classification of flowering plants. New York: Columbia University Press.

Tutel, B. 1982. Comparison of the taxonomy and leaf anatomy of the genus *Biebersteinia* with the other genera of Geraniaceae of Turkey. Istanbul Cnicersitesi Fen Fakultesi Mecmuasi. Sen B 4748: 51–86.

Tzakou, O., Yannitsaros, A., Vassiliades, D. 2001. Investigation of the $C_{16:3}/C_{18:3}$ fatty acid balance in leaf tissues of *Biebersteinia orphanidis* Boiss. (Biebersteiniaceae). Biochem. Syst. Ecol. 29: 765–767.

Vassiliades, D., Yannitsaros, A. 2000. Orphanides's best discovery. Bot. Chron. 13: 241–248.

Xu Langran, Vassiliades, D.D. 2008. Biebersteiniaceae. In: Wu Zhengyi, Raven, P.H., Hong Deyuan (eds.) Flora of China, vol. 11, pp. 31–32. Beijing: Science Press, St. Louis: Missouri Bot. Gard.

Zhang, X.F., Hu, B.L., Zhou, B.N. 1995. Studies on the active constituents of Tibetan herb *Biebersteinia heterostemon* Maxim. Acta Pharmaceutica Sin. 30: 211–214.

Burseraceae

Burseraceae Kunth (1824), nom. cons.

D.C. Daly, M.M. Harley, M.-C. Martínez-Habibe, and A. Weeks

Trees or shrubs, sometimes rupicolous, very rarely scandent or epiphytic, with schizogenous resin canals in most vascularized tissues. Leaves alternate, rarely stipulate or pseudostipulate, imparipinnately compound or infrequently unifoliolate or apparently simple, rarely bipinnate or hetero-trifoliolate, often the petiole pulvinate at base, the rachis sometimes winged, often the lateral or at least the terminal petiolules pulvinulate at apex; leaflets (sub-)opposite (to subalternate in *Boswellia*). Inflorescences (pseudo)terminal and/or axillary, rarely cauliflorous, structurally thyrsoid but sometimes appearing spicate, fasciculate, or racemose. Flowers actinomorphic, unisexual or less often bisexual; hypanthium sometimes present; perianth 2-whorled and 3–5(6)-parted; calyx partially synsepalous, valvate (less often imbricate or apert), sometimes partly accrescent in fruit; petals distinct or less often ± fused, induplicate-valvate, imbricate in the middle (*Canarium*), or rarely imbricate (*Boswellia*); androecium (meta-)obdiplostemonous, rarely haplostemonous (the antepetalous series missing); filaments usually distinct or rarely basally to mostly connate (some *Canarium*), anthers sometimes continuous with the filaments (not sagittate), dehiscence longitudinal and introrse or latrorse; disk intrastaminal, rarely extrastaminal (*Triomma*, *Aucoumea*), sometimes adnate to the receptacle or apparently absent (taxa with hypanthium); gynoecium syncarpous and 2–5(6–12)-carpellate and -locular, in staminate flowers reduced or less often rudimentary or absent, sometimes disk and pistillode replaced by a parenchymatous ovariodisk; ovary superior but sometimes semi-inferior within a hypanthium; most locules abortive in most genera; ovules 2 per locule, epitropous, collateral or (*Beiselia*) superposed, attached laterally or (sub-)apically; style 1, erect, sometimes shortly branched near the apex, stigmatal lobes usually as many as carpels but sometimes stigma capitate. Fruit a fleshy or dry compound drupe or pseudocapsule; each seed contained in a pyrene, the pyrenes distinct and the fruit berry-like (e.g., *Garuga*), or fused and indehiscent (most Canarium Alliance taxa), or separated by a columella and dehiscent and the pyrene (sub-)alate or (partly) enveloped in a pseudaril. Seeds exalbuminous, the embryo minute and straight; cotyledons entire and plano-convex or lobed and contortuplicate, rarely transversely twice-folded. Germination epigeal or hypogeal, cryptocotylar or phanerocotylar, with the first eophylls opposite or alternate, simple or trifoliate or pinnately compound.

Approximately 700 species in 19 genera in the tropics and subtropics, represented by few taxa in some warm temperate areas.

Vegetative Morphology. Burseraceae are entirely woody. Most taxa are small to large trees, but species in Boswellia and most Bursera Alliance genera can be shrubby, pachycaulous, or caudiciform under arid conditions; some of these can be effectively lithophytic. *Dacryodes scandens* is probably sarmentose, and *Bursera standleyana* is a small, epiphytic cloud forest tree.

Buttresses are often present in moist forest taxa; these are (sometimes branched) plank buttresses, rarely flying buttresses, and/or stilt roots. The bark is diverse; the color is usually gray or brown, but can be red, yellow, or green (e.g., many *Bursera* spp.) or even blue (some *Commiphora*). It is close in most genera, but can be papery and exfoliating (e.g., *Bursera* subgen. *Bursera*); it can be thin and smooth (most *Protium*) to thick and deeply fissured (e.g., *Tetragastris*).

Resin ducts are associated with virtually all vascularized tissues. The resin can be clear, translucent, milky, or rarely reddish, and the consistency can be watery or gummy; the clear and translucent resins usually have a strong terpenoid odor. It can dry crystalline and powdery, or solid and translucent, or rubbery and opaque. In most species the resin is flammable when dry, in some when fresh as well.

Most taxa are unarmed, but some *Commiphora* and very few *Bursera* have spines that terminate short shoots, sometimes on adventitious shoots from the trunk and boughs; axillary spines have been reported on seedlings of *Protium javanicum* (Lam 1932b), and observed on *Protium melinonis.* The persistent swollen petiole bases of *Beiselia* are often cuspidate.

Cataphylls are sometimes present (some *Bursera* subg. *Elaphrium*). The leaves are evergreen or less often deciduous (*Beiselia,* most Bursera and Boswellia Alliance taxa) and alternate, often conferted at branch apices. Most are imparipinnate, but a few *Bursera* spp. are bipinnate, and a number of genera have unifoliolate species that may appear simple in genera lacking pulvinuli. The petiole is often basally pulvinate, the rachis sometimes winged (petiole distally winged in *Ambilobea*). Most Protium and Canarium Alliance taxa have a pulvinulus at least on the terminal petiolule, and usually at both ends of the lateral petiolules. The leaflets are (sub)opposite (to subalternate in *Boswellia*); in *Garuga* only, stipels can occur at the base of lateral petiolules. The leaflet apex can be gland-tipped (e.g., some *Canarium*). The margin is entire or variously crenate or toothed, these rarely gland-tipped.

Stipules–here meaning laminar to linear structures fully developed before leaf expansion, unlike the leaflets, and usually associated with the point of insertion of a leaf on a stem, but often inserted on the petiole at or near the base–occur in most *Canarium* and some *Garuga.* Pseudostipules–here meaning reduced leaflets that develop with the leaf, and are usually modified basal leaflets or occurring near the petiole base–are found in some *Canarium* (Lam 1932b) and *Pseudodacryodes.*

Leaflet architecture (following Ellis et al. 2009) is moderately diverse. Primary venation is always pinnate, the secondary fabric most often (festooned-)brochidodromous or less often craspedodromous (*Beiselia,* some Bursera Alliance taxa), semi-craspedodromous (e.g., *Crepidospermum*) or eucamptodromous (some Canarium Alliance taxa). Intersecondaries are sometimes present, also usually epimedial tertiaries. The intercostal tertiary fabric is variously percurrent, random-reticulate, or admedially ramified (sometimes composite admedial). In *Trattinnickia* sect. *Burserifoliae,* the areoles define papilla-filled abaxial laminar crypts. Freely ending veinlets are usually dendritic, rarely dichotomous, and most often 0–4-branched (most Protium Alliance taxa), rarely highly branched (some Protium Alliance taxa), sometimes ending in tracheoid idioblasts (many arid-zone Bursera Alliance taxa) or highly branched sclereids (e.g., *Aucoumea,* some Protium Alliance taxa).

Punctate leaflets characterize *Protium* sect. *Icicopsis* but are not restricted to that section. Asperous leaflets are found in some Canarium Alliance taxa, while hairy-tuft domatia occur in the secondary vein axils of some *Haplolobus.*

Trichomes are discussed under Vegetative Anatomy.

Vegetative Anatomy. Literature on Burseraceae anatomy is scant, especially considering the significance of the family's resins and gum-resins. These are produced in the schizogenous canals that are present in the phloem of virtually all vascularized tissues, but they are obtained usually via controlled wounding of the bark, or sometimes by harvesting resin lumps produced by bark-boring weevils (e.g., Plowden et al. 2002). Most pioneering work on the anatomy of the Burseraceae was conducted by Solereder (1908) and Guillaumin (1909), later reiterated by Metcalfe and Chalk (1950); except where otherwise noted, the descriptions of anatomical characters are from Solereder. The work of later authors is largely descriptive and part of regional surveys (e.g., Barajas-Morales and Gómez 1989); very little is diagnostic. Leaf anatomy is especially under-investigated; therefore, the taxonomic distribution of anatomical characters reported in the literature needs to be vetted.

Primary axis (Gum-)resin ducts form from the procambium in the phloem region or in the cambial zone toward the phloem (e.g., *Commiphora mukul*; Setia et al. 1977). Pachycaulous *Bursera* species can have stems with a chlorophyll-rich periderm (Gibson 1981). Cork usually arises in the sub-epidermis of the young stem. The primary cortex has sclerenchymatous idioblasts in a number of genera (with sclerenchymatous fibers in *Dacryodes*). The pericycle is characterized by either arcs or composite continuous rings of sclerenchyma. The pith is homogeneous or heterogeneous, and usually lignified.

Leaf Vascular bundles in the petiole usually form a ring, or an arc in at least some *Commiphora*. Medullary vascular bundles have been observed in the petioles of *Protium* sect. *Icicopsis* (Solereder 1908), and some Bursera Alliance taxa and *Garuga*, while inverted bundles occur in the medulla in most Canarium Alliance taxa (Lam 1932b).

One of the most characteristic leaf features of most Protium and Canarium Alliances taxa is the presence of a pulvinulus on at least the terminal petiolule. Guillaumin (1909) observed vascular bundles scattered throughout the pulvin(ul)i of Burseraceae, with the vascular bundles re-coalescing in a ring or a cylinder at either end. Typically in other groups, the vascular bundles in pulvini are compactly grouped in the center, and surrounded by parenchyma (Esau 1965).

The leaf epidermis (especially the upper epidermis) has mucilage cells. The mesophyll usually has one palisade layer (sometimes two in *Protium*); the palisade cells sometimes are mucilaginous (Paleotropical *Protium*). The lower epidermis can have groups of silicified cells (*Protium*). Dendritic crystals can be found in the epidermis and hypodermis (e.g., *Commiphora angolensis*). Vertically transcurrent vascular bundles have been observed in *Dacryodes* and *Protium*. Laminar crypts apparently filled with papillae occur on the abaxial leaf surface in one section of *Trattinnickia* (Daly 1999). Leaflets are amphistomatic in some *Tetragastris* and *Commiphora* (also *Bursera hindsiana*; Gibson 1981); stalked stomates have been observed in *Santiria*. The nature and taxonomic distribution of leaflet punctations in the family are worth investigating. To Solereder (1908), these corresponded to large solitary crystals in the palisade layer in Asian *Protium*; we have observed them in *Protium* sect. *Icicopsis*, while the sunken punctate 'glands' in some sect. *Pepeanthos* must be anatomically distinct.

Wood Vessels are small to medium, solitary or in multiples of 2–3 (4–5) cells. Perforations are simple (scalariform in *Beiselia*; Forman et al. 1989). Intervascular pitting is alternate, with large hexagonal borders, the pits to ray cells and parenchyma simple (scalariform in *Beiselia*; Forman et al. 1989), and usually unilaterally compound. Thin-walled tyloses occur in many genera, occasionally with starch grains, crystals, or gums. The parenchyma is paratracheal, sheathing the vessels, and usually scant; when vasicentric, it is usually in one row, but biseriate in some Canarium Alliance taxa; occasionally it is diffuse. Broken bands or uniseriate terminal bands are reported in several genera. Rays are uniseriate in some species, up to six cells wide in others; they are heterogeneous, with 1–3 rows of square or upright marginal cells, often with rhomboidal crystals or druses (*Haplolobus*; Chattaway 1955). Intercellular resin canals occur in secondary rays in some species. Fibers are septate; the pits are simple, small and slit-like to almost round, often in radial rows; sometimes with a mucilaginous layer.

Among others, Webber (1941) and Kryn (1952) commented on the close similarities in the anatomy of the Burseraceae and Anacardiaceae, because both have radial intercellular canals, similar ray-vessel pitting, and septate fibers, and they lack banded parenchyma other than terminal. They also commented on the relative uniformity of wood and stem anatomy in the family. In their atlas of Amazonian woods, Détienne and Jacquet (1983) wrote a joint key to the woods of some Burseraceae and Anacardiaceae species, although they were able to distinguish several species of *Protium* based on the presence or absence of silica crystals in the fibers, and on whether the rays are uniseriate or biseriate.

Trichomes Trichomes of Burseraceae exhibit greater diversity than most other anatomical features of the family. They are unicellular or multicellular, sessile or stalked, rarely fasciculate (*Santiria*; Lam 1932b), sometimes markedly retrorse (e.g., abaxial side of corolla in *Canarium* and *Trattinnickia*), non-glandular or glandular, sometimes snail-shaped (several clades), malpighiaceous (e.g., Central American *Protium*), long and dendritic (some Paleotropical *Dacryodes*), peltate scales, or stellate (some Canarium Alliance taxa, these stipitate in *Dacryodes normandii*; Aubréville 1962). The capitate glands often have a several-celled stalk and 4-celled head (*Garuga*); they are branched in *Commiphora merkeri* (van der Walt and van der Schijff 1969).

Inflorescences. Inflorescences are often initiated with leaf flush or, in the case of most arid-zone taxa, beforehand. They are usually axillary, but may be truly terminal (some *Canarium*) or pseudoterminal (e.g., *Trattinnickia*).

Inflorescence architecture is consistently thyrsoid, i.e., a panicle of cymes, but a great deal of morphological diversity is achieved through the suppression or proliferation of axes or the pedicels, resulting in pseudoracemes, pseudofascicules, or pseudospikes, the latter characteristic of some lineages (e.g., *Protium* sect. *Icicopsis*).

Sexual dimorphism is common particularly in the Canarium Alliance; usually the male inflorescences are longer and more laxly branched, with more (and smaller) flowers.

Inflorescence bracts may be caducous to persistent, and they range from highly reduced to foliose (many *Canarium*); in *Rosselia* the bracts resemble reduced leaflets (although more ovate).

Flower Structure. Burseraceae flowers are actinomorphic and unisexual or less often bisexual. In dioecious taxa, flowers are usually structurally bisexual, but with either a reduced pistillode lacking stigmas or reduced staminodes lacking pollen. Some taxa show marked sexual dimorphism; the staminate flowers can be larger in *Commiphora* (Voigt et al. 2005), while in *Canarium* the pistillate ones are much larger.

A hypanthium is characteristic of *Garuga* and *Scutinanthe* (Fig. 15D, H, I), but perigynous species occur also in *Bursera*, *Commiphora*, and *Boswellia*. The perianth is always 2-whorled and can be 3–5(6)-parted. The calyx is partially synsepalous (rarely divided to the base) and lobed (rarely truncate), and the aestivation is usually valvate (apert in *Bursera*, imbricate in *Boswellia*, *Commiphora*, and *Haplolobus*); in some *Canarium* it is enlarged in fruit.

The petals are distinct or less often partly to mostly fused. The aestivation can be valvate, induplicate-valvate (including all Protium Alliance) or valvate at the apex and imbricate along the sides (many Canarium Alliance taxa), rarely imbricate (*Boswellia*). In most taxa there is an inflexed apiculum (Fig. 15). Petal color ranges from pale green to (greenish) yellow, white, cream, pink, and red (the latter most often in Canarium Alliance).

The androecium is usually obdiplostemonous (carpels are antepetalous), but the two whorls often appear in a single cycle (meta-obdiplostemony; Lam 1931, 1932a); some taxa in six genera in which the antepetalous series is missing are haplostemonous. The filaments are usually distinct, rarely basally to mostly connate and forming a tube (some *Canarium*, Fig. 15W, X), and the anthers are sometimes continuous with the filaments (not sagittate; e.g., some *Dacryodes*), the dehiscence longitudinal and introrse or latrorse.

The disk is usually intrastaminal and annular (Fig. 15), less often adnate to the receptacle or apparently absent (some taxa with a hypanthium, Fig. 15E), very rarely extrastaminal (*Triomma*, Fig. 15O, Q; *Aucoumea*). Sometimes the disk and pistillode are replaced by a parenchymatous ovariodisk (*Tetragastris*, some *Canarium* and *Protium*).

In staminate flowers the pistillode is variously reduced: most often resembling a reduced pistil and parenchymatous or with locules and rudimentary ovules, capped by a style (Fig. 15C); or consisting of a parenchymatous cylinder (e.g., *Protium* sect. *Sarcoprotium*); or obsolete (Fig. 15L).

The gynoecium is syncarpous, 2–6-locular and rarely (*Beiselia*) 10–12-locular (Fig. 16), the ovary usually superior (perianth epigynous when a hypanthium is present, e.g., *Garuga*), the placentation axile. The common style is apical, erect, sometimes shortly branched near the apex, the stigmas or stigmatal lobes usually as many as carpels, but sometimes capitate.

In most Burseraceae, the gynoecium is synascidiate at least to the midpoint of the ovary, and symplicate up to the base of the distinct stigmas, but in *Beiselia* the synascidiate region encompasses the whole ovary and extends beyond the locules, and the symplicate zone below the postgenitally united distinct carpel tips is short. In the flower of *Beiselia*, the notable massive remnant of the floral apex is uplifted to the base of the distinct carpel tips, as is the case in Anacardiaceae and *Kirkia* (Bachelier and Endress 2009).

The ovules are usually bitegmic, slightly campylotropous, and have a longer and thicker inner integument and an S-shaped or zigzag micropyle, traits shared with *Kirkia*, Meliaceae, Simaroubaceae, and Rutaceae. Apparently unitegmic ovules are known from some *Canarium*, *Commiphora*, and *Santiria* (Bachelier and Endress 2009).

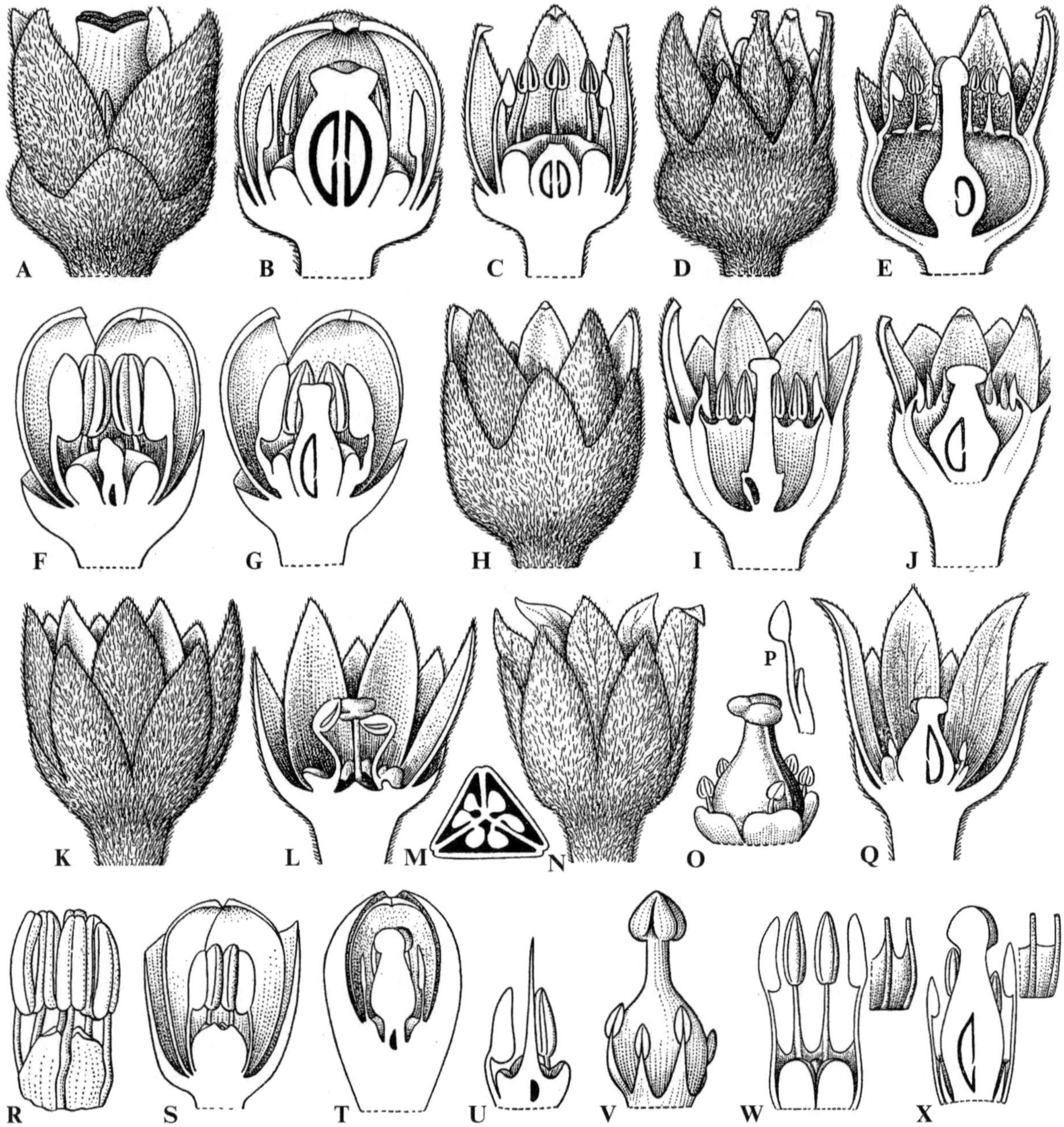

Fig. 15. Flower structure in Burseraceae. **A–C** *Protium connarifolium*. **A, B** Female flower and longitudinal section. **C** Male flower, longitudinal section. **D, E** *Garuga floribunda*, flower and longitudinal section. **F, G** *Haplolobus floribundus*, male and female flowers in longitudinal section. **H–J** *Scutinanthe brunnea*, male flower and male and female flowers in longitudinal section. **K–Q** *Triomma malaccensis*. **K** Male flower. **L** Ditto, longitudinal section. **M** Transversal section of ovary. **N** Female flower. **O** Ovary, staminodes, and extrastaminal disk. **P** Tangential section of staminode and disk. **Q** Female flower, longitudinal section. **R–T** *Canarium maluense*. **R** Androecium and ovariodisk. **S** Male flower, longitudional section. **T** Female flower, longitudinal section. **U, V** *Canarium caudatum*. **U** ♂ Ovariodisk with stamens and rudimentary style. **V** ♀ Androecium and gynoecium. **W, X** *Canarium asperum*. **W** Androecium and disk of male flower, longitudinal section, with detail of staminal tube. **X** Androecium, disk, and gynoecium of female flower, longitudinal section. (Lam 1932a; drawn by Darmosudiro)

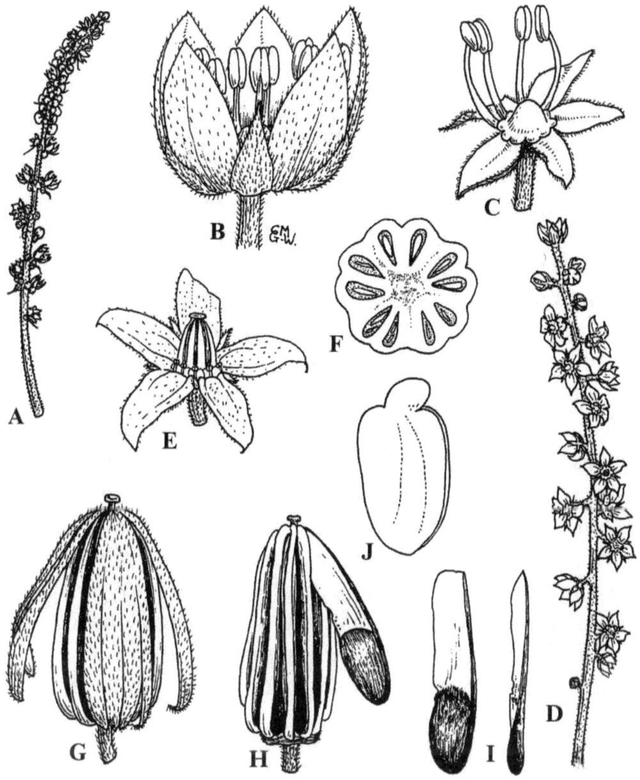

Fig. 16. Burseraceae. *Beiselia mexicana*. **A** Male inflorescence. **B** Male flower. **C** Male flower with petals and most stamens removed to show the ovariodisk. **D** Female inflorescence. **E** Female flower. **F** Transverse section of ovary. **G** Fruit with pericarp splitting. **H** Flanged columella of fruit with one pyrene. **I** Pyrenes from side and abaxial views. **J** Embryo. (Forman et al. 1989; drawn by C. Grey-Wilson)

Additional details on the floral anatomy of the family can be found in Narayana (1959, 1960) and Bachelier and Endress (2009).

FRUIT AND SEED, GERMINATION. Burseraceae fruits are variations on the drupe, and the unit of dispersal is the pyrene, either simple or compound. Functionally, the fruits are either dehiscent pseudocapsules or more often compound drupes, the latter either indehiscent or dehiscent septicidally and acropetally via valves.

Pseudocapsules occur in four different lineages; they are lobed or winged structures with coriaceous or dry valves, releasing dry, compressed, usually winged (except *Ambilobea*) pyrenes separated by a persistent columella with as many lobes as the number of valves; each pyrene is compressed and/or winged in a plane perpendicular to radii of the columella, except in *Beiselia* where the plane is parallel to the columella lobes.

The compound drupes of Burseraceae are of four types: (1) Fruits with fleshy to leathery valves, pyrene 1 (including 1–2 connate or connivent abortive locules, rarely these developing), basally attached to the receptacle, and invested with a fleshy and brightly colored pseudaril that usually only partly covers the pyrene (*Bursera*, *Commiphora*). (2) Fruits with fleshy to leathery (rarely sublignified) valves, pyrenes 1–5, separated by a columella, enveloped in a sweet pulpy white or rarely red pseudaril, on dehiscence tenuously suspended from the fruit apex by an inverted V-shaped structure (Protium Alliance). (3) Indehiscent fruits with the exocarp thin and dry; the mesocarp fleshy and oily or resinous or rarely dry; pyrene (endocarp) of 2–3 bony or cartilaginous (papery in *Haplolobus*), connate locules, in most of the genera the undeveloped locules highly reduced and compressed and forming an often separable articulated plate on a margin of the developed locule (most Canarium Alliance taxa). (4) A berry-like indehiscent fruit in which the 1–5 bony pyrenes are distinct and embedded in a sweet fleshy pulp (*Garuga*).

There is disagreement about the homology of the pseudaril in the anatomy of the first two types. van der Walt (1975) considered it a zone of the mesocarp, while other authors associate it with the endocarp; further developmental studies are needed.

The endocarp in all but one genus has a zone of weakness (usually at the apex) where it splits open during germination; in *Canarium*, the endocarp opens via an operculum or lid (Hill 1933, 1937).

The testa in Burseraceae is thin but sometimes with irregular thickenings and then infolded with contortuplicate cotyledons, a morphology sometimes referred to as a "labyrinth seed" (van Heel 1970). The seed is exalbuminous, the embryo straight, and the cotyledons account for most of the seed volume. Cotyledon morphology in Burseraceae ranges from entire and plano-convex and straight or variously curved, to entire and plicate or contortuplicate, to pinnately or palmately divided and folded, to 3-lobed and transversely reverse-folded. Correspondingly, germination and seedling morphology in the family are remarkably diverse; germination can be epigeal or hypogeal and cryptocotylar or phanerocotylar, and the first eophylls can be opposite or alternate, simple or trifoliate or pinnately compound, the margin entire or less often toothed.

Phytochemistry. Overviews: Hegnauer (1964, 1989), Khalid (1983). The strong diversification of its terpenoid chemistry is the family's most characteristic trait. The resins generally consist of a mixture of a volatile and a non-volatile fraction (Boer and Ella 2000). In the volatile fraction, monocyclic and bicyclic monoterpenes are equally important constituents of oleoresins. The volatile sequiterpenes are of more restricted occurrence, but are strongly diversified in *Canarium* and *Commiphora.* Macrocyclic cembranoid diterpenes occur in the gum-resins of *Commiphora* and the frankincense of some *Boswellia* (Khalid 1983), while triterpenes are frequent and highly diversified in the family, always with compounds from the ursane and oleanane series (Lima et al. 2004). The sterol guggulsterone and guggulsterols derived from the crude drug 'guggul' (*Commiphora mukul*) are reported to have anti-inflammatory activity (Burris et al. 2005). Other remarkable compounds include ellagic acid from the wood, proanthocyanins from leaves, tannins, and the biflavonoid amentoflavone from a group of flavonoid compounds that elsewhere in Sapindales are known only from Anacardiaceae (Hegnauer 1964, 1989).

Terpenoids (mostly volatile mono- and sesquiterpenes) serve in communication and defense, for example, as attractants for pollinators and seed dispersers, competitive phytotoxins, antibiotics, and as herbivore repellents and toxins (e.g., Langenheim 1994; Becerra 2003; Fine et al. 2005). The most important function of volatiles is in intra- and interspecific communication; characteristic blends are also synthesized de novo in response to wounding, discriminating between mechanical and herbivore-induced wounding (Piechulla and Pott 2003); it is significant that culturally and economically valued Burseraceae resins are obtained by repeated wounding (e.g., Atal et al. 1975).

Given the family's importance in traditional medicine throughout the tropics, the Burseraceae resins have been the subject of numerous studies of their chemistry and pharmacology for anticancer, antioxidant, antimicrobial, analgesic, anti-inflammatory, anti-cholesterol, cercaricidal, acaricidal, insecticial, and repellent properties; most of these studies have focused on several species of the frankincense and myrrh genera *Boswellia* and *Commiphora* (e.g., Strappaghetti et al. 1982; Dolara et al. 1996; Rahman et al. 2008; Mothana et al. 2009), while more recently, Neotropical species of *Bursera*, *Protium*, and *Trattinnickia* have been examined for similar applications (e.g., Aregullin et al. 2002; Oliveira et al. 2005; Robles et al. 2005; Pontes et al. 2007; Aragão et al. 2008), as well as the monotypic African *Aucoumea* (Liang et al. 1989).

In commerce, the soft resins like the Manila elemis obtained from several Asian *Canarium* species contain a volatile (essential oil) fraction forming up to 30% or more of the total mass; they are soluble in alcohol and ether. Currently, Manila elemi is used mainly for fragrance applications (soaps and perfumes), and occasionally as an ingredient in lacquers and varnishes to give them toughness and elasticity. In contrast, hard resins like that from *Canarium hirsutum* are suitable for spirit varnish; they are soluble in benzene, chloroform, kerosene, petroeum ether, and turpentine (Boer and Ella 2000).

Pollination and Breeding Systems. Burseraceae are predominantly dioecious. *Boswellia* and *Garuga* have perfect flowers, as do some *Dacryodes* and *Bursera.* Monoecy is rare (e.g., *Commiphora monoica*; Vollesen 1989), as is selfing (e.g., *C. samharensis* subsp. *terebinthina*; Gillett 1991). Non-pseudogamous apomixis has been observed in the ethnobotanically and commercially notable *Commiphora wightii* (Gupta et al. 1996). In Mexico, *Bursera morelensis* flowers that were not pollinated formed parthenocarpic fruits, and it has been suggested that this might function as a deceit strategy against insect seed predators (Ramos-Ordoñez et al. 2008).

Dioecy has been associated with the woody habit and entomophily (e.g., Matallana et al. 2005), and specifically with small generalist insect pollination (e.g., Bawa and Opler 1975); the relatively scant pollination literature on the family suggests that this holds true for most Burseraceae regardless of habitat, and examples are *Protium divaricatum* and *Commiphora leptophloeos* (diverse small insects; van Dulmen 2001 and Machado and Lopes 2004, respectively), *Protium spruceanum* (*Trigona* sp. and *Apis mellifera*; Vieira and Carvalho 2008), and *Trattinnickia peruviana* (small bees; van Dulmen 2001). In eastern South Africa, 28 insect species were observed visiting the flowers of *Commiphora harveyi* (Voigt et al. 2005).

Flowers in Burseraceae tend to be short-lived (e.g., *Bursera simaruba*; Primack 1985). Both "Big Bang" and "trapline" flowering occur in the family (see Gentry 1974), and members of the same clade tend to show the same syndrome (e.g., trapline in *Protium* sect. *Icicopsis* vs. Big Bang in sect. *Sarcoprotium*).

Sexual dimorphism in Burseraceae flowers is usually revealed only through flower dissection, because the non-functional organs are usually reduced rather than obsolete, and notable size differences between genders are found in few groups (e.g., most *Canarium* and less so in *Protium* sect. *Icicopsis*).

The pollinator reward is nectar, produced by the usually annular disk. For example, for *Protium spruceanum* in Central Brazil both flower sexes produced relatively abundant nectar (~4 μl), with an average 30% concentration of sucrose equivalents (Vieira and Carvalho 2008).

Gene flow in Burseraceae is poorly known. In a dry forest in Puerto Rico, potential for long-distance pollen movement was high, and between 45% and 100% of seedlings were sired from outside their stand, but there were few pollen donors per maternal tree and seed abortion was high (Dunphy and Hamrick 2007).

Hybridization has been implicated repeatedly in *Bursera*, not only to explain morphological intermediates but also as a mechanism of speciation (e.g., Rzedowksi and Ortiz 1988); recent molecular genetic work (Weeks and Simpson 2004) confirmed hybridization in this genus, and a number of problematic intermediates in *Protium* (e.g., Daly 2007) will likely prove to be hybrids as well.

KARYOLOGY. There are few chromosome counts in the literature for Burseraceae. Chromosome numbers for nine genera have been reported.

- Beiselia Alliance—*Beiselia*: $n = 13$ (Forman et al. 1989).
- Protium Alliance—*Protium serratum* (as *Bursera serrata*): $n = 11$ (Gillett 1980).
- Bursera Alliance—*Bursera* and *Commiphora*: $n = 12$ (Fedorov 1969).
- Boswellia Alliance—*Boswellia*: $n = 11$ (Gill et al. 1979, 1990; Thulin and Warfa 1987); *Garuga*: $n = 11$ (Singhal and Gill 1984).
- Canarium Alliance—*Canarium ovatum*: $n = 23$ (Villegas and Coronel 1979); *Canarium* (3 spp.): $n = 24$ (Chen 1993); *Dacryodes*: $n = 23$ (Raven 1975); *Haplolobus*: $n=22$ (Oginuma et al. 1999).

Chromosome numbers are as yet unknown for *Aucoumea*, *Crepidospermum*, *Pseudodacryodes*, *Rosselia*, *Santiria*, *Scutinanthe*, *Tetragastris*, *Trattinnickia*, and *Triomma*.

PALYNOLOGY (by M.M. Harley). An account of the pollen morphology of Burseraceae was given by Harley et al. (2005), who proposed 14 different pollen types for the family. The palynology of the Protium Alliance was published by Harley and Daly (1995), and the pollen morphology of African Burseraceae was treated by Harley and Hall (1999).

Pollen grains usually isopolar or, rarely, sub-isopolar (*Canarium gracile*, Fig. 17Q, R) or heteropolar (*Scutinanthe brunnea*). Aperture system tricolporate, rarely tricolporate lobed-vestibulate (*Scutinanthe brunnea*, Fig. 17O) or triporate-projectate (*Canarium gracile*, Fig. 17Q, R); aperture disposition usually angulaperturate (e.g., *Bursera kerberi*, Fig. 17B and *Aucoumea klaineana*, Fig. 17T), occasionally fossaperturate, or rarely planaperturate (*Rosselia bracteata*, Fig. 17G).

Pollen shape either suboblate (e.g., *Canarium gracile*, Fig. 17Q, R, *Scutinanthe brunnea*, Fig. 17O); oblate-spheroidal to prolate-spheroidal (Fig. 17A, C, E, F, I, M, N, S), notably in Bursera Alliance; subprolate to prolate, typical of most members of the Protium Alliance (Fig. 17J, L, P), *Boswellia* (Boswellia Alliance) and most members of Canarium Alliance; or, rarely, almost per-prolate (e.g., *Protium aracouchini*, P/E 1.95). Outline in polar view circular, rounded-triangular (Fig. 17K), triangular (Fig. 17T), triangular-lobed (typical of *Bursera*, Fig. 17B) or, rarely, hexagonal (e.g., *Commiphora sarcopoda*) or triangular projectate (*Canarium gracile*).

Polar length ranges from ca. 17–83 μm, and equatorial width from ca. 14–73 μm. Very large pollen grains (>55 μm) are found mainly in *Garuga* and *Boswellia* (both Boswellia Alliance), while very small pollen grains (<20 μm) occur occasionally in *Bursera* (Bursera Alliance),

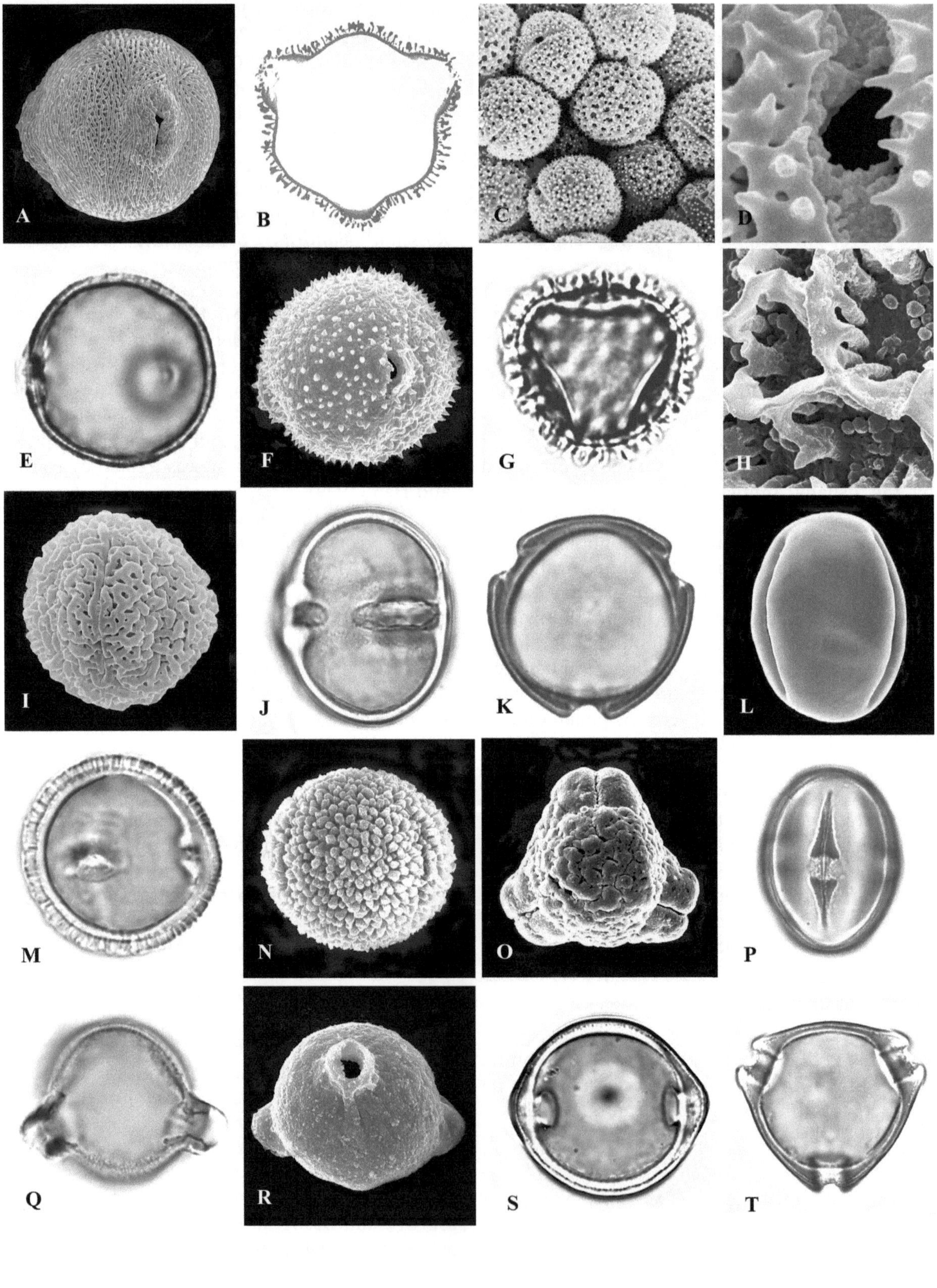
A
B
C
D
E
F
G
H
I
J
K
L
M
N
O
P
Q
R
S
T

infrequently in *Protium* (Protium Alliance), and occasionally in *Canarium*, *Dacryodes*, and *Santiria* (Canarium Alliance).

The colpi may be very short, typically in the Bursera Alliance (*Aucoumea*, Fig. 17S; *Bursera*, Fig. 17A) but also occurring rarely in the Canarium Alliance (*Triomma*, Fig. 17F; *Scutinanthe brunnea*, Fig. 17O) as well as *Rosselia*. Long or very long (Fig. 17L) colpi are typical of most members of the Protium and Canarium Alliances, also of *Boswellia* (Fig. 17P).

Endoapertures usually lalongate (Fig. 17M) or per-lalongate, notably in *Trattinnickia* (Canarieae) and *Crepidospermum* (Protieae: Fig. 17J), circular, notably in Bursera and Boswellia Alliances (Fig. 17D, E, S), subcircular or, rarely, lolongate (a few species of *Commiphora*); in light microscopy (LM) costae may appear indistinct, distinct, or very pronounced. Costae absent or, more frequently, present and either polar-oriented and sickle-shaped (Fig. 17M), circular (e.g., *Triomma*, Fig. 17E) or, less frequently, meridionally extended parallel to colpus margins and tapered toward apices (Boswellia Alliance: *Boswellia* and *Garuga*, Fig. 17P).

Exine tectate or semitectate and either psilate (Fig. 17L), psilate-perforate, foveolate, low relief rugulate, scabrate-rugulate, perforate-rugulate, striate, striate-perforate, striate-reticulate (Fig. 17A), or (very rare) striate-reticulate with long, frequently hooked, striate spines (*Ambilobea madagascariensis*), spinulose (Fig. 17F), micro-reticulate spinulose, reticulate or foveolate-spinulose (Fig. 17C, D), supra-reticulate clavate (*Canarium oleiferum*, Fig. 17N), coarsely reticulate with shallow lumina and psilate rounded muri (*Protium robustum*, Fig. 17I), or very coarsely reticulate with deep lumina and psilate sharply angular muri (*Rosselia bracteata*, Fig. 17H).

Apocolpial wall thickness 1–4.5 μm; thicker walls tend to be associated with more elaborate tectum types or larger pollen (e.g., *Boswellia*), exceptionally the psilate-perforate pollen of *Aucoumea klaineana* has a thick walled apocolpium (Fig. 17S). Infratectum columellate or granular, or occasionally granular-columellate. Apocolpial ultrastructure: between taxa the thicknesses of the ectexine layers (tectum, infratectum and foot layer) vary relative to each other (cf. Fig. 17B, G, M, S). Endexine generally thin or absent in mesocolpial areas (Fig. 17B, K, T), present and thicker in apertural areas, even thicker in endoapertural region, rarely present at apocolpia.

The 14 pollen types described for Burseraceae in Harley et al. (2005) are based on combinations of shape, size, and aperture and exine morphology. All but one are basically tricolporate and oblate spheroidal to prolate, all characteristics frequently encountered in other eudicotyledonous families. Nevertheless, nine pollen types are unique to Burseraceae.

The pollen of *Rosselia*, originally described in Forman et al. (1994), is unique within the family and unusual outside the family as well, superficially resembling some *Ruellia* (Acanthaceae) in its reticulum and the granulae in the bases of the lumina. *Canarium gracile* and *Scutinanthe brunnea* represent extremely unusual pollen aperture morphologies.

The combination of tricolporate pollen with a distinctly striate exine, especially frequent in tribe Canarieae, is not particularly common among colporate eudicot pollen, but it is often encountered in Sapindalean families. It is common in Anacardiaceae (Erdtman 1952; Baksi 1976; Ibe and Leis 1979) and found in *Brucea* (Simaroubaceae, Durdana Yunus and Nair 1988–1989). On

Fig. 17. Burseraceae. Pollen grains. **A** *Bursera confusa*. Equatorial view. SEM ×1,750. **B** *Bursera kerberi*. Equatorial plane. TEM ×1,750. **C** *Commiphora africana* var. *tubak*. Group of pollen grains. SEM ×850. **D** *Commiphora incisa*. Close up of colporus. SEM ×5,000. **E** *Triomma malaccensis*. Polar plane. LM, mid focus ×1,300. **F** *Triomma malaccensis*. Equatorial view. SEM ×1,400. **G** *Rosselia bracteata*. Equatorial plane. LM high-mid focus ×1,150. **H** *Rosselia bracteata*. Close-up of exine. SEM ×6,000. **I** *Protium robustum*. Equatorial view. SEM ×1,500. **J** *Crepidospermum rhoifolium*. Polar plane. LM mid focus ×950. **K** *Protium tenuifolium*. Equatorial plane. LM mid focus ×1,500. **L** *Protium copal*. Equatorial view. SEM ×950. **M** *Canarium schweinfurthii*. Polar plane. LM ×1,000. **N** *Canarium oleiferum*. Equatorial view. SEM ×1,000. **O** *Scutinanthe brunnea*. Polar view of colpus 'side' of a heteropolar grain. SEM ×900. **P** *Garuga pinnata*. Polar plane. LM ×700. **Q** *Canarium gracile*. Polar plane. LM ×1,150. **R** *Canarium gracile*. Oblique equatorial view. SEM ×1,150. **S** *Aucoumea klaineana*. Polar plane. LM ×825. **T** *Aucoumea klaineana*. Equatorial plane. LM ×825. (Photos M.M. Harley and U. Song)

the other hand, in Rutaceae the exine more often tends to be reticulate-striate rather than purely striate (e.g., Erdtman 1952; Morton and Kallunki 1993; Grant et al. 2000), and in Sapindaceae the ectexine is frequently striate, although here the colpi tend to be longer than in Canarieae or, quite often, may be syn- or para-syncolporate (Muller and Leenhouts 1976; Van der Ham 1995, and references cited therein). In Meliaceae, few taxa have striate pollen, and the only genus with pollen resembling Burseraceae is *Aglaia* ('Protium type' of Harley and Daly 1995; see, for example, Tissot et al. 1994).

Dispersal. Burseraceae display two basic dispersal syndromes, zoochory and anemochory. Winged pyrenes have arisen independently in at least four distinct lineages: *Beiselia*, *Aucoumea* (Bursera Alliance), *Boswellia*, and *Triomma* (Canarium Alliance). Hydrochory may occur in the few floodplain species but has not been recorded.

The reward for dispersers differs among lineages. Protium Alliance fruits are dehiscent and their pseudarils provide a low-energy resource (e.g., 6.5 kJ/pseudaril in *Tetragastris panamensis*; see Howe 1980). In the Bursera Alliance, *Bursera* is also dehiscent, but in contrast, the caloric content of *Bursera simaruba* and *B. hindsiana* pseudarils is very high (32.8±1.3 and 27.4±0.8 kJ/g, respectively; Bates 1992). Of 34 zoochorous tree species studied in SE Madagascar, *Canarium* (Canarieae Alliance) fruits were the second largest, third heaviest, second highest in protein, and third highest in fat (Bollen et al. 2004).

Dispersers can be diverse among or even within species. In Gabon, *Dacryodes* species are dispersed by birds, monkeys, and elephants, while *Santiria trimera* is dispersed by birds and elephants (Gautier-Hion et al. 1985). In southern Africa, *Commiphora edulis* fruits are ingested by birds, rodents, and baboons (van Wyk and van Wyk 1997). The fruits or fruit pulp of *Dacryodes*, *Santiria*, and *Canarium* are ingested by lowland gorillas and chimpanzees in Gabon (Tutin et al. 1994), and by bonobos in Zaire (Idani et al. 1994). At least eight species of lemurs ingest or eat the pulp of *Canarium* species on Madagascar (e.g., Britt 2000).

In the South Pacific, *Canarium* fruits do not float but rather some species are dispersed by fruit pigeons, which ingest the fruits whole and only later disgorge the pyrenes elsewhere, often effecting dispersal among islands; pyrenes obtained from the crops of these pigeons measured up to 3 × 2.5 cm (Guppy 1906). In Amazonia and the Guianas, bats are reported as dispersers of several *Protium* species (Charles-Dominique 1991; Parrado-Rosselli 2005; Lobova et al. 2009), all of which have green or whitish fruits.

Many species of Burseraceae show a 'generalist strategy', but few visitor species are effective or significant dispersers. In Panama, 26 species of birds were observed feeding on the fruits of *Bursera simaruba*, but only some of these (notably oropendulas and caciques) swallowed the pyrene + pseudaril and later regurgitated the pyrene some distance away (Trainer and Will 1984). In Panama, 25 species were observed visiting *Tetragastris panamensis* fruits, and a variety of mammals and birds removed the pyrenes, but monkeys were the primary dispersal agents and the howler monkey *Alouatta* alone was responsible for 74% of seed removal; seeds from their feces were viable (Howe 1980).

The 'quality' of dispersal is extremely important; for example, in Mexico, only two bird species ingested the pyrene + pseudaril of *Bursera fagaroides*, but those passed by *Vireo griseus* did not germinate, whereas 17% of those ingested by *Dumetella carolinensis* germinated (Ortiz-Pulido and Rico-Graya 2006).

The exposed pyrene + pseudaril of the Bursera and especially of Protium Alliances is tenuously attached (at base and apex, respectively) and eventually falls if not dispersed; that of *Tetragastris panamensis* can hang suspended for 72 h before dropping to forest floor (Howe 1980). Primary dispersal of *Commiphora guillauminii* is effectively carried out by the parrot *Coracopsis nigra*, and secondary dispersal by the ant *Aphaenogaster swammerdami* (Böhning-Gaese et al. 1999). It is much rarer to find fallen intact fruits of indehiscent Canarium Alliance, but in Thailand seven mammal species were observed consuming fallen fruits of *Canarium euphyllum* (Kitamura et al. 2006).

Although many Burseraceae would be considered generalists in regard to dispersers, there are many cases of strong (inter-)dependence. For the oil-bird *Steatornis* on Trinidad, Burseraceae comprised more than 25% of fruits consumed and was the second most important family after palms; *Dacryodes* fruits often outnumbered all other species in samples (Snow 1962). In Thailand, *Canarium euphyllum* seeds were dispersed by the pigeon *Ducula badia* and four hornbill species; these frugivorous species often have low tolerance to negative human impacts, and their loss would severely compromise the tree's successful regeneration (Kitamura et al. 2006). In Sonora, Mexico, the migratory gray vireo's 5-month winter diet consists almost entirely of *Bursera microphylla* fruits, and there is strong overlap between the vireo's winter range and that of *B. microphylla* (Bates 1992).

On Madagascar, large-fruited *Canarium* trees are highly dependent on large-bodied lemurs, not only for effective dispersal away from the mother tree but also for germination (e.g., Birkinshaw and Colquhoun 2004). Conversely, in one area *Canarium* fruits accounted for 92% of the feeding records of *Varecia variegata* (Britt 2000), which ingests and passes whole fruits (Simons 1991), and at another site Sanford's lemurs chased crowned lemurs from fruiting *Canarium* trees (Freed 2007).

One phenomenon that occurs across several lineages in Burseraceae is clump-dispersal, usually by frugivores that defecate seeds in masses, often corresponding to unexpectedly high seedling survival, considering density-dependent mortality and seed-shadow effects. In Panama, 24% of pyrenes of *Tetragastris panamensis* were dropped in fecal clumps, and droppings of howler monkey, coatimundi, and white-faced monkey contained 1–60, 6–45, and 4–9 pyrenes, respectively (Howe 1989). In Colombian Amazonia, there was a relatively high proportion of viable seeds of *Dacryodes chimantensis* under and near the parent tree crown; similarly, overall seed predation of *Protium paniculatum* fruits was lower than that for many other tropical species, and there was high survivorship capacity close to the parent tree (<10–15 m), suggesting a need for only occasional longer-distance dispersal to initiate new 'patches' (Parrado-Rosselli 2005).

In contrast, the increase in the probability of becoming established as a seedling away from parent trees due to dispersal was 80 times for the Malagasy *Commiphora guillauminii*, whose seedlings were found mostly under and close to the nearest *Commiphora* tree due to a depauperate avian frugivore community, compared to only six times higher for South African *C. harveyi*, whose seedlings were found relatively far from the nearest *Commiphora* tree (median = 21 m) in a region richer in avian frugivores.

In a northeastern Brazilian forest, leaf-cutter ants piled pyrenes of *Protium heptaphyllum* in large numbers 3.4 ± 2.2 m from parent trunks, subsequently cutting and harvesting the epicotyls of most seedlings, but ultimately effecting short-distance dispersal (Silva et al. 2007).

Paleobotany (by A. Weeks). Fossil evidence suggests that all major clades of the Burseraceae had evolved and were distributed across at least the Northern Hemisphere by the end of the Late Oligocene approximately 23 Ma. Well-preserved, three-dimensional fossil fruits and pyrenes attributable to the Burseraceae have been found in the London Clay of southern England, a deposit that spans the Early Eocene to Early Oligocene (56.5–29.3 Ma; Reid and Chandler 1933). *Bursericarpum aldwickense* Chandler is assignable to extant Protium Alliance on the basis of the number of pyrenes per fruit (Chandler 1961; Harley and Daly 1995), and dates to the Early Eocene, 56.5–50 Ma. *Protocommiphora europea* Reid & Chandler (Reid and Chander 1933) has a bifacial pyrene and is assignable to *Commiphora* or *Bursera* subg. *Elaphrium. Protocommiphora* fossils span the Early Eocene (56.5–50 Ma) and Late Eocene (38.6–35.4 Ma) of the London Clay. Fossil endocarps from the Czech Republic attributed to *Canarium* date to the Late Oligocene, 29.3–23.3 Ma (Gregor and Goth 1979). These fossil endocarps contain three fused locules, and have only one fertile locule per fruit.

Besides fossil casts, several impression fossils of putative Burseraceae are available, most of which do not expand the temporal or spatial range set by the fossil fruits and pyrenes, except for fruit and leaf impressions of *Bursera* reported from the Eocene Green River Flora of Colorado and Utah, and leaf impressions of *Canarium* that have been reported from the Early Oligocene of Japan (*C. ezoanum* Tanai; Tanai 1970), but the latter cannot be assigned with confidence to any particular lineage within the Burseraceae.

Phylogeny (by A. Weeks and D.C. Daly). Our knowledge about the evolutionary history of

Burseraceae lineages remains far from complete despite the insights gained from recently published phylogenies; the following reflects current consensus regarding major lineages.

Molecular phylogenetic studies have confirmed the monophyly of many taxonomic groups originally defined by morphological characters, but have also uncovered relationships not previously hypothesized. As an example of the latter, Thulin et al. (2008) determined that the sole Malagasy *Boswellia* species not only constituted a new genus, *Ambilobea madagascariensis*, but also that it is sister to the Canarium Alliance rather than nested with Boswellia Alliance s.s. members (*Garuga* + African *Boswellia*).

The monophyly of the Burseraceae has been repeatedly confirmed (e.g., Weeks et al. 2005; Thulin et al. 2008), as has the position of *Beiselia* as basal and sister to the rest of the family, beginning with Clarkson et al. (2002). The Protium Alliance (not including *Trattinnickia*) has comprised the best-supported clade, but the most recent molecular phylogenies of the group show *Tetragastris* and *Crepidospermum* as nested within *Protium* (Fine et al. 2005).

The extreme polyphyly of Bursereae subtribe Boswelliinae sensu Daly (Harley and Daly 1995) demonstrates our lack of knowledge about morphological synapomorphies for some lineages within the family. Current consensus recognizes tribe Beiselieae and a Bursera Alliance clade composed of *Aucoumea* + Bursereae subtribe Burserinae (*Bursera* and *Commiphora*) sister to a clade containing tribe Canarieae sister to *Boswellia* and *Garuga*, with *Triomma* in the Canarieae or sister to the Protium Alliance (Weeks et al. 2005; Thulin et al. 2008). The morphological basis for some re-alignments is not clear, and in fact examining morphological character evolution in a phylogenetic context will be crucial for re-circumscribing taxonomic groups within the family.

Like *Protium*, the majority of the other widespread, species-rich genera are para- or polyphyletic as currently circumscribed. *Canarium* sect. *Canariellum* lacks stipules and had been recognized by Engler as a distinct genus; the two species sampled in a molecular phylogenetic study appear to be more closely related to other Canarieae members *Trattinnickia* and *Santiria* (Weeks 2009). Several studies have suggested that *Bursera* subg. *Elaphrium* is sister to *Commiphora* + *B.* subg. *Bursera* (Weeks 2003; Weeks and Simpson 2007; Thulin et al. 2008), although subsequent investigations support the monophyly of *Bursera* (S. Magallón, pers. comm. 2009).

Several problematic lineages, including the genera *Rosselia*, *Scutinanthe*, and *Pseudodacryodes*, are not yet represented in molecular phylogenetic studies because the plants are so difficult to access. Indeed, the number of important genera and the range of phylogenetic data yet to be sampled, combined with the surprising outcomes from work to date, suggest the study of Burseraceae phylogeny will remain a dynamic field for the foreseeable future.

ECONOMIC BOTANY. The economic botany and ethnobotany of the Burseraceae are remarkable for the long history, great diversity, and convergence of uses of this family throughout the tropics. Few products of the Burseraceae are important in international commerce today, but locally the family is almost always of high cultural and often economic importance where it occurs. The family is best known for frankincense (*Boswellia* spp.), myrrh and Balm of Gilead (both *Commiphora* spp.), and other species of Biblical fame (e.g., Zohary 1982), but the family's importance in ancient trade routes far predates Christian times. Around 2,800 B.C., King Isesi of Egypt sent expeditions to Eritrea for incense; the Egyptian queen Hatshepsut introduced *Boswellia* and *Commiphora* plants from the same region in the 15th Century B.C. Later, Alexander the Great sent Anaxicrates to Yemen and Oman for the same purpose in the 4th Century B.C. During the 1st Century A.D., the same period that Pliny and Dioscorides wrote about these ancient commodities, it is estimated that Rome imported 2,500–3,000 tons of frankincense and 450–600 tons of myrrh (Groom 1981).

Use of Burseraceae resins in the New World resins predates European contact. For example, the burseraceous copals of the Aztecs were of tremendous ritual significance (Stacey et al. 2006), and use of Burseraceae resin to caulk canoes was observed in the Brazilian Amazon in the 17th Century (Cristóvão de Lisboa 1968).

Surprisingly, the most ancient known uses of Burseraceae appear to be from Melanesia, where cultivated or managed *Canarium* species are ubiquitous around coastal villages and are used for many purposes including their edible seeds. Widespread archaeological evidence from up to 14,000 years B.P. suggests that *Canarium* may be the oldest domesticated species in Melanesia (McClatchey et al. 2006).

The ancient human history of Burseraceae led to much confusion about the botanical sources (e.g., Lindley 1853; Marchand 1867–1868), but these are mostly resolved (e.g., Tucker 1986; Thulin and Warfa 1987; Thulin and Claeson 1991). The domestication and introduction of Burseraceae has taken place several times, including large plantations of the Mexican *Bursera delpechiana* in India before synthetic linalool was developed.

There has been remarkable convergence of uses for Burseraceae throughout the tropics. They are used for caulking and making canoes, illumination, glazes or varnishes, firewood, repellant or insecticide, treating inflammation and rheumatism, treating pulmonary disorders and skin ailments, living fences (*Bursera* and *Commiphora*), edible seeds and (parboiled) mesocarp (*Canarium* and *Dacryodes*) and important rituals in localities as far-flung as West Africa (e.g., Raponda-Walker and Sillans 1961; Arbonnier 2002), the Philippines (e.g., Quisumbing 1951; Coronel 1996), and Amazonia (e.g., Grimes et al. 1994; Machado et al. 2003; pers. obs.).

Equally impressive is the number of uses for a single species. The ethnobotany of southern Oman alone records 43 distinct uses for *Boswellia sacra* and similar use diversity for several other Burseraceae (Miller and Morris 1988). Additional uses for Burseraceae include arrow poisons (Thulin 1993), fine carvings (Peters et al. 2003), and treatment for venereal diseases (Arbonnier 2002).

In stark contrast to ancient times, the Burseraceae are of little current importance in international trade, an exception being okoumé (*Aucoumea klaineana*), the most important timber species of the Congo basin (e.g., Ruiz Pérez et al. 2005). Still, some species are of widespread use, such as *Dacryodes edulis*, the most widely consumed fruit in Central Africa (Baxter 2006), and many local economies throughout the tropics depend heavily on Burseraceae species, such as a lowland Quichua community in western Amazonia where one-half the market value of their non-timber forest products is accounted for by four *Protium* species whose resin is sold primarily for pottery glazing and incense, while a total of ten Burseraceae species are used for medicine, food, crafts, timber, and rituals (Grimes et al. 1994). The cultural value of Burseraceae in tropical communities cannot be overemphasized.

Key to the Genera

1. Branches and trunk armed with laterally compressed, cuspidate protuberances formed by persistent swollen petiole bases; secondary leaflet venation craspedodromous, fruit a pseudocapsule with 10(12) narrow valves dehiscing to release as many pyrenes separated by a columella, the pyrenes compressed and distally winged, the wings parallel to radii of the fruit axis. Mexico **1. *Beiselia***
- Trunk unarmed except sometimes with spiny short shoots (some Bursera Alliance taxa); secondary leaflet venation diverse, only rarely craspedodromous; fruit drupaceous or, if a pseudocapsule, releasing 3–5 pyrenes compressed perpendicular to radii of the fruit axis 2
2. Stipels often present at base of lateral petiolules; flowers functionally bisexual; a hypanthium present, the disk adnate to the receptacle, with 10 glandular lobes between the filament bases; fruit indehiscent; pyrenes 1–5, distinct. Asia to W Pacific region **6. *Garuga***
- Stipels absent; flowers unisexual or, if bisexual and a hypanthium present (*Boswellia*), the disk intrastaminal; fruit dehiscent or, if indehiscent, fruit containing a single 2–3-locular pyrene 3
3. Inflorescence bracts foliose, broadly ovate and cordate, persistent, much larger than the flowers, with festooned-brochidodromous secondary venation; sepals reflexed; petals erect and dorsally keeled; fruit pyriform and long-acuminate, indehiscent. Papua New Guinea **19. *Rosselia***
- Inflorescence bracts not much larger than flowers; sepals usually not reflexed, petals not dorsally keeled; fruit dehiscent if pyriform 4
4. Pulvinulus present at least on terminal petiolule; plants evergreen 5
- Pulvinuli lacking altogether; plants evergreen or more often deciduous 14
5. Disk extrastaminal, of (4)5 unlobed or bilobed glands; fruit a dry or leathery pseudocapsule, essentially pyriform (ovate in profile) or turbinate, at least slightly winged or angled, dehiscing to release pyrenes compressed in plane perpendicular to radius of fruit and at least slightly winged 6
- Disk intrastaminal and annular or adnate to receptacle; fruit fleshy and drupaceous, or indehiscent if dry, neither turbinate nor pyriform, variously rounded or lobed but not angled or winged; pyrenes not markedly compressed or winged 7
6. Several long parallel intersecondary veins present on each leaflet; stamens 5, the filaments glabrous; ovary 3-locular, stigma lobes 3; fruit markedly

3-winged, broadly ovate in profile; pyrenes 3, seed toward base of pyrene. Asia **19. *Triomma***

– Short epimedial tertiary veins present on leaflets; stamens 10, the filaments densely pubescent; ovary 5-locular, stigma capitate and unlobed; fruit narrowly turbinate; pyrenes 5, seed toward distal end of pyrene. W Africa **9. *Aucoumea***

7. Perianth 4–5-parted (rarely a few 3-parted flowers in some *Protium*); fruit dehiscent by 1–5 valves, with 1–5 unilocular pyrenes separated by a columella; each pyrene enveloped in a spongy, white (rarely red), sweet pseudaril 8

– Perianth 3-parted (*Scutinanthe* with 5-parted perianth but hypanthium present and ovary trilocular); fruit indehiscent, with a single compound 2–3-locular pyrene, usually 1–2 locules abortive 9

8. Resin clear; pulvinulus present on distal end of terminal petiolule but not on laterals; petals connate at least ½ their length (some of these usually persisting in fruit); on stamens and sometimes staminodes the anthers continuous with the filaments (base entire, not sagittate); disk and pistillode fused to form a parenchymatous ovariodisk; cotyledons plano-convex and straight. Neotropical **3. *Tetragastris***

– Resin clear or milky; pulvinulus present on distal end of terminal and lateral petiolules (if absent from laterals, then resin milky and/or the leaflets serrulate at least at apex); petals distinct or irregularly fused less than ½ their length; anthers usually sagittate; disk and pistillode usually distinct; cotyledons variously configured. Pantropical **2. *Protium***

9. Perianth 5-parted (calyx persisting on fruit), hypanthium present; fruit maturing yellow; pericarp fleshy and sweet; cotyledons entire and contortuplicate. Paleotropical **15. *Scutinanthe***

– Perianth 3-parted, hypanthium absent (rare in *Canarium*); mature fruit variously colored but not yellow; pericarp consisting of a thin exocarp and a (sometimes thin) oil-rich, resinous mesocarp; cotyledons variously configured 10

10. Petals partially connate, aestivation induplicate-valvate; pyrene bony and tuberculate, 2–3-locular and with 2–3 equal lobes, 2–3-horned at apex. Neotropical **14. *Trattinnickia***

– Petals distinct; fruit globose, (ob)ovoid, ellipsoid, or oblong; pyrene bony to papery, smooth or very rarely tuberculate, not lobed (some *Canarium* trigonous) nor horned; tropics 11

11. Leaves with a pair of linear to foliose stipules on petiole at or near base (rarely absent); leaflet secondary venation brochidodromous or eucamptodromous; mesocarp of fruit often thick; pyrene bony, abortive locule(s) not visible as an articulated plate on the fertile locule; fruit radially (sub)symmetrical, an operculum developing at apex of each fertile locule and visible on pyrene surface. Paleotropical **10. *Canarium***

– Leaves exstipulate; leaflet secondary venation brochidodromous; mesocarp usually thin; pyrene cartilaginous to papery, without axial intrusions; aborted locule(s) occurring as a 1- or 2-parted articulated plate or shield on pyrene; opercula absent 12

12. Leaflet tertiary and quaternary veins with marked admedial branching; fruit strongly oblique and style (or style scar) strongly excentric (usually close to 90° from vertical); pyrene thinly cartilaginous; cotyledons plano-convex and folded. Paleotropical **13. *Santiria***

– Leaflet tertiary and quaternary veins without marked admedial branching (except some *Dacryodes*); fruit symmetrical or only slightly oblique and style not strongly excentric; pyrene thinly cartilaginous or papery, when plano-convex, cotyledons straight 13

13. Domatia sometimes present in abaxial axils of leaflet secondary veins; intersecondary veins absent (long perpendicular epimedial tertiaries present instead); fruit maturing red, 3-locular, only slightly wrinkled when dry (mesocarp very thin); pyrene papery; cotyledons plano-convex and straight. Malesia to W Polynesia **12. *Haplolobus***

– Domatia lacking; intersecondary veins usually present; fruit variously colored, 2–3-locular, markedly wrinkled when dry; pyrene cartilaginous; cotyledons palmatifid and folded or contortuplicate. Pantropical **11. *Dacryodes***

14. Spacing between leaflet pairs gradually decreasing toward leaf base and leaflet size progressively smaller; calyx divided to base; fruit indehiscent, compressed oblong-ellipsoid, a longitudinal depression on both sides of the fruit marking the 2 locules. C Africa **17. *Pseudodacryodes***

– Leaflets without such a reduction series toward leaf base; calyx usually not divided completely to base; fruit dehiscent, (ob)ovoid to turbinate or pyriform, without longitudinal depressions 15

15. Perianth (4–)5-parted, petals usually clawed at base; fruit a turbinate to pyriform, dry or leathery pseudocapsule; pyrenes dorsiventrally compressed in a plane perpendicular to radius of fruit 16

– Perianth 3–5-parted, petals not clawed at base; fruit a usually ovoid dehiscent drupe (nuculanum); pyrenes not dorsiventrally compressed 17

16. Petiole with a distal pair of wings or dilations; flowers unisexual, perianth aestivation valvate; ovary 3-locular; pyrenes without wings, adnate to interior of fruit valve at dehiscence (this comprising the unit of dispersal). Madagascar **16. *Ambilobea***

– Petiole without distal wings or dilations; flowers bisexual, perianth aestivation imbricate; ovary (2)3 (4–6)(8)-locular; pyrenes often at first fringed by a fragile temporary wing, distinct from fruit valve but tenuously attached to the placenta near the apex. Paleotropical **5. *Boswellia***

17. Most surfaces with snail-shaped glands; flowers always unisexual, often isostemonous; perianth 5-parted; ovary 5-locular and stigmas (or stigma

lobes) 5; pyrenes 1–5, covered with a white, spongy, sweet pseudaril; pyrenes after dehiscence tenuously suspended from fruit apex by a V-shaped structure; cotyledons plano-convex and uncinately curved to form a J- or U-shaped structure. Neotropical **4. *Crepidospermum***

- Snail-shaped glands absent; flowers sometimes perfect, very rarely isostemonous; perianth 3–5-parted; ovary 2–3-locular and stigma lobes 2–3; pyrenes 1(2), rarely 3, basifixed, pseudaril fleshy and bright red, yellow, or orange; cotyledons palmatifid and folded or contortuplicate. Neo- or Paleotropical 18

18. Branches often spinose due to modified short shoot apices; calyx cupular to bell-shaped, lobes closed in bud; antepetalous stamens usually much shorter than antesepalous ones; pseudaril rarely covering pyrene entirely, its arms or lobes (when present) occurring on sutures as well as on 1 or both faces. Paleotropics, some Neotropical **8. *Commiphora***

- Branches only rarely spinose; calyx saucer-shaped to shallowly cupular, the lobes open in bud; stamens in two (sub)equal series; arillate structures, when not covering pyrene entirely, with arms on sutures but never on faces. North and tropical America **7. *Bursera***

Alliances and Genera of Burseraceae

I. Beiselia Alliance

Beiselieae Thulin, Beier & Razafim., Nord. J. Bot. 26: 226–227 (2008).

Ovules 2, superposed. Gynoecium and fruit 10 (–12)-locular; each locule of the pseudocapsule releasing one winged, radially flattened pyrene.

1. *Beiselia* Forman Fig. 16

Beiselia Forman, Kew Bull. 42: 262 (1987).

Dioecious trees or shrubs; trunk armed with cuspidate protuberances formed by persistent petiole bases. Leaves imparipinnate, deciduous, exstipulate, leaflet secondary venation craspedodromous, opposed tertiaries freely branching and meeting only at higher ranks; lateral leaflets sessile, coarsely dentate to slightly serrate. Inflorescences axillary, pseudoracemose. Flowers pedicellate; perianth 5(6)-parted; calyx apert in bud, divided almost to base; corolla valvate; petals distinct; stamens in two series; filaments distinct and inserted at base of intrastaminal annular disk; in male flowers disk and pistillode fused to form an ovariodisk; in female flowers staminodes significantly reduced; gynoecium 10 (–12)-locular and slightly 10(–12)-lobed; stigma subsessile, discoid; ovules 2, superposed. Fruit a pseudocapsule, ovoid-oblong, septicidal, each of 10(12) narrow valves dehiscing acropetally to release a pyrene separated by the columella, the pyrenes flattened parallel to radii of the fruit axis and distally winged. Cotyledons flat, entire, accumbent, radicle apical and small. Germination epigeal, cotyledons opposite, entire, first eophylls simple. $n = 13$.

A single sp., *B. mexicana* Forman, endemic to dry forests in Michoacán, Mexico.

II. Protium Alliance

Protieae Marchand, Adansonia 8: 17–71 (1867–1868).

Leaves exstipulate, pulvinulus present at least at apex of the terminal petiolule (except *Crepidospermum*). Flowers (3)4–5-merous; corolla aestivation induplicate-valvate; disk intrastaminal; staminate flowers with pistillode or ovariodisk. Fruit a compound dehiscent drupe; pericarp fleshy to coriaceous or rarely sublignified, septicidally and acropetally dehiscent by 1–5 valves (= number of pyrenes developed); locules separated by a columella; pyrenes 1–5, each enveloped in a pseudaril and tenuously suspended from the fruit apex by an inverted V-shaped structure; $n = 11$.

2. *Protium* Burm. f. Fig. 15A–C

Protium Burm. f., Fl. Indica: 88 (1768); Swart, J.J., A monograph of the genus *Protium* and some allied genera (Burseraceae). Gouda: Drukkerij Koch en Knuttel (1942).
Icica Aubl. (1775).
Marignia Commers. ex Kunth (1824).
Icicopsis Engl. (1874).

Small to large trees, rarely shrubs, often buttressed. Pulvinulus always present at apex of terminal petiolule and usually at both ends of laterals; rachis dilated in *P. beandou*; leaflets entire or rarely (sub)serrate. Flowers pedicellate or sometimes (sub)sessile, 4–5-parted (some male flowers on each inflorescence 3-parted in sect. *Sarcoprotium*), obdiplostemonous (antesepalous series only in *P. divaricatum* subsp. *divaricatum*); sepals partly fused; petals distinct, rarely fused irregularly near base; pistillode usually present in male flowers, usually containing locules and reduced ovules, sometimes reduced to a parenchymatous cylinder or the disk and pistillode fused to form an ovariodisk; disk instrastaminal (when not fused to pistillode), sometimes pubescent; ovary 4–5-locular; style sometimes with 4–5 short lateral branches; stigmas sometimes spiculate. Fruit obliquely ovoid to subglobose or ellipsoid, with cartilaginous to sublignified pericarp, maturing red or less often green, orange, or brown, the valves white or pale red within; pyrenes cartilaginous to bony or rarely papery, enveloped in a spongy pseudaril. Testa often infolded with cotyledons; cotyledons lobed and contortuplicate or simply plano-convex or broadly 3-lobed plus laterally reverse 2-plicate and uncinately folded. Germination patterns and seedlings diverse. $n = 11$.

As presently circumscribed, ca. 180 spp. in S Mexico and the Neotropics, tropical China, Madagascar, Mauritius, India, the Malesian region.

3. *Tetragastris* Gaertn.

Tetragastris Gaertn., Fruct. Sem. Pl. 2: 130, t. 109, Fig. 5 (1802); Daly, Kew Bull. 45: 179–194 (1990), rev.
Hedwigia Sw. (1788).

Trees. Bark usually thick and fissured. Leaves with pulvinulus always present at apex of terminal petiolule but absent from lateral petiolules; leaflet margin entire, freely ending veinlets highly branched. Flowers 4–5-merous, diplostemonous, the two series of subequal or unequal length; calyx partially synsepalous but sinuses often split to unequal depth at anthesis; petals fused at least ½ their length; filaments in male flowers strap-shaped and continuous with the anthers; disk and pistillode replaced by a continuous (rarely articulated) ovariodisk; in female flowers the anthers sagittate, intrastaminal annular disk present. Fruit obliquely ovoid to subglobose, maturing red or less often pale green, the valves pale red within, pseudaril white, pyrenes cartilaginous. Testa papery; cotyledons plano-convex, straight or rarely uncinately folded. Germination epigeal and phanerocotylar; first eophylls opposite, simple.

Nine species in Hispaniola, Puerto Rico, and Central America south to Bolivia and Brazil.

4. *Crepidospermum* Hook.f.

Crepidospermum Hook.f. in Benth. & Hook.f., Gen. Pl. 1: 325 (1862); Daly, Kew Bull. 57: 471–477 (2002).
Hemicrepidospermum Swart (1942).

Trees or shrubs, sometimes buttressed. Leaves usually clustered distally on branchlets; petiolules lacking pulvinuli; leaflet margin serr(ul)ate, rarely (*C. cuneifolium*) the teeth glandular; freely ending veinlets highly branched. Inflorescences sometimes pseudospicate. Flowers 5-merous, obdiplostemonous or haplostemonous; petals distinct; disk usually glabrous; pistillode in male flowers either containing locules and reduced ovules or sometimes reduced to a parenchymatous cylinder; pistil 5-locular; style unbranched, stigmas 5. Fruit obliquely ovoid, maturing orange to (yellowish) red, valves white or pale red within, pericarp thinly cartilaginous, pseudaril white, pyrenes cartilaginous. Testa papery, cotyledons plano-convex and uncinately folded to form a 'J'- or 'U'-shaped structure. Germination phanerocotylar, first eophylls opposite, simple, margin crenate.

Seven species, moist forests of northern South America.

III. Boswellia Alliance

Pulvinuli lacking on petiolules; leaflet margin often crenate or serrate. Flowers bisexual, 5-merous, obdiplostemonous; petals distinct; filaments often pubescent; pollen grains large (>55 μm), endoapertures with distinct, meridionally extended, tapering costae; style usually longer than ovary. Cotyledons palmatifid.

5. *Boswellia* Roxb. ex Colebr.

Boswellia Roxb. ex Colebr., Asiatic Res. 9: 379 (1807); Thulin et al., Kew Bull. 42: 487–500 (1987).

Shrubs or trees, sometimes rooted on stone cliffs via swollen 'holdfasts' at base of trunk; outer bark variously colored, usually papery and peeling; resin watery to milky, sometimes aromatic (e.g., frankincense spp.). Leaves deciduous, usually clustered at ends of thick short shoots, imparipinnate or rarely tri- or unifoliolate or heterotrifoliolate; petiole and rachis sometimes winged; caducous stipules sometimes present; leaflets usually subsessile, and entire, serrate, or less often crenate or dentate; freely ending veinlets 1–2-dichotomously branched. Inflorescences produced with or before leaves, axillary or (pseudo)terminal, often pseudoracemose or pseudospicate. Perianth 5-parted, imbricate; calyx shallowly cupular; petals usually clawed at base; stamens in two series, subequal or heteromorphic, inserted on outside of disk or below its base; disk intrastaminal, annular, glabrous, rarely adnate to receptacle to form a short hypanthium; gynoecium (2)3(4–6)(8)-locular, sessile; stigma usually 3-lobed. Fruit a turbinate to pyriform pseudocapsule, usually trigonous, septicidal, valves dehiscing acropetally to release pyrenes separated by a persistent columella with as many short lateral septal wings as locules; pyrenes tenuously attached to the placenta near apex, essentially obovoid but variously compressed dorsally (perpendicular to radii of the fruit axis), often slightly horned on extremes and/or sides, often at first fringed by a fragile temporary wing. Cotyledons folded and palmately 3- or 5-lobed or palmatifid, or 3-lobed with outer lobes bifid. $n = 22$.

About 20 species in dry areas from Ivory Coast to Arabia and India and S to NE Tanzania.

6. *Garuga* Roxb. Fig. 15D, E

Garuga Roxb., Pl. Corom. 3: 5, t. 208 (1811); Leenhouts, Fl. Males. I, 5: 215–218 (1956).

Shrubs or trees, sometimes buttressed, bark smooth to reticulate-fissured or shed in large irregular plates; resin cream-colored or pinkish; vascular bundles in pith of branchlets lacking. Leaves usually deciduous, clustered at branchlet ends, sometimes with caducous stipules, imparipinnate, toward leaf base leaflets progressively smaller; stipels often present, inserted at base of lateral petiolules; leaflet margin serrate or irregularly crenate-serrate, rarely subentire. Inflorescences axillary, usually anticipating leaf flush. Flowers 5-merous, perigynous, calyx and receptacle forming a campanulate to urceolate hypanthium; calyx valvate; petals distinct, induplicate-valvate; stamens in two subequal or unequal series; filaments inserted on margin of receptacle; anther thecae divergent at base; disk of 10 glandular lobes adnate to receptacle between filament bases; gynoecium (4)5-locular, substipitate; stigma capitate, (4)5-lobed. Fruit a fleshy, compound drupe, globose to obliquely pyriform but laterally compressed and slightly 2–5-lobed, the pericarp fleshy and sweet, enveloping 1–5 distinct, gibbous, bony, tuberculate, 1-seeded pyrenes. Cotyledons palmatifid, contortuplicate. Germination phanerocotylar, first eophylls (sub)opposite, trifoliolate, margin at least sometimes serrate. $n = 13$.

About four spp. distributed in continental SE and S Asia, N and E Malaysia, Java, Borneo, N and W Australia, Philippines, New Guinea, W Pacific region.

IV. Bursera Alliance

Leaves exstipulate. Petals distinct. Pollen with very short colpi. Fruits dehiscent, the valves dehiscing acropetally.

7. *Bursera* Jacq.

Bursera Jacq. ex L., Sp. Pl. ed. 2, 1: 471 (1762), nom. cons.; Rzedowski et al., Acta Bot. Mex. 70: 85–111 (2005).
Terebinthus P. Browne (1756).
Elaphrium Jacq. (1760).

(Polygamo-)dioecious shrubs to large trees, sometimes caudiciform, rarely lithophytic or epiphytic; bark papery and peeling or close; resin clear or translucent; branchlets usually unarmed. Cataphylls sometimes present; leaves usually clustered near branchlet apices, usually deciduous, imparipinnate or sometimes unifoliolate, rarely bipinnate; rachis often winged; petiolules lacking pulvinuli; leaflets entire, crenate, or dentate, freely ending veinlets often with tracheoid idioblasts. Inflorescences axillary or subterminal, sometimes pseudospicate, pseudofasciculate, or pseudoracemose. Flowers unisexual or less often bisexual, in subg. *Bursera* male flowers (3–4)5-merous, female and bisexual flowers 3-merous; in subg. *Elaphrium* flowers 4-merous; receptacle flat to saucer-shaped, rarely cup-shaped, sometimes forming a hypanthium in male flowers; calyx usually divided almost to base, apert; corolla induplicate-valvate at apex but slightly imbricate along sides, usually with inflexed apiculum; stamens in two usually subequal series; pistillode greatly reduced, with rudimentary locules and ovules, or sometimes spindle-shaped and parenchymatous, occasionally obsolete; disk intrastaminal, usually annular, sometimes adnate to hypanthium (when latter present); gynoecium 3-carpellate (subg. *Bursera*) or 2-carpellate (subg. *Elaphrium*); style usually short, stigmas or stigma lobes 3–2. Fruit a septicidally dehiscent compound drupe with fleshy to coriaceous pericarp, ellipsoid to (ob)ovoid or subglobose, often slightly oblique; at dehiscence 2–3 valves (= number of locules) fall away, exposing the basifixed compound pyrene(s); pyrene cartilaginous to bony, either enveloped by a brightly colored, fleshy, arillate structure (subg. *Bursera*), or the arillate structure partially covering the pyrene from the base and extending arms or lobes up the sutures (most of subg. *Elaphrium*); 1(–3) locule (s) developing a seed, the other(s) abortive, these not separable from fertile locule but often distinguishable by a ridge or furrow. Testa membranaceous, cotyledons palmately 3-lobed or usually multilobed, folded. Germination epigeal, phanerocotylar, first eophylls alternate and simple. $n = 12$.

Approx. 100 species, SW U.S.A., Mexico (center of diversity), Central America, Caribbean, Colombia, Venezuela, Trinidad, Guyana, SW Ecuador, NW Peru, extreme N Brazil (Roraima).

8. *Commiphora* Jacq.

Commiphora Jacq., Pl. Hort. Schoenbr. 2: 66, t. 249 (1797); Gillett, J.B., Burseraceae, in: Flora of Tropical East Africa (1991), rev., and in Kew Bull. 34: 569–587 (1980), neotrop. spp.; Rzedowski & Palacios-Chávez, Taxon 34: 207–210 (1985), Mexican spp.
Balsamodendrum Kunth (1824).
Neomangenotia J. Leroy (1976).

(Polygamo-)dioecious shrubs or trees, sometimes caudiciform, sometimes rupicolous; bark close or more often papery and peeling; inner bark usually green; resin usually aromatic and drying gummy; branches often armed, the spines modified axillary shoots or shoot apices. Leaves usually clustered near branchlet apices, deciduous, exstipulate, often flushing during or after anthesis, imparipinnate or sometimes tri- or unifoliolate, rarely hetero-trifoliolate; rachis terete (rarely winged); petiolules lacking pulvinuli; leaflets entire, crenate, or toothed, sometimes lobed, intersecondary veins frequent and parallel to secondaries. Inflorescences axillary or subterminal, sometimes pseudospicate or pseudofasciculate or dichasial, or reduced to 1–3-flowered cymes, rarely with large foliose bracts. Flowers unisexual or rarely some flowers bisexual, 4(5)-parted; receptacle flat to funnel-shaped, sometimes with the calyx forming a hypanthium in male flowers; calyx cupular to tubular, usually divided less than halfway to base, imbricate; corolla (induplicate-) valvate at apex but often imbricate along sides, usually with inflexed apiculum; stamens in two strongly unequal series, rarely haplostemonous, inserted at base of or on top of intrastaminal disk, sometimes filaments adnate to a hypanthium; intrastaminal annular disk distinct, sometimes adnate to a hypanthium, rarely pubescent; pistillode with reduced locules and ovules, or rudimentary, or obsolete, rarely fused with disk to form an ovariodisk; gynoecium 2(3)-carpellate, style short, stigma capitate. Fruit a compound drupe with fleshy to coriaceous pericarp, septicidally dehiscent by 2 (4, very rarely 3) valves, exposing a basifixed compound pyrene, this either enveloped by a brightly colored, fleshy, arillate structure, or the arillate structure partially covers the pyrene from the base and extends lobes or linear arms up the faces and/or sutures (usually both); 1(2) locule(s) developing a seed and 1(2) usually abortive; apex of pyrene sometimes with two apical pits between the locules (these probably assisting the absorption of water before germination). Testa membranaceous; cotyledons palmately 3-lobed or multilobed and flat or folded. Germination phanerocotylar, the first eophylls opposite and simple. $n = 13$.

Approx. 185 spp., Africa, Madagascar, Arabia, peninsular India, Sri Lanka, Vietnam, Pakistan, Iran, Venezuela, C and NE Brazil, E Bolivia.

9. *Aucoumea* Pierre

Aucoumea Pierre, Bull. Soc. Linn. Paris 157: 1241–1242 (1896).

Large dioecious trees, often buttressed, bark scaly, shed in large, thick, elongate plates; resin sticky, drying grayish-white. Leaves evergreen, exstipulate, imparipinnate; petiolules pulvinate at both ends; leaflets entire; freely ending veinlets highly branched. Inflorescences axillary or terminal. Flowers 5-parted; calyx valvate, lobes divided nearly to base; petals distinct, brown-tomentose, subterete at base; disk extrastaminal, a circle of 5 discrete bilobed glands alternating with petals; stamens in two unequal series; filaments distinct and densely pubescent; pistillode reduced but containing 5 locules each with 2 reduced ovules; staminodes shorter than pistil; pistil 5-locular and 5-lobed, the stigma capitate, unlobed. Fruit a turbinate pseudocapsule, septicidal, the 5 coriaceous valves dehiscing to release 5 crustaceous pyrenes separated by a persistent columella; pyrenes ovoid but compressed perpendicular to radii of the fruit axis, marginal wing with a long basal extension, oblanceolate in dorsal view. Cotyledons foliose, orbicular, cordate, entire but folded in seed, with brochidodromous venation. Germination epigeal, phanerocotylar, first eophylls alternate, simple. $n = 13$.

One sp., *A. klaineana* Pierre, in W Africa (Gabon, People's Republic of Congo, Cameroon, Equatorial Guinea).

V. Canarium Alliance

Canarieae Engl., Bot. Jahrb. Syst. 48: 444 (1912).

Trees or shrubs, dioecious (some *Dacryodes* hermaphrodite), leaves imparipinnate (rarely unifoliolate in *Santiria*). Fruit an indehiscent drupe or less often a pseudocapsule.

10. *Canarium* Stickm. Figs. 15R–X, 18

Canarium Stickm., Herb. Amb.: 10 (1754) (quoad *Cenarium*); Leenhouts, Blumea 9: 275–475 (1959), rev.
Pimela Lour. (1790).
Canariellum Engl. (1896).

Small to large, usually dioecious trees, rarely shrubby or scandent, buttresses often present; bark smooth, rough, scaly, or scalloped, often fissured, sometimes densely lenticellate; resin translucent, watery or sticky, drying white or dark brown or blackish. Leaves imparipinnate, evergreen, usually with linear to foliose stipules on petiole toward or at base or less often on branch near petiole base; pseudostipules sometimes present; pulvinuli usually present on terminal and lateral petiolules; leaflets entire to serrate or dentate; tertiary veins usually mixed opposite-alternate percurrent. Inflorescence axillary or (pseudo)terminal, sometimes pseudospicate, pseudoracemose, or fasciculate. Flowers 3-parted, unisexual, rarely the plants monoecious; calyx and receptacle cupular to funnel-shaped; calyx synsepalous but variously divided, valvate, in

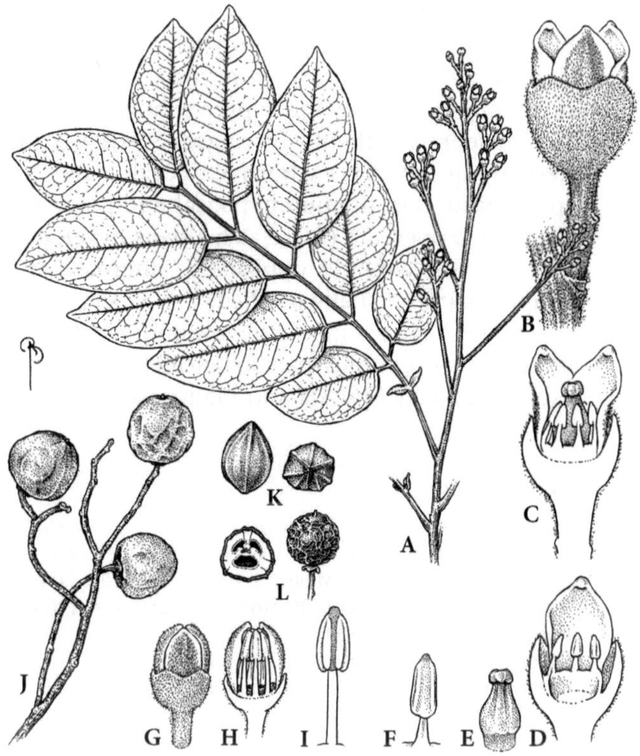

Fig. 18. Burseraceae. *Canarium globosum* Daly (ined.). **A** Flowering branchlet, the leaf with pseudostipules. **B** Female flower. **C** Ditto, with calyx lobe and petal removed, showing staminodes inserted on rim of disk. **D** Ditto, with pistil removed. **E** Pistil. **F** Staminode, adaxial view. **G** Male flower. **H** Ditto, with calyx lobe and petal removed and showing stamens inserted at base of disk. **I** Stamen. **J** Part of infructescence. **K** Pyrene in lateral and apical views. **L** Fruit (dry), at left in cross-section. (Reproduced with permission of the artist Bobbi Angell)

fruit usually persistent and enlarged; petals distinct, valvate at apex but imbricate in middle, the apex inflexed-apiculate; stamens in two subequal or less often unequal series, rarely the antepetalous series obsolete; filaments sometimes partly or mostly connate to form a staminal tube; stamens inserted at base of disk, sometimes on the disk near the base, rarely on the rim; disk usually annular but variously shaped, sometimes pubescent, rarely adnate to a shallow hypanthium in male flowers; pistillode usually ± reduced, sometimes fused with disk to form an ovariodisk, rarely the disk, stamens and ovary on a short androgynophore; staminodes usually only slightly reduced; ovary 3-locular; style short; stigma subcapitate or 3-lobed, or 3 separate subsessile stigmas. Fruit an indehiscent compound drupe, (ob)ovoid to ellipsoid, oblong, (sub)globose, or fusiform, often slightly trigonous, rarely slightly oblique; exocarp thin; mesocarp fleshy, oily-resinous, usually thick; pyrene bony, smooth to rugose, rarely slightly 3-winged. Seeds 1–2 (3) developing, other(s) abortive, cotyledons 3+ -palmatifid. Germination hypogeal and phanerocotylar via opercula (these as many as developed locules); first eophylls opposite or alternate and simple, margin usually toothed.

About 120 spp. in mostly lowland, wet to less often semi-deciduous paleotropical tropical forests, principally Malesia and Madagascar.

11. *Dacryodes* Vahl

Dacryodes Vahl, Skr. Naturhist.-Selsk. 6: 115 (1810); Cuatrecasas, Tropical Woods 106: 46–65 (1957); Neotropical spp.; Leenhouts, Fl. Males. ser I, 5: 219–229 (1956); Daly & Martínez-Habibe, Brittonia 54: 266–274 (2003).
Pachylobus G. Don (1832).
Canarium sect. *Africana* Engl. in A. DC. & C. DC. (1883).
Canarium sect. II, ser. 4 *Tenuipyrena* Engl. in A. DC. & C. DC. (1883), pro parte.
Curtisina Ridl. (1920).
Hemisantiria H. J. Lam (1929).

Dioecious or hermaphrodite trees (one sp. scandent), sometimes with plank buttresses; bark smooth or rough due to scales or raised lenticels, relatively thin and shed in irregular plates; resin clear or somewhat milky. Stellate hairs sometimes present. Leaves evergreen, exstipulate; pulvinuli usually apparent on terminal and lateral petiolules; leaflet margin entire; intersecondary veins usually present,. Inflorescences axillary or sometimes (pseudo)terminal. Flowers unisexual or less often bisexual; perianth 3-parted, calyx (shallowly) cupular to urceolate, truncate or 3-lobed, sometimes divided nearly to base; petals distinct, valvate at apex, on sides valvate (American species) or slightly imbricate, with an inflexed apiculum; stamens in two subequal series, inserted at base of disk or on the side or rim; anthers continuous with the filaments or sagittate; pistillode either reduced but provided with locules and ovules, or a variously shaped parenchymatous structure, or fused with the disk to form an ovariodisk; otherwise disk annular; pistil 2–3-locular; style short and thick, stigma 2–3-lobed. Fruit an indehiscent, compound drupe, oblique-ovoid to ellipsoid, rarely globose, exocarp thin and drying wrinkled; mesocarp thin and oily-resinous; endocarp (thinly) cartilaginous, smooth, not lobed; pyrenes 1(2)-seeded, the 1–2 aborted locules strongly compressed, forming an articulated, separable plate on periphery of pyrene. Testa smooth, cotyledons palmate or up to 13-palmatifid; folded or contortuplicate. Germination usually epigeal and phanerocotylar, rarely hypogeal and cryptocotylar, the first eophylls opposite or alternate and simple or trifoliate.

About 70 species in the moist tropics, usually lowland but some montane.

12. *Haplolobus* H. J. Lam Fig. 15F, G

Haplolobus H. J. Lam, Ann. Jard. Bot. Btzg. 42: 25 (1931); Leenhouts, Blumea 20: 283–310 (1972).
Canarium L. (1759), pro parte.
Santiria sect. *Anisandra* Laut. (1920).

Dioecious trees; bark grayish, scaly. Leaves imparipinnate, evergreen, exstipulate; inverted vascular bundles present in pith of petioles; pulvinuli usually present at both ends of petiolules; leaflets entire, sometimes domatiiferous; epimedial tertiaries perpendicular to the midvein, long, and basiflexed, freely ending veinlets with tracheoid idioblasts. Inflorescences axillary or pseudoterminal, rarely pseudospicate, male inflorescences generally larger than female. Flowers 3-parted; calyx cupular but 3-lobed and divided up to half its length, imbricate; petals distinct, with inflexed apiculum, valvate at apex but imbricate at middle; stamens in 2(1) series, inserted below or on base of intrastaminal annular disk, rarely on the rim; disk often persistent in fruit; pistillode reduced and parenchymatous; pistil 3-locular; style short, stigma 3-lobed or subglobose. Fruit an indehiscent, compound drupe, ellipsoid to ovoid or subglobose; calyx persistent; exocarp dry and thin, drying rather smooth; mesocarp thin; pyrenes papery, 1(2)-seeded, the 1–2 aborted locules strongly compressed, forming an articulated, separable plate on periphery of pyrene. Seed testa thin; cotyledons plano-convex, straight. Germination hypogeal, cryptocotylar; first eophylls opposite, simple.

About 16 spp., in lowland to montane forests, from Borneo and Celebes through Moluccas and New Guinea to W Polynesia.

13. *Santiria* Blume

Santiria Blume, Mus. Bot. Ludg. Bat. 1: 209, f. 40 (1850); Leenhouts, Fl. Males. I, 5: 229–238 (1956).
Trigonochlamys Hook.f. (1860).
Icicaster Ridl. (1917).

Dioecious trees, sometimes with plank buttresses or stilt roots; bark usually densely lenticellate; resin white to translucent. Leaves evergreen, exstipulate, imparipinnate or rarely unifoliolate, pulvinulus evident on terminal petiolule and usually on laterals; leaflets entire; intercostal tertiary veins alternate-percurrent and admedially ramified. Inflorescences axillary or pseudoterminal, paniculate; male inflorescences generally longer than female. Receptacle shallowly concave; perianth 3-parted; calyx divided to varying depths; petals distinct, valvate at apex but imbricate along sides; stamens in 2 subequal or unequal series (sometimes the antepetalous series obsolete or vestigial), inserted on disk or at its base or rarely on the rim; anthers continuous with filaments or sagittate; pistillode reduced but containing reduced locules; disk intrastaminal, annular; gynoecium 2–3-locular, style short and thick, stigma (sub)sessile, 3-lobed or truncate. Fruit a compound, indehiscent drupe, small (less than 2.5 cm long), oblique and ovoid or ellipsoid, rarely globose or acuminate; stigma markedly excentric, offset often more than 90° from vertical; calyx usually persistent; exocarp thin and almost smooth when dry; mesocarp thin; endocarp thinly cartilaginous; fruit 1–3-seeded, aborted locules strongly compressed, forming small, articulated, detachable plates on periphery of pyrene between the fruit base and excentric apex. Cotyledons 5-lobed or -fid, folded or contortuplicate. Germination epigeal or hypogeal; first eophylls opposite and simple or pinnate.

About 22 spp., in W Africa, Malay Peninsula, Borneo, Sumatra, Philippines, Moluccas, Celebes, New Guinea.

14. *Trattinnickia* Willd.

Trattinnickia Willd., Sp. Pl. 4: 887 (1805); Daly, Kew Bull. 54: 129–137 (1999).

Dioecious trees, rarely shrubby; outer bark thin and smooth in some taxa, in most of sect. *Rhoifoliae* rough and shallowly to deeply fissured and shed in irregular plates; hoop marks frequently present; resin clear and watery or oily (rarely reported as white), slightly sticky, drying yellowish or white and hard, flammable in some species. Leaves evergreen, exstipulate, imparipinnate, rachis usually diamond-shaped in cross-section; pulvinuli usually apparent on lateral petiolules and almost always on the terminal ones, often galled; usually one or both leaflet surfaces asperous; leaflets entire; sometimes (sect. *Burserifoliae*) the vein areoles on the abaxial side defining papilla-filled laminar crypts. Inflorescences axillary or (pseudo)terminal. Perianth 3-parted; calyx cupular but 3-lobed, valvate, sometimes invested with retrorse hairs abaxially, not persisting in fruit; corolla of 3 partly to mostly connate petals, fleshy, induplicate-valvate, with an inflexed apiculum, both surfaces always with at least some retrorse hairs; stamens of two subequal series; filaments distinct, usually strap-shaped, anthers often continuous with filaments; stamens inserted at or near base of disk; staminodes reduced, inserted on disk rim; disk usually annular, thin in female flowers; pistillode reduced to a conical parenchymatous structure either surmounting the disk or fused with disk to form an ovariodisk; gynoecium 2–3-locular; style short, stigma 1, capitate, 2–3-lobed. Fruit a compound, indehiscent drupe, depressed-globose to depressed-ovoid; mesocarp oily and resinous; pyrene bony, 2–3-locular and 2–3-lobed, tuberculate, with sutures between the lobes but the carpels not separable, 2–3-horned at apex. Germination hypogeal, phanerocotylar; cotyledons palmatifid; first eophylls alternate, simple.

Approx. 14 spp., Costa Rica south to Brazil and Bolivia (excluding the Caribbean).

15. *Scutinanthe* Thwaites Fig. 15H–J

Scutinanthe Thwaites in Hook., Kew J. 8: 266, t. 8 (1856); Leenhouts, Fl. Males. I, 5: 246–249 (1956).
Santiriopsis Engl. (1890).

Dioecious trees; outer bark smooth and with hoop marks, sometimes scaly, resin translucent to cream-colored. Pith of branchlets and of petioles without vascular bundles. Leaves evergreen, exstipulate, imparipinnate; pulvinuli apparent on terminal and lateral petiolules; leaflets entire; composite admedial tertiary veins often present, freely ending veinlets 4+-branched. Inflorescences axillary. Perianth 5-parted, valvate; receptacle cupular, forming a hypanthium with the calyx; calyx lobed, with an inflexed apiculum; petals distinct, with inflexed apiculum; stamens in two subequal series; filaments flattened and deltate, apparently connate at base but inserted on rim of disk; pollen heteropolar, tricolporate, lobed-vestibulate; disk adnate to receptacle and not easily distinguished from it; pistillode subcylindrical, provided with reduced locules and ovules; stigma capitate or subpeltate; ovary ovoid, 3-locular; stigma subpeltate, sometimes slightly 3-lobed. Fruit a compound, indehiscent drupe, ellipsoid to ovoid and slightly oblique; 5-parted calyx persistent; pericarp fleshy, edible; pyrene bony, 1-seeded; the two aborted locules strongly compressed, forming an articulated, non-detachable plate on periphery of pyrene. Cotyledons entire, contortuplicate. Germination epigeal, phanerocotylar; first eophylls simple and alternate.

Two spp., in Sri Lanka, Malay Peninsula, Borneo, and SE Sumatra.

16. *Ambilobea* Thulin, Beier & Razafim.

Ambilobea Thulin, Beier & Razafim., Nord. J. Bot. 26: 223 (2008).
Boswellia madagascariensis Capuron, Adansonia II, 2: 268 (1962).

Dioecious trees or shrubs; bark gray with longitudinal fissures; resin fragrant. Leaves deciduous, clustered at ends of branches, exstipulate, imparipinnate; petiole with a distal pair of decurrent wings or dilations; leaflets shallowly serrulate. Inflorescences produced with the leaves, axillary, dimorphic, the staminate ones lax pedunculate dichasial cymes, the pistillate ones few-flowered pseudoracemes or panicles. Perianth (4)5-parted, valvate in bud; receptacle and calyx cup-shaped; petals with an acute inflexed apiculum; stamens in two subequal series; pollen exine striate-reticulate with long, frequently hooked, striate spines; disk intrastaminal, annular; pistillode reduced, depressed-globose, with style but stigmas undeveloped; ovary 3-locular, style thick and short, stigma capitate, slightly 3-lobed. Fruit a 3-locular pseudocapsule, narrowly pyriform; central columella 3-winged; pyrenes usually 3, bony, triangular and compressed perpendicular to radii of the fruit axis, adnate to interior of fruit valve at dehiscence. Testa thin, cotyledons divided and contortuplicate.

One sp., *A. madagascariensis* (Capuron) Thulin, Beier & Razafim., endemic to northernmost Madagascar, in dry deciduous forests, on limestone cliffs and rocks, or on sand.

17. *Pseudodacryodes* Pierlot

Pseudodacryodes Pierlot, Bull. Jard. Bot. Natl. Belg. 66: 175 (1997).

Dioecious trees. Bark gray, finely reticulate-fissured. Leaves exstipulate; toward leaf base the interjuga gradually shortening and leaflet size progressively smaller, basal pair (pseudostipules) close to leaf insertion point; pulvinuli absent; margin entire; intersecondary veins perpendicular to midvein present. Inflorescences terminal and pseudoracemose. Perianth 3-parted, aestivation valvate; calyx divided to base; petals with an inflexed apiculum; male flowers unknown; staminodes in 2 subequal series, inserted at base of intrastaminal annular disk; ovary 2-locular, ovoid; style short, stigma capitate and 2-lobed. Fruit a compound, indehiscent drupe, slightly compressed oblong-ellipsoid, a longitudinal depression on both sides of the fruit marking the 2 locules; mesocarp thin; endocarp thinly cartilaginous; one seed developing in each of the (sub) equal locules. Cotyledons entire, folded.

One sp., *P. leonardiana* Pierlot, in the People's Republic of Congo.

18. *Triomma* Hook.f. Fig. 15K–Q

Triomma Hook.f., Trans. Linn. Soc. 23: 171 (1860); Leenhouts, Fl. Males. I, 5: 218–219 (1956).

Dioecious large trees, tall plank buttresses present; bark scaly or smooth. Pith of branchlets without vascular bundles; resin translucent, aromatic. Leaves evergreen, imparipinnate, exstipulate; pulvinulus present at apex of terminal petiolule; leaflets entire; several long parallel inter-secondary veins present on each leaflet, freely ending veinlets highly branched. Inflorescences (pseudo)terminal. Perianth (4)5-parted, densely brown-pubescent; calyx divided to near base; petals distinct, corolla induplicate-valvate, with inflexed apiculum; disk extrastaminal, of 5 discrete lobes; stamens 5, antesepalous, inserted inside base of disk lobes, filaments sometimes twisted so anther dehiscence appearing extrorse; pollen endoaperture circular, exine spinulose, tectate, apocolpial endexine present; pistillode greatly reduced and parenchymatous; ovary 3-locular, triangular; style short and thick, stigma capitate and 3-lobed. Fruit a pseudocapsule, obturbinate to pyriform in profile, 3-winged, the 3 woody valves dehiscing to release 3 pyrenes separated by a persistent columella, the pyrenes strongly compressed and broadly winged in plane perpendicular to radii of fruit axis, the wings rounded at base; seed located toward distal end of pyrene. Cotyledons shallowly 5-lobed and folded or contortuplicate. Germination epigeal, phanerocotylar, first eophylls alternate, simple, margin serrate.

One sp., *T. malaccensis* Hook.f., on the Malay Peninsula, Borneo, and Sumatra, and Bihar in India.

UNALIGNED:

19. *Rosselia* Forman

Rosselia Forman, Kew Bull. 49: 601–621 (1994).

Dioecious trees; bark dark brown. Leaves congested near branch apices, imparipinnate, 1–3-jugate; terminal leaflet cordiform and resembling the inflorescence bracts; pulvinuli lacking; leaflets entire; tertiary veins alternate-percurrent but also forming composite admedial tertiaries parallel to the secondaries. Inflorescences axillary and terminal, elongate and narrow, the primary axis bearing spaced peduncled cymes, each arising from the axil of a foliaceous, cordate bract. Flowers 3-parted; calyx lobes valvate, divided nearly to base and reflexed; corolla of 3 distinct, navicular, dorsally keeled petals, valvate at apex but imbricate in middle, apex usually with inflexed apiculum; stamens equal and apparently in one series; disk and pistillode in male flowers fused to form a depressed-globose ovariodisk; in female flowers staminodes reduced and disk intrastaminal, annular; ovary 3-locular, subglobose, style long, stigma with 3 subglobose lobes. Fruit a compound indehiscent drupe, pyriform and acuminate, slightly trigonous, drying slightly wrinkled; endocarp bony, one locule developing, the 2 aborted locules compressed, in cross-section visible on periphery of pyrene and forming two articulated but nonseparable plates with short, pointed, distinct apical extensions (horns). Cotyledons apparently lobed or palmatifid and laterally reverse 2-plicate.

One sp., *R. bracteata* Forman, apparently endemic to Rossel Island in the Louisiade Archipelago, Papua New Guinea.

Selected Bibliography

Aragão, G.F., Cunha, M.C., Nogueira, P., Gomes, T.L., de Barros, V. 2008. Analgesic and anti-inflammatory activities of the isomeric mixture of alpha- and beta-amyrin from *Protium heptaphyllum* Aubl. March. [sic] J. Herbal Pharmacother. 7(2): 31–47.

Arbonnier, M. 2002. Arbres, arbustes et lianes des zones sèches d'Afrique de l'Ouest. Montpelier: CIRAD/MNHN/UICN.

Arctander, S. 1960. Perfume and flavor materials of natural origin. Arctander, Elizabeth, NJ.

Aregullin, M., Gompperb, M.E., Rodrigueza, E. 2002. Sesquiterpene lactone in *Trattinnickia aspera*. Biochem. Syst. Ecol. 30: 187–188.

Arora, R.B., Taneja, V., Sharma, R.C., Gupta, S.K. 1972. Anti-inflammatory studies on a crystalline steroid isolated from *Commiphora mukul*. Indian J. Med. Res. 60 6: 929–931.

Atal, C.K., Gupta, O.P., Afaq, S.H. 1975. *Commiphora mukul*: source of guggal in Indian systems of medicine. Econ. Bot. 29: 209–218.

Aubréville, A. 1962. Burséracées. In: Flore du Gabon. Vol. 3, pp. 65–89. Paris: Muséum Nationale d'Histoire Naturelle.

Bachelier, J.B., Endress, P.K. 2009. Comparative floral morphology and anatomy of Anacardiaceae and Burseraceae (Sapindales), with a special focus on gynoecium structure and evolution. Bot. J. Linn. Soc. 159: 499–571.

Baksi, S.K. 1976. Pollen morphology of the genera *Gluta* Linnaeus and *Melanorrhoea* Wallich Anacardiaceae. In: Ferguson, I.K., Muller, J. (eds.) The evolutionary significance of the exine. Linnean Society Symposium Series I. London: Academic Press, pp. 379–405.

Barajas-Morales, J., Gómez, C.L. 1989. Anatomía de maderas de México: especies de una selva baja caducifolia. México City: Universidad Nacional Autónoma de México.

Basak, R.F. 1968. Studies on the pollen morphology of Simaroubaceae. Bull. Bot. Surv. India 1967 (9): 63–67.

Bates, J.M. 1992. Frugivory on *Bursera microphylla* Burseraceae by wintering gray vireos *Vireo vicinior*, Vireonidae in the coastal deserts of Sonora, México. Southwest. Nat. 37 3: 252–258.

Bawa K.S., Opler, P.A. 1975. Dioecism in tropical forest trees. Evolution 29: 167–179.

Baxter, J. 2006. Blue fruit and novel tastes: Selling agroforestry delicacies in central Africa. Geneflow 2006: 9.

Becerra, J.X. 2003. Evolution of Mexican *Bursera* Burseraceae inferred from ITS, ETS, and 5S nuclear ribosomal DNA sequences. Mol. Phylogenet. Evol. 26: 300–309.

Birkinshaw, C., Colquhoun, I.C. 2004. Lemur food plants. In: Goodman, S., Benstead, J. (eds.) Natural history of Madagascar. Chicago: University of Chicago Press, pp. 1207–1220.

Boer, E., Ella, A.B. 2000. Plants producing exudates. Plant Resources of South-East Asia no. 18. Leiden: Backhuys.

Boesewinkel, F.D., Bouman, F. 1984. The seed: structure. In: Johri, B.M. (ed.) Embryology of angiosperms. Heidelberg: Springer, pp. 567.

Böhning-Gaese, K., Gaese, B.H., Rabemanantsoa, S.B. 1999. Importance of primary and secondary seed dispersal in the Malagasy tree, *Commiphora guillaumini* [sic]. Ecology 80: 821–832.

Bollen, A., Van Elsacker, L., Ganzhorn, J.U. 2004. Tree dispersal strategies in the littoral forest of Sainte Luce, SE-Madagascar. Oecologia 139: 604–616.

Britt, A. 2000. Diet and feeding behavior of the black-and-white ruffed lemur *Varecia variegata variegata* in the Betampona Reserve, Eastern Madagascar. Folia Primatol. 71: 133–141.

Burris, T.P., Montrose, C., Houck, K.A., Osborne, H.E., Bocchinfuso, W.P., Yaden, B.C., Cheng, C.C., Zink, R.W., Barr, R.J., Hepler, C.D., Krishnan, V., Bullock, H.A., Burris L.L., Galvin, R.J., Bramlett, K., Stayrook, K.R. 2005. The hypolipidemic natural product guggulsterone is a promiscuous steroid receptor ligand. Mol. Pharmacol. 67: 948–954.

Case, R.J., Tucker, A.O., Maciarello, M.J., Wheeler, K.A. 2003. Chemistry and ethnobotany of commercial incense Copals, Copal Blanco, Copal Oro, and Copal Negro, of North America. Econ. Bot. 57 2: 189–202.

Chander, R., Khanna, A.K., Kapoor, N.K. 1998. Lipid lowering activity of Guggulsterone from *Commiphora mukul* in hyperlipaemic rats. Phytother. Res.10 6: 508–511.

Chandler, M.E.J. 1961. The Lower Tertiary floras of southern England. London: British Museum of Natural History.

Charles-Dominique, P. 1991. Feeding strategy and activity budget of the frugivorous bat *Carollia perspicillata* (Chiroptera: Phyllostomidae) in French Guiana. J. Trop. Ecol. 7: 243–256.

Chattaway, M.M. 1955. Crystals in woody tissues; Part I. Trop. Woods 102: 55–74.

Chen, R.Y. 1993. Chromosome atlas of Chinese fruit trees and their close wild relatives. Chromosome Atlas of Chinese Principal Economic Plants. Beijing: International Academic Publishers, vol. 1, 626 pp.

Clarkson, J.J., Chase, M.W., Harley, M.M. 2002. Phylogenetic relationships in Burseraceae based on plastid *rps*16 intron sequences. Kew Bull. 57: 183–193.

Coronel, R.E. 1996. Pili Nut - *Canarium ovatum* Engl. Rome: International Plant Genetic Resources Institute IPGRI, 57 pp.

Cristóvão de Lisboa, F. 1968. História dos animais e árvores do Maranhão. Universidade Federal do Paraná, Conselho de Pesquisas, Curitiba.

Daly, D.C. 1989. Studies in Neotropical Burseraceae II. Generic limits in Neotropical Protieae and Canarieae. Brittonia 41: 17–27.

Daly, D.C. 1990. The genus *Tetragastris* and the forests of Eastern Brazil. Studies in Neotropical Burseraceae III. Kew Bull. 45: 179–194.

Daly, D.C. 1999. Notes on *Trattinnickia*, including a synopsis in eastern Brazil's Atlantic forest complex. Studies in Neotropical Burseraceae IX. Kew Bull. 54 1: 129–137.

Daly, D.C. 2002. *Crepidospermum atlanticum* sp. nov., a genus new to eastern Brazil's Atlantic forest complex. Studies in Neotropical Burseraceae X. Kew Bull. 57: 471–477.

Daly, D.C. 2007. A new section of *Protium* Burm. f. from the Neotropics. Studies in Neotropical Burseraceae XIV. Brittonia 59: 1–24.

Daly, D.C., Martínez H., M.C. 2003. Notes on *Dacryodes* Vahl, including a new species from the Rio Negro basin. Studies in Neotropical Burseraceae XI. Brittonia 54: 266–274.

Détienne, P., Jacquet, P. 1983. Atlas d'identification des bois de l'Amazonie et des régions voisines. Nogent-sur-Marne: Centre Technique Forestier Tropical.

Dolara, P., Luceri, C., Ghelardini, C., Monserrat, C., Aiolli, S., Luceri, F., Lodovici, M., Menichetti, S., Romanelli, M.N. 1996. Analgesic effects of myrrh. Nature 379: 29.

Dunphy, B.K., Hamrick, J.L. 2007. Estimation of gene flow into fragmented populations of *Bursera simaruba* Burseraceae in the dry-forest life zone of Puerto Rico. Am. J. Bot. 94: 1786–1794.

Durdana Yunus, Nair, P.K.K. 1988–1989. Pollen morphology of Indian Geraniales: a research monograph. New Delhi: Today and Tomorrow's Printers & Publishers.

Ellis, B., Daly, D.C., Hickey, L., Johnson, K., Mitchell, J., Wilf, P., Wing, S. 2009. Manual of leaf architecture. Ithaca: The New York Botanical Garden and Comstock Publishing Associates, Cornell University Press.

Engler, A. 1931. Burseraceae. In: Engler & Prantl, Die natürlichen Pflanzenfamilien, 2nd edn, 19a: 405–456. Leipzig: W. Engelmann.

Erdtman, G. 1952. Pollen morphology and plant taxonomy. Stockholm: Almqvist and Wiksell.

Esau, K. 1965. Plant anatomy. New York: Wiley.

Fedorov, A.A. (ed.) 1969. Chromosome numbers of flowering plants (in Russian). Leningrad: Nauka.

Fine, P.V., Daly, D.C., Villa, G., Mesones, I., Cameron, K.M. 2005. The contribution of edaphic heterogeneity to the evolution and diversity of Burseraceae trees in the western Amazon. Evolution 59: 1464–1478.

Fine, P.V.A., Miller, Z.J., Mesones, I., Irazuzta, S., Appel, H.M., Stevens, M.H.H., Sääksjärvi, I., Schultz, J.C., Coley, P.D. 2006. The growth-defense trade-off and habitat specialization by plants in Amazonian forests. Ecology 87: S150–S162.

Forman, L.L., Brandham, P.E., Harley, M.M., Lawrence, T.J. 1989. *Beiselia mexicana* (Burseraceae) and its affinities. Kew Bull. 44: 1–31.

Forman, L.L., van der Ham, R.W.J.M., Harley, M.M., Lawrence, T.J. 1994. *Rosselia*, a new genus of Burseraceae from the Louisiade Archipelago, Papua New Guinea. Kew Bull. 49: 601–621.

Freed, B.Z. 2007. Polyspecific associations of Crowned Lemurs and Sanford's Lemurs in Madagascar. In: Gould, L., Sauther, M. L. (eds.) Developments in primatology: progress and prospects. New York: Springer.

Gautier-Hion, A., Duplantier, J.M., Quris, R., Feer, F., Sourd, C., Decoux, J.P., Dubost, G., Emmons, L., Erard, C., Hecketsweiler, P., Moungazi, A., Roussilhon, C., Thiollay, J.M. 1985. Fruit characters as a basis of fruit choice and seed dispersal in a tropical forest vertebrate community. Oecologia 65: 324–337.
Gentry, A.H. 1974. Flowering phenology and diversity in tropical Bignoniaceae. Biotropica 6: 64–68.
Gibson, A.C. 1981. Vegetative anatomy of *Pachycormus* Anacardiaceae. Bot. J. Linn. Soc. 83: 273–284.
Gill, B.S., Bir, S.S., Singhal, V.K. 1979. IOPB chromosome number reports LXIV. Taxon 28: 403.
Gill, B.S., Singhal, V.K., Bedi, Y.S., Bir, S.S. 1990. Cytological evolution in the woody taxa of Pachmarhi Hills. J. Cytol. Genet. 25: 308–320.
Gillett, J.B. 1980. *Commiphora* (Burseraceae) in South America and its relationship to *Bursera*. Kew Bull. 34: 569–587.
Gillett, J.B. 1991. Burseraceae. In: Polhill, R.M. (ed.) Flora of tropical East Africa. Rotterdam: A.A. Balkema.
Goodman, S.M., Sterling, E.J. 1996. The utilization of *Canarium* (Burseraceae) seeds by vertebrates in the Réserve Naturelle Intégrale d'Andringitra, Madagascar. In: Goodman, S.M. (ed.) A floral and faunal inventory of the Eastern slopes of the Reserve Naturelle Intégrale d'Andringitra, Madagascar: with reference to elevational variation. Fieldiana Zool. 85: 83–89.
Grant, M., Blackmore, S., Morton, C. 2000. Pollen morphology of the subfamily Aurantoideae (Rutaceae). Grana 39: 8–20.
Gregor, V.J., Goth, K. 1979. Erster Nachweis der Gattung *Canarium* Stickman 1759 Burseraceae im europaischen Alttertiar. Stuttgarter Beitrage Naturkunde Serie B, Geol. Palaeontol. 47: 1–15.
Grimes, A., Loomis, S., Jahnige, P., Burnham, M., Onthank, K., Alarcón, R., Palacios, W., Cuenca, C., Martínez, C., Neill, D., Balick, M., Bennett, B., Mendelsohn, R. 1994. Valuing the rain forest: the economic value of nontimber forest products in Ecuador. Ambio 23: 405–410.
Groom, N. 1981. Frankincense and myrrh: a study of the Arabian incense trade. Longman: London.
Guillaumin, A. 1909. Recherches sur la structure et le développement des Burseracées. Applications à la systématique. Ann. Sci. Nat. IX, Bot. 10, 4: 201–301.
Guppy, H.B. 1906. Observations of a naturalist in the Pacific between 1896 and 1899. Vol. II. Plant-Dispersal. London: Macmillan.
Gupta, P., Shivanna, K.R., Mohan-Ram, H.Y. 1996. Apomixis and polyembryony in the guggul plant, *Commiphora wightii*. Ann. Bot. 78: 67–72.
Harley, M.M., Daly, D.C. 1995. Burseraceae Kunth, Protieae March. em. Engl. World Pollen Spore 20: 1–44.
Harley, M.M., Hall, D. 1999. Pollen morphology of the African Burseraceae and related genera. Palaeoecol. Africa Surrounding Islands 26: 225–242.
Harley, M.M., Song, U., Banks, H. 2005. Pollen morphology and systematics of Burseraceae. Grana 44: 282–289.
Hegnauer, R. 1964. Chemotaxonomie der Pflanzen. Vol. 3. Basel: Birkhäuser.
Hegnauer, R. 1989. Chemotaxonomie der Pflanzen. Vol. 8. Basel: Birkhäuser.
Hill, A.W. 1933. The method of germination of seeds enclosed in a stony endocarp. Ann. Bot. 47: 873–887.
Hill, A.W. 1937. The method of germination of seeds enclosed in a stony endocarp. II. Ann. Bot. 51: 239–256.
Howe, H.F. 1980. Monkey dispersal and waste of a neotropical fruit. Ecology 6: 944–959.
Howe, H.F. 1989. Scatter- and clump-dispersal and seedling demography: hypothesis and implications. Oecologia 79: 417–426.
Ibe, R.A., Leis, R.A. 1979. Pollen morphology of the Anacardiaceae of northeastern North America. Bull. Torrey Bot. Club 106: 140–144.
Idani, G., Kuroda, S., Kano, T., Asato, R. 1994. Flora and vegetation of Wamba Forest, Central Zaire with reference to bonobo *Pan paniscus* foods. Tropics 3 (3/4): 309–332.
Khalid, S.A. 1983. Chemistry of the Burseraceae. In: Waterman, P.G., Grundon, M.F. (eds.) Chemistry and chemical taxonomy of the Rutales. New York: Academic Press.
Kitamura, S., Suzuki, S., Yumoto, T., Poonswad, P., Chuailua, P., Plongmai, K., Maruhashi, T., Noma, N., Suckasam, C. 2006. Dispersal of *Canarium euphyllum* Burseraceae, a large-seeded tree species, in a moist evergreen forest in Thailand. J. Trop. Ecol. 22: 137–146.
Kryn, J.M. 1952. The anatomy of the wood of the Anacardiaceae and its bearing on the phylogeny and relationships of the family. Ph.D. Thesis, Ann Arbor, MI: University of Michigan.
Lam, H.J. 1931. Beiträge zur Morphologie der dreizähligen Burseraceae-Canarieae. Ann. Jard. Bot. Buitenzorg 42: 25–56, t. v–vii.
Lam, H.J. 1932a. Beiträge zur Morphologie der Burseraceae, insbesondere der Canarieae. Ann. Jard. Bot. Buitenzorg 42: 97–226, t. xi–xvi.
Lam, H.J. 1932b. The Burseraceae of the Malay Archipelago and Peninsula. Bull. Jard. Bot. Buitenzorg III, 12: 281–561.
Langenheim, J.H. 1994. Higher plant terpenoids: a phytocentric overview of their ecological roles. J. Chem. Ecol. 20 6: 1223–1280.
Langenheim, J.H. 2003. Plant resins: chemistry, evolution, ecology, and ethnobotany. Portland, OR: Timber Press.
Leenhouts, P.W. 1956. Burseraceae. Flora Malesiana, ser. I(5): 209–296.
Leenhouts, P.W. 1959. Revision of the Burseraceae of the Malaysian area in a wider sense Xa. *Canarium* Stickm. Blumea 9(2): 275–475.
Leenhouts, P.W. 1972. A revision of *Haplolobus* (Burseraceae). Blumea 20: 283–310.
Liang, G.Y., Gray, A.I., Patalinghug, W.C., Skelton, B.W., Waterman, P.G., White, A.H. 1989. Chemistry of the Burseraceae. XI. X-Ray-Diffraction studies on some tirucall-7-ene triterpenes from *Aucoumea klaineana* (Burseraceae). Aust. J. Chem. 42 7: 1169–1175.
Lima, M.P., Braga, P.A.C., Macedo, M.L., Silva, M.F.G.F., Ferreira, A.G., Fernandes, J.B., Vieira, P.C. 2004. Phytochemistry of *Trattinnickia burserifolia*, *T. rhoifolia*, and *Dacryodes hopkinsii*: chemosystematic implications. J. Braz. Chem. Soc. 15: 385–394.

Lindley, J. 1853. The vegetable kingdom, 3rd edn. London: Bradbury & Evans.

Lobova, T.A, Geiselman, C.K., Mori S.A. 2009. Seed dispersal by bats in the Neotropics. New York: The New York Botanical Garden Press.

Lobreau-Callen, D., Oltmann, O., Straka, H. 1975. Fam. 99–110. In: Straka, H. (ed.) Palynologia Madagassica et Mascarenica. Pollen Spores 17: 11–57.

MacGinitie, H.D. 1941. A Middle Eocene Flora from the Central Sierra Nevada. Contributions to Paleontology. Carnegie Institution of Washington Publications no. 534.

MacGinitie, H.D., 1953. Fossil plants of the Florissant beds, Colorado. Carnegie Inst. Washington 599: 1–188.

MacGinitie, H.D., 1969. The Eocene Green River flora of northwestern Colorado and northeastern Utah. Univ. Calif. Publ. Geol. Sci. 83: 1–140.

Machado, I.C., Lopez, A.V. 2004. Floral traits and pollination systems in the Caatinga, a Brazilian tropical dry forest. Ann. Bot. 94 3: 365–376.

Machado, L.B., Zoghbi, M.G.B., Andrade, E.A.H. 2003. Seasonal variation in the composition of the essential oils from leaves, thin branches and resin of *Protium spruceanum* (Benth.) Engl. Flavour Fragr. J. 18: 338–341.

Marchand, N.L. 1867–1868. Recherches sur l'organisation des Burseracées. Adansonia 8: 17–71.

Matallana, G., Wendt, T., Araújo, D.S.D., Scarano, F.R. 2005. High abundance of dioecious plants in a tropical coastal vegetation. Am. J. Bot. 92: 1513–1519.

McClatchey, W., Sirikolo, M.Q.B. Jr., Kaleveke, L., Pitanapi, C. 2006. Differential conservation of two species of *Canarium* (Burseraceae) among the Babatana and Ririo of Lauru Choiseul, Solomon Islands. Econ. Bot. 60: 212–226.

McGarvey, D.J., Croteau, R. 1995. Terpenoid metabolism. Plant Cell 7: 1015–1026.

McVaugh, R., Rzedowski, J. 1965. Synopsis of the genus *Bursera* L. in western Mexico, with notes on the material of *Bursera* collected by Sessé & Mociño. Kew Bull. 18: 317–382.

Metcalfe, C.R., Chalk, L. 1950. Anatomy of the dicotyledons. Vol. 1. Oxford: Oxford University Press.

Miller, A.G., Morris, M. 1988. Plants of Dhofar: the southern region of Oman, traditional, economic and medicinal uses. Oman: Office of the Adviser for Conservation of the Environment, Diwan of Royal Court.

Morton, C.M., Kallunki, J.A. 1993. Pollen morphology of the subtribe Cuspariinae (Rutaceae). Brittonia 45: 286–314.

Mothana, R.A., Lindequist, U., Gruenert, R., Bednarski, P.J. 2009. Studies of in vitro anticancer, antomicrobial and antioxidant potentials of selected Yemeni medicinal plants from the island Socotra. Complementary and Alternative Medicine 9: 1472–6882.

Muller, J., Leenhouts, P.W. 1976. A general survey of pollen types in Sapindaceae in relation to taxonomy. In: Ferguson, I.K., Muller, J. (eds.) The evolutionary significance of the exine. Linnean Society Symposium Series I. London: Academic Press, pp. 407–445.

Narayana, L.L. 1959. Microsporogenesis and female gametophyte in *Boswellia serrata* Roxb. Curr. Sci. 28: 77–78.

Narayana, L.L. 1960. Studies in Burseraceae – I. J. Indian Bot. Soc. 39: 204–209.

Oginuma, K., Damas, K., Tobe, H. 1999. A cytology of some plants from Papua New Guinea: additional notes. Acta Phytotaxon. Geobot. 50: 43–50.

Oliveira, F.A., Chaves, M.H., Almeida, F.R.C., Lima, R.C.P. Jr., Silva, R.M., Maia, J.L., Brio, G.A.A.C., Santo, F.A., Rao, V.S. 2005. Protective effect of a- and b-amyrin, a triterpene mixture from *Protium heptaphyllum* Aubl. March. [sic] trunk wood resin, against acetaminophen-induced liver injury in mice. J. Ethnopharmacol. 98: 103–108.

Ortiz-Pulido, R., Rico-Graya, V. 2006. Seed dispersal of *Bursera fagaroides* Burseraceae: the effect of linking environmental factors. Southwest. Nat. 51: 11–21.

O'Sullivan, J. 1983. Structural diversity and distribution of lignans in the Rutales. In: Waterman, P.G., Grundon, M.F. (eds.) Chemistry and chemical taxonomy of the Rutales. London: Academic Press, pp. 267–279.

Parrado-Rosselli, A. 2005. Fruit availability and seed dispersal in terra firme rain forests of Colombian Amazonia. Ph.D. Thesis, Amsterdam: University of Amsterdam.

Pernet, R. 1972. Phytochimie des Burseracées. Lloydia 35: 280–287.

Peters, C.M., Purata, S.E., Chibnik, M., Brosi, B.J., Lopez, A.M., Ambrosio, M. 2003. The life and times of *Bursera glabrifolia* H.B.K. Engl. in Mexico: a parable for ethnobotany. Econ. Bot. 57 4: 431–141.

Piechulla, B., Pott, M.B. 2003. Plant scents—mediators of inter-and intraorganismic communication. Planta 217: 687–689.

Plowden, C., Uhl, C., Assis de Oliveira, F. 2002. Breu resin harvest by Tembé Indians and its dependence on a bark-boring beetle. In: Stepp, J.R., Wyndham, F.S., Zarger, R.K. (eds.) Ethnobiology and biocultural diversity. Athens, GA: University of Georgia Press.

Pontes, W.J.T., Oliveira, J.C.S., Camara, C.A.G., Lopes, A.C.H.R., Gondim, M.G.C., Oliveira, J.V., Schwartz, M.O.E. 2007. Composition and acaricidal activity of the resin's essential oil of *Protium bahianum* Daly against two spotted spider mite *Tetranychus urticae*. J. Essent. Oil Res. 19: 379–383.

Primack, R.B. 1985. Longevity of individual flowers. Annu. Rev. Ecol. Syst. 16: 15–37.

Provan, G.J., Gray, A.I., Waterman, P.G., 1987. Monoterpene-rich resins from some Kenyan Burseraceae. Flavour Fragr. J. 2: 115–118.

Quisumbing, E. 1951. Medicinal plants of the Philippines. Technical Bulletin no. 16, Department of Agriculture and Natural Resources, Manila.

Rahman, M.M., Garvey, M., Piddock, L.J.V., Gibbons, S. 2008. Antibacterial terpenes from the oleo-resin of *Commiphora molmol* Engl. Phytother. Res. 22: 1356–1360.

Ramos, M.F.S., Guimarães, A.C., Siani, A.C. 2003. Volatile monoterpenes from the oleoresin of *Trattinnickia rhoifolia*. Biochem. Syst. Ecol. 31: 309–311.

Ramos-Ordoñez, M.F., Márquez-Guzmán, J., Arizmendi, M.C. 2008. Parthenocarpy and seed predation by insects in *Bursera morelensis*. Ann. Bot. 102: 713–722.

Raponda-Walker, A., Sillans, R. 1961. Les plantes utiles du Gabon. Paris: Ed. Paul Lechevalier.

Raven, P.H. 1975. The bases of angiosperm phylogeny: cytology. Ann. Missouri Bot. Gard. 62: 724–764.
Reid, E.M., Chandler, M.E.J. 1933. The London Clay flora. London: British Museum Natural History.
Robles, J., Torrenegra, R., Gray, A.I., Piñeros, C., Ortíz, L., Sierra, M. 2005. Triterpenos aislados de corteza de *Bursera graveolens* Burseraceae y su actividad biológica. Rev. Bras. Farmacogn. 15 4: 283–286.
Rüdiger, A.L., Siani, A.C., Veiga, V.F. 2007. The chemistry and pharmacology of the South America genus *Protium* Burm.f. Burseraceae. Pharmacog. Rev. 1: 93–104.
Ruiz Pérez, M., Ezzine de Blas, D., Nasi, R., Sassen, M., Sayer, J. Angoué, C., Gami, N., Ndoye, O., Ngono, G., Nguinguiri, J.C., Nzala, D., Toirambe, B., Yalibanda, Y. 2005. Who is logging the Congo? ITTO Tropical Forest Update 14 4: 3–6.
Rzedowski, J., Ortiz, E., 1988. Estudios quimiotaxonómicos de *Bursera* Burseraceae. II. Una especie nueva de origen híbrido de la Barranca de Tolantongo, estado de Hidalgo. Acta Bot. Mex. 1, pp. 11–19.
Rzedowski, J., Medina Lemos, N., Calderón de Rzedowski, G. 2005. Inventario del conocimiento taxonómico, así como de la diversidad y del endemismo regionales de las especies mexicanas de *Bursera* (Burseraceae). Acta Bot. Mex. 70: 85–111.
Setia, R.C., Parthasarahaty, M.V., Shah, J.J. 1977. Development, histochemistry and ultrastructure of gum-resin ducts in *Commiphora mukul* Engl. Ann. Bot. 41: 999–1004.
Siani, A.C., Garrido, I.S., Monteiro, S.S., Carvalho, E.S., Ramos, M.F.S. 2004. *Protium icicariba* as a source of volatile essences. Biochem. Syst. Ecol. 32: 477–489.
Silva, P.D., Leal, I.R., Wiurth, R., Tabarelli, M. 2007. Harvesting of *Protium heptaphyllum* Aubl. March. [sic] Burseraceae by the leaf-cutting ant *Atta sexdens* L. promotes seed aggregation and seedling mortality. Rev. Bras. Bot. 30: 553–560.
Simons, H. 1991. Social organization and ecology of Black and White Ruffed Lemurs *Varecia variegata rubra* in lowland rain forest, Nosy Mangabe, Madagascar. Ph.D. Thesis, New Haven, CT: Yale University.
Singhal, V.K., Gill, B.S. 1984. SOCGI plant chromosome number reports - II. J. Cytol. Genet. 19: 115–117.
Snow, D.W. 1962. The natural history of the Oilbird, *Steatornis caripensis*, in Trinidad, W.I. Part 2: Population, breeding ecology and food. Zoologica, N.Y. 47: 199–221.
Solereder, H. 1908. Systematic anatomy of the dicotyledons. Translated by L.A. Boodle & F.E. Fritsch, revised by D.H. Scott. Oxford: Clarendon Press.
Stacey, R.J., Cartwright, C.R., McEwan, C. 2006. Chemical characterization of ancient Mesoamerican "copal" resins: preliminary results. Archaeometry 48: 323–340.
Strappaghetti, G., Corsane, S., Craveiro, A., Proietti, G. 1982. Constituents of essential oil of *Boswellia frereana*. Phytochemistry 21: 2114–2115.
Swart, J.J. 1942. A monograph of the genus *Protium* and some allied genera Burseraceae. Gouda: Drukkerij Koch en Knuttel.
Tanai, T., 1970. The Oligocene Xoras from the Kushiro Coal Field, Hokkaido, Japan. J. Fac. Sci. Univ. Tokyo Sect. 2, Ser. IV. 14: 383–514.
Thulin, M. 1993. Burseraceae. In: Thulin, M. (ed.) Flora of Somalia, vol. 2. Kew: Royal Botanic Gardens, pp. 183–229.
Thulin, M., Claeson, P. 1991. The botanical origin of scented myrrh bissabol or habak hadi. Econ. Bot. 45: 487–494.
Thulin, M., Warfa, A.M. 1987. The frankincense trees *Boswellia* spp., Burseraceae of northern Somalia and southern Arabia. Kew Bull. 42: 487–500.
Thulin, M., Beier, B.A., Razafimandimbison, S.G., Banks, H.I. 2008. *Ambilobea*, a new genus from Madagascar, the position of *Aucoumea*, and comments on the tribal classification of the frankincense and myrrh family. Nord. J. Bot. 26: 218–229.
Tissot, C., Chikhi, H., Nayar, T.S. 1994. Pollen of the wet evergreen forests of the western Ghats, India. Institut Français de Pondichéry, Publications du Département d'Écologie 35: 1–133.
Toledo, C.A. 1982. El género *Bursera* Burseraceae en el estado de Guerrero. Tesis, Facultad de Ciencias, Universidad Nacional Autónoma de México.
Trainer, J.M., Will, T.C. 1984. Avian feeding methods on *Bursera simaruba* Burseraceae fruits in Panama. The Auk 101: 193–195.
Tucker, A.O. 1986. Frankincense and myrrh. Econ. Bot. 40: 425–433.
Tutin, C.E.G., White, L.J.T., Williamson, E.A., Fernandez, M., McPherson, G. 1994. List of plant species identified in the northern part of the Lopé Reserve, Gabon. Tropics 3: 249–276.
Van der Ham, R.W.J.M. 1995. Palynology. In: Mabberley, D.J., Pannell, C.M., Sing, A.M. (eds.) Meliaceae. Flora Malesiana I, 12, part 1: 6–7. Foundation Flora Malesiana.
van der Walt, J.J.A. 1975. The fruit of *Commiphora*. Boissiera 24a: 325–330.
van der Walt, J.J.A., van der Schijff, H.P. 1969. The anatomy of the petiole as an aid to the identificaiton of South African *Commiphora* species. Kirkia 9: 95–107.
van Dulmen, A. 2001. Pollination and phenology of flowers in the canopy of two contrasting rain forest types in Amazonia, Colombia. Plant Ecol. 153: 73–85.
van Heel, W.A. 1970. Some unusual tropical labyrinth seeds. Proc. Kon. ned. Akad. Wet. C 73: 288–301.
van Wyk, B., van Wyk, P. 1997. Field Guide to Trees of Southern Africa. Cape Town: Struik.
Vieira, F.A., Carvalho, D. 2008. Genetic structure of an insect-pollinated and bird-dispersed tropical tree in vegetation fragments and corridors: implications for conservation. Biodivers. Conserv. 17: 2305–2321.
Villegas, V.N., Coronel, R.E. 1979. Note: Cytology of Philippine Fruits. 1. Pili (*Canarium ovatum* Engl.) and barobo (*Diplodiscus paniculatus* Turzc.) Philipp. Agric. 63 2: 174–178.
Voigt, F.A., Jung, S., Farwig, N., Böhning-Gaese, K. 2005. Low fruit set in a dioecious tree: pollination ecology of *Commiphora harveyi* in South Africa. J. Trop. Ecol. 21: 179–188.
Vollesen, K. 1989. Burseraceae. In: Hedberg, I., Edwards, S. (eds.) Flora of Ethiopia. Vol. 3. The National Herbarium, Biology Department, Science Faculty, Addis Ababa University, Ethiopia and The

Department of Systematic Botany, Uppsala University, Sweden, pp. 422–477.
Webber, I.E. 1941. Systematic anatomy of the woods of the "Burseraceae". Lilloa 6: 441–465.
Weeks, A. 2003. The molecular systematics and biogeography of the Burseraceae. Ph.D. Thesis, Austin, TX: The University of Texas at Austin.
Weeks, A. 2009. Evolution of the pili nut genus *Canarium* L., Burseraceae and its cultivated species. Genet. Resour. Crop Evol. 56: 765–781.
Weeks, A., Simpson, B.B. 2004. Molecular genetic evidence for interspecific hybridization among endemic Hispaniolan *Bursera* Burseraceae. Am. J. Bot. 91: 967–984.
Weeks, A., Simpson, B.B. 2007. Molecular phylogenetic analysis of *Commiphora* Burseraceae yields insight on the evolution and historical biogeography of an "impossible" genus. Mol. Phylogenet. Evol. 42: 62–79.
Weeks, A., Daly, D.C., Simpson, B.B. 2005. The phylogenetic history and biogeography of the frankincense and myrrh family (Burseraceae) based on nuclear and chloroplast sequence data. Mol. Phylogen. Evol. 35: 85–101.
Zohary, M. 1982. Plants of the Bible. Cambridge: Cambridge University Press.

Coriariaceae

Coriariaceae DC. (1824), nom. cons.

K. Kubitzki

Erect or scandent rhizomatous shrubs, subshrubs, or rarely perennial herbs, evergreen or summergreen; the roots harbouring actinorhizal symbionts; stems with corky lenticels, sharply quadrangular, long and short, eventually arching to pendulous, sometimes with clusters of axillary inflorescence buds produced on main axes. Leaves opposite, more rarely in trimerous whorls, simple, entire, subsessile; venation palmate, veins 3(5); stipules minute, deciduous. Inflorescences many-flowered racemes, terminating main or lateral axes, often hidden under densely leafy stems. Flowers hermaphrodite or functionally unisexual, markedly protogynous, regular, small, 5-merous, diplostemonous; sepals distinct, quincuncially arranged; petals distinct, with open or valvate aestivation, keeled ventrally, smaller than sepals at anthesis but later accrescent and becoming fleshy, protruding between the carpels; stamens distinct or the antepetalous ones adnate to petals; anthers large, basifixed or slightly dorsifixed, tetrasporangiate, dithecal, dehiscing longitudinally, introrse; ovary superior, 5- or more rarely 10-carpellate and -locular, superior, synascidiate for two thirds of its length, proximally plicate; each carpel with a long slender stylodium that is papillose and stigmatic over its entire surface; ovule 1 per locule, pendulous, anatropous, bitegmic, crassinucellar, the micropyle directed upwards; placentation median, apical. Fruit of distinct, small, purple to black hard-walled nuts, enclosed by the persistent, fleshy petals. Seeds slightly compressed, endosperm scant or 0; embryo straight, oily. $x=10$.

A single genus with about 17 disjunct spp., in Europe, South, East and Southeast Asia, the south-eastern Pacific, and Central and South America.

Morphology and Anatomy. Leaves and branches of *Coriaria* are usually arranged to form flattened, frond-like sprays. Stipules are minute; they have long been overlooked (Weberling 1955). At the nodes, *Coriaria* has distinctive trichomoids (half hair, half scale, fide L. Croizat, pers. comm. by M. Heads), shared with *Corynocarpus*. Most southern hemisphere species such as *C. ruscifolia*, *C. microphylla*, *C. arborea*, *C. papuana*, and most species from New Zealand are evergreen phanerophytes; *C. plumosa*, *C. pottsiana* and *C. angustissima* from the subalpine belt of New Zealand are low subshrubs that die down to ground level every winter. In all these species the racemes are usually produced in pairs in the axils or are terminal on shoots arising from the axils of leaves of the current year's growth. *Coriaria terminalis*, which grows in the northern hemisphere where its distribution overlaps in the eastern Himalaya with that of *C. nepalensis*, is a hemicryptophyte that produces racemes terminal on stems developed in the current season. While *C. terminalis*, like all southern hemisphere species, lacks bud scales at the base of lateral branches, these are present in all other northern hemisphere species.

The northern hemisphere species *C. myrtifolia*, *C. intermedia* and *C. japonica* are deciduous phanerophytes that flower on old wood (Fig. 19A). Their racemes are few-flowered but aggregated in axillary clusters of up to 12 per node subtended by prominent bud-scales (Thompson and Gornall 1995). *Coriaria nepalensis* and *C. duthiei* have terminal racemes.

Coriaria harbours endosymbiontic bacteria (*Frankia*) in the roots, in which nodule lenticels allow the passage of oxygen into the nodule tissue. In this, *Coriaria* resembles *Datisca*, the only other genus with actinorhizal symbiosis in Cucurbitales (Swensen 1996).

Young stems have a large pith. Vessel elements are short but wide, with simple perforation plates, lateral wall pits are circular to

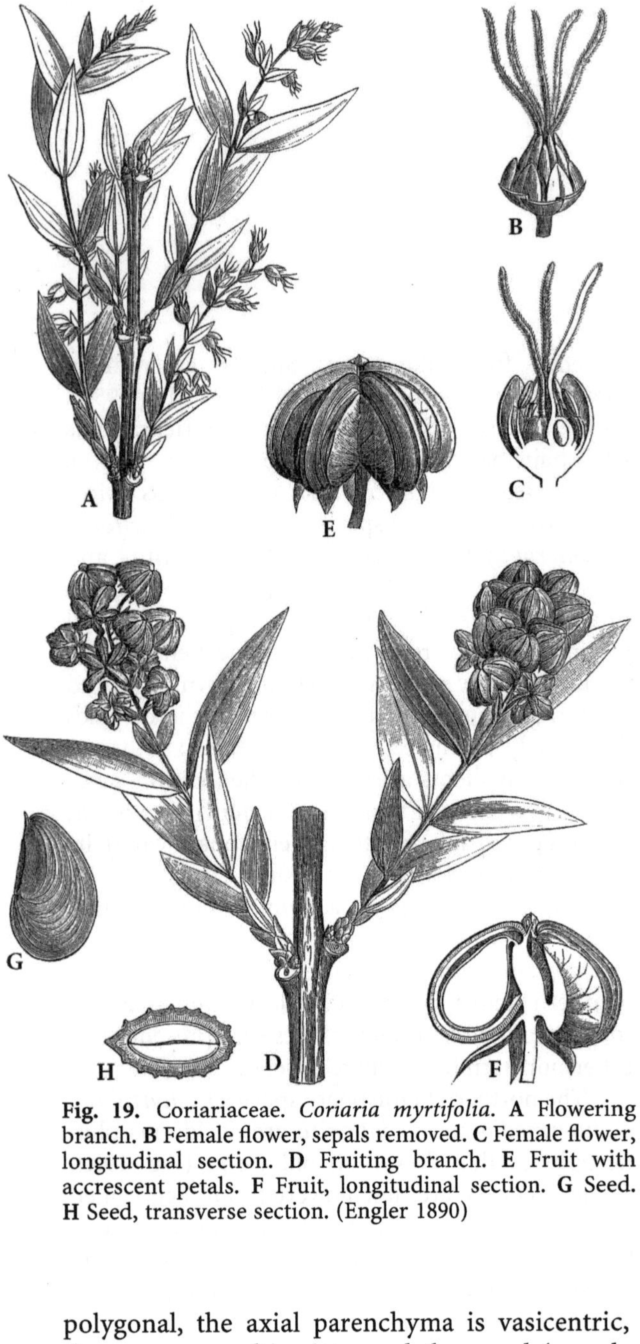

Fig. 19. Coriariaceae. *Coriaria myrtifolia.* **A** Flowering branch. **B** Female flower, sepals removed. **C** Female flower, longitudinal section. **D** Fruiting branch. **E** Fruit with accrescent petals. **F** Fruit, longitudinal section. **G** Seed. **H** Seed, transverse section. (Engler 1890)

polygonal, the axial parenchyma is vasicentric, the rays are multiseriate, and the wood (vessels and axial parenchyma) is storied. Within Cucurbitales, *Coriaria* shares a special form of axial parenchyma and Homogeneous Type II rays with *Corynocarpus* and Cucurbitaceae (Carlquist 1985; Carlquist and Miller 2001).

The gynoecium has often been interpreted as (half-)apocarpous because the dorsal bulging of the carpels gives an apocarpous impression, until Guédès (1971) showed that it is fully syncarpous; a compitum is present (Matthews and Endress 2004). The stylodia are unifacial; they lack a ventral slit, and the pollen tube transmission tissue does not contact with the morphological surface, except at the stigma.

EMBRYOLOGY. Mature pollen grains are 2- and 3-celled. The ovule is anatropous, bitegmic and crassinucellate with a dorsal raphe. The micropyle is formed by the inner integument alone. The embryo sac is of the Polygonum type. The endosperm is Nuclear but is consumed by the embryo and persists as a single layer, so that the seed is practically exalbuminous (Sharma 1968).

POLLEN MORPHOLOGY. Pollen grains are suboblate or rarely oblate, brevi-(2)3(4)-colporate with lalongate ora or 3(4)-zonoporate, peritreme or slightly goniotreme; the exine is tectate-baculate, verrucate and microperforate (Praglowski 1970).

POLLINATION AND SEX EXPRESSION. *Coriaria* is wind pollinated, which agrees with the well-developed, abundantly papillose stylodia and the anthetically inconspicuous petals, which in fruit form a fleshy layer around the hard carpels. Self compatibility, which is reported for many species, seems to be the general condition.

The racemes in all southern hemisphere species and in *C. terminalis* usually exclusively bear hermaphrodite flowers, which are strongly protogynous. The stigmas are receptive for about 14 days and there is an overlap of 1–2 days during which the stamens are exserted and dehisce while the stigmas are still able to receive pollen. All inflorescences on a given branch, but not on the entire plant, develop synchronously (hemi-synchronous type).

In the northern hemisphere species, racemes are produced in nodal clusters in which there is a preponderance of male racemes over hermaphrodite or female ones. Within each nodal cluster, male racemes generally flower first, with a flowering phase of 2–3 months, which is followed, with little or no overlap, by the hermaphrodite or female racemes, which last 6–8 weeks. The hermaphrodite or female flowers of the northern hemisphere species also develop synchronously on a branch and, in contrast to the southern hemisphere species, also synchronously on all branches of a plant. Thomson and Gornall (1995), from which these observations are taken, argue that the southern species represent the more basal condition and that vegetative and reproductive traits of the northern species have evolved in connection with the transition to a more seasonal climate.

Fruit and Seed. The accrescent petals close over the nutlets and become fleshy, thus transforming the fruit into a drupe-like structure (Fig. 19E), in some cases purple-black, in others (*C. japonica*) red or (*C. terminalis*) orange (Ridley 1930). The nutlets of *Coriaria* have a particularly thick and hard pericarp, which has been studied by Tobe et al. (1992) and apparently replaces the reduced seed coat. A well-stratified pericarp structure is shared by the species from New Zealand, South and Central America, and the Asian *C. terminalis*; four other Asian species, *C. nepalensis*, *C. sinica*, *C. intermedia* and *C. japonica*, have a more specialised structure, in which the outer zone of the mesocarp is reduced in favour of a thickened zone of the middle mesocarp. *Coriaria myrtifolia* and *C. japonica* exhibit further specialised traits.

The seeds are small, compressed, reddish-brown, exalbuminous or retain a trace of endosperm (Fig. 19G, H); the outer epidermis of the testa consists of cubic cells with thickened walls, filled with tannin; the mesophyll and inner integument are thin-walled; the tegmen is crushed (Corner 1976).

Phytochemistry. *Coriaria* leaves, twigs and fruits contains large amounts of gallic and ellagic acid and tannins based on them. A black pigment extracted from the fruits has been or still is used for dying. Toxic bitter principles of the coriamyrtin type have been isolated from the leaves, stems and fruits. These are sequiterpenoid bitter compounds structurally similar to picrotoxin and acting like it, producing spastic convulsions (Hegnauer 1964, 1989).

Dispersal. Ridley (1930) has assembled several records from various parts of the World according to which *Coriaria* fruits are being eaten by birds. A convincing proof of the dispersability of its diaspores is provided by its occurrence on remote and geologically young volcanic islands such as Tahiti and Raoul Island of the Kermadecs (Carlquist 1985).

Distribution and Habitats. *Coriaria* species are colonizers with a preference for disturbed ground such as road cuts, dry river beds, or lava flows. In the tropics they grow at middle or high elevation (in the Himalayas up to 3,000 m, in Colombia to 3,200 m), whereas in the southern temperate zone they grow in the lowland, as in Chile, but also at higher altitudes, as in New Zealand, where the genus occurs with eight (or more?) species from the coastal to the subalpine belt beyond the tree line. On the steep, wet forest of the Southern Alps of New Zealand, patches of *C. arborea* form the first real forest cover on the constantly occurring landslides (M. Heads, pers. comm. March 2009).

The global distribution of *Coriaria* is highly scattered into numerous smaller and larger disjunct areas. In South America, *C. ruscifolia* grows in central and southern Chile and adjacent Argentina from 33°N to 41°S, and *C. microphylla* on the western slope of the Andes from Peru (15°N) northwards to Venezuela and Panama and further to Mexico. *C. myrtifolia* is confined to the western Mediterranean (Europe and North Africa), Italy and Greece. *C. terminalis* grows in the Himalayas and in Tibet; *C. duthei* is restricted to Pakistan and India; *C. nepalensis* and *C. sinica* range from Nepal to western China, *C. japonica* grows in Japan, and *C. intermedia* is restricted to Taiwan and the Philippines. *C. papuana* occurs in eastern New Guinea, and eight species are recognised from New Zealand (Allan 1961). A possibly hitherto unnamed species is known from Tahiti, the Kermadecs and from New Caledonia (Yokoyama et al. 2000).

Good (1930) was one of the first to point to the remarkably disjunct distribution of the genus, which he thought to be due to vicariance and later influences by climatic changes in the Tertiary and Pleistocene. Yokoyama et al. (2000) tried to reconstruct the possible evolution of the genus in space and time on the basis of a DNA sequence analysis, in which the northern hemisphere species (including *C. terminalis*) form a clade sister to the southern hemisphere species.

Fossils. Apart from several records of fossil leaves ascribed to *Coriaria*, a flowering branchlet of *Coriaria* is known from the Oligocene of France (Saporta 1865), and *Coriaria* seeds were recorded from the Oligocene to Pliocene of Europe (Gregor 1980).

INFRAGENERIC RELATIONSHIPS. Both fruit structure and growth form agree in demonstrating a separation of extant species into a southern and a northern species group, with the northern *C. terminalis* showing strong affinities with the southern group. In the DNA sequence analysis of Yokoyama et al. (2000), however, *C. terminalis* is included in the northern group.

AFFINITIES. The position of *Coriaria* has been uncertain and unstable, until DNA sequence studies resolved the genus as sister to *Corynocarpus* within Cucurbitales (Zhang et al. 2006). Despite the differences in the pollination mode of the two genera, both share several traits such as stipules, wide wood rays, ellagic acid, displostemony, superior ovaries, the single, median, pendant ovule, and S_oP_o sieve element plastids (Behnke 1981).

USES. The "fruits" (see above) are used for dying; those of *C. terminalis* are said to be eaten, whereas those of other species are very toxic, those of *C. ruscifolia* being used in southern Chile to make rat poison.

Only one genus:

***Coriaria* L.** Fig. 19

Coriaria L., Gen. pl. ed. 5: 459 (1754); Allan, Fl. New Zealand 1: 300–305 (1961); Skog, Rhodora 74: 242–253 (1972), rev., New World.

Selected Bibliography

Allan, H.H. 1961. Coriariaceae. In: Flora of New Zealand, vol. 1. Wellington: R.E. Owen.

Behnke, H.-D. 1981. Sieve element characters. Nord. J. Bot. 1: 381–400.

Carlquist, S. 1985. Wood anatomy of Coriariaceae: phylogenetic and ecological implications. Syst. Bot. 10: 174–183.

Carlquist, S., Miller, R.B. 2001. Wood anatomy of *Corynocarpus* is consistent with cucurbitalean placement. Syst. Bot. 26: 54–65.

Corner, E.J.H. 1976. The seeds of the dicotyledons. Cambridge: Cambridge University Press.

Engler, A. 1890. Coriariaceae. In: Engler & Prantl, Natürl. Pflanzenfam. III, 5: 128–129.

Garg, M. 1981. Pollen morphology and systematic position of *Coriaria*. Phytomorphology 30: 5–10.

Good, R.D'O. 1930. The geography of the genus *Coriaria*. New Phytol. 29: 170–198.

Gregor, H.-J. 1980. Seeds of the genus *Coriaria* Linné (Coriariaceae) in the European Neogene. Tert. Res. 3: 61–69.

Guédès, M. 1971. Carpel peltation and syncarpy in *Coriaria ruscifolia* L. New Phytol. 70: 213–227.

Hegnauer, R. 1964. Chemotaxonomie der Pflanzen, Bd. 3. Basel: Birkhäuser.

Hegnauer, R. 1989. Chemotaxonomie der Pflanzen, Bd. 8. Basel: Birkhäuser.

Matthews, M.L., Endress, P.K. 2004. Comparative floral structure and systematics in Cucurbitales (Corynocarpaceae, Coriariaceae, Tetramelaceae, Datiscaceae, Begoniaceae, Cucurbitaceae, Anisophylleaceae). Bot. J. Linn. Soc. 145: 129–185.

Mauritzon, J. 1936. Zur Embryologie der Berberidaceen. Acta Horti Gothoburg. 11: 1–18.

Oginuma, K., Nakata, M., Suzuki, M., Tobe, H. 1991. Karyomorphology of *Coriaria* (Coriariaceae): taxonomic implications. Bot. Mag. Tokyo 104: 297–308.

Poole, A.L., Adams, N.M. 1964. Trees and shrubs of New Zealand. Wellington: R.E. Owen.

Praglowski, J. 1970. 1. Coriariaceae. In: Erdtman, G. (ed.) World Pollen Flora. Copenhagen: Munksgaard.

Ridley, H.N. 1930. The dispersal of plants throughout the World. Ashford: Reeve.

Saporta, G. de. 1865. Etudes sur la végétation du Sud-Est de la France à l'époche tertiaire. Ann. Sci. Nat. Bot. V, 4: 5–264, 13 pl.

Sharma, V.K. 1968. Floral morphology, anatomy, and embryology of *Coriaria nepalensis* Wall. with a discussion of the interrelationships of the family Coriariaceae. Phytomorphology 18: 143–153.

Skog, L.E. 1972. The genus *Coriaria* (Coriariaceae) in the Western Hemisphere. Rhodora 74: 242–253.

Susuki, M., Yoda, K. 1986. Comparative wood anatomy of *Coriaria* in East Asia. J. Jap. Bot. 61: 289–296 and 333–342.

Swensen, S.M. 1996.The evolution of actinorhizal symbioses: evidence for multiple origins of the symbiotic association. Am. J. Bot. 83: 1503–1512.

Thomson, P.N., Gornall, R.J. 1995. Breeding systems in *Coriaria* (Coriariaceae). Bot. J. Linn. Soc. 117: 293–304.

Tobe, H., Suzuki, M., Fukuhara, T. 1992. Pericarp anatomy and evolution in *Coriaria* (Coriariaceae). Bot. Mag. Tokyo 105: 289–302.

Weberling, F. 1955. Die Stipularbildungen der Coriariaceae. Flora 142: 629–630.

Yokoyama, J., Suzuki, M., Iwatsuki, K., Hasebe, M. 2000. Molecular phylogeny of *Coriaria*, with special emphasis on the disjunct distribution. Mol. Phylogen. Evol. 14: 11–19.

Zhang, L.-B., Simmons, M.P., Kocyan, A., Renner, S.S. 2006. Phylogeny of the Cucurbitales based on DNA sequences of nine loci from three genomes: implications for morphological and sexual system evolution. Mol. Phylogen. Evol. 39: 305–322.

Corynocarpaceae

Corynocarpaceae Engler (1897), nom. cons.

K. Kubitzki

Evergreen, glabrous trees; branches often in pseudowhorls. Leaves alternate, more or less crowded towards the ends of the shoots, simple, entire, shining; stipules intrapetiolar, subtending the axillary bud, caducous. Inflorescences terminal or sometimes axillary umbellate thyrses. Flowers regular, hermaphrodite or plants gynodioecious, 5-merous, diplostemonous, with a shallow receptacular cup; sepals distinct, unequal, quincuncially arranged, fleshy, persistent, petaloid; petals distinct, cochlear; staminodes 5, antesepalous, small, petal-like, with pointed tip, irregularly dissected, with a large, cushion-like nectary on the ventral base; fertile stamens 5, attached to base of petals; anthers dorsifixed, tetrasporangiate and dithecal, dehiscing lengthwise, introrse; gynoecium (pseudo-?)monomerous, ascidiate; ovary superior; style solitary (occasionally 2 stylodia present); stigma capitate; ovule 1 per locule, pendulous, anatropous, bitegmic, crassinucellar; placentation median, apical. Fruit a drupe. Seed with membranous testa, exalbuminous, very poisonous; the embryo large, straight, containing oil and starch; cotyledons plano-convex. $n = 23$.

A single genus with six spp. native to New Guinea, the Solomon Islands, Vanuatu, New Caledonia, eastern Australia, and New Zealand including Raoul Island and Chatham Islands.

Morphology and Anatomy. For the distinctive "trichomoids" at the nodes shared with *Coriaria*, see under Coriariaceae. Nodes are trilacunar. Stomates are paracytic. The vessel elements are short with simple perforations; the imperforate tracheary elements have simple or faintly bordered pits; wood rays are notably broad, pluriseriate, heterocellular, the axial parenchyma is vasicentric in broad paratracheal strips; the cambium is storied.

The gynoecium of *Corynocarpus* has been viewed either as truly unicarpellate or as pseudomonomerous, i.e. reduced from a bicarpellate condition; in the latter case, the second stylodium that sometimes can be observed (Fig. 20E) is taken as indicative of a second carpel. The ovary is thick-walled and ascidiate; the stigma appears unifacial; the style has a ventral slit that is postgenitally fused in the inner region of the style; the pollen tube transmitting tissue has no contact with the outer surface (Philipson 1987; Matthews and Endress 2004).

Pollen Morphology. Erdtman (1952) characterised the strange pollen grains of *Corynocarpus* as "bilateral (10.5×22.5×16 µm), 2-colporate (?), very flattened (?)". After a careful study of the pollen morphology of *C. laevigata*, Novicke and Skvarla (1983) designated the colpi as meridionally oriented and at right angle to the equator only in order to be able to describe the pollen grains. Under this assumption, the grains are weakly heteropolar, with a rounded and a flattened pole, in equatorial view on the mesocolpium elliptic, in aperture-centred equatorial view subtriangular, 2-colporate, with each colpus consisting of a short, well-defined part on the rounded pole and a long and poorly defined part on the flattened pole, and with an ovoid endoaperture; the exine sculpture psilate and scabrate, the tectum underlain by granular/subcolumellate material and a thick endexine.

Embryology. The ovules are bitegmic, crassinucellate, anatropous, and the micopyle is formed by both integuments (Matthews and Endress 2004). The endosperm is Nucleate (Davis 1966), and the seeds are possibly pachychalazal (Corner 1976).

Karyology. All species, except the tetraploid *C. rupestris*, are diploid, with n=23 (Dawson 1997).

Fig. 20. Corynocarpaceae. *Corynocarpus laevigata.* **A** Flowering branch. **B** Flower bud. **C** Perianth unrolled with androecium, the staminodes with basal glands. **D** Perianth of another flower with only 2 staminodial glands. **E** Distylous flower in longitudinal section. **F** Unripe fruit. **G** Fruit, vertical section. (Engler 1897; drawn by J. Tegetmeyer)

Phytochemistry. Ellagic acid and tannins (but no proanthocyanins) are known from the bark of *C. laevigatus*, the "karaka" tree. Its seeds are toxic for humans and dogs, while pigs and cattle are less affected. The toxic principle karakin is probably a glucose ester of nitropropionic acid (Hegnauer 1964, 1989).

Sexual Dimorphism. *Corynocarpus laevigatus* is gynodioecious, and flowers on male plants produce pollen and each has a fully developed ovule; fruit set on them is low. Female plants have smaller flowers with functional gynoecia and stamens with empty anthers; their fruit set is ample (Garnock-Jones et al. 2007). Other *Corynocarpus* species might also show sexual dimorphism.

Dispersal. The large, orange, fleshy, aromatic drupes (Fig. 20F) suggest dispersal by strong animals. For *C. similis*, there is a report of dispersal by bats (Wheatley 1992), and in New Zealand native pigeon have been observed feeding on the fruit and thus transporting the seeds over short distances before dropping them (Stevenson 1978). Since the fruits are eaten by indigenous people of Vanuato, the Solomon Islands, and New Zealand, these also contribute to localised dispersal.

Fossils. Pollen of *Corynocarpus* from the Quaternary (?) of New Zealand (Mildenhall 1980) and *Corynocarpus laevigatus*-like fruits from the Miocene of New Zealand (Campbell 2002) are the only fossil records of the genus known to me.

Distribution and Distributional History. *Corynocarpus* is distributed in tropical and warm-temperate regions of the south-western Pacific. The most widely distributed species, *C. similis*, ranges from the Bismarck Archipelago, New Ireland, New Britain, and the Solomon Islands to Vanuato. *Corynocarpus dissimilis* is endemic to New Caledonia, and *C. laevigatus* is found in New Zealand including the Chatham and Kermadec Islands. *Corynocarpus cribbianus* grows in New Guinea and north-eastern Queensland, while *C. rupestris* occurs in eastern Australia. In a gene sequence analysis, Wagstaff and Dawson (2000) found *C. similis* in a basal position and hypothesized that from a Palaeotropical centre the more austral regions were reached in two tracks, one extending through New Guinea to Australia (*C. cribbianus*, *C. rupestris*), and another via New Caledonia to New Zealand (*C. dissimilis* and *C. laevigatus*).

Affinities. Gene sequence analyses resolve *Corynocarpus* as member of the Cucurbitales clade, where it is sister to *Coriaria*. For the morphological similarity between the two genera, see under Coriariaceae.

Uses. The prune-like fruits of *C. similis* and *C. laevigata*, which raw are very toxic, used to be eaten by indigenous people for their fleshy mesocarp and the seeds after removal of the toxins. This was accomplished by prolonged steaming and soaking. The trunks of the trees are well known for the glyphs Maori people cut upon them on Chatham Island, and have been also used for making canoes.

Only one genus:

Corynocarpus J.R. Forst. & G. Forst. Fig. 20

Corynocarpus J.R. Forst. & G. Forst., Char. Gen. Pl.: 31, t. 16 (1776).

Description as for family.

Selected Bibliography

Campbell, J.D. 2002. Angiosperm fruit and leaf fossils from Miocene silcrete, Landslip Hill, northern Southland, New Zealand. J. R. Soc. New Zealand 32: 149–154.

Corner, E.J.H. 1976. The seeds of the dicotyledons. Cambridge: Cambridge University Press.

Davis, G.L. 1966. Systematic embryology of angiosperms. New York: Wiley.

Dawson, M.I. 1997. Chromosome numbers in *Corynocarpus* (Corynocarpaceae). New Zealand J. Bot. 35: 255–258.

Engler, A. 1897. Corynocarpaceae. In: Die natürl. Pflanzenfam., Nachtr. u. Register zu Teil II–IV. Leipzig: W. Engelmann.

Erdtman, G. 1952. Pollen morphology and plant taxonomy. Angiosperms. Stockholm: Almqvist and Wiksell.

Garnock-Jones, P.J., Brockie, R.E., FitzJohn, R.G. 2007. Gynodioecy, sexual dimorphism and erratic fruiting in *Corynocarpus laevigatus* (Corynocarpaceae). Austr. J. Bot. 55: 803–808.

Hegnauer, R. 1964. Chemotaxonomie der Pflanzen, Bd. 3. Basel: Birkhäuser.

Hegnauer, R. 1989. Chemotaxonomie der Pflanzen, Bd. 8. Basel: Birkhäuser.

Krause, J. 1942. Corynocarpaceae. In: Engler, A., Prantl, K. (eds.) Die natürl. Pflanzenfam., 2nd edn, 20b. Leipzig: W. Engelmann.

Matthews, M.L., Endress, P.K. 2004. Comparative floral structure and systematics in Cucurbitales (Corynocarpaceae, Coriariaceae, Tetramelaceae, Datiscaceae, Begoniaceae, Cucurbitaceae, Anisophylleaceae). Bot. J. Linn. Soc. 145: 129–185.

Mildenhall, D.C. 1980. New Zealand Late Cretaceous and Cenozoic plant biogeography: a contribution. Palaeogeogr Palaeoclimatol Palaeoecol 31: 197–233.

Narayana, L.L., Satyanarayana, P., Raju, V.S., Prakasa Rao, P.S. 1986. The floral anatomy of *Corynocarpus laevigatus* (Corynocarpaceae). Phytomorphology 36: 325–329.

Novicke, J.W., Skvarla, J.J. 1983. Pollen morphology and the relationships of the Corynocarpaceae. Taxon 32: 176–183.

Patel, R.N. 1975. Wood anatomy of the dicotyledons indigenous to New Zealand. 9, *Corynocarpus*. New Zealand J. Bot. 13: 19–29.

Philipson, W.R. 1987. *Corynocarpus* J.R. and G. Forst. - an isolated genus.Bot. J. Linn. Soc. 95: 9–18.

Stevenson, G. 1978. Botanical evidence linking the New Zealand Maoris with New Caledonia and the New Hebrides. Nature 276: 704–705.

Wagstaff, S.J., Dawson, M.I. 2000. Classification, origin, and patterns of diversification of *Corynocarpus* (Corynocarpaceae) inferred from DNA sequences. Syst. Bot. 25: 134–149.

Wheatley, J.I. 1992. A guide to the common trees of Vanuato. Port Vila, Vanuato: Department of Forestry.

Cucurbitaceae

Cucurbitaceae Durande (1782), nom. cons.

H. Schaefer and S.S. Renner[1]

Tendril-bearing monoecious or dioecious climbers or trailers, rarely without tendrils, herbaceous annual vines or woody perennial lianas, exceptionally trees (*Dendrosicyos*), often with tuberous roots or rootstocks or with leafless and ± succulent stems. Shoots usually angulate, herbaceous, woody, or succulent, with bicollateral vascular bundles. Leaves spiral, estipulate, petiolate, membranaceous or succulent, simple and entire or palmately or pedately lobed, or palmati- or pedati-compound, the margins with small hydathodes (or glandular teeth); tendrils usually one per node. Inflorescences generally racemes, thyrses, panicles, fascicles, rarely spikes or umbels. Flowers with calyx and corolla, unisexual (very rarely bisexual), epigynous, usually pentamerous and actinomorphic, rarely zygomorphic; corolla aestivation valvate, contort, quincuncial; petals (3–)5(–10), entire, 2-lobed or fringed, rarely with a basal scale; nectary formed by mesophyll tissue or hairs; some species with floral oil glands; androecium of 3–5 alternipetalous stamens; thecae 1 or 2, distinct or connate along their filaments and/or anthers; stamens often joined or connate in two pairs, filaments or stamens sometimes connate into a central column; anthers typically basifixed and thecae longitudinally dehiscent, straight or variously bent or folded, rarely forming a horizontal ring; gynoecium with (1–)3(–5) carpels, inferior or semi-inferior; stylodia distinct or connate into a single style; stigma entire, lobulate, or divided; placentation parietal; ovules anatropous, bitegmic, and crassinucellar, horizontal, pendent, oblique, or ascendent. Fruit many-seeded, rarely 1-seeded, usually a soft-shelled or hard-shelled berry, less often a capsule, rarely samaras or achenes. Seeds with an exotestal seed coat, sometimes flattened, winged, or enclosed by mucilaginous tissue, exalbuminous; embryo straight, usually oily, with large, flat cotyledons. Producing oxygenated tetracyclic triterpenoids (cucurbitacins) with bitter taste and purging or abortive effect.

The family contains 97 genera and 940–980 species. Its distribution is essentially tropical and subtropical, with relatively few species reaching the temperate regions of the world. The aerial parts of all species are sensitive to frost.

Vegetative Morphology. Most Cucurbitaceae are annual or perennial herbs with relatively thin roots and shoots. Many perennial species have tuberous roots or pachypodia, and their herbaceous shoots die and re-grow in an annual cycle (hemicryptophytic life form). Pachypodia can reach well over a meter in diameter (e.g., in *Baijiania yunnanensis*). The family also includes a few shrub species (*Acanthosicyos horridus*, Fig. 29, *Corallocarpus glomeruliflorus*, *Momordica spinosa*) and lianas with woody, perennial stems up to 10 cm across (*Alsomitra*, *Bayabusua*, *Coccinia grandis*, *Siolmatra*, *Zanonia*; Carlquist 1992). The sole tree in the family, *Dendrosicyos socotranus* (Fig. 27), results from an extreme case of pachycauly, with stems up to 1 m in diameter (Olson 2003).

Phyllotaxy in Cucurbitaceae is spiral. Only the first leaves following the cotyledons may be opposite or rarely verticillate (*Bayabusua*). The cotyledons are often large, fleshy, ovate to elongate, green, and long-lived. Epigeal germination is the rule, hypogeal germination an exception (Zimmermann 1922; Parfitt et al. 1990). Cucurbitaceae leaves lack stipules and usually have

[1]We thank C. Jeffrey, R. Pozner, R.W.J.M. van der Ham, C.B. Mennes, W.J.J.O. de Wilde, R. Wunderlin, R. Lira Saade, and I. Telford for comments and corrections.

well-developed petioles; sessile or subsessile leaves occur in a few species of *Coccinia*, *Momordica*, *Sicyos*, *Solena*, and *Cephalopentandra ecirrhosa*. The blade is simple or lobed, less often 3-foliolate, rarely up to 9-foliolate (*Momordica enneaphylla*), the venation is palmate. As is typical of climbers, leaf shape can vary strikingly along shoots of the same plant and between individuals growing under different environmental conditions (Jones 1993; Pozner 1998a, b). Probracts (usually foliar structures at the base of the peduncles, see Fig. 30D) occur in some inflorescences, and these bracts often bear extrafloral nectaries (Fig. 31B, H; see also Sexual Strategies, Pollination, and Herbivores). The indumentum of Cucurbitaceae is often pubescent or prickly, with trichomes that have calcified cell walls and cystolith-bearing bases; glandular hairs are also common, as are foliar nectaries and mucilage glands.

Most Cucurbitaceae have tendrils, and this ancestral condition presents a clear morphological synapomorphy for the family; evolutionarily, these tendrils are modified shoots (Kumazawa 1964; Lassnig 1997; Gerrath et al. 2008). Whether tendrils are simple or divided and how they coil is taxonomically useful (Fig. 21A–C illustrates the main types). Tendrils that coil below the tendril branching point are found in the more basal clades (see below), and have traditionally been referred to as zanonioid (from Zanonieae). In a few species, the tendrils have been transformed into thorns (*Acanthosicyos horridus*, Fig. 29, *Citrullus naudinianus*, *Momordica spinosa*) or were lost (*Citrullus ecirrhosus*, *Dendrosicyos socotranus*, *Ecballium elaterium*, *Melothria campestre*, *Cucumis messorius*, *Trochomeria polymorpha*). Sometimes, the tendrils form adhesive pads similar to those of *Parthenocissus* in the Vitaceae (*Alsomitra macrocarpa*, *Bayabusua clarkei*, *Neoalsomitra sarcophylla*, *Polyclathra cucumerina*, *Trichosanthes cucumerina*). The first tendrils usually appear on the 4th to 6th node of a seedling (Zimmermann 1922).

ANATOMY. The wood anatomy of Cucurbitaceae reflects the climbing habit of most species (Carlquist 1992), with a wide vessel diameter compensating for a small transsectional area of secondary xylem. In some groups, parenchyma is abundant, with the vessels sheathed in thick-walled libriform fibers and vasicentric tracheids (Zimmermann 1922; Carlquist 1992). Uniseriate rays are absent. Large primary rays separate perennial vascular bundles, or large secondary rays are initiated within vascular bundles (Schweingruber et al. 2010). Successive cambia occur in *Bryonia* and *Ecballium*, and some genera have medullary phloem and bicollateral vascular bundles (Schweingruber et al. 2010). The phloem is mostly simply structured. Sieve tubes and

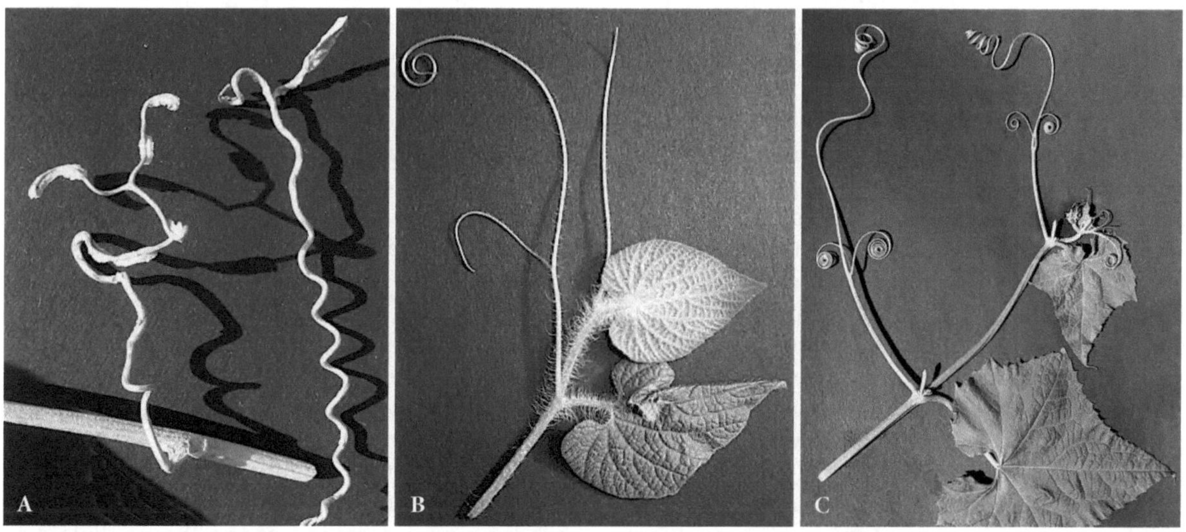

Fig. 21. The main tendrils type of Cucurbitaceae. **A** *Neoalsomitra sarcophylla* (Gomphogyneae, apically 2-fid (=zanonioid) tendril). **B** *Thladiantha dubia* (Thladiantha Clade, simple or 2-fid tendril). **C** *Luffa cylindrica* (Sicyoeae). (Photos G. Hausner)

parenchyma are difficult to distinguish. Raphids occur in the phloem of at least *Ecballium* and *Cucurbita pepo*. Calcium oxalate crystals as well as cystoliths, small calcium carbonate bodies of variable shape, are abundant in *Telfairia* (Okoli and McEuen 1986) and also occur in other genera, especially near the bases of hairs. At least *Cucurbita* contains crystalline silica grains (SiO_2), and in the fossil record, where they are called phytoliths, their shapes can document the presence of particular domesticated forms (Piperno et al. 2002; Piperno and Stothert 2003).

INFLORESCENCES AND FLORAL STRUCTURE. Flowers are generally borne in few- to many-flowered racemes, thyrses, panicles, or fascicles, rarely in spikes or umbels. The family's basic inflorescence type seems to be monotelic, and cases of "open" inflorescences are therefore best described as racemiform, spiciform or umbelliform cymes. About 50% of the c. 960 species are monoecious, and in these species, inflorescences can be cosexual or unisexual. The flowers of Cucurbitaceae are usually unisexual. Bisexual flowers are exceedingly rare. They regularly occur in *Schizopepon bryoniifolius* (Fukuhara and Akimoto 1999; Akimoto et al. 1999) and *Zehneria hermaphrodita* (de Wilde and Duyfjes 2006a; see also Sexual Strategies, Pollination, and Herbivores). Floral symmetry is mostly actinomorphic. Zygomorphy has evolved only in a few species, for example, in *Gerrardanthus* (Fig. 25) and *Xerosicyos*. Male and female perianths are usually similar; occasionally, they differ in size and exceptionally also in shape (*Momordica, Peponopsis*). At about day six of floral development (at least in *Cucumis*), either the stamen primorida or the carpel primordia begin to expand rapidly, while the primorida of the other sex are arrested (Kater et al. 2001). In female flowers, the aborted stamens are visible as staminodes. Analysis of cucumber floral homeotic mutants suggests that the inhibition of stamens or pistils depends on whorl position, not specific sexual organ identity (Kater et al. 2001). The calyx and corolla in Cucurbitaceae are usually pentamerous. Sepal aestivation is valvate or open, exceptionally overlapping (*Luffa*; Pozner 1998a). The corolla consists of ± connate or distinct petals highly variable in size, shape, and consistence, though rather uniform in color (usually white, yellow, or orange).

The hypanthium in Cucurbitaceae derives from the expansion of sepal and petal bases, and their receptacular insertion area (Leins and Galle 1971; Pozner 1993a, b). In male flowers, stamens arise at the bottom of the hypanthium or are inserted at different levels on the hypanthium wall, with the stamen bases then contributing to the hypanthium. These "appendicular" hypanthia may take a wide range of shapes: from flat and patelliform (*Cyclanthera*) to long and tubular (*Ceratosanthes*; Vogel 1981a; Pozner 2004). The hypanthium floor typically bears a nectary, which can be mesenchymal (most genera) or trichomatous (Sicyoeae; Vogel 1981a, 1997; Pozner 2004). In female flowers, the hypanthium includes the ovary wall, producing a "hypanthial gynoecium" (Leins et al. 1972). The gynoecia consist of 1–5 carpels, the 3-carpellate condition being the most common.

A taxonomically useful character is the number of stylodia: Gomphogyneae, Triceratieae, and Zanonieae usually have three (sometimes two or five) distinct stylodia (e.g., Figs. 23F, 24G,25F). The more derived clades have a single style with 2–3(–5) stigmas (e.g., Fig. 30F), which can be enlarged to mimic an androecium, probably to attract pollen-seeking bees (Dukas 1987; Rust et al. 2003). Stigma shape is diverse and taxonomically useful. Ovules are anatropous and bitegmic, their number ranging between 1 to several hundreds. Placentae are typically large, and ovule numbers can be huge (Matthews and Endress 2004). In the more derived clades, the ovules are embedded within individual chambers (ovular or seminal chambers) formed by hypanthial-carpellary-placentary tissue (Pozner 1994). Ovule orientation in the *Actinostemma* Clade, Triceratieae, Gomphogyneae, and Zanonieae (as well as in the derived clade Sicyoeae) is mostly pendent. Ovule orientation in the remaining clades is typically horizontal, horizontal in the upper and middle region of the ovary, and erect at the base (*Cucurbita*) or entirely erect (*Cayaponia*). However, ovule orientation is not known for many groups.

A family-wide evolutionary trend is an increasing fusion of neighboring stamens and an enlargement of the pollen-producing space through sigmoid coiling of the thecae. Five distinct, bithecal stamens may represent the ancestral state, which survives in a few unrelated groups (*Anisosperma* and some *Telfairia*; see Fig. 26C). Five distinct monothecal stamens, as in *Fevillea*, occur only rarely, while androecia with three stamens (two 2-thecous, one 1-thecous, e.g., Fig. 30B), four stamens (via the loss of one), or two stamens (below) evolved repeatedly and are common. Cucurbit flowers initiate five distinct stamen primordia, even those with highly connate stamens (Matthews and Endress 2004, and references therein), and it is assumed that the 3-merous androecia evolved from the fusion of two pairs of monothecal stamens, leaving one monothecal unpaired stamen (Eichler 1875). An analogous process may be observed in some genera with postgenital fusion of stamens, such as *Cucurbitella* (Pozner 1993b, 1994, 1998a), and double vascular bundles in filaments also fit with such an interpretation (*Thladiantha*; Vogel 1990; *Momordica charantia*; Deshpande et al. 1986). Another type of fusion involves the filaments, which may form a central column (in the unrelated genera *Echinopepon*, *Frantzia*, *Gynostemma*, *Hanburia*, *Ibervillea*, *Marah*, *Penelopeia*, and *Sicyos*). Not all filament columns are homologous: some derive from staminal filaments, while others are a receptacle expansion (Imaichi and Okamoto 1992). Striking is the independent evolution of circular anther heads in the unrelated genera *Cyclanthera*, *Cyclantheropsis*, and *Penelopeia*. Stamens of Cucurbitaceae often bear connective trichomes, which may arise in marginal rows, resulting in fringed connectives (*Cucumis*, *Melothria*), or else they may produce sticky secretions that agglutinate pollen grains (Zimmermann 1922; Vogel 1981b). Connective outgrowths are usually apical, and may be taxonomically useful, although this needs critical study (*Cucumis*, *Citrullus*).

Karyology. Chromosome numbers are available for at least 141 species from about half of the 97 genera (Beevy and Kuriachan 1996; Index to Plant Chromosome Numbers, http://mobot.mobot.org/W3T/Search/ipcn.html), mostly those of economic importance. Reported haploid (gametophytic) chromosome numbers range from 7 to 24, with $x = 12$ a prevalent number (Beevy and Kuriachan 1996). In the relatively basal Gomphogyneae, reported numbers for *Gynostemma* are $n = 11, 22, 33, 44, 66, 88$ (Gao et al. 1995), for *Gomphogyne* $n = 16$ (Thakur and Sinha 1973), and for *Hemsleya* $n = 14$ (Samuel et al. 1995). *Actinostemma* has $2n = 16$ (Probatova and Rudyka 1981), and the more derived clades, such as the *Thladiantha*, *Siraitia*, and *Momordica* Clades, have $n = 9$ in *Thladiantha*, $n = 12$ in *Siraitia*, $n = 16$ in *Baijiania* (Li et al. 1993), or $n = 11$ and 14 in *Momordica* (Beevy and Kuriachan 1996). *Telfairia* has $x = 12$ (Okoli 1987). Bryonieae have $x = 9$ (*Ecballium*) or 10 (*Bryonia*) (Volz and Renner 2008). Schizopeponeae have $x = 10$ in *Schizopepon* (Nishikawa 1981) and $x = 11$ in *Herpetospermum* (Thakur and Sinha 1973). In the Sicyoeae (as defined here), *Luffa* has $n = 13$ (Whitaker 1933; Samuel et al. 1995), *Hodgsonia* $x = 9$ (Chen 1993), *Trichosanthes* $x = 11$ or 12 (Beevy and Kuriachan 1996), *Echinopepon* and *Sicyos* $n = 12$ (Ward and Spellenberg 1988; Turala-Szybowska 1990), *Marah* $n = 15$ (Parfitt et al. 1990), and *Cyclanthera* and *Echinocystis* $n = 16$ (Samuel et al. 1995; Gervais et al. 1999). The few counted species of Coniandreae have $n = 13$ (*Corallocarpus*, *Kedrostis*; Beevy and Kuriachan 1996) or $n = 14$ (*Apodanthera*; Ward 1984). Benincaseae may have a base number of $n = 12$, as reported for eight of their genera (Beevy and Kuriachan 1996), but there is also much polyploidy and aneuploidy (Thakur and Sinha 1973; Beevy and Kuriachan 1996). The Cucurbiteae may have fixed polyploidy, with $n = 20$ (*Sicana*; Mercado and Lira Saade 1994; *Cucurbita*; Samuel et al. 1995).

Pollen Morphology. (contributed by C.B. Mennes and R.W.J.M. van der Ham). The morphology of the pollen grains is known for all but one very rare genus (*Tumamoca*). Pollen of Cucurbitaceae is tectate to intectate, and grains are shed as monads, rarely as tetrads (*Borneosicyos*, *Gurania*, *Psiguria*). Pollen grain size can reach up to very large (to 200 μm; some *Cayaponia* and *Polyclathra*; Khunwasi 1998; Barth et al. 2005), but most species have large grains (50–100 μm in diam.). When describing the pollen of individual genera for this treatment, we have applied Erdtman's (1952) subdivision: 10–25 μm = small, 25–50 μm = medium-sized, 50–100 μm = large, 100–200 μm = very large. Pollen of the phylogenetically early-branching *Actinostemma* Clade, the Triceratieae, Gomphogyneae, and Zanonieae is always tricolporate and, different from the more derived clades, usually small to medium-sized, with diameters less than 40 μm. The exine in these clades is usually striate, although *Gerrardanthus* (Zanonieae) has reticulate exines, and *Alsomitra macrocarpa* (Gomphogyneae) a perforate to indistinctly rugulate exine (van der Ham 1999). Nevertheless, striate-reticulate exines are also found in a few genera in more derived clades, such as *Kedrostis* in the Coniandreae and *Peponium*, *Scopellaria*, and *Papuasicyos* (including *Urceodiscus*) in the Benincaseae (Duyfjes et al. 2003). Clades V–XV in Fig. 22 usually have pollen with reticulate or echinate exines, and porate, colporate or colpate apertures. Thus, Coniandreae (including *Bambekea* and *Eureiandra*) and Benincaseae tend to have reticulate 3-colporate pollen, and Cucurbiteae echinate 3-porate to periporate pollen. An African member of Benincaseae with unusual 6-aperturate pollen is *Zehneria peneyana* (van der Ham and Pruesapan 2006). Echinate exines also occur in *Benincasa* (including *Praecitrullus*) and *Diplocyclos* (Benincaseae). Other examples of parallel evolution are the pollen tetrads in *Gurania* and *Psiguria* (Coniandreae) and in *Borneosicyos* (Benincaseae; van der Ham and van Heuven 2003). Another striking case of morphological similarity (or parallel evolution) is the intectate, gemmate pollen with 3 operculate pori that is found in the Asian Schizopeponeae and in the African *Cephalopentandra ecirrhosa* (Benincaseae) (van der Ham and Mennes, unpubl. data). Pollen of Sicyoeae is rather heterogeneous, as expected in an old and species-rich group. Thus, *Trichosanthes* has 3 (4)-colporate to 3(4)-porate pollen (Pruesapan and van der Ham 2005), and the New World *Linnaeosicyos* and Sicyoeae have 4- to 16-colporate or -colpate pollen (Stafford and Sutton 1994; Schaefer et al. 2008a). Such New World sicyoid pollen is also known as *Hexacolpites echinatus* from the Oligocene of Cameroon (Salard-Cheboldaeff 1978; Muller 1985).

Embryology. Important aspects of floral structure and embryology are unknown for most Cucurbitaceae, especially the early-branching lineages (Matthews and Endress 2004). In all species studied, ovules are anatropous, bitegmic, and crassinucellar (Johri et al. 1992; Matthews and Endress 2004). Typically, ovules of Cucurbitaceae develop a nucellar beak: a more or less cylindrical protuberance of the nucellus that fills the micropylar channel, and in some cases also contacts the epidermis of the ovular chamber (Pozner 1993a). The outer integument has a vascular bundle running from the funiculus around the chalaza and reaching the opposite, distal end of the integument. The outer integument, and particularly its outer epidermis, forms the testa (resulting in the family's characteristic exotestal seed coat; Johri et al. 1992). The inner integument is typically 2-layered (but thicker around the micropyle) and disintegrates by the time of fertilization. Megaspores are arranged in a linear tetrad (Johri et al. 1992), and megaspore selection is usually by postmeiotic competition (Pozner 1994). As far as known, embryo sacs are of the Polygonum type, except in *Benincasa hispida*, which can have a Polygonum type or an Allium type embryo sac (Chopra and Basu 1965). The endosperm is of the nuclear type (Johri et al. 1992). Shortly after fertilization, the lower part of the embryo sac of many (most?) cucurbit genera forms a chalazal endosperm haustorium, which can be coenocytic or become cellular (Chopra 1955; Chopra and Basu 1965; Chopra and Seth 1977). The haustorium functions until the growing embryo reaches the heart-shaped stage and then disorganizes or is pressed to the base of the endosperm. In the studied genera, the endosperm is completely consumed during embryo development and the mature seeds are non-endospermic (Chopra 1955). Embryo development follows the onagrad, asterad, or solanad type (Johri et al. 1992). Embryos are straight and have flat cotyledons.

The anthers are bi- or tetrasporangiate, and in *Cucumis sativus* and *Echinocystis lobata* both conditions may occur in the same flower (Davis 1966). Anther wall development corresponds to Davis's basic type (Davis 1966; Johri et al. 1992), although the pattern of cell layer segregation is unstable (Pozner 1993b). The endothecium develops fibrous thickenings, and anther dehiscence has some variations particularly in species with tightly folded thecae (Pozner 1993b). The 1–3 middle layers are ephemeral, and the tapetum is glandular, with uni- to multinucleate cells. The microspore mother cells after meiosis undergo simultaneous cytokinensis, and the microspore tetrads are tetrahedral. Pollen grains are 2- or

3-celled when shed (Johri et al. 1992). The pollen tube is persistent within the nucellar beak after fertilization, and in some species it may grow as a swollen structure the function of which is unclear (Pozner 1993a).

FRUITS AND SEEDS. The morphology of cucurbit fruits and seeds is highly variable and often useful for identifying genera. Fruits are typically many-seeded, the ancestral condition in the clade formed by Begoniaceae, Cucurbitaceae, Datiscaceae, and Tetramelaceae (Zhang et al. 2006). One-seeded fruits evolved in *Hodgsonia* (1–3 seeds per pyrene), *Sicyos*, and *Sicydium*. Berries are the most common fruit type, and they can be hard-shelled, then called gourd or pepo (*Citrullus, Cucumis, Cucurbita*), or leathery with a fibrous mesocarp (*Sicyos*). Especially the commercially important species often have hard-shelled berries that can reach huge dimensions (to 1 m diameter in *Cucurbita pepo*). In the seasonally dry habitats, where most of these species occur, hard-shelled water-storing fruits allow for prolonged protected seed maturation, which continues even after the remainder of the vegetative shoot has mostly dried out and died off. Dehiscent berries that expose seeds surrounded by a showy, fleshy, arilloid jacket characterize *Momordica*, and explosive fruits *Ecballium* and *Hanburia*. In the latter, the seeds are ejected while the fruit stays on the plant, whereas in *Ecballium* and some species of *Cucumis* the mature fruits separate from the peduncle and eject the seeds by elastic contraction. Capsules are less common (e.g., Figs. 23G, 25G), and they may open apically by 3-radiate slits (*Gerrardanthus, Siolmatra*), or the upper part may fall off to release the seeds (pyxidium; *Actinostemma, Echinopepon, Luffa*). Achenes are found in *Sicydium*, samaras in *Cyclantheropsis, Pteropepon*, and *Pseudosicydium*, and geocarpic fruits evolved independently in a few species of *Echinopepon, Kedrostis*, and *Cucumis*. Vivipary occurs in chayote, *Sicyos edule*, when the testa does not differentiate sclerenchymatous layers, and the epidermis of both the testa and cotyledons differentiate as haustorial epitelium (Giusti et al. 1978).

Seed morphology is extremely variable, and a few seed shapes are unique to Cucurbitaceae, which may permit the assignment of fossil seeds to particular genera (cf. Fossils and Biogeography). Seeds of fleshy fruits may be globose, ovoid, pyriform (*Halosicyos*), falcate (*Abobra*), compressed (most cases), or even winged (*Cyclanthera* p.p.), usually surrounded by an arilloid jacket derived from the closest carpellary tissue (the ovular or seminal chamber) around the ovule. That arilloid jacket is usually fleshy, hyaline (*Cucumis*), green (*Cucurbitella*), yellow or red (*Momordica*) and sticky with mucilage, which contributes to seed dispersal by adhesion (see Dispersal). Seeds from dry, dehiscent fruits do not have an arilloid jacket. They may be more or less globose (*Echinopepon* p.p.), compressed (*Luffa*), or frequently winged. Seeds with wings predominate in Triceratieae, Gomphogyneae, and Zanonieae. The wings can be huge and unilateral (*Gerrardanthus, Neoalsomitra, Zanonia*), bilateral (*Siolmatra*), or peripheral (*Alsomitra*). The testa can be smooth, tuberculate, or scrobiculate, and it can bear spongy outgrowth (*Apodanthera*) or hairs (some species of *Cucumis, Melothria, Indomelothria, Tecunumania*, and *Zehneria*). Seeds of *Ecballium* contain mucilage in their testa cells that rapidly hydrates, surrounding seeds with a jelly coat. The testa can also be thin and delicate, especially in the 1-seeded undehiscent fruits (*Pteropepon*), or it can be hard and highly lignified (*Cayaponia*). The tegmen is always thin and delicate (see Embryology).

SEXUAL STRATEGIES, POLLINATION, AND HERBIVORES. Throughout the evolution of Cucurbitaceae, there have been numerous changes between dioecy and monoecy (Roy and Saran 1990; Zhang et al. 2006; Kocyan et al. 2007; Volz and Renner 2008; Schaefer and Renner 2010a), and the phylogenetic distribution of monoecy and dioecy on the family phylogeny suggests that dioecy may be the ancestral condition. A cucurbit, *Bryonia dioica*, was the first experimental system for the genetic analysis of the inheritance of sex in any organism (Correns 1903, 1907; Rheinberger 2000), and it was from the sex ratios of the offspring from reciprocal pollinations between this species and the monoecious *B. alba* that Correns inferred that half the pollen grains of *B. dioica* must carry a "female tendency," the other half a "male tendency." Correns's results were confirmed in a series of later studies that also inferred XY sex determination in *Bryonia*, with the male the heterogametic sex. While the chromosomes of *Bryonia* are not morphologically differentiated, those of *Coccinia grandis* are. Male individuals of this species have a pair of different-sized chromosomes, with one, interpreted as the Y-chromosome, 2.5-times longer than its homolog and all autosomes (Bhaduri and Bose 1947; N. Holstein and S. Renner, pers. obs.). Very few Cucurbitaceae have functional bisexual flowers, and these may occur only in some populations (see Inflorescences and Floral Structure). Deviations from pure monoecy (every individual with functional male and female flowers) and pure dioecy (every individual either male or female) have been reported (Morimoto et al. 2004), but there is surprisingly little fieldwork on the role of such deviations (Schaefer and Renner 2010a). In the cultivated species, especially of *Cucumis*, hormone application has permitted the planting of large fields of female (gynoecious) individuals of *C. sativus*. All species of Cucurbitaceae that have been investigated in this regard are self-compatible, fitting with little heterosis being known in the family (Gusmini and Wehner 2008; cf. Phytochemistry and Economic Importance).

Pollen of Cucurbitaceae is usually covered with a thick layer of oily yellow to orange-colored pollenkitt (Zimmermann 1922; Vasil 1960), and pollen-foraging bees are the predominant pollinators of Cucurbitaceae, with several clades more or less oligolectic on cucurbits (below). Cucurbit flowers usually open early in the morning, and anthers usually dehisce hours before the flowers open; evening or nocturnal flowering are less common, but occur in bat- and hawkmoth-pollinated species.

About 86 species of *Momordica*, *Thladiantha*, and a few other genera have oil-secreting trichomes on the petal bases, and are pollinated by specialized oilbees of the genus *Ctenoplectra* (Vogel 1990; H. Schaefer and S. Renner, unpubl. data); over the course of its evolution, *Ctenoplectra* has broadened its host spectrum from *Momordica* to the unrelated clades *Thladiantha*, *Siraitia*, and *Telfairia*. The squash bees *Xenoglossa* and *Peponapis* are specialized on the extremely coarse pollen of *Cucurbita*, and *Andrena florea* depends on the pollen and nectar of *Bryonia*. Hawkmoth pollination evolved independently in several genera that all have pale, usually fragrant flowers opening at night, often with nectar in elongated receptacle-tubes: *Cayaponia*, *Dieterlea fusiformis*, *Hodgsonia*, some *Lagenaria*, some *Momordica* species, *Peponium*, *Selysia*, *Trichosanthes*, *Tricyclandra*, and *Trochomeria*. The conspicuously fringed petals of several of these species (*Hodgsonia*, *Linnaeosicyos*, *Telfairia*, *Tricyclandra*, most *Trichosanthes*) likely are an adaptation to moth pollination (Vogel 1954; Endress and Matthews 2004). For *Lagenaria siceraria*, pollination by the sphingids *Agrius convolvuli* and *Hippotion celerio* has been confirmed by field observations in Kenya (Morimoto et al. 2004). The Indonesian cucurbit *Bayabusua* with purple-red, hairy, fleshy flowers of unknown scent may be adapted to pollination by flies. Bird pollination is characteristic for *Gurania* and *Psiguria*; their bright orange to red flowers, often in dense heads or umbels, attract straight-billed hummingbirds. Species in these genera are also visited by butterflies of the genus *Heliconius* that depend on their pollen for protein (Murawski and Gilbert 1986). Bat pollination has been reported for species in the South American genera *Calycophysum*, *Cayaponia*, and *Cionosicys* (Vogel 1958, 1969; Sazima et al. 1996), which form a clade, and a phylogeny of *Cayaponia* implies that bat pollination may be ancestral in this genus, with bee pollination evolving repeatedly as species entered more open habitats (Duchen and Renner 2010).

Cucurbits often attract nectary-tending ants with extrafloral nectaries on bracts, petioles, leaf bases, or flower buds (Zimmermann 1922; Okoli and Onofeghara 1984; Ilyas 1992; Agarwal and Rastogi 2008). Other insects feed on cucurbit shoots, leaves, and flowers. From Tanzania, Zimmermann (1922) lists species of *Orthezia* and *Helopeltis* (Hemiptera); *Epilachna* species (Coccinellidae), chrysomelid beetles, and gall-inducing Curculionidae. Several species of the ladybird beetle genus *Epilachna* (Coccinellidae) also specialize on cucurbits, with larvae and adults both feeding on the leaves of their hosts: *E. borealis*, the squash ladybird beetle, feeds on *Cucurbita*, *E. argus*, the bryony ladybird beetle, mainly on *Bryonia* and *Ecballium*, and *E. chrysomelina*, the melon ladybird beetle, mainly on *Citrullus lanatus*. Cucumber beetles or rootworms (Chrysomelidae, Luperini) feed on leaves and pollen of several cucurbit genera, and the larvae of some species feed on cucurbit roots (Metcalfe 1986; Gillespie et al. 2003). Larvae of the melon fly *Bactrocera cucurbitae* (Diptera, Tephritidae), a native of India, Southeast Asia, New Guinea, and Australia (introduced in Hawaii, Egypt, Kenya, and Tanzania), develop in the fruits and fleshy stems of many species, causing considerable economic damage (Heppner 1989). Whiteflies (Aleyrodidae) feed on most species of the family and can be a big problem in cultivation. The most common whitefly species on Cucurbitaceae seem to be the polyphagous *Aleurodicus dispersus*, *Bemisia tabaci*, and *Trialeurodes vaporariorum* (Evans 2007). In the neotropics, larvae of *Blepharoneura*, a genus of tephritid fruit flies, feed within the flowers or fruits of Cucurbitaceae, especially *Gurania* and *Psiguria* (Condon et al. 2008). Snails and slugs also feed on many cultivated cucurbit species.

Dispersal. Dispersal is mostly by animals, more rarely by wind or gravity (ballistic). Wind dispersal of fruits or seeds occurs in *Alsomitra*, *Bayabusua*, *Neoalsomitra*, *Pseudosicydium*, *Pteropepon*, *Siolmatra*, *Zanonia*, and *Sicyos*. Typical bird fruits, i.e., red small fleshy berries, are common in *Bryonia*, *Cucumis*, *Diplocyclos*, and *Zehneria*. Larger fruits may be swallowed entirely by large birds, such as ostriches, emus, and bustards (*Cucumis*, *Austrobryonia*). Others are picked open by birds that feed on the pulp, sometimes also the seeds, e.g., *Dieterlea* (Lott 1986). Large seabirds, such as albatrosses, shearwaters and storm petrels, nest in habitats were *Sicyos* occurs and probably disperse the seeds, which can be glandular sticky or ornamented with retrorse barbs. The pigeon *Zenaida maculata* eats, and occasionally disperses, the seeds of the *Cayaponia* species that occurs on the island of Fernando de Noronha (Ridley 1930), and similar occasional dispersal on or in birds likely explains the presence of cucurbits on other ocean islands. Mammals are also important dispersal agents of Cucurbitaceae, although they probably destroy most seeds. Spider monkeys (*Ateles*) feed on *Cayaponia* (Link and Di Fiore 2006), and rodents burry and disperse *Marah* seeds (Borchert 2004). The Maned wolf (*Chrysocyon brachyurus*) in Goias, Brazil, occasionally feeds on *Cayaponia espelina* fruits (Rodrigues et al. 2007). The geocarpic fruits of *Cucumis humifructus* are dug out and eaten by the aardvark (*Orycteropus afer*), which apparently also disperses the seeds (Meeuse 1962). Fruit bats and flying foxes feed on, and disperse, species of *Guriana*, *Coccinia*, and probably quite a few other cucurbits (Condon and Gilbert 1988; Medellín and Gaona 1999; Elangovan et al. 2001). Herbivorous fishes in Suriname feed on the fruits of *Cayaponia cruegeri* (herbarium specimen label N.M. Heyde 469, Herbarium Utrecht), and there is evidence of *C. cruegeri* plants occurring on floating mats in the Suriname River, already close to the Atlantic ocean (herbarium specimen label J. van Donselaar 3854, Herbarium Utrecht). In *Luffa* and *Cayaponia*, the fruit veins persist as a spongy web enclosed in the papery or leathery exocarp, which enables the fruit to float for many days in fresh or salty water (Ridley 1930). Other genera (*Hodgsonia*, *Fevillea*, *Sicana*) evolved large fleshy buoyant fruits apparently adapted to water dispersal. These observations fit well with the inferred transoceanic dispersal of several Cucurbitaceae between South America to Africa, and between the Malesian region or India and Madagascar and Africa (Schaefer et al. 2009).

Phytochemistry and Economic Importance. Probably the most characteristic chemicals are cucurbitacins, saponins, especially triterpenesaponins, and non-proteinogenic aminoacids (Hegnauer 1964, 1989). Cucurbitacins are a group of bitter triterpenes confined mainly to the seeds of Cucurbitaceae (Chen et al. 2005). Biologically, they are effective herbivore deterrents, although certain chrysomelid beetles are adapted to, and even require, these substances (e.g., Metcalfe 1986; Gillespie et al. 2003). Cucurbitacins are effective in slowing or stopping cell division, and there is much research on their medical uses, with hundreds of papers just in the past few years. Cucurbits that have been studied usually also contain saponins, e.g., *Bryonia dioica* (Oobayashi et al. 1992), *Gynostemma pentaphylla*, *Hemsleya chinensis*, and many others. The cucurbitane-type triterpene glycoside constituents of various *Siraitia* (especially *S. grosvenorii*) are the source of plant-derived sweeteners, which may become commercially important in the future.

Numerous species of Cucurbitaceae have economic importance, usually as vegetables. The cucurbit crops that are grown most commonly are cucumber, melon, and watermelon. The number of tons of pickling cucumbers, fresh-market cucumbers, muskmelons, honeydew melons, and watermelons produced in the U.S. alone is staggering (see U.S. Department of Agriculture, National Agricultural Statistics Service online). However, since cucumber, melon, and watermelon show almost no heterosis effects (Gusmini and Wehner 2008), increases in yield in the past 55 years are probably mostly due to improved agricultural practices, and less so to breeding of F1 hybrids. Indeed, yields in processing cucumber seem to have reached a plateau (Gusmini and Wehner 2008), and breeders now place their hopes in fully sequenced genomes, such as that of a Chinese fresh-market type of *C. sativus* (Huang et al. 2009), a North American pickling type, and an isogenic gynoecious breeding line. More locally important Cucurbitaceae crops include squash or pumpkin (*Cucurbita maxima*), zucchini (*C. pepo* subsp. *pepo*), bitter gourd (*Momordica charantia*), waxgourd (*Benincasa hispida*), chayote (*Sicyos* (*Sechium*) *edule*), casa banana or casbanan (*Sicana odorifera*), and *Coccinia grandis*. Medicinally used species are *Cayaponia racemosa*, *Fevillea cordifolia*, and species of *Bryonia* (see under these genera). *Luffa* are also used as sponges (see under that genus), and *Lagenaria siceraria*, the bottle gourd, as vessels in African and Asian cultures (Whitaker and Davis 1962; Erickson et al. 2005; Clarke et al. 2006). Finally, the leaves and shoots of many species are boiled and eaten as a vegetable in both Africa and Asia (e.g., Okoli 1984), and many species play a role in folk medicine.

Relationships to Other Cucurbitales and Within-family Relationships. Molecular data place the Cucurbitaceae in a polytomy with Begoniaceae, Datiscaceae, and Tetramelaceae (Zhang et al. 2006), a clade supported by shared inferior ovaries and parietal placentation. The precise family relationships at this writing (2010) are unresolved. Their tendrils readily distinguish Cucurbitaceae from their closest relatives, and the family's monophyly is well supported by molecular data (Zhang et al. 2006; Kocyan et al. 2007; Schaefer et al. 2009). Molecular phylogenies that include all genera (Fig. 22; except *Khmeriosicyos*) reveal five well-supported clades, namely, (1) a group of five to six Asian genera including *Alsomitra*, *Bayabusua*, and *Neoalsomitra*, which corresponds to tribe Gomphogyneae of Bentham & Hooker (1867); (2) a group of one African genus and five Neotropical genera, including *Fevillea* and *Sicydium*, which corresponds to tribe Triceratieae of A. Rich. (1845); (3) a group of four or five genera from Madagascar, continental Africa, Asia, and South America, corresponding to tribe Zanonieae of Bentham and Hooker (1867); (4) a clade consisting of the Asian *Actinostemma*; and (5) a group of c. 100 genera comprising our clades V–XV, traditionally ranked as subfamily Cucurbitoideae of Kosteletzky (1833). Earlier classification schemes for the family often allocated our clades II–V to a subfamily variously called Zanonioideae (Benth. & Hook.f.) Luerss. or Nhandiroboideae (Kosteletzky 1833; Jeffrey 1980, 1990, 2005). However, Nhandiroboideae is an illegitimate name (see under the genus *Fevillea*), and Zanonioideae (Benth. & Hook.f.) Luerss. is a taxonomic synonym of Fevilleoideae Burnett (the latter validly published by Burnett 1835, p. 756). Neither morphological nor molecular data support the division of Cucurbitaceae into more or less equivalent, large clades that might be ranked as subfamilies, and we are therefore not using this rank. The morphological characters thought to distinguish Zanonioideae from Cucurbitoideae, namely, striate pollen, winged seeds, and "zanonioid" tendrils (tendrils in which the lower section is capable of curving), all occur also in other clades of Cucurbitaceae, and molecular phylogenies clearly show that Zanonioideae were a paraphyletic assembly.

Fossils and Biogeography. The fossil record of Cucurbitaceae and indeed of the order Cucurbitales is sparse (Zhang et al. 2006 give a brief review). The oldest fossils are seeds from the Uppermost Paleocene and Lower Eocene London Clay (65 Ma) that, based on their shape and testa morphology, represent Cucurbitaceae (Chandler 1964; Collinson et al. 1993). The earliest pollen of Cucurbitaceae is *Hexacolpites echinatus* from the Oligocene of Cameroon (Salard-Cheboldaeff 1978; accepted by Muller 1985); these grains under the light microscope are hexacolpate or stephanocolpate, and resemble polycolpate pollen of New World Sicyoeae. Leaves from the North American Paleocene, described as *Vitis lobata* (Knowlton) Brown and mentioned as possibly

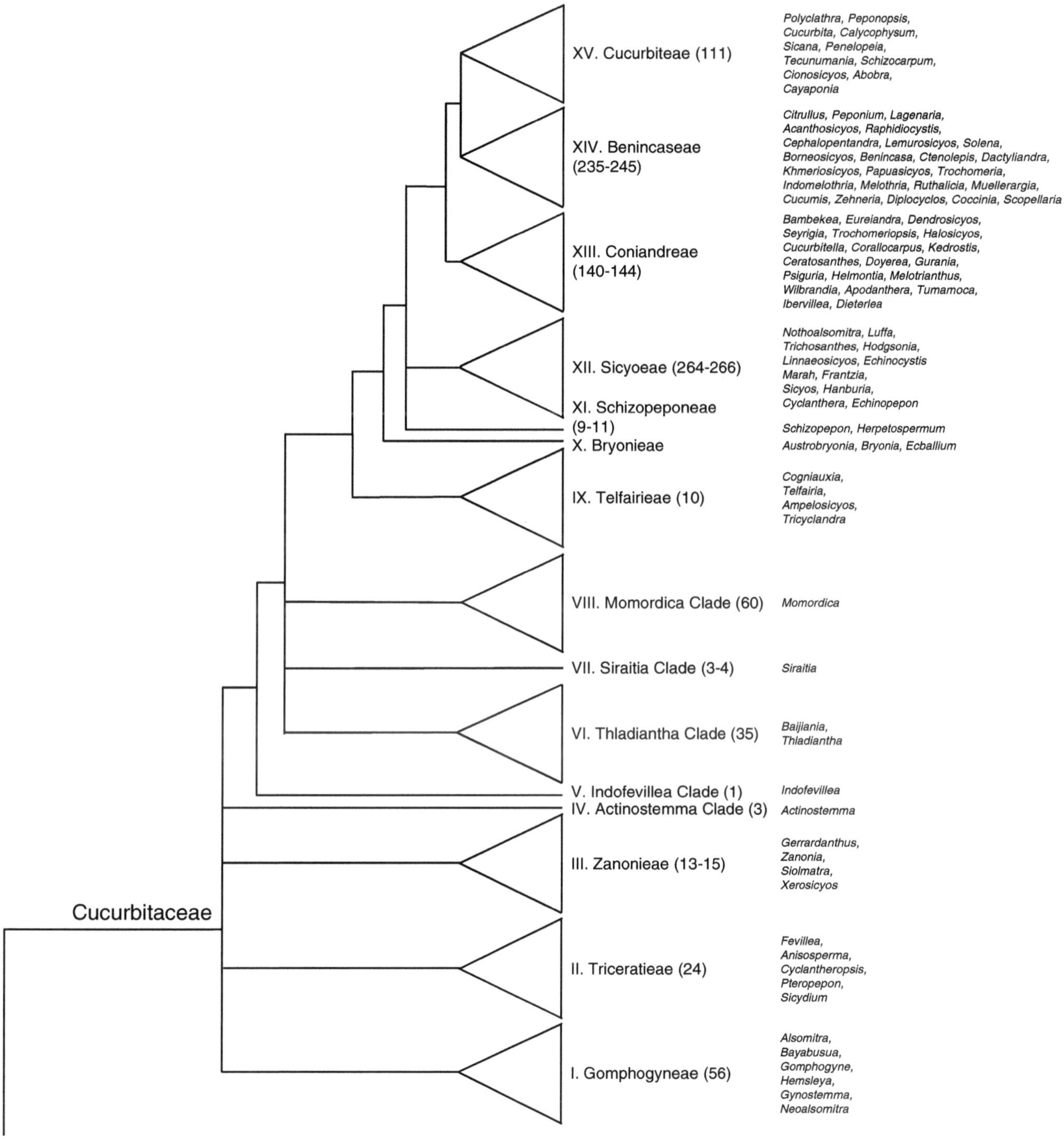

Fig. 22. Phylogenetic relationships among the genera and tribes of Cucurbitaceae as resolved by chloroplast and nuclear data (Kocyan et al. 2007; Schaefer et al. 2009). In parentheses, the respective species numbers. Formal names of all clades are proposed in Schaefer and Renner (in press)

cucurbitaceous in Raven and Axelrod (1974), probably represent Vitaceae, not Cucurbitaceae (R. Burnham and S. Renner, based on images of the type material, August 2005). Seeds of various species of *Cucurbitospermum* have been described from the Early Miocene (17.8 Ma) sites of Rusinga Island in Lake Victoria, Kenya (Chesters 1957; Collinson et al. 2009). *Bryonia*-like seeds from fossil beds at Tambov, Western Siberia (Dorofeev 1963, 1988) date to the Lower Sarmat, 15–13 Ma ago.

Subfossil records of *Cucurbita pepo* have been dated to 8,000–7,000 B.C. at Guilá Naquitz, and to about 7,000–6,500 B.C. at Ocampo Cave, Tamaulipas (Smith 1997), those of *C. moschata* in the northern Peruvian Andes to up to 9,200 B.P. (Dillehay et al. 2007), and phytoliths from Early Holocene domesticated *Cucurbita* are known from Southwest Ecuador (Piperno and Stothert 2003). *Lagenaria siceraria* rind fragments from Mesoamerican archaeological deposits have been radiocarbon-dated to 10,000 B.P., indicating that the bottle gourd was present in the Americas as a domesticated plant by that time (Erickson et al. 2005).

Based on outgroup comparison, Cucurbitaceae originated in Asia sometime in the Late Cretaceous (Schaefer et al. 2009). The five deepest evolutionary divergences in the family all date to the Late Cretaceous, 70–80 Ma. Two of these ancient clades (the Gomphogyneae and the *Actinostemma* Clade) are now almost restricted to Asia. A third, the Triceratieae, is mainly Neotropical, except for a small African presence, *Cyclantheropsis*, with two species in Africa and one in Madagascar. The ancestors of the Triceratieae probably were more widely distributed in the Laurasian tropics and reached the American continent by dispersing across a still narrow Atlantic. *Cyclantheropsis* most likely results from a back dispersal from South America to Africa in the middle Eocene. The ancestors of the fourth early-diverging clade, the Zanonieae, apparently reached the African continent early, and from there dispersed to Madagascar (the early Eocene *Xerosicyos* lineage). Later, in the Oligocene, at least two long-distance dispersal events brought the *Siolmatra* lineage to America, and the *Zanonia* lineage back to tropical Asia. Clades V–XV, finally, diversified partly in Asia (e.g., *Thladiantha*, *Siraitia*, *Trichosanthes*), partly in Africa (e.g., *Momordica*, *Cucumis*, *Coccinia*, *Kedrostis*). Other examples of transoceanic dispersal are known from *Cayaponia* and *Luffa* (see under those genera). Dispersal from Africa to Asia occurred in *Coccinia*, *Corallocarpus*, *Kedrostis*, and *Momordica*.

The native European cucurbit flora belongs to a single clade, Bryonieae, comprising *Bryonia*, with 10 species, and its monotypic sister *Ecballium*. The sister group of both is the Australian genus *Austrobryonia*, with the split between the two clades dating to 36 (50–24) million years ago (Schaefer et al. 2008b). The remaining cucurbit species occurring in Europe are the result of recent introductions (*Echinocystis lobata*, *Sicyos angulatus*, *Thladiantha dubia*), or casual escapes from cultivation (*Citrullus lanatus*, *Cucumis melo*, *C. sativus*, *Cucurbita pepo*). The native African Cucurbitaceae, most of which belong to clades V–XV, evolved from five successful dispersals from Asia to Africa, and two from America to Africa (in *Melothria* and *Cayaponia*). The famous cucumber tree, *Dendrosicyos socotranus*, endemic on Socotra some 350 km from the Arabian peninsula, diverged from its closest relative 34 (47–22) Ma, while the Socotra archipelago is only some 10 million years old. *Dendrosicyos* therefore seems to be an island relict of an old lineage of Coniandreae that went extinct on the mainland. Madagascar has 50 native species Cucurbitaceae that are currently classified in 16 genera. Based on molecular sequence data, it appears that this diversity evolved from 13 ancestral lines that reached Madagascar from the African mainland. Using Madagascar as a steppingstone, one of these clades, *Peponium*, later reached the Seychelles (Schaefer et al. 2009).

South America has about 360 species of Cucurbitaceae that descend from just a few transoceanic dispersal events, mostly from Africa to South America. These events involved the ancestors of the Cucurbiteae, lineages of the Sicyoeae, part of the Coniandreae, and *Melothria*, *Lagenaria*, and *Luffa* (see under these genera). For *Melothria*, it appears that its ancestors came across the Pacific, since the sister group of *Melothria*, *Indomelothria*, is endemic in Southeast Asia. North American cucurbits descend from seven expansions of Central and South American lineages that occurred at widely different times (Schaefer et al. 2009). The indigenous Australian Cucurbitaceae flora, finally, consists of 30 species in 12 genera of which two are endemic, *Nothoalsomitra*, a single liana species of Queensland's humid rainforests, and *Austrobryonia*, four species of trailers or creepers in the dry regions of (mostly) Central Australia (Schaefer et al. 2008b).

Key to the Genera

1. Tendrils absent 2
- Tendrils present 14
2. Trees or (sub)shrubs or erect herbs, not climbing or trailing 3
- Herbaceous trailers or creepers 11
3. Trees to 6 m tall with large trunks (to 1 m across). Socotra **45. *Dendrosicyos***
- Shrubs or subshrubs or erect herbs 4
4. Shrubs or subshrubs 5
- Erect or prostrate herbs 7
5. Plants not spiny. Africa, Madagascar, and Asia **50. *Corallocarpus***
- Plants spiny 6
6. Tendrils transformed into c. 1 cm long, straight spines; leaves reduced to small scale-like, ovate, c. 2 mm long bracts; thecae flexuous. Southern Africa **66. *Acanthosicyos***
- On older stems, the bases of the tendrils thickened and transformed into a pair of straight or curved, rather blunt, 0.4–3 cm long spines; leaves well-developed, petiolate, 2–7 by 2–5 cm, broadly ovate; thecae curved. East and Northeast Africa **21. *Momordica*** (*M. spinosa*, *M. macrocarpa*)
7. Fruit a small subglobose berry, to 2 cm long 8
- Fruit a large gourd-like pepo, >2 cm long, ripening green or yellow 9
8. Receptacle-tube cylindrical, 10–18 mm long with conical nectary; leaves entire or 3-lobed. South and East Africa **77. *Trochomeria*** (*T. polymorpha*)
- Receptacle-tube shallowly saucer-shaped, 0.5–1 mm long; leaves deeply palmately dissected. East Africa **82. *Cucumis*** (*C. messorius*)
9. Ripe fruit expelling seeds explosively. Mediterranean region and North Africa **28. *Ecballium***
- Fruit indehiscent 10
10. Fruit <10 cm; anthers distinct. Africa **82. *Cucumis*** (*C. canoxyi*, *C. reticulatus*, *C. rigidus*)
- Fruit >10 cm; anthers connate into a central head. America, introduced in Africa, Europe, Asia, Australia **89. *Cucurbita*** (*C. pepo* cultivars)
11. Plant spiny. Southern Africa **63. *Citrullus*** (*C. naudinianus*)
- Plant not spiny 12
12. Plant with long underground branches and subterranean fruits. Southern Africa **51. *Kedrostis*** (*K. psammophila*)
- Plant creeping and fruiting above ground 13
13. Thecae flexuous, glabrous. Southern Africa **63. *Citrullus*** (*C. ecirrhosus*)
- Thecae straight, fringed with hairs. Brazil **79. *Melothria*** (*M. campestre*)
14. Tendrils in groups of 5–8 per node, simple; fruit geocarpic, maturing below ground. Southern Africa **82. *Cucumis*** (*C. humifructus*)
- Tendrils solitary (1 or rarely 2 per node), simple or 2–8-fid with a basal, unbranched part; fruit usually maturing above ground (but see 45. *Echinopepon*) 15
15. Tendrils simple, not 2-fid or multi-fid with basal, unbranched part 16
- Tendrils 2–8-fid (often only at apex, which might be lost in herbarium material) 107
16. Thecae fringed with hairs 17
- Thecae glabrous (or hairs minute) 20
17. Stamens 3; anthers all 2-thecous. Africa, Asia, Australia, and Pacific Islands **83. *Zehneria***
- Stamens 3; two anthers 2-thecous, one 1-thecous18
18. Testa covered by long appressed hairs. South and Central America (naturalized in Asia) **79. *Melothria***
- Testa glabrous 19
19. Fruit small, gourd-like, up to 2.5 cm long, ornamented with long, soft bristles. Madagascar, Indonesia, Northeastern Australia **81. *Muellerargia***
- Fruit a globose to ellipsoid, up to 20 cm long, smooth pepo. Tropical Africa and South/Central America **79. *Melothria***
20. Ovules few, pendent; pollen small to medium-sized, striate 21
- Ovules many, horizontal; pollen mostly medium-sized to large, reticulate, perforate, gemmate or echinate/baculate, very rarely striate-reticulate (*Dactyliandra*, *Papuasicyos*) 23
21. Fruit indehiscent, globose. South America **11. *Sicydium***
- Fruit dehiscent. Asia 22
22. Fruit dehiscing into 3 valves; seeds winged **13. *Zanonia***
- Fruit operculate; seeds winged or unwinged **16. *Actinostemma***
23. Petals fringed 24
- Petals not fringed 27
24. Thecae circular. Madagascar **25. *Tricyclandra***
- Thecae straight or folded, not circular 25
25. Stamens 5; anthers all 1-thecous; thecae triplicate. Madagascar **24. *Ampelosicyos***
- Stamens 3; two anthers 2-thecous, one 1-thecous 26
26. Anthers connate into a central head; filaments distinct. Hispaniola **35. *Linnaeosicyos***
- Anthers distinct. Asia **33. *Trichosanthes***
27. Pollen echinate or perforate 28
- Pollen reticulate or striate-reticulate 36
28. Filaments connate into a central column 29
- Filaments distinct, sometimes very short or absent 33
29. Thecae connate into a horizontal, ring-like structure. South and Central America **41. *Cyclanthera***
- Thecae distinct or connate into a central head-like structure 30
30. Fruits fleshy, unarmed, indehiscent, 1-seeded. Central America **39. *Sicyos***
- Fruit ± dry, seeds few to many 31

31. Fruit globose, smooth, 5–7 cm in diam., indehiscent. Hispaniola **92. *Penelopeia***
- Fruit dehiscent, ± setose or prickly. America 32
32. Fruit operculate; seeds relatively small, ± compressed **42. *Echinopepon***
- Fruit not operculate; seeds large, ± globose **37. *Marah***
33. Fruit a large, fleshy, indehiscent pepo 34
- Fruit a small dry or fleshy berry 35
34. Petals yellow. America, introduced in Africa, Europe, Asia, Australia **89. *Cucurbita***
- Petals greenish white. Central America **95. *Cionosicys***
35. Fruit a fleshy berry, ripening red. Argentina and Uruguay **96. *Abobra***
- Fruit a dry berry with a firm, thin wall, ripening green, red, brown or black; seeds in loose cellular pulp. America **97. *Cayaponia***
36. Pollen striate-reticulate 37
- Pollen reticulate or gemmate 38
37. Stigma 3-lobed. Africa and Asia **74. *Dactyliandra***
- Stigma-lobes feather-like divided. New Guinea **76. *Papuasicyos***
38. Stamens 5 39
- Stamens 2–3 45
39. Thecae triplicate/sinuate 40
- Thecae straight or ± curved 41
40. Sepals > petals. Indonesia **83. *Zehneria*** (*Z. macrosepala*)
- Sepals < petals. Africa and Socotra **44. *Eureiandra***
41. Plant densely black- or reddish-glandular hairy. Tropical Africa **20. *Siraitia*** (*S. africana*)
- Plant glabrous or hairy but not black- or reddish-glandular 42
42. Seeds pear-shaped to subglobose. Africa, Madagascar, Asia 43
- Seeds ovate-oblong, ± compressed. Asia 44
43. Fruit operculate, the basal part of the fruit green, expanded into a cup, the upper part red **50. *Corallocarpus***
- Fruit indehiscent or opening by valves, ripening entirely orange to red **51. *Kedrostis***
44. Petals small, to 5 mm long, cream-colored or white **19. *Baijiania***
- Petals >5 mm long, yellow **18. *Thladiantha***
45. Stamens 2 46
- Stamens 3 51
46. Adult plants usually leafless with green, succulent stems. Madagascar **46. *Seyrigia***
- Adult plants with well-developed leaves; stems not succulent 47
47. Sepals showy, orange to red, sepals > petals. Tropical America **54. *Gurania***
- Sepals green or dark-colored, sepals < petals 48
48. Flowers small; petals inconspicuous, c. 3 mm long, yellowish-green. Tropical America **56. *Helmontia***
- Flowers medium-sized; petals >5 mm long, yellow, orange or white 49
49. Male flowers often subtended by a ± orbicular bract, often 1–3 of the petals with an incurved basal scale. Africa and Asia, introduced in Australia and America **21. *Momordica***
- Male flowers not subtended by bracts; petal scales absent 50
50. Pollen often in tetrads; woody or less often herbaceous climber; petals orange, red or pink. Tropical and subtropical America **55. *Psiguria***
- Pollen in monads; herbaceous climber or trailer; petals greenish-white to yellow. Southern US to Argentina **59. *Apodanthera***
51. Pollen in tetrads. Indonesia **71. *Borneosicyos***
- Pollen in monads 52
52. Male flowers often subtended by a ± orbicular bract, often 1–3 of the petals with an incurved basal scale. Africa and Asia, introduced in Australia and America **21. *Momordica***
- Male flowers not subtended by an orbicular bract, petal scales absent 53
53. Petiole base with suborbicular ciliate bract 54
- Petiole base without ciliate bract 55
54. Thecae linear, straight; petals ± 1 mm long. Africa, Madagascar, and Asia **73. *Ctenolepis***
- Thecae triplicate; petals larger. Africa and Madagascar **77. *Trochomeria***
55. Thecae straight or slightly curved (sometimes apically hooked) 56
- Thecae strongly curved, duplicate, triplicate or flexuous 81
56. Petals 2-furcate to deeply 2-fid. Central and South America 57
- Petals entire 59
57. Filaments very short, distinct **52. *Ceratosanthes***
- Filaments longer, connate into a central column 58
58. Flowers small, opening during the day **61. *Ibervillea***
- Flowers medium-sized to large, fragrant, opening at night **62. *Dieterlea***
59. Fruit a large, hard-shelled pepo, to 20 cm long; thecae fringed with hairs. Tropical Africa and America **79. *Melothria***
- Fruit smaller, a fleshy berry or gourd; if large (*Cucumis melo*), then thecae not hairy 60
60. Fruit ornamented 61
- Fruit ± smooth 62
61. Fruit with long, soft bristles. Madagascar, Northern Australia, and Indonesia **81. *Muellerargia***
- Fruit with dense to scattered fleshy spines, pustules or tubercles that end in a hyaline bristle. Africa, Asia, naturalized in America, Australia, and the Pacific Islands **82. *Cucumis***
62. Stamens inserted near the base or halfway up the receptacle-tube 63
- Stamens inserted in the upper half or near the mouth of the receptacle-tube 64
63. Anthers all 2-thecous; leaves petiolate, triangular to ± ovate, entire to 3-lobed. Africa, Asia, Australia, and Pacific Islands **83. *Zehneria***

- Two anthers 2-thecous, one 1-thecous; leaves shortly petiolate to sessile, base cordate or hastate. Asia **70. *Solena***
64. Filaments connate into a central column. Mexico and Southern US 65
- Filaments distinct or very short to absent 66
65. Fruit a fusiform or ellipsoid berry, shortly rostrate, 6–15 cm long and 3–6 cm in diam. **62. *Dieterlea***
- Fruit a globose berry, c. 1 cm in diam., glabrous, with remains of flower **60. *Tumamoca***
66. Stigma 3-lobed, long-hairy. Southeast Asia **78. *Indomelothria***
- Stigmas 1–5, entire or lobed, if 3-lobed, then glabrous, papillose or short-hairy
67. Stigmas 2. Central and South America 68
- Stigmas 3–5 69
68. Fruit fleshy, indehiscent, ovoid to ellipsoid, ± rostrate, 1–7 cm long, ripening green or red to brown often with white stripes or spots **59. *Apodanthera***
- Fruit an ovoid-conical berry, c. 2 cm long and 1.5 cm in diam., sessile in the leaf axils, rostrate **58. *Wilbrandia***
69. Stigmas 4–5 70
- Stigmas 3 71
70. Testa smooth, chocolate-brown, often with distinct, ivory-colored margin. Southern US to Argentina **59. *Apodanthera***
- Testa smooth, margin ± distinct, not winged. South America **49. *Cucurbitella***
71. Two anthers 2-thecous, one 1-thecous 72
- All anthers 2-thecous 75
72. Testa covered by long appressed hairs. America, introduced in Asia **79. *Melothria***
- Testa glabrous or rarely puberulent 73
73. Testa light-colored, ± yellowish. Africa, Asia, Australia, introduced in America **82. *Cucumis***
- Testa brown 74
74. Testa finely sculptured, no distinct margin. Madagascar **47. *Trochomeriopsis***
- Testa smooth, often with distinct, ivory-colored margin. Southern US to Argentina **59. *Apodanthera***
75. Anthers 2 76
- Anthers 3 77
76. Testa chocolate-brown. Southern US to Argentina **59. *Apodanthera***
- Testa pale brown or yellowish. Africa, Asia, Australia **83. *Zehneria***
77. Seeds tumid to globose 78
- Seeds compressed 79
78. Disk in male flowers urceolate, ± connate with base of tube. New Guinea **76. *Papuasicyos***
- Disk in male flowers ± globose, distinct. Africa, Asia, Australia **83. *Zehneria***
79. Testa covered by long appressed hairs. America, introduced in Asia **79. *Melothria***
- Testa glabrous 80
80. Stigmas hairy. Southeast Asia **86. *Scopellaria***
- Stigmas glabrous or papillose but not hairy. Africa, Asia, Australia **83. *Zehneria***
81. Thecae strongly curved or duplicate 82
- Thecae triplicate or flexuous 91
82. Filaments inserted near the base or in the lower half of the receptacle-tube 83
- Filaments inserted halfway up or in the upper half of the tube 84
83. Leaves petiolate, triangular to ± ovate, entire to 3-lobed; disk in male flowers globose, entire or 3-parted. Africa, Asia, Australia, and Pacific Islands **83. *Zehneria***
- Leaves shortly petiolate to sessile, base cordate or hastate; disk in male flowers 3–4-lobed, conspicuous, carnose. Asia **70. *Solena***
84. Receptacle-tube elongate, tubular to cylindrical, often dilated at the apex. Asia, Australia, introduced in Africa and the Neotropics **33. *Trichosanthes***
- Receptacle-tube broadly campanulate or ± cylindrical but not elongated 85
85. Fruit laterally compressed, shortly rostrate. Argentina **48. *Halosicyos***
- Fruit ellipsoid to oblong or subglobose, not laterally compressed, sometimes rostrate 86
86. Seeds pear-shaped, slightly compressed, reddish brown, with distinct pale brown margin. Central America **53. *Doyerea***
- Seeds compressed, pale or dark-colored 87
87. Testa verrucous. Brazil **57. *Melothrianthus***
- Testa smooth or finely scrobiculate 88
88. Testa finely scrobiculate. New Guinea **76. *Papuasicyos***
- Testa smooth 89
89. Testa brown. Southern US to Argentina **59. *Apodanthera***
- Testa pale yellowish or cream-colored 90
90. Staminodes forming a ring; seeds with distinct margin. Madagascar **73. *Ctenolepis***
- Staminodes distinct; seeds without distinct margin or rarely margin thickened. Australia **26. *Austrobryonia***
91. Filaments inserted halfway up or in the upper half of the tube 92
- Filaments/stamens inserted near the base or in the lower half of the receptacle-tube 94
92. Seeds with broad, flattened margin; leaves pedately 3–7-lobed, drying black. Tropical West Africa **80. *Ruthalicia***
- Seeds without distinct margin or margin narrow; leaves unlobed or palmately 3–5-lobed, usually drying green 93
93. Seeds tumid, subglobose, or asymmetrically ovoid. Africa and Socotra **44. *Eureiandra***
- Seeds ovate or elliptic, small to medium-sized, globose or lenticular compressed. Africa, Asia, Australia, introduced in America **82. *Cucumis***
94. Fruit densely brown-setose. Tropical Africa and Madagascar **67. *Raphidiocystis***
- Fruit ± glabrous or sparsely setose or hairy or with prominent spines but not brown-setose 95

95. Fruit hairy 96
- Fruit ± glabrous 98
96. Testa blackish, smooth, not winged, no distinct margin. Africa and Madagascar **64. *Peponium***
- Testa brown or grayish-brown, without distinct margin or with dentate or narrow, corky margin 97
97. Testa brown; leaves petiolate, palmately 3–5-lobed, the lobes lobulate-dentate. Madagascar **69. *Lemurosicyos***
- Testa grayish-brown; leaves shortly petiolate to sessile, the blade ovate or elliptic, margin entire, base cordate or hastate. Asia **70. *Solena***
98. Leaves very shortly petiolate to sessile, ± amplexicaul 99
- Leaves with distinct petioles 102
99. Seeds slightly compressed to ± globose 100
- Seeds strongly compressed 101
100. Testa grayish-brown, sometimes with narrow, corky margin. Asia **70. *Solena***
- Testa whitish, without distinct margin. Africa and Madagascar **77. *Trochomeria***
101. Testa black, verrucous. Africa **68. *Cephalopentandra***
- Testa pale, smooth to fibrillose. Africa, Asia, introduced in Australia and America **85. *Coccinia***
102. Seeds subglobose, ovoid or ellipsoid, tumid; testa smooth, hard, whitish. Africa and Madagascar **77. *Trochomeria***
- Seeds compressed 103
103. Testa bright brown, finely grooved, with broad, grooved, crenulate-tuberculate margin. Cambodia **75. *Khmeriosicyos***
- Testa without distinct margin or margin not grooved, crenulate-tuberculate 104
104. Receptacle-tube elongated, ± cylindrical. Africa and Madagascar **64. *Peponium***
- Receptacle-tube campanulate or turbinate 105
105. Flowers small, in axillary, racemose panicles or sub-umbellate fascicles. Europe, Northern Africa, Canaries, Central Asia **27. *Bryonia***
- Flowers medium-sized to large, if small then solitary (rarely in few-flowered fascicles). Africa and Asia, introduced in Australia and America 106
106. Thecae triplicate; style filiform; fruit small and globose or cylindrical and to 30 cm long, baccate, usually glabrous **85. *Coccinia***
- Thecae flexuous; style short, columnar; fruit large, globose or oblong, glabrous or covered with prominent spines **63. *Citrullus***
107. Tendrils 2-fid 108
- Tendrils 3–8-fid 175
108. Basal part of the tendrils sensitive and thus tendrils coiling above and below the bifurcation 109
- Basal part of the tendrils insensitive, not coiling 127
109. Sepals and petals 4; leaves entire or 3-lobed, orbicular or flat, succulent or non-succulent, perennial or deciduous. Madagascar **15. *Xerosicyos***
- Sepals and petals 5, or sepals 3–4 and petals 5 110
110. Sepals 3–4 and petals 5 111
- Sepals and petals 5 113
111. Stamens 3, two anthers 2-thecous, one 1-thecous; thecae vertical; ovules many per locule; seeds with a butterfly-shaped wing, expanded laterally and divaricate. South East Asia to New Guinea **1. *Alsomitra***
- Stamens 5, distinct; thecae horizontal; ovules 2 per locule 112
112. Leaves entire; seeds with a chalazal wing. Indomalesia **13. *Zanonia***
- Leaves compound, palmate or pedate; seeds in the median position of an encircling wing expanded along the chalaza-micropyle axis. South America **14. *Siolmatra***
113. Filaments connate into a central column 114
- Filaments ± distinct 117
114. Thecae 2–3, horizontal; fruit a samara, indehiscent 115
- Thecae 5, vertical; fruit a dry achene, indehiscent or a subglobose capsule, dehiscent 116
115. Thecae 2, semicircular, forming together a split ring. Africa and Madagascar **9. *Cyclantheropsis***
- Thecae 3, straight, forming the sides of an equilateral triangle. South America **10. *Pteropepon***
116. Leaves cordate; ovule 1 per ovary; fruit dry, globose, indehiscent; seed subspherical. Mexico **11. *Sicydium***
- Leaves compound, pedate, 3–7-foliolate; stylodia 3; ovules more than 1 (usually 2–4) per ovary; fruit a subglobose capsule (3-valvate at the apex), dehiscent; seeds not compressed, unwinged or winged. Asia and Indomalesia **5. *Gynostemma***
117. Stamens 1, anther 2-thecous. South America **10. *Pteropepon***
- Stamens 2–5, all anthers 1-thecous, distinct or two anthers 2-thecous, one 1-thecous (sometimes the 1-thecous or one of the 2-thecous anthers reduced or aborted) 118
118. Stamens 3, two anthers 2-thecous, one 1-thecous; ovary 1-locular; ovule 1 per ovary; fruits indehiscent, baccate; seed subspherical 119
- Stamens 4 and 1 staminode or stamens 5 121
119. Fruit a cylindrical-clavate capsule, c. 20 cm long, opening with 3 valves; seeds with broad circular membraneous wing, c. 5 cm in diam., testa finely verrucous, dull brown, margin coarsely 8–9-spined. Malaysia **2. *Bayabusua***
- Fruit indehiscent, much smaller. South and Central America 120
120. Fruit a large, fibrous samara with 1 continuous, encircling wing or small, membranaceous, with two lateral wings **10. *Pteropepon***
- Fruit baccate, globose, fleshy or fibrous, not winged **11. *Sicydium***
121. Stamens 4 and 1 staminode 122
- Stamens 5 123
122. Corolla slightly to strongly zygomorphic; ovary 3-locular at the apex, 1-locular at the base; ovules many per locule; fruit a capsule (apically 3-valvate); seeds clearly winged, fusiform. Africa **12. *Gerrardanthus***
- Corolla actinomorphic; ovary 1-locular; ovules 1 per ovary; fruit a samara; seed compressed;

pericarp (wing included) fibrous or membranaceous. South America **10. *Pteropepon***
123. Petals lanceolate, long acuminate; corolla actinomorphic. Asia **16. *Actinostemma***
– Petals ovate or oblong, never long-acuminate; corolla not actinomorphic 124
124. Fruit operculate or capsule or opening by longitudinal splits. America 125
– Fruit not operculate. Asia or Africa 126
125. Fruit operculate (rarely a capsule); leaves pedately 3–7-lobed or 3–5-foliolate **7. *Fevillea***
– Fruit opening by longitudinal splits; leaves simple, entire **8. *Anisosperma***
126. Fruit fleshy 127
– Fruit dry 129
127. Petals fringed. Tropical Africa and Madagascar **23. *Telfairia***
– Petals entire 128
128. Plant glabrous or hairy but not glandular. Asia **19. *Baijiania***
– Plant ± glandular hairy. Tropical Africa or Asia **20. *Siraitia***
129. Fruit indehiscent, 20–30 cm long; seeds large, 3.5–4 cm long. Asia **17. *Indofevillea***
– Fruit apically 3-valvate, to 8 cm long; seeds small 130
130. Seeds with wing on the chalazal end; flowers actinomorphic. China, Indomalesia, and Australia **6. *Neoalsomitra***
– Seeds unwinged or wing encircling the seed, uniform in width or extended along the chalaza-micropyle axis. Asia 131
131. Annual, herbaceous climbers or trailers, to 5 m long with fibrous roots **3. *Gomphogyne***
– Mostly perennial and tuberous climbers or trailers **4. *Hemsleya***
132. Filaments connate into a central column 133
– Filaments distinct or connate to pairs only 140
133. Male flowers 4-merous, female flowers 3-merous. Central America **39. *Sicyos***
– All flowers 5-merous 134
134. Thecae connate into a horizontal, ring-like structure. South and Central America **41. *Cyclanthera***
– Thecae distinct or connate into a central head-like structure 135
135. Fruits fleshy, unarmed, indehiscent, 1-seeded. Central America **39. *Sicyos***
– Fruit ± dry, if fleshy, then seeds few to many 136
136. Fruit indehiscent, smooth. Hispaniola **92. *Penelopeia***
– Fruit dehiscent, ± setose or prickly 137
137. Fruit operculate; seeds relatively small, ± compressed. America **42. *Echinopepon***
– Fruit not operculate 138
138. Seeds large, ± globose. North and Central America **37. *Marah***
– Seeds small, ± compressed 139
139. Fruits dry, indehiscent. America, Pacific Islands, Australia, introduced in Africa **39. *Sicyos***
– Fruits ± fleshy, opening explosively, solitary. Tropical America **40. *Hanburia***
140. Pollen baculate or echinate 141
– Pollen reticulate or striate 148
141. Fruit indehiscent 142
– Fruit dehiscing 146
142. Fruit fleshy 143
– Fruit dry 145
143. Fruit a small red, ovoid to globose, fleshy berry, c. 1 cm long and 1 cm in diam. Argentina and Uruguay **96. *Abobra***
– Fruit a fleshy pepo, >5 cm long 144
144. Testa densely appressed hairy, pale yellowish-brown. Central America **93. *Tecunumania***
– Testa glabrous, smooth, cream-colored or black. America, introduced in Africa, Europe, Asia, Australia **89. *Cucurbita***
145. Stamens inserted in the center of the flower. Seeds compressed, ovoid or less often ± triangular or dagger-shaped and apically tricornute; tests not banded. Tropical and subtropical America **97. *Cayaponia***
– Stamens inserted near the mouth of the receptacle-tube. Seeds ovoid, compressed; testa brown or banded crosswise with light and dark stripes. Mexico and Guatemala **94. *Schizocarpum***
146. Fruit ± ribbed, 5–8 cm long, seeds 6, 12 or c. 48, oblong or obovate, compressed; testa smooth, margin obtuse, not winged. Asia **30. *Herpetospermum***
– Fruit ± rounded, ellipsoid to pear-shaped, rostrate or not, smooth 147
147. Seeds broadly ovate, compressed, contracted at base; testa brown, narrowly winged. Central America **87. *Polyclathra***
– Seeds ovoid, compressed, not contracted; testa brown or banded crosswise with light and dark stripes, margin with or without wing. Mexico and Guatemala **94. *Schizocarpum***
148. Petals fringed 149
– Petals not fringed 152
149. Seeds small. Asia to Australia **33. *Trichosanthes***
– Seeds large 150
150. Anthers connate into a central globose head. Asia **34. *Hodgsonia***
– Anthers distinct 151
151. Stamens 5, anthers all 1-thecous. Madagascar **24. *Ampelosicyos***
– Stamens 3–5, one or several anthers 2-thecous. Tropical Africa and Madagascar **23. *Telfairia***
152. Stamens 2 153
– Stamens 3 or 5 154
153. Male flowers often subtended by a ± orbicular bract, often 1–3 of the petals with an incurved basal scale; receptacle-tube broadly campanulate. Africa and Asia, introduced in Australia and America **21. *Momordica***
– Male flowers long pedunculate but without orbicular sheathing bract; receptacle-tube elongate-cylindrical. Peru **59. *Apodanthera***
154. Stamens 5 155
– Stamens 3 158

155. Seeds pear-shaped to subglobose. Africa, Madagascar, Asia 156
- Seeds ± compressed 157
156. Fruit operculate, the basal part green, expanded into a cup, the upper part red **50. *Corallocarpus***
- Fruit indehiscent or opening by valves, ripening entirely orange to red **51. *Kedrostis***
157. Fruit a ± fleshy berry. Asia **18. *Thladiantha***
- Fruit dry with seeds in fibrous tissue. Africa, Arabia, Asia, Australia, America **32. *Luffa***
158. Male flowers often subtended by a ± orbicular bract, often 1–3 of the petals with an incurved basal scale; receptacle-tube broadly campanulate. Africa and Asia, introduced in Australia and America **21. *Momordica***
- Male flowers without prominent, ± orbicular sheathing bract and without petal scales 159
159. Thecae straight or ± curved 160
- Thecae duplicate or triplicate or convoluted 162
160. Seeds 1–3, pendent, ovate, compressed; testa brown, ± sculptured, margin irregularly dentate, not winged. Asia **29. *Schizopepon***
- Seeds usually more than 3; testa yellowish or brown, ± smooth, margin sometimes distinct but not dentate 161
161. Stamens inserted near mouth of receptacle-tube; seeds often chocolate-brown with distinct ivory-colored margin. America **59. *Apodanthera***
- Stamens inserted halfway up the receptacle-tube; seeds yellowish or brown but not with distinct, ivory-colored margin. Africa, Asia, Australia, introduced in America **82. *Cucumis***
162. Thecae duplicate. Tropical Africa **22. *Cogniauxia***
- Thecae triplicate, flexuose or convoluted 163
163. Petioles with two, ± conspicuous apical glands. Africa, introduced in Asia and America **65. *Lagenaria***
- Petioles not with paired glands 164
164. Receptacle-tube elongate, tubular to cylindric 165
- Receptacle-tube short, broad, shallow 166
165. Stamens inserted halfway up the receptacle-tube. Asia, Australia, introduced in Africa and America **33. *Trichosanthes***
- Stamens inserted near the mouth of the tube. Africa and Madagascar **64. *Peponium***
166. Fruits in clusters or racemes 167
- Fruit solitary (rarely 2–3) 169
167. Fruits up to 8 in racemes, globose, c. 2.5 cm across, with strong gourd-like odor, style and calyx rests persistent on fruit. Tropical West Africa **43. *Bambekea***
- Fruits in clusters of 2–6, globose, ellipsoid or ovoid, usually <2 cm across, style and calyx not persistent on fruit 168
168. Stamens inserted near the mouth of the receptacle-tube; fruits ripening bright red with silvery white stripes or marks. Africa, Asia, Australia **84. *Diplocyclos***
- Stamens inserted halfway up the receptacle-tube; fruits ripening yellow, orange, red or greenish to brownish, often with longitudinal pale stripes. Africa, Asia, Australia, introduced in America **82. *Cucumis***
169. Fruit dry with seeds embedded in fibrous tissue. Africa, Asia, Australia, America **32. *Luffa***
- Fruit ± fleshy 170
170. Fruit a small, baccate, globose, ellipsoid or ovoid berry with white pulp, ripening bright red with silvery white stripes or marks. Africa, Asia, Australia **84. *Diplocyclos***
- Fruit a ± large pepo or gourd, if berry, then not with white pulp and bright red pericarp 171
171. Fruit at first hispid, later glabrous, dark green and covered with white wax; seeds many, compressed, smooth, white with thick margin. Asia, Australia, Pacific Islands, introduced in Africa **72. *Benincasa***
- Fruit not hispid when young, not covered with white wax when older 172
172. Stamens inserted near the mouth of the receptacle-tube. Australia **31. *Nothoalsomitra***
- Stamens inserted at the base of the tube or halfway up 173
173. Stamens inserted halfway up the tube. Africa, Asia, Australia, introduced in America **82. *Cucumis***
- Stamens inserted at the base of the tube. Africa, Asia, introduced in Australia and America 174
174. Style filiform; fruit small and globose or cylindrical and to 30 cm long, baccate, usually glabrous **85. *Coccinia***
- Style short, columnar; fruit large, globose or oblong, glabrous or covered with prominent spines **63. *Citrullus***
175. Petals fringed 176
- Petals not fringed 177
176. Seeds small. Asia, Australia, introduced in Africa and America **33. *Trichosanthes***
- Seeds large. Asia **34. *Hodgsonia***
177. Petals 6. North America **36. *Echinocystis***
- Petals 3–5 178
178. Male flowers 4-merous, female flowers 3-merous. Central America **39. *Sicyos***
- All flowers 5-merous 179
179. Pollen reticulate 180
- Pollen echinate, baculate or perforate 183
180. Receptacle-tube elongated, cylindrical 181
- Receptacle-tube (broadly) campanulate 182
181. Thecae triplicate; stamens inserted halfway up the tube. Asia **33. *Trichosanthes***
- Thecae straight or ± curved; stamens inserted near the mouth of the tube. America **59. *Apodanthera***
182. Fruit ± fleshy, indehiscent. Africa, Asia, introduced in Australia and America **63. *Citrullus***
- Fruit dry with fibrous tissue, operculate. Africa, Asia, Australia, America **32. *Luffa***
183. Filaments connate into a central column 184
- Filaments distinct 190

184. Thecae connate into a horizontal, ring-like structure. South and Central America **41. *Cyclanthera***
- Thecae distinct or connate into a central head-like structure 185
185. Fruit dry, globose, smooth, 5–7 cm in diam., indehiscent. Hispaniola **92. *Penelopeia***
- Fruit ± fleshy, if dry, then not globose 186
186. Fruit 1-seeded, fleshy, medium-sized to large. Central America 187
- Fruit few- to many-seeded, if 1-seeded, then fruit dry or small 188
187. Nectaries in open pouches. Mexico **39. *Sicyos***
- Nectaries often with umbrella-like covering (not in *F. tacaco* and *F. talamancensis*!). Costa Rica, Nicaragua, Panama **38. *Frantzia***
188. Fruit explosively dehiscent **40. *Hanburia***
- Fruit indehiscent or operculate 189
189. Fruits indehiscent, winged or 3–4-angled, small ovoid to fusiform, armed with retrorse barbs or unarmed, glabrous or villous, sometimes enclosed by a subtending leaf; seed solitary, ovoid or tumid to compressed. America, Pacific Islands, Australia, introduced in Africa **39. *Sicyos***
- Fruits operculate, echinate, rostrate, if indehiscent, then subterranean (to 3.5 cm deep), on 6–9 cm long peduncle (*E. arachoidea*); seeds solitary or few, quadrangular or angular-ovoid, compressed. America **42. *Echinopepon***
190. Fruit dehiscent, dry or fleshy 191
- Fruit indehiscent 193
191. Fruit a fleshy pepo, splitting into three carpellar segments at maturity **88. *Peponopsis***
- Fruit ± dry at maturity 192
192. Fruit splitting into several irregular segments. Central America **87. *Polyclathra***
- Fruit splitting into 3 valves. Asia **30. *Herpetospermum***
193. Fruit a dry berry 194
- Fruit a medium-sized to large, fleshy pepo 195
194. Flowers large, solitary in the axils; peduncle of male flowers to 30 cm long, to 15 cm in female; receptacle-tube campanulate to urceolate, ± inflated. South America **90. *Calycophysum***
- Flowers small, in racemes, panicles, pairs or solitary, male and female often coaxial; receptacle-tube campanulate, not inflated. America **97. *Cayaponia***
195. Stamens inserted close to the mouth of the receptacle-tube; fruit a globose, ellipsoid or cylindrical pepo, smooth, to 60 cm long. Central America **91. *Sicana***
- Stamens inserted near the base of the receptacle-tube 196
196. Receptacle-tube and corolla ± campanulate or funnel-shaped. America, introduced in Africa, Europe, Asia and Australia **89. *Cucurbita***
- Receptacle-tube flat, saucer-shaped, villous; corolla flat, villous outside, smooth inside. Asia **72. *Benincasa* (*B. fistulosa*)**

Genera of Cucurbitaceae

I. Tribe Gomphogyneae Benth. & Hook.f. (1867).

Tendrils apically 2-fid (rarely simple). Stamens 3 or 5. Fruit a capsule or berry. Seeds 1 to many, often winged.

1. *Alsomitra* (Blume) Spach

Alsomitra (Blume) Spach, Hist. Nat. Vég. Phan. 6: 187 (1838); Duyfjes & W.J. de Wilde, Proc. 4th Int. Flora Males. Symp. 1998, Kuala Lumpur: 101–105 (1998).
Macrozanonia Cogn. (1893).

Dioecious, perennial, woody liana, 30–50 m long, the stem up to 15 cm in diam. Leaves broadly ovate to rounded-ovate, entire, rarely 3-lobed, to 16 cm long; petiole with basal ring-shaped callus; young plants with c. 1 cm long, oblong, hastate, ± auriculate, distichous leaves; tendrils with elongated, not peltate adhesive pads. Male and female flowers in panicles or racemes; receptacle-tube campanulate; calyx first completely connate with minute 5-lobed orifice, at anthesis tearing into (2)3 (4) irregular parts; petals narrowly elliptic, acute, papillose at apex; stamens 3, inserted near the mouth of the tube; filaments short; all anthers 2-thecous or two anthers 2-thecous and one 1-thecous; thecae straight, vertical, papillose hairy; pollen small (polar axis 19–26 μm, equatorial axis c. 21 μm), 3-colporate, perforate to indistinctly rugulate (Khunwasi 1998; van der Ham 1999); pistillodes 3, minute; ovary ± ellipsoid; placentae 3, apical; ovules many; stylodia 3; stigmas fleshy, 2-lobed. Fruit a large, ovoid-globose to ovoid-cylindrical capsule, 20–25 cm in diam., dehiscent by an apical 3-radiate slit into 3 valves, ripening brown. Seeds many, compressed, suborbicular to elliptic, 25–30 by 20–23 mm; testa smooth, margin with large, membranous wing, laterally expanded, butterfly-like, 10–12 cm wide; cotyledons remaining in the seed during germination.

One species, *A. macrocarpa* (Blume) M. Roem., in Thailand, Malaysia, Indonesia, Philippines, and New Guinea; in tropical lowland riverine forests on rich clay soil; flowering Dec.–Jan., fruiting March–June (in Java).

2. *Bayabusua* W.J. de Wilde

Bayabusua W.J. de Wilde, Sandakania 13: 1 (1999).

Dioecious, perennial, woody lianas, 20–40 m long. Leaves broadly cordate-ovate, entire; young plants with 4-verticillate basal leaves; tendrils to 15 cm long, inserted axillary, always with circular, peltate adhesive pads. Male flowers in erect, axillary spikes or racemes; female flowers solitary or in small groups; receptacle-tube shallowly cup-shaped; sepals almost distinct, broadly obtuse-triangular; corolla rotate; petals broadly obovate, almost distinct, reflexed, purple-red, densely covered with multicellular hairs; stamens 3, inserted near the center of the tube; filaments distinct; two anthers 2-thecous (the thecae halfway connate), one 1-thecous, creamy white, somewhat fleshy; thecae straight, oblong; pollen medium-sized (polar axis c. 34 μm, equatorial axis c. 21 μm), 3-colporate or partly syncolporate, striate (van der Ham 1999); ovules many. Fruit a cylindrical-clavate capsule, c. 20 cm long, opening with 3 valves, ripening brown. Seeds many, compressed; testa finely verrucous, dull brown, margin coarsely 8–9-spined, with broad circular membraneous wing, c. 5 cm in diam.

One species, *Bayabusua clarkei* (King) W.J. de Wilde, endemic in Peninsular Malaysia; in lower montane forest (200–800 m a.s.l.), extremely rare; flowering in Feb., June, Aug.; fruiting Dec.–Feb.

3. *Gomphogyne* Griff.

Gomphogyne Griff., Account Bot. Coll. Cantor: 26 (1845); de Wilde, Duyfjes & van der Ham, Thai For. Bull. (Bot.) 35: 45–68 (2007).

Dioecious, annual, herbaceous climbers or trailers, to 5 m long; roots fibrous. Leaves simple or pedately 5–(7–9)-foliolate, petiolulate, ovate to subcircular. Male flowers in racemes or thyrses, female flowers in racemes or fascicles (rarely solitary), often with 1–2 small tendrils on the peduncle close to the flowers; receptacle-tube saucer-shaped, reduced; sepals long-triangular; corolla rotate; petals long-acuminate, white; stamens 5, inserted near the center of the tube; filaments short, distinct, diverging; anthers all 1-thecous; thecae straight or ± curved; pollen medium-sized (polar axis 33–40 μm, equatorial axis 21–32 μm), 3-colporate, striate (Khunwasi 1998; de Wilde et al. 2007a); ovary turbinate or subclavate, 3-locular at apex and 1-locular at base; ovules few; stylodia 3, short; stigmas 2-fid. Fruit foveolate. Seeds 1–9, (little) compressed, ellipsoid; testa thick, black, sparsely irregularly verrucose, unwinged. $n = 16$ (Thakur and Sinha 1973).

Two species, *G. cissiformis* Griff. and *G. nepalensis* W.J. de Wilde & Duyfjes, in Asia; mountain slopes, evergreen and deciduous forest, or open scrub.

4. *Hemsleya* Cogn. ex F. B. Forbes & Hemsl.

Hemsleya Cogn. ex F. B. Forbes & Hemsl., J. Linn. Soc. Bot. 23: 490 (1888); D.-Z. Li, Systematics and evolution of *Hemsleya* (Cucurbitaceae). Kunming: Yunnan Sc. Tech. Pr. (1993).

Dioecious, usually perennial and tuberous climbers or trailers. Leaves pedately (3–)5–9(–11)-foliolate, rarely simple. Male flowers in thyrses, female flowers in racemes; receptacle-tube rotate; sepals oblong or lanceolate; corolla very variable in form; petals membranaceous, oblong or ovate, white to deep orange-brown; stamens 5; filaments short, distinct; anthers all 1-thecous; pollen medium-sized (polar axis 33–40 μm, equatorial axis 21–29 μm), 3-colporate, (indistinctly) striate (Khunwasi 1998; de Wilde et al. 2007a); ovary 3-locular at apex and 1-locular at base; placentae 3; ovules many; stylodia 3, short; stigmas 2-lobed. Fruit a clavate-cylindrical to globose capsule, opening apically triradiately. Seeds compressed; testa hard with (or rarely without) an encircling woody (rarely membraneous) wing, uniform in width or expanded along the chalaza-micropyle axis. $n = 14$ (Samuel et al. 1995).

About 30 species mostly in China, a few in the Himalaya, Indochina, Eastern Malesia.

5. *Gynostemma* Blume

Gynostemma Blume, Bijdr.: 23 (1825); W.J. de Wilde & Duyfjes, Blumea 52: 263–280 (2007).
Pestalozzia Zoll. & Moritzi (1846).
Trirostellum Z. P. Wang & Q. Z. Xie (1981).

Dioecious or monoecious, small herbaceous or woody climbers with or without tuberous rootstock. Leaves 3–9-foliolate (rarely simple), leaflets petioluled, margin dentate. Male flowers in panicles, female flowers in fascicles; receptacle-tube reduced, saucer-shaped; sepals triangular; corolla rotate; petals long triangular, greenish white, subulate; stamens 5, inserted near the base of the tube; filaments connate into a central column; anthers 1-thecous, connate into a central head;

thecae straight, short-ellipsoid; pollen small to medium-sized (polar axis 20–35 μm, equatorial axis 15–23 μm), 3-colporate, striate (Khunwasi 1998; de Wilde et al. 2007a); ovary subglobose, (2)3–5-locular; ovules 2 per locule; stylodia (2)3 or 5, short; stigmas 2-fid. Fruit a dry berry or capsule, (sub)globose, to 10 mm in diam., opening apically triradiately. Seeds 1–5, ovoid or subtriangular; testa verrucous, not winged or with a narrow encircling wing. $n = 11$ (Gao et al. 1995).

About 10 species in India, Sri Lanka, China, Taiwan, Japan, Indomalesia, New Guinea; in moist forests, thickets, and meadows.

6. **Neoalsomitra** Hutch. Fig. 23

Neoalsomitra Hutch., Ann. Bot. (London) II, 6: 97 (1942); W.J. de Wilde & Duyfjes, Blumea 48: 99–121 (2003).

Dioecious (rarely monoecious), perennial (rarely annual), herbaceous to woody climbers, to 30 m long, with or without tuberous rootstock and in a few species with conspicuously swollen base (pachypodium); the lower parts of the stem in *N. schefferiana* (Cogn.)Hutch. ornamented with hard, green, 1–2(–4) cm long thorns. Leaves lobed or 3–5-foliolate, subcircular. Inflorescences many-flowered, paniculate. Flowers in panicles or racemes; receptacle-tube cup-shaped; sepals distinct; corolla rotate or cup-shaped; petals very short-connate at base, yellowish or greenish; stamens 5, inserted centrally; filaments distinct or ± connate; anthers all 1-thecous; thecae straight; pollen small to medium-sized (polar axis 17–36 μm, equatorial axis 17–35 μm), 3-colporate, striate (Khunwasi 1998; van der Ham 1999); ovary cylindrical-clavate, 3-locular at the apex and unilocular at the base; ovules 5(–10) per placenta; stylodia 3, short; stigmas reniform; staminodes 0 or 5. Fruit a cylindrical-clavate capsule, solitary or in groups, to 8 cm long, glabrous or pubescent, apex truncate, opening triradiately. Seeds compressed, mostly horned or star-shaped; testa finely tubercled or smooth with narrow or broad margin, with membranous translucent wing.

About 12 species, NE India, Malesia, S China, New Guinea, Australia, Fiji; in humid ravines, deciduous forest, lowland savannah and dry forest, primary evergreen forest, coastal rainforest, riverbanks.

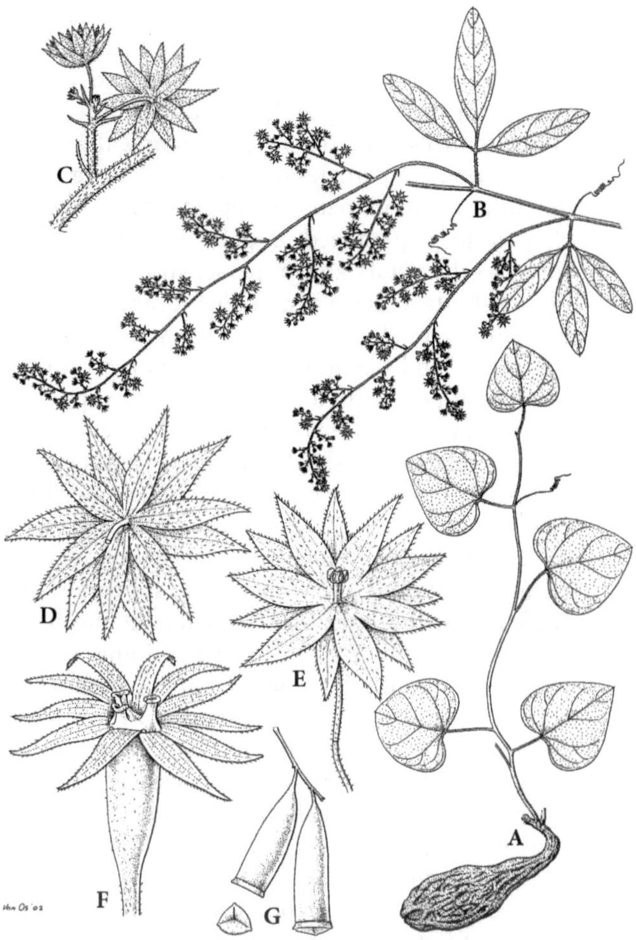

Fig. 23. Cucurbitaceae. *Neoalsomitra angustipetala*. **A** Sprouted tuber with simple eophylls. **B** Male inflorescence. **C** Portion of female inflorescence. **D, E** Male flower. **F** Female flower. **G** Fruits. (de Wilde and Duyfjes 2003; artist J. van Os)

II. Tribe Triceratieae A. Rich. (1845).

Fevilleeae Benth. & Hook.f. (1867).

Leaves simple. Tendrils simple or apically 2-fid. Stamens 1–5. Fruit a pepo, samara or achene. Seeds solitary or 10–15, often with narrow, wing-like margin.

7. ***Fevillea*** L

Fevillea L., Sp. Pl. 2: 1013 (1753) and Gen. Pl., ed. 5: 443 (1754); G.L. Robinson & R.P. Wunderlin, Sida 21: 1971–1996 (2005).
Nhandiroba Plum. ex Adans., nom. illegit.
Hypanthera Silva Manso (1836).

Dioecious, perennial, woody climbers, to several meters long. Leaves pedately 3–7-lobed or 3–5-foliolate, with 2 glands at the distal end of the petiole, or marginal glands at the tip of the main veins; tendrils short or very long, apically 2-fid. Male flowers in panicles or fascicles, female flowers solitary or in pairs; receptacle-tube saucer-shaped to cup-shaped; sepals ± 2 mm long, fused to the petals above; petals suborbicular or oblong-hastate, greenish, yellow, orange, or dull brown, ± 4 mm long, the lower margin fused with the sepals and extending a glandular protuberance, each petal with a median uncinate appendage; stamens 5, inserted near the center of the flower; filaments short, distinct; anthers all 2-thecous; thecae straight, vertical; pollen medium-sized (polar axis 27–33 μm, equatorial axis c. 21–34 μm), 3-colporate, (coarsely) striate (Khunwasi 1998); ovary obconical, subtrigonous, 3-locular at the apex and 1-locular at the base; placentae 3; ovules 4 per locule; stylodia 3; stigmas 2-fid; staminodes 5 or 0. Fruit a ± globose or ovate-oblong pepo or capsule, 3.5–16 cm long, 3–13 cm in diam., indehiscent or circumscissile dehiscent along calyx scar, ripening mottled green, brown or reddish. Seeds 10–17, orbicular, ± compressed, large, to 6×2 cm and 3–9 g dry weight, or much smaller (in *F. anomalosperma* M. Nee c. 1×1×0.2 cm), oil-rich; testa pale brown, smooth to striate-verrucous or pustulate, often with narrow, wing-like margin.

Eight species, from Southern Mexico to Northern Argentina, also in the Caribbean; canopy plant in moist or wet forests.

Fevillea cordifolia L. is widely cultivated for the oil-rich seeds, and this and *F. trilobata* L. have been used for centuries by indigenous South Americans as candles, purgative, and antidote for several kinds of poisoning (Gentry and Wettach 1986). Host of the fungus *Uromyces novissimus* Speg. (Monoson and Rogers 1978).

8. *Anisosperma* Silva Manso

Anisosperma Silva Manso, Enum. Subst. Brazil.: 38 (1836).

Dioecious, perennial, woody climber, to several meters long, with caudex to 5 cm in diam. Leaves unlobed, simple, narrowly ovate, acuminate; tendrils stout, long, apically 2-fid. Male flowers in axillary fascicles, female flowers solitary or in pairs; receptacle-tube saucer-shaped to cup-shaped; sepals ± 2 mm long, fused to the petals above; petals oblong-hastate, greenish-white, ± 4 mm long, the lower margin fused with the sepals, each petal with a median uncinate appendage; stamens 5, inserted near the centre of the flower; filaments short, distinct; anthers all 2-thecous; thecae straight, vertical; pollen medium-sized (polar axis 29 μm, equatorial axis 31 μm), 3-colporate, striate (Khunwasi 1998); ovary fully inferior. Fruit ovoid or oblong, subtrigonous, and shortly apiculate, dehiscent by longitudinal splits. Seeds c. 15–20, suborbicular, compressed, 3.5–3.5 cm long, 3–4 cm wide, and c. 1.5 cm thick; testa yellowish brown, striate-verrucous, with broad wing-like margin.

One species, *A. passiflora* (Vell.) Silva Manso, from gallery forest in Brazil.

Molecular phylogenetic data suggest that this monotypic genus is sister to *Fevillea* (Nee et al. 2010).

9. *Cyclantheropsis* Harms Fig. 24

Cyclantheropsis Harms, Bot. Jahrb. Syst. 23: 167 (1896).

Dioecious, perennial, herbaceous to softly woody climber to 5 m long, with tuberous rootstock. Leaves entire or 3–5-lobed, the blade ovate, base cordate, apically acute; tendrils apically 2-fid. Flowers small; male flowers in axillary panicles, female flowers 3–6, in thyrses or monochasia; receptacle-tube saucer-shaped; sepals triangular, 0.5–1 mm; corolla regular; petals triangular, c. 1 mm, distinct, greenish-yellow; stamen 1, central; thecae 2, horizontal, semicircular, at the top of the column, forming a split ring; pollen medium-sized (polar axis 29–34 μm, equatorial axis 24–27 μm), 3-colporate, striate (Khunwasi 1998); ovary compressed; placenta 1, apical; ovule 1; stylodia 3, short; stigmas 2-fid; staminodes 3, small. Fruits 1–4, an elliptic compressed samara, to 55 by 22 mm, indehiscent, ripening brown. Seed solitary, elliptic and compressed, to 11 mm long; testa pale brown, slightly rough; germination hypogeal (Zimmermann 1922).

Three species, two in East and South tropical Africa, and one endemic in Madagascar; in lowland evergreen forest, deciduous forest, and bushland.

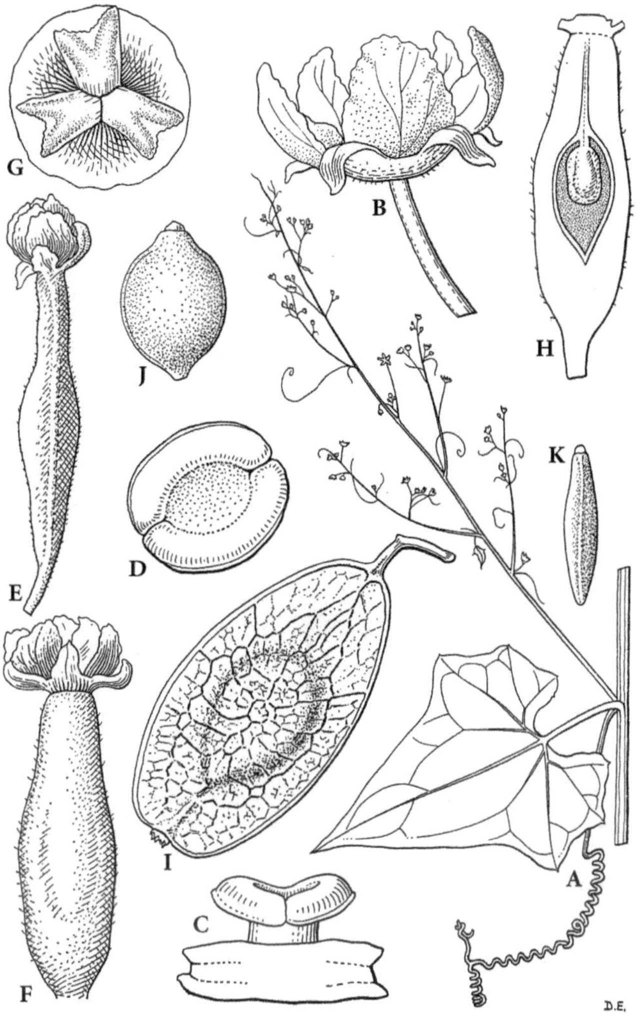

Fig. 24. Cucurbitaceae. *Cyclantheropsis parviflora.* **A** Node with branched tendril and inflorescence. **B** Male flower. **C** Staminal column and disk. **D** Anther thecae, in plan. **E** Female flowers, side view. **F** Same, face view. **G** Stylodia. **H** Ovary, median longitudinal section. **I** Fruit. **J** Seed, face view. **K** Same, side view. (Jeffrey 1967)

10. *Pteropepon* (Cogn.) Cogn.

Pteropepon (Cogn.) Cogn., Pflanzenreich IV. 275 (Heft 66): 260 (1916).
Sicydium sect. *Pteropepon* Cogn. (1878).
Pseudosicydium Harms (1927).

Monoecious or dioecious, perennial, herbaceous or woody climbers with tuberous roots. Leaves entire or 3–5-lobed, the blade subdeltoid, subcircular subcordate or subtruncate with 2 lateral punctate glands just above the insertion of the petiole at the base; tendrils apically 2-fid. Flowers small; male flowers in panicles or thyrses, female flowers solitary or few in monochasia or panicles; receptacle-tube saucer-shaped; sepals linear, or deltoid to lanceolate; corolla rotate; petals ovate-lanceolate, connate at the base, whitish, greenish or yellowish; stamens 1–3, inserted at the base of the tube; filaments short, recurved, distinct or absent or connate into a central column; anthers all 1-thecous or two anthers 2-thecous, one 1-thecous or one anther 2-thecous and one 1-thecous or only one 2-thecous anther; thecae straight, horizontal; pollen medium-sized (polar axis 33–35 μm, equatorial axis 27–32 μm), 3-colporate, striate (Khunwasi 1998); ovary oblong, strongly compressed subtrigonous, 1-locular with an apical placenta; ovule 1; stylodia 3; stigmas 3, 2-fid; staminodes reduced. Fruit a large, fibrous samara with 1 continuous, encircling wing or small, membranaceous, with 2 lateral wings, solitary or 2–3, compressed. Seed solitary, pendent, compressed; testa verrucous or scrobiculate.

Five species in South America, two of them endemic in Argentina (Martínez Crovetto 1952), one in Peru, and two in Brazil; in humid ravines close to rivers, primary rainforest, and secondary scrub.

11. *Sicydium* Schlechtend.

Sicydium Schlechtend., Linnaea 7: 388 (1832).
Triceratia A. Rich. (1845).
Chalema Dieterle (1980).

Dioecious or rarely monoecious, perennial or annual, herbaceous climber with tuberous or fibrous roots. Leaves (sub)cordate, tip acuminate; tendrils simple or apically 2-fid. Flowers small, in panicles; receptacle-tube saucer-shaped; corolla rotate; petals ovate-lanceolate to triangular, whitish-greenish; stamens 3 or 5, inserted near base of the tube; filaments very short, distinct or connate into a central column; anthers all 1-thecous or two anthers 2-thecous, one 1-thecous; thecae straight; pollen small to medium-sized (polar axis 17–37 μm, equatorial axis 17–29 μm), 3-colporate, striate (Khunwasi 1998; Lira Saade et al. 1998); staminodes 0 or 3; ovary ovoid, 1-locular; ovule 1; stylodia 3, linear; stigmas linear or punctiform. Fruit baccate, globose, indehiscent, fleshy or fibrous, ripening black or a dry, globose achene, c. 3 mm in diam. Seed solitary, brownish, (sub)globose or compressed; testa rugose-verrucous, no distinct margin.

About seven species in Central to tropical South America, and the Caribbean and Mexico (Lira Saade 1995, 2004a, b); in disturbed tropical and deciduous forest and along rivers, in dry forest and among shrubs of coastal lowlands; flowering and fruiting all year.

III. Tribe Zanonieae Benth. & Hook.f. (1867).

Tendrils 2-fid, rarely simple. Stamens 4–5. Fruit a dry capsule. Seeds few, winged.

12. *Gerrardanthus* Harv. ex Hook.f. Fig. 25

Gerrardanthus Harv. ex Hook.f. in Benth. & Hook.f., Gen. Pl. 1: 820, 840 (1867).
Atheranthera Masters (1871).

Herbaceous to softly woody climbers to 15 m long, with tuberous rootstocks, often as partly exposed pachypodia (to 1.8 m in diam.). Leaves petiolate, ovate-cordate, unlobed or palmately 3–5-lobed; tendrils apically 2-fid. Male flowers in axillary panicles, female flowers solitary or in small groups; receptacle-tube broad, saucer-shaped; sepals 5, small, ovate or triangular; corolla regular to strongly zygomorphic; petals 5, distinct, unequal, yellowish to orange or brown; stamens 5 (one often reduced to a staminode), inserted near the center of the tube; filaments distinct; anthers all 1-thecous, 2 pairs and 1 single; thecae straight, horizontal; pollen medium-sized (polar axis 42–50 μm, equatorial axis 41–52 μm), 3-colporate, reticulate (Khunwasi 1998); ovary 3-sided, 3-locular at the apex and 1-locular at the base; placentae 3; ovules several, pendent; stylodia 3, divergent; stigmas reniform, 2-lobed; staminodes 5. Fruit obconic-cylindric, 3-sided, dehiscing by an apical triradiate slit, ripening pale yellow. Seeds fusiform, the body to 2 cm long; testa (pale) brown, with distal, membranous, to 2 cm long wing; germination epigeal (Zimmermann 1922).

Three to five species in tropical Africa, and two species in South Africa; in lowland rainforest, deciduous bushland, and wooded grassland (Crouch et al. 1999).

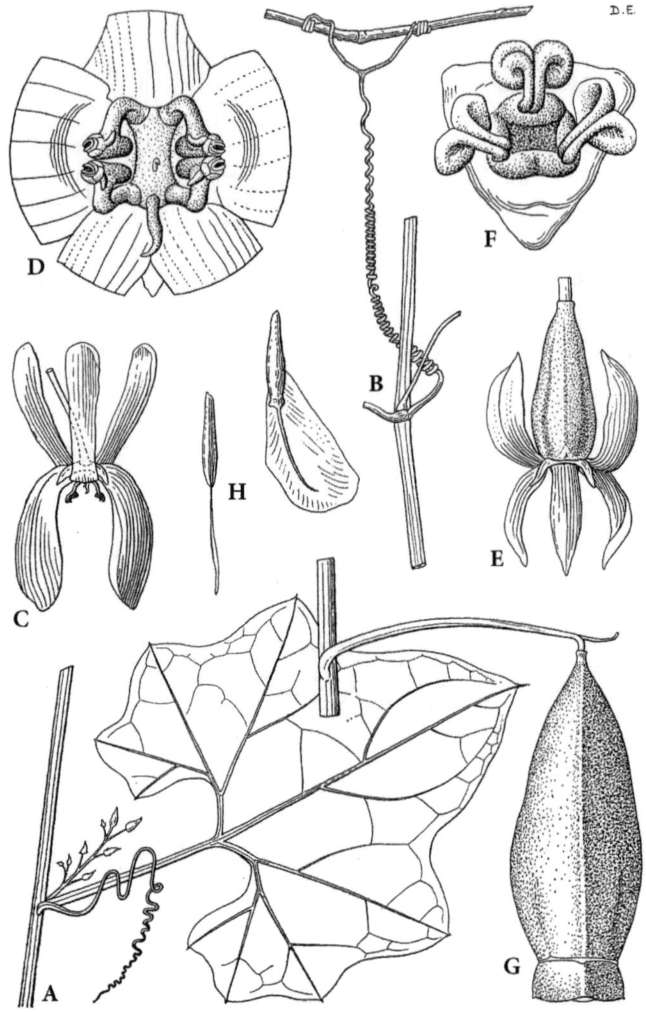

Fig. 25. Cucurbitaceae. *Gerrardanthus lobatus*. **A** Flowering node. **B** Apically bifid tendril. **C** Male flower. **D** Stamens, in plan. **E** Female flower. **F** Stylodia. **G** Fruit. **H** Seeds, side and face view. (Jeffrey 1967)

13. *Zanonia* L.

Zanonia L., Sp. Pl. 2: 1028 (1753) & Gen. Pl., ed. 5: 454 (1754); W.J. de Wilde & Duyfjes, Blumea 52: 281–290 (2007).

Woody climber to 15 m tall. Leaves simple, shortly petiolate, ovate-oblong, coriaceous; tendrils apically 2-fid (or rarely simple) in juvenile plants with apical adhesive pads of to 5 mm length. Male flowers in pendent, many-flowered, to 60 cm long panicles, female flowers in to 40 cm long racemes; receptacle-tube saucer-shaped; sepals 3–4 (but calyx morphologically 5-merous), c. 2 mm long; petals 5, distinct, fleshy, 2.5–7 mm long, cream-colored; stamens 5, inserted centrally; filaments distinct, short, thick; anthers all 1-thecous; thecae horizontal; pollen medium-sized (polar axis c. 40 μm, equatorial axis c. 29 μm), 3-colporate, striate (Khunwasi 1998); ovary clavate, 3-locular at the apex and 1-locular at base; placentae 3; ovules 2 per locule, pendent; stylodia 3, short; stigmas 2-fid; staminodes very small or absent. Fruit to 10×5 cm, elongate-cylindrical, claviform, apex truncate, pendent, dehiscent by a 3-radiate, apical slit into 3 valves. Seeds few, ovate, compressed; testa smooth, leathery winged, the wing to 8×2 cm.

One species, *Z. indica* L., with two subspecies in Cambodia, India, Sri Lanka, Bhutan, Indonesia, Laos, Thailand, Malesia, Myanmar, Vietnam, Southern China, Philippines, and New Guinea; forest edges, riversides, open forest on mountain slopes, 0–2,300 m; flowering and fruiting collections throughout the year.

14. *Siolmatra* Baill.

Siolmatra Baill., Bull. Mens. Soc. Linn. Paris 1: 458 (1885); Robinson & Wunderlin, Sida 21: 1961–1969 (2005).

Woody climbers. Leaves compound, petiolate, the petiole with a basal ring-shaped callus, the blade 3-palmate or 5-pedate, lateral leaflets asymmetrical; tendrils apically 2-fid. Flowers in panicles; receptacle-tube saucer-shaped; sepals 3; corolla contorted; petals 5, unguiculate, obovate, whitish; stamens 5, distinct, inserted at the base of the tube; filaments short; anthers all 1-thecous; thecae straight, horizontal; pollen small to medium-sized (polar axis 24–28 μm, equatorial axis 23–26 μm), 3-colporate, striate (Khunwasi 1998); staminodes absent; ovary obconical, subtrigonous, 3-locular at the apex and 1-locular at the base; placentae 3; ovules 2, pendent; stylodia 3; stigmas 2-lobed, reniform. Fruit nodding, obconical-subtrigonous, apex truncate, to 3 cm long, dehiscent by a 3-radiate, apical slit into 3 short teeth. Seeds strongly compressed, ellipsoid; testa yellowish brown, finely verrucous, margin with membranous wing, expanded along the chalaza-micropyle axis, c. 2 cm long.

Two species endemic in the Amazon basin (Brazil, Peru, Bolivia); in tropical rainforest.

15. *Xerosicyos* Humbert

Xerosicyos Humbert, Compt. Rend. Hebd. Séances Acad. Sci. 208: 220 (1939).
Zygosicyos Humbert (1945).

Herbaceous to woody climbers or shrubs (*X. danguyi* Humbert) with rootstock or large, partially exposed pachypodium, to 1 m in diam. and 1 m high (*X. pubescens* Keraudr.). Leaves simple, shortly petiolate, 3-lobed or 3-foliolate; tendrils apically 2-fid, glabrous or pubescent. Male flowers in sessile or pedunculate fascicle, female flowers solitary or in pedunculate fascicles; receptacle-tube reduced, flat; sepals 4, small, lanceolate-deltoid; corolla irregularly contorted, rotate or zygomorphic; petals 4, distinct, (long) lanceolate, yellow or yellowish-green; stamens 4, inserted near the mouth of the tube; filaments distinct or connate in pairs in the lower half or united into a central column but distinct at the tip; anthers 4, all 1-thecous; thecae reniform, horizontal; pollen small to medium-sized (polar axis 24–30 μm, equatorial axis 16–23 μm), 3-colporate, striate (Keraudren 1968); ovary obconical, 2-locular at the apex and 1-locular at the base; placentae 2; ovules 2 per locule, ± pendent; stylodia 2, straight, ± divergent; stigma ± 2-lobed or horse-shoe-shaped; staminodes 4. Fruit obconical, compressed, 2–3 cm long, apically dehiscent by the ventral carpellar suture, ripening yellow. Seeds 4, elliptical, compressed, c. 1 cm long; testa smooth, pale brown, narrowly winged.

Five species endemic in Madagascar; in xerophilous forest and bushland (Rauh 1996; Eggli 1998).

IV. Actinostemma Clade

Tendrils 2-fid, rarely simple. Stamens 5 or 6. Fruit a dry pyxidium. Seeds few.

16. *Actinostemma* Griff.

Actinostemma Griff., Account Bot. Coll. Cantor: 24 (1845).
Mitrosicyos Maximowicz (1859).
Pomasterion Miquel (1865).
Bolbostemma Franquet (1930).

Herbaceous climbers with fibrous roots or underground stolons bearing crowded scales at the tip.

Leaves hastate-cordate, entire or 3–5-lobed, base often with glands. Flowers unisexual or rarely bisexual; receptacle-tube cup -shaped; sepals linear-lanceolate; petals ovate-lanceolate or ovate, long caudate-acuminate; stamens 5 (rarely 6) or 2 pairs and a single stamen, inserted on the base of the tube; filaments distinct, short; anthers all 1-thecous; thecae straight; pollen small to medium-sized (polar axis c. 26–40 μm, equatorial axis c. 21–42 μm), 3-colporate, striate (Khunwasi 1998); ovary subglobose or ovoid, semi-inferior, verrucous; placentae 1 or 3; ovules 2–4; style 1, short; stigmas 2 or 3, reniform; staminodes 0–5. Fruit ovoid, conical, the upper part falling off at maturity, glabrous or echinate. Seeds ovate compressed; margin denticulate, with short chalazal wing or not winged. $n = 8$ (Probatova and Rudyka 1981).

Three species, two in China and one, *A. tenerum* Griff., widespread in Russia, India, China, Taiwan, Korea, Laos, Vietnam, Japan (Franquett 1930; Ali Khan 2002); in grassland and thickets near open water.

V. Indofevillea Clade

17. *Indofevillea* Chatterjee

Indofevillea Chatterjee, Nature 158: 345 (1946) & Kew Bull. 2: 119 (1947).

Woody climber, to several meters long. Leaves leathery, broadly ovate-cordate, entire, to 20 cm long; tendrils 2-fid, 20–30 cm long. Flowers in axillary panicles; sepals ovate-lanceolate, to 6 mm long; petals ovate-lanceolate, to 4 mm long; stamens 5, inserted near the base of the tube; filaments very short, four in pairs, one distinct; anthers all 1-thecous; thecae reniform, hairy; pollen medium-sized (polar axis c. 49 μm, equatorial axis c. 53 μm), 3-colporate, reticulate (Khunwasi 1998); ovary ellipsoid to oblong. Fruits 3–6 in clusters, c. 30 cm long, oblong, dry, indehiscent with thick woody pericarp. Seeds many, compressed, unwinged, ovoid, 3.5–4 cm long, c. 2 cm broad, 0.5 cm thick; testa yellowish, smooth, faintly winged.

One species, *I. khasiana* Chatterjee, in NE India, Bhutan, Tibet; in open forest on mountain slopes; flowering and fruiting in August.

VI. Thladiantha Clade

Tendrils simple or 2-fid. Stamens 5, inserted near the mouth of the tube; filaments short, two pairs connate at the base, one solitary; anthers all 1-thecous. Fruits fleshy, indehiscent. Seeds many, unwinged.

18. *Thladiantha* Bunge

Thladiantha Bunge, Enum. Pl. China Bor. 29. (1833).

Herbaceous climbers, with tuberous roots. Leaves entire or pedately 3–7-foliolate, the base cordate. Flowers medium-sized; receptacle-tube shortly campanulate or cup-shaped; sepals linear, lanceolate, ovate-lanceolate or oblong; corolla campanulate; petals yellow, entire, oblong, broadly ovate or obovate; thecae straight; pollen large (polar axis 52–79 μm, equatorial axis 64–71 μm), 3-colporate, reticulate (Khunwasi 1998); ovary ovate, oblong or fusiform, smooth or verrucous; stigmas 3, 2-lobed, reniform. Fruit smooth or verrucous, ribbed or not ribbed. Seeds horizontal; testa brown or blackish. $n = 9$ (Li et al. 1993).

About 30 species in China, Taiwan, Tibet, India, Korea, Japan, Thailand, Vietnam, Indonesia, Philippines, New Guinea, one species, *T. dubia* Bunge, naturalized in Europe and N America; in montane forest and bushland, tropical rainforest, riverine forest, on disturbed and cultivated ground. For floral biology and pollination, see Vogel (1990).

19. *Baijiania* A. M. Lu & J. Q. Li

Baijiania A. M. Lu & J. Q. Li in J. Q. Li, Acta Phytotax. Sin. 31: 50 (1993); W.J. de Wilde & Duyfjes, Blumea 48: 279–284 (2003).
Sinobaijiania C. Jeffrey & W.J. de Wilde; W.J. de Wilde & Duyfjes, Blumea 51: 494–498 (2006).

Woody or herbaceous climber to 6 m long, with small or very large spherical tubers, often as partly exposed pachypodia. Leaves ovate-cordate or ovate-oblong, unlobed or 2–3-lobed, to 30 cm long; tendrils apically 2-fid (rarely simple), to 25 cm long. Flowers small; receptacle-tube shallow, cup-shaped; sepals short, triangular-linear or triangular-ovate; petals elliptic, rounded, cream-colored or greenish white, to 5 mm long; thecae ± curved; pollen medium-sized (polar axis 24–25 μm, equatorial axis 26–27 μm), 3-colporate, reticulate (Zhang and Lu 1989); disk at the base of the tube conspicuous, 3-parted; ovary ellipsoid; style c. 2 mm long; stigma 3-lobed, the lobes notched; staminodes 5, 2 pairs and 1 solitary. Fruit solitary or 2–3, subglobose or cylindric, 3–4.5 cm in diam., soft hairy or ± glabrous, ripening orange. Seed ± ovate or ovate-oblong, rounded, c. 5 mm in diam. $n = 16$ (*B. yunnanensis* (A.M. Lu & Zhi Y. Zhang) A.M. Lu & J.Q. Li).

Five species, in China, Taiwan, Thailand, and Borneo (Sabah, SE Kalimantan, Sarawak); along forest margins, in primary or moderately disturbed forest.

VII. Siraitia Clade

20. *Siraitia* Merr.

Siraitia Merr., Pap. Michigan Acad. Sci. 19: 200 (1934); J.Q. Li, Acta Phytotax. Sin. 31: 45–55 (1993); J.W. de Wilde & Duyfjes, Blumea 51: 409–503 (2006).
Microlagenaria (C. Jeffrey) A.M. Lu & J.Q. Li (1993).

Herbaceous climbers to 7 m long, with spherical, enlarged tubers. Leaves entire and unlobed or palmately 3–5-lobed, blade ovate-cordate, sparsely dentate; plant covered with black, blackish-brown or yellowish glandular hairs; tendrils apically 2-fid. Male flowers to 50 in racemes or panicles, female flowers solitary or fasciculate; receptacle-tube short, campanulate; sepals entire, linear to lanceolate or triangular; petals distinct, rounded, lanceolate or obovate-lanceolate, cream-colored, 1–3 with an incurved basal scale; stamens 5, distinct or 2 pairs and one single, inserted near the base of the tube; filaments distinct; anthers all 1-thecous; thecae straight, curved or triplicate; pollen medium-sized (polar axis 35–51 μm, equatorial axis 36–54 μm), 3-colporate, reticulate (Zhang and Lu 1989; Khunwasi 1998); ovary ovoid, hairy; style stout, apex 3-lobed; stigma 2-lobed; ovules many, horizontal; staminodes 5. Fruit (sub)globose or cylindric, fleshy, indehiscent, tomentose, ripening yellow. Seeds few, ± compressed, subovoid or ovoid or oblong; testa pale brown or yellowish, rarely with 2 longitudinal parallel central ridges (*S. africana*) margin unwinged or with 2 or 3 corky wings. $n = 14$ (Li et al. 1993).

Three or four species in India, Indonesia, Peninsular Malaysia, Thailand, South and Southwest China; forest on mountain slopes, riversides or thickets.

Molecular phylogenetic data indicate that *S. africana* (C. Jeffrey) A. M. Lu & J. Q. Li in Southern Tanzania (2 localities) and Southeast Nigeria (1 locality) indeed is closely related to the Asian species; on lake shores and in thickets at low altitudes; not recollected since the 1960s. Locally used as a source of cucurbitane glycosides as a natural sweetener.

VIII. Momordica Clade

21. *Momordica* L.

Momordica L., Sp. Pl.: 1009 (1753).
Dimorphochlamys Hook.f. (1867).
Raphanocarpus Hook.f. (1871).
Raphanistrocarpus (Baill.) E.G.O. Müll. & Pax (1889).
Calpidosicyos Harms (1923).

Herbaceous or woody climber or trailer to 15 m long (rarely small shrubs) with fibrous or woody, sometimes tuberous and greatly enlarged root or rootstock. Leaves entire or pedately 3–7–15-foliolate, often with discoidal glands/nectaries; tendrils simple or apically 2-fid (rarely paired at the nodes and spinose); probract absent or sessile, orbicular (*M. calantha* Gilg). Male flowers solitary or in umbels, racemes, fascicles or pseudopanicles; female flowers solitary; receptacle-tube short, broad, ± campanulate; sepals entire; corolla rotate, campanulate-urceolate or zygomorphic; petals distinct, entire, white, yellow, cream-colored or greenish, usually with black center, 1–3 with an incurved scale inside; stamens 3 or 2, inserted in the lower half of the tube; filaments distinct; two anthers 2-thecous, one 1-thecous or one 3-thecous and one 2-thecous; thecae arcuate, duplicate or triplicate; pollen large (polar axis 65–73 μm, equatorial axis 68–79 μm), 3-colporate, reticulate (Keraudren 1968; Khunwasi 1998); ovary smooth, ribbed, tuberculate or papillose; ovules few to many, horizontal or pendent or erect; stigma 3-lobed; staminodes 5. Fruit small to large, fusiform or ovoid-ellipsoid or globose, usually spiny, tuberculate, winged or ridged, indehiscent or dehiscent by 3 valves or irregularly. Seeds few to several, yellow, brown or black, often with white, yellow or red arilloid, medium-sized to large, subglobose to compressed; testa smooth or variously sculptured, margin often grooved; germination epigeal or hypogeal (Zimmermann 1922; Schaefer, unpubl. data). $n = 11$ or 14 (Beevy and Kuriachan 1996).

About 60 species in tropical and subtropical Africa, Arabia, (sub)tropical Asia, Malesia and Northeastern Australia (Schaefer and Renner 2010a; H. Schaefer, monograph in preparation); two species, *M. charantia* L. and *M. balsamina* L., naturalized in the Americas and most of the Pacific islands; in tropical rainforest, deciduous forest and bushland, savannah and semi-deserts. Host of the fungi *Puccinia cucumeris* Henn., *P. vanderystii* Henn., and *P. momordicae* Kalchbr. and Cooke (Berndt 2007). For floral biology and pollination, see Vogel (1990).

IX. Tribe Telfairieae Arn. (1841).

Tendrils simple or 2-fid. Stamens 3 (rarely 5). Fruit fleshy, ellipsoid or pear-shaped. Seeds 5–200; testa unwinged.

22. *Cogniauxia* Baill.

Cogniauxia Baill., Bull. Mens. Soc. Linn. Paris 1: 423 (1884).
Cogniauxella Baill. (1884).

Dioecious, herbaceous climber or trailer, to several meters long. Leaves simple, the blade ovate-cordate, entire or 3–5-lobed, to 18 cm long; tendrils 2-fid. Flowers large, showy, 7–8 cm in diam.; male flowers in racemes, female flowers solitary; receptacle-tube elongated, dilated at the apex; sepals triangular; petals distinct, ± asymmetric, obovate, yellow to orange; stamens 3 (rarely 5), inserted near the mouth of the tube; filaments distinct; two anthers 2-thecous, one 1-thecous; thecae duplicate; pollen medium-sized to large (polar axis c. 60 μm, equatorial axis 48–51 μm), 3-colporate, irregularly reticulate (Khunwasi 1998); ovary narrowly oblong; placentae 3; ovules many, horizontal; style short, fleshy; stigmas 2-lobed; staminodes 5. Fruit ovoid, shortly rostrate, fleshy, smooth, to 15 cm long and 8 cm in diam., ripening red. Seeds compressed, to 2 cm long, with acuminate apex and almost 2-lobed base; testa brown, smooth.

Two species in tropical Africa (Gabon, Cameroon, Congo, Angola); along forest margins and roadsides, also in secondary forest.

23. *Telfairia* Hook. Fig. 26

Telfairia Hook., Bot. Mag.: 2751 (1827).
Joliffia Bojer ex Delile (1827).

Dioecious (rarely monoecious fide Akoroda et al. 1990), large, woody liana, to 30 m long, with strong, fleshy, ± tuberous roots. Leaves petiolate, pedately (3–)5–7-foliolate, leaflets ± elliptic; tendrils apically 2-fid; probracts tubular, with nectaries. Flowers large, showy, sweet-scented, diurnal; male flowers in racemes, female flowers solitary (rarely in pairs); receptacle-tube short, campanulate; sepals triangular-acuminate, dentate; petals c. 2 cm long, distinct, white, purplish or pink, fringed; stamens 5 or 3, inserted halfway up the tube; filaments distinct; anthers five, 2-thecous or three (two 4-thecous, one 2-thecous); thecae ± straight; pollen large (polar axis 60–82 μm, equatorial axis 50–70 μm), 3-colporate, reticulate (Keraudren 1968; Khunwasi 1998); ovary ribbed; ovules many, horizontal; stigmas 3. Fruit fleshy, ellipsoid, to 60 cm long and 25 cm in diam., weighing to 12 kg, ribbed, with or without an expanded, basal collar, with white waxy surface when young, dehiscing apically by 8–10 longitudinal valves. Seeds up to 200, large, broadly ovate, 3.5–5 cm in diam., ± compressed, in yellowish-white pulp, covered by fibrous sheath, containing edible oil; testa yellowish to pale or dark reddish brown, smooth or ± verrucous; germination hypogeal (Zimmermann 1922), seeds often germinate within the fruit (Akoroda et al. 1990). $n=12$ in *T. occidentalis* Hook.f. (Okoli 1987).

Three species in tropical Africa; in lowland rainforest, often cultivated in villages and nearby for the leaves and oily seeds (Okoli and Mgbeogu 1983).

24. *Ampelosicyos* A. Thouars

Ampelosicyos A. Thouars, Hist. Vég. Isles Austr. Afrique 68 (1808), as *Ampelosycios*.
Delognaea Cogn. (1884).

Monoecious (*A. meridionalis* Keraudren perhaps dioecious), herbaceous to woody climbers or trailers, some (or all) with partly exposed, tuberous rootstock of to 30 cm in diam. Leaves simple, petiolate, 3–5-foliolate, the leaflets oblong-lanceolate, entire or deeply lobed; tendrils simple or 2-fid, long, glabrous or (partly) pubescent.

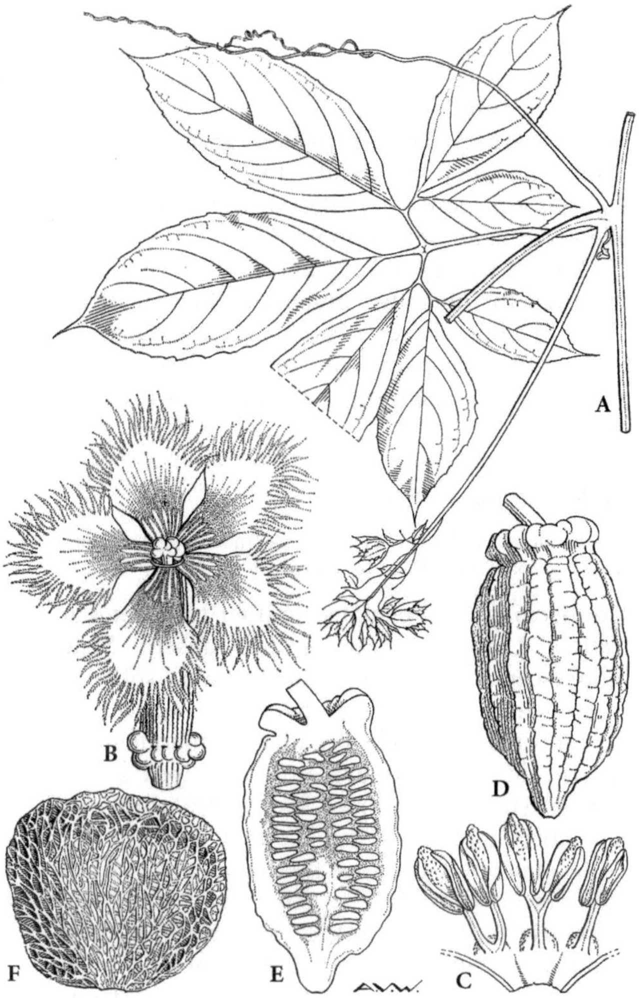

Fig. 26. Cucurbitaceae. *Telfairia pedata.* **A** Node with male inflorescence. **B** Female flower. **C** Stamens. **D** Fruit. **E** Fruit, longitudinally sectioned. **F** Seed with endocarpic fibrous sheath. (Jeffrey 1967)

Flowers medium-sized to large; male flowers in small racemes or solitary, female flowers solitary; receptacle-tube elongate, cylindrical, apically expanded, to 20 cm long and 2 mm in diam. near the base, apically expanded to to 12 mm; sepals triangular-dentiform, about 2 mm long; petals distinct, to 3 cm long, ± ellipsoidal, white to yellowish, margin with to 2 cm long fringes; stamens 5, inserted near the mouth of the tube; filaments very short; anthers all 1-thecous; thecae triplicate; pollen medium-sized to large (polar and equatorial axes c. 46–66 μm), 3-colporate, (micro)reticulate to striate-reticulate (Keraudren 1968); ovary smooth; placentae 3; ovules many, horizontal; stigmas 3, 2-lobed. Fruit ± pear-shaped, 7–15 cm long, fleshy, smooth, indehiscent, yellow with whitish pulp. Seeds horizontal, bean-shaped, large, to 3 cm long and 2 cm in diam., rich in oil; testa pale cream-colored, smooth, no distinct margin.

Three species endemic in Madagascar (Keraudren-Aymonin 1971); in forest remnants, highly endangered.

25. *Tricyclandra* Keraudren

Tricyclandra Keraudren, Bull. Soc. Bot. France 112: 327 (1966).
Odosicyos Keraudren (1981).

Dioecious, herbaceous climber or trailer to several meters long, with large tuberous root (weighing several kg) or subterranean, tuberous rootstock, 1 m or more in diam. Leaves simple, petiolate, the blade pedately 3–5-lobed; tendrils simple, long. Flowers medium-sized, opening at night, with light, pleasant fragrance; male flowers in pedunculate, elongated racemes, female flowers unknown; receptacle-tube funnel-shaped, elongate; sepals triangular, 1 mm long; petals ± connate, white to cream, apically divided into long fringes; stamens 3, inserted in the upper half of the tube; filaments distinct, very short; anthers all 1-thecous; thecae circular; pollen (*T. leandrii* Keraudr.) large (polar axis c. 55 μm, equatorial axis c. 55 μm), 3-porate, verrucate (Keraudren 1968). Fruit ± ellipsoid, fibrous pepo, 3 cm long, 2 cm in diam., or 12–15 cm long and 5–8 cm in diam., indehiscent, rostrate, smooth. Seeds 5–6 or 15–25, obovate, asymmetrical, compressed; testa pale brown, smooth.

Two species endemic in Madagascar; in dry bush and deciduous forest; highly endangered in their natural range but often cultivated in Europe as ornamental plants; flowering Sept.–Nov.

X. Tribe Bryonieae Dumort. (1827).

Tendrils simple or absent. Stamens 3. Fruit a small or oblong berry, indehiscent or watery and ejecting seeds by elastic contraction. Seeds few to many, compressed, unwinged.

26. *Austrobryonia* H. Schaef.

Austrobryonia H. Schaef. in Schaefer et al., Syst. Bot. 33: 126 (2008).

Monoecious, perennial or annual trailers with woody rootstock. Leaves ± ovate, unlobed or shallowly 3-, 5- or 7-lobed; tendrils simple. Flowers solitary or in fascicles; receptacle-tube broadly campanulate; sepals small, narrow-triangular; corolla ± rotate; petals ovate, yellow-green or yellow; stamens inserted halfway up the tube; filaments distinct, short; two anthers 2-thecous, one 1-thecous; thecae curved; pollen prolate, 3-colporate (Jeffrey 1969); ovary subglobose or ellipsoidal; ovules many, horizontal; disk annular; style very short or stigma ± sessile; stigma 2- or 5-lobed, the lobes capitate or linear, papillose; staminodes 3. Fruit a globose or ellipsoidal, 10–35 mm long berry, ripening green to yellow. Seeds few to several, ovate, compressed; testa smooth, pale, margins sometimes thickened.

Four species endemic in dry regions of Central and Western Australia (molecular phylogeny: Schaefer l.c.); on clay soils of river flood plains, waterhole and dam margins and swales in dunefields, grasslands on cracking clay, grassy woodlands on red earth.

27. *Bryonia* L.

Bryonia L., Sp. Pl.: 1012 (1753); Jeffrey, Kew Bull. 23: 441–461 (1969).

Monoecious or dioecious, perennial climbers or trailers with fleshy or woody rootstock. Leaves ovate-cordate to triangular, entire to 5-lobed; tendrils simple. Flowers small, in axillary, racemose panicles or sub-umbellate fascicles; receptacle-tube shortly campanulate; corolla almost rotate; petals connate at base, greenish-white; stamens inserted near the rim of the tube; filaments short, distinct; two anthers 2-thecous, one 1-thecous; thecae triplicate; pollen medium-sized to large (polar axis c. 42–60 μm, equatorial axis c. 35–52 μm), 3-colporate, reticulate (Khunwasi 1998); ovary globose; style elongate, 3-fid; stigmas 2-lobed; ovules many, horizontal; staminodes 3–5 or absent. Fruit a fleshy berry, smooth, ripening green with pale stripes, red or black, when mature separating from the peduncle and in *B. verrucosa* ejecting the seeds by elastic contraction. Seeds few, compressed, ellipsoid; testa brown, smooth, no distinct margin. Usually $n = 10$ (Volz and Renner 2008).

Ten species in Europe, North Africa, Canary Islands, Central Asia (Volz and Renner 2009); forest margins and disturbed ground, semi-deserts and dry bushland. Host of the fungus *Puccinia isiacae* Winter (Berndt 2007) and of the oligolectic sand bee *Andrena florea*.

Medicinal uses of *Bryonia* have been recorded for over two millennia (Renner et al. 2008). Today, there is a considerable market for *Bryonia* preparations, mostly for homeopathic medicine, although effectiveness remains contested.

28. *Ecballium* A. Rich.

Ecballium A. Rich. in Bory de St.-Vincent, Dict. Class. Hist. Nat. 6: 19 (1824), nom. cons.

Monoecious or dioecious, annual to perennial trailers. Leaves cordate; tendrils absent. Male inflorescence a raceme, female flowers solitary; receptacle-tube short-campanulate; sepals linear-lanceolate; corolla broadly campanulate or almost rotate; petals ovate-oblong, apex acute, yellow; stamens inserted near the center of the tube; filaments short, distinct; two anthers 2-thecous, one 1-thecous; thecae reflexed; pollen medium-sized to large (polar axis c. 67 μm, equatorial axis c. 38 μm), 3-colporate, reticulate (Khunwasi 1998); ovary oblong, hispid; style short; stigmas 3, 2-lobed; ovules many, horizontal; staminodes 3. Fruits oblong, hispid, scabrous, watery, when mature separating from the peduncle and contracting at the base, ejecting the seeds by elastic contraction. Seeds many, oblong, compressed; testa pale yellow to brown, smooth, narrowly marginate. $2n = 18$ (Slavik et al. 1993).

One species, *E. elaterium* (L.) A. Rich., with a monoecious and a dioecious subspecies (Costich and Galán 1988; Costich and Meagher 1992). Mediterranean to North Africa and Southwest Asia; on disturbed ground. Pollination biology: Dukas (1987) and Rust et al. (2003).

XI. Tribe Schizopeponeae C. Jeffrey (1964).

Herpetospermeae (C. Jeffrey) C. Jeffrey, Bot. Zhurn. 90: 333 (2005).

Tendrils 2 or 3-fid. Stamens 3. Fruit indehiscent or 3-valved. Seeds 1–48, unwinged.

29. *Schizopepon* Maxim.

Schizopepon Maxim., Mém. Sav. Étr. Acad. St. Pétersbourg 9: 110 (1859).

Dioecious or monoecious climbers; rarely flowers bisexual. Leaves ovate-cordate or hastate, usually 5–7-lobed; tendrils 2-fid. Flowers small; male flowers usually in racemes, female flowers solitary or few in a raceme; receptacle-tube cupular or campanulate; sepals lanceolate or subulate; petals white, ovate; stamens inserted at the base of the tube; filaments short, distinct or connate; two anthers 2-thecous, one 1-thecous; thecae straight; pollen (*S. longipes* Gagnep.) medium-sized (polar axis c. 43 μm, equatorial axis c. 47 μm), 3-colporate, reticulate (Khunwasi 1998); ovary ovate or conical, 3-locular; ovule pendent, one per locule; style short; stigmas 3(–5), slightly expanded, 2-lobed; hermaphrodite individuals produce solitary perfect flowers from leaf axils; each hermaphrodite flower has three stamens, a short style with a 3-lobed stigma, and a triangular hypogenous ovary. Fruits small, ovate or conical, smooth or punctate, apex acute or long-acuminate, 3-valved or indehiscent. Seeds 1–3, pendent, ovate, compressed; testa brown, ± sculptured, margin irregularly dentate. $n = 10$ (Nishikawa 1981).

Six to eight species in Russia, India, Myanmar, China, and Japan (Lu 1985); in river valleys, thickets, forests, on roadsides and mountain slopes up to 3,000 m; flowering and fruiting May–Nov. Details on floral biology: Akimoto et al. (1999) and Fukuhara and Akimoto (1999).

30. *Herpetospermum* Wall.

Herpetospermum Wall. ex Benth. & Hook.f., Gen. 1: 834 (1867).
Edgaria C.B. Clarke (1876).
Rampinia C.B. Clarke (1876), nom. illegit.
Warea C.B. Clarke (1876), nom. illegit.
Biswarea Cogn. (1882).

Dioecious climbers with spreading roots. Leaves ovate-cordate, to 15 cm long, 5–7-lobed or unlobed, margin entire or irregularly dentate; tendrils 2(–3)-fid. Flowers medium-sized, showy, fragrant; male flowers in racemes (rarely solitary), female flowers solitary; receptacle-tube narrowly tubular below, dilated above and broadly campanulate; sepals linear to subulate; corolla broadly campanulate to rotate; petals connate at the base, entire, elliptic, yellow; stamens inserted in the upper half of the tube; filaments distinct; two anthers 2-thecous, one 1-thecous; thecae straight, duplicate or triplicate; pollen very large (polar axis 108–110 μm, equatorial axis 111–134 μm), 3-porate, baculate/gemmate (Khunwasi 1998); ovary oblong to narrowly ovoid, 3-locular; ovules 1–6 or 16 per locule, pendent or ± horizontal; stigmas 3, dilated; staminodes 3 or absent. Fruit dry, fibrous, broadly oblong to ellipsoid-fusiform, ± ribbed, 5–8 cm long, apically dehiscing into 3 valves. Seeds 6, 12 or c. 48, oblong or obovate, compressed; testa smooth, margin obtuse. $n=11$ in *H. darjeelingensis* (Thakur and Sinha 1973).

Three species in India, Myanmar, Nepal, Tibet, China (Yunnan); among shrubs and on riverbanks; flowering July–October.

XII. Tribe Sicyoeae Schrad. (1838).

Tendrils simple or 2–8-fid. Stamens 2–5. Fruit fleshy or dry, indehiscent, explosively dehiscent or operculate. Seeds solitary or few to many.

31. *Nothoalsomitra* I. Telford

Nothoalsomitra I. Telford, Fl. Australia 8: 388, 172 (1982).

Dioecious, perennial, herbaceous climbers with woody base, to several meters long. Leaves pedately 3-foliolate, the leaflets ± equal, ovate to lanceolate, to 11 cm long; tendrils 2-fid. Flowers small; male flowers in racemes, female flowers solitary; receptacle-tube long and deeply campanulate; sepals 5, triangular, to 2 mm long; petals 5, to 6 mm long, white-tomentose outside, yellow inside; stamens 3, inserted near the mouth of the tube; filaments distinct, relatively long; two anthers 2-thecous, one 1-thecous, distinct but appressed into a central head; thecae flexuose, triplicate; pollen large (polar axis c. 60 μm, equatorial axis c. 63 μm), 3-colporate, reticulate (Khunwasi 1998); ovary ellipsoidal; ovules many, horizontal; style short, thick; stigmas 3, the lobes spreading, flexuose; staminodes 3.

Fruit fleshy, ellipsoidal, 8–12 cm long and 4–5 cm in diam., glabrous, indehiscent, ripening variegated green to yellowish. Seeds many, ovoid, 11–13 by 7–9 mm, ± tumid, truncate; testa brown, smooth, no distinct margin.

One species, *N. suberosa* (Bailey) I. Telford, endemic to subtropical E Australia; in rainforest and humid *Eucalyptus* forest.

32. *Luffa* Mill.

Luffa Mill., Gard. Dict. Abridg. ed. 4 (1754).
Trevouxia Scopoli (1777).
Turia Forssk. (1775), vide I. Friis, Taxon 33: 666 (1984).

Monoecious or dioecious (*L. echinata*), herbaceous climbers or trailers to 15 m long. Leaves simple, the blade ovate-cordate, palmately 3–5-lobed; tendrils apically 2–6-fid; probract small, lingulate. Flowers large; male flowers in racemes, female flowers solitary; receptacle-tube campanulate; sepals 3 or 5, entire; petals 5, distinct, entire, yellowish-white to golden yellow; stamens 5, inserted near the mouth of the tube; filaments distinct; anthers all 1-thecous or two 2-thecous and one 1-thecous; thecae convoluted; pollen (very) large (polar axis 70–110 μm, equatorial axis 70–110 μm), 3-colporate, perforate to reticulate (Khunwasi 1998); ovary smooth, ribbed, tuberculate or spiny; ovules many, horizontal; stigmas 3, 2-lobed; staminodes 5. Fruit dry with fibrous tissue, subglobose to cylindrical, beaked, smooth, ribbed or ± spiny, ripening brown, operculate (a pyxidium). Seeds many, oblong-elliptic, compressed; testa smooth, blackish, with or without a narrow, ± distinct membraneous border and 2 oblique bumps above hilum on each face. $n = 13$ in *L. acutangula* (L.) Roxb., *L. aegyptiaca* Mill. and *L. operculata* L. (Dutt and Roy 1971; Heiser and Schilling 1988; Heiser et al. 1988; Singh 1991; Beevy and Kuriachan 1996).

Five or seven species: four in Africa, Asia, Australia, and Polynesia, one or three in Central and South America; on riverbanks, along forest margins, and on disturbed ground.

Loofah sponges constitute an important biodegradable and renewable resource, and demand is rising, along with an interest in producing large acreages of *Luffa* in regions with a long growing season and warm temperatures, such as the southeastern United States.

33. *Trichosanthes* L.

Trichosanthes L., Sp. Pl.: 1008 (1753).
Anguina Mill. (1755).
Cucumeroides Gaertner (1791).
Involucraria Ser. (1825).
Gymnopetalum Arn. in R. Wight (1840).
Tripodanthera M. J. Roem. (1846).
Scotanthus Naudin (1862).
Eopepon Naudin (1866).
Platygonia Naudin (1866).

Dioecious or rarely monoecious, annual or perennial, herbaceous climbers, some with woody rootstock. Leaves simple, the blade entire or palmately 3–7(–9)-lobed, rarely compound, 3–5-foliate, margin usually denticulate; tendrils 2–5-fid (rarely simple), sometimes with apical adhesive pads; probract often present. Flowers medium-sized to large, mostly fragrant and opening at night; male flowers usually bracteate, in racemes, rarely solitary and coaxillary with a raceme, female flowers solitary (rarely in racemes); receptacle-tube elongate, tubular to cylindric, often dilated at the apex; sepals 5, entire, serrate or laciniate, triangular to lanceolate; petals 5, long-fimbriate less often entire, white, rarely pink or red; stamens 3, inserted halfway up the tube; filaments very short, distinct; two anthers 2-thecous, one 1-thecous; thecae triplicate; pollen medium-sized to (very) large (polar axis 32–98 μm, equatorial axis 34–125 μm), 3(4)(col)porate, psilate, perforate, rugulate, verrucate or (micro) reticulate (Khunwasi 1998; Pruesapan and van der Ham 2005); ovary ovoid or fusiform, glabrous to villous; placentae 3; ovules many, horizontal, ± pendent; style slender to filiform; stigmas 3, entire or 2-fid. Fruit fleshy, pulpy, globose, ovoid to ellipsoid or fusiform, indehiscent, usually glabrous and smooth, sometimes ribbed, ripening orange to red (rarely metallic blue). Seeds many, oblong or ovate, sagittate, or ± rounded, 1-loculed, compressed or 3-loculed, turgid, the two lateral locules empty; testa ± smooth, yellowish to white, black or dark brown, with or without distinct margin; germination epigeal. $n = 11$ or 12 (Beevy and Kuriachan 1996), up to $2n = 88$ in *T. kirilowii* Maxim.

About 100 species in India, China, Taiwan, Japan, Southeast Asia, New Guinea, Northeast Australia (de Wilde and Duyfjes 2004, 2006b); in humid forest; one species, *T. cucumerina* L. var. *anguina* (L.) Haines, cultivated in tropical regions of Africa, Asia, Central and South America. Host of the fungi *Puccinia gymnopetali-wightii* T.S. Ramakr., Srinivasan and Sundaram, and *Uredo trichosanthis* (Berndt 2007).

Molecular phylogenetic data indicate that *Gymnopetalum* is nested inside *Trichosanthes*, a genus that itself is polyphyletic (Schaefer et al. 2008a) and in need of re-evaluation.

34. *Hodgsonia* Hook.f. & Thomson

Hodgsonia Hook.f. & Thomson, Proc. Linn. Soc. London 2: 257 ('1853', 1854); W.J. de Wilde & Duyfjes, Blumea 46: 169–179 (2001).

Dioecious, perennial, woody liana, to 30 m long, stems to 7 mm in diam. Leave simple, petiolate (to 8 cm long), the blade subcircular, palmately 3–5-lobed, to 25 cm in diam.; tendrils 2–3-fid; probract thorn-like, c. 5 mm long. Flowers large, fragrant, opening at night; male flowers in bracteate, pedunculate racemes, female flowers solitary (rarely in short racemes); receptacle-tube elongate, to 12 cm long, apically dilated into a shallow cup; sepals 5, small (1–4 mm long); corolla rotate; petals 5, distinct, cuneate, to 5 cm long, white to yellowish, long-fimbriate with 5–15 cm long, spiraling or straight threads; stamens 3, inserted in the upper half of the tube; filaments distinct, short; two anthers 2-thecous, one 1-thecous, connate into a globose head; thecae duplicate; pollen (*H. macrocarpa* Cogn.) very large (polar axis c. 158 μm, equatorial axis 148 μm), 3-colporate, coarsely reticulate (Khunwasi 1998); disk 3-parted, free or joined to base of tube; ovary subglobose, secondarily 3-carpellate, secondarily 6-locular; placentae 6, parietal; ovules 6 or 12 in 6 collateral pairs, erect or pendent; style filiform; stigma large, obconical, 3-lobed; staminodes absent. Fruit a large, pulpy drupe, hard-walled, smooth or shallowly 6–12-grooved, depressed globose, to 25 cm in diam., with 6 large, simple or compound, ± ovoid, veined pyrenes. Seeds 1–3 per pyrene, compressed, large, corky, containing edible oil; testa thin. $n = 9$ (Chen 1993).

Two species in Northeast India, Bhutan, South China, Myanmar, Laos, Cambodia, Vietnam, Thailand, Malaysia, Indonesia; in lowland and lower montane forest, on riverbanks; sometimes cultivated for the seeds.

35. *Linnaeosicyos* H. Schaef. & Kocyan

Linnaeosicyos H. Schaef. & Kocyan in Schaefer et al., Syst. Bot. 33: 349–355 (2008).

Dioecious, perennial climber or trailer to 6 m long, with fleshy rootstock. Leaves simple, the blade reniform to suborbicular, entire to deeply 3-lobed, the upper side distinctly pustulate with short trichomes on whitish-gray, discoidal, multicellular, cystolith-bearing hairbases; tendrils simple, to 12 cm long. Flowers solitary; receptacle-tube broadly campanulate, in buds to 20 mm long, glabrous; sepals 5, narrow-triangular, c. 10 mm long; petals 5, ovate, 30 mm×12 mm, white with green veins, fimbriate; stamens 3, inserted 10 mm below the mouth of the receptacle-tube; filaments distinct, c. 1 mm long, glabrous; two anthers 2-thecous, one 1-thecous, connate into a head, c. 9 mm long; thecae triplicate; pollen reticulate, 4-colporate, c. 30 μm in diam. (Schaefer et al. 2008a); ovary ellipsoidal, c. 25 mm long; placentae 3; ovules numerous; stigma 3-lobed, the lobes capitate; staminodes minute. Fruit turbinate to ellipsoidal, green, pendent, 8–12 cm long, 3–4 cm diam. Seeds in soft, whitish pulp, many (several hundreds), linear-oblong, compressed; testa yellowish-brown, margin distinct, flat.

One species, *L. amara* (L.) H. Schaef. & Kocyan, endemic in Hispaniola (Dominican Republic); among cacti in dry thickets and in dry forests from sea level to 300–400 m; flowering December to May, ripe fruits in April, June, and October.

Molecular sequence data show that this species is the sister to all other New World Sicyoeae (Schaefer et al. 2008a).

36. *Echinocystis* Torr. & A. Gray

Echinocystis Torr. & A. Gray, Fl. N. Am. 1: 542 (1840), nom. cons.
Pseudoechinopepon (Cogn.) Cockerell (1897).

Monoecious, annual, herbaceous climber, to several meters long. Leaves simple, 5-lobed; tendrils 3–5-fid. Flowers small; male flowers in racemes, female flowers solitary (rarely pairs), coaxillary with the male raceme; receptacle-tube flat; sepals 6; corolla rotate; petals 6, white; stamens 3, inserted near the center of the tube; filaments very short; thecae triplicate; pollen large (polar axis c. 54 μm, equatorial axis c. 60 μm),

5-colporate, perforate-rugulate (Khunwasi 1998); ovary globose, echinate; placentae 2; style very short; stigma capitate. Fruit an ovoid, fleshy pepo, echinate with slender spines, apically dehiscent. Seeds 4, compressed; testa pale brown; germination epigeal. n=16 (Samuel et al. 1995; Gervais et al. 1999).

One species, *E. lobata* (Michx.) Torr. & A. Gray, in Eastern North America (Stocking 1955); in thickets, along roadsides, and in other disturbed areas.

37. *Marah* Kellogg

Marah Kellogg, Proc. Calif. Acad. Sci. 1: 38 (1854); l.c. ed. 2, 1: 37 (1873); S. T. Dunn, Bull. Misc. Inf. (Royal Gardens, Kew) 1913(4): 145–153.
Megarrhiza Torr. & A. Gray (1860–1861).

Monoecious (sometimes temporarily dioecious), perennial, herbaceous climber or trailer with (very) large tuberous rootstocks, often as partly exposed pachypodia. Leaves petiolate, the blade round, cordate, palmately 3-9-lobed; tendrils simple or 2-3-fid. Male flowers in racemes or panicles (or solitary), female flowers solitary, often coaxillary with male; receptacle-tube campanulate to saucer-shaped; sepals 5, filiform or absent; corolla 3–15 mm wide (wider in female), cup-shaped to rotate; petals 5, oblong to lanceolate, white or cream to yellowish green; stamens 3 (rarely 4), inserted near the center of the tube; filaments connate into a central column; anthers twisted together; thecae flexuose; pollen large (polar axis 61–92 μm, equatorial axis 54–88 μm), 4–5-colporate, perforate-rugulate (Khunwasi 1998); ovary ovoid to globose, rostrate, glabrous or setose; placentae 2–4; style short; stigma 1, ± hemispherical, 2–5-lobed; ovules 1–8 per locule, erect; staminodes 0–3. Fruit a dry, round, ovate, or oblong capsule, irregularly dehiscent, ± symmetric, 3–6 cm in diam., ± prickly or setose, sometimes rostrate, glabrous or tomentose. Seeds 4–30, ± globose, turgid, to 3.5 cm in diam.; testa smooth, yellowish to gray, margin not distinct; germination hypogeal; n=15 (Parfitt et al. 1990).

About seven species in the Western and Southeastern US (Washington to California, Arizona, New Mexico) and Mexico; in moist canyons and scrubland; flowering in February–June.

38. *Frantzia* Pittier

Frantzia Pittier, Contr. U.S. Natl. Herb. 13(4): 127–128 (1910).
Polakowskia Pittier (1910).

Monoecious, perennial, herbaceous climbers, to several meters long, with tuberous roots; tendrils 3- to 5-fid. Leaves simple, long-petiolate, the blade palmately lobed or angulate. Flowers small; male flowers in racemes, female flowers solitary or 2–5, often coaxillary with male inflorescence; receptacle-tube semi-globose, with 10 pouch-like nectaries at the base, some species with umbrella-like covering over the nectaries; sepals 5, triangular or thick and rounded; corolla rotate; petals 5, ovate-lanceolate, apex acute, white; stamens 3, inserted on the base of of the tube; filaments connate into a central column; two anthers 2-thecous, one 1-thecous or anthers connate into a subglobose head; thecae flexuous; pollen large (polar axis 74–77 μm, equatorial axis 85–88 μm), 7–10-colpate, echinate (Khunwasi 1998); ovary fusiform, setose; placenta 1; style 1, short; stigma 3–5-lobed, lobes reflexed; ovule 1, pendent. Fruit medium-sized, 3–6 cm long, fleshy, ovoid to fusiform, indehiscent, sulcate at the apex, sparsely spiny along the ridges or at the apices or glabrous, ripening green, yellowish or purple. Seed solitary, ovate, compressed, pendent, woody, germinating within the fruit. Chromosome numbers are $n = 12$ in *F. villosa* Wunderlin and $n = 14$ in *F. venosa* L. D. Gomez (Mercado and Lira Saade 1994).

About five species in Central America (Costa Rica, Nicaragua, Panama); in forest and secondary scrub (Wunderlin 1976); one species, *F. tacaco* (Pittier) Wunderlin, is a widely cultivated vegetable.

Molecular data (Sebastian et al. 2010 and unpubl. data) suggest that a monophyletic *Frantzia* minimally includes *F. pittieri* (Cogn.) Pittier, *F. tacaco*, *F. talamancensis* Wunderlin, *F. venosa* L. D. Gómez, and *F. villosa* Wunderlin.

39. *Sicyos* L.

Sicyos L., Sp. Pl. 2: 1013 (1753).
Sicyoides Mill. (1754).
Sechium P. Browne (1756), nom. cons.
Bryoniastrum Heist. ex Fabr. (1759).
Chayota Jacq. (1780).
Sicyosperma A. Gray (1853).
Microsechium Naudin (1866).

Sechiopsis Naudin (1866); D. M. Kearns. Syst. Bot. 17: 395–408 (1992), rev.
Pterosicyos Brandegee (1914).
Ahzolia Standl. & Steyerm. (1944).
Anomalosicyos Gentry (1946).
Sicyocaulis Wiggins (1970).
Skottsbergiliana H. St. John (1974).
Parasicyos Dieterle (1975).
Sicyocarya (A. Gray) H. St. John (1978).
Sarx H. St. John (1978).
Cladocarpa (H. St. John) H. St. John (1978).
Costarica L.D. Gómez (1983).

Monoecious, annual or perennial, herbaceous climbers or trailers, to 10 m long, with fibrous to tuberous roots or woody rootstocks. Leaves simple, petiolate (rarely sessile), blade angulate or lobed, rarely suborbicular; tendril (2)3–5(6)-fid, rarely simple, with long stout peduncle. Flowers small (even minute), white, greenish or yellow; male flowers in racemes or panicles, female flowers solitary, in small racemes, umbels or dense capitula of 3–40 (rarely solitary or pairs), usually coaxillary with the male flowers, sometimes enclosed in a pair of dentate bracts; receptacle-tube cup-shaped to broadly campanulate, sometimes pitted with nectariferous foveolae (pouches); sepals 5 (rarely 3–4), very small; corolla rotate; petals 5 (rarely 3–4), basally connate, white or yellowish-green; stamens (2)3(–5), inserted near the base of the tube; filaments more or less connate into a central column; anthers sessile; thecae sigmoid, flexuous, or straight; pollen medium-sized to large (polar axis 31–92 μm, equatorial axis 34–110 μm), 6–12-colpate, echinate (Khunwasi 1998); ovary ovoid, fusiform, angular, or rarely winged, ± pubescent; ovule 1, pendent, reflexed; style slender or fleshy; stigmas 2–3, ± dilated, often reflexed. Fruits fleshy or dry, indehiscent, clustered in capitula, sometimes enclosed by a subtending leaf or bracts, small or up to 20 cm long (*S. edulis*), ovoid to fusiform, armed with retrorse barbs or unarmed, occasionally a winged samara, glabrous or villous. Seed solitary, tumid to compressed; testa smooth, no distinct margin. $n = 12$ in *Sicyos angulatus* L. (Turala-Szybowska 1990) and *S. nihoaense* H. St. John (Carr 1985), $n = 12$, 13, or 14 in *S. edule* Jacq. (Beevy and Kuriachan 1996), $n = 14$ in *Microsechium compositum* Donn. Sm., $n = 14$ in *M. hintonii* (Paul G. Wilson) C. Jeffrey, and $n = 15$ in *Sechium chinantlense* (Mercado and Lira 1994).

About 75 species, mostly from Mexico to Argentina, Hawaii, North America (2 species), Australia, New Zealand, Norfolk and Lord Howe Islands, Galapagos (*Sicyocaulis pentagonus* Wiggins, Sta. Cruz, Isabela; *S. villosus* Hook.f., Floreana, known only from the type collection by Darwin and apparently extinct), *S. polyacanthus* Cogn. in Africa (introduced); forest margins, hillsides, clearings, roadsides, pastures, seabird colonies.

According to molecular phylogenetic results (Sebastian et al. 2010, and unpubl. data), all the above-listed monotypic or small genera are nested among species of *Sicyos*, including *Sechium* P. Browne, *Sechiopsis* Naudin (including *Pterosicyos* Brandegee, as suggested by Kearns 1992), *Sicyosperma* A. Gray, *Sicyocaulis* Wiggins, *Parasicyos* Dieterle, and *Costarica* L. D. Gómez, which we therefore synonymize here and with the required formal transfers in a forthcoming paper.

40. *Hanburia* Seem.

Hanburia Seem., Bonplandia 6: 293 (1858).
Elateriopsis A. Ernst (1873).
Nietoa Seem. ex Schaffner (1876).

Monoecious, perennial, herbaceous climber, to 15 m long. Leaves simple, the blade broadly ovate to cordate or pentagonal, entire or 3–7-lobed, some species with discoidal glands at the base of the leaf; tendrils 2–5-fid (rarely simple), sometimes with adhesive disks. Flowers medium-sized to large, some species with vanilla-scent; male flowers in pedunculate racemes, female flowers solitary; receptacle-tube short, urceolate-cylindrical or campanulate; sepals 5, short, ± triangular or linear to subulate; corolla campanulate; petals 5(6), triangular, 5–30 mm long, yellow or (greenish-)white; stamens 3–5, inserted near the base of the tube; filaments connate into a central column; anthers connate into a central, ± globose head, all 1-thecous; thecae triplicate or convolute; pollen (very) large (polar axis 88–122 μm, equatorial axis 82–130 μm), 4–7-colporate, perforate-rugulate (Khunwasi 1998); ovary ovoid to subglobose or oblique, mostly rostrate, hispid; ovules several, erect to ascendent; style elongate; stigma peltate. Fruit fleshy, 11–14 cm long, 7.5 cm in diam., setose, rostrate, ± asymmetrically marsupiform, explosively dehiscent. Seeds few, large, circular, to 2–4 cm in diam., or ovate to pear-shaped, compressed, angularly lobed, in white, spongy pulp; testa black to gray, smooth or minutely rugulate, margin distinct.

Seven species in Central to tropical South America; in primary and disturbed rainforest, deciduous forest, and cloud forest.

41. *Cyclanthera* Schrad.

Cyclanthera Schrad., Index Sem. Gött. 1831: 2 (1831).
Discanthera Torr. & A. Gray (1840).
Rytidostylis Hook. & Arn. (1840).
Pseudocyclanthera Mart. Crov. (1954).
Cremastopus Paul G. Wilson (1962).

Monoecious, annual or perennial, herbaceous climbers, to 10 m long, sometimes with woody base. Leaves simple or pedately 3–7-foliolate, the blade lanceolate to orbicular, entire or 3–9-lobed; tendrils simple or 2-fid (rarely to many-fid), to 30 cm long. Flowers small to medium-sized; male flowers in racemes or panicles, female flowers usually solitary (rarely in groups of 2–3), often coaxillary with the male inflorescence; receptacle-tube cup-shaped, cupular or elongate-tubular; sepals 5, subulate, dentiform to filiform or absent; corolla rotate; petals 5(–10), yellow, white or greenish, united at the base only, (ovate-)triangular to lanceolate, usually acute; stamens 3, fused, inserted in the center of the tube; filaments united into a short or elongate, bottle-shaped central column; anthers connate into a globose head; thecae united into a horizontal, flat or 10-folded ring, opening by a continuous split; pollen medium-sized to (very) large (polar axis 58–107 μm, equatorial axis 45-167 μm), 4–11-colporate, perforate-rugulate (Khunwasi 1998); ovary ovoid to oblique, hirsute, echinate or setose, ± rostrate, 1-locular; ovules few to several, ascendent; style very short or elongate, slender; stigma large, subglobose or spherical. Fruit (oblique) ovoid to triangular or reniform, ± fleshy, setose, setiform or echinate, rarely glabrous, explosively dehiscent, rarely indehiscent. Seeds solitary or many, compressed, angled, 2-lobed at apex and base, ± turtle-shaped; testa crustaceous, ± verrucouse. $n = 16$ (Samuel et al. 1995, Gervais et al. 1999).

About 40 species in Southwestern USA, Mexico, Central and South America, one species extending into the Galapagos archipelago; roadsides, forest clearings, on riverbanks and cultivated ground, hedges, tropical deciduous forest, in humid lowland forest, dry xeric forest, and montane cloud forest. *Cyclanthera pedata* (L.) Schrad. is cultivated in Asia.

Molecular phylogenetic data indicate that *Rytidostylis* and *Pseudocyclanthera* are nested inside *Cyclanthera*.

42. *Echinopepon* Naudin

Echinopepon Naudin, Ann. Sci. Nat. Bot. V, 6: 17 (1866).
Brandegea Cogn. (1890).
Vaseyanthus Cogn. (1891).
Apatzingania Dieterle (1974).

Monoecious, annual or perennial, herbaceous climbers, to 5 m long, with fibrous roots or woody rootstock. Leaves simple, the blade thin, angulate-cordate, often palmately 3–5(–9)-lobed or dissected, margin entire or denticulate; tendrils simple or 2–3-fid. Flowers small; male inflorescence a raceme or panicle, female flowers mostly solitary; receptacle-tube cup-shaped to urceolate or shallowly campanulate; sepals 5, small or minute, green; corolla rotate or campanulate; petals 5, ovate-triangular, white or cream-colored, sometimes knobby-glandular (*E. insularis*); stamens 3–5; filaments connate into a central column; anthers distinct, all 2-thecous; thecae straight, curved or duplicate; pollen (very) large (polar axis 58–168 μm, equatorial axis 78–168 μm), 5–14-colpate or colporate, some pantocolpate-inaperturate, sometimes the colpi with distinct margins, perforate (some weakly verrucate or micro-reticulate) (Khunwasi 1998); ovary conic to ovoid, ± rostrate; placentae 1–2; ovules 1–5 per locule, erect to ascending (rarely horizontal or pendent); style short; stigma fleshy, subglobose. Fruits ovoid or ellipsoid, operculate (a pyxidium), glabrous or hairy, often conspicuously echinate, rostrate, rarely dry, subterranean (to 3.5 cm deep), on 6–9 cm long peduncle (*E. arachnoideus*). Seeds solitary or few, quadrangular or angular-ovoid, compressed; testa smooth, rugose or sculptured, no distinct margin. $n = 12$ (Ward and Spellenberg 1988).

About 20 species, Southern United States to Northern Argentina (Gentry 1950; Dieterle 1974; Monro and Stafford 1998); in forest clearings, semi-deserts and ravines, on hillsides, roadsides, sand dunes and seaside gravel shores, some are weeds of cultivated ground.

XIII. Tribe Coniandreae Endl. ex M. Roem. (1846).

Tendrils simple or 2–3-fid, rarely absent. Stamens 2, 3 or 5. Fruit fleshy, indehiscent. Seeds few or many, unwinged.

43. *Bambekea* Cogn.

Bambekea Cogn., Bull. Jard. Bot. État 5: 115 (1916).

Dioecious, perennial, woody climber or trailer, to 15 m long, with large tuberous rootstock, to 25 cm in diam. and 1 m long. Leaves broadly ovate, entire to palmately 3–5(–7)-lobed; tendrils 2-fid, to 20 cm long. Flowers in racemes, often with a coaxillary solitary flower; receptacle-tube very short, broad, shallow; sepals small; corolla rotate; petals entire, distinct, yellowish to orange; male petals 6 mm long, 4.5 mm broad; stamens 5, inserted near the mouth of the tube; filaments distinct; anthers all 1-thecous; thecae triplicate; pollen medium-sized (polar axis c. 38 μm, equatorial axis c. 40 μm), 3-colporate, reticulate (Khunwasi 1998); female flowers with sepals to 7 mm long, petals to 12 mm long, 5 mm broad; ovary ellipsoid to cylindrical; ovules many, horizontal; style fleshy, 7–8 mm long; stigmas globular; staminodes 5. Fruits to 8 in racemes, globose, c. 2.5 cm across, with strong gourd-like odor, style and calyx rests persistent on fruit. Seeds many, c. 5 mm long; testa smooth.

One species, *B. racemosa* Cogn., in tropical Central and West Africa (Nigeria, Gabon, Ivory Coast, Cameroon, Congo); in lowland secondary rainforest.

44. *Eureiandra* Hook.f.

Eureiandra Hook.f., Gen. Pl. 1: 826 (1867).

Dioecious, perennial (rarely annual?), herbaceous to ± woody climbers with tuberous rootstocks (to 20 cm in diam.). Leaves entire or palmately 3–5-lobed; tendrils simple. Flowers medium-sized to large, often on leafless shoots; male flowers in short, pedunculate fascicles or solitary, female flowers solitary; receptacle-tube narrowly campanulate and apically dilated, short; sepals lanceolate, acute, often acuminate, finely pubescent; petals distinct, to 3.5 cm long, cream-colored to orange-yellow (rarely white), obovate, rounded, apiculate; stamens 5 or 3, inserted about halfway up the tube; filaments distinct or 2 pairs connate; anthers all 1-thecous or two 2-thecous and one 1-thecous; thecae triplicate, glabrous or minutely to conspicuously hairy; pollen large (polar axis 54–79 μm, equatorial axis 63–79 μm), 3-colporate, reticulate (Khunwasi 1998); ovary ellipsoid-cylindrical, ± rostrate; ovules many, horizontal; stigma 3-lobed; staminodes 3–5. Fruit to 13 cm long and 8 cm in diam., ellipsoid or cylindrical, rostrate, fleshy, indehiscent, ripening orange to red. Seeds ovate to ± globose; testa blackish, smooth or fibrillose, margin not distinct or narrow.

About eight species in tropical and subtropical Africa, one species, *E. balfourii* Cogn., endemic in Socotra; in woodland and wooded grassland, coastal forests.

45. *Dendrosicyos* I.B. Balfour Fig. 27

Dendrosicyos I.B. Balfour, Proc. R. Soc. Edinburgh 11: 513 (1882).

Monoecious, perennial tree with a bloated trunk, 3 (–6) m high and to 1 m in diam. and few, thick, pendent branches. Leaves ovate-cordate, deeply pedately 4–6-lobed, prickly, the margin serrate, with characteristic, unpleasant odor; tendrils

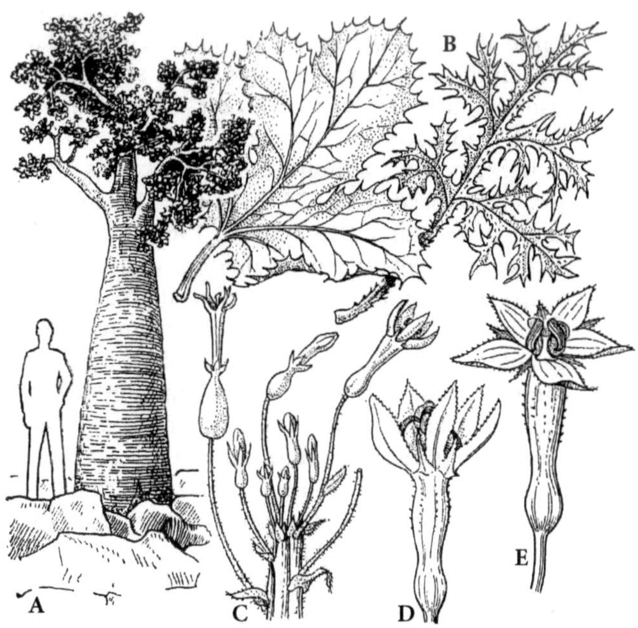

Fig. 27. Cucurbitaceae. *Dendrosicyos socratanus*. **A** Habit. **B** Lower (left) and upper (right) leaves. **C** Flowering shoot. **D** Male flower. **E** Female flower. (Takhtajan 1981)

absent. Flowers medium-sized (2.5 cm in diam.), in pendent axillary fascicles; receptacle-tube funnel-shaped; sepals lanceolate; petals linear-lanceolate, yellow, tomentose; stamens 3, inserted on mouth of the tube, exserted; thecae straight; pollen large (polar axis c. 69 μm, equatorial axis c. 71 μm), 3-colporate, reticulate (Khunwasi 1998); ovary ovoid, smooth, rostrate; style long, slender; stigmas 3, 2-lobed; staminodia 5. Fruit cylindric, to 4 cm long, rostrate, glandular setose, ripening orange. Seeds c. 6 mm long, compressed, in orange pulp.

One species, *D. socotranus* I. B. Balfour, endemic in Socotra; in shrubland communities. For morphology: Olson (2003), biogeography: Schaefer et al. (2009).

46. *Seyrigia* Keraudren

Seyrigia Keraudren, Bull. Soc. Bot. France 107: 299 (1961).

Dioecious, perennial climber with ± succulent, much-branched, (3–)5-angled stems, to 3 m high, the roots with potato-shaped tubers. Leaves mostly absent in adult plants or very small, caducuous, 3–5-lobed; tendrils simple, partly pubescent, to 20 cm long. Flowers small; male flowers in pedunculate, ± condensed racemes, female flowers solitary or in pairs; receptacle-tube cup- to funnel-shaped; sepals 2 mm long, triangular; petals elliptic-lanceolate, 1–3 mm long, yellowish-white; stamens 2, inserted near the mouth of the tube; filaments short, pubescent; anthers 2-thecous; thecae straight; pollen medium-sized (polar axis 42–46 μm, equatorial axis 25–31 μm), 3-colporate, (micro)reticulate (Keraudren 1968); pistillode cup-shaped; ovary ovate-oblong; placentae 2; ovules horizontal, 2 per locule; stigmas lobed; staminodes 2. Fruit fleshy, c. 2 cm long, ovoid, apiculate to rostrate, glabrous, indehiscent, ripening brilliant red. Seeds 2–4, 6–7 mm long, 4 mm large, in transparent arilloid and red pulp; testa brown, slightly sculptured, no distinct margin. $n = 13$ in *S. bosseri* Keraudren, *S. multiflora* Keraudren, *S. humbertii* Keraudren and *S. gracilis* Keraudren (Keraudren 1968).

Six species endemic in South and Southwest Madagascar; in xerothermic forest and bushland.

47. *Trochomeriopsis* Cogn.

Trochomeriopsis Cogn., Monogr. Phan. 3: 661 (1881).

Dioecious, perennial, herbaceous climber or trailer, to several meters long. Leaves entire, 5-lobed, or 3–5-foliolate; tendrils simple. Flowers medium-sized to large; male flowers in panicles, female flowers solitary, in pairs or racemes; receptacle-tube elongate; sepals small; petals sublinear, 5–6 cm long, yellowish-green; stamens 3, sessile, inserted at the mouth of the tube; filaments very short or absent; two anthers 2-thecous, one 1-thecous; thecae straight; pollen medium-sized (polar axis c. 37 μm, equatorial axis c. 33 μm), 3-colporate, reticulate (Keraudren 1968) or medium-sized (polar axis c. 49 μm, equatorial axis c. 56 μm), 3-colporate, reticulate according to Khunwasi (1998); ovary subcylindrical; placentae 3; style short; stigmas 3; staminodes 3. Fruit a cylindrical, smooth, fleshy berry, to 12 cm long and 3.5 cm in diam., ripening red, tinged with yellow. Seeds 20–30, with red arilloid, triangular, to 8 mm long; testa pale brown or blackish, finely sculptured, no distinct margin. $n=12$, with diploid, triploid and tetraploid individuals (Keraudren 1968).

One species, *T. diversifolia* Cogn., endemic in Madagascar; in dry *Euphorbia* bushland and dry forest, on sand dunes; flowering and fruiting all year.

48. *Halosicyos* Mart. Crov.

Halosicyos Mart. Crov., Bol. Soc. Argent. Bot. 2: 84 (1947).

Dioecious, perennial, herbaceous climber or trailer with large woody rootstock. Leaves ± circular, finely dissected, the lobes 10 mm long and 2 mm broad; tendrils simple, filiform. Flowers small; male flowers in racemes, female flowers solitary; receptacle-tube subcylindrical, the throat long-hairy; sepals dentiform; corolla rotate; petals subspathulate, 3.5–4 mm long, green; stamens 3, inserted near the mouth of the tube; filaments distinct; two anthers 2-thecous, one 1-thecous, coherent; thecae curved; pollen 3-colporate, reticulate (Pozner 1998a); ovary oblong, compressed; placentae 2; style elongate, c. 4 mm long; stigmas 2; staminodes 5. Fruit 7–8 mm long, laterally compressed, glabrous, shortly rostrate, ripening red. Seeds 4–8, pyriform, 5–5.5 by ± 2 by ± 1.5 mm; testa sculptured or rugose, slightly winged.

One species, *H. ragonesei* Mart. Crov., endemic in Central Argentina; on sandy soil and in halophilous bushland on the border of salinas.

49. *Cucurbitella* Walp.

Curcubitella Walp., Repert. Bot. Syst. 6: 50 (1846) = *Cucurbitella* Walp. corr. Walpers (1847); R. Pozner, Ann. Missouri Bot. Gard. 85: 425–439 (1998).
Prasopepon Naudin (1866).

Monoecious or dioecious, perennial, herbaceous climbers or trailers, to several meters long, with tuberous roots. Leaves entire, dissected, or palmately 3-7-lobed; tendrils simple. Flowers small; male flowers in racemes (rarely solitary), female flowers solitary; receptacle-tube campanulate; corolla imbricate; petals connate in the lower half, orange to yellow; stamens 3, inserted near the mouth of the tube; two anthers 2-thecous, one 1-thecous; filaments distinct, short, hirsute; thecae straight; pollen large (polar axis 59–67 μm, equatorial axis 58–71 μm), 3-colporate, finely reticulate (Khunwasi 1998); ovary oblong, pubescent; placentae 5; ovules horizontal, many; style columnar; stigmas 5, 2-fid; staminodes 0. Fruit a globose berry, ripening greenish with white spots or lines. Seeds many, ovate, compressed, in green, sticky pulp (arilloid jacket); testa smooth, brown, margin ± distinct.

One variable species, *Cucurbitella asperata* (Gillies ex Hook. & Arn.) Walp., in Argentina, Bolivia, Brazil, Uruguay, Paraguay (Pozner 1998b); in dry bushland, along roadsides, and on disturbed ground.

50. *Corallocarpus* Welw. ex Hook.f.

Corallocarpus Welw. ex Hook.f., Gen. Pl. 1: 831 (1867); Jeffrey, Kew Bull. 30: 485–491 (1975).
Phialocarpus Deflers (1895), pro parte, vide *Kedrostis*
Gijefa (M. Roem.) O. Kuntze (1903 ('1904')).

Monoecious, perennial, herbaceous climbers or trailers with tuberous rootstock or small shrubs (*C. glomeruliflorus* Schweinf. ex Deflers). Leaves ovate to reniform-cordate, palmately 3-5-lobed to finely dissected; tendrils simple (rarely 2-fid or absent). Flowers small, greenish-yellow; male flowers in small, pedunculate racemes, female flowers solitary or fasciculate, often coaxillary with the male raceme; receptacle-tube campanulate; sepals small; corolla rotate; petals united at base; stamens 5, inserted in the mouth of the tube; filaments distinct, short; anthers all 1-thecous, sometimes in two pairs with one single, appearing as two 2-thecous and one 1-thecous; thecae straight; pollen medium-sized to large (polar axis 51–75 μm, equatorial axis 49–73 μm), 3-colporate, finely reticulate to microreticulate (Khunwasi 1998); ovary smooth; ovules few to many, horizontal; staminodes 5 or absent; stigmas 2(3)-lobed. Fruit a small berry, to 2 cm long, ovoid, ellipsoid, often rostrate, operculate, the basal part green, expanded into a cup, the upper part red, solitary or in small groups. Seeds few to several, small, pear-shaped (rarely subglobose). $n = 13$ in *C. epigaeus* (Rottler) Hook.f. (Beevy and Kuriachan 1996).

Thirteen species in mainland Africa, two endemic in Madagascar, two in Arabia, India, and Pakistan; rainforest margins, wooded grasslands, deciduous and evergreen bushland.

51. *Kedrostis* Medik.

Kedrostis Medik., Philos. Bot. 2: 69 (1791).
Coniandra Schrad. ex Eckl. & Zeyh. (1836).
Cyrtonema Schrad. ex Eckl. & Zeyh. (1836).
Rhynchocarpa Schrad. ex Endl. (1839).
Achmandra Arn. (1840). ('*Aechmandra*', 1841), as to type *Aechmandra rostrata* (Rottler) Arn. = *Kedrostis rostrata* (Rottler) Cogn.
Pisosperma Sonder (post 15 Oct. 1862).
Cerasiocarpum Hook.f. (1867).
Toxanthera Hook.f. (1883).
Phialocarpus Deflers (1895), pro parte, vide *Corallocarpus*

Monoecious or dioecious, perennial, herbaceous climbers or trailers to 2 m long with tuberous rootstock, or woody climbers or subshrubs to 12 m long, with thick basal stems and coral-like bark or tuberous herb creeping with white underground branches each with several subsidiary tubers and only very short, leaf-bearing aboveground twigs (*K. psammophila* P. Bruyns). Leaves entire (rarely 3-foliolate), ± ovate or hastate, ± palmately lobed to deeply dissected into lanceolate segments; tendrils simple or 2-fid or absent (*K. psammophila*). Flowers small to large; male flowers in pedunculate racemes, female flowers solitary, paired or in small fascicles, in *K. psammophila* arising from subterranean stems; receptacle-tube shortly campanulate, in female *K. psammophila* initially horizontal then erect, narrowly cylindrical and solid, 35–70 mm long and 2–3 mm in diam., mostly subterranean; sepals small, ± lanceolate; corolla rotate; petals ± ovate, 1.5–8 mm long, in female *K. psammophila* 10–12 mm long, united at base, greenish to yellow; stamens 5, inserted near the mouth of the tube; filaments short; anthers all 1-thecous, two in pairs and one solitary or the pairs connate (two 2-thecous and one 1-thecous); thecae ± straight; pollen medium-sized (polar axis 53–91 μm, equatorial axis 48–95 μm), 3-colporate, perforate to (striate) reticulate (Khunwasi 1998); ovary ovoid, smooth or finely papillate; ovules horizontal; style slender; stigmas 2(3)-fid; staminodes (3–)5. Fruits solitary or in clusters, baccate, fleshy, subglobose, ovoid-rostrate, conical or fusiform, to 9 cm long and 3 cm in diam., indehiscent or opening by valves, ripening orange to red, rarely subterranean and ripening white (in *K. psammophila*). Seeds 1–10, small, tumid, asymmetrically pear-shaped to subglobose; testa smooth. $n = 12$ in *K. elongata* Keraudren (Keraudren 1968) and $n = 13$ in *K. foetidissima* (Jacq.) Cogn. (Beevy and Kuriachan 1996).

About 20 species in tropical and subtropical Africa and Arabia, six species in Madagascar, and four species in Asia (India, Sri Lanka, W Malesia); in deciduous bushland, thickets, woodland, lowland rainforest, and semi-desert grassland. Host of the fungus *Puccinia arbor-miraculensis* R. Berndt (Berndt 2007).

52. *Ceratosanthes* Adans.

Ceratosanthes Adans., Fam. Pl. 2: 139, 535 (1763).

Dioecious or monoecious (*C. hilariana* Cogn.), perennial, herbaceous climber, to 5 m long, with large tuberous rootstock. Leaves ovate, pentagonal to reniform, palmately 3–5-lobed, in flower, sometimes reduced or caducous; tendrils simple, filiform, short. Flowers small, opening at night; male flowers in long pedunculate racemes, female flowers solitary or in fascicles of 2–4; receptacle-tube elongate, cylindrical, apically expanded; sepals lanceolate, ± 2 mm long; petals cream-colored, 2-fid in the apical half, ± 10 mm, usually involute; stamens 3, inserted near the mouth of the tube; filaments very short, distinct; two anthers 2-thecous, one 1-thecous; thecae straight; pollen medium-sized (polar axis 50–60 μm, equatorial axis 53–63 μm), (3)4-colporate, irregularly reticulate (Khunwasi 1998); ovary globose to fusiform; placentae 2; ovules many, horizontal; stigmas 2, 2-fid. Fruit an ovoid-oblong berry, to 4 cm long and 2 cm in diam., rostrate, smooth, glabrous, indehiscent, ripening green or red, often with white spots. Seeds many, ovoid to subglobose, tumid; testa smooth, pale, with distinct margin.

Four species, Central America to northern Argentina; semi-arid plains and mountain slopes, roadsides, cultivated ground.

53. *Doyerea* Grosourdy

Doyerea Grosourdy, Med. Bot. Criollo 1(2): 338 (1864).
Anguriopsis J.R. Johnst. (1905).

Dioecious, perennial, woody climber to 6 m long, with thick, trunk-like base, to 15 cm high. Stems scandent, zigzag and conspicuously compressed. Leaves rounded-cordate, unlobed or 3-lobed (sometimes to almost 3-foliolate), often with prominent, marginal callosities; tendrils simple, woody, with thickened base that persists as conical spur-like structure. Inflorescence in short, sessile, few- to 40-flowered racemes, to 3 per axil; flowers small, in dense clusters; receptacle-tube turbinate-campanulate; sepals valvate, acute, to 1 mm long; corolla rotate; petals yellowish-green; stamens 3, inserted near the mouth of the tube; filaments short or absent; two anthers 2-thecous, one 1-thecous; thecae curved; pollen small (polar axis c. 43 μm, equatorial axis c. 41 μm), 3-colporate, reticulate (Khunwasi 1998); ovary ellipsoidal; placentae 2; ovules 4–6 per locule; style thick, simple, apically shortly 2-fid; stigmas penicillate-fringed, 2-lobed; staminodes 3. Fruits ellipsoid or oblong, fleshy, rostrate, 1–3 cm long, indehiscent, thin-walled, ripening red with white spots. Seeds 8–15, pear-shaped, slightly compressed, to 3–4 mm long; testa reddish brown, with distinct pale brown margin.

One species, *D. emetocathartica* Grosourdy, Caribbean, Central America, Venezuela, Guyanas, Brazil; at low altitudes in dry thickets and woodlands or on rocky slopes; flowering and fruiting Jun.–Dec. Host of the fungus *Uromyces corallocarpi* Dale (Monoson and Rogers 1978).

54. *Gurania* (Schltdl.) Cogn. Fig. 28

Gurania (Schltdl.) Cogn., Bull. Soc. R. Bot. Belg. 14: 239 (1875).
Dieudonnaea Cogn. (1875).
Ranugia (Schltdl.) T. Post & O. Kuntze (1903 ('1904')).

Appearing dioecious, but almost certainly monoecious, with plants first male, then female, perennial, herbaceous or woody climber, to 15 m or more in length. Leaves ovate-cordate, simple, unlobed or palmately lobed or 3-foliolate; tendrils simple. Flowers small or medium-sized; male flowers in fascicles or racemes (to 120 per raceme), female flowers solitary, in small groups or pendulous racemes; receptacle-tube urceolate to cylindrical, bright orange to red; sepals mostly prominent, fleshy, often enlarged, shiny orange to red, sometimes with green or yellow tip, rarely small, ± triangular, reflexed; petals inconspicuous, lanceolate, erect, fleshy, orange or yellowish-green; stamens usually 2, inserted halfway up or near the mouth of the tube; filaments short, distinct; anthers 2-thecous; thecae straight or curved to ± convolute; pollen mostly in tetrads (these 149–174 μm in diam.), the monads medium-sized to large (polar axis 51–88 μm, equatorial axis 62–111 μm), 3-porate, perforate, reticulate or psilate and baculate (Khunwasi 1998); ovary cylindrical, smooth; placentae 2; ovules many, horizontal; stigmas 2. Fruits fleshy, to 7 cm long and 2–3 cm in diam., cylindrical to pear-shaped, indehiscent, ripening (yellowish-)green. Seeds many, in yellow pulp, oblong-elliptic, ± compressed; testa smooth, gray to blackish, sometimes slightly marginate.

About 37 species in Central to tropical South America; in tropical forest. Host of the fungi *Passalora guraniae* R. Kirschner and *Stenella praelonga* (Syd.) U. Braun (Kirschner and Piepenbring 2006). For sexual strategy, see Condon and Gilbert (1988).

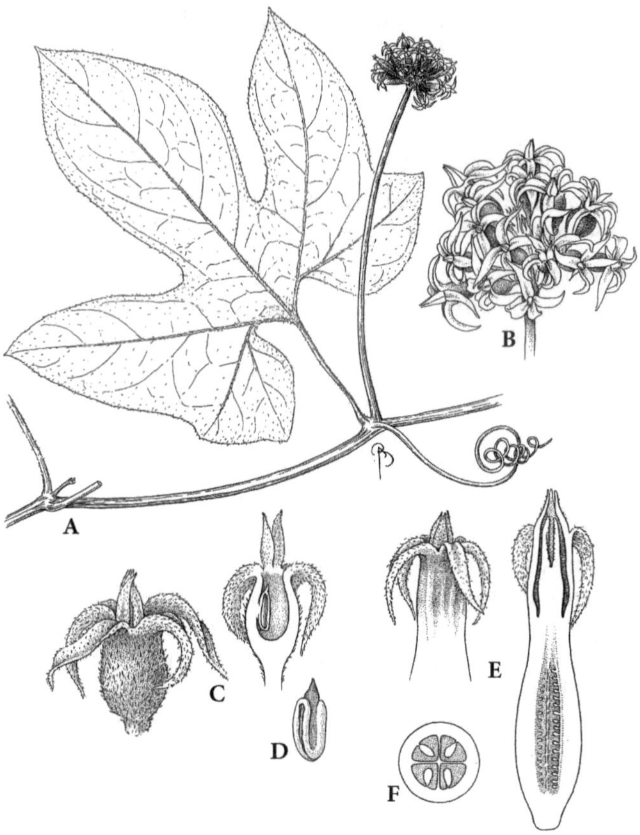

Fig. 28. Cucurbitaceae. *Gurania subumbellata*. **A** Node with female inflorescence and tendril. **B** Male inflorescence; calyx more conspicuous than corolla. **C** Male flowers. **D** Anther with connective. **E** Female flowers. **F** Transverse section of ovary showing locules and placentation. (Reproduced with permission of the artist Bobbi Angell)

55. *Psiguria* Neck. ex Arn.

Psiguria Neck. ex Arn., J. Bot. (Hooker) 3: 274–275 (1841).
Anguria Jacq. (1760), nom. illegit.

Appearing dioecious, but almost certainly monoecious, with plants first male, then female, perennial, herbaceous or woody climber, to 10 m or more in length. Leaves simple or 3-foliolate, entire or palmately 3–5-lobed; tendrils simple. Flowers small to large; male flowers in racemes or axillary spikes, female flowers solitary or in groups of 2–5; receptacle-tube urceolate to cylindrical, green; sepals small, ± triangular, green; petals broad, spreading, red, orange, or pink; stamens 2, inserted halfway up the tube; filaments distinct; anthers 2-thecous; thecae duplicate or rarely straight; pollen in monads or tetrads (152–185 μm in diam.), the monads medium-sized to large (polar axis 77–80 μm, equatorial axis 98–126 μm), 3–6-porate, verrucate, perforate or psilate (Khunwasi 1998); ovary oblong, smooth; placentae 2; ovules many, horizontal; stigmas 2. Fruits fleshy, 3–8 cm long and 2–3 cm in diam., ellipsoid to oblong, smooth, indehiscent, ripening yellowish-green or black, sometimes striped. Seeds many, to 11 mm long, oblong-elliptic, ± compressed; testa smooth, gray, no distinct margin.

About 6–12 species in Central to tropical South America; in tropical forest. Host of the fungi *Uromyces poliotelis* Syd. and *U. anguriae* H.S. Jack. and Holw. (Monoson and Rogers 1978). For a molecular phylogeny, see Steele et al. (2010).

56. *Helmontia* Cogn.

Helmontia Cogn., Bull. Soc. R. Bot. Belg. 14: 239 (1875).

Appearing dioecious, but almost certainly monoecious, with plants first male, then female, perennial, herbaceous or woody climbers, to several meters long. Leaves simple or 3-foliolate, unlobed or palmately lobed; tendrils simple. Flowers small; male flowers many (to 120 per raceme), in racemes or umbels, female flowers in pendulous racemes; receptacle-tube obconic to cylindrical; sepals small, ± triangular, reflexed; petals distinct, c. 3 mm long, yellowish-green; stamens 2, inserted near the mouth of the tube; filaments distinct; anthers 2-thecous; thecae straight; pollen simple (not in tetrads), medium-sized (polar axis 51–69 μm, equatorial axis 62–74 μm), 3-porate, perforate or reticulate (Khunwasi 1998). Fruits fleshy, ovoid to ellipsoid, indehiscent, ripening yellowish-green. Seeds many, seeds oblong-elliptic, ± compressed; testa smooth.

Two to four species in Guyana, Venezuela, and Brazil; in tropical forest.

Molecular data suggest eventual inclusion in *Psiguria/Gurania* (Kocyan et al. 2007).

57. *Melothrianthus* Mart. Crov.

Melothrianthus Mart. Crov., Notul. Syst. (Paris) 15: 58 (1954).

Dioecious, herbaceous climber or trailer. Leaves unlobed, lanceolate with subcordate to sagittate base; tendrils simple. Flowers small; male flowers in corymbs, bracteate, female flowers solitary; receptacle-tube narrowly campanulate; sepals lanceolate; petals connate at base, oblong-lanceolate, entire; stamens 3, inserted near the mouth of the tube; filaments absent; two anthers 2-thecous, one 1-thecous; thecae curved; pollen medium-sized (polar axis c. 43 μm, equatorial axis c. 45 μm), 3-colporate, perforate (Khunwasi 1998); ovary lanceolate; placentae 2; ovules many, horizontal; style robust, inserted in the center of a circular disk; stigma solitary, 2-lobed, fleshy. Fruit oblong, slightly pubescent. Seeds compressed, oblong; testa verrucous, no distinct margin.

One species, *M. smilacifolius* (Cogn.) Mart. Crov., endemic in Brazil; growing in humid places. Possibly 2–3 additional as yet undescribed species.

58. *Wilbrandia* Silva Manso

Wilbrandia Silva Manso, Enum. Subst. Brazil.: 30 (1836).

Monoecious or dioecious, perennial climber or trailer with woody rootstock. Leaves entire to palmately 3–7-lobed; tendrils simple, circinate. Flowers small; male flowers in racemes or spikes, female flowers solitary or in dense axillary clusters; petals oblong to lanceolate, papillose, white; stamens 3, inserted in the upper third of the tube; filaments very short, distinct; two anthers 2-thecous, one 1-thecous, coherent into a central head; thecae straight; pollen medium-sized (polar axis c. 52–53 μm, equatorial axis c. 54–56 μm), 3-colporate, reticulate (Khunwasi 1998); ovary ovoid-oblong, rostrate; placentae 2 or 3; ovules many, horizontal; style 2–3 mm long; stigmas 2 or 3, entire or 2-fid. Fruit an ovoid-conical berry, c. 2 cm long and 1.5 cm in diam., sessile in the leaf axils, rostrate. Seeds many, ovate to oblong, compressed, c. 5 mm long; testa with distinct margin.

Five species in South America (Brazil, Paraguay, Argentina; Martínez Crovetto 1946); in rainforest and secondary scrub.

59. *Apodanthera* Arn.

Apodanthera Arn., J. Bot. (Hooker) 3(21): 274 (1841).
Guraniopsis Cogn. (1908).

Monoecious or dioecious (*A. congestiflora*?, *A. hatschbachii*?, *A. succulenta*?, *A. villosa*?), herbaceous climbers or trailers to 5 m long, some with succulent stems, with perennial rootstock (to 1 m long and 10 cm in diam.). Leaves simple or 5-foliolate (*A. fasciculata* Cogn.), often 3–5 (–9)-lobed, often undulate and with nasty odor; tendrils simple or 2–3-fid. Flowers small to medium-sized; male inflorescence a pedunculate raceme, female flowers usually solitary, in some species long-pedunculate; receptacle-tube elongated, cylindrical; sepals linear; corolla to 6 cm across; petals (almost) distinct, greenish-white or yellow; stamens 2–3, inserted halfway up or near the mouth of the tube; filaments short and distinct or absent; two anthers 2-thecous, one 1-thecous or two 2-thecous; thecae straight or ± curved; pollen medium-sized to large (polar axis 47–103 μm, equatorial axis 52–104 μm), 3-colporate, reticulate (Khunwasi 1998); ovary ovoid or oblong; ovules numerous, horizontal; style columnar; stigmas (2–)3(–5), U-shaped. Fruit fleshy, indehiscent, ovoid to ellipsoid, ± rostrate, 1–7 cm long, ripening green or red to brown often with white stripes or spots, edible. Seeds few to many, ovoid, compressed; testa smooth, chocolate-brown (all species?), often with distinct, ivory-colored margin. n = 14 (Ward 1984).

About 16 species in America (Texas to Argentina); on roadsides and cultivated ground, in bushland and Andean grasslands (Martínez Crovetto 1956).

The seeds of *A. aspera* Cogn. have been used as oil-rich food in Mexico since pre-colonial times; remains have been found in the caves of Tehuacán, Puebla and Guilá Naquitz, Oaxaca, Mexico (Lira Saade 2004a, b). *Apodanthera sagittifolia* (Griseb.) Mart. Crov. differs from the rest of the genus in the presence of long hairs at the base of the filaments (Jeffrey 1978a, b), and does not group with *A. mandonii* Cogn. in molecular phylogenetic analyses (Schaefer et al. 2009). The genus is in need of revision.

60. *Tumamoca* Rose

Tumamoca Rose, Contr. U.S. Natl. Herb. 16: 21 (1912); D. M. Kearns, Madroño 41: 23–29 (1994).

Monoecious, perennial, herbaceous to ± woody climber or trailer with a bundle of tuberous roots (each to 15 cm in diam.). Leaves pedately 3-lobed, the lobes 2–4 cm long, divided into narrow, obtuse segments; tendrils simple, short. Flowers small, opening at night; male flowers in racemes, female flowers solitary; receptacle-tube elongate, narrowly cylindrical, c. 1 cm long; sepals triangular, minute; corolla rotate; petals narrowly linear, 4–6 mm long, pale yellow; stamens 3, inserted in the upper half of the tube; two anthers 2-thecous, one 1-thecous; pollen unknown; ovary globose to fusiform; ovules many, horizontal; staminodes 3. Fruit a globose berry, c. 1 cm in diam., glabrous, with remains of flower, ripening red (rarely yellow). Seeds 2-several, obovoid, 7–8 mm long, truncate at the apex; testa black, tuberculate-rugose, no distinct margin.

Two species in Arizona (near Tucson) and Mexico (Sonora); in semi-desert and xeric bushland; extremely rare; flowering June–September.

61. *Ibervillea* Greene

Ibervillea Greene, Erythea 3(5): 75 (1895).
Maximowiczia Cogn. (1881), nom. illegit.

Dioecious, annual or perennial climbers, to 3 m long, with large tuberous rootstocks, partly exposed as fleshy pachypodia. Leaves sublobate to 3–5-lobed, the lobes often dissected, to 10 cm long and 6 cm broad; tendrils simple. Flowers small, opening during the day; male flowers in racemes or fascicles (rarely solitary), female flowers solitary; receptacle-tube narrowly campanulate; sepals small, acute; corolla narrowly campanulate; petals 4–5 mm long, emarginate to 2-furcate, united near base, yellowish; stamens 3, inserted near the mouth of the tube; filaments connate into a central column; two anthers 2-thecous, one 1-thecous; thecae straight; pollen medium-sized (polar axis 56–68 μm, equatorial axis 56–67 μm), 3-colporate, reticulate (Khunwasi 1998); ovary ovoid to fusiform; placentae 3(–5); ovules many, horizontal; style columnar; stigmas 3–5; staminodes 3–5 (or 0). Fruit a fleshy, indehiscent, globose, ovoid or ellipsoid berry, 1.5 to 4 cm in diam., ripening orange to red. Seeds many, irregularly ovoid, scarcely compressed, in orange-red pulp; testa verrucous, transversely ridged or ± smooth, the margins raised.

Seven to eight species, Texas to Guatemala; semi-deserts, grassy plains, swampy woodlands, thorn-forest, margins of cultivated land; flowering June–Nov. For detailed morphological work, see Kearns (1994).

62. *Dieterlea* E. J. Lott

Dieterlea E. J. Lott, Brittonia 38: 407 (1986).

Dioecious, perennial, woody climber to 12 m long, with very large, tuberous rootstock. Leaves ovate-cordate to reniform, unlobed or ± 3–5-lobed; tendrils simple. Flowers large, in *D. fusiformis* E.J. Lott strongly fragrant and opening at night; male flowers in racemes, female flowers solitary; receptacle-tube narrowly cylindrical, 2–4.5 cm long; sepals distinct or united at base; petals distinct, 1.5–2.5 cm long, entire or apically 2-fid, white or pale yellow; stamens 3, inserted on mouth of the tube; filaments distinct; two anthers 2-thecous, one 1-thecous; thecae straight; pollen (*D. fusiformis*) large (polar axis c. 67 μm, equatorial axis c. 62 μm), 4-colporate, reticulate (Khunwasi 1998); ovary cylindrical, glabrous, 4-to 5-locular; ovules many, horizontal; stigmas 4–5, 2-lobed; staminodes 5 (rarely 3–4). Fruit fusiform or ellipsoid, shortly rostrate, 6–15 cm long and 3–6 cm in diam., indehiscent, ripening yellow to red. Seeds many, turgid, in red pulp; testa smooth, dark gray, margins raised, convex.

Three species in Mexico; tropical deciduous forest, dry rocky slopes.

Molecular data (Kocyan et al. 2007) suggest inclusion in *Ibervillea* but a broader analysis is needed.

XIV. Tribe Benincaseae Ser. (1825).

Dioecious or monoecious, annual or perennial, herbaceous or woody climbers or trailers, rarely shrubs. Leaves simple or 3–7-lobed, rarely absent; tendrils simple, 2–5-fid or absent. Sepals 5; petals 5; stamens (2)3–(5), two anthers 2-thecous, one 1-thecous, less often all 2-thecous or all 1-thecous; pollen mostly 3-colpate, reticulate. Fruit usually indehiscent, fleshy, medium-sized to large.

63. *Citrullus* Schrad. ex Eckl. & Zeyh.

Citrullus Schrad. ex Eckl. & Zeyh., Enum. Pl. Afr. Austral: 279 (1836), nom. cons., *Cucurbita citrullus* L., typ. cons.

Monoecious or dioecious, annual or perennial, herbaceous trailers to 6 m long, some with tuberous root to 1 m long. Leaves simple, petiolate, the blade rounded or broadly to triangular-ovate, palmately 3–5-lobed, the segments lobulate or dissected; tendrils 2–3-fid, simple or absent, in *C. naudinianus* Hook.f. spiniform. Flowers solitary (rarely fasciculate), axillary; receptacle-tube broadly campanulate; sepals 5, narrow; corolla rotate or broadly campanulate, medium-sized; petals 5, yellow to white, ovate-oblong, united at base; stamens 3, inserted near the base of the tube; filaments distinct, short; two anthers 2-thecous, one 1-thecous, distinct or slightly coherent; thecae flexuous; pollen medium-sized (polar axis 43–59 µm, equatorial axis 41–56 µm), 3-colporate, irregularly reticulate (Khunwasi 1998); ovary ovoid; placentae 3; style short, columnar; stigmas 3, thick, reniform, ± 3-lobed; ovules many, horizontal; staminodes 3, setiform or ligulate. Fruit large, globose or oblong, fleshy or dry, indehiscent, glabrous or covered with prominent spines. Seeds many, oblong, compressed; testa pale yellowish, blackish or brown, smooth, with or without distinct margin. $n = 11$ in *C. lanatus* (Thunb.) Mansf. and *C. colocynthis* Pangalo (Beevy and Kuriachan 1996).

Four species in the eastern Mediterranean region, North and tropical Africa, and western Asia (Fursa 1972a, b); in semi-deserts and xeric bushlands, on sand dunes and other disturbed ground; one species, *C. lanatus*, widely cultivated and a casual or locally naturalized in many parts of the tropics and subtropics. Host of the fungi *Puccinia citrulli* Syd., P. Syd. & Butler and *P. citrullina* Ragunathan & K. Ramakr. (Berndt 2007).

64. *Peponium* Engl.

Peponium Engl. in Engler & Prantl, Nat. Pflanzenfam., Nachtr.: 318 (1897).
Peponiella O. Kuntze (1898).

Dioecious, perennial, herbaceous climbers or trailers to 8 m long, some with tuberous rootstock. Leaves simple, petiolate, blade ± ovate-cordate, unlobed or palmately 3–5-lobed; tendrils 2-fid (rarely simple); probract oblanceolate to obovate, to 18 mm long. Flowers to 8 cm in diam., some or all species sweet-scented and opening in the evening (Zimmermann 1922); male flowers in pedunculate racemes or solitary, often raceme and solitary flower in the same axil, female flowers solitary; receptacle-tube elongated, ± cylindrical; sepals 5, linear-lanceolate; petals 5, entire, distinct, obovate, to 50 mm long and 40 mm broad, white or yellow; stamens 3, inserted near the mouth of the tube; filaments distinct; anthers all 2-thecous or two 2-thecous and one 1-thecous, connate into a central head; thecae triplicate; pollen large in continental African species (polar axis 90–98 µm, equatorial axis 92–102 µm), 3-colporate, striate (Page and Jeffrey 1975; Khunwasi 1998), in the nine Madagascan endemics medium-sized to large (polar and equatorial axis 38–82 µm), 3-colporate with very short narrow colpi, reticulate-rugulate, rugulate or striate-rugulate (Keraudren 1968); ovary ellipsoid, ± hairy; ovules many, horizontal; stigma 3-lobed; staminodes 3. Fruit fleshy, indehiscent, subglobose to ellipsoid, ± rostrate, ripening orange or red. Seeds many, elliptic, compressed; testa blackish, smooth, no distinct margin; germination epigeal (Zimmermann 1922). $n = 12$ in *P. betsiliense* Keraudr. (Keraudren 1968).

About 20 species, ten in Madagascar, one in the Seychelles and Aldabra, six in tropical Africa, and three in South Africa; in rainforest, wood-, bush- and grassland, often near open water.

65. *Lagenaria* Ser.

Lagenaria Ser., Mém. Soc. Phys. Genève 3: 26 (1825).
Adenopus Benth. (1849).
Sphaerosicyos Hook.f. (1867).

Monoecious, annual or perennial, herbaceous climbers. Leaves simple, petiolate, petioles with a pair of glands near the apex; leaf-blade ovate-cordate or reniform; probracts and bracts absent or small, tubular (*L. sphaerica* E. Mey.); tendrils 2-fid. Flowers large, solitary, sweet-scented and opening in the evening or in the morning (Zimmermann 1922), pollinated by hawkmoths (e.g., *Agrius convolvuli*, *Hippotion celerio*), moths (Noctuidae), butterflies (e.g., *Gorgyra johnstoni*), and honeybees (Morimoto et al. 2004); receptacle-tube narrowly campanulate or funnel shaped (male flower) or cupular (female flower); sepals 5; petals 5, oblong-obovate, apex retuse, white; stamens 3, inserted on the receptacle-tube; filaments distinct; two anthers 2-thecous, one 1-thecous; thecae triplicate or much convoluted; pollen large (polar axis 62–77 μm, equatorial axis 69–81 μm), 3-colporate, perforate (Khunwasi 1998), in *L. sphaerica* smaller (polar axis 60 μm, equatorial axis 60 μm), 3-colporate, reticulate (Keraudren 1968); ovary ovate or cylindric; placentae 3; ovules numerous, horizontal; stylodia short; stigmas 3, 2-lobed; staminodes 3. Fruits fleshy, medium-sized to large, globose, bottle-shaped or ellipsoid, indehiscent, ripening green, often flecked with white, woody when mature. Seeds numerous, obovate, compressed; testa with distinct margin; germination epigeal (Zimmermann 1922). $n = 11$ in *L. siceraria* (Molina) Standl. (Keraudren 1968; Beevy and Kuriachan 1996).

Six species in tropical Africa and Madagascar; along forest margins and on disturbed ground, usually near water; one species, *L. siceraria*, is cultivated throughout the tropics and subtropics and in warm temperate regions.

DNA sequence analysis of archaeological bottle gourd specimens and comparison with modern Asian and African landraces reveal detailed patterns of natural and anthropogenic dispersal (Erickson et al. 2005; Clarke et al. 2006).

66. *Acanthosicyos* Welw. ex Hook.f. Fig. 29

Acanthosicyos Welw. ex Hook.f., Gen. Pl. 1: 824 (1867).

Dioecious, perennial shrub, to 1 m high with woody, robust, angled branches. Leaves reduced to small scale-like ovate, c. 2 mm long, bracts; tendrils transformed into c. 10 mm long, straight spines. Flowers medium-sized, tomentose, ± sessile; male flowers solitary or in small fascicles, female flowers solitary; receptacle-tube turbinate; sepals 5, short, unequal (2 suborbicular to obcordate, 3 ovate); petals 5, connate at the base, broadly ovate, entire, yellow; stamens 3, inserted near the mouth of the tube; filaments distinct, very short, fleshy; two anthers 2-thecous, one 1-thecous; thecae flexuous; pollen large (polar axis c. 68 μm, equatorial axis c. 70 μm), 3-colporate, reticulate (Khunwasi 1998). Fruit globose, verrucous, indehiscent, edible, ripening orange. Seeds many, oblong, tumid, 14–15 mm long and 9–11 mm broad; testa cream-colored, crustaceous, no distinct margin.

One species, *A. horridus* Welw., in Angola, Namibia, Botswana, and Republic of South Africa; in sandy places like dunes and riverbanks, where it forms stands of several meters in diam.

Citrullus naudinianus (Sond.) Hook.f. was transferred to *Acanthosicyos* by Jeffrey (1962) based on biochemical and seed coat characters, a view not accepted by Fernandes et al. (1986), Marticorena (1963), and Fursa (1972a, b), and also contradicted by molecular data (Schaefer et al. 2009).

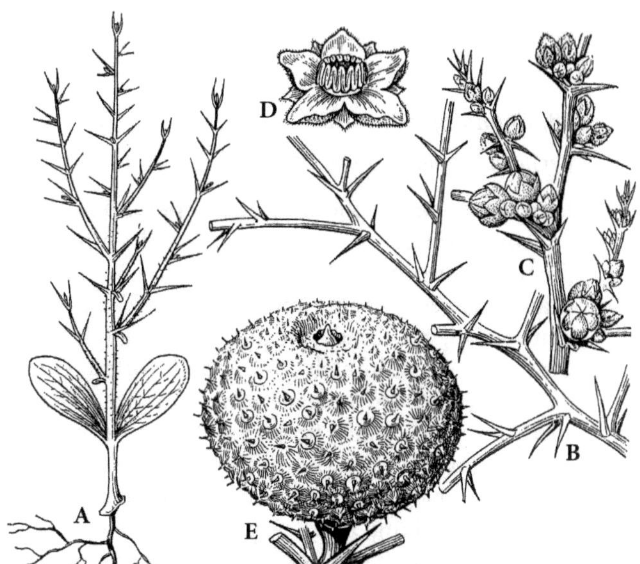

Fig. 29. Cucurbitaceae. *Acanthosicyos horridus*. **A** Seedling. **B** Branch of adult plant. **C** Branch with flower buds. **D** Male flower. **E** Fruit. (Takhtajan 1981)

67. *Raphidiocystis* Hook.f

Raphidiocystis Hook.f. in Benth. & Hook., Gen. 1: 828 (1867); Keraudren & Jeffrey, Bull. Jard. Bot. Nat. Belg. 37: 319–328 (1967), rev.

Monoecious, perennial, herbaceous climbers to 6 m long. Leaves simple, blade ovate-cordate, margin entire or sinuate; tendrils simple; pro-bract elliptic, hooded. Flowers in racemes, often male and female coaxillary or solitary; receptacle-tube short, obconic or cylindrical; sepals 5, triangular-lanceolate, in some species pinnately divided and convoluted, to 10 mm long and broad; corolla campanulate to urceolate; petals 5, entire, reflexed, yellow to orange; stamens 3, inserted in the lower half of the tube; filaments long; anthers all 2-thecous; thecae triplicate; pollen large (polar axis 56–73 μm, equatorial axis 44–70 μm), 3-colporate, reticulate (Keraudren 1968; Khunwasi 1998); ovary ellipsoid, densely bristly; ovules many, horizontal; style columnar; stigmas 3, lobed; staminodes 3. Fruits fleshy, ellipsoid to spherical, to 7 cm long, densely brown-setose, ripening reddish, dehiscing into 10 longitudinal valves and extruding the seeds in red pulp at the apex (Zimmermann 1922). Seeds many, broadly ovate, c. 5 mm long and 4 mm broad, compressed; testa smooth.

Five species, four throughout tropical Africa and one, *R. brachypoda* Baker, endemic in Madagascar; in lowland rainforest, often in clearings or along rivers.

68. *Cephalopentandra* Chiov.

Cephalopentandra Chiov., Fl. Somalia 1: 187 (1929).

Dioecious (rarely monoecious?), perennial, herbaceous climber to 2 m long, with tuberous, partly above-ground rootstock, to 20 cm in diam. Leaves simple, subsessile, semi-amplexicaul, crisp, the blade elliptic-cordate, (slightly) pinnately or palmately 5–7-lobed (often *Quercus*-like), to 9 cm long; tendrils simple. Flowers medium-sized; male flowers solitary or paired, female flowers solitary; receptacle-tube cylindrical to campanulate; sepals 5, triangular-lanceolate, 2–3 mm long; petals 5, partly connate, yellow or cream-colored with green veins; stamens 3, all 2-thecous; filaments distinct, inserted in the lower half of the tube; thecae triplicate; pollen large (polar axis 82–84 μm, equatorial axis 79–80 μm), 3-porate, (baculate?)/gemmate (Aloyshina 1971); ovary ellipsoid, smooth, hairy; ovules many, horizontal. Fruit ellipsoid, attenuate at the ends, baccate, smooth, to 8 cm long, 4 cm in diam., ripening red. Seeds many, compressed, pear-shaped; testa black, verrucous.

One species, *C. ecirrhosa* (Cogn.) C. Jeffrey, in Northeastern tropical Africa (Kenya, Uganda, Somalia, Ethiopia); deciduous *Acacia* woodland and bushland of lowland semi-deserts to 1,000 m.

69. *Lemurosicyos* Keraudren

Lemurosicyos Keraudren, Bull. Soc. Bot. France 110: 405 (1964).

Monoecious, annual, herbaceous climber or trailer, to 5 m long. Leaves simple, petiolate, palmately 3–5-lobed, the lobes lobulate-dentate to 7 cm long; tendrils simple. Male flowers in racemes, female flowers solitary or in pairs, often coaxillary with the male raceme; receptacle-tube campanulate; sepals 5, small; petals 5, 3–5 mm long, oblong-lanceolate, white; stamens 3, inserted in the lower half of the tube; filaments distinct; two anthers 2-thecous, one 1-thecous, coherent, forming a central globose head; thecae triplicate; pollen large (polar axis 65–70 μm, equatorial axis 52 μm), 3-colporate, reticulate (Keraudren 1968); ovary oblong, pubescent; placentae 3; ovules many, horizontal; style slender; stigmas 3; staminodes 3. Fruit a fleshy berry, oblong to pear-shaped, hairy, ripening scarlet. Seeds many (c. 25–30), oblong, c. 10 mm long and 5 mm broad, in yellow pulp; testa brown, with dentate margin. $n = 12$ (Keraudren 1968).

One species, *L. variegata* (Cogn.) Keraudren, endemic in Madagascar.

70. *Solena* Lour.

Solena Lour., Fl. Cochinch. 477, 514 (1790), nom. cons.; W.J.J.O. de Wilde & Duyfjes, Blumea 49: 69–81 (2004), rev.
Karivia Arn. (1840).
Melothria sect. *Solena* (Lour.) Cogn. (1881) p.p.

Monoecious or dioecious, perennial, herbaceous or ± woody climber or trailer, 2–6 m long, with tuberous roots. Leaves simple, shortly petiolate to sessile, the blade ovate or elliptic, very variable, base cordate or hastate, to 22 cm long; tendrils simple, glabrous; probract very small, linear or absent. Flowers small; male flowers in condensed racemes, female flowers solitary, sometimes coaxillary with male raceme; receptacle-tube campanulate; sepals 5, ± subulate, minute; petals 5, distinct, triangular, yellow or yellowish-white; stamens 3, inserted near the base of the tube; filaments distinct, long; two anthers 2-thecous, one 1-thecous; thecae straight, duplicate or triplicate; pollen medium-sized to large (polar axis c. 63 μm, equatorial axis c. 49 μm), 3-colporate, (perforate-)verrucate (*S. heterophylla* Lour., *S. umbellata* (Klein ex Willd.) W.J. de Wilde & Duyfjes) or reticulate (*S. amplexicaulis* (Lam.) Ghandi) (van der Ham, pers. comm.); disk 3–4-lobed, conspicuous, carnose; ovary oblong, glabrous or hairy; ovules few to several, horizontal; staminodes 3 (rarely 4). Fruit fleshy, oblong or ovoid, attenuate at both ends, glabrous or hairy, indehiscent, ripening yellow or red. Seeds few to 20, slightly compressed to ± globose; testa smooth, grayish-brown, sometimes with narrow, corky margin. $n=12$ or 24 in *S. amplexicaulis* (Beevy and Kuriachan 1996).

Three species in Afghanistan, India, Myanmar, Sri Lanka, Malesia, Vietnam, Cambodia, and China; in thickets, on roadside slopes; flowering and fruiting all year. Host of the fungus *Pseudocercospora solenae-heterophyllae* (R.K. Verma and Kamal) U. Braun (Kirschner and Piepenbring 2006).

71. *Borneosicyos* W.J. de Wilde

Borneosicyos W.J. de Wilde, Reinwardtia 11: 224 (1998).

Dioecious, perennial, herbaceous climber, to 12 m long and stems to 1 cm in diam., roots unknown. Leaves simple, petiolate, the blade ovate-oblong, entire; probract oblong, small; tendrils simple; receptacle-tube shallow. Flowers small; male flowers in racemes or panicles, female flowers solitary or in racemes coaxillary with a single female flower; sepals 5, triangular, minute; petals 5, elliptic, entire, distinct, pale yellow; stamens 3, inserted toward the base of the tube; filaments distinct, short; two anthers 2-thecous, one 1- or 1.5-thecous, distinct but appressed into a central head; thecae triplicate; pollen in tetrads, these c. 85 μm in diam., the monads medium-sized (polar axis c. 44 μm, equatorial axis c. 59 μm), 3-colporate, micro-reticulate-gemmate (van der Ham and van Heuven 2003); ovary cylindrical-oblong; ovules few, horizontal; style cylindrical 9–10 mm long; stigmas 3, heart-shaped, papillose; staminodes 4–5. Fruit solitary, oblong, 8–10 cm long and 4–4.5 cm in diam., glabrous, indehiscent, ripening red. Seeds few, subglobose, c. 10 by 8–9 by c. 4 mm; testa smooth, faintly winged.

One species, *B. simplex* W.J. de Wilde (and possibly another undescribed species; de Wilde et al. 2003), endemic in Borneo (Sarawak, Sabah); in tropical montane primary forest, 1,000–1,800 m a.s.l.; flowering Jul., Dec., fruiting Jul., Oct.

72. *Benincasa* Savi

Benincasa Savi, Biblioth. Ital. 9: 158 (1818).
Camolenga T. Post & O. Kuntze (1903 ('1904')).
Praecitrullus Pangalo (1944).

Monoecious, annual, herbaceous climber or trailer with hollow stems. Leaves simple, petiolate, reniform-ovate, 5–11-lobed or -angled, deeply cordate; tendrils 2-fid or 3–4(5)-fid. Flowers solitary, medium-sized; receptacle-tube broadly campanulate or flat, saucer-shaped, villous; sepals 5 (rarely 6), short, triangular; corolla rotate, flat, villous outside, smooth inside; petals 5 (rarely 6), connate at base, obovate, entire, (sulfur-)yellow; stamens 3, inserted at the base of the tube; filaments short, distinct; two anthers 2-thecous, one 1-thecous (rarely three 2-thecous); thecae triplicate; pollen large (polar axis 51–64 μm, equatorial axis 58–70 μm), 3-colporate, reticulate in *B. hispida* (Thunb.) Cogn., baculate in *B. fistulosa* (Khunwasi 1998); ovary globose to ovoid; placentae 3; ovules many, horizontal; style short, thick; stigmas 1–3, undulate; staminodes 3. Fruit baccate, indehiscent, oblong-terete, at first hispid, later glabrous, ripening light or dark green, in *B. hispida* covered with white wax. Seeds many, oblong, compressed; testa smooth, black or white, with thick margin. $n = 12$ (Beevy and Kuriachan 1996).

Two species: *Benincasa hispida*, native to New Caledonia, New Ireland, New Guinea, tropical NE Australia; along forest margins and in secondary scrub; cultivated throughout the tropics (Marr et al. 2007). A second species,

B. *fistulosa* (= *Praecitrullus fistulosus*), in India and Pakistan is apparently known in cultivation only.

Nicolson and Fosberg (2004: 382–383) have documented that the name *Benincasa hispida* (Thunb.) Cogn. does not need to be replaced by *Cucurbita pruriens* Solander in ms because its basionym (*Cucurbita hispida*) was first validly published by Thunberg in July 1784.

73. *Ctenolepis* Hook.f. Fig. 30

Ctenolepis Hook.f., Gen. Pl. 1: 832 (1867).
Blastania Kotschy & Peyritsch (1867).
Zombitsia Keraudren (1963).

Monoecious or dioecious, perennial, herbaceous or woody climber or trailer. Leaves simple, petiolate, entire to palmately 3–5-lobed; probract stipuliform, fimbriate; tendrils simple, to 30 cm long. Flowers small, greenish or yellow; male flowers in pedunculate racemes, female flowers solitary or in groups of 4–6, often coaxillary with male raceme; receptacle-tube obconic or campanulate; sepals 5, narrowly triangular, to 3 mm long; corolla rotate; petals 5, ± 1 mm long, united at base, yellowish, reflexed or not; stamens 3, inserted in lower half or near mouth of the tube; filaments distinct; two anthers 2-thecous, one 1-thecous, distinct, exserted or all 2-thecous (*C. lucorum*); thecae short, straight or duplicate and coherent in the center of the flower (*C. lucorum*); pollen medium-sized (polar axis 45–52 μm, equatorial axis 45–53 μm), 3-colporate, (irregularly) reticulate (Khunwasi 1998); ovary ellipsoid, smooth; ovules few or many, horizontal; style cylindric; stigma 2- or 3-lobed. Fruit a berry, small or to 5 cm in diam., ± globose, smooth, ripening red. Seeds mostly 2 or many (*C. lucorum*), ovoid, plano-convex or compressed with distinct margin; testa smooth, cream-colored. $n = 12$ in *C. garcinii* (Burm. f.) C. B. Clarke (Beevy and Kuriachan 1996). Three species: *C. cerasiformis* (Stocks) Hook.f., widespread from tropical and subtropical Africa to India and Pakistan, *C. garcinii*, endemic in India and Sri Lanka; on lake shores and river margins, in wood- and grassland, mostly at low altitudes.

One species, *C. lucorum*, endemic in Madagascar; in lowland forest. Host of the fungus *Puccinia ctenolepidis* Ramachar and Bagyanar (Berndt 2007).

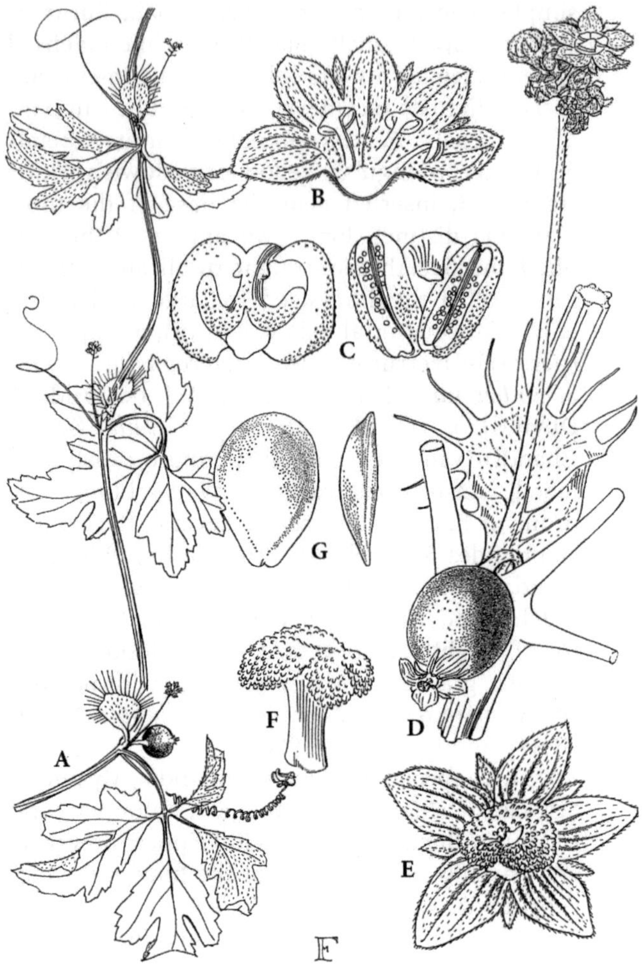

Fig. 30. Cucurbitaceae. *Ctenolepis cerasiformis*. **A** Habit. **B** Male flower. **C** Bithecate anther, ventral and dorsal view. **D** Flowering node with male inflorescence, young fruit, and stipuliform probract. **E** Female flower in plan. **F** Style with stigmas. **G** Seed, face and lateral views. (Jeffrey 1978b)

74. *Dactyliandra* (Hook.f.) Hook.f.

Dactyliandra (Hook.f.) Hook.f., Fl. Trop. Afr. 2: 557 (1871).

Monoecious or dioecious, perennial, herbaceous climbers or trailers. Leaves simple, petiolate, unlobed to palmately 5-lobed, with stipuliform bract at petiole base; tendrils simple. Flowers small, yellow; male flowers in pedunculate racemes, female flowers mostly solitary; receptacle-tube campanulate; sepals 5, triangular, small; corolla rotate; petals 5, distinct, ± 2 mm long; stamens 3, inserted near base of the tube; filaments distinct; two anthers 2-thecous, one 1-thecous, distinct; thecae duplicate; pollen large (polar axis 59–73 μm, equatorial axis 62–65 μm), 3-colporate, striate-reticulate (Khunwasi 1998); ovary ovoid; placentae 3; ovules many, horizontal; style 1; stigma 3-lobed; staminodes absent. Fruit a globose berry, ripening red. Seeds few, compressed, oblong to pear-shaped, ± angular; testa smooth, thick.

Two species: *D. welwitschii* Hook.f. in the Namib desert (Namibia, Angola) and the Thar desert (India and Pakistan), and *D. nigrescens* C. Jeffrey, endemic in Kenya; in dry bushland. Host of the fungus *Puccinia antennata* R. Berndt & A. Rössel (Berndt 2007).

75. *Khmeriosicyos* W.J. de Wilde & Duyfjes

Fig. 31

Khmeriosicyos W.J. de Wilde & Duyfjes, Blumea 49: 441 (2004).

Monoecious, herbaceous climber or trailer, possibly with tuberous root. Leaves simple, petiolate, the blade pentagonal, deeply 3(–5)-lobed, to 7 cm long; tendrils simple; probract conspicuous, 2–5 mm long, with several glands. Flowers small; male flowers in long-pedunculate racemes, female flowers solitary, often coaxillary with male raceme but flowering earlier; receptacle-tube shallow, cup-shaped, 5-ribbed; sepals 5, small, linear; petals 5, ovate-elliptic, 6–8 mm long; stamens 3, inserted halfway up the tube; filaments very short, distinct; two anthers 2-thecous, triplicate, one 1-thecous, distinct but appressed into a central head; pollen medium-sized (polar axis 52 μm, equatorial axis 51 μm), 3-colporate, reticulate (de Wilde et al. 2004). Fruit an ovoid, scabrous berry, c. 3 cm long and 2 cm in diam., pulpy. Seeds many, 8.5–9 mm long and 4.5–5 mm broad, strongly compressed; testa bright brown, finely grooved, with broad, grooved crenulate-tuberculate margin.

One species, *K. harmandii* W.J. de Wilde & Duyfjes, endemic in Northern Cambodia, known only from the type collection; in Dipterocarp forest.

76. *Papuasicyos* Duyfjes

Papuasicyos Duyfjes in Duyfjes, van der Ham & de Wilde, Blumea 48: 123–128 (2003).
Urceodiscus W.J. de Wilde & Duyfjes (2006).

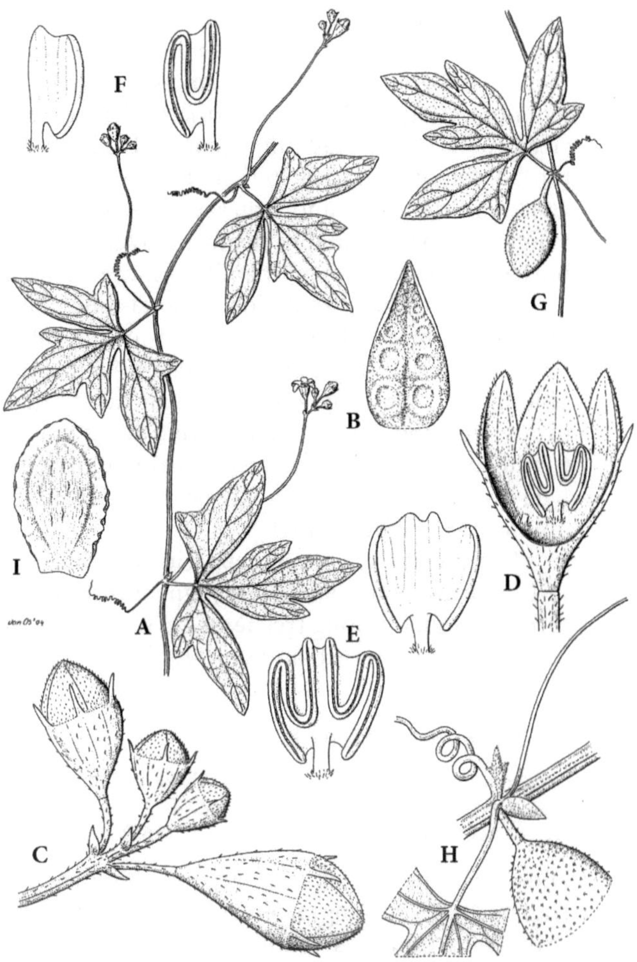

Fig. 31. Cucurbitaceae. *Khmeriosicyos harmandii*. **A** Branch with male inflorescences. **B** Glandular probract. **C** Tip of male inflorescence. **D** Opened male bud. **E** Bithecate stamen, ventral and dorsal view. **F** Monothecate stamen, dorsal and ventral view. **G** Node with fruit. **H** Ditto, enlarged, showing persistent probract. **I** Seed. (de Wilde et al. 2004; drawn by J. van Os)

Monoecious, annual to perennial, herbaceous climbers, to 6 m long. Leaves simple, petiolate, the blade entire or lobed, elliptic to hastate; tendrils simple. Flowers small to medium-sized, to 20 mm in diam.; male flowers in pedunculate racemes and solitary, female flowers solitary; often a single male or female flower coaxillary with a male raceme; receptacle-tube shallow, cup-shaped or urceolate to campanulate; corolla 10–15 mm in diam.; sepals 5, minute; petals 5, distinct, entire, cream-colored, yellow or orange, aestivation imbricate; stamens 3, distinct, inserted halfway up the tube; filaments long, slender, rarely short; anthers all 2-thecous, distinct but often appressed into a subglobose head; thecae straight, curved or sigmoid; pollen [*P. papuana* (Cogn.) Duyfjes, *P. belensis*] medium-sized (polar axis 31–38 μm, equatorial axis 32–38 μm), 3-colporate, striate-reticulate (Duyfjes et al. 2003; van der Ham and Pruesapan 2006); ovary subglobose or ellipsoid to fusiform; ovules many, horizontal; style 3-lobed, papillose-hairy or with forked and feather-like divided stigma-lobes; staminodes 3, inserted near the mouth of the tube. Fruit an edible, juicy berry, globose to ellipsoid-oblong, to 3 cm long and 1.5 cm in diam., glabrous, ripening scarlet to glossy red. Seeds many, tumid to globose, ovoid; testa cream-colored to pale brown, finely scrobiculate or foveolate, margin narrow or not distinct.

About eight species endemic in New Guinea; in (disturbed) montane *Nothofagus* forest, along forest margins, among tree ferns, in low scrub, in lowland swamp forest, and on river banks.

77. *Trochomeria* Hook.f.

Trochomeria Hook.f. in Benth. & Hook., Gen. 1: 822 (1867).
Heterosicyos Welw. ex Benth. & Hook. (1867).

Dioecious, perennial, herbaceous climbers or trailers to 2.5 m long or erect herb without tendrils, with tuberous (edible?) rootstock. Leaves simple, subsessile to petiolate, linear, elliptic, ovate-cordate, or sagittate, 3-lobed or palmately 3–5-lobed, often with ciliate stipuliform bract at petiole base; tendrils simple or absent. Flowers medium-sized, often on leafless stems, opening in the evening (Zimmermann 1922), exceptionally flowers bisexual (Jeffrey 1967: 89); male flowers in pedunculate clusters (rarely simple), female flowers solitary (rarely paired); receptacle-tube cylindrical, elongated; sepals 5, minute; corolla rotate; petals 5, ± distinct, often 10 mm long, triangular to linear, spreading, greenish or lemon-yellow sometimes with yellow papillae; stamens 3, inserted in the upper half of the tube; filaments distinct; two anthers 2-thecous, one 1-thecous, united into an oblong head; thecae triplicate; pollen large (polar axis 62–83 μm, equatorial axis 62–83 μm), 3-colporate, perforate or reticulate (Khunwasi 1998); ovary ovoid to subglobose, rostrate; ovules horizontal; stigma 3-lobed; staminodes 3. Fruit a fleshy, ellipsoid to subglobose, ± beaked, small to medium-sized berry, indehiscent, ripening scarlet. Seeds few, subglobose, ovoid or ellipsoid, tumid; testa smooth, hard, whitish (rarely pitted); germination hypogeal (Zimmermann 1922).

Eight species in tropical and subtropical Africa; in deciduous forest and ± dry bushland. Host of the fungus *Puccinia trochomeriae* Cooke (Berndt 2007).

78. *Indomelothria* W.J. de Wilde & Duyfjes

Indomelothria W.J. de Wilde & Duyfjes, Blumea 51: 5–9 (2006).

Monoecious, perennial, herbaceous climber, to 5 m long. Leaves simple, entire or shallowly lobed; probract absent; tendrils simple. Male flowers in pedunculate racemes; receptacle-tube campanulate to urceolate; sepals 5, minute, linear; corolla to 10 mm in diam.; petals 5, small, white (or yellow?), aestivation valvate; stamens 3, inserted in the upper half of the tube; filaments distinct, short, thick; two anthers 2-thecous, one 1-thecous; thecae straight (or slightly curved); pollen (*I. chlorocarpa* W.J. de Wilde & Duyfjes) medium-sized (polar axis c. 42 μm, equatorial axis c. 39 μm), 3-colporate, reticulate (van der Ham and Pruesapan 2006); disk free, ± 3-parted. Female flowers solitary, often coaxillary with male raceme; ovary narrowly ellipsoid; stigma 3-lobed, long-hairy; staminodes 0; disk annular, slightly 3-lobed. Fruit solitary, narrowly ellipsoid to fusiform, to 7 cm long on short pedicel, glabrous, smooth, ripening green. Seeds many, in pulp, compressed, ovate-elliptic, with dense appressed hairs, no distinct margin.

Two species in Southeast Asia (Myanmar, Thailand, Malesia (Sumatra, Borneo, Java)); forest margins, marshland, disturbed ground.

79. *Melothria* L.

Melothria L., Sp. Pl.: 35 (1753).
Diclidostigma Kunze (1844).
Landersia Macfadyen (1850).
Melancium Naudin (1862).
Cucumeropsis Naudin (1866).
Posadaea Cogn. (1890).

Monoecious, annual or perennial, herbaceous climbers or trailers to 10 m long, often with perennial rootstock. Leaves simple, petiolate, entire or palmately lobed, often with unpleasant odor; tendrils simple, rarely 2-fid, solitary, rarely 2 per node; probract absent. Flowers small; male flowers in pedunculate racemes or umbels, female flowers solitary (rarely in groups of 2–3), usually coaxillary with male; receptacle-tube campanulate to cylindrical; sepals 5, short, dentiform; corolla rotate; petals 5, entire, connate at the base, yellow, less often white (*M. dulcis* Wunderlin, *M. warmingii* Cogn.); stamens 3, inserted in the upper half of the tube; filaments distinct, short, slender; two anthers 2-thecous, one 1-thecous (rarely all 2-thecous); thecae ± straight, fringed with hairs; pollen medium-sized to large (polar axis 36–67 μm, equatorial axis 36–69 μm), 3-colporate, (micro)reticulate (Khunwasi 1998); disk globose or depressed, entire or 3-lobed; ovary smooth, ovoid to fusiform; style short, surrounded at base by an annular disk; placentae 3; ovules many, horizontal; stigmas 3, 2-lobed or 1, 3-lobed; staminodes 3 or absent. Fruit a small or up to 20 cm long berry, on long pedicel, fleshy, indehiscent, smooth, globose or ellipsoid, ripening cream with green stripes, yellow, orange, reddish or purple-blackish, edible. Seeds many, compressed, ovoid, with or without arilloid; testa smooth, ivory-colored, covered by long appressed hairs, no distinct margin. $n=12$ in *M. sphaerocarpa* (= *Cucumeropsis mannii*) (Osuji et al. 2006).

Melothria has about 12 species in tropical Central and South America, one species, *M. mannii*, in West tropical Africa and in Central and tropical South America, one species, *M. pendula* L., naturalized in Asia; roadsides and cultivated ground, arid plains, clearings, forest margins, grass- or woodlands. Host of the fungus *Uromyces novissimus* Speg. (Monoson and Rogers 1978).

Molecular data from plastid and nuclear loci indicate that *Melancium*, *Cucumeropsis*, and *Posadaea* are nested inside *Melothria*, and that there is next to no genetic difference between South American and African individuals (Schaefer and Renner 2010b).

80. *Ruthalicia* C. Jeffrey

Ruthalicia C. Jeffrey, Kew Bull. 15: 360 (1962).

Dioecious, perennial, herbaceous climber. Leaves simple, petiolate, the blade broadly ovate, pedately 3–7-lobed; tendrils simple. Flowers large, showy; male flowers in racemes, female flowers axillary, solitary or in pairs or few-flowered fascicles; receptacle-tube funnel-shaped, apically broadened; sepals 5, triangular; corolla rotate; petals 5, distinct or connate at the base, to 1.5 cm long, yellow; stamens 3, inserted near the mouth of the tube; filaments distinct. 2 anthers 2-thecous, one 1-thecous; thecae triplicate; pollen large (polar axis 52–80 μm, equatorial axis 53–78 μm), 3-colporate, reticulate (Khunwasi 1998); ovary ellipsoid; style fleshy, c. 8 mm long; ovules many, horizontal; stigmas 3; staminodes 3. Fruit globular to ellipsoid, to 4 cm long and 2.5 cm in diam., smooth, ripening red. Seeds many, in yellowish pulp, compressed, ellipsoid to oblong; testa slightly sculptured, with broad, flattened margin; germination epigeal (Zimmermann 1922).

Two species in tropical West Africa; along forest margins, in secondary forest and bushland.

81. *Muellerargia* Cogn.

Muellerargia Cogn., Monogr. Phan. 3: 630 (1881).

Monoecious, perennial?, herbaceous climber or trailer, to 2 m long. Leaves simple, petiolate; blade entire, ovate, triangular, or 3–5-lobed; tendrils simple; probract reniform or suborbicular. Male flowers in (umbellate) racemes, female flowers solitary, coaxillary with the male raceme; receptacle-tube campanulate, 2.5 mm long; sepals 5, triangular, small; corolla rotate; petals 5, (ob) ovate, 0.7 mm long; stamens 3, inserted near base or halfway up the tube; filaments very short to absent; two anthers 2-thecous, one 1-thecous; thecae straight and apically hooked; pollen (*M. jeffreyana* Keraudr.) medium-sized (polar axis 50 μm, equatorial axis 46–48 μm), 3-colporate, finely reticulate (Keraudren 1968); ovary ovoid or ellipsoid, setose; placentae 2–3; ovules many, horizontal; style short; stigmas ± 2-lobed, forming a central globular body; staminodes 3 or 0. Fruit fleshy, ovoid, to 2.5 cm long and 1–2 cm in diam., ± rostrate, ornamented with long, soft bristles, indehiscent or dehiscing through ruptured attachment of pedicel. Seeds many, oblong, compressed, 8–10 mm long; testa smooth or finely pitted, pale whitish, margin slightly thickened.

Two species, one endemic in Madagascar, and one collected a few times in tropical northern Australia and Timor, and the Lesser Sunda islands (Telford 1989); in forest remnants; highly endangered; flowering Feb. to May.

This ancient clade, with a highly disjunct range and just two surviving species, is the sister group to *Cucumis* (Renner and Schaefer 2008).

82. *Cucumis* L.

Cucumis L., Sp. Pl.: 1011 (1753); H. Schaefer, Blumea 52: 165–177 (2007).
Melo Mill. (1754).
Mukia Arn. (1840).
Oreosyce Hook. (1871).
Dicoelospermum C.B. Clarke (1879). ('*Dicaelospermum*', correction T. Post and O. Kuntze (1903 ('1904')).
Hymenosicyos Chiov. (1911).
Cucumella Chiov. (1929).
Myrmecosicyos C. Jeffrey (1962).

Monoecious or dioecious, annual or perennial, small to medium-sized, herbaceous or woody climbers or trailers, with fibrous roots or perennial rootstock (rarely tubers). Leaves simple, petiolate, the blade unlobed or palmately lobed; tendrils solitary or rarely in groups of 5–8, simple, rarely absent (exceptionally 2-fid). Male flowers solitary or in few-flowered groups, female flowers solitary or in fascicles of 2–6, usually separate from male flowers; flowers small to medium-sized; receptacle-tube funnel-shaped to campanulate or shallowly saucer-shaped; sepals 5 (rarely 4), small, long-triangular to filiform; petals 5, elliptic or (ob)ovate, distinct or united at base, yellow; stamens 3, inserted halfway up the tube; filaments short, glabrous; two anthers 2-thecous, one 1-thecous; thecae lateral, straight (sometimes apically hooked) or triplicate, rarely horizontal, arcuate and slightly coherent [*C. messorius* (C. Jeffr.) Ghebret. & Thulin], glabrous or fringed with minute hairs; pollen medium-sized to large (polar axis 22–61 μm, equatorial axis 49–80 μm), 3-(col) porate, (micro)reticulate-perforate (Khunwasi 1998); disk obconic or depressed globose, basal, free from the tube, rarely indistinct; ovary hairy, globose to oblong; ovules several to many, horizontal; style terete, thick, glabrous; stigma entire, sublobate or 3-lobed, the lobes carnose, papillose, often with 1–9 finger-like projections on the margin; staminodes 3 or 0. Fruit solitary or clusters of 2–6, fleshy, (sub)globose or ellipsoid, cylindrical, (ob)ovoid, or spindle-shaped, smooth and glabrous or pubescent or with dense to scattered fleshy spines, pustules or tubercles, sometimes beaked or fusiform, ripening yellow, orange, red or greenish to brownish, often with longitudinal pale stripes (rarely maturing underground), indehiscent (rarely expelling seeds explosively). Seeds few to many, ovate or elliptic, small to medium-sized, globose or lenticular compressed; testa light-colored, smooth or ornamented, glabrous or rarely puberulent, margin often distinct, usually not winged; germination epigeal. $n = 7$ or 12, with polyploids and aneuploids also reported (Beevy and Kuriachan 1996).

About 55 species in Africa, Asia and Australia; in semi-deserts and savannas, dry bushland and along forest margins, often on disturbed or cultivated ground. Host of the fungus *Puccinia cucumeris* Henn. (Berndt 2007).

Cucumis melo L. and *C. sativus* L. are among the World's most important vegetable crops, and there are three fully sequenced *C. sativus* genomes, namely, that of a Chinese fresh market cucumber (Huang et al. 2009), a North American pickling type, and an isogenic gynoecious breeding line, making the cucumber the sixth flowering plant to have been completely sequenced. The genus has also been the study system for one of the first biosystematic studies ever, that of Charles Naudin (1859), who over many years performed controlled crossings among species he had in cultivation in Paris. Molecular data have revealed that the former genus *Mukia* (de Wilde and Duyfjes 2007c) comprised a mix of species not closely related to each other (P. Sebastian, H. Schaefer, I. Telford, and S. S. Renner, unpubl. data).

83. *Zehneria* Endl.

Zehneria Endl., Prodr. Fl. Norfolk.: 69 (1833).
Pilogyne Eckl. ex Schrad. (1835).
Anangia W.J. de Wilde & Duyfjes, Reinwardtia 12(3): 219 (2006).
Neoachmandra W.J. de Wilde & Duyfjes, Blumea 51(1): 12 (2-3, 13; figs. 1c,2c) (2006).

Dioecious or monoecious, annual or perennial, herbaceous climbers or trailers, to 10 m long, some with tuberous roots. Leaves simple, petiolate, triangular to ± ovate, entire to 3–5-lobed, rarely 5-foliolate; tendrils simple; probract linear, minute, caducous or absent. Flowers small (rarely medium-sized), mostly unisexual but in one species bisexual (*Z. hermaphrodita* W.J. de Wilde & Duyfjes); male flowers solitary or 2–8 per node or in pedunculate, crowded racemes, female flowers solitary or in small groups. In monoecious species, female flowers coaxillary with male raceme or mixed racemes with flowers of both sexes; sepals 5, minute, triangular to narrowly elliptic, rarely much longer than the petals (*Z. macrosepala*); petals 5, distinct, white or cream-colored (rarely yellow?); stamens 3 or 5 (rarely 2), inserted near the base or in the upper half of the tube; filaments distinct, long and slender or short; anthers all 1-thecous or 2-thecous; thecae lateral, straight or curved to sinuate, often fringed with hairs; pollen medium-sized to large (polar axis 28–73 µm, equatorial axis 29–73 µm), 3-colporate, (micro)reticulate (rarely micro-reticulate-perforate) (Khunwasi 1998; van der Ham and Pruesapan 2006); disk globose, entire or 3-lobed; ovary globose to ellipsoid; stigma 3-lobed or style 3-parted with 2-lobed stigmas, ± hairy; staminodes 3 or 0. Fruit a pulpy berry, solitary or in fascicles on short pedicel, globose to ellipsoid or fusiform, with pitted pericarp, to 7 cm long, ripening green, white, pale yellowish, orange, red or blackish. Seeds few to many, compressed (rarely globose), ovate to elliptic; testa pale, smooth, glabrous or sometimes hairy (at the ends or throughout), margin narrow or indistinct; germination epigeal (Zimmermann 1922). $2n = 48$ in *Z. maysorensis* Arn. (Beevy and Kuriachan 1996).

About 60 species in tropical and subtropical Africa, five endemic in Madagascar, the rest from India and China to Northern Australia and the Pacific Islands; on disturbed ground, along forest margins, and in clearings, scrubland, grassland, on riverbanks, and in mangroves. Host of the fungi *Puccinia arisanensis* Hirats.f. and Hashioka, *P. melothriicola* (*Uredo melothriae* (Henn.) R. Berndt), *P. hieroglyphica*, *P. rhytidioderma* R. Berndt (*Uromyces zehneriae*), and *Uromyces cantonensis* (Berndt 2007).

Molecular data show that the type species of the recently described genera *Anangia* and *Neoachmandra* are nested within *Zehneria* (Schaefer et al. 2009). *Zehneria peneyana* (Naudin) Schweinf. & Asch. (*Pilogyne peneyana* Naudin) with 6-porate or 6-brevicolporate pollen (Keraudren 1968), which has not yet been sequenced for any locus, may represent a separate lineage. The generic name *Pilogyne* (type species *Pilogyne suavis* Schrad. from South Africa) has recently been taken up for about 20 species until now placed in *Zehneria* (de Wilde and Duyfjes 2009).

84. *Diplocyclos* (Endl.) T. Post & O. Kuntze

Diplocyclos (Endl.) T. Post & O. Kuntze, Lex.: 178 ('*Diplocyclus*'). (1903 ('1904')).
Ilocania Merr. (1918).

Monoecious, herbaceous climbers, to 6 m long. Leaves simple, petiolate, broadly ovate, palmately 5-lobed; tendrils 2-fid; probract c. 3 mm long, with nectaries. Flowers small, fasciculate, often male and female together in the axils; receptacle-tube broadly campanulate; sepals 5, triangular-dentiform, to 2 mm long; corolla broadly campanulate; petals 5, ovate, white to greenish-yellow; stamens 3, inserted near the mouth of the tube; filaments distinct, short; two anthers 2-thecous, one 1-thecous; thecae triplicate; pollen large (polar axis 57–107 µm, equatorial axis 63–104 µm), 3-(col)porate, echinate and reticuloid (Khunwasi 1998); ovary globose or ovate; placentae 3; ovules few, horizontal; style slender; stigmas 3, 2-lobed; staminodes 3. Fruit solitary or in clusters of 2–5, baccate, globose or ovoid, ripening bright red with silvery white stripes or marks. Seeds to 6 mm long; testa slightly scorbiculate, strongly winged; germination epigeal (Zimmermann 1922). $n = 12$ in *D. palmatus* (L.) C. Jeffrey (Beevy and Kuriachan 1996).

Four species in tropical and subtropical Africa, Asia, Australia; rainforest clearings and margins, secondary growth.

85. *Coccinia* Wight & Arn.

Coccinia Wight & Arn., Prodr. Fl. Ind. Orient. 1: 347–348 (1834).
Cephalandra Eckl. & Zeyh. (1836).
Physedra Hook.f. (1867).
Staphylosyce Hook.f. (1867).

Dioecious, perennial, herbaceous to woody climbers or trailers to 10 m long, with tuberous roots reaching up to 10 kg. Leaves simple, blade angled, cordate, or deeply lobed; tendrils simple or 2-fid; probracts and bracts variable, often with nectaries. Flowers with petals 1.4–5.5 cm long; male flowers solitary, clustered or in racemes, female flowers solitary or in racemes; receptacle-tube short, campanulate or turbinate; sepals 5, entire, usually small and dentiform; corolla short-tubulate, to 8 cm in diam.; petals 5, connate, white, salmon, yellow or orange; stamens 3, inserted at the base of the tube; filaments connate apically or completely into a central column, rarely distinct; anthers all 2-thecous, less often two 2-thecous, one 1-thecous and forming a central head; thecae triplicate; pollen large (polar axis 58–92 μm, equatorial axis 35–92 μm), 3-colporate, reticulate (Khunwasi 1998); ovary ovoid, oblong or linear, smooth; placentae 3; ovules many, horizontal; style filiform; stigmas 3, 2-lobed; staminodes 3, oblong or subulate. Fruit fleshy, small and globose, ovoid, or cylindrical and up to 30 cm long, baccate, indehiscent, ripening orange to red, sometimes with green and white spots. Seeds many, ovate to pear-shaped, compressed; testa fibrillose, marginate; germination epigeal (Zimmermann 1922). $2n = 22 +$ XY sex chromosomes in *Coccinia grandis* (L.) Voigt (Bhaduri and Bose 1947; Bhar and Datta 1982).

About 30 species in tropical and subtropical Africa, one species, *C. grandis* (L.) Voigt, also in Asia and naturalized on the American and Australian continents; in rainforest, cloud forest, deciduous bushland, riverine forests, and semi-desert shrubland/bushland, rarely on sand dunes. Host of several fungi, including *Puccinia windhoekensis* Mennicken, Maier & Oberw., *P. cucumeris* Henn., *P. physedrae* Syd., *P. cephalandrea* Thümen, and *P. cephalandrae-indicae* Syd. & P. Syd. (Berndt 2007).

The tubers, leaves, and fruits of several species are edible, but *C. trilobata* fruits appear to be poisonous. Male individuals of *C. grandis* have a pair of different-sized chromosomes, interpreted as an X- and a Y-chromosome (Bhaduri and Bose 1947; N. Holstein and S. Renner, pers. obs.).

86. *Scopellaria* W.J. de Wilde & Duyfjes

Scopellaria W.J. de Wilde & Duyfjes, Blumea 51: 297 (2006).
Scopella W.J. de Wilde & Duyfjes (2006).

Monoecious, annual, biennial or perennial, herbaceous climbers or trailers, to 6 m long. Leaves simple, ovate, angular, or 3–5-lobed; tendrils simple, hairy throughout their length; probract absent. Flowers small to medium-sized, to 10 mm in diam.; sepals 5, minute, linear; petals 5, distinct, ovate-elliptic, yellow, aestivation imbricate; receptacle-tube campanulate; male flowers in short, pedunculate, crowded racemes, female flowers solitary (or pairs), coaxillary with male racemes; stamens 3, inserted near the mouth of the tube; filaments distinct, long, slender; anthers all 2-thecous; thecae lateral, straight; pollen [*S. marginata* (Blume) W.J. de Wilde & Duyfjes] medium-sized (polar axis c. 40 μm, equatorial axis c. 47 μm), 3-(brevi)colporate, irregularly striate-reticulate (van der Ham and Pruesapan 2006); disk free, globose; ovary globose to ellipsoid; stigma 3-lobed, hairy; staminodes 3; disk annular, free. Fruit a smooth berry, solitary (or in pairs) on ± long pedicel, juicy or pulpy, globose or ellipsoid to fusiform, to 3 cm long, ripening red. Seeds 1–40, to 6 mm long, compressed, ovate-elliptic; testa scorbiculate, pale, with distinct margin.

Two species from Southern China to Indonesia and the Philippines; in thickets, along forest margins and roadsides.

XV. Tribe Cucurbiteae Dumort. (1827).

Tendrils simple or 2–7-fid. Stamens (2)3(4). Fruits small, dry, indehiscent, or medium-sized to large pepos, or dry and splitting into several valves. Seeds one to many, unwinged, less often narrowly winged.

87. *Polyclathra* Bertol.

Polyclathra Bertol., Novi Comment. Acad. Sci. Inst. Bononiensis 4: 438 (1840); Jeffrey, Kew Bull. 25: 196–198 (1971).
Pentaclathra Endl. (1842).
Pittiera Cogn. (1891 ('1892')).
Roseanthus Cogn. (1896).

Monoecious, annual, herbaceous climbers or trailers, to several meters long. Leaves cordate, entire or ± 3–5-lobed; tendrils (2–)4(–6)-fid, to 20 cm long, with stout peduncle and (not only apically) adhesive pads. Flowers large, showy, solitary, axillary, the male on very long, the female on short stalks, opening at night; receptacle-tube campanulate to obconic-cylindric, much shorter in the female than in the male flowers; sepals triangular, narrower in the female flowers then in the male; petals white, broadly rounded; stamens inserted halfway up the tube; filaments long, distinct; two anthers 2-thecous, one 1-thecous, connate into a central head; thecae triplicate; pollen very large (polar axis 176–180 μm, equatorial axis 176–180 μm), pantoporate, echinate (Khunwasi 1998); ovary ellipsoid; ovules many, horizontal; style slender, elongated; stigmas 3, deeply 2-lobed. Fruit a dry berry, medium-sized, ellipsoid to oblong, rounded at the apex, green with white or yellow marks, the pericarp splitting into several irregular segments at maturity, exposing the seeds. Seeds many, broadly ovate, contracted at base, compressed; testa brown, narrowly winged.

Fide Kearns (1992), six species in Mexico, Costa Rica, Guatemala, Panama, and Nicaragua but only one formally described; in tropical or oak-pine forests; flowering and fruiting I–IV. Host of the fungus *Uromyces novissimus* Speg. (Monoson and Rogers 1978).

88. *Peponopsis* Naudin

Peponopsis Naudin, Ann. Sci. Nat., Bot. IV, 12: 88 (1859); Jeffrey, Kew Bull. 25: 194–196 (1971).

Dioecious, perennial, ± woody climber, 8–10 m long. Leaves broadly ovate-cordate, unlobed to 3–5-lobed, 10–18 cm long, usually with a few disk glands near the leaf base; tendrils multifid, densely short-villous, with apical adhesive pads. Flowers medium-sized, solitary, axillary; receptacle-tube obconic-tubular, apically expanded; sepals of male flowers ovate-lanceolate and to 2 cm long, of female flowers triangular-lanceolate, acute, 6–7 mm long; corolla broadly campanulate; petals connate halfway, broadly obovate-oblong, rounded, to 4 cm long, white to greenish-white; stamens inserted near the base of the tube; filaments distinct, 5 mm long; two anthers 2-thecous, one 1-thecous; thecae much convoluted; pollen large (polar axis and equatorial diameter 78–82 μm), 3-porate, echinate (R. van der Ham, pers. comm., 9 Dec. 2009); ovary ovoid, ± glabrous; placentae 3; ovules many, horizontal; style 12–14 mm long; stigmas 3, papillose, oblong to ovate-oblong; staminodes 3–4. Fruit a fleshy, subglobose pepo, 8–10 cm in diam., splitting into three carpellar segments at maturity, exposing the seeds. Seeds many (c. 200 per fruit), ovate-oblong, 4–9 by 2–5 mm, compressed; testa greenish to gray, finely perforate, margin narrowly winged.

One species, *P. adhaerens* Naudin, endemic in Mexico (Querétaro, Hidalgo, Puebla, and Veracruz); extremely rare, in pine forest 800–1,500 m a.s.l.; flowering and fruiting Jun.–Oct.

Molecular data indicate that this is the sister species to *Cucurbita* (Schaefer et al. 2009).

89. *Cucurbita* L.

Cucurbita L., Sp. Pl.: 1010 (1753), nom. cons.
Melopepo Mill. (1754).
Pepo Mill. (1754).
Ozodycus Raf. (1832).
Sphenantha Schrad. (1838).
Mellonia Gasp. (1847).

Monoecious, annual or perennial, herbaceous climbers or trailers to 6 m long, with hollow stems and fibrous or fleshy roots. Leaves ovate-cordate to suborbicular, pedately 3–5(–7)-lobed, sometimes with a nasty odor; tendrils 2- to 7-fid, rarely simple (absent in some cultivars). Flowers large, solitary in axils, some fragrant; male flowers with campanulate or elongated receptacle-tube; sepals lanceolate or foliaceous at the apex; corolla campanulate; petals ± connate, yellow; stamens inserted at the base of the tube; filaments usually distinct, short and fleshy; two anthers 2-thecous, one 1-thecous, connate into a central head; thecae reflexed; pollen very large (124–154 μm in diam.), pantoporate, echinate (Khunwasi 1998); ovary oblong, globose, cylindric or pear-shaped, constricted at apex; placentae 3–5; ovules many, horizontal; stylodia short, thick, united into a column; stigmas usually 3, 2-fid; staminodes 3, short-triangular. Fruit a large, fleshy, indehiscent pepo of variable form and color, with woody or corky peduncle, smooth or ribbed, interior tissue soft, fibrous, white, yellow or orange, sweet or bitter. Seeds many, ovate to elliptic, strongly compressed; testa smooth, cream-colored or black, with or without distinct margin. $n = 20$ in *C. maxima*, *C. digitata* A. Gray, *C. foetidissima* H.B. & K., and *C. palmata* S. Wats. (McKay 1931; Beevy and Kuriachan 1996).

About 15 wild species in tropical and subtropical America (Sanjur et al. 2002; M. Nee, pers. comm., Feb. 2010) and five domesticated ones cultivated worldwide. Whether the domesticated species should continue to be ranked as species or as forms of their wild progenitors is a matter of opinion. Disturbed places, humid ravines, floodplains, tropical deciduous forest, grasslands, deserts, rocky hillsides, oak- and pine-oak forests.

Subfossil records of *Cucurbita pepo* L. and *C. moschata* Duchesne ex Poir. from Central

America and the northern Andes indicate that squashes are among the oldest neotropical domesticated plants (Smith 1997; Piperno and Stothert 2003; Dillehay et al. 2007).

90. *Calycophysum* Triana

Calycophysum Triana, Nuev. Jen. Esp. 20 (1854 [1855]). [as "*Calycophisum*"]; Pittier, H. Contr. US. Natl. Herb. 20: 487–490 (1922); Jeffrey, Kew Bull. 25: 192–194 (1971).
Edmondia Cogn. (1881) (non Cassini 1818), nom. illegit.
Bisedmondia J. Hutch. (1967).

Monoecious, herbaceous (or woody) climber, root not tuberous, to 10 m long. Leaves ovate-cordate, entire or palmately 3–5–7-lobed; tendrils 3–6-fid, with apical, adhesive pads. Flowers bat-pollinated, large, solitary in the axils; peduncle of male flowers to 30 cm long, to 15 cm in female; receptacle-tube campanulate to urceolate, ± inflated; sepals large, ovate-lanceolate; corolla rotate to tubular-campanulate; petals white or yellowish-green; stamens inserted in the upper half of the tube; filaments distinct; anthers ± distinct, two 2-thecous, one 1-thecous; thecae duplicate; pollen (very) large (polar axis 90–168 μm, equatorial axis 91–169 μm), 3-porate, echinate (Khunwasi 1998); ovary ellipsoid; placentae 3; ovules horizontal, numerous; staminodes 3; stylodia fleshy; stigmas 3. Fruit a smooth, elliptic pepo, green with darker green stripes, ripening yellow to orange. Seeds many, compressed, irregularly elliptical, with arilloid; testa dark brown, irregularly structured.

Five species from Venezuela, Colombia, Ecuador, Peru, Bolivia; in Andean cloud forests and lowland rainforest.

91. *Sicana* Naudin

Sicana Naudin, Ann. Sci. Nat., Bot. IV, 18: 180 (1862).

Monoecious, annual or perennial, herbaceous climber, to 15 m long. Leaves ovate to sub-orbiculate, palmately 3–9-lobed, often with diskoidal glands at the leaf base, to 24 cm long; tendrils 3–5-fid, with apical adhesive pads. Flowers solitary, medium-sized to large, showy; receptacle-tube obconical or campanulate; sepals triangular-lanceolate, reflexed or less often ascendent; corolla campanulate, divided in the upper ¼; petals yellow; stamens inserted close to the mouth of the tube; filaments short, distinct or ± connate; two anthers 2-thecous, one 1-thecous; thecae duplicate; pollen large (polar and equatorial axes 79–98(120?) μm), (3?)6–8-pantoporate, echinate (Marticorena 1963; Ayala-Nieto et al. 1988); ovary elliptical; placentae 3; ovules many, horizontal; style short; stigmas 3; staminodes 0, 3 or 5. Fruit a globose, ellipsoid or cylindrical pepo, smooth, to 60 cm long. Seeds many, compressed, ovate, with distinct margin, sometimes narrowly winged. $n = 20$ in *S. odorifera* (Vell.) Naudin (Mercado and Lira Saade 1994).

About four species in the Caribbean Islands and Central America (Lira Saade 1991); in rainforest and secondary scrub; one species, *S. odorifera* Naudin, widely cultivated as a vegetable.

Sicana sphaerica Hook.f. may belong in a different genus (C. Jeffrey, pers. comm., 2008).

92. *Penelopeia* Urb.

Penelopeia Urb., Repert. Spec. Nov. Regni Veg. 17: 8 (1921).
Anacaona A. H. Liogier (1980).

Monoecious or dioecious, perennial, herbaceous to woody climber, to 10 m long. Leaves triangular, pedately 3–5-lobed; tendrils simple or 2–3-fid. Flowers small to medium-sized. Male flowers solitary or in axillary fascicles, female flowers solitary; receptacle-tube shortly cup-shaped to campanulate; sepals triangular, 4–10 mm long, connate in the lower half or distinct; corolla ± cylindrical; petals 5 (rarely 4), connate in lower half or distinct, yellow or pale green; filaments connate into a central column; anthers connate into a central head; thecae triplicate; pollen large (polar axis and equatorial diameter 65–75 μm in *P. sphaerica*), 3(4)-porate, echinate (R. van der Ham, pers. comm., 9 Dec. 2009); ovary ovoid; stigmas 3; placentae 3; ovules many, horizontal. Fruit globose, smooth, 5–7 cm in diam., indehiscent, ripening yellowish. Seeds many, elliptical.

Two species endemic in Hispaniola; in humid montane forest.

Molecular data (Kocyan et al. 2007) show that the two species are each others' closest relative, arguing for placing them in a single genus, rather than two genera, since this creates a more informative classification.

93. *Tecunumania* Standl. & Steyerm.

Tecunumania Standl. & Steyerm., Publ. Field Mus. Nat. Hist., Bot. 23(2): 96–97 (1944).

Dioecious, herbaceous climber. Leaves ovate-cordate and palmately 3–7-lobed, to 17×16 cm, often with tuft of hair and patelliform glands at leaf base; tendrils 2-(4-)fid. Flowers large, solitary; receptacle-tube campanulate; sepals linear, to 2 cm long; corolla rotate; petals obovate, connate at base, yellow; stamens inserted near the base of the tube; filaments distinct but closely appressed; anthers 2-thecous, connate into a globose head; thecae triplicate; pollen large (polar axis c. 97 μm, equatorial axis c. 102 μm), 3-porate, echinate (Khunwasi 1998); ovules many, horizontal; style elongated; stigmas 3, 2-lobed; staminodes 3. Fruit fleshy, subglobose, 7–8 cm long, indehiscent, ripening dark green. Seeds many, 6–7 mm long and 4–5 mm broad, compressed; testa pale yellowish-brown, densely appressed hairy, no distinct margin.

One species, *T. quetzalteca* Standl. & Steyerm., in Costa Rica, Mexico, Guatemala; in wet montane forest.

Molecular data do not yet firmly resolve the placement of the genus relative to *Schizocarpum*.

94. *Schizocarpum* Schrad.

Schizocarpum Schrad., Index Sem. Gött. 1830: 4 (1830); Jeffrey, Kew Bull. 25: 198–200 (1971).

Monoecious, annual or perennial, herbaceous climbers, to 7 m long. Leaves entire or 3–5-lobed; tendrils 2-fid. Flowers solitary, large, showy; receptacle-tube elongated, subcylindric at base, expanded distally; sepals reflexed or ascending, linear, ovate or triangular; corolla campanulate; petals connate in the lower half, triangular, yellow with dark central spot inside; stamens (2–)3(–4), inserted near the mouth of the tube; filaments distinct; anthers connate into a central ovoid head; thecae triplicate; pollen very large (polar axis 119–125 μm, equatorial axis 119–125 μm), pantoporate, echinate (Khunwasi 1998); ovary ovoid to fusiform, often rostrate; placentae 3; ovules 12 to many, ascending; style slender; stigmas 3, fleshy; staminodes 3. Fruit a dry, ± woody capsule, ellipsoid to pear-shaped, smooth, rostrate, indehiscent or dehiscing into three lobes from the apex downward, each with two rows of seed chambers, followed by the abscission of the pericarp. Seeds ovoid, compressed, 8–10 mm long; testa smooth, brown or banded crosswise with light and dark stripes, margin with or without wing.

Eleven species in Mexico and Guatemala; in oak or pine forest, tropical deciduous forest, disturbed ground, coastal plains; flowering and fruiting Jun.–Nov.

95. *Cionosicys* Griseb.

Cionosicys Griseb. Fl. Brit. W. I.: 288 (1860); *Cionosicyos* [orth. var.] Hook.f. (1867); Jeffrey, *Cionosicyos*, Kew Bull. 25: 200–201 (1971).

Monoecious, perennial herbaceous or woody climber, to several meters long. Leaves large, coriaceous, ovate to roundish, entire or 3-lobed; tendrils simple, stout. Flowers large, solitary, axillary; receptacle-tube turbinate (male) or cup-shaped (female); sepals ovate-lanceolate; corolla funnel-shaped to rotate; petals ovate-oblong, connate at the base, greenish-white; stamens inserted at the base of the tube; filaments distinct; anthers connate into a central column; thecae triplicate; pollen very large (polar axis 103–144 μm, equatorial axis c. 105–146 μm), 3-porate, echinate (Khunwasi 1998); placentae 3; ovules many, ascending to horizontal; stigmas strongly papillose. Fruit a large, ovoid, fleshy, hard-shelled pepo, to 10 cm long, glabrous, ripening yellowish. Seeds many, elliptic, tumid, 1–1.8 cm long; testa dark brown to black.

Four (or five) species in Central America, Cuba, and Jamaica; along forest margins and in montane forest.

96. *Abobra* Naudin

Abobra Naudin, Rev. Hort. 1862: 111 (1862).

Dioecious, perennial, herbaceous trailer, to 7 m long, with fleshy rootstock. Leaves small, palmately 5-lobed to dissected, to 12 cm long and broad; petioles 1–4 cm long; tendrils simple or 2-fid. Flowers small with strong odor; male flowers solitary or in pedunculate racemes, female flowers solitary; receptacle-tube cup-shaped; sepals short; corolla rotate; petals ovate-lanceolate, greenish-white; stamens inserted near the mouth of the tube; filaments short, distinct; two anthers 2-thecous, one 1-thecous; thecae triplicate; pollen large (polar axis c. 68 μm, equatorial axis c. 71 μm), 3-porate, margin distinct, echinate (Khunwasi 1998); ovary globose; placentae 3; ovules 6, erect; style slender; stigmas 3 (rarely 4), linear; staminodes 3. Fruit a fibrous red berry, with a firm, thin wall, c. 1 cm in diam. Seeds 3–6, 7–8 by 2–3 by ± 1.5 mm, slightly compressed (falcate); testa smooth, green or brownish.

One species, *A. tenuifolia* (Gillies in Hook.) Cogn., in Brazil, Argentina, and Uruguay; in xeric bushland and on dry soil. Host of the fungus *Uromyces novissimus* Speg. (Monoson and Rogers 1978).

Molecular data indicate that this is the closest relative of *Cayaponia*.

97. *Cayaponia* Silva Manso

Cayaponia Silva Manso, Enum. Subst. Braz.: 31 (1836), nom. cons.; Jeffrey, Kew Bull. 25: 201–234 (1971).
Alternasemina Silva Manso (1836).
Dermophylla Silva Manso (1836).
Perianthopodus Silva Manso (1836).
Arkezostis Raf. (1836 [1838]).
Trianosperma (Torr. & A. Gray) Mart. (1843).
Allagosperma M. Roem. (1846).
Cionandra Griseb. (1860).
Antagonia Griseb. (1874).
Selysia Cogn. (1881); Jeffrey, Kew Bull. 25: 234–236 (1971).

Monoecious or rarely dioecious, herbaceous or suffrutescent, often much-branched climbers or creepers, to 20 m long, with perennial roots. Leaves ovate-cordate, simple, palmately 3–7-lobed or entire, less often 3–5-foliolate, the blade often decurrent on the petiole, often with disk-shaped glands on the abaxial side; tendrils 2–7-fid (rarely simple), sometimes with apical adhesive pads. Flowers small, in racemes, panicles, pairs or solitary, male and female often coaxial, opening at night; receptacle-tube shortly campanulate to cylindrical; petals yellow, white or light green; stamens inserted near the base of the tube; filaments distinct; two anthers 2-thecous, one 1-thecous or all 2-thecous, usually coherent in a head; thecae duplicate or triplicate; pollen large to very large (polar axis 61–208 μm, equatorial axis 68–196 μm), 3–6-porate to 4–10-pantoporate, echinate with short pila between the spines (Khunwasi 1998; Barth et al. 2005); ovary globose, ovoid or oblong or ellipsoid-rostrate; placentae 1–3; ovules 2–30, ascending; style erect, linear, inserted on a basal nectary; stigmas 3, dilated, reflexed, strongly papillose or smooth; staminodes 3, minute; disk 3-lobed or ring-like. Fruits small, dry, indehiscent, smooth, with a firm, thin wall, ripening green, red, brown or black, often with very bitter taste. Seeds 1–30, in loose cellular pulp, erect, irregularly ovate or oblong, ± compressed, sometimes ± triangular or dagger-shaped, truncate and apically tricornute; testa brown, rigid and smooth, woody, no distinct margin.

About 50–59 species, most of them in tropical South America, few in Central America, Mexico and Southern US; one species endemic in Fernando de Noronha Island; one species, *C. africana* (Hook.f.) Exell, in West and Central Africa (São Tomé, Senegal, Guinea Bissau, Ivory Coast, Ghana, Cameroon, Gabon, Congo) and on Madagascar; one species, *C. martiana* (Cogn.) Cogn., introduced to Indonesia (Java); along forest margins, in clearings, and riverine forest. Host of the fungi *Uromyces pentastriatus* Viegas, *U. novissimus* Speg., *U. ratus* H. S. Jack. & Holw. (Monoson and Rogers 1978), *Passalora cayaponiae* (F. Stevens & Solheim) U. Braun & Crous, and *Stenella praelonga* (Syd.) U. Braun (Kirschner and Piepenbring 2006).

The genus appears to have repeatedly switched from ancestral pollination by bats to pollination by bees (Duchen and Renner 2010).

Selected Bibliography

Agarwal, V.M., Rastogi, N. 2008. Deterrent effect of a guild of extrafloral nectary-visiting ant species on *Raphidopalpa foveicollis*, a major insect pest of sponge gourd, *Luffa cylindrica*. Entomologia Experimentalis et Applicata 128: 303–311.

Akimoto, J., Fukuhara, T., Kikuzawa, K. 1999. Sex ratio and genetic variation in a functionally androdioecious species, *Schizopepon bryoniaefolius* (Cucurbitaceae). Am. J. Bot. 86: 880–886.

Akoroda, M.O., Ogbechie-Odiaka, N.I., Adebayo, M.L., Ugwo, O.E., Fuwa, B. 1990. Flowering, pollination and fruiting in fluted pumpkin (*Telfairia occidentalis*). Sci. Hort. 43: 197–206.

Ali Khan, A. 2002. *Actinostemma tenerum* Griff., Cucurbitaceae, a new phytogeographic record from Aligarh, Uttar Pradesh. J. Bombay Nat. Hist. Soc. 99: 365–366.

Ayala-Nieto, M.L., Lira Saade, R., Alvarado J.L. 1988. Morfología polínica de las Cucurbitaceae de la Península de Yucatán, Mexico. Pollen Spores 30: 5–28.

Barth, O.M., Pinto da Luz, C.F., Gomes-Klein, V.L. 2005. Pollen morphology of Brazilian species of *Cayaponia* Silva Manso (Cucurbitaceae, Cucurbiteae). Grana 44: 129–136.

Beevy, S.S., Kuriachan, P. 1996. Chromosome numbers of South Indian Cucurbitaceae and a note on the

cytological evolution in the family. J. Cytol. Genet. 31: 65–71.
Berndt, R. 2007. A global survey of *Puccinia*-rust on Cucurbitaceae. Mycol. Progress 6: 151–178.
Bhaduri, P.N., Bose, P.C. 1947. Cyto-genetical investigations in some common cucurbits with special reference to fragmentation of chromosomes as a physical basis of speciation. J. Genet. 48: 237–256.
Bhar, N.C., Datta, K.B. 1982. Cytomorphological studies in diploid and induced tetraploid of *Coccinia grandis* (L.) Voigt. Proc. Ind. Sci. Congress Ass. 69(3/6): 216.
Borchert, M. 2004. Vertebrate seed dispersal of *Marah macrocarpus* (Cucurbitaceae) after fire in the Western transverse ranges of California. Ecoscience 12: 463–471.
Burnett, G.T. 1835. Outlines of Botany, including a general history of the vegetable kingdom in which plants are arranged according to the system of natural affinities. J. Churchill. London.
Carlquist, S. 1992. Wood anatomy of selected Cucurbitaceae and its relationship to habit and systematics. Nord. J. Bot. 12: 347–355.
Carr, G.D. 1985. Additional chromosome numbers of Hawaiian flowering plants. Pacific Sci. 39: 302–306.
Chandler, M.E.J. 1964. The Lower Tertiary floras of southern England IV: a summary and survey of findings in the light of recent botanical observations. London: British Museum (Natural History), xii+151 pp.
Chen, J.C.C., Chiu, M.H., Nie, R.L., Cordell, G.A. & Qiu, S. X. 2005. Cucurbitacins and cucurbitane glycosides: structures and biological activities. Nat. Prod. Rep. 22: 386–399.
Chen, R.Y. 1993. Chromosome atlas of Chinese fruit trees and their close wild relatives 1. Beijing: International Academic Publishers.
Chesters, K.I.M. 1957. The Miocene flora of Rusinga Island, Lake Victoria, Kenya. Palaeontographica 101B: 30–71.
Chopra, R.N. 1955. Some observations on endosperm development in the Cucurbitaceae. Phytomorphology 5: 219–230.
Chopra, R.N., Basu, B. 1965. Female gametophyte and endosperm of some members of the Cucurbitaceae. Phytomorphology 15: 217–223.
Chopra, R.N., Seth, P.N. 1977. Some aspects of endosperm development in Cucurbitaceae. Phytomorphology 27: 112–115.
Clarke, A.C., Burtenshaw, M.K., McLenachan, P.A., Erickson, D.L., Penny D. 2006. Reconstructing the origins and dispersal of the Polynesian bottle gourd (*Lagenaria siceraria*). Mol. Biol. Evol. 23: 893–900.
Collinson, M.E., Boulter, M.C., Holmes, P.R. 1993. Magnoliophyta (Angiospermae). In: Benton, M.J. (ed.), The Fossil Record 2. London: Chapman and Hall, pp. 809–841, 864.
Collinson, M.E., Andrews, P., Bamford, M.K. 2009. Taphonomy of the early Miocene flora, Hiwegi Formation, Rusinga Island, Kenya. J. Human Evol. 57: 149–162.
Condon, M.A., Gilbert, L.E. 1988. Sex expression of *Gurania* and *Psiguria* (Cucurbitaceae): Neotropical vines that change sex. Am. J. Bot. 75: 875–884.
Condon, M.A., Scheffer, S.J., Lewis, M.L., Swensen, S.M. 2008. Hidden Neotropical diversity: greater than the sum of its parts. Science 320: 928–931.
Correns, C. 1903. Über die dominierenden Merkmale der Bastarde. Ber. Deutschen Bot. Ges. 21: 133–147.
Correns, C. 1907. Die Bestimmung und Vererbung des Geschlechtes, nach Versuchen mit höheren Pflanzen. Verhandlungen Ges. deutscher Naturforscher Ärzte 1907: 794–802.
Costich, D.E., Galán, F. 1988. The ecology of the monoecious and dioecious subspecies of *Ecballium elaterium* (L.) A. Rich. (Cucurbitaceae) I. Geographic distribution and its relationship to climatic conditions in Spain. Lagascalia 15 (suppl.): 697–710.
Costich, D.E., Meagher, T.R. 1992. Genetic variation in *Ecballium elaterium*: breeding system and geographic distribution. J. Evol. Biol. 5: 589–601.
Crouch, N., Prentice, C., Smith, G.F., Symmonds, R. 1999. South Africa's rarest caudiciform cucurbit, *Gerrardanthus tomentosus*. Bradleya 17: 95–100.
Davis, G.L. 1966. Systematic embryology of the Angiosperms. New York: Wiley.
Deshpande, P.K., Bhuskute, S.M. Makde, K.H. 1986. Microsporogenesis and male gametophyte in some Cucurbitaceae. Phytomorphology 36: 145–150.
de Wilde, W.J.J.O., Duyfjes, B.E.E. 1999. *Bayabusua*, a new genus of. Cucurbitaceae. Sandakania 13: 1–13.
de Wilde, W.J.J.O., Duyfjes, B.E.E. 2003. Revision of *Neoalsomitra* (Cucurbitaceae). Blumea 48: 99–121.
de Wilde, W.J.J.O., Duyfjes, B.E.E. 2004.The genus *Trichosanthes (Cucurbitaceae)* in Sabah. Sandakania 14: 5–32.
de Wilde, W.J.J.O., Duyfjes, B.E.E. 2006a. Redefinition of *Zehneria* and four new related genera (Cucurbitaceae), with an enumeration of the Australasian and Pacific species. Blumea 51: 1–88.
de Wilde, W.J.J.O., Duyfjes, B.E.E. 2006b. Review of the genus *Gymnopetalum* (Cucurbitaceae). Blumea 51: 281–296.
de Wilde, W.J.J.O., Duyfjes, B.E.E. 2006c. *Scopellaria*, a new genus name in Cucurbitaceae. Blumea 51: 297–298.
de Wilde, W.J.J.O., Duyfjes, B.E.E. 2006d. The subtribe Thladianthinae (Cucurbitaceae) in Indochina and Malesia. Blumea 51: 493–518.
de Wilde, W.J.J.O., Duyfjes, B.E.E. 2007a. *Gynostemma* (Cucurbitaceae) in Thailand and Malesia. Blumea 52: 263–280.
de Wilde, W.J.J.O., Duyfjes, B.E.E. 2007b. Diversity in *Zanonia indica* (Cucurbitaceae). Blumea 52: 281–290.
de Wilde, W.J.J.O., Duyfjes, B.E.E. 2007c. *Mukia* Arn. (Cucurbitaceae) in Asia, in particular in Thailand. Thai Forest Bull. (Bot.) 34: 38–52.
de Wilde, W.J.J.O., Duyfjes, B.E.E. 2009. Miscellaneous Southeast Asian Cucurbit news II. Reinwardtia 12: 405–414.
de Wilde, W.J.J.O., Duyfjes, B.E.E., van der Ham, R.W.J.M. 2003. *Borneosicyos simplex* (Cucurbitaceae) a veritable rare plant peculiar of Kinabalu Park. Flora Malesiana Bull. 14: 33–42.
de Wilde, W.J.J.O., Duyfjes, B.E.E., van der Ham, R.W.J.M. 2004. *Khmeriosicyos*, a new monotypic genus of Cucurbitaceae from Cambodia. Blumea 49: 441–446.
de Wilde, W.J.J.O., Duyfjes, B.E.E., van der Ham, R.W.J.M. 2007a. Revision of the genus Gomphogyne (Cucurbitaceae). Thai For. Bull. (Bot.) 35: 45–68.

de Wilde, W.J.J.O., Duyfjes, B.E.E., van der Ham, R.W.J.M. 2007b. *Borneosicyos simplex* (Cucurbiaceae), a veritable rare plant peculiar to Kinabalu Park. Flora Malesiana Bull. 14 (1 & 2): 33–42.
Dieterle, J.V.A. 1974. A new geocarpic genus from Mexico: *Apatzingania* (Cucurbitaceae). Brittonia 26: 129–132.
Dillehay, T.D., J. Rossen, T.C. Andres, Williams, D.E. 2007. Preceramic adoption of peanut, squash, and cotton in Northern Peru. Science 316: 1890–1893.
Dorofeev, P.I. 1963. The Tertiary floras of western Siberia (in Russian). Moskva: Izd. Akad. Nauk SSSR, p. 287.
Dorofeev, P.I. 1988. Miozäne Floren des Bezirks Tambov. Moskva: Izd. Akad. Nauk SSSR.
Duchen, P., Renner, S.S. 2010. The evolution of *Cayaponia* (Cucurbitaceae): repeated shifts from bat to bee pollination and long-distance dispersal to Africa 2-6 million years ago. Am. J. Bot.
Dukas, R. 1987. Foraging behavior of three bee species in a natural mimicry system: Female flowers which mimic male flowers in *Ecballium elaterium* (Cucurbitaceae). Oecologia 74: 256–263.
Dutt, B., Roy, R.P. 1971. Cytogenetic investigations in Cucurbitaceae. I. Interspecific hybridization in *Luffa*. Genetica 42: 139–156.
Duyfies, B.E.E., van der Ham, R.W.J.M., de Wilde, W.J.J.O. 2003. *Papuasicyos*, a new genus of Cucurbitaceae. Blumea 48: 123–128.
Eggli, U. 1998. First report of female flowers for *Xerosicyos pubescens* Keraudren (Cucurbitaceae). Cact. Succ. J. (USA) 70: 40–41.
Eichler, A.W. 1875. Blüthendiagramme 1. Leipzig.
Elangovan, V., Marimuthi, G., Kunz, T.H. 2001. Temporal patterns of resource use by the short-nosed fruit bat, *Cynopterus sphinx* (Megachiroptera: Pteropodidae). J. Mammalogy 82: 161–165.
Erdtman, G. 1952. Pollen morphology and plant taxonomy. Angiosperms. Stockholm: Almqvist and Wiksell.
Erickson, D.L., Smith, B.D., Clarke, A.C., Sandweiss, D.H., Tuross, N. 2005. Asian origin for a 10,000-year-old domesticated plant in the Americas. Proc. Natl. Acad. Sci. USA 102: 18315–18320.
Evans, G.A. 2007. Host plant list of the whiteflies (Aleyrodidae) of the World. www.sel.barc.usda.gov:591/1WF/WhiteflyHost.pdf
Fernandes, A., Mendes, E.J., Jeffrey, C., Fernandes, R.B. 1986. Cucurbitaceae. Flora de Moçambique, vol. 80. Lisboa: Junta de Investigações Científica Tropical, Centro de Botânica.
Franquet, M.R. 1930. L'*Actinostemma paniculatum* Maxim. ex Cogn. doit constituer un genre nouveau de Cucurbitacées. Bull. Mus. Hist. Nat. Paris, Ser. 2, 2: 324–328.
Fukuhara, T., Akimoto, J. 1999. Floral morphology and vasculature of *Schizopepon bryoniaefolius* (Cucurbitaceae). Acta Phytotax. Geobot. 50: 59–73.
Fursa, T.B. 1972a. On the taxonomy of genus *Citrullus* Schrad. Bot. Zhurn. 57: 31–41.
Fursa, T.B. 1972b. On the evolution of the genus *Citrullus* Schrad. Bot. Zhurn. 57: 1365–1372.
Gao X.F., Chen S.K., Gu Z.J., Zhao, J.Z. 1995. A chromosomal study on the genus *Gynostemma* (Cucurbitaceae) (in Chinese). Acta Bot. Yunnanica 17: 312–316.
Gentry, A. 1950. Taxonomy and evolution of *Vaseyanthus*. Madroño 10: 142–155.
Gentry, A.H., Wettach, R.H. 1986. *Fevillea* - a new oil seed from Amazonian Peru. Econ. Bot. 40: 177–185.
Gerrath, J.M., Guthrie, T.B., Zitnak, T.A., Posluszny, U. 2008. Development of the axillary bud complex in *Echinocystis lobata* (Cucurbitaceae): interpreting the cucurbitaceous tendril. Am. J. Bot. 95: 773–781.
Gervais, C., Trahan, R., Gagnon, J. 1999. IOPB chromosome data 14. Newslett. Int. Org. Plant Biosyst. 30: 10–15.
Gillespie, J.J., Kjer, K.M., Duckett, C.N., Tallamy, D.W. 2003. Convergent evolution of cucurbitacin feeding in spatially isolated rootworm taxa (Coleoptera: Chrysomelidae; Galerucinae, Luperini). Mol. Phylogenet. Evol. 29: 161–175.
Giusti, L., Resnik, M., Ruiz, T. del V., Grau, A. 1978 Notas acerca de la biologia de *Sechium edule* (Jacq.) Swartz (Cucurbitaceae). Lilloa 35: 5–13.
Gusmini, G., Wehner, T.C. 2008. Fifty-five years of yield improvement for cucumber, melon, and watermelon in the United States. Hort. Technol. 18: 9–12.
Hegnauer, R. 1964. Chemotaxonomie der Pflanzen. Vol. 3. Basel: Birkhäuser.
Hegnauer, R. 1989. Chemotaxonomie der Pflanzen. Vol. 8 (Nachträge zu Band 3 und 4). Basel: Birkhäuser.
Heiser, C.B., Schilling, E.E. 1988. Phylogeny and distribution of *Luffa* (Cucurbitaceae) Biotropica 20: 185–191.
Heiser, C.B., Schilling E.E., Dutt, B. 1988. The American species of *Luffa* (Cucurbitaceae). Syst. Bot. 13: 138–145.
Heppner, J.B. 1989. Larvae of fruit flies. V. *Dacus cucurbitae* (Melon Fly) (Diptera: Tephritidae). Florida Dept. Agric. & Consumer Services, Division of Plant Industry. Entomology Circular 315: 1–2.
Huang, S., Zhang, Z., Gu, X. et al. (biology analysis group). 2009. The genome of the cucumber, *Cucumis sativus* L. Nature Genetics 41: 1275–1281.
Ilyas, M.H.M. 1992. Studies on the extrafloral nectaries in some members of Cucurbitaceae. Acta Botanica Indica 20: 116–119
Imaichi, R., Okamoto, K. 1992. Comparative androecium morphogenesis of *Sicyos angulatus* and *Sechium edule* (Cucurbitaceae). Bot. Mag. Tokyo 105: 539–548.
Jeffrey, C. 1962. Notes on Cucurbitaceae, including a proposed new classification of the family. Kew Bull. 15: 337–371.
Jeffrey, C. 1967. Cucurbitaceae. In: Milne-Redhead, E., Polhill, R.M. (eds.) Flora of Tropical East Africa.
Jeffrey, C. 1969. The genus *Mukia* in Asia, Malesia and Australasia. Hooker's Icon. Pl. 5, 7, 3: 1–12.
Jeffrey, C. 1978a. Further notes on Cucurbitaceae: IV. Some New World taxa. Kew Bull. 33: 347–380.
Jeffrey, C. 1978b. Cucurbitaceae. In: Flora Zambesiaca 4: 414–499.
Jeffrey, C. 1980. A review of the Cucurbitaceae. Bot. J. Linn. Soc. 81: 233–247.
Jeffrey, C. 1990. Appendix: an outline classification of the Cucurbitaceae. In: Bates, D.M., Robinson, R.W., Jeffrey, C. (eds.) Biology and utilization of the Cucurbitaceae. Ithaca, NY: Comstock Publication Associates, Cornell University Press, pp. 449–463.
Jeffrey, C. 2005. A new system of Cucurbitaceae. Bot. Zhurn. 90: 332–335.

Johri, B.M., Ambegaokar, K.B., Srivastava, P.S. 1992. Comparative embryology of angiosperms. Vol. 1. Berlin: Springer.

Jones, C.S. 1993. Heterochrony and heteroblastic leaf development in different subspecies of *Cucurbita argyrosperma* (Cucurbitaceae). Am. J. Bot. 80: 778–795.

Kater, M.M., Franken, J., Carney, K.J., Colombo, L., Angenent, G.C. 2001. Sex determination in the monoecious species cucumber is confined to specific floral whorls. Plant Cell 13: 481–493.

Kearns, D.M. 1992. A revision of *Polyclathra* (Cucurbitaceae). In: Biosystematics of Mexican Cucurbitaceae. Ph.D. Thesis, University of Texas.

Kearns, D.M. 1994. The genus *Ibervillea* (Cucurbitaceae): an enumeration of the species and two new combinations. Madroño 41: 13–22.

Keraudren, M. 1968. Recherches sur les cucurbitacées de Madagascar. Mém. Mus. Hist. Nat. B 16: 122–330.

Keraudren-Aymonin, M. 1971. La survivance des *Ampelosicyos* (Cucurbitacees) a Madagascar. Bull. Soc. Bot. France 118: 281–286.

Khunwasi, C. 1998. Palynology of the Cucurbitaceae. Doctoral Dissertation Naturwiss. Fak., University of Innsbruck.

Kirschner, R., Piepenbring, M. 2006. New species and records of cercosporoid hyphomycetes from Panama. Mycol. Progress 5: 207–219.

Kocyan, A., Zhang, L.-B., Schaefer, H., Renner, S.S. 2007. A multi-locus chloroplast phylogeny for the Cucurbitaceae and its implications for character evolution and classification. Mol. Phylogenet. Evol. 44: 553–577.

Kosteletzky, V.F. 1833. Allgemeine medizinisch-pharmazeutische Flora. Vol. 6. Prag: Borrosch & Andre.

Kumazawa, M. 1964. Morphological interpretations of axillary organs in the *Cucurbitaceae*, Phytomorphology 14: 287–298.

Lassnig, P. 1997. Verzweigungsmuster und Rankenbau der Cucurbitaceae. Trop. Subtrop. Pflanzenwelt 98. Akad. Wissensch. Lit., Mainz, F. Steiner, Stuttgart, pp. 1–156.

Leins, P., Galle, P. 1971. Entwicklungsgeschichtliche Untersuchungen an Cucurbitaceen-Blüten. Österr. Bot. Z. 119: 531–548.

Leins, P., Merxmüller, H., Sattler, R. 1972. Zur Terminologie interkalarer Becherbildungen in Blüten. Ber. Deutsch. Bot. Ges. 85: 294.

Li, J.Q., Wu, Z.Y., Lu, A.-M. 1993. Cytological observation on the plants of Thladianthinae (Cucurbitaceae). Acta Bot. Yunnanica 15: 101–104.

Link, A., Di Fiore, A. 2006. Seed dispersal by spider monkeys and its importance in the maintenance of neotropical rain-forest diversity. J. Trop. Ecol. 22: 235–246.

Lira Saade, R. 1991. Observaciones en el género *Sicana* (Cucurbitaceae). Brenesia 35: 19–59.

Lira Saade, R. 1995. A new species of *Sicydium* Schlechtendal (Cucurbitaceae: Zanonioideae, Zanonieae, Sicydiinae) for the Flora Mesoamericana. Novon 5: 284–286.

Lira Saade, R. 2004a. El género *Sicydium* (Cucurbitaceae, Zanonioideae, Sicydiinae) en México. Acta Bot. Mex. 68: 39–64.

Lira Saade, R. 2004b. Cucurbitaceae de la Peninsula de Yucatán. Taxonomía, florística y etnobotánica. Etnoflora Yucatanense 22. Mérida, Yucatán, México: Universidad Autónoma de Yucatán/CONACyT.

Lira Saade, R., Alvarado J.L., Ayala-Nieto, M.L. 1998. Pollen morphology in *Sicydium* (Cucurbitaceae, Zanonioideae). Grana 37: 215–221.

Lu, A. 1985. Studies on the genus *Schizopepon* Max. (Cucurbitaceae). Acta Phytotax. Sin. 23: 106–120.

Lu, A., Huang, L., Chen, S., Jeffrey, C. 2009. Flora of China. Vol. 19. Cucurbitaceae (draft). http://flora.huh.harvard.edu/china/mss/volume19/Cucurbitaceae-MO_reviewing.htm

Marr, K.L., Xia, Y.-M., Bhattarai, N.K. 2007. Allozymic, morphological, phenological, linguistic, plant use, and nutritional data of *Benincasa hispida* (Cucurbitaceae). Econ. Bot. 61: 44–59.

Marticorena, C. 1963. Material para una monografia de la morfología del polen de Cucurbitaceae. Grana Palynol. 4: 78–91.

Martínez Crovetto, R. 1946. Nota taxonómica sobre *Wilbrandia sagittifolia* Griseb. (Cucurbitaceae). Bol. Soc. Argent. Bot. 1: 312–317.

Martínez Crovetto, R. 1952. El género *Pteropepon* (Cucurbitaceae) en la República Argentina. Bol. Soc. Argent. Bot. 4: 177–182.

Martínez Crovetto, R. 1956. Especies nuevas o críticas del género *Apodanthera* (Cucurbitaceae) II. Bol. Soc. Argent. Bot. 6: 94–97.

Matthews, M.L., Endress, P.K. 2004. Comparative floral structure and systematics in Cucurbitales (Corynocarpaceae, Coriariaceae, Tetramelaceae, Datiscaceae, Begoniaceae, Cucurbitaceae, Anisophylleaceae). Bot. J. Linn. Soc. 145: 129–185.

McKay, J.W. 1931. Chromosome studies in the Cucurbitaceae. Univ. Calif. Publ. Bot. 16: 339-350.

Medellín, R.A., Gaona, O. 1999. Seed dispersal by bats and birds in forest and disturbed habitats of Chiapas, Mexico. Biotropica 31: 478–485.

Meeuse, A.D.J. 1962. The Cucurbitaceae of Southern Africa. Bothalia 8: 1–111.

Mercado, P., Lira Saade, R. 1994. Contribution al conocimiento de los numeros chromosomicos de los generos *Sechium* P. Br. y *Sicana* Naudin (Cucurbitaceae). Acta Bot. Mex. 27: 7–13.

Metcalfe, R.L. 1986. Coevolutionary adaptations of rootworm beetles (Coleoptera: Chrysomelidae) to cucurbitacins. J. Chem. Ecol. 12: 1109–1124.

Monoson, H.L., Rogers, G.M. 1978. Species of *Uromyces* that infect New World Cucurbitaceae. Mycologia 70: 1144–1150.

Monro, A.K., Stafford, P.J. 1998. A synopsis of the genus *Echinopepon* (Cucurbitaceae: Sicyoeae), including three new taxa. Ann. Missouri Bot. Gard. 85: 257–272.

Mori, S.A., Cremers, G., Gracie, C.A., de Granville, J.-J., Heald, S.V., Hoff, M., Mitchell, J.D. 2002. Guide to the vascular plants of Central French Guiana. Part 2. Dicotyledons. Mem. N.Y. Bot. Gard. 76 (2).

Morimoto, Y., Gikungu, M., Maundu, P. 2004. Pollinators of the bottle gourd (*Lagenaria siceraria*) observed in Kenya. Int. J. Trop. Insect. Science 24: 79–86.

Muller, J. 1985. Significance of fossil pollen for angiosperm history. Ann. Missouri Bot. Gard. 71: 419–443.

Murawski, D.A., Gilbert, L.E. 1986. Pollen flow in *Psiguria warscewiczii*: a comparison of *Heliconius* butterflies and hummingbirds. Oecologia 68: 161-167.

Naudin, C.V. 1859. Essaie d'une monographie des espèces et des variétés du genre *Cucumis*. Ann. Sci. Nat., Bot. sér. 4, 11: 5–87.

Nee, M., Schaefer, H., Renner, S.S. 2010. The relationship between *Anisosperma* and *Fevillea* (Cucurbitaceae), and a new species of *Fevillea* from Bolivia. Syst. Bot. 34: 704–708.

Nicolson, D., Fosberg, R. 2004. The Forsters and the botany of the second Cook expedition. Rugell: Gantner Verlag.

Nishikawa, T. 1981. Chromosome counts of flowering plants of Hokkaido (5). Rep. Taisetsuzan Inst. Sci. 16: 45–53.

Okoli, B.E. 1984. Wild and cultivated cucurbits in Nigeria. Econ. Bot. 38: 350–357.

Okoli, B.E. 1987. Morphological and cytological studies in *Telfairia* Hooker (Cucurbitaceae). Feddes Repertorium. 98: 505–508.

Okoli, B.E., McEuen, A.R. 1986. Calcium-containing crystals in *Telfairia* Hooker (Cucurbitaceae). New Phytol. 102: 199–207.

Okoli, B.E., Mgbeogu, C.M. 1983. Fluted Pumpkin, *Telfairia occidentalis*: West African vegetable crop. Econ. Bot. 37: 145–149.

Okoli, B.E., Onofeghara, F.A. 1984. Distribution and morphology of extrafloral nectaries in some Cucurbitaceae. Bot. J. Linn. Soc. 89: 153–164.

Olson, M.E. 2003. Stem and leaf anatomy of the arborescent Cucurbitaceae *Dendrosicyos socotrana* with comments on the evolution of pachycauls from lianas. Plant Syst. Evol. 239: 199–214.

Oobayashi, K., Yoshikawa, K., Arihara, S. 1992. Structural revision of bryonoside and structure elucidation of minor saponins from *Bryonia dioica*. Phytochemistry 31: 943–946.

Osuji, J.O., Okoli, B.E., Heslop-Harrison, J.S.P. 2006. Cytology and molecular cytogenetics of *Cucumeropsis mannii* Naudin: implications for breeding and germplasm characterization. Int. J. Bot. 2: 187–192.

Page, J.S., Jeffrey, C. 1975. A palyno-taxonomic study of African *Peponium* (Cucurbitaceae). Kew Bull. 30: 495–502.

Parfitt, B.D., Pinkava, D.J., Rickel, D., Fillipi, D., Eggers, B., Keil, D.J. 1990. Documented chromosome numbers 1990: 1. Miscellaneous North American vascular plants. Sida 14: 305–308.

Piperno, D.R., Stothert, K.E. 2003. Phytolith evidence for Early Holocene *Cucurbita* domestication in Southwest Ecuador. Science 299: 1054–1057.

Piperno, D.R., Holst, I., Wessel-Beaver, L., Andres, T.C. 2002. Evidence for the control of phytolith formation in *Cucurbita* fruits by the hard rind (Hr) genetic locus: archaeological and ecological implications. Proc. Natl. Acad. Sci. 99: 10923–10928.

Pozner, R. 1993a. Sistemas reproductivos en Cucurbitaceae Argentinas. Ph.D. Thesis, Buenos Aires University, Argentina.

Pozner, R. 1993b. Androsporangio, androsporogénesis y androgametogénesis en *Cayaponia citrullifolia, Cayaponia bonariensis* y *Cucurbitella duriaei* (Cucurbitaceae). Darwiniana 32: 109–123.

Pozner, R. 1994. Rudimento seminal y ginosporogénesis en *Cucurbitella duriaei* y *Cayaponia bonariensis* (*Cucurbitaceae*). Kurtziana 23: 55–72.

Pozner, R. 1998a. Revisión del género *Cucurbitella* (Cucurbitaceae). Ann. Missouri Bot. Gard. 85: 425–438.

Pozner, R. 1998b. Cucurbitaceae. In: Hunziker, A.T. (ed.) Flora Fanerogámica Argentina, fascicle 53, family 275, pp. 1–58. Córdoba: Pugliese Siena.

Pozner, R. 2004. A new species of *Echinopepon* from Argentina and taxonomic notes on the subtribe Cyclantherinae (Cucurbitaceae). Syst. Bot. 29: 599–608.

Probatova, N.S., Rudyka, E.G. 1981. Chromosome numbers of some vascular plant species from the Soviet Far East. Izvestiya Sibirskogo Otdeleniya Akad. Nauk SSSR, Ser. Biol. Nauk 2 (10): 77–81.

Pruesapan, K., van der Ham, R.W.J.M. 2005. Pollen morphology of *Trichosanthes* (Cucurbitaceae). Grana 44: 75–90.

Rauh, W. 1996. Observations complementaires sur *Xerosicyos pubescens* (Cucurbitaceae) de Madagascar. Bull. Mus. Natl. Hist. Nat., B, Adans. 18: 161–166.

Raven, P.H., Axelrod, D.I. 1974. Angiosperm biogeography and past continental movements. Ann. Missouri Bot. Gard. 61: 539–637.

Renner, S.S., Schaefer, H. 2008. Phylogenetics of *Cucumis* (Cucurbitaceae) as understood in 2008. In: Pitrat, M. (ed.) Cucurbitaceae 2008. Proc. IXth EUCARPIA Meeting Genetics and Breeding of Cucurbitaceae, Avignon (France), 21–24 May 2008, pp. 53–58. https://w3.avignon.inra.fr/dspace/handle/2174/236

Renner, S.S., Schaefer, H., Kocyan, A. 2007. Phylogenetics of *Cucumis* (Cucurbitaceae): Cucumber (*C. sativus*) belongs in an Australian/Asian clade far from African melon (*C. melo*). BMC Evol. Biol. 7: 58.

Renner, S.S., Scarborough, J., Schaefer, H., Paris, H.S., Janick, J. 2008. Dioscorides's bruonia melaina is *Bryonia alba*, not *Tamus communis*, and an illustration labeled bruonia melaina in the *Codex Vindobonensis* is *Humulus lupulus* not *Bryonia dioica*. In: Pitrat, M. (ed.) Cucurbitaceae 2008. Proc. IXth EUCARPIA Meeting Genetics and Breeding of Cucurbitaceae, Avignon (France), 21–24 May 2008, pp. 273–280. https://w3.avignon.inra.fr/dspace/handle/2174/218

Rheinberger, H.J. 2000. Mendelian inheritance in Germany between 1900 and 1910. The case of Carl Correns (1864–1933). C. R. Acad. Sci. Ser. III. Sci. Vie/Life Sci. 323: 1089–1096.

Ridley, H.N. 1930. The dispersal of plants throughout the World. Ashford: L. Reeve.

Rodrigues, F.H.G., Hass, A., Lacerda, A.C.R., Grando, R.L. S.C., Bagno, M.A., Bezerra, A.M.R., Silva, W.R. 2007. Feeding habits of the maned wolf (*Chrysocyon brachyurus*) in the Brazilian Cerrado. Mastozoología Neotropical 14: 37–51.

Roy, R.P., Saran, S. 1990. Sex expression in the Cucurbitaceae. In: Bates, D.M., Robinson, R.W., Jeffrey, C. (eds.) Biology and utilization of the Cucurbitaceae. Ithaca: Comstock Publ. Associates, Cornell University Press, pp. 251–268.

Rust, R.W., Vaissière, B.E., Westrich, P. 2003. Pollinator biodiversity and floral resource use in *Ecballium elaterium* (Cucurbitaceae), a Mediterranean endemic. Apidologie 34: 29–42.

Salard-Cheboldaeff, M. 1978. Sur la palynoflore Maestrichtienne et Tertiaire du bassin sédimentaire littoral du Cameroun. Pollen Spores 20: 215–260.
Samuel, R., Balasubramaniam, S., Morawetz, W. 1995. The karyology of some cultivated Cucurbitaceae of Sri Lanka. Ceylon J. Sci., Biol. Sci. 24: 17–22.
Sanjur, O.I., Piperno, D.R., Andres, T.C., Wessel-Beaver, L. 2002. Phylogenetic relationships among domesticated and wild species of *Cucurbita* (Cucurbitaceae) inferred from a mitochondrial gene: implications for crop plant evolution and areas of origin. Proc. Natl. Acad. Sci. 99: 535–540.
Sazima, M., Buzato, S., Sazima, I. 1996. *Cayaponia cabocla* (Curcubitaceae) parece uma espécie quiróptero͏́fila. In: VLVII Congresso Nacional de Botânica, 1996, Nova Friburgo, 1996, pp. 407–408.
Schaefer, H., Kocyan, A., Renner, S.S. 2008a. *Linnaeosicyos* (Cucurbitaceae): a new genus for *Trichosanthes amara*, the Caribbean sister species of all Sicyoeae. Syst. Bot. 33: 349–355.
Schaefer, H., Heibl, C., Renner, S.S. 2009. Gourds afloat: a dated phylogeny reveals an Asian origin of the gourd family (Cucurbitaceae) and numerous oversea dispersal events. Proc. R. Soc. B 276: 843–851.
Schaefer, H., Renner, S.S. 2010a. A three-genome phylogeny of *Momordica* (Cucurbitaceae) suggests seven returns from dioecy to monoecy and recent long-distance dispersal to Asia. Mol. Phylogen. Evol. 54: 553–560.
Schaefer, H., Renner, S.S. 2010b. A gift from the New World? The West African crop *Cucumeropsis mannii* and the American *Posadaea sphaerocarpa* (Cucurbitaceae) are the same species. Syst. Bot.
Schaefer, H., Renner, S.S. Phylogenetic relationships in the order Cucurbitales and a new classification of the gourd family (Cucurbitaceae). Taxon.
Schaefer, H., Telford, I.R.H., Renner, S.S. 2008b. *Austrobryonia* (Cucurbitaceae), a new Australian endemic genus, is the closest living relative to the Eurasian and Mediterranean *Bryonia* and *Ecballium*. Syst. Bot. 33: 125–132.
Schweingruber, F.H., Börner, A., Schulze, E.-D. 2010. Stem anatomy of herbs, shrubs and trees; an ecological approach. Heidelberg: Springer.
Sebastian, P., Schaefer, H., Renner, S.S. 2010. Darwin's Galapagos gourd: providing new insights 175 years after his visit. J. Biogeogr. 37: 975–980.
Sebastian, P.M., Schaefer, H., Telford, I.R.H., Renner, S.S. 2010. Cucumber and melon have their wild progenitors in India, and the sister species of *Cucumis melo* is from Australia. Proc. Nat. Acad. Sc. 107: 14269–14273.
Singh, B.P. 1991. Interspecific hybridization in between New and Old-World species of *Luffa* and its phylogenetic implication. Cytologia 56: 359–365.
Slavik, B., Jarolimova, V., Chrtek, J. 1993. Chromosome counts of some plants from Cyprus. Candollea 48: 221–230.
Smith, B.D. 1997. The initial domestication of *Cucurbita pepo* in the Americas 10,000 years ago. Science 276: 932–934.
Stafford, P.J., Sutton, D.A. 1994. Pollen morphology of the Cyclantherinae C. Jeffr. (tribe Sicyoeae Schrad., Cucurbitaceae) and its taxonomic significance. Acta Bot. Gallica 141: 171–182.
Steele, P.R., Friar, L.M., Gilbert, L.E., Jansen, R.K. 2010. Molecular systematics of the Neotropical genus *Psiguria* (Cucurbitaceae): implications for phylogeny and species identification. Am. J. Bot. 97: 156–173.
Stocking, K.M. 1955. Some considerations of the genera *Echinocystis* and *Echinopepon* in the United States and northern Mexico. Madroño 13: 84–100.
Takhtajan, A.L. (ed.) 1981. Flowering plants, vol. 5 (2). Moscow: Proswjeschtschenie.
Telford, I.R.H. 1989. Rediscovery of *Muellerargia timorensis* (Cucurbitaceae). Austral. Syst. Bot. Soc. Newslett. 59: 4.
Thakur, G.K., Sinha, B.M.B. 1973. Cytological investigation in some cucurbits. J. Cytol. Gen. 7/8: 122–130.
Turala-Szybowska, K. 1990. Further studies in chromosome numbers of Polish angiosperms 23. Acta Biologica Cracoviensia, Series Botanica 32: 172, 179–180.
van der Ham, R.W.J.M. 1999. Pollen morphology of *Bayabusua* (Cucurbitaceae) and its allies. Sandakania 13: 17–22.
van der Ham, R.W.J.M., Pruesapan, K. 2006. Pollen morphology of *Zehneria* s. l. (Cucurbitaceae). Grana 45: 241–248.
van der Ham, R.W.J.M., van Heuven, B.J. 2003. A new type of Old World Cucurbitaceae pollen. Grana 42: 88–90.
Vasil, I.K. 1960. Studies on pollen germination of certain Cucurbitaceae. Am. J. Bot. 47: 239–247.
Vogel, S. 1954. Blütenbiologische Typen als Elemente der Sippengliederung, dargestellt anhand der Flora Südafrikas. Bot. Stud. 1. Jena: Fischer.
Vogel, S. 1958. Fledermausblumen in Südamerika. Ein Beitrag zur Kenntnis des chiropterophilen Stiltypus. Österreichische Bot. Zeitschr. 104: 491–530.
Vogel, S. 1969. Chiropterophilie in der neotropischen Flora. Neue Mitteilungen II. Flora 158: 185–222.
Vogel, S. 1981a. Trichomatische Blütennektarien bei Cucurbitaceen. Beitr. Biol. Pfl. 55: 325–353.
Vogel, S. 1981b. Die Klebstoffhaare an den Antheren von *Cyclanthera pedata* (*Cucurbitaceae*). Pl. Syst. Evol. 137: 291–316.
Vogel, S. 1990. Ölblumen und Ólsammelnde Biennen dritte Folge: *Momordica*, *Thladiantha* in die Ctenoplectridae. Trop. Subtrop. Pflanzenwelt 73: 1–181.
Vogel, S. 1997. Remarkable nectaries: structure, ecology, organophyletic perspectives I. Substitutive nectaries. Flora 192. 305–333.
Volz, S.M., Renner, S.S. 2008. Hybridization, polyploidy, and evolutionary transitions between monoecy and dioecy in *Bryonia* (Cucurbitaceae). Am. J. Bot. 95: 1297–1306.
Volz, S.M., Renner, S.S. 2009. Phylogeography of the ancient Eurasian medicinal plant genus *Bryonia* (Cucurbitaceae) inferred from nuclear and chloroplast sequences. Taxon 58: 550–560.
Ward, D.E. 1984. Chromosome counts from New Mexico and Mexico. Phytologia 56(1): 55–60.
Ward, D.E., Spellenberg, R. 1988. Chromosome counts of angiosperms from New Mexico and adjacent areas. Phytologia 64: 390–398.
Whitaker, T.W. 1933. Cytological and phylogenetic studies in the Cucurbitaceae. Bot. Gaz. 94: 780–790.
Whitaker, T.W., Davis, G.N. 1962. Cucurbits – Botany, cultivation and utilization. London: Leonard Hill.

Wunderlin, R.P. 1976. Two new species and a new combination in *Frantzia* (Cucurbitaceae). Brittonia 28: 239–244.
Zhang, Z.-Y., Lu, A.-M. 1989. Pollen morphology of the subtribe Thladianthinae (Cucurbitaceae) and its taxonomic significance. Cathaya 1: 23–36.
Zhang, L.-B., Simmons, M.P., Kocyan, A., Renner, S.S. 2006. Phylogeny of the Cucurbitales based on DNA sequences of nine loci from three genomes: implications for morphological and sexual system evolution. Mol. Phylogen. Evol. 39: 305–322.
Zimmermann, A. 1922. Die Cucurbitaceen. Beiträge zur Anatomie, Physiologie, Morphologie, Biologie, Pathologie und Systematik. Vols. 1 and 2. Jena: Fischer.

Datiscaceae

Datiscaceae Bercht. & J.Presl (1820), nom. cons.
Tetramelaceae Airy Shaw (1965).

S. Swensen and K. Kubitzki

Dioecious tall soft-wooded, buttressed trees or (andro)dioecious robust perennial actinorhizal herbs; cork cambium initially superficial. Leaves spiral, petiolate, simple and heart-shaped or imparipinnate to pinnatifid, entire or dentate, estipulate. Inflorescences thyrsoid, long, pendant, terminal or terminal and axillary spikes or (males only) thyrses, or compound, contracted thyrses. Flowers subsessile or shortly petiolate; male ones: calyx tube very short or 0, with 3–10 lobes; petals 0 or (*Octomeles*) small and greenish; stamens 4–15(25); filaments very short or elongate; anthers basifixed, bilocular, dehiscing longitudinally; vestigial gynoecium sometimes present; female and bisexual flowers: ovary inferior, unilocular with 3–8 longitudinal parietal or protruding-diffuse placentae; carpels forming a roof over the ovary on the rim of which the calyx tube with the widely separated stylodia is inserted; calyx lobes 3–8, short; petals 0; stamens, if present, few; stylodia short and broad with a distinct stigma or elongate and bifid and stigmatic throughout; ovules in the single cavity 20–100, pendulous to horizontal, anatropous, bitegmic, crassinucellate. Fruit capsular, dehiscing either apically or laterally. Seeds very numerous, minute; endosperm scant or 0; embryo straight. $x = 11, 23$.

Three genera with four spp., from S Asia to the SW Pacific region, and SE Europe and USA (California) and Mexico (Baja California).

Morphology and Anatomy. A comprehensive study of the vegetative and reproductive morphology and anatomy of the family was undertaken by Davidson (1973, 1976), and in a comparative study of the floral structure of Cucurbitales important information on *Octomeles* and *Datisca* was provided by Matthews and Endress (2004). The nodular roots of *Datisca* contain *Frankia* endosymbionts, which are similar to those known from *Coriaria* (Swensen 1996). Nodes in *Octomeles* and *Tetrameles* are trilacunar (with double strands in the lateral gaps in *Octomeles*), and those of *Datisca* are 3-trace, unilacunar. The wood of *Octomeles* and *Tetrameles* appears very specialised; it is storied and consists only of large vessel elements with simple perforation plates, thin-walled libriform fibres, paratracheal axial parenchyma, and rays. The inflorescences of Datiscaceae are specialised thyrses. Spike-like contracted thyrses or compound spikes are characteristic of *Octomeles* and *Tetrameles* (Fig. 32A; Davidson 1973) and, in *Datisca*, Wydler (1878) and Himmelbaur (1909) found compound, contracted thyrses (Fig. 32E). A peculiar feature of Datiscaceae and related families is the strong development of the carpels on their ventral side, by which an extensive "roof" is formed over the ovary locule, which forms the bottom of the perianth tube and on which the sepals and the stylodia are inserted. Inside the floral tube of male *Octomeles* flowers, histologically distinct tissue has been interpreted as a nectary (Davidson 1973; Matthews and Endress 2004). In *Octomeles* and *Datisca*, there is an indication of the presence of a compitum. In contrast to previous concepts, Matthews and Endress (2004) describe the ovary of *Octomeles* as septate with protruding-diffuse placentation, which may also be true for *Tetrameles*; *Datisca* has parietal placentation.

Embryology. Pollen of *Datisca* is shed in the binucleate state (Davidson 1973). In *Datisca canabina* the embryo sac is formed according to the Allium Type. The ovules are bitegmic, anatropus and crassinucellate. The endosperm is

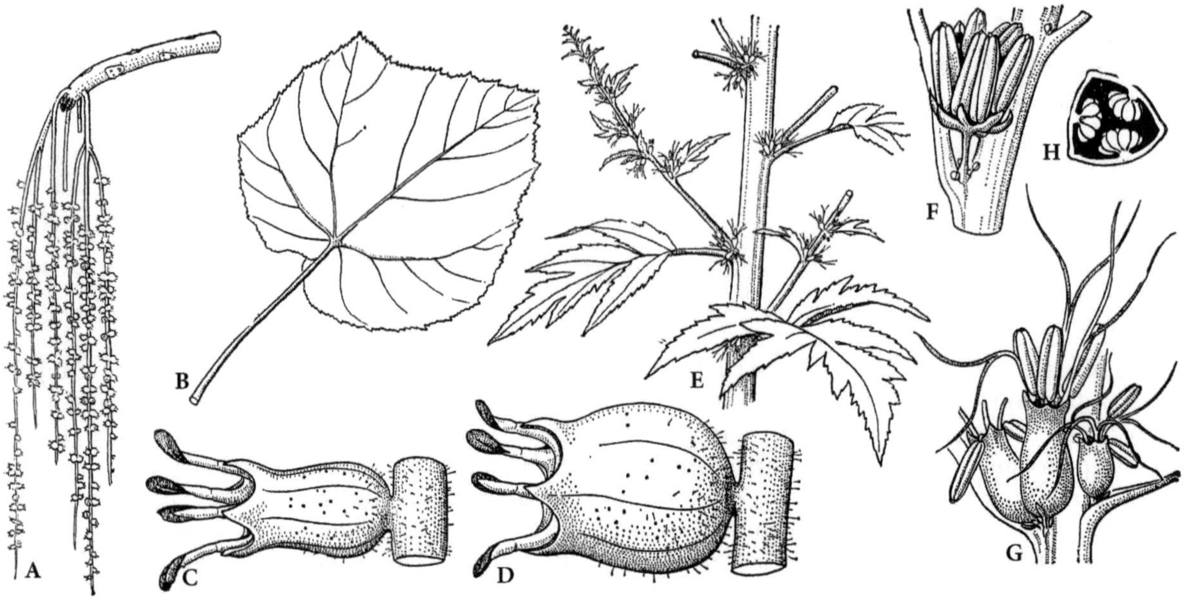

Fig. 32. Datiscaceae. **A–D** *Tetrameles nudiflora*. **A** Female inflorescence. **B** Leaf. **C** Female flower. **D** Fruit. **E–H** *Datisca glomerata*. **E** Part of flowering shoot. **F** Male flower. **G** Bisexual flower. **H** Transverse section of ovary. (Takhtajan 1981)

nuclear, later becoming cellular (Himmelbaur 1909; Boesewinkel 1984).

Pollen Morphology. Pollen grains of Datiscaceae are relatively small, spheroidal, tricolporate, tectate-columellate, those of *Octomeles* 10–12 μm long, psilate and in monads, and those of *Datisca* 13–16 μm long, rugulate and in loose tetrads (Davidson 1973).

Pollination. Both the generally psilate, relatively small pollen and, in *Tetrameles*, the abundance of staminate flowers aggregated at the tips of leafless twigs might indicate anemophily, but the probable presence of nectaries in *Octomeles* (life observations are unknown to us) would militate against this. The small seeds are clearly wind-disseminated, which enables the plants to occupy transient habitats such as river margins and cleared sites.

Reproductive Systems. *Datisca glomerata* is one of the two known flowering plant species which have a functionally androdioecious breeding system which may have evolved from a dioecious precursor (Liston et al. 1989; Zhang et al. 2006), and not from hermaphroditism, as often is postulated for the evolution of androdioecy. High outcrosssing rates, along with significant levels of inbreeding depression, have been observed in populations of *D. glomerata* and are consistent with theoretical models proposed to explain the maintenance of males in androdioecious populations (Fritsch and Rieseberg 1992; Rieseberg et al. 1993). The other three species of Datiscaceae are strictly dioecious.

Fruit and Seed. In the *Octomeles* fruit, the exocarp splits longitudinally from the base and is shed apically along with the calyx tube, whereas the endocarp splits from the apex into hard pale-brown valves which persist as a campanulate basket-like structure. The fruits of *Tetrameles* and *Datisca* split on the distal end, with the valves curving inwards to leave an apical pore through which the minute seeds probably are shaken out. Seeds in Datiscaceae come close to dust seeds, they are produced in very large quantities and hardly exceed the length of 1 mm (*Datisca*: 0.8–0.9×0.3×0.4 mm; *Octomeles* 0.75×1.0×0.2 mm). The seeds are exotestal, with polygonal lignified testa cells, the remaining seed coat tissue and nucellus being obliterated. A rim which surrounds the hilar region is lifted during germination without a special zone of rupture, and functions as an operculum. Both the thin endosperm and the straight cylindric embryo are rich in oil and aleurone but free of starch (Boesewinkel 1984).

Phytochemistry. Glucosides of the simple flavonoids quercetin and kaempferol are found in all species of the family, whereas both species of *Datisca* contain glucosides of unusual flavonols, galangin (5, 7-dihyroxyflavonol), 7-O-methylgalangin, datiscetin (3, 5, 7, 2'-tetrahydroxyflavonol) and 7-O-methyldatiscetin (Bohm 1988). *Datisca glomerata* contains various cucurbitacins; alkaloids have not been found (Hegnauer 1989).

Fossils. Fossil wood from the Intertrappean beds near Nagpur in Central India has been described as *Tetrameleoxylon prenudiflora* and tentatively included in Datiscaceae by Lakhanpal and Verma (1965). If confirmed, it would indicate the presence of the family on the Deccan Peninsula at the early Eocene.

Family Status and Affinities. Traditionally, Datiscaceae have been circumscribed to include the herbaceous *Datisca* along with the woody *Octomeles* and *Tetrameles* (e.g. Lindley 1846; Warburg 1895). Families such as Cucurbitaceae, Loasaceae and Begoniaceae were considered as the closest relatives of Datiscaceae and, in the Englerian system, were included in the broadly construed Parietales (Gilg 1925). The great phenetic differences between the herbaceous and woody genera within the family led Airy Shaw (1965) to propose, for the latter, the new family Tetramelaceae. Gene sequence analyses such as those of Swensen et al. (1994), Swensen (1996), Swensen et al. (1998), Wagstaff and Dawson (2000) and Goodall-Copestake et al. (2009) showing Datiscaceae (s.l.) to be polyphyletic were contradictory in regard to its closest relatives, and usually lacked full statistical support. Even the most comprehensive analysis of Cucurbitales to date based on sequence data of nine loci from the three genomes (Zhang et al. 2006) presented Datiscaceae (s.str.) in a tetratomy with Tetramelaceae, Begoniaceae and Cucurbitaceae. Therefore, the traditional wide family concept, based on perspicacious observations of early botanists, is followed here in consonance with the careful observations by Davidson (1973) (see also Brummitt 2007); note that the splitting off of a monotype (*Datisca*) from a taxon which would continue to be its closest relative (Tetramelaceae) is avoided because it would mean loss of phylogenetic information.

In contrast, the position of Datiscaceae within Cucurbitales and close to Begoniaceae and Cucurbitaceae is well supported by morphological (Matthews and Endress 2004) and molecular studies (Chase et al. 1993; Zhang et al. 2006; and many others).

Distribution and Habitats. *Datisca* has an intercontinentally disjunct distribution: *D. glomeratus* is native to California and adjacent northern Baja California, whereas *D. cannabina* is distributed in C and SW Asia and the W Mediterranean. Axelrod (1975) hypothesized that such Mediterranean-American disjuncts achieved their present distribution in the Paleogene. An analysis of isozymes in *Datisca* (Liston et al. 1989) has demonstrated that, in spite of the lack of any substantial morphological differentiation between the two disjunct species, there is a considerable molecular-based divergence between them. In fact, these species exhibit one of the lowest known genetic identity values for congeneric species, which provides evidence for an ancient origin of their disjunct distribution. Among populations of *D. cannabina*, Mediterranean populations show a higher identity with *D. glomerata* than Asian populations, which may support a possible dispersal route across the Atlantic Ocean.

Octomeles is found from Sumatra through all Malesia (except for Java and the Lesser Sunda Islands) to New Guinea and the Solomon Islands, where it occurs up to c. 800 m above sea level, but mostly in riverine forests on alluvial ground. *Tetrameles* is distributed from the Western Ghats of India through Malesia to Australia (Queensland), with a preference for limestone substrates, and always is bound to a pronounced seasonal climate, which may explain its absence from W Malesia (most of Sumatra, Borneo, Philippines). Datiscaceae are probably not overly long-lived members of seral vegetation, but colonize open and/or disturbed habitats, thereby producing large numbers of offspring.

Uses. The roots and leaves of *Datisca cannabina* are an old oriental dye source; the pigment is datiscetin-3-rutinoside (datiscin). The timber of both genera is soft and perishable, but in spite of this, that of *Octomeles*, traded as "binuang", is used for cabinet making, furniture and in general constructions, whereas that of *Tetrameles*, "thitpok", is resistant to attack by marine boring organisms and therefore used in naval constructions.

Key to the Genera

1. Robust herbs; leaves imparipinnate or pinnatifid **3. *Datisca***
- Large trees; leaves simple, ± heart-shaped to slightly palmately lobed 2
2. Evergreen; leaves with pocket domatia; inflorescences solitary, axillary; male flowers 6–8-merous, with petals **1. Octomeles**
- Deciduous; domatia 0; inflorescences clustered at the apex of defoliate twigs; male flowers apetalous, 4–5-merous **2. *Tetrameles***

Genera of Datiscaceae

1. ***Octomeles*** Miq.

Octomeles Miq., Fl. Ned. Ind., Erste bijv.: 336 (1861); Steenis, Fl. Males. I, 4: 382–384 (1953).

Evergreen fast-growing, buttressed trees. Leaves 5–9-veined, beneath with many pocket domatia in the axils of the veins and on the veins. Inflorescences axillary pendant spikes. Flowers subsessile; males 6–8-merous, with campanulate calyx tube; calyx lobes short, erect; petals greenish, narrowly triangular, exceeding the sepals; stamens antesepalous; anthers C-shaped, basifixed, introrse; female flowers: (5)6–8-merous; ovary cylindric to barrel-shaped, well demarcated from campanulate calyx tube; petals 0; stylodia short and broad, lacking a ventral slit; stigmas capitate. Fruit throwing off the irregularly torn exocarp and calyx; endocarp splitting from apex downwards, persistent. Seeds spindle-shaped, very numerous.

A single sp., *O. sumatrana* Miq., from Sumatra through all Malesia (Java and the Lesser Sunda Islands excepted) to New Guinea and Solomon Islands.

2. ***Tetrameles*** R.Br. Fig. 32A–D

Tetrameles R.Br. in D. Denham & H. Clapperton, Narr. Trav. Afr., Bot. Append.: 230 (1826); Steenis, Fl. Males. I, 4: 385–387 (1953).

Tall deciduous trees. Leaves 3–7-veined. Inflorescences pendant, fascicled at defoliate branch tips, males mostly elongate-thyrsoid, females spicate or few-branched. Flowers 4(5)-merous, apetalous, crowded or slightly spaced, subsessile; male flowers: calyx tube short, the lobes oblong to triangular; stamens antesepalous; anthers medifixed; female flowers: ovary globose; calyx tube slightly demarcated from ovary; calyx lobes triangular; stylodia elongate, with obliquely clavate stigma. Fruit dehiscing from distal end, the valves curving inwards to leave an apical pore; seeds with loose wing-like testa. $n = 23$.

A single sp., *T. nudiflora* R.Br., from W India and Sri Lanka through Nepal and Yunnan to SE Asia, Malesia and Queensland.

Datisca L. Fig. 32E–H

Datisca L., Sp. pl.: 1037 (1753).

Dioecious or androdioecious, perennial, glabrous, actinorhizal (*Frankia* symbiosis) herbs. Leaves serrate, imparipinnate to deeply pinnatifid, the upper ones successively reduced to ternate or deeply incised to unlobed simple leaves or bracts. Inflorescence a thyrsoid with dense axillary fascicles, or uppermost flowers solitary. Flowers apetalous; male flowers: floral base flat, thin; calyx tube very short, with 3–4(5) small lobes; stamens 6–15(25), the anthers subsessile, oblong; female and hermaphrodite flowers: calyx lobes 3–8, small; stamens (in bisexual fl.) 2–4; gynoecium 3–4(5)-carpellary; ovary oblong to cylindric, with 6 longitudinal ridges; stylodia 3–4(5), deeply bifid, flexuous, the branches papillate-stigmatic throughout. Fruit capsular, dehiscing distally with the valves curving inwards to leave an apical pore. Seeds numerous, with loose testa. $n = 11$.

Two spp., one from the eastern Mediterranean to Central Asia, and the other in California and northern Baja California.

Selected Bibliography

Airy Shaw, H.K. 1965. Diagnoses of new families, new names, etc., for the seventh edition of Willis's "Dictionary". Kew Bull. 18: 249–273.

Axelrod, D.I. 1975. Evolution and biogeography of Madrean-Tethyan sclerophyll vegetation. Ann. Missouri Bot. Gard. 62: 280–334.

Boesewinkel, F.D. 1984. Ovule and seed structure in Datiscaceae. Acta Bot. Neerl. 32: 417–425.
Bohm, B.A. 1988. Flavonoid systematics of the Datiscaceae. Biochem. Syst. Ecol. 16: 151–155.
Brummitt, R. 2007. Datiscaceae. In: Heywood, V.H., Brummitt, R.K., Culham, A., Seberg, O., Flowering plant families of the world. Buffalo, NY: Firefly.
Chase, M.W., Soltis, D.E., Olmstead, R.G., Morgan, D., Les, D.H. and 37 further authors. 1993. Phylogenetics of seed plants: an analysis of nucleotide sequences from the plastid gene *rbc*L. Ann. Missouri Bot. Gard. 80: 528–580.
Davidson, C. 1973. An anatomical and morphological study of Datiscaceae. Aliso 8: 49–110.
Davidson, C. 1976. Anatomy of xylem and phloem of Datiscaceae. Nat. Hist. Mus. Los Angeles County, Contr. Sci. 280: 1–28.
Fritsch, P., Rieseberg, L.H. 1992. High outcrossing rates maintain male and hermaphrodite individuals in populations of the flowering plant *Datisca glomerata*. Nature 359: 633–636.
Gilg, E. 1925. Datiscaceae. In: Engler, A., Prantl., K., Die natürl. Pflanzenfam, 2nd edn, vol. 21. Leipzig: Engelmann, pp. 543–547.
Goodall-Copestake, W.P., Harris, D.J., Hollingsworth, P.M. 2009. The origin of a mega-diverse genus: dating *Begonia* (Begoniaceae) using alternative datasets, calibrations, and reloxed clock methods. Bot. J. Linn. Soc. 159: 363–380.
Hegnauer, R. 1989. Chemotaxonomie der Pflanzen. Vol. 8. Basel: Birkhaeuser.
Himmelbaur, W. 1909. Eine blütenmorphologische und embryologische Studie über *Datisca cannabina* L. Sitzungsber. Kais. Akad. Wiss., Math.-Nat. Cl. II, Abt. 1, 118: 91–113.
Lakhanpal, R.N., Verma, J.K. 1965. Fossil wood of *Tetrameles* from the Deccan Intertrappean beds of Mohgaonkalan, Madhya Pradesh. Paleobotanist 14: 209–213.
Lindley, J. 1846. The vegetable kingdom, vol. 1. London: Bradbury & Evans.
Liston, A., Rieseberg, L.H., Elias, T.S. 1989. Morphological stasis and molecular divergence in the intercontinental disjunct genus *Datisca* (Datiscaeae). Aliso 12: 525–542.
Matthews, M.L., Endress, P.K. 2004. Comparative floral structure and systematics in Cucurbitales (Corynocarpaceae, Coriariaceae, Tetramelaceae, Datiscaceae, Begoniaceae, Cucurbitaceae, Anisophylleaceae). Bot. J. Linn. Soc. 145: 129–185.
Rieseberg, L.H., Philbrick, C.T., Pack, P.E., Hanson, M.A., Fritsch, P. 1993. Inbreeding depression in androdioecious populations of *Datisca glomerata* (Datiscaceae). Am. J. Bot. 80: 757–762.
Steenis, C.G.G.J. van 1953. Datiscaceae. In: Flora Malesiana I, 4: 382–387.
Swensen, S.M. 1996. The evolution of actinorhizal symbioses: evidence for multiple origins of the symbiotic association. Am. J. Bot. 83: 1503–1512.
Swensen, S.M., Mullin, B.C., Chase, M.W. 1994. Phylogenetic affinities of Datiscaceae based on an analysis of nucleotide sequences from the plastid *rbc*L gene. Syst. Bot. 19: 157–168.
Swensen, S.M., Luthi, J.N., Rieseberg, L.H. 1998. Datiscaceae revisited: monophyly and the sequence of breeding system evolution. Syst. Bot. 23: 157–169.
Takhtajan, A.L. (ed.) 1981. Flowering plants, vol. 5 (2). Moscow: Proswjeschtschenie.
Wagstaff, S.J., Dawson, M.I. 2000. Classification, origin, and patterns of diversification of *Corynocarpus* (Corynocarpaceae) inferred from DNA sequences. Syst. Bot. 25: 134–149.
Warburg, O. 1895. Datiscaceae. In: Engler, A., Prantl, K., Die natürl. Pflanzenfam. III, 6a: 150–155. Leipzig: W. Engelmann.
Wydler, H. 1878. Zur Morphologie, hauptsächlich der dichotomen Blüthenstände. Jahrb. wiss. Bot. 11: 313–379.
Zhang, Li-Bing, Simmons, M.P., Kocyan, A., Renner, S.S. 2006. Phylogeny of the Cucurbitales based on DNA sequences of nine loci from three genomes: implications for morphological and sexual systems evolution. Mol. Phylogen. Evol. 39: 305–322.

Kirkiaceae

Kirkiaceae Takht. (1967).

A.N. Muellner

Trees or shrubs. Leaves alternate; imparipinnate, estipulate. Inflorescences axillary or ramiflorous, compound thyrsoids with the cymes dichasial and in higher-branching orders monochasial. Flowers mostly tetramerous, haplostemonous, functionally unisexual, but morphologically bisexual; plants monoecious, polygamomonoecious or dioecious; sepals distinct or very shortly connate at base, triangular, erect; petals distinct, linear and acute; stamens distinct, alternipetalous, inserted outside and beneath disk, in male flowers usually about as long as the petals, in female flowers much reduced; filaments filiform and often broadened slightly towards the base; anthers dorsifixed; disk annular, usually fleshy; gynoecium of 4 or 8 carpels, much reduced in male flowers; ovary superior, syncarpous, synascidiate, narrowly ampulliform, on short gynophore; ovules 1 per loculus; stylodia 4 or 8, slender, proximally contiguous, distally connate; the 4 stigmas capitate, coherent. Fruit a schizocarp, woody and dry, breaking up into 4 or (7)8 single-seeded mericarps, these remaining suspended from the apex of the columella and carrying on their dorsal side the portion of the stylodium to which they were attached. Seeds almost as large as the mericarps, slender, broader and rounded at one end and pointed at the other end, slightly curved, triangular in cross section; endosperm 0; embryo curved; cotyledons fleshy and even.

A unigeneric family with six species in eastern tropical Africa, South Africa and Madagascar.

Vegetative Morphology. The *Kirkia* species are small to medium-sized deciduous trees, up to 20 m high, and shrubs. The bark is silvery grey, grey, with or without black spots, or yellowish with small black or grey scattered spots, fissured or smooth with salmon-pink lenticels (becoming fissured with age; Stannard 1981; Immelman 1984). The branchlets are almost at right-angles to the main axis, or form an acute angle with the main axis.

The leaves are deciduous, up to 40 cm long, mostly crowded at the ends of the shoots, and have 2–30 or more pairs of opposite to subopposite or alternate leaflets (Stannard 1981; Schatz 2001). The rhachis is sometimes winged, and the leaflets are sessile or very shortly petiolulate.

Vegetative Anatomy. In some species, the lower epidermis of the leaves is papillose. *Kirkia wilmsii* has a papillose epidermis, *K. acuminata* not (Jadin 1901). The stomata are paracytic (?). The heart wood is pale brown or green-brown, hard, with an odour reminiscent of honey (the latter in *Kirkia leandrii*; Schatz 2001; Brink 2007). In *Kirkia acuminata*, the heart wood has a dark brown veining; the sapwood is yellow-white or pale grey and up to 7.5 cm wide. The grain is usually straight, locally interlocked, the texture is fine (Brink 2007). The vessel elements have simple perforations; the lateral pitting is alternate. Fibres have simple pores and are septate. The rays are heterogeneous; the axial parenchyma is vasicentric.

Inflorescences. Inflorescences are axillary or ramiflorous (the latter only in *Kirkia leandrii*), borne below the subterminal cluster of expanding leaves. Inflorescences are compound thyrsoids with the cymes dichasial and in higher-branching orders monochasial (Fig. 39A). Inflorescences are shorter, as long as or longer than the subtending leaves, 4–26 cm long. They have leaf-like, lanceolate caducous bracts 1.0–6.5 cm long on the peduncle, and minute prophylls on the inflorescence branches (Stannard 1981).

Flower Structure. Flowers in *Kirkia* are functionally unisexual. In general, the male flowers have large anthers, long filaments, but the gynoecium is reduced to no more than a small pyramid of tissue in the centre of a well-developed, nectiferous disk. In female flowers the anthers are reduced and sterile with short filaments (Fig. 33B, C). Plants can change from male to female overnight (Immelman 1984). The plants are monoecious or polygamomonoecious; *Kirkia leandrii* is dioecious (Heywood et al. 2007).

Most of the following information on flower structure is derived from Jadin (1901), Stannard (1981) and Bachelier and Endress (2008). The flowers are mostly tetramerous, but pentamerous or hexamerous flowers are found on lower-order axes of the inflorescence, and trimerous flowers on higher-order axes. The flowers are relatively small, i.e. less than 1 cm in diameter. They have long, jointed pedicels and a broad floral base. The stamens are alternipetalous and the carpels are antepetalous. A short floral cup is formed by congenitally connate petal and stamen bases. The sepals are distinct or very shortly connate at base. The petals expand between the sepal margins with a dorsal bulge. In contrast to the sepals, their aestivation is basally open but is imbricate further up. The stamen filaments lack a ligulate appendage. The thick and lobed intrastaminal nectary disk expands only late in development.

Fig. 33. Kirkiaceae. *Kirkia acuminata*. **A** Flowering branch. **B** Male flower. **C** Female flower. **D** Gynoecium after dehiscence of the style. **E** Vertical section of one carpel. **F** Fruit. **G** Fruit showing suspension of mericarps from a central carpophore. (Wild and Phipps 1963)

The carpels are closed entirely by postgenital fusion. They are usually unilocular, but *K. wilmsii* has two radially disposed locules, a most unusual condition. The syncarpous ovary has a short stalk (gynophore). Above the ovary the gynoecium is apocarpous. The stylodia are contiguous, form a short conical stylar part and are distally postgenitally connate for half of their length. They form an oblique and flattened four-lobed receptive plate (stigmatic head) in which each lobe corresponds to the tip of a carpel. The stylodia are plicate and have ventral slits extending from the stigma down to the ovary. The united stigmas form an external compitum. The stigmatic surface has unicellular (spherical) and uniseriate multicellular (moniliform) papillae and is covered with secretion.

Embryology. The carpels are uniovulate. The ovule is long and cylindrical. It is crassinucellar, bitegmic, epitropous and slightly campylotropous with only the very base of the nucellus and embryo sac curved. The ovule fills the locule and the micropyle is contiguous with the placenta (Bachelier and Endress 2008).

Although the carpels are uniovulate, at least in *Kirkia wilmsii* they have two axile and almost collateral placentae in the uppermost part of the locule. The second placenta slightly protrudes in such a way that it resembles a second ovule aborting early in development. Behind the second placenta and towards the centre of the gynoecium, there is a small gap. This may correspond to the “inner locule” described in other *Kirkia* species (Capuron 1961; Bachelier and Endress 2008).

Pollen Morphology. The pollen grains are subspheroidal, relatively large (long axis c. 42 μm), tricolporate-syncolpate, crassisexinous, grossly reticulate (curvimurate; muri simplibaculate), with circular ora (Erdtman 1952, with illustration).

Fruit and Seed. The schizocarp beaks into 4 or 7–8 mericarps, which are the dispersal units. They develop from the outward-bulging dorsal region of the carpels and detach from the central part of the synascidiate ovary that develops into the "carpophore" (Engler 1931; Stannard 1981) or "central column" (Capuron 1961). The exocarp is made up of 2–4 layers of thin-walled parenchyma with an outer epidermis covered with thick cuticle. Up to eight layers of larger thin-walled parenchyma compose the mesocarp, with centrally located vascular bundles that have a sclereid bundle external to the phloem. Along the inner boundary with the endocarp is a region, 2–4 cells thick, of lightly thickened cells with very prolific crystals. The endocarp is composed of a broad zone of dense, heavily lignified, elongated sclereids, with some tangentially oriented. The inner epidermis is strongly lignified (Fernando and Quinn 1992). The testa is very thin; endosperm is lacking, and the embryo is slightly curved (Takhtajan 1997).

Phytochemistry. In hydrolysed leaf-extracts of *Kirkia wilmsii*, quercetin, caffeic acid, ellagic acid and a large quantity of gallic acid were found to be present, but no myricetin, cyanidin and delphinidin (Nooteboom 1967). More recently, 3,3'-di-*O*-methylellagic acid 4-*O*-β-D-xylopyranoside was isolated from *Kirkia acuminata* roots (Diakanamwa et al. 1991).

From the stem bark and wood of *Kirkia acuminata* and *K. wilmsii*, Mulholland et al. (2003) isolated lignans, nor-carotinoids, an isocoumarin and a flavonol, but no typical simaroubaceous metabolites such as limonoids, quassinoids, or quinolone alkaloids. These findings further substantiate the recognition of Kirkiaceae as distinct from Simaroubaceae.

Systematics and Affinities. *Kirkia* was described by Oliver (1868a, b), who included it in Simaroubaceae (as Simarubeae). Engler (1896) placed *Kirkia* in Simaroubaceae as well and established the monotypic tribe Kirkieae in Simarouboideae. Later, Engler (1931) added three new species to the genus and raised Kirkieae to subfamilial level. Based on a similar fruit structure but double the number of carpels, Capuron (1961) described a monotypic genus *Pleiokirkia*, endemic to Madagascar and considered to be close to *Kirkia*. The close relationship between *Kirkia* and *Pleiokirkia* was also supported by fruit anatomy (Fernando and Quinn 1992). *Pleiokirkia* was included in *Kirkia* by Stannard (2007).

The affinities of *Kirkia* within Simaroubaceae remained uncertain for a long time. Potential relationships of *Kirkia* with other sapindalean families were never suggested, although Oliver (1868b) mentioned that it could be a Burseracea. Molecular phylogenetic studies showed that Simaroubaceae are an artificial taxon made up partly of components currently placed in Sapindales and Malpighiales (Irvingiaceae), or unplaced in malvids (Picramniaceae; Fernando et al. 1995; Stevens 2001 onwards). Fernando et al. (1995), based on *rbc*L data, suggested a position of *Kirkia* near Simaroubaceae. New studies based on the same DNA region with an extended sapindalean taxon sampling (Muellner et al. 2007) suggest a position of Kirkiaceae close to the Anacardiaceae-Burseraceae clade. Comparative morphological studies on Kirkiaceae (Bachelier and Endress 2008) and Anacardiaceae and Burseraceae (J.B. Bachelier and P.K. Endress, unpubl. data) revealed a combination of features that appears to be unique for these taxa. These include a pronounced convex remnant of the floral apex on top of the syncarpous and entirely synascidiate ovary, and the almost complete absence of a symplicate zone in the gynoecium, as in *Beiselia* (Burseraceae) and Spondioideae-Anacardiaceae and that may be synapomorphic for the three families. A number of other features of Kirkiaceae, such as anthers with pseudopit, campylotropous ovules, antitropous curvature of ovules, inner integument thicker than outer (Endress and Stumpf 1991; Endress and Matthews 2006), and the tendency to form gynoecia with an increased number of carpels, may be synapomorphic at the level of Sapindales or even malvids (Endress and Matthews 2006).

Distribution and Habitats. *Kirkia* occurs from Ethiopia and Somalia south to the Transvaal in South Africa, and in Namibia, and a single species (*K. leandrii*) in wet-central Madagascar

(Bemaraha). All species are usually in dry habitats, often on limestone hills or open woodland (Heywood et al. 2007).

Economic Importance and Uses. *Kirkia acuminata* is planted as a live fence in southern tropical Africa, and provides poles, planks, household utensils (bowls, spoons), carts, musical instruments, tourist items, veneer and plywood (Brink 2007; Heywood et al. 2007). In South Africa, the wood is made into furniture and floor blocks. The wood is also considered suitable for light construction, flooring, vehicle bodies, cabinet work, interior trim, agricultural implements, boxes and crates, core stock, matches, toys and novelties, turnery, hardboard and particle board, and as pulpwood. In Malawi, the wood is made into charcoal (Brink 2007). The bark fibres of *Kirkia acuminata* are woven into cloth (Brink 2007); the bark of *K. wilmsii* is used to make cord (Heywood et al. 2007). The seeds and leaves of *Kirkia acuminata* are browsed by livestock. The swollen roots of *Kirkia acuminata*, and the roots of *K. tenuifolia* in East Africa and of *K. wilmsii* in the Transvaal (tubers up to 30 cm across) are chewed to quench thirst in times of drought (Brink 2007; Heywood et al. 2007). In Zimbabwe, an infusion of the bark of *Kirkia acuminata* is taken against vomiting and abdominal pain. An infusion of the root is taken to treat cough. The fruit sap is applied on wounds and as an antidote on snake bites (Brink 2007). Pulverized roots are a remedy for toothache (Brink 2007; Heywood et al. 2007).

Only one genus:

Kirkia **Oliv.** Fig. 33

Kirkia Oliv., Fl. trop. Afr. 1: 310 (1868); Stannard, Kew Bull. 35: 829–839 (1989), and ibid 62: 151–152 (2007), rev. *Pleiokirkia* Capuron (1961).

Description as for the family.

Selected Bibliography

Bachelier, J.B., Endress, P.K. 2008. Floral structure of *Kirkia* (Kirkiaceae) and its position in Sapindales. Ann. Bot. 102: 539–550.

Bakker, F.R., Vassiliades, D.D., Morton, C., Savolainen, V. 1998. Phylogenetic relationships of *Biebersteinia* Stephan (Geraniaceae) inferred from *rbc*L and *atp*B sequence comparisons. Bot. J. Linn. Soc. 127: 149–158.

Brink, M. 2007. *Kirkia acuminata* Oliv. [Internet] Record from Protabase. Louppe, D., Oteng-Amoako, A.A., Brink, M. (eds.) PROTA (Plant Resources of Tropical Africa/Ressources végétales de l'Afrique tropicale), Wageningen, The Netherlands. http://database.prota.org/search.htm

Capuron, R. 1961. Contributions à l'étude de la flore forestière de Madagascar. III. Sur quelques plantes ayant contribué au peuplement de Madagascar. Adansonia II, 1: 65–92.

Diakanamwa, C., Diallo, B., Vanhaelen-Fastre, M. 1991. 3,3'-Di-O-methylellagic acid 4-O-â-D-xylopyranoside from *Kirkia acuminata* roots. Fitoterapia 62: 87–88.

Endress, P.K., Matthews, M.L. 2006. First steps towards a floral structural characterization of the major rosid subclades. Pl. Syst. Evol. 260: 223–251.

Endress, P.K., Stumpf, S. 1991. The diversity of stamen structures in "lower" Rosidae. Bot. J. Linn. Soc. 107: 217–293.

Engler, A. 1896. Simarubaceae. In: Die natürl. Pflanzenfam. III, 4: 202–230. Leipzig: W. Engelmann.

Engler, A. 1897. V – Neue Arten aus Transvaal. Notizbl. Königl. Bot. Gart. Mus. Berlin 2: 25–26.

Engler, A. 1931. Simaroubaceae. In: Die natürl. Pflanzenfam., ed. 2, 19a: 359–405. Leipzig: W. Engelmann.

Erdtman, G. 1952. Pollen morphology and plant taxonomy. Stockholm: Almqvist & Wiksell.

Fernando, E.S., Quinn, C.J. 1992. Pericarp anatomy and systematics of the Simaroubaceae sensu lato. Austral. J. Bot. 40: 263–289.

Fernando, E.S., Gadek, P.A., Quinn, C.J. 1995. Simaroubaceae, an artificial construct: evidence from *rbc*L sequence variation. Am. J. Bot. 82: 92–103.

Heywood, V.H., Brummitt, R.K., Culham, A., Seberg, O. 2007. Flowering plant families of the World. Richmond Hill, Ontario, Canada: Firefly Books.

Immelman, K.L. 1984. Flowering in *Kirkia wilmsii* Engl. Bothalia 15: 151–152.

Jadin, F. 1901. Contribution à l'étude des Simaroubacées. Ann. Sci. Nat., Bot. Sér. 8: 201–304.

Muellner, A.N.,Vassiliades, D.D., Renner, S.S. 2007. Placing Biebersteiniaceae, a herbaceous clade of Sapindales, in a temporal and geographic context. Pl. Syst. Evol. 266: 233–252.

Mulholland, D.A., Cheplogoi, P., Crouch, N.R. 2003. Secondary metabolites from *Kirkia acuminata* and *Kirkia wilmsii* (Kirkiaceae). Biochem. Syst. Ecol. 31: 793–797.

Nooteboom, H.P. 1967. The taxonomic position of Irvingioideae, *Allantospermum* Forman and *Cyrillopsis* Kuhlm. Adansonia II, 7: 161–168.

Oliver, D. 1868a. Flora of tropical Africa. London: L. Reeve.

Oliver, D. 1868b. *Kirkia acuminata*. Hooker's Icones Plantarum, III, 11: 26–27, plate 1066.

Schatz, G.E. 2001. Generic Flora of the Trees of Madagascar. Royal Botanic Gardens, Kew & Missouri Botanical Garden, St Louis.

Stannard, B.L. 1981. A revision of *Kirkia* (Simaroubaceae). Kew Bull. 35: 829–839.

Stannard, B.L. 2007. The inclusion of *Pleiokirkia* in *Kirkia* (Kirkiaceae), and corresponding combination. Kew Bull. 62: 151–152.

Stevens, P.F. 2001 onwards. Angiosperm phylogeny website, Version 8, June 2007 (and more or less continuously updated since). http://www.mobot.org/MOBOT/research/APweb/

Takhtajan, A. 1967. Systema et Phylogenia Magnoliophytorum (in Russian). Moscow: Nauka.

Takhtajan A. 1997. Diversity and classification of flowering plants. New York, NY: Columbia University Press.

Wild, H., Phipps, J.B. 1963. Simaroubaceae. In: Flora Zambesiaca 2(1), pp. 210–220.London: Crown Agents.

Meliaceae

Meliaceae Juss. (1789, 'Melieae'), nom. cons.

D.J. Mabberley

Trees, treelets, often pachycaul or, more rarely, shrubs or suckering shrublets, monopodial or sympodial, rarely with *Terminalia* branching (*Vavaea*), dioecious (though sometimes, at least, 'male' trees occasionally producing bisexual flowers), polygamous, monoecious or with all flowers bisexual; indumentum of simple, bifid or stellate hairs or stellate or peltate scales or sometimes mixtures of these, sometimes with small glands; buds naked or with scale-leaves. Leaves exstipulate (occasionally pseudostipules present), in spirals, rarely decussate, pinnate, sometimes with a terminal 'bud', i.e. pseudogemmula, trifoliolate, with a single blade (simple or unifoliolate) or rarely bipinnate (*Melia*); rachis very rarely winged; leaflets usually entire, rarely lobed or serrate (or spinous), sometimes with minute black glandular dots. Inflorescences axillary, supra-axillary, ramiflorous, cauliflorous to ground level or rarely epiphyllous (*Chisocheton*), thyrsoid, racemose or spicate, sometimes reduced to fascicles or solitary flowers. Flowers bisexual and/or more usually, unisexual, with well-developed rudiments of opposite sex; calyx usually ± lobed, sometimes with distinct sepals, these occasionally in spirals and transitional to bracts (*Dysoxylum*), sometimes truncate or closed in bud and circumscissile at base at anthesis; petals 3–7(–14) in 1 (rarely in a spiral (*Chisocheton*) to give up to 2 apparent) whorls, green, white, cream, pink to claret and violet or yellow (*Aglaia*); stamens usually partially or completely united by a tube with or without lobes; anthers 3–10(–30) in 1 or, rarely, 2 or more whorls, sometimes locellate, at tips of filaments or at the margin of the tube or within its throat; nectary disk around ovary, cushion-like, tubular or 0; ovary (1)2–6(–20)-locular, each locule with 1–many ovules; ovules pendulous, epitropous, anatropous and bitegmic, less often hemianatropous, campylotropous, or orthotropous, and unitegmic; stylehead discoid to capitate. Fruit a capsule, berry or drupe. Seed with fleshy aril or sarcotesta or a combination of these or winged and these attached to a woody columella, or with corky outer layers, or very rarely without any of these, endosperm usually 0; cotyledons collateral, superposed or, rarely, oblique, emergent or not at germination, when scale-leaves are sometimes produced before first foliage leaves, which can be opposite or in spirals, simple or pinnate with later ones simple to bipinnate. $2n=16$–c. 360.

A family of 50 genera and c. 575 spp., throughout the tropics and subtropics, with weak representation in temperate zones.

Vegetative Morphology[1]. In terms of 'architecture', most Meliaceae correspond to the models of Corner (unbranched, inflorescences lateral) and related structures with this pattern variously reiterated (model of Champagnat) in the scheme of Hallé and Oldeman (1970), from pachycaul to leptocaul, while *Vavaea* has *Terminalia* branching (model of Aubréville) and *Xylocarpus* architecture corresponds to the model of Rauh.

In subfam. Cedreloideae, the buds are surrounded by small subulate scales, which are only sporadically found in subfam. Melioideae. The form of the young emergent leaves is a useful character in the recognition of the sections of *Dysoxylum* (Mabberley in Mabberley et al. 1995: 63). The mature leaves are usually in spirals, though apparently always decussate in *Capuronianthus* and some species of *Turraea* (Mascarenes) and *Dysoxylum*.

[1]General chapters based on Pennington and Styles (1975, 1981) and Mabberley et al. (1995), unless otherwise stated.

Most species have pinnate leaves, those in *Melia* always bipinnate, a feature occurring in juveniles (and very occasionally adults, as in some *Chukraria tabularis*) in other genera. Sometimes the leaflets of pinnate leaves are deeply lobed in mature foliage as well as in juveniles; this is characteristic of some *Dysoxylum* species brought into cultivation because of their 'ferny' foliage, but also in some Mascarene *Turraea*. In most species of *Turraea* besides all of *Calodecaryia*, *Humbertioturraea*, *Nymania* and *Vavaea*, the leaves are simple; unifoliolate leaves are found in some species of *Aglaia*, *Malleastrum*, *Reinwardtiodendron*, *Trichilia* and *Walsura*, exclusively paripinnate ones in *Anthocarapa*, *Khaya*, *Neobeguea*, *Owenia* and *Xylocarpus*. The leaves of most species of *Chisocheton* and *Guarea* have apical 'pseudogemmulae', which are crozier-like buds of undeveloped leaflets, from which leaflets unfold at intervals, such that the leaves develop over several seasons and have annual rings in the ageing petioles. In some species in both of these genera, however, the pseudogemmula is effectively inactive and may fall off before any second 'flush', the leaf appearing paripinnate, whereas some species in these genera have long imparipinnate leaves. In species of *Cabralea*, *Dysoxylum* and *Ruagea*, the development of the apical leaflets may be delayed, but there is a terminal leaflet and no resting meristem in the form of a pseudogemmula.

'Pseudostipules' occur sporadically in the family, notably in South American species of *Trichilia*, but also some species of *Chisocheton*.

Vegetative Anatomy. The wood anatomy of the family is well documented because of the economic significance of the family (see Baas in Mabberley et al. 1995: 5 for a review). The limits of the family, particularly the distinction from Rutaceae in referring *Chloroxylon*, *Flindersia* and *Ptaeroxylon* to the latter, were first realized through a study of wood anatomy (Kribs 1930).

The association of non-septate fibres with terminal parenchyma bands helps to characterize the tribes 'Turraeeae', Melieae, 'Trichilieae' (except *Cipadessa*) and Sandoriceae, while sepate fibres in combination with the absence or only sporadic occurrence of marginal parenchyma characterize the other tribes recognized here. Many individual genera and species overlap in their wood-anatomical range, however, so that only a few genera can be recognized by their wood structure alone: *Aphanamixis*, *Astrotrichilia*, *Azadirachta*, *Cipadessa*, *Ekebergia*, *Melia*, *Nymania*, *Owenia*, *Sandoricum*, *Turraeanthus* and *Vavaea*. It is of interest that several of them have the habitats the most different from the bulk of the family (see below).

An indumentum of stellate hairs or scales is the only type found in *Aglaia*, *Astrotrichilia*, *Lepidotrichilia*, *Melia* and *Pterorhachis* (but occurs sporadically in other genera too).

Inflorescences. Although basically thyrsoid, the inflorescences are sometimes reduced to spikes or solitary flowers. Sometimes, as in some South American *Guarea* species, inflorescences show indeterminate growth (see also under *Melia* below), like the leaves. In *Chisocheton*, inflorescences are borne in axillary, supra-axillary or ramiflorous positions (Fisher and Rutishauser 1990). In each of these places and also on the bole, the inflorescences may be borne on dwarf shoots with reduced leaves. Such shoots are also known in *Melia* (q.v.), where they continue growth after fruit set. There are also species with epiphyllous inflorescences borne on the adaxial surface of the leaf rachis, their vascular supplies being connected directly with that of the adjacent rachis, there being no adnation or other 'fusion'. In *C. tenuis* there are also vegetative buds, their origin like that of the epiphyllous inflorescences in this species, and *C. pohlianus* being considered heterotopic or, indeed, homoeotic.

Floral Morphology. Flowers vary in size from the long tubular ones to 10 cm or more in length in some *Turraea*, to those 0.05 mm across in some species of *Aglaia*. Male flowers are in general smaller than female ones. Flowers are globular in, e.g. *Aphanamixis* and *Aglaia*, tubular in many other genera. Usually the corolla is white or pinkish and often very highly scented. Distinct sepals are found in, e.g. *Ruagea* and *Cabralea*, but also in some *Dysoxylum* in which they grade into bracts. The corolla rarely has more than one whorl (some *Chisocheton*) and sometimes the petals are united by a basal tube, which sometimes also bears the staminal tube.

Generally the stamens are united by a basal tube (e.g. Figs. 34B, 35D), though filaments are distinct in *Cedrela* and *Toona*, while some species of *Walsura* and *Trichilia* have flattened filaments. There is a gradation from those with a basal tube and otherwise distinct filaments to those with tubes where the anthers are borne along the rim or inserted within it: the genus *Trichilia* shows the complete range. The position of the anthers is often a character of generic value, being inserted at the end of the filaments or, as in *Dysoxylum*, on the margin of the tube with the anthers completely included or partially exserted. However, the positioning does vary within some genera such as *Aglaia* and *Owenia*. The apex of the filaments or margin of the tube often bear appendages, simple or variously lobed. In *Sphaerosacme* and *Reinwardtiodendron*, the anthers are in two distinct whorls.

Nectary disks are common between the ovary base and the androecium, being long cylinders enveloping the ovaries in, for example, *Dysoxylum* but, in some genera like *Aglaia*, *Aphanamixis* and *Lansium*, they are absent. In *Guarea* the ovary is borne on an expanded stipitate disk and, in *Toona* and *Cedrela*, such a structure supports both the gynoecium and androecium. In nearly all Melioideae, each ovary locule has one or two collateral or superposed ovules, in most Cedreloideae they have 3 to many biseriate ones. The styleheads vary from the pinhead-shaped ones in many species of *Aglaia* to the flattened discoid ones of most Cedreloideae and those with large stigmatic lobes in *Azadirachta* and *Melia*; in *Turraea*, the stylehead forms a receptaculum pollinis topped with an apical stigmatic zone, pollen being deposited on the stylehead before the flower opens.

Pollen Morphology. The pollen grains are nearly always isopolar, radially symmetrical, suboblate to subprolate monads (van der Ham in Mabberley et al. 1995: 8). Rarely are they oblate or prolate and only in some *Dysoxylum* are they shed in rhomboidal tetrads. They are tri-, tetra- or penta-colporate, the second being the most frequent condition. In 'Turraeeae' (though not *Munronia*) and in Vavaeeae, they are usually tri-colporate; pollen in most genera has distinctly thickened ecto- and endo-aperture margins. Little is recorded of exine structure. There are no significant pollen differences between the subfamilies.

Karyology. There is a wide range of somatic chromosome number, from $2n = 16$ (*Sandoricum koetjape*) to c. 360 in some African *Trichilia*, with polyploid series in some genera, e.g. *Aphanamixis* and *Aglaia*, and also within species, e.g. *Chisocheton cumingianus*. The base number is probably $x = 6, 7$ (see Khosla and Styles 1975, and Mabberley et al. 1995: 8).

Pollination. Most species appear to be insect-pollinated, the agents possibly being bees, sting-less sweatbees or syrphids in those cases examined (*Aglaia* spp., *Xylocarpus* spp.), while some species are strongly scented particularly in the evening, which, with their white flowers, suggests moth pollination, e.g. *Chisocheton* and *Dysoxylum* spp. of Asia, as recorded in neotropical *Cedrela* and *Guarea* spp. Some species of these two genera have long flagelliform inflorescences suggesting bat pollination, though the Bornean species are known to be visited by spiderhunters; whether these birds are efficacious pollinators is not known.

Reproductive Systems. As it seems that male flowers may fall before females or bisexuals, it is often difficult in the absence of field studies to ascertain the true sexual arrangements of any particular species. Even dioecious species occasionally produce bisexual flowers, as was shown for *Dysoxylum spectabile* in New Zealand (Braggins et al. 1999). Apomixis is recorded in cultivated clones of *Lansium domesticum*.

Fruits and Seeds. There is a wide range of fleshy fruits, from the brightly coloured capsules of *Aphanamixis* and *Dysoxylum* to the dull berries of many *Aglaia* and *Lansium* species. The fruit is most commonly a capsule, loculicidal in Melioideae, septifragal in Cedreloideae (and one species of *Walsura*). Cedreloid capsules are usually woody with a ridged or angled columella (e.g. Fig. 34C), rudimentary in *Schmardaea*, *Carapa* and *Xylocarpus*, and bearing the seeds: dehiscence is basal, apical or both. Capsules of Melioideae have no columella and are fleshy, membranous, leathery or even woody. Indehiscent fruits include berries and drupes, the endocarp of those in *Astrotrichilia* and *Owenia* being very thick and woody.

By contrast with the relative uniformity of the flowers, the seeds of Meliaceae are some of the most diverse and intricate in structure so far investigated (Cheek 1989). The seeds are usually pendulous and epitropous (but apotropous in the Australian *Synoum*) in relation to the placenta. They are usually anatropous (but hemi-anatropous in most 'Turraeeae' and *Cipadessa*), occasionally orthotropous as in all *Chisocheton* and some *Guarea* spp., campylotropous in, e.g. *Nymania*, usually with a distinct funicle and raphe. Integument vascularization occurs in a few bitegmic genera, e.g. *Chisocheton* and *Heynea* (vascularized tegmen), *Dysoxylum* and *Trichilia* (vascularized testa) and is usual in pachychalazal seed-coats. In the 'Guareeae', arillate seeds characterize genera with bitegmic seeds, sarcotestal ones the unitegmic genera, while both conditions are found in *Chisocheton* and *Dysoxylum*, the genus with the greatest variation in seed anatomy investigated by Cheek.

The single most taxonomically useful layer of the seed-coat in the majority of genera is the exotegmen: dimensions of the fibres and their wall characters but also the number of cell layers. However, in *Dysoxylum*, for example, the type of seed appendage and its organization are more helpful in recognizing species groupings because the exotegmen is reduced and poorly developed there. From his survey of the family, Cheek concludes that the ancestral Meliaceous seed was comparatively large and bitegmic, with a small chalaza and a funicular-raphe-aril. From such can be derived the unitegmic sarcotestal state and then unitegmic 'pseudoarillate' seen in many species.

Dispersal. The fruits of species of Cedreloideae are dry dehiscent capsules, the winged seeds of the tall emergent and canopy trees being wind-dispersed, the irregular corky-coated ones of *Xylocarpus* being dispersed by saltwater in which they float just below the surface. The indehiscent fleshy fruits of *Azadirachta indica* are known to be dispersed by bats and baboons in Africa, where it is introduced, those of *Melia azedarach* by birds. Those species of *Aglaia*, *Chisocheton*, *Dysoxylum* and *Aphanamixis* with dehiscent capsules and arillate or otherwise fleshy seeds are also known to be taken by birds, apparently attracted by the contrasting colours of seeds and/or pericarp, while those species with indehiscent fruits, at least in *Aglaia* and *Reinwardtiodendron*, are eaten by primates which pass the seed divested of its fleshy layers. The bird and primate 'syndromes' are associated in a sample of *Aglaia* spp. at least, with characteristic chemistry as might be expected, those taken by birds being high in lipid, those by primates higher in free carbohydrates, but it is clear that few if any tree species are restricted to any one species, or indeed group, of vertebrate dispersers. Moreover, as there are no native primates in some parts of the ranges of certain species—*Aglaia elaeagnoidea* in New Caledonia, for example—it must be concluded that bats or birds are active dispersal agents even for those species which are thought to be dispersed by primates or other terrestrial mammals elsewhere.

In New Guinea, birds of paradise are the only known dispersal agents for *Chisocheton lasiocarpus* seeds. Those of the Malesian *Dysoxylum angustifolium* and possibly *Sandoricum borneense* are believed to be dispersed by fish, like those of the neotropical *Guarea guidonia* (Pennington and Styles 1981: 11). This 'syndrome' has yet to be analyzed chemically.

The inflated fruits of *Nymania capensis* serve as dispersal units in southern Africa.

Phytochemistry. This is reviewed by Hegnauer in Mabberley et al. (1995: 8). Characteristic of the order Sapindales is the synthesis of nortriterpenoids derived from tetracyclic triterpenes known as protolimonoids. Those of Meliaceae are unique to the family and are of great interest because of their insecticidal qualities.

Meliaceae also differ from the closely allied Rutaceae in both not accumulating large amounts of steam-volatile phenylpropanoids in their essential oils and in the absence of family-specific alkaloids. In Meliaceae true saponins are rare, but saponin-like glycosides are common. Some species, notably of *Azadirachta* and *Dysoxylum*, have characteristic onion- or garlic-like smells due to the presence of sulphur-containing volatiles. More widespread are mono- and sesqui-terpenoids in idioblasts, some of which are insecticidal.

Subdivision and Relationships Within the Family. The family is divisible into two subfamilies (Muellner et al. 2003, 2006, 2008a)—Melioideae and Cedreloideae (monophyletic sister groups), both pantropical, with the former much larger in terms of numbers of genera and species. Subfam. Melioideae has been divided into eight tribes (Pennington in Pennington and Styles 1975; Leroy 1976) with 36 genera. All those recognized tribes are represented in tropical Asia, but only two ('Guareeae' and 'Trichilieae') are pantropical, while another two are restricted to the Old World ('Turraeeae' and Melieae), the remaining three being restricted to Indomalesia and the western Pacific (Vavaeeae, Aglaieae and Sandoriceae). Trichilieae, Guareeae and Aglaieae have been treated as one tribe by some workers, and molecular work (Muellner and Mabberley 2008; Muellner et al. 2008a) shows that Turraeeae are best combined with Trichilieae, though several genera formerly assigned to the latter are excluded from this pantropical grouping. This modified Trichilieae grouping is sister to one (pantropical) comprising Aglaieae (monophyletic), Guareeae, Vavaeeae (unigeneric) and *Lepidotrichilia* (formerly in Trichilieae). Sandoriceae (unigeneric) are sister to this combined group + *Walsura* (formerly Trichilieae), with *Quivisianthe* (formerly considered to comprise the unigeneric Quivisianthoideae of Madagascar) sister to that. Melieae (monophyletic) are sister to that grouping + *Astrotrichilia* (formerly Trichilieae).

Affinities. These are fully discussed by Pennington and Styles (1975), who argue for the integrity of the family, which is, nevertheless, closely allied to Rutaceae, some of which (*Cneorum* and *Harrisonia*) have very similar limonoids, the oxidised terpenoids which characterize the family, and provide some of the most significant plant products in terms of potential insecticides. DNA work has confirmed that the family is most closely allied to Rutaceae and Simaroubaceae, which with Sapindaceae form the core group of Sapindales (Muellner et al. 2003).

Distribution and Habitats. Muellner et al. (2006) argue that the family is of West Gondwanan origin. At the generic level, the family demonstrates some remarkable transoceanic affinities: Asiatic and Malesian *Toona* with neotropical *Cedrela*; *Dysoxylum* and *Chisocheton* of Indomalesia are apparently close to *Guarea* of the neotropics and Africa; *Xylocarpus* of East Africa to the Pacific is closely allied to *Carapa* of the Neotropics to West Africa, while the species *Carapa procera* is found on both sides of the Atlantic. *Naregamia* has one species in India, one in Africa. No genus is pantropical and Madagascar is remarkable for having no fewer than seven endemic genera (*Astrotrichilia*, *Calodecaryia*, *Capuronianthus*, *Humbertioturraea*, *Malleastrum*, *Neobeguea* and *Quivisianthe*). *Nymania* is found only in southern Africa, *Schmardaea* in Andean cloud forest, *Sphaerosacme* in the Himalaya, but all these are monospecific.

Meliaceae are very common trees of the canopy and understorey of lowland primary forest, notably in Malesia, making up to 17% of all trees over 10 cm bole diameter in the forests of Sumatra, for example, and being absent from only the driest zones, though in Australia they are found even there, as species of *Owenia*. They are represented by species of *Xylocarpus* on rocky shores and in mangrove swamps. They are poorly represented at higher altitudes, although some *Dysoxylum* and *Toona sinensis* are sometimes conspicuous in lower montane forest in Asia, *Ruagea* spp. in America, and *Schmardaea* restricted to the cloud forest of the Andes. The family is represented in freshwater swamp forest in Borneo by *Sandoricum borneense* and *Chisocheton amabilis* and includes some species restricted to limestone, like *C. ruber* of Sarawak, while *Walsura monophylla* is restricted to ultramafics in the Philippines. Along rivers in west Malesia are a number of rheophytic species of *Aglaia* and *Dysoxylum angustifolium*. A few species are tolerant of more open conditions and will colonize large gaps in forest or are frequently encountered in secondary forest, e. g. *Toona* spp., and *Chukrasia tabularis* which colonizes even bare ground along road cuttings in the Malay Peninsula and is a naturalized weedy tree in parts of tropical Australia.

In Africa, the commercially significant cedreloid Meliaceae are emergent trees, some of the biggest in the continent and, in Java, 58-m-tall specimens of the melioid *Dysoxylum mollissimum* were some of the island's tallest trees, but these have long been removed, like the commercial mahoganies, *Swietenia* spp., of South America, now severely depleted, and the red cedar, *Toona ciliata* in north-eastern Australia.

Silviculture and Pests. The most widely grown for timber are Cedreloideae, the neotropical *Swietenia mahagoni*, *S. macrophylla*, and the Asiatic *Toona ciliata* and *Chukrasia tabularis* as well as Old World Melioideae, *Azadirachta indica* and *Melia azedarach*, all of which have also been grown as shade or avenue trees; less often seen are species of *Entandrophragma* and *Khaya*, Cedreloideae from Africa. The major problem besetting plantation forestry of Meliaceae is the attacks of the moths (Lepidoptera, Pyralidae) of the genus *Hypsipyla*, larvae of which burrow into young plants and seedlings, causing their collapse and death, though Melioideae seem to be rarely attacked—a good argument for their promotion as plantation trees.

The shoot-borer is perhaps one of the most economically important insect pests in tropical forestry. In the Neotropics, the moths appear to be attracted by a chemical, possibly an alkaloid or limonoid, allied to substances which, in species from other parts of the world, are toxic to their local moths. Thus, *Toona ciliata* is resistant to attack in Costa Rica and this resistance may be transferred to susceptible *Cedrela odorata* if a scion of this is grafted on to a stock of *Toona*. *Chukrasia tabularis*, on the other hand, is not resistant in Costa Rica. There have been many attempts at biological control of the moths, for externally applied insecticides have little effect and systemic ones are expensive. Mixed and enrichment planting with non-susceptible species has been shown to reduce damage and there are possible advances to be made in breeding resistance to attack (Newton et al. 1993).

Palaeobotany. Fossils with features which, if they were found in a living plant, would place them in the concept of the modern Meliaceae are known from the Upper Cretaceous, some of the earliest ones being referred to the genus *Guarea* (certainly by the Oligocene), while others are referred to *Cedrela* and the closely allied Indomalesian *Toona* from the Eocene of the northern hemisphere (wood of *Cedreloxylon* allied to both of which being known from the Pliocene of China as well as the Eocene and Miocene of central Europe, and possibly the Eocene of North America; Cheng et al. 2006) and, among others, to the African *Entandrophragma* from the Miocene of Kenya, the tropical Old World *Melia* from the Miocene of Poland and Washington State, USA, and *Trichilia* from the Lower Miocene, Cameroon. The volcanic basalt deposits of the Cretaceous/Tertiary boundary, the Deccan Intertrappean beds of Mandla District, have yielded fossils referred to *Aglaioxylon* and *Heyneoxylon*, allied to *Aglaia* and *Heynea* respectively, while '*Carapa* spp.' (i.e. *Xylocarpus* spp.?) are known from the so-called peat bed near Calcutta, 22°08'N (see Mabberley et al. 1995: 2–3 and Muellner et al. 2006 for summaries of the earlier literature; see also Gregory et al. 2009).

Economic Importance and Conservation. The timbers of certain Meliaceae are some of the most sought after in the world, such that natural stands have been much depleted and serious conservation measures have been proposed for wild mahoganies (*Swietenia* spp.) in tropical America. The original 'mahogany' of the furniture-makers Hepplewhite and Chippendale was *S. mahagoni*, allowing the construction of more graceful and woodworm-proof furniture than could the oak and walnut previously used in Europe: this species has suffered severe genetic erosion and most 'mahogany' (if Meliaceous at all!) seen today is derived from *S. macrophylla* introduced to the Old World, probably from Honduras, in 1876 and described as a new species from material cultivated in India.

The other important timbers are also generally Cedreloideae, notably toon, *Toona ciliata*, from India to Australia ('red cedar'), where most of it has been long cut out, having been the most desirable timber on that continent. Others include the neotropical *Cedrela odorata* and species of the African genera *Entandrophragma* (*sapele, utile*), *Khaya* (African mahogany) and *Lovoa* (Nigerian golden walnut).

The locally important fruit trees of Malesia, *Lansium domesticum* (langsat) and *Sandoricum koetjape* (sentul) exist in a number of forms, wild, cultivated and naturalized, though they are not grown on a commercial plantation scale, those reaching markets being largely those selected from village trees. Seeds of a number of species of *Chisocheton* and *Aphanamixis* yield an oil which has been used as an illuminant.

The bitterness of the barks of Meliaceae has long been known and they have been used in medicine, some being eagerly sought by Europeans in the eighteenth century. *Soymida febrifuga* from India was much prized but barks of other genera have been used, while the whole plant of *Munronia pinnata* is an important item of materia medica in southern Asia. The bark and, indeed, the leaves of *Azadirachta indica*, the neem, are powerful insecticides and this tree has a host of uses including planting in the reclamation of derelict land: it is perhaps one of the most all-round useful trees of Asia. Its young shoots are used as a vegetable (sadao) and sold in markets, as in Australia.

The triterpenoids which are responsible for the insecticidal (and molluscicidal, e.g. *Khaya* spp. bark and seeds) properties have aroused considerable commercial interest and have been examined in a number of genera for their use as biological pesticides. The biological activities of these compounds, including insect antifeedant and growth-regulating properties, medicinal effects in humans and other animals, as well as antifungal, bactericidal and antiviral activity, are reviewed by Champagne et al. (1992).

The genera are arranged according to Muellner et al. (2008a) and Muellner and Mabberley (2008), though there is more work to be done, and the tribes recognized by Pennington and Styles (1975) cannot be satisfactorily remodelled as yet (but see above); thus, only those which are clearly monophyletic are dealt with formally below.

Key to the Genera

1. Loculi 1- or 2-ovulate (rarely with 1 or 2 additional vestigial ovules) 2
– Loculi multiovulate 58
2. Leaves decussate, capsule with partial septifragal dehiscence (Madagascar) **12. *Capuronianthus***
– Leaves almost always in spirals, very rarely decussate (in which case capsule with loculicidal dehiscence) 3
3. Fruit dry, seed winged (Madagascar) **18. *Quivisianthe***
– Fruit fleshy or leathery; seed unwinged 4
4. Leaves bipinnate **6. *Melia***
– Leaves pinnate to simple 5
5. Leaves all simple (or unifoliolate) 6
– Leaves pinnate to trifoliolate 18
6. Hairs simple only 7
– Indumentum of at least some stellate hairs and/or stellate or peltate scales 16
7. Fruit a capsule 8
– Fruit a berry or, at least, indehiscent 12
8. Filaments 0 or less than half length of basal tube 9
– Filaments usually united only at base, at most with a tube shorter than them 11
9. Leaves simple (Old World) **34. *Turraea***
– Leaves unifoliolate, with distinct articulation (New World) 10
10. Anthers inserted within the staminal tube **44. *Guarea***
– Anthers on rim of staminal tube or tips of filaments **28. *Trichilia***
11. Capsule inflated (southern Africa) **32. *Nymania***
– Capsule not inflated (Madagascar) **36. *Calodecaryia***
12. Anthers in 2 whorls of 5 (Indomalesia) **48. *Reinwardtiodendron***
– Anthers in 1 whorl 13
13. Flowers conspicuous; trees with *Terminalia* branching (Malesia, Pacific) **23. *Vavaea***
– Flowers small; small trees or shrubs 14
14. Filaments almost distinct (Philippines, Sri Lanka) **20. *Walsura***
– Filaments very short, atop a tube (Madagascar) 15
15. Petals imbricate **35. *Humbertioturraea***
– Petals valvate **30. *Malleastrum***
16. Hairs mixed simple and stellate, suckering shrublets to 1 m **24. *Munronia***
– Indumentum of stellate hairs and/or stellate or peltate scales; suckers 0 17
17. Nectary disk 0 (Indopacific) **50. *Aglaia***
– Nectary disk present (elsewhere) **28. *Trichilia***
18. Leaves all trifoliolate 19
– Leaves pinnate 22
19. Indumentum of stellate hairs and/or stellate or peltate scales **50. *Aglaia***
– Hairs simple 20
20. Fruit a capsule 21
– Fruit a drupe (Malesia) **19. *Sandoricum***
21. Flowers solitary or paired **33. *Naregamia***
– Flowers in thyrses **28. *Trichilia***
22. Leaves with pseudogemmula (apical bud) 23
– Leaves without pseudogemmula 24
23. Disk stipitate, seeds anatropous (Neotropics) **44. *Guarea***
– Disk 0 to patelliform, seeds orthotropous (Indomalesia) **39. *Chisocheton***
24. Leaves with scales and/or stellate hairs 25
– Leaves with simple hairs 32
25. Shrublets with large white flowers and capsular fruits (Indomalesia) **24. *Munronia***
– Trees or treelets 26
26. Nectary disk present 27
– Disk 0; scales often present 30
27. Fruit a drupe (Madagascar) **17. *Astrotrichilia***
– Fruit a capsule 28
28. Rachis winged (west Africa) **29. *Pterorhachis***
– Rachis unwinged 29
29. Nectary disk tubular (Indomalesia, Pacific) **40. *Dysoxylum***
– Nectary disk a fleshy annulus **28. *Trichilia***
30. Petals valvate (Africa, Madagascar) **22. *Lepidotrichilia***
– Petals imbricate (Indomalesia, Pacific) 31

31. Female inflorescence and infructescence long spicate, scales 0 **46. *Aphanamixis***
– Not this combination of characters **50. *Aglaia***
32. Leaves paripinnate 33
– Leaves imparipinnate 41
33. Stigma with conspicuous lobes; fruit a drupe **15. *Azadirachta***
– Stigma unlobed 34
34. Fruit a capsule 37
– Fruit indehiscent 35
35. Fruit a drupe (Australia) **31. *Owenia***
– Fruit a berry (Indomalesia) 36
36. Anthers in 1 whorl of 10; berries on branches and trunk **49. *Lansium***
– Anthers in 2 whorls of 5; berries on axillary infructescences **48. *Reinwardtiodendron***
37. Anthers on rim of staminal tube or tip of filaments **28. *Trichilia***
– Anthers inserted within throat of staminal tube, at most partially exserted 38
38. Staminal tube cyathiform 39
– Staminal tube cylindrical to patelliform 40
39. Seeds united by joint raphe-arils (Australia) **38. *Synoum***
– Seeds sarcotestal (Malesia to Pacific) **37. *Anthocarapa***
40. Nectary disk present (Indomalesia, Pacific) **40. *Dysoxylum***
– Nectary disk absent (west Africa) **42. *Turraeanthus***
41. Leaf rachis swollen at insertion of leaflets **20. *Walsura***
– Leaf rachis not swollen thus 42
42. Fruit a berry 43
– Fruit a capsule or drupe 44
43. Petals valvate (Madagascar) **30. *Malleastrum***
– Petals imbricate (Malesia) **25. *Pseudoclausena***
44. Stigma with conspicuous lobes; fruit a 1(2)-seeded drupe **15. *Azadirachta***
– Stigma without such lobes; fruit a capsule or drupe with 2–5 (6) pyrenes 45
45. Staminal tube deeply cleft 46
– Staminal tube not deeply cleft 48
46. Corolla valvate; fruit a 5- or 6-pyrened drupe (Indomalesia) **26. *Cipadessa***
– Corolla imbricate 47
47. Fruit a drupe (Africa) **27. *Ekebergia***
– Fruit a 1-seeded capsule 48
48. Abaxial leaf-surface glaucous, rachis contracted at leaflet attachments when dried (Indomalesia) **21. *Heynea***
– Leaves different (not Indomalesia) **28. *Trichilia***
49. Nectary disk 0 50
– Nectary disk present 52
50. Petals 3 (Indomalesia) **46. *Aphanamixis***
– Petals 4 or more 51
51. Staminal tube globose (Himalaya) **47. *Sphaerosacme***
– Staminal tube cylindrical (west Africa) **42. *Turraeanthus***
52. Ovary 1-locular with parietal placentae (west Africa) **41. *Heckeldora***
– Ovary with more than 1 locule 53
53. Nectary disk tubular (to subannular) (Indomalesia, Pacific) **40. *Dysoxylum***
– Nectary disk stipitate to cyathiform 54
54. Leaves with pellucid lines or dots (Neotropics) 55
– Pellucid lines or dots 0 56
55. Stylehead discoid **45. *Cabralea***
– Stylehead capitate **28. *Trichilia***
56. Stylehead capitate **28. *Trichilia***
– Stylehead discoid 57
57. Calyx lobes imbricate, sarcotesta basal, swollen (Neotropics) **43. *Ruagea***
– Calyx lobes open, sarcotesta different **44. *Guarea***
58. Stamens 5 59
– Stamens 8–10 60
59. Seeds winged below, attached to distal end of columella (Neotropics) **2. *Cedrela***
– Seeds winged at both ends or if with one wing attached towards base of columella (Asia) **1. *Toona***
60. Leaves (1)2–4(5)-jugate (maritime Old World) **14. *Xylocarpus***
– Leaves with more leaflets, forest trees 61
61. Seeds unwinged **13. *Carapa***
– Seeds winged 62
62. Capsule globose to trigonous, not or scarcely longer than broad; seeds winged all round 63
– Capsule elongate, at least twice as long as broad; seeds with terminal wing or wings at both ends 64
63. Capsule ± globose with 4–6 valves remaining joined to one another **3. *Khaya***
– Capsule ± trigonous, the valves falling separately (Madagascar) **4. *Neobeguea***
64. Seed with a wing at both ends (India, Sri Lanka) **5. *Soymida***
– Seed with a terminal wing only 65
65. Staminal tube margin entire to crenulate; capsule with 60–100 terminally winged seeds per locule **7. *Chukrasia***
– Staminal tube with appendages or distinct lobes (rarely entire); seeds <20 per locule 66
66. Seeds attached by seed-end towards apex of columella 67
– Seeds attached by the wing-end towards apex of columella 68
67. Leaflets entire; capsule pendent (Africa) **6. *Entandrophragma***
– Leaflets dentate to serrate or undulately lobed; capsule erect 69
68. Capsule claviform or oblong (Africa) **9. *Pseudocedrela***
– Capsule ellipsoid or fusiform (Andes) **8. *Schmardaea***
69. Capsule erect, woody, ovoid or obovoid to oblong; flowers mostly 5-merous (Neotropics) **10. *Swietenia***
– Capsule pendent, subwoody, ellipsoid to tetragonal; flowers 4-merous (Africa) **11. *Lovoa***

Genera of Meliaceae

I. Subfam. Cedreloideae Arn. (1832).

Swietenioideae Kostel. (1836).
Lovooideae Kribs (1930).
Capuronianthoideae T.D. Penn. (1975).

Monoecious or dioecious trees; hairs usually simple; buds nearly always protected by scale-leaves. Leaves pinnate, spirally arranged. Loculi nearly always with 3 or more ovules, these biseriate; stylehead discoid or very rarely capitate. Fruit a septifragal capsule with a central columella. Seeds winged or capsule subwoody or leathery with rudimentary columella and seeds unwinged with a woody or corky outer layer.

1. ***Toona*** (Endl.) M. Roem.

Toona (Endl.) M. Roem., Fam. Nat. Syn. Monogr.1: 131, 139 (1846); Edmonds, Fl. Males. 12: 358–371 (1995); Peng Hua & Edmonds, Fl. China 11: 112–115 (2008).
Cedrela L. sect. *Toona* Endl. (1840).

Deciduous or semi-evergreen trees to 50 m, monoecious. Leaves usually paripinnate; leaflets entire to serrate, domatia usually present. Inflorescences much-branched thyrses. Flowers unisexual, rarely bisexual; calyx 5(6)-lobed or 5(6) sepals distinct, imbricate to cupulate in bud; petals 5(6), distinct, imbricate (quincuncial), usually adnate to pulvinate androgynophore (disk); stamens 5(6), distinct, arising from androgynophore, sometimes alternating with 1–5 filamentous staminodes; ovary 5-locular, each locule with 6–10 ovules; stylehead discoid, usually 5-rayed. Fruit a woody septifragal capsule; valves opening from apex; columella softly woody, 5-angled, extending to capsule apex. Seeds winged at both ends, when attached distally, or at one end, when attached by seed-end to proximal part of the columella; endosperm residual; cotyledons collateral, flattened, leaf-like; radicle laterally exserted; germination phanerocotylar. $2n = 46, 52, 56$.

Four or five spp. from eastern Pakistan to southern China and eastern Australia. *T. ciliata* M.J. Roem. (toon), almost throughout the range, was the most important cabinet timber (red cedar) in Australia.

2. ***Cedrela*** P. Br.

Cedrela P. Br., Civ. Hist. Nat. Jamaica: 158, t. 10, Fig. 1 (1756); T.D. Pennington & A.N. Muellner, Monogr. *Cedrela* (2010).

Deciduous trees. Leaves usually paripinnate; leaflets entire. Inflorescence a thyrse. Calyx deeply lobed to cup-shaped; petals 5, distinct, longer than calyx in bud, imbricate and adnate to long columnar andogynophore for up to half their length; stamens 5, distinct, adnate to andogynophore proximally; staminodes 0; ovary 5-locular, each locule with 8–14 ovules; stylehead discoid with glandular stigmatic papillae. Fruit a woody septifragal capsule opening from apex with 5 valves; columella woody, broadly winged, extending to apex of capsule. Seeds with terminal wing attached by seed end to distal part of columella; endosperm residual; cotyledons collateral, leaf-like. Germination phanerocotylar; eophylls opposite, trifoliolate; leaflets sinuate, entire. $2n = 50, 56$.

About 17 spp. in the Neotropics. One of the world's most important timber tree genera, though greatly attacked by *Hypsipela* shoot-borer larvae; *C. odorata* L. (Spanish cedar, invasive in the Galapagos) and lesser quality *C. fissilis* Vell. are often sold mixed together. Timber for all uses, formerly much for cigar-boxes.

3. ***Khaya*** A. Juss.

Khaya A. Juss., Bull. Sci. Nat. Géol. 23: 238 (1830); White & Styles, Fl. Trop. E. Afr. Meliaceae: 46–49 (1991).

Large trees, monoecious. Leaves paripinnate. Flowers in large thyrses; calyx lobed almost to base, lobes suborbicular, imbricate; petals 4 or 5, distinct, contorted; staminal tube urceolate, margin with 8–10 suborbicular emarginate to irregularly lobed overlapping appendages alternating with included anthers or antherodes; nectary disk in male flowers cushion-shaped, united to base of pistillode but free from base of androecium, smaller in females; ovary 4- or 5-locular, each locule with 12–18 ovules; stylehead discoid with crenulate margin, upper surface with 4 or 5 radiating stigmatic ridges. Fruit an erect subglobose, woody septifragal capsule opening by 4 or 5(6) valves from the apex, the valves remaining joined at the base, often with rough fibrous strands at margins; columella not extending to capsule apex, with 4 or 5(6) sharp tough woody ridges; seed-scars white, conspicuous. Seeds 8–18 per locule, broadly transversely ellipsoid or suborbicular, narrowly winged all round. $2n = 50$.

About 5 spp. in Africa, 1 in Madagascar. Major timber trees (African mahogany) used as a *Swietenia* substitute; some with locally important medicinal bark.

4. *Neobeguea* J.-F. Leroy

Neobeguea J.-F. Leroy, J. Agric. Trop. Bot. Appl. 17: 232 (1970) & Adansonia II, 16: 174, t. 1–3 (1976).

Deciduous trees. Indumentum of simple hairs or dendroid glands. Leaves usually paripinnate, rarely imparipinnate; leaflets entire, crenulated, dentate or denticulate. Inflorescences thyrses, sometimes appearing terminal. Calyx 4-lobed; petals 4(5), distinct, contorted or imbricate, reflexed at anthesis; staminal tube cupular to urceolate, margin with 8–16 ± bifid appendices; anthers 8(10) included to weakly exserted; nectary disk cushion-shaped in male flowers, much reduced in females; ovary (2)3(4)-locular, each locule with 4–6 ovules; stylehead discoid. Fruit a septifragal capsule, ± trigonous, 3-valved, woody, valves falling individually; columella woody, 3-angled, reaching apex of capsule. Seeds 3 or 4 per locule, subcircular, flattened and broadly winged around the margin; endosperm residual; embryo with collateral flattened cotyledons; radicle superior. $2n = 50, 52$.

Three spp., Madagascar.

5. *Soymida* A. Juss.

Soymida A. Juss., Bull. Sci. Nat. Géol. 23: 238 (1830); Mabb., Fl. Ceylon 9: 293–296 (1995).

Deciduous tree. Leaves paripinnate, glabrous; leaflets entire. Thyrses of unisexual flowers, with well-developed vestiges of opposite sex, in the most terminal axils. Sepals 5, ± distinct, imbricate; petals 5, contorted, spreading at anthesis; staminal tube shortly cylindrical to urceolate, margin with 10 bifid spreading lobes; anthers inserted between teeth of the lobes; nectary disk in male flowers shallowly patelliform, united with base of pistillode, in females an obscure swelling at ovary base; ovary 5-locular, each locule with 12–16 ovules; style 0, stylehead discoid, 5-lobed. Capsule ± ovoid, woody, septifragal, erect, opening from apex, the 5 valves separating into two layers: columella woody with 5 distinct ridges and reaching to capsule apex, seed-scars inconspicuous. Seeds unequally winged at both ends, attached by longer wing to distal end of columella; endosperm present; embryo with flattened foliaceous cotyledons; radicle obliquely superior. Germination phanerocotylar, cotyledons becoming green; eophylls opposite or in spirals, simple with minute serrate margin, later trifoliolate. $2n = 56$.

One sp., *S. febrifuga* (Roxb.) A. Juss., C, W and S India, Sri Lanka; locally important medicinal bark, tanbark and timber.

In many respects *Soymida* closely resembles *Khaya*, notably in leaf characters and flowers (petals erect in *Khaya* at anthesis; staminal tube lobes rounded or irregularly lobed) and seeds (winged equally all round in *Khaya*); the capsule in *Khaya* has a perceptibly bilamellate structure but does not separate into two distinct layers.

6. *Entandrophragma* C.DC. Fig. 34

Entandrophragma C.DC., Bull. Herb. Boissier 2: 582, t. 21 (1894); White & Styles, Fl. Trop. E. Afr. Meliaceae: 49–56 (1991).
Heimodendron Sillans (1953).

Large trees, dioecious. Leaves paripinnate, sometimes appearing imparipinnate. Inflorescences large thyrses. Calyx cupuliform, margin entire or with 5 acute lobes with open aestivation; petals 5, distinct, contorted; staminal tube cupular to urceolate, margin entire to lobed with 10 shortly stalked anthers or antherodes on margin or lobes; appendages 0; nectary disk cushion-shaped, united with base of ovary or pistillode but free from androecium and connected to it by 10 or 20 ridges or partitions; ovary 5-locular, each locule with 4–12 ovules; stylehead discoid, with 5 radiating stigmatic lobes. Fruit a pendulous, elongate woody septifragal capsule, opening by 5 valves from the apex or base, or from both simultaneously; columella softly woody, extending to apex of capsule, 5-angled or 5-ridged, deeply indented with the imprints of seeds. Seeds 3–9 per locule, each with a terminal wing, attached by the seed end to distal part of columella and winged towards base of capsule. $2n = 36, 72$.

Eleven spp., tropical Africa. Many important timbers, including sapele (*E. cylindricum* (Sprague) Sprague) with trees to 55 m or more, and utile (*E. utile* (Dawe & Sprague) Sprague).

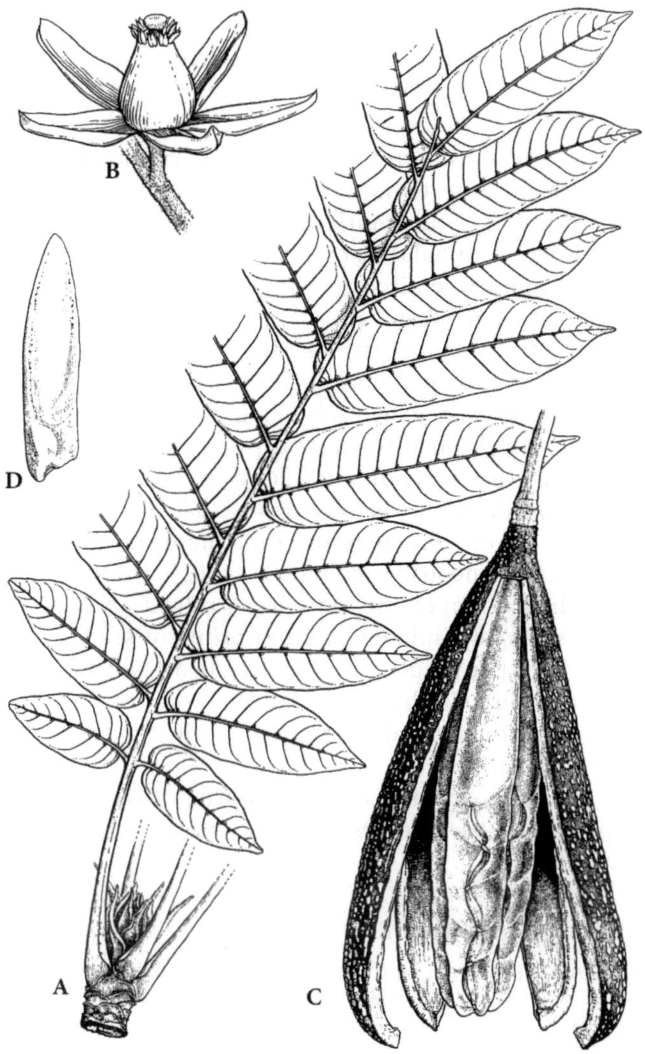

Fig. 34. Meliaceae. *Entandrophragma utile.* **A** Leaf and apical bud. **B** Flower. **C** Capsule, one valve removed. **D** Seed. (White and Styles 1991; drawn by Janet Dyer)

7. *Chukrasia* A. Juss.

Chukrasia A. Juss., Bull. Sci. Nat. Géol. 23: 239 (1830); Kalangire et al., Austr. J. Bot. 50: 319–330 (2002).

Deciduous trees. Leaves paripinnate with terminal leaflet represented by a spike, imparipinnate and bipinnate with incised or lobed leaflets in juveniles, rarely retained at maturity. Thyrses axillary, often subterminal appearing terminal. Calyx 4- or 5-lobed; petals 4 or 5, distinct, contorted and much longer than calyx in bud; staminal tube cylindrical, somewhat narrowing distally, margin entire to crenulate; anthers attached to margin; nectary disk obscure to narrowly cushion-shaped; ovary flask shaped, 3–5-locular, each loculus with numerous ovules; stylehead capitate with 3–5 stigmatic ridges. Capsule ovoid or ellipsoid, woody, opening by 3–5 values from the apex, the valves splitting into an outer and inner bifid layer; columella with 3–5 sharply angled ridges, extending to apex of capsule; seed-scars conspicuous. Seeds 60–100 per locule, wings terminal, arranged laterally in tiers in two ranks; endosperm present; cotyledons subcircular; radicle obliquely exserted. $2n = 26$.

One somewhat variable sp., *C. tabularis* A. Juss., from India and Sri Lanka, eastwards through tropical Asia to western Malesia, perhaps divisible into ecological races or even ecospecies (Kalangire et al. 2002). Valuable timber (*chickrassy*, *yinma* or Chittagong wood).

8. *Schmardaea* H. Karst.

Schmardaea H. Karst., Fl. Columb. 1: 187, t. 93 (1861); Styles, Fl. Neotrop. 28: 387–389 (1981).

Tree or shrub, deciduous. Leaves usually imparipinnate; leaflets with crenate to coarsely dentate margins. Flowers unisexual, in axillary thyrses. Calyx lobed almost to base, lobes 4, imbricate; petals 4, linear, contorted; staminal tube with 8 appendages; anthers included or shortly exserted, connective extended into long thread-like appendage; nectary disk ridged, short-stipitate; ovary 4-locular, each locule with up to 25 ovules; stylehead narrowly discoid with 4 stigmatic lobes. Capsule septifragal, erect, opening from apex with 4 valves; inner layers of valves linked by fibrous network; columella 0. Seeds up to 20 per locule with terminal wing, attached to dissepiments; endosperm in thick layer, embryo with thin collateral cotyledons; radicle laterally exserted.

One sp., *S. microphylla* (Hook.) C. Muell., Andes from Venezuela to Peru, up to 2,700 m in cloud forest.

9. *Pseudocedrela* Harms

Pseudocedrela Harms in Engl., Bot. Jahrb. Syst. 22: 153 (1895); White & Styles, Fl. Trop. E. Afr. Meliaceae: 56–58 (1991).

Tree, monoecious. Leaves paripinnate. Calyx (4)5-lobed almost to base, lobes ovate to suborbicular; petals (4) 5 distinct, slightly contorted, boat-shaped; staminal tube urceolate, margin with (8) 10 bifid reflexed lobes with anthers between them; nectary disk annular surrounding ovary base; ovary 4- or 5-locular, each locule with 4–6 ovules; stylehead discoid, upper surface with 4 or 5 radiating stigmatic ridges. Fruit an erect elongate woody septifragal capsule, opening from the apex by 5 divergent valves remaining attached at the base, connected by a fibrous network; columella woody extending to capsule apex, sharply 4- or 5-angled, indented with imprints of the seeds. Seeds 4 or 5 per locule, winged, attached by the seed-end to distal part of columella. $2n = 56$.

One sp., *P. kotschyi* (Schweinf.) Harms, tropical Africa. Timber valuable.

10. *Swietenia* Jacq.

Swietenia Jacq., Enum. Syst. Pl.: 4, 20 (1760); Lamb, Mahogany of trop. America (1966); Styles, Fl. Neotrop. 28: 389–406 (1981).

Deciduous trees. Leaves almost always paripinnate; leaflets entire. Flowers unisexual in axillary thyrses; calyx 5-lobed to half way, lobes rounded to obtuse, imbricate; petals (4)5, contorted; staminal tube cup-shaped to urceolate, with 8–10 partially exserted anthers (antherodes) and 8–10 acuminate appendages; nectary disk annular; ovary (4)5(6)-locular, each locule with 9–16 ovules; stylehead discoid with (4)5 stigmatic lobes. Capsule septifragal, oblong to ovoid, erect, the 5 valves opening from base or from base and apex simultaneously, separating into two layers, the outer thick and woody; columella the length of the capsule, 5-angled. Seeds 9–16 per locule, hanging by wing-end from distal part of columella; endosperm present; embryo with thin cotyledons. $2n = 48, 54, 56$.

Three spp., the true mahoganies (see Lamb 1966), tropical America extending to southern Florida. *S. mahagoni* (L.) Jacq. (Caribbean) and *S. macrophylla* King are widely planted for timber (Spanish mahogany), the latter perhaps now the most valuable timber in South America, because the former has been so reduced in quantity and quality through over-exploitation and genetic erosion.

11. *Lovoa* Harms

Lovoa Harms in Engl. & Prantl, Nat. Pflanzenfam. III, 4: 307 (1896); White & Styles, Fl. Trop. E. Afr., Meliaceae: 58–60 (1991).

Large trees, monoecious. Leaves paripinnate. Flowers in large thyrses; calyx lobed almost to base, lobes 2 + 2, imbricate; petals 4, distinct, imbricate; staminal tube cupular or shortly cylindrical, margin entire or with paired deltate-acuminate appendages alternating with 8 anthers; nectary disk broadly cushion-shaped, enveloping base of ovary or pistillode but free from androecium; ovary 4-locular, each locule with 4–6(–8) locules; stylehead discoid or capitate, obscurely 4-lobed. Fruit a pendulous, elongate, tetragonal or ellipsoid, thinly woody septifragal capsule, dehiscing from the apex or from the apex and base simultaneously; columella softly woody, extending to apex of capsule, 4-ridged, each ridge shallowly indented with imprints of seeds. Seeds c. 2 per locule, winged, attached to distal part of columella by wing-end, the body of the seed hanging towards the apex of the capsule. $2n = 50, 52, 56$.

Two spp., tropical Africa. Important timbers, especially *L. trichilioides* Harms (Nigerian golden walnut).

12. *Capuronianthus* J.-F. Leroy Fig. 35

Capuronianthus J.-F. Leroy, Compt. Rend. Hebd. Séances Acad. Sci. 247: 1374 (1958) & Adansonia II, 16: 176–181 (1976).

Trees, monoecious. Buds naked. Leaves pinnate, decussate. Flowers in thyrses or short racemes; calyx deeply lobed or of 4 distinct sepals, lobes imbricate; petals 4(5), distinct, generally contorted, more rarely imbricate; staminal tube urceolate or shortly cylindrical, margin with 8 appendices alternating with anthers inserted on margin; nectary disk present; ovary locules with 2 superposed ovules plus 1 or 2 vestigial ones; style very short or 0, stylehead capitate, minute. Fruit a 3- or 4-valved capsule dehiscing irregularly or not at all; epicarp and mesocarp fibrous, indehiscent or with partial irregularly septifragal dehiscence; endocarp cartilaginous. Seeds 1 or 2 per valve, unwinged; testa thick; endosperm 0; cotyledons plano-convex, collateral to oblique; radicle superior or adaxial. $2n = 58$.

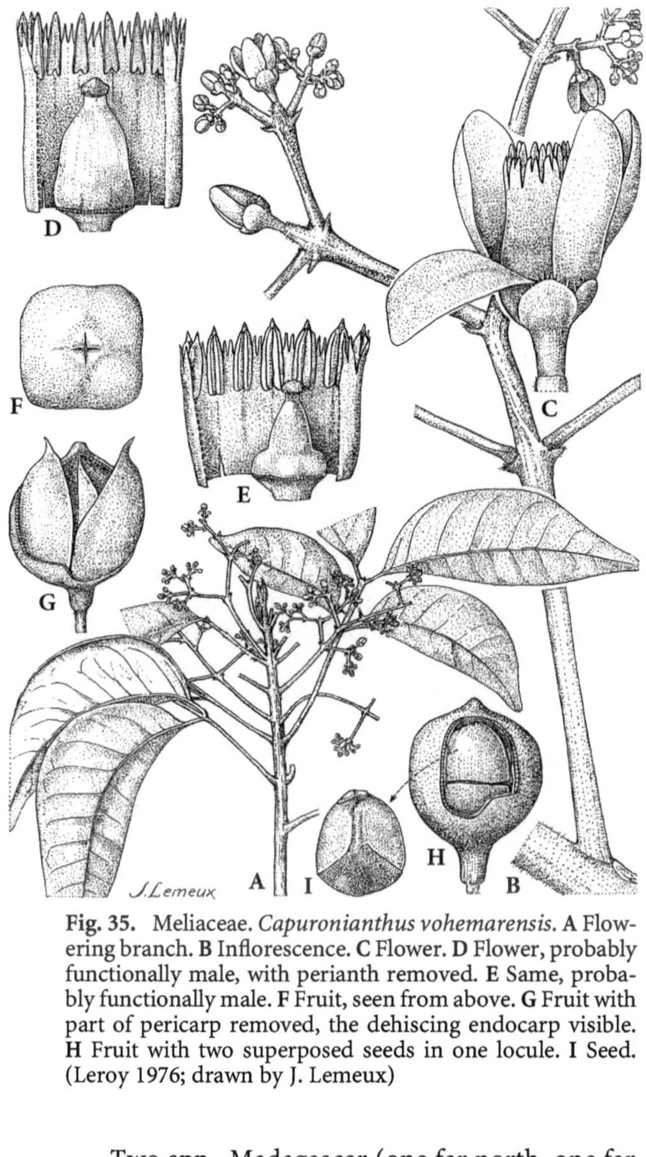

Fig. 35. Meliaceae. *Capuronianthus vohemarensis.* **A** Flowering branch. **B** Inflorescence. **C** Flower. **D** Flower, probably functionally male, with perianth removed. **E** Same, probably functionally male. **F** Fruit, seen from above. **G** Fruit with part of pericarp removed, the dehiscing endocarp visible. **H** Fruit with two superposed seeds in one locule. **I** Seed. (Leroy 1976; drawn by J. Lemeux)

Two spp., Madagascar (one far north, one far south).

Although the genus has recently been afforded subfamilial rank, Muellner et al. (2003) argue for its inclusion in Cedreloideae, supporting earlier anatomical work, for the seed-coat is like that of *Carapa* (Cheek 1989).

Genera 13 and 14 make up:

Tribe Xylocarpeae Blume (1825).

Carapeae Harms (1896).

Trees. Flowers without a gynophore; petals distinct; staminal tube urceolate or cupular, margin lobed; anthers 8–10. Capsule ± woody or leathery with a rudimentary columella. Seeds unwinged, with a corky or woody outer layer; cotyledons large, united; endosperm 0.

13. *Carapa* Aubl.

Carapa Aubl., Hist. Pl. Guian. Franç. 2, suppl.: 32, t. 387 (1775); Styles, Fl. Neotrop. 28: 406–418 (1981).

Deciduous or semi-evergreen trees, monoecious. Leaves almost always paripinnate. Flowers in large, erect, axillary thyrses, sometimes subterminal and appearing terminal; calyx 4- or 5-lobed almost to base, the lobes rounded or ovate, imbricate; petals 4 or 5(6), imbricate; staminal tube cupular to cylindrical or urceolate with 8–10 entire to lobed appendages alternating with sessile included anthers (antherodes); nectary disk cushion-shaped, free of staminal tube; ovary 4- or 5(6)-locular, partly sunk in disk, each locule with (2)3–8 ovules; stylehead discoid. Capsule septifragal, subglobose to cylindrical, pendulous, opening from apex and base simultaneously by 4 or 5 leathery valves, columella disintegrating. Seeds 8–35, angular, with outer surface rounded; sarcotesta woody, thick; endosperm 0; embryo with large united cotyledons. $2n = 58$.

Two variable spp. (possibly divisible into more) in tropical America and Africa, *C. procera* DC. in both. Timber valuable; seed oil (andiroba) locally medicinal.

14. *Xylocarpus* Koenig Fig. 36

Xylocarpus Koenig, Naturforscher 20: 2 (1784); Mabb., Malays. For. 45: 448–450 (1982), rev., Fl. Ceylon 9: 296–300 (1995) & Fl. Males. 12: 371–380 (1995).

Semi-evergreen maritime trees. Twigs lenticellate, cicatrose. Leaves paripinnate with (1)2–4(5) pairs leaflets, entire, glabrous. Flowers unisexual in short axillary thyrses; calyx 4-lobed to about the middle, valvate; petals 4, contorted and much longer than the calyx in bud; staminal tube margin with 8 suborbicular, retuse or shallowly and irregularly divided lobes; anthers 8, included; nectary disk cushion-shaped, beneath or surrounding and united with ovary, red; ovary 4(5)-locular, each locule with 3 or 4 (–6) ovules; style short, stylehead discoid, its margin crenellate and its upper surface with four radiating stigmatic grooves. Fruit a large pendulous subspherical capsule, tardily dehiscing by 4(5) leathery valves from apex. Seeds 5–20, large, irregularly tetrahedral or pyramidal, outermost surface convex, attached to central columella, with aerenchymatous (?sarcotestal) coat; embryo with the radicle lying above the hilum; germination cryptocotylar, the numerous subulate cataphylls verticillate or in spirals; true leaves simple, entire, later ones trifoliolate (rarely first ones thus). $2n = 52$.

Three spp. throughout the coastal regions of the Old World tropics from E Africa to the W Pacific in mangrove swamps and coastal woodlands on rock and other substrates. The wood and bark are locally important for construction, and both tanning and medicine.

Fig. 36. Meliaceae. *Xylocarpus granatum*. **A** Flowering branch. **B** Flower in longisection. **C** Unripe fruit and seed. **D** Bark. (Mabberley et al. 1995; drawn by Rosemary Wise)

II. Subfam. Melioideae Arn. (1832).

Trichilioideae Kostel. (1836).
Quivisianthoideae T.D. Penn. & Styles (1975).

Buds usually naked; plants dioecious, polygamous or with bisexual flowers entirely; loculi 1-, 2- or 3-ovular with vestigial third ovule, very rarely multi-ovulate; ovules collateral or superposed. Fruit a columella-less fleshy or leathery capsule, berry or drupe. Seed usually with a fleshy testa or aril or combination of these, very rarely (*Quivisianthe*) winged; endosperm present or absent. Pantropical.

Genera 15 and 16 make up:

Tribe Melieae DC. (1824).

Trees, polygamous; hairs stellate and/or simple. Leaves pinnate or bipinnate. Flowers bisexual or male; staminal tube cylindrical, lobed; nectary disk annular; stylehead 3–8-lobed. Fruit a drupe. Seed exarillate; endosperm thin; cotyledons collateral; radicle superior, exserted.

15. *Azadirachta* A. Juss.

Azadirachta A. Juss., Bull. Sci. Nat. Géol. 23: 236 (1830); Mabb., Fl. Ceylon 9: 280–284 (1995) & Fl. Males. 12: 337–343 (1995).

Trees; hairs simple; buds thinly encrusted with resin. Leaves pinnate with 2 pairs of glands at base of petiole. Flowers bisexual and male on same individual (polygamous); calyx 5-lobed to proximal half, the lobes imbricate; petals 5, distinct, imbricate; staminal tube cylindrical, slightly expanded at mouth, margin (8–)10-lobed, the lobes rounded, truncate, emarginate or bifid; anthers (8–)10, glabrous, inserted at base of and opposite lobes; nectary disk annular, united with base of ovary; ovary 3-locular, each locule with 2 collateral ovules; stylehead with apical swollen torus with 3 acute, partially united papillose stigmatic lobes. Fruit a 1(2)-seeded drupe; endocarp thin, cartilaginous. Seed ovoid, distally pointed; testa thin, membranous with small adaxial sarcotesta; cotyledons plano-convex, collateral; radicle superior, short projecting from cotyledons. Germination phanerocotylar; eophylls opposite, trifoliolate, leaflets deeply incised or pinnatifid. $2n = 28, 30$.

Two spp. native to Indomalesia, though one, *A. indica* A. Juss., the neem, is widely cultivated in warm countries throughout the world and is naturalized in some of them, colonizing deforested land.

Formerly confused with *Melia*, *Azadirachta* differs in its simple indumentum, pinnate leaves, collateral ovules, 3-lobed stylehead and 1(2)-seeded drupes. The wood is readily distinguished in that that of *Melia* is ring porous.

In West Africa, where the tree has come to dominate large areas of the savanna, the seeds are dispersed by fruit bats and baboons, after passage through which their germination is enhanced. Held sacred by the Hindus, the neem is potentially one of the most important of all tropical seasonal forest trees, having proved to be very adaptable and able to withstand arid conditions. It can be grown in impoverished soil and is a fast-growing source of fuelwood. In Central America, it is now being planted as a substitute for *Swietenia* mahogany. It is also widely planted as a windbreak, shade and avenue tree, the world's biggest plantation being of 50,000 trees in the plains of Saudi Arabia, planted to shade the two million or so Muslim pilgrims camping there annually for 'Haj' rites. It is a soil ameliorant and is potentially a source of many valuable by-products: it has been an ingredient of soaps, toothpaste and lotions in commerce for decades. The seeds contain some 40% oil by weight and this bitter material has been used in lamps and as a lubricant and has potential as a fuel-source, the mesocarp being a promising substrate for the production of methane gas. Neem cake is an excellent fertilizer and the leaves and twigs are used as a mulch in Asia. The bark produces a valuable gum and tannin worth exploiting. The pressed leaves have long been put in books to ward off insects, the repellent being the limonoid azadirachtin, which is in the seeds as well as the leaves; house-sparrows in India incorporate leaves into their nests, perhaps reducing parasite loads. Azadirachtin is absorbed by plants and acts as a systemic insecticide so efficient that Japanese beetles and other insects, even including the desert locust, will starve rather than eat plants treated with it. Of the five limonoids known from the tree, deacetylaxadirachnol (salannin) is as potent as azadirachtin in inhibiting ecdysis in tobacco budworm. Neem seed powder with carbofuran greatly reduces leaf-hoppers and rice tungro virus in rice. The leaves, bark and seed oil have been used in the treatment of a wide range of ailments, including malaria, eczema, dysentery and ulcers, but particularly effective as a parasiticide for skin diseases such as scabies. Neem oil also has significant post-coital contraceptive action (see Mabberley et al. 1995: 342–343).

16. *Melia* L.

Melia L., Sp. Pl.: 384 (1753); Harms in Engl. & Prantl, Nat. Pflanzenfam., ed. 2, 19b1: 99–102, t. 26, 27 (1940); Mabb., Gdns' Bull. Sing. 37: 49–62 (1984) & Fl. Ceylon 9: 274–279 (1995) & Fl. Males. 12: 329–336 (1995).
Antelaea Gaertn. (1788).

Trees, occasionally flowering precociously as shrublets; hairs simple and stellate-tufted. Leaves 2(3)-pinnate. Inflorescence thyrsoid, axillary. Calyx 5(6)-lobed to near base, lobes somewhat imbricate; petals 5(6), distinct, imbricate; staminal tube narrowly cylindrical, slightly expanded at mouth, 10(12)-ribbed, with 10 or 12 truncate, bifid or 4-fid filiform lobes; anthers 10(12), inserted at margin of or just within tube, alternating with or opposite lobes; nectary disk small, surrounding base of ovary; ovary 4–8-locular, each locule with 2 superposed ovules; stylehead capitate to coroniform with 4–8 short, erect or incurved stigmatic lobes. Drupe 3–8-locular; endocarp thick, bony, deeply dimpled at base and apex; loculi 1 (2)-seeded. Seed oblong, laterally compressed; testa leathery sometimes slightly swollen and fleshy around hilum; embryo embedded in oily endosperm; cotyledons flat. Germination phanerocotylar; eophylls opposite, pinnatisect or trifoliolate. $2n = 28$.

Two or three spp., one, *M. azedarach* L. (white cedar, Persian lilac), in Indomalesia and one or possibly two in S tropical Africa. Forms of *M. azedarach* are widely cultivated and naturalized throughout the warm parts of the world.

The 'inflorescences' of *M. azedarach* are often short shoots with terminal buds. The true inflorescences are borne in the axils of rudimentary leaves and, after fruit abscission, the terminal bud may grow out into a leafy shoot. The fruits, which have been long used for beads, are toxic to man, some 6–8 considered a fatal dose, and also to pigs but apparently not to birds, sheep or goats: the toxic principles are the limonoids, meliatoxins A1, A2, B1, B2. The wood of different forms of the species has been used for furniture and light construction and for the production of fibreboard. It has been used as a fast-growing coffee-shade and it has been alleged that fruit trees grown under it remain relatively free from aphids. Indeed, a decoction of the fruits has long been used as an insecticide for plants in India and fruits or leaves have been placed with dry fruit, clothing and in books to keep insects away. A glycopeptide, meliacin, isolated from the leaves and roots is responsible for inhibition of in-vitro replication of various DNA and RNA viruses, e.g. poliovirus, herpes simplex Type I (see Mabberley 1995: 336).

17. *Astrotrichilia* (Harms) T.D. Penn.

Astrotrichilia (Harms) J.-F. Leroy ex T.D. Penn., Blumea 22: 477–478 (1975); J.-F. Leroy & Lescot, Bull. Mus. Natl. Hist. Nat. IV, 18: 20–34 (1996).
Trichilia P. Br. sect. *Astrotrichilia* Harms (1896).

Trees or shrubs. Indumentum of stellate hairs, rarely bifid, mixed with simple hairs and small glandular hairs. Leaves pinnate to trifoliolate. Flowers unisexual in axillary thyrses; calyx with 5 distinct lobes, imbricate or open in bud; petals 5, distinct, imbricate; staminal tube cyathiform or shortly cylindrical, margin entire or with 10(11) dentiform appendages bearing anthers; nectary disk patelliform or 0; ovary 2–5-locular, each locule with 2–5 superposed, rarely collateral, ? orthotropous ovules; stylehead discoid. Fruits drupes with ± persistent calyx, 1–3-locular; pericarp thick, resinous; endocarp chartaceous. Seeds one per locule, exarillate; endosperm thick; embryo with thin flat collateral cotyledons, radicle small, abaxial.

Twelve spp., Madagascar.

18. *Quivisianthe* Baill.

Quivisianthe Baill. in Grandidier, Hist. Phys. Madagascar 33, 3 [fasc. 34], Atlas 2: ad t. 251 (1894); T.D. Penn., Blumea 22: 508–509 (1975).

Trees; hairs simple; buds naked. Leaves pinnate, in spirals. Flowers usually unisexual (trees dioecious), in thyrses; calyx 5-lobed; petals distinct, valvate; staminal tube urceolate to shortly cylindrical, margin entire or subentire; anthers 5 inserted on the margin; nectary disk annular to patelliform or 0 or a short stipe; ovary 3- or 4-locular, each locule with (1)2 ovules; stylehead obscurely 3-lobed. Fruit a loculicidal capsule, 3(4)-angled, 3(4)-valved. Seeds (1)2 per locule; testa dry with apical wing attached to placenta; endosperm present; embryo with lateral flat cotyledons; radicle superior, exserted.

Two spp., Madagascar.

Although recently assigned to a subfamily of its own, on DNA grounds (Muellner et al. 2003) the genus appears close to *Ekebergia* ('Trichilieae'). The bitegmic seeds are like those of Cedreloideae.

19. *Sandoricum* Cav.

Sandoricum Cav., Diss. 7: 359 (1789); Mabb., Blumea 31: 146–151 (1985), rev. & Fl. Males. 12: 344–353 (1995).

Trees; hairs simple. Leaves trifoliolate. Flowers in axillary thyrses; calyx ± truncate to shallowly 4- or 5-lobed; petals (4)5, distinct, imbricate; staminal tube cylindrical, ribbed distally; margin with 5 or 10 short lobes; anthers 10, glabrous, included; nectary disk tubular, free, margin coarsely toothed; ovary slightly sunk in receptacle, 4- or 5-locular, each locule with 2 collateral ovules; stylehead with 4- or 5-lobed stigma. Fruit a drupe, 1–5-locular, pyrenes 1(or 2)-seeded; outer mesocarp rather dry-fleshy or soft and fibrous, inner mesocarp fleshy or spongy-fibrous; endocarp thin, cartilaginous. Seeds exarillate, pachychalazal, kidney-shaped, laterally compressed with thin sarcotesta; endosperm 0; cotyledons thick, plano-convex, collateral, radicle apical, extending to surface or slightly exserted. Germination phanerocotylar; eophylls trifoliolate, opposite. $2n = 16, 22, 44, 64$.

Five spp., all but one, *S. koetjape* (Burm.f.) Merr., restricted to W Malesia, where the cultivated forms of *S. koetjape* (santol) may have arisen, though wild relations appear to be native as far east as New Guinea. All five are wild in Borneo, to which three are restricted.

The fruit (where known as mature) of all species is edible and the timber of some value, the bark used for tannin and medicine locally, some of the triterpenoids extracted from the stems having been shown to have significant cytotoxic activity against cultured cancer cells and to be insect antifeedants.

20. *Walsura* Roxb.

Walsura Roxb., Fl. Ind. 2: 386 (1832); T. Clark, Blumea 38: 247–291 (1994), rev.; Mabb., Gdns' Bull. Sing. 55: 195–199 (2003).
Napeodendron Ridl. (1920).

Pachycaul to leptocaul trees; hairs simple and/or bifid. Leaves imparipinnate or unifoliolate, leaflets opposite. Inflorescences axillary thyrses of hemaphrodite and male flowers. Calyx 5-lobed; petals 5, distinct, imbricate to valvate; androecium of distinct filaments or a 10-lobed tube with truncate to weakly bifid lobes; nectary disk annular; ovary 2 (or incompletely 4?)-locular, each locule with 2 collateral ovules; stylehead capitate to cylindrical, sometimes with 2 apical lobes. Fruit a 1- or 2(?4)-seeded berry or 1- or 2-seeded weakly dehiscent capsule; pericarp with thin layer of sclerenchyma, the locules separated by a thin septum. Seeds pre-raphe-funicular-arillate. $2n = 28$.

Sixteen spp., from Sri Lanka to the Himalaya and Indochina through Malesia to New Guinea.

21. *Heynea* Roxb.

Heynea Roxb. in Curt., Bot Mag. 41: t. 1738 (1815); Mabb., Fl. Males. 12: 41–44 (1995); Peng Hua & Mabb., Fl. China 11: 120–121 (2008).
Walsura Roxb. sect. *Heynea* (Roxb.) Harms (1896), nom. illeg.
Ailantopsis Gagnep. (1944).
Picroderma Gagnep. (1944).

Trees; twig pith vessel-less; hairs simple. Leaves imparipinnate; rachis compressed, not swollen at points of attachment of leaflets; abaxial surface of leaflets papillate, glandular. Inflorescences corymbose cymes with long peduncles. Calyx 4- or 5-lobed, the lobes imbricate; petals 4 or 5, ± imbricate; androecium with cylindrical staminal tube to 1/3 length with 8 or 10 filaments with bifid apices; nectary disk annular; ovary 2- or 3-locular, each locule 2-ovulate; stylehead 2- or 3-lobed. Fruit a capsule; pericarp with sclereids. Seeds 1 or 2, pre-raphe-funicular-arillate. $2n = 28$.

Two spp., E and SE Asia.

22. *Lepidotrichilia* (Harms) T.D. Penn.

Lepidotrichilia (Harms) J.-F. Leroy ex T.D. Penn., Blumea 22: 473–475 (1975); J.-F. Leroy & Lescot, Bull. Mus. Natl. Hist. Nat. Paris IV, 18, B Adansonia: 7–13 (1996).
Trichilia sect. *Lepidotrichilia* Harms (1896).

Trees; hairs simple and stellate. Leaves imparipinnate. Flowers unisexual, rarely bisexual, in axillary thyrses; calyx 5(6)-lobed; petals 5, distinct, valvate; staminal tube entire or with distinct filaments, cylindrical; appendages filiform to lanceolate; anthers 10, inserted between appendices; nectary disk 0; ovary 2–5-locular, each locule with 1 ovule; stylehead capitate with 3–5 stigmatic lobules. Fruit a 2–5-locular drupe; endocarp membranous or cartilaginous. Seeds 1–5, oblong to globose, exarillate, testa (?pachychalaza) membranous or cartilaginous; endosperm 0; embryo with collateral or oblique plano-convex cotyledons; radicle superior or abaxial, included to slightly exserted. $2n = 38$.

Four spp., 1 in tropical E Africa, 3 in Madagascar.

23. *Vavaea* Benth.

Vavaea Benth. in Hook., Lond. J. Bot. 2: 212 (1843); T.D. Penn., Blumea 17: 351–366 (1969), rev. & 22: 464–466 (1975).
Lamiofrutex Lauterb. (1924).

Trees or treelets with sympodial *Terminalia* branching. Hairs simple. Leaves simple. Flowers mostly bisexual; calyx 4 or 5(7)-lobed, with open rarely imbricate aestivation; petals (3)4–6, distinct, imbricate, rarely contorted; staminal tube cylindrical or cyathiform; anthers 9–23, attached at ends of filaments, filaments partly distinct; nectary disk patelliform or cyathiform, united to base of tube, or forming androecial ribbing, or 0; ovary 2–6-locular, locules with 1 or with 2(3) collateral ovules or with 4–10 in 2 rows. Fruit a berry with fleshy to woody pericarp. Seeds 1–3 (7) ovoid or plano-convex, with thin sarcotesta; thin endosperm sometimes present; cotyledons plano-convex, collateral; radicle superior, small, included or extending to the surface. Germination cryptocotylar, eophylls opposite, simple.

Four spp. from Sumatra eastwards to tropical Australia, Micronesia, Melanesia and Polynesia, with one restricted to Fiji and two to New Guinea.

24. *Munronia* Wight — Fig. 37

Munronia Wight, Ic. Pl. Ind. Orient. 1, 5: [1] (1838); Harms in Engl. & Prantl, Nat. Pflanzenfam., ed. 2, 19b1: 91–93 (1940); Mabb., Fl. Males. I, 12: 30–34 (1995); Peng Hua & Mabb., Fl. China 11: 118–119 (2008).

Unbranched or sparsely branched shrublets, sometimes suckering with apparently short-lived shoots, sometimes possibly dioecious; hairs simple and bifid or stellate. Leaves imparipinnate to simple, margins often crenate to serrate. Flowers bisexual, solitary or in few-flowered thyrses, pseudopedicellate, usually white; calyx 5-lobed to near base, lobes often somewhat foliaceous; petals 5, valvate to imbricate, adnate to staminal tube basally, rarely becoming free later; staminal tube narrowly cylindrical or weakly obconical, the margin with 10 entire or bilobed appendages or, rarely, with 10 reflexed filiform appendages recurved some distance below margin; anthers 10, pubescent, inserted on tube rim, alternating with appendages, connective often produced apically forming an appendage, which is rarely filiform; nectary disk 0 or tubular, free or united with base of staminal tube; ovary (3) 5-locular, each loculus with 1 or 2 superposed ovules; stylehead capitate, sometimes with 5 stigmatic lobes. Fruit a 5-valved loculicidal capsule, each locule with 1 or 2 seeds. Seed plano-convex with a bony tegmen with an outgrowth enveloping a thick sarcotesta (Cheek 1989). $2n = 50$.

About four spp., tropical Asia and subtropical China, two extending eastwards to Timor.

25. *Pseudoclausena* T. Clark

Pseudoclausena T. Clark, Blumea 38: 291–295 (1994) & Fl. Males. 12: 55–57 (1995).

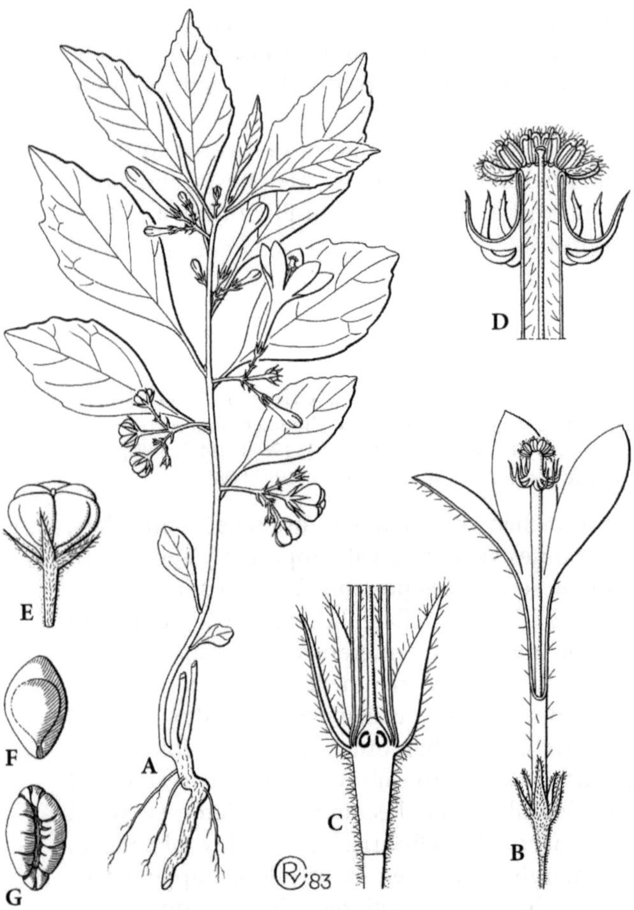

Fig. 37. Meliaceae. *Munronia humilis*. **A** Habit. **B** Flower, half corolla removed. **C** Ovary, vertical section. **D** Androecium, vertical section. **E** Fruit. **F, G** Seeds. (Mabberley et al. 1995; drawn by R. van Crevel)

Tree; hairs simple. Leaves imparipinnate. Inflorescences thyrses of bisexual or male flowers. Calyx 5-lobed; petals distinct, imbricate; androecium ± cylindrical, each filament linear with bifid apex; nectary disk 0; ovary 4- or 5-locular, each locule with 1 ovule. Fruit a 1- or 2-seeded berry, asymmetric, shortly beaked. Seed ellipsoid, exarillate, possibly pachychalazal.

One variable sp., *P. chrysogyne* (Miq.) T. Clark, Indochina to west New Guinea, in wet evergreen forest.

26. *Cipadessa* Blume

Cipadessa Blume, Bijdr.: 162 (1825); Mabb., Fl. Ceylon 9: 270–274 (1995) & Fl. Males. 12: 57–60 (1995).

Small tree; hairs simple. Leaves imparipinnate. Inflorescences polygamous, axillary thyrses. Calyx 5-lobed in apical half; petals 5(6), distinct, valvate; filaments alternatively long and short, connate near base, terminated by a pair of narrowly lanceolate to filiform or erose appendages; anthers 10, acute, inserted between the appendages, pubescent; nectary disk patelliform; ovary 5(6)-locular; each locule with 1(2) collateral ovules; style short, stout, columnar, swollen stylehead with 5(6) stigmatic lobes. Fruit a globular drupe with 5(6) pyrenes, each 1 (2)-seeded; endocarp bony. Seed orange-segment-shaped, exarillate; testa thin, membranous; embryo embedded in endosperm; cotyledons thin, flat, collateral; radicle superior, long exserted. $2n = 28, 56$.

One sp., *C. baccifera* (Roth) Miq., tropical and subtropical Asia from Nepal, India and Sri Lanka to S China, Indochina and C Malesia.

The rest of the genera (see introductory remarks) make up the remodelled:

Tribe Trichilieae DC. (1824), including a modified Turraeeae Harms (1896, including Nymanieae F. White (1986)), Guareeae T.D. Penn. (1975) and Aglaieae Blume (1825).

Trees, shrubs or suffrutices; hairs simple or stellate, rarely peltate scales. Leaves usually pinnate, rarely trifoliolate, unifoliolate or simple. Indumentum of simple or stellate hairs. Plants dioecious or flowers bisexual; staminal tube usually complete or filaments partially connate, rarely distinct; anthers nearly always inserted apically on margin of staminal tube or on filaments; disk small, usually annular or patelliform, rarely 0. Fruit a capsule, berry or drupe. Seed usually arillate or with sarcotesta, mostly without endosperm; cotyledons usually collateral.

27. *Ekebergia* Sparrm.

Ekebergia Sparrm., Kongl. Vetensk. Akad. Handl. 40: 282, t. 9 (1779); White & Styles, Fl. Trop. East Africa, Meliaceae: 37–39 (1991).

Trees or shrubs, dioecious; hairs simple. Leaves imparipinnate. Inflorescences of contracted thyrses. Calyx (4)5(6)-lobed in distal half; petals (4)5 distinct, imbricate; staminal tube cup-shaped, with very short distinct filaments, appendages 0; anthers usually 10, inserted on rim; nectary disk in male flowers annular or patelliform, partly united with base of staminal tube and ovary, in females a small swelling at ovary base; ovary 2–5(6)-locular, each locule with 2 superposed ovules; style short and stout; stylehead capitate with 2–5 indistinct lobes. Fruit a drupe with 2–4(–6), 1(2)-seeded pyrenes. $2n = 46, 50$.

Four spp., tropical Africa.

28. *Trichilia* P. Br.

Trichilia P. Br., Civ. Hist. Nat. Jamaica: 278 (1756), nom. cons.; de Wilde, Meded. Landbouw. Wageningen 68-2(1968), reg. rev.; T.D. Penn., Fl. Neotrop. 28: 25–233 (1981), reg. rev.
Pseudobersama Verdc. (1956).
Burseranthe Rizz. (1974).

Trees and treelets, usually dioecious, rarely polygamous. Buds usually naked, rarely subtended by a cluster of small scale-leaves; hairs simple, less often malpighiaceous, forked or stellate hairs, or with peltate scales. Leaves pinnate, less often unifoliolate, trifoliolate or rarely digitate; leaflets sometimes glandular-punctate and -striate. Infloresences thyrses, rarely fasciculate or corymbose or few-flowered racemes. Calyx usually (3)4–6-lobed, rarely sepals distinct, aestivation usually open; petals (3)4 or 5(6), distinct or partially united, imbricate, rarely quincuncial; staminal tube cyathiform, urceolate or shortly cylindrical, margin usually toothed or lobed, rarely stamens distinct; anthers (4)5–10 (–12) inserted between teeth or lobes or apical on distinct filaments; nectary disk usually a fleshy annulus around ovary base; ovary 2- or 3(–5)-locular, locules with 1 or 2 collateral or less often superposed ovules; stylehead usually capitate, with or without lobes. Fruit a 2- or 3(4 or 5 [*Pseudobersama*])-valved loculicidal capsule with leathery or woody valves and 1 or 2 seeds per locule. Seed ± plano-convex, with partial to complete raphe-funicular aril or sarcotesta (pachychalaza), usually without endosperm; embryo usually with plano-convex collateral cotyledons; radicle superior, usually included. $2n = 24, 28, 46, 50, 92$, c. 360.

About 70 spp. in mainly lowland tropical America, 18 in Africa, 6 in Madagascar. Frequent understorey trees in Amazonia.

Pseudobersama is included here with a little hesitation: it has five-lobed fruits but otherwise seems indistinct (cf. Cheek 1992). Asiatic species recently included are now referred back to *Heynea*.

29. *Pterorhachis* Harms

Pterorhachis Harms in Engl., Bot. Jahrb. Syst. 22: 155 (1895); T.D. Penn., Blumea 22: 471–472 (1975).

Small tree, dioecious; hairs stellate. Leaves imparipinnate or trifoliolate, rachis winged. Flowers in axillary thyrses. Calyx 5-lobed to near base, aestivation ± imbricate; petals 5, distinct, imbricate; staminal tube urceolate to cylindrical, with distinct filaments as long as it, each terminated by (3)4–6(10) unequal filiform appendages (fewest in females), the outer pair longer than the rest; anthers (9)10 inserted just below appendages, connective aciculate; nectary disk a swollen annulus; ovary 3(4)-locular, each locule with 1 or 2 superposed ± orthotropous ovules; stylehead capitate with 3(4) minute obtuse lobes. Capsule loculicidal, 3(4)-valved, each locule with 1 or 2 seeds. Seed plano-convex with flattened base; testa leathery with aril surrounding basal part; endosperm 0; embryo with plano-convex collateral cotyledons, radicle superior. $n = 14$.

One (*P. zenkeri* Harms) or possibly two spp., Cameroon and Gabon.

30. *Malleastrum* (Baill.) J.-F. Leroy

Malleastrum (Baill.) J.-F. Leroy, J. Agric. Trop. Bot. Appl. 11: 128–149 (1964); Leroy & Lescot, Bull. Mus. Natl. Hist. Nat., Paris IV, 18: 13–20 (1996).
Cipadessa Bl. sect. *Malleastrum* Baill. (1874).

Trees, shrubs or shrublets; hairs simple. Leaves imparipinnate to trifoliolate or unifoliolate; rachis sometimes winged. Flowers bisexual or unisexual, in small axillary thyrses; calyx 4- or 5-lobed or -dentate; petals (3) 5 or 6, distinct, valvate; staminal tube entire, cyathiform to shortly cylindrical, to filaments almost distinct; appendices (8–)10, narrowly lanceolate or filiform, ± bifid; anthers (8–)10, alternating with appendices; nectary disk annular or 0; ovary of 1–3(–5)-locular, each locule with 2 superposed or slightly oblique ovules; stylehead capitate with small stigma with apical depression. Fruit a berry, often asymmetric with 1–3 (5) locules and thin pericarp. Seeds 1 per locule, oblong or spheroid, exarillate, pachychalaza thin, membranous, sometimes thicker and hard; endosperm 0; embryo with plano-convex cotyledons, collateral to oblique; radicle superior, usually not exserted.

Twenty-three spp., Madagasacar. Some locally medicinal.

31. *Owenia* F. Muell.

Owenia F. Muell. in Hook., Kew J. Bot. 9: 303 (1857); T.D. Penn., Blumea 22: 478–479 (1975); Mabb., Telopea 8: 47–48 (1998).

Deciduous trees with resinous buds and often milky sap, dioecious; hairs simple. Leaves paripinnate. Flowers in axillary thyrses; calyx of 5 distinct imbricate sepals; petals 5, distinct, imbricate; staminal tube irregularly lobed or with distally distinct filaments, each with 2 apical appendages; anthers 10, glabrous; nectary disk obscure, annular; ovary 2–4-locular, each locule 1-ovulate; stylehead conical. Drupe 2–4-locular; mesocarp fleshy to leathery; endocarp woody, hard. Seed ovoid to orange-segment-shaped, with thin sarcotesta. $2n = 28$.

Five spp., Australia, particularly drier regions.

32. *Nymania* Lindb.

Nymania Lindb., Not. Sällsk. Fauna Fl. Fenn. Förh. 9: 290 (1868, 'Trans. Nov. Scand.'); White, Bothalia 16: 146–148 (1986).
Aitonia Thunb. (1781), non *Aytonia* Forst. & Forst.f. (1776).

Shrub; hairs simple. Leaves simple, in fascicles on short lateral shoots. Flowers bisexual, axillary, solitary; calyx 4-lobed to near base; petals 4, distinct, imbricate; filaments 8 or 9, curved, with basal tube, anthers glabrous, versatile; nectary disk thin, partly united with staminal tube; ovary 4(5)-locular, each locule with 2 collateral campylotropous ovules; style long with minute capitate stylehead. Capsule loculicidal, thinly membranous, each locule with 1 (or 2) seeds. Seed reniform, with fleshy sarcotesta around hilum. $2n = 40$.

One sp., *N. capensis* (Thunb.) Lindb., Namibia and South Africa.

The inflated pink to purple capsular fruits are blown about and break up, thereby scattering the seeds. The placement in 'Turraeeae' is confirmed by the presence of the limonoid nymania-1, typical of *N. capensis*, in *Turraea obtusifolia* Hochst. (Fraser et al. 1995).

33. *Naregamia* Wight & Arn.

Naregamia Wight & Arn., Prodr. Fl. Pen. Ind. Or. 1: 116 (1834), nom. cons.; Harms in Engl. & Prantl, Nat. Pflanzenfam., ed. 2, 19bI: 90–91 (1940); T.D. Penn., Blumea 22: 454–455 (1975).
Nelanaregam Adans. (1763), nom. rejic.

Small shrublets, sometimes geoxylic; hairs simple. Leaves trifoliolate, leaflets entire to weakly lobed, rachis broadly winged. Flowers bisexual, axillary, solitary or paired; calyx 4- or 5-lobed to middle; petals 4 or 5, distinct, imbricate; staminal tube narrowly cylindrical, margin entire to weakly crenulated; anthers 8 or 10 with connective produced apically to form an appendage up to as long as anther; nectary disk absent; ovary 3- or 4-locular, each locule with 2 collateral ovules; style slender; stylehead discoid to obconical. Fruit a 3- or 4-valved loculicidal capsule, each locule with 1 or 2 seeds. Seed curved, truncate at both ends; testa bony, thick; aril small. $2n = 46$.

Two spp., one on west coast of India, one in Angola.

Cheek (1996) combined this with *Turraea*, but molecular analyses (Muellner et al. 2003, 2006) support earlier authors' view as to its distinctness.

34. *Turraea* L.

Turraea L., Mant. Alt.: 150 (1771); Harms in Engl. & Prantl, Nat. Pflanzenfam., ed. 2, 19b1: 85–90 (1940); Mabb., Fl. Males. I, 12: 24–27 (1995), reg. rev.; Mabb. & Cheek, Taxon 41: 541–545 (1992), typification.

Small trees or shrubs; hairs simple. Leaves simple entire, sometimes sinuate or crenate (*T. pulchella* (Harms) T.D. Penn. (South Africa) with imparipinnate or trifoliolate leaves). Flowers bisexual, in axillary or apparently terminal, fascicles or thyrses, or solitary; calyx cupuliform, (3) 4 or 5 (6)-lobed or -toothed; petals (3) 4 or 5, distinct, imbricate or contorted; staminal tube cylindrical or rarely cyathiform, complete or filaments less than half as long as basal tube; margin shallowly lobed or entire or terminated by simple or bilobed appendages, as many as or twice as many as the anthers; anthers (7–)10(20); nectary disk small or 0; ovary (3)4–10(20)-locular; loculi biovulate; style usually expanded near apex forming a *receptaculum pollinis* surmounted by a discoid stigmatic area. Fruit a (3)4–10(–20)-valved, loculicidal capsule; loculi 1- or 2-seeded. Seed plano-convex, raphe-funicular-arillate. $2n = 36, 46, 50$.

About 60 spp., in Africa, Madagascar (31), the Mascarenes (8) and one in Indomalesia, extending from India, through Malesia to tropical Australia. The African species are found in a variety of forest types and bush vegetation, often markedly seasonal.

The formerly enigmatic *Turraea breviflora* Ridl., found in a few localities in the Malay Peninsula, is now referred to *Munronia* (Muellner and Mabberley 2008).

35. *Humbertioturraea* J.-F. Leroy

Humbertioturraea J.-F. Leroy, Compt. Rend. Hebd. Séances Acad. Sci. D, 269: 2311, figs D, H, P, S & 2322 (1969).

Small trees, shrubs or shrublets; hairs simple. Leaves simple, entire. Flowers bisexual in small axillary fascicles, sometimes 1- or 2-flowered. Calyx cupuliform, (4)5-lobed; petals (4)5, distinct, imbricate; staminal tube cylindrical, margin subentire, appendices alternating with anthers, emarginate to bifid; anthers 10–12, sessile or shortly stalked, connective developed into apical mucro; nectary disk small, ± united with base of tube, margin often lobed; ovary 10–14-locular, each locule with 2 superposed ovules; stylehead enlarged into a *recaptaculum pollinis*, with apical stigmatic surface. Fruits indehiscent, 10–14-locular. Seeds plano-convex, with thick woody, glabrous testa; sarcotesta adaxial, hilum large.

Ten spp., Madagascar.

36. *Calodecaryia* J.-F. Leroy

Calodecaryia J.-F. Leroy, J. Agric. Trop. Bot. Appl. 7: 379–382, Fig. 1–4 (1960).

Small trees or shrubs; hairs simple. Leaves simple, entire. Flowers bisexual, in small axillary thyrses or fascicles; calyx 4- or 5-lobed to middle; petals 4 or 5, distinct, contorted; stamens 8–10, united only at base or almost distinct; anthers basifixed; nectary disk annular to patelliform; ovary 4- or 5-locular, each locule with 2 collateral campylotropous ovules; style straight; stylehead capitate to discoid, usually with 4 or 5 small stigmatic lobes. Fruit a 2–5-valved loculicidal capsule, each locule with 1 or 2 plano-convex to reniform seeds; testa thick, tough, sarcotesta surrounding hilum. $2n = 36$.

Two spp., Madagascar.

37. *Anthocarapa* Pierre

Anthocarapa Pierre, Fl. For. Cochinch. 5: ad t. 343 (1897); Mabb., Fl. Nouv.-Caléd. 15: 70–74 (1988) & Fl. Males. 12: 133–136 (1995), reg. revs.

Trees; hairs simple. Leaves paripinnate. Flowers unisexual (trees dioecious or monoecious), in axillary or supra-axillary thyrses. Calyx 4- or 5-lobed; petals (4)5, distinct, imbricate at least at apices; staminal tube ± urceolate, margin crenulate; anthers 10, inserted within tube; nectary disk obscure, thick, fleshy, annular or patelliform in male flowers, small, annular at the base of ovary in females; ovary (2)3(4)-locular, locules uniovulate, placentation axile; stylehead discoid. Fruit a 2- or 3(4)-valved capsule, (tardily) dehiscent. Seed with non-vascularized sarcotesta; embryo with thick collateral cotyledons; radicle superior, included.

One (*A. nitidula* (Benth.) Mabb.) or possibly two spp., from Philippines to New Caledonia.

38. *Synoum* A. Juss.

Synoum A. Juss., Bull. Sci. Nat. Géol. 23: 237 (1830); T.D. Penn., Blumea 22: 499–500 (1975); Mabb., Telopea 8: 48 (1998).

Tree; hairs simple. Leaves imparipinnate. Flowers unisexual, in axillary thyrses. Calyx deeply 4- or 5-lobed, extended into pseudopedicel; petals 4 or 5, distinct, imbricate; staminal tube cyathiform, margin irregularly and shallowly lobed; anthers

8 or 10, partly exserted; nectary disk obscure. Capsule loculicidal, 2- or 3-valved, locules usually 2-seeded. Seeds collateral, pendulous, epitropous, united by joint raphe-arils partly enveloping them; embryo with thick plano-convex, collateral cotyledons, radicle superior. $2n = 84$.

One sp., *S. glandulosum* (Sm.) A. Juss., NE Australian rain forests.

39. *Chisocheton* Blume Fig. 38

Chisocheton Blume, Bijdr.: 168 (1825); Mabb., Bull. Br. Mus. Nat. Hist. Bot. 6: 301–386 (1979) & Gdns' Bull. Sing. 55: 189–195 (2003).

Megaphyllaea Hemsl. (1887).

Trees pachycaul to leptocaul, sometimes unbranched, sometimes laticiferous or myrmecophilous, very rarely foetid, dioecious or polygamous; indumentum usually of simple, rarely of 4-stellate, hairs, sometimes irritant, with small glandular hairs. Leaves pinnate and pseudogemmulate or imparipinnate, very rarely paripinnate. Inflorescence paniculate to thyrsoid or with long peduncle and congested racemose, axillary to supra-axillary, ramiflorous or rarely borne on congested cauliflorous branches, or epiphyllous (New Guinea). Flowers sometimes with elongated receptacle (pseudopedicel); calyx ± cupuliform, usually obscurely 3–6-lobed; petals (3)4–6(–14) in 1(2) whorls, distinct, imbricate, quincuncial or alternative, often merely at apices, or valvate, rarely weakly united below or with base of staminal tube; staminal tube cylindrical, margin entire, crenate or with 4–10(–30) emarginate, truncate or narrowly lanceolate 2- or 3-fid lobes; anthers (3)4–10(–30), usually attached within the tube, alternating with lobes, usually locellate; nectary disk usually 0, less often stipitate, annulate or patelliform, occasionally lobed; ovary 2–8-locular, each locule with 1 or 2 collateral or superposed orthotropus ovules; stylehead clavate or discoid. Fruit a 2–5(–8)-valved capsule, the valves 1(2)-seeded. Seeds obovoid-spheroid to scutelliform or orange-segment-shaped, variously arillate or sarcotestal, orthotropous, with large chalaza; cotyledons collateral, oblique or superposed. $2n = 46, 92$.

Fifty-three spp., from Assam and tropical China throughout Malesia SE to N New South Wales and Vanuatu.

Fig. 38. Meliaceae. *Chisocheton macrophyllus.* **A** Leaf and inflorescence. **B** Young leaves. **C** Leaflet. **D** Flower, vertical section. **E** Fruit. (Mabberley et al. 1995; drawn by Rosemary Wise)

40. *Dysoxylum* Blume

Dysoxylum Blume, Bidjr.: 172 (1825); Harms in Engl. & Prantl, Nat. Pflanzenfam., ed. 2, 19b1: 160–166 (1940); Mabb., Fl. Nouv.-Caléd. 15: 23–70 (1988), Fl. Ceylon 9: 241–248 (1995), Fl. Males. 12: 61–133 (1995), Telopea 10: 725–729 (2004), Tree Fl. Sabah & Sarawak 6: 146–176 (2007) & Fl. China 11: 125–129 (2008).

Piptosaccos Turcz. (1858).

Pseudocarapa Hemsl. (1884).

Meliadelpha Radlk. (1890).

Trees or shrubs, often very pachycaul, dioecious, more rarely with bisexual flowers; hairs simple, very rarely stellate. Leaves in spirals, rarely opposite, pinnate, occasionally with tardily developed apical leaflets, very rarely with winged rachis. Inflorescences thyrses to racemose or spicate, sometimes reduced to fascicles or solitary flowers, axillary, ramiflorous or cauliflorous, sometimes with conspicuous bracteoles and these sometimes transitional to distinct sepals. Calyx of distinct sepals or, more usually, a 3–5(6)-lobed tube; petals 3–6, distinct or adnate to base of staminal tube; staminal tube cylindrical to urceolate, margin entire, lobed or tipped with 6–10(–13) appendages; anthers 6–16, within throat of tube, pollen very rarely shed in tetrads; nectary disk free, tubular though sometimes short or even subannular, margin sometimes lobed; ovary 2–6-locular, each locule with 1 or 2 seeds; stylehead capitate to discoid. Fruit a 2–6-valved capsule, each valve with 1 or 2 seeds. Seeds anatropous, very variable, usually with aril or sarcotesta.

About 80 spp., from India and Sri Lanka (3) to S China (11), Indochina, throughout Malesia (including Christmas Island, 1) to the Pacific S to Australia (14), New Caledonia (9), Norfolk Is. (1), Lord Howe Is. (1, endemic), New Zealand (1, endemic) and E to Niue (1), the greatest distribution of any Indopacific genus in the family with high levels of endemism in New Guinea (16 of 28; cf. Borneo with 2 of 22, Malay Peninsula with 2 of 17), Fiji (7 of 9) and New Caledonia (8 of 9), demonstrating a distinctly austral richness by comparison with its ally *Chisocheton* and the other large genus in Malesia, *Aglaia*.

41. *Heckeldora* Pierre.

Heckeldora Pierre, Bull. Mens. Soc. Linn. Paris 2: 1286 (1897); de Wilde, Blumea 52: 179–199 (2007), rev.
Guarea F. Allam. sect. *Heckeldora* (Pierre) Harms (1940).

Small trees, dioecious; hairs simple. Leaves imparipinate. Flowers in narrow axillary thyrses or racemes; calyx shallowly 3–5-lobed; petals 4, distinct, imbricate; staminal tube cylindrical, margin with shallow rounded lobes; anthers 8, inserted within tube and sometimes partly exserted; nectary disk long, slender, at apex forming collar beneath constricted ovary base; ovary 1-locular with 2(3) parietal placentas, each with 2 collateral orthotropous ovules; stylehead discoid, flat or with slight central depression. Fruit a narrow rostrate 2–4-seeded berry, with constrictions between seeds. Seed with unitegmic vascular sarcotesta (pachychalazal); embryo with thick plano-convex collateral cotyledons, radicle superior, extending to surface.

Six closely allied spp., tropical W Africa.

42. *Turraeanthus* Baill.

Turraeanthus Baill., Adansonia 11: 261 (1874); T.D. Penn., Blumea 22: 493–494 (1975).

Trees or treelets, dioecious; hairs simple. Leaves impari- or paripinnate. Calyx almost entire; petals 4 or 5(6), valvate, united to half way or more with the staminal tube; staminal tube cylindrical, margin crenate or shallowly lobed; anthers 8–12 in a single whorl within the throat; nectary disk 0; ovary 4- or 5-locular, each locule with 2 superposed or oblique ovules; stylehead discoid with a central depression. Fruit a leathery 3–5-valved loculicidal capsule. Seeds with sarcotestal seed-coat (pachychalazal?); embryo with thick plano-convex, superposed cotyledons, radicle included. $2n =$ c. 280.

Three spp., tropical W Africa. Wood used in cabinet-making.

43. *Ruagea* H. Karst.

Ruagea H. Karst., Fl. Columb. 2: 51, t. 126 (1863); T.D. Penn., Fl. Neotrop. 28: 242–255 (1981), rev.
Guarea sect. *Ruagea* (H. Karst.) C. DC. (1878).

Trees and treelets, dioecious; bud-scales rare. Leaves pinnate, sometimes with limited apical growth; leaflets not glandular-punctate or -striate. Inflorescences paniculate, axillary or in axils of fallen bud-scales. Sepals 5, distinct, quincuncial; petals 5, distinct, quincuncial; staminal tube cylindrical, cyathiform or urceolate, margin shallowly lobed; anthers (7–)10(11), inserted within throat; nectary disk usually short, rarely 0; ovary (2)3(4)-locular, each locule with (1)2 superposed ovules; stylehead discoid. Fruit a 2- or 3-valved loculicidal capsule, each locule with 1(2) superposed seeds; endocarp thin, cartilaginous. Seed with fleshy raphe-chalazal aril; embryo with thick plan-convex, collateral cotyledons; radicle apical.

Five poorly defined spp. of montane rain forest and cloud forest from Guatemala, Costa Rica, Panama and south to Peru.

44. *Guarea* F. Allam.

Guarea F. Allam. in L., Mant. Pl.: 150 (1771), nom. cons.; Pennington, Fl. Neotrop. 28: 255–359 (1981).
Samyda L. (1753), nom. rejic.
Leplaea Vermoesen (1921).
Urbanoguarea Harms (1937).

Trees and treelets, dioecious. Bud-scales 0; hairs simple. Leaves usually pinnate with terminal pseudogemmula, very rarely unifoliolate. Leaflets sometimes glandular-punctate and -striate. Inflorescence an axillary to cauliflorous panicle, raceme or spike. Calyx with almost entire margin to 3–7-lobed, aestivation open; petals (3)4–6(7), distinct, usually valvate; staminal tube cylindrical, sometimes contracted at throat, margin entire to lobed; anthers (7)8–12(–14) inserted within throat, alternate with lobes; nectary disk short to long-stipitate usually forming a collar around ovary base; ovary 2–10(–14)-locular, each locule with 1 or 2 superposed anatropous to orthotropous ovules; stylehead discoid. Fruit a 2–10(–14)-valved loculicidal capsule, each locule with 1 or 2 seeds; pericarp leathery or woody, endocarp thin, cartilaginous. Seed often orange-segment-shaped with thin fleshy sarcotesta; endosperm 0; embryo with thick plano-convex usually superposed cotyledons, radicle abaxial. $2n = 72$.

About 40 spp., tropical America and five in tropical Africa. Some promising timbers.

45. *Cabralea* A. Juss.

Cabralea A. Juss., Bull. Sci. Nat. Géol. 23: 237 (1830); T.D. Penn., Fl. Neotrop. 28: 234–242 (1981).

Tree or treelet; hairs simple. Leaves usually pinnate; leaflets glandular-punctate and -striate. Inflorescences of panicles, axillary, rarely borne on branches or trunk. Flowers usually bisexual; calyx of 5 ± distinct usually imbricate or quincuncial sepals; petals 5, distinct; staminal tube cylindrical, margin with (9)10(–12) appendages; anthers alternating with appendages, within throat; nectary disk cyathiform; ovary (4)5-locular, semi-inferior, each locule with 2 superposed ovules; stylehead discoid. Fruit a 4- or 5-valved loculicidal capsule, each locule with 1 or 2 seeds, held together by entwined funicular-raphe arils; embryo with thick plano-convex cotyledons; radicle apical.

One variable sp., *C. canjerana* (Vell.) Mart., from Costa Rica through tropical South America to northern Argentina.

46. *Aphanamixis* Blume

Aphanamixis Blume, Bijdr.: 165 (1825); Mabb., Blumea 31: 136–140 (1985), rev. & Fl. Males. 12: 187–194 (1995). ? *Ricinocarpodendron* Boehm. (1760).

Trees or pachycaul treelets with cicatrose twigs; hairs simple, rarely basally bifid and stellate. Leaves imparipinnate, leaflets opposite. Inflorescences axillary to supra-axillary, male flowers (smallest) in panicles, female and bisexual in long spikes or racemes, rarely panicles. Calyx deeply 5-lobed, lobes imbricate; petals 3, imbricate, united with staminal tube basally; staminal tube globose to deeply cyathiform; anthers 3–8, glabrous, inserted within tube; ovary 3(4)-locular, each locule with (1)2 collateral to superposed ovules; style stout; stylehead conical to truncate, 3-angled or with impressions of anthers. Fruit a 2- or 3(4)-valved, loculicidal capsule, loculi 1- or 2-seeded. Seeds arillate, cotyledons plano-convex, collateral (?) united; radicle small, superior, included. $2n = 36, 76$, c. 150.

Three very closely related spp. in Indomalesia from Ceylon and India to Bhutan, tropical China and Indochina, throughout Malesia, to the Solomon Is. Local timber, oil for soap-making and medicaments; leaf extracts are effective anti-feedants.

Except for the apparent unity of the cotyledons (a feature not investigated in all *Aglaia* spp. so far), there is no other macroscopic character which separates *Aphanamixis* from *Aglaia* absolutely. At the microscopic level, only one *Aglaia* sp. has the 4-colporate pollen grains found in *Aphanamixis* and the wood of the latter differs from that of the *Aglaia* species formerly included in *Amoora* in having confluent and banded paratracheal parenchyma (Pennington in Pennington and Styles 1975).

47. *Sphaerosacme* Royle

Sphaerosacme Wall. ex Royle, Ill. Bot. Himal. Mts: 142 (1835); Pennington, Blumea 22: 488–489 (1975); Grierson [& Mabb.], Fl. Bhutan 2: 35–36 (1991).

Tree, dioecious; hairs simple. Leaves imparipinnate. Flowers in axillary to supra-axillary panicles with racemose branches. Calyx 4- or 5-lobed, lobes imbricate to open; petals 5, distinct, imbricate; staminal tube almost globose, margin undulate; anthers 10, in 2 alternate whorls of 5, the upper partly exserted; ovary 5-locular, each locule with 1 ovule; style 0, stylehead pileate, 5-lobed. Fruit a 2–5-lobed loculicidal capsule. Seed partly or completely enclosed in aril;

embryo with plano-convex, collateral, united cotyledons; radicle small, superior, included.

One sp., *S. decandra* (Wall.) T.D. Penn., Himalaya.

48. *Reinwardtiodendron* Koord.

Reinwardtiodendron Koord., Meded. s'Lands Plantentuin 19: 389 (1898); Mabb., Blumea 31: 144–146 (1985), rev. & Fl. Males. 12: 322–328 (1995); Peng Hua & Mabb., Fl. China 11: 124–125 (2008).
Aglaia Lour. sect. *Lansium* (Corrêa) Kosterm. (1966), p.p.

Trees; hairs simple. Leaves pinnate or unifoliolate, the leaflets alternate, the most apical on one side appearing terminal; petiolules often swollen; domatia frequently present. Inflorescences spikes or basally branched panicles of spikes. Flowers bisexual, yellow; calyx deeply 5-lobed, the lobes orbicular, imbricate; petals 5, distinct from each other but united with staminal tube at base; staminal tube globose to ovoid, with an undulate to toothed margin; anthers 10 in 2 whorls of five, glabrous, the upper ones partly exserted, the lower alternating with the upper and completely included, their connectives extended to form a short acute appendage; ovary 5-locular, each locule with one ovule; style very short, with a small capitate or pileate, obscurely lobed apex. Fruit a 1–5-seeded berry. Seed apparently sarcotestal; embryo with thick plano-convex, superposed, free cotyledons, radicle included.

Probably seven spp., restricted to Indomalesia, one, *R. anaimalaiense* (Bedd.) Mabb., in the western Ghats of India, the rest occurring in Malesia, with one extending to southern China.

Reinwardtiodendron closely resembles *Aglaia* in pollen and secondary xylem as well as overall facies but it differs in its simple indumentum, the two whorls of anthers with appendages and the 5-locular ovary. It is closest to *Lansium* which shares the indumentum and the leaflet form and has similar venation but differs in its single whorl of 10 stamens without appendages, the ramiflorous to cauliflorous inflorescences and the seeds which are arillate and pachychalazal.

49. *Lansium* Corrêa

Lansium Corrêa, Ann. Mus. Hist. Nat. Paris 10: 157 (1807); Mabb., Blumea 31: 140–143 (1985), rev. & Fl. Males. 12: 314–322 (1995).
Aglaia Lour. sect. *Lansium* (Corrêa) Kosterm. (1966).

Trees; hairs simple. Leaves pinnate, the leaflets subopposite to alternate, to most apical on one side appearing terminal; petiolules pulvinate at base. Inflorescences spikes, racemes or more rarely basally branched panicles with spicate or racemose branches, borne on twigs, branches or bole. Flowers unisexual (tree dioecious) and bisexual, these larger than male ones; calyx deeply 5-lobed, the lobes imbricate; petals 5, free from each other but united with staminal tube in proximal third to half, imbricate; staminal tube globose to cyathiform, margin ± undulate; anthers (8)10 in one whorl inside the throat of the tube, their tips not or slightly exserted, without appendages; ovary 3–5-locular, each locule with one ovule; style long and broad-columnar, its flanks ribbed with the impressions of the surrounding anthers. Fruit a 1–5-seeded berry. Seed usually arillate, aril completely enveloping seed; embryo with thick plano-convex, superposed free cotyledons, radicle included. $2n = 144$.

Three spp., the genus possibly being the only one restricted to Malesia, but planting of the important local fruit tree *L. domesticum* Corrêa (langsat) elsewhere, which may be native in southern Thailand in any case, has obscured this.

50. *Aglaia* Lour. Fig. 39

Aglaia Lour., Fl. Cochinch.: 173 (1790), nom. cons.; Pannell, Kew Bull. Add Ser. 16 (1992), rev.
Amoora Roxb. (1820).

Trees, dioecious. Indumentum of stellate hairs or scales or of peltate scales. Leaves imparipinnate. Flowers (males smaller than females) in axillary thyrses, male thyrses larger than female. Calyx (2)3–5-lobed, aestivation open or imbricate; petals usually 3 or 5; anthers sessile, inserted within staminal tube; ovary with (1)2 or 3 locules; stigma sessile, either ovoid with 2 or 3 apical lobes or depressed-globose. Fruits dehiscent with 3 locules or indehiscent with with 1 or 2 locules. Seeds 0 or 1 per locule; aril completely surrounding rest of seed to vestigial. $2n=40, 92$.

About 120 spp., Indomalesia to Australia and E to Samoa. As currently understood (with *Amoora*, etc. included), paraphyletic with regard to *Lansium* and *Reinwardtiodendron* (Muellner et al. 2008b). Some locally important fruit trees and timbers.

Fig. 39. Meliaceae. *Aglaia silvestris*. **A** Leaf and male inflorescence. **B** Infructescence with young fruits. **C** Male flower longitudinally sectioned. **D** Mature fruits. **E** Peltate scale. (Mabberley et al. 1995; drawn by Rosemary Wise)

Selected Bibliography

Braggins, J.E., Large, M.F., Mabberley, D.J. 1999. Sexual arrangements in kohekohe (*Dysoxylum spectabile*, Meliaceae). Telopea 8: 315–324.

Champagne, D.E., Koul, O., Idsman, M., Scudder, G.E., Towers, G.H.N. 1992. Biological activity of the limonoids from the *Rutales*. Phytochemistry 31: 3877–3894.

Cheek, M.R. 1989. The systematic seed anatomy of the Meliaceae. D. Phil. Thesis, Bodleian Library, Oxford.

Cheek, M.R. 1992. The wood anatomy of *Pseudobersama mossambicensis* and *Trichilia capitata* (Meliaceae) compared. Kew Bull. 47: 753–758.

Cheek, M.R. 1996. The identity of *Naregamia* Wight & Arn. (*Meliaceae*). Kew Bull. 51: 716.

Cheng, Y.-M., Ferguson, D.K., Li, C.-S., Jiang, X.-M., Wang, Y.-F. 2006. *Cedreloxylon cristalliferum*, a new record of angiosperm wood of Pliocene age from Yunnan, China. IAWA J. 27: 145–152.

Fisher, J. B., Rutishauser, R. 1990. Leaves and epiphyllous shoots in *Chisocheton* (Meliaceae): a continuum of woody leaf and stem axes. Can. J. Bot. 68: 2316–2328.

Fraser, L.-A, Mulholland, D.A., Taylor, D.A.H. 1995. The chemotaxonomic significance of the limonoid, nymania-1, in *Turraea obtusifolia*. S. Afr. J. Bot. 61: 281–282.

Gregory, M., Poole, I., Wheeler, E.A. 2009. Meliaceae. IAWA J., Suppl. 6: 99–101.

Hallé, F., Oldeman, R.A.A. 1970. Essai sur l'architecture et la dynamique de croissance des arbres tropicaux. Paris: Masson.

Harms, H. 1940. Meliaceae. In: Engler, A., Prantl, K., Die natürlichen Pflanzenfamilien, ed. 2, 19b I: 1–172. Leipzig: W. Engelmann.

Khosla, P.K., Styles, B.T. 1975. Karyological studies and chromosomal evolution in Meliaceae. Silvae Genet. 24: 75–83.

Kribs, D.A. 1930. Comparative anatomy of the woods of the Meliaceae. Am. J. Bot. 17: 724–738.

Leroy, J.-F. 1976. Essais de taxonomic syncrétique 1. Etude sur les Meliaceae de Madagascar. Adansonia II, 16: 167–203.

Mabberley, D.J. 1979. The species of *Chisocheton* (Meliaceae). Bull. Br. Mus. Nat. Hist. Bot. 6: 301–386.

Mabberley, D.J. 1988. Meliaceae. In: Morat, P., MacKee, H.S. (eds.) Flore de la Nouvelle-Calédonie et Dépendances 15, pp. 17–89.

Mabberley, D.J. 1995. Meliaceae [Aglaia by C.M. Pannell]. In: Dassanayake, M.D. (ed.) Revised Handbook of the Flora of Ceylon 9, pp. 228–300.

Mabberley, D.J., Pannell, C.M., Sing, A.M. 1995. Meliaceae. Flora Malesiana 12: 1–407.

Muellner, A.N., Mabberley, D.J. 2008. Phylogenetic position and taxonomic disposition of *Turraea breviflora* (Meliaceae), a hitherto enigmatic species. Blumea 53: 607–616.

Muellner, A.N., Samuel, R., Johnson, S.A., Cheek, M., Pennington, T.D., Chase, M.W. 2003. Molecular phylogenetics of Meliaceae (Sapindales) based on nuclear and plastid DNA sequences. Am. J. Bot. 90: 471–480.

Muellner, A.N., Savolainen, V., Samuel, R., Chase, M.W. 2006. The mahogany family "out-of-Africa": divergence time estimation, global biogeographic patterns inferred from plastid *rbcL* DNA sequences, extant, and fossil distribution of diversity. Mol. Phylog. Evol. 40: 236–250.

Muellner, A.N., Samuel, R., Chase, M.W., Coleman, A., Stuessy, T.F. 2008a. An evaluation of tribes and generic relationships in Melioideae (Meliaceae) based on nuclear ITS ribosomal DNA. Taxon 57: 98–108.

Muellner, A.N., Pannell, C.M., Coleman, A., Chase, M.W. 2008b. The origin and evolution of Indomalesian, Australasian and Pacific island biotas: insights from Aglaieae (Meliaceae, Sapindales). J. Biogeogr. 35: 1769–1789.

Newton, A.C., Baker, P., Ramnarine, S., Mesén, J.F., Leakey, R.R.B. 1993. The mahogany shoot borer: prospects for control. For. Ecol. Manag. 57: 301–328.

Pannell, C.M. 1992. A taxonomic monograph of the genus *Aglaia* Lour. (Meliaceae). Kew Bull. Add. Ser. 16.

Pennington, T.D., Styles, B.T. 1975. A generic monograph of the Meliaceae. Blumea 22: 419–540.

Pennington, T.D., Styles, B.T. [Swietenioideae] 1981. Meliaceae. Flora Neotropica Monograph 28.
Scott, A.J. 1997. Méliacées. In: Bosser, J., Cadet, T., Gueho, J., Marais, W. (eds.) Flore des Mascareignes 69–79, pp. 1–17.
Smith, A.C. 1985. Meliaceae. Flora vitiensis nova 3: 527–578.
Styles, B.T., Khosla, P.K. 1976. Cytology and reproductive biology of Meliaceae. In: Burley, J., Styles, B.T. (eds.) Tropical trees: variation, breeding and conservation. London: Academic Press, pp. 61–67.
White, F., Styles, B.T. 1991. Meliaceae. Flora of Tropical East Africa. Rotterdam: Balkema.

Myrtaceae

Myrtaceae Juss., Gen.: 322 (1789) ('Myrti').

Peter G. Wilson

Trees or shrubs, frequently with conspicuous oil glands. Leaves opposite, alternate (disjunct-opposite or spiral), rarely whorled, simple, entire, pinnately veined (triplinerved in a few genera). Stipules absent or minute and inconspicuous. Indumentum absent or of simple, thick-walled, unicellular hairs; other variants, such as biramous (two-armed), stellate, infundibular or multicellular hairs, are less common and occur in a few genera only. Inflorescences mostly determinate, terminal or axillary, panicles, thyrsoids, metabotryoids, botryoids, dichasia, triads or single flowers, sometimes further condensed into conflorescences of various types. Flowers mostly bisexual, occasionally unisexual, predominantly actinomorphic, 4-5-merous (very rarely 6–12-merous); hypanthium present, fused in varying degrees to the ovary; sepals free, occasionally fused into a calyptra, imbricate, mostly persistent; petals imbricate, usually distinct and caducous, occasionally persistent, sometimes fused as part of a calyptra, sometimes cohering and falling as a unit at anthesis; stamens usually multiseriate, numerous and free, sometimes in clusters or fused into fascicles opposite the petals, rarely reduced to be equal to or less than the number of perianth segments; anthers dorsifixed and versatile, less commonly basifixed and not versatile; dehiscing by slits, rarely by pores or valves; ovary usually inferior to half-inferior (superior in Psiloxyloideae) with 1–5 carpels, rarely more; placentation parietal, axile or basal; style terminal or set into a pit on the ovary summit, rarely gynobasic; stigma small or capitate, rarely lobed; ovules 2–many, anatropous, hemitropous or campylotropous, integuments two, rarely one. Fruit a dry, loculicidal capsule, a dry indehiscent fruit, or a one- to many-seeded fleshy berry (rarely drupe-like). Seeds one to many; endosperm mostly absent; embryo straight, curved or coiled; cotyledons flat or variously folded, thin (occasionally broad and leafy) or thick and fleshy, the latter sometimes fused.

As understood here, the Myrtaceae include *Psiloxylon* and *Heteropyxis*, referred to the monogeneric families Psiloxylaceae and Heteropyxidaceae by some workers. The family has a predominantly tropical to southern temperate distribution but with relatively low representation in the African region. The family includes a total of around 142 genera and in excess of 5,500 species.

Vegetative Morphology. Myrtaceae are all woody plants. The majority of species are trees of closed or open forests or woodlands; trees in closed forest communities are commonly buttressed. Bark type is variable but one noteworthy type is smooth bark that is shed annually; this is apparently a plesiomorphic character-state that has a scattered occurrence throughout the family, in both capsular and non-capsular taxa, and is found in other members of the Myrtales (e.g. *Lagerstroemia*, Lythraceae). In addition to trees, there are also many taxa that are shrubs or subshrubs. There are a few species that produce flowers and fruits when only a few centimetres in height, including some low-growing to prostrate shrubs that sometimes produce adventitious roots. These are primarily restricted to shrub communities in arid regions, cool temperate areas and heathland. Of particular note are the tall, shrubby eucalypt species known as 'mallees' that have multiple stems arising from an underground stem sometimes called a 'lignotuber', the two southern African species of *Eugenia* and a number of Brazilian species from various genera found in Cerrado vegetation (*Psidium*, *Campomanesia* and *Eugenia*) that are rhizomatous subshrubs; these all appear to be adaptations to fire-prone habitats. The genus *Metrosideros* commonly has tree species that produce large numbers of adventitious roots, and certain of these have been recorded as beginning their lives as strangling epiphytes. Also, a number of species of *Metrosideros* sens. lat. from New Zealand and Papua New Guinea are root-climbing lianes. There is only a single mangrove in the family, the monotypic *Osbornia*, a shrub or small tree that lacks pneumatophores.

There is a rather strong correlation between leaf morphology and habitat. In rainforest species, the leaves are usually broad and mesophyllous, sometimes with acuminate tips; in drier forests and woodlands they are often broad and leathery (sclerophyllous); shrubby species of arid region or coastal heath communities often have smaller and narrower leaves, rarely subterete or scale-like. The leaves are always simple, the margins are mostly entire, rarely crenulate and only occasionally with non-vascularised, tooth-like projections of the leaf margin, and some members of the Chamelaucieae have an expanded and dissected cuticle that can make leaves appear denticulate or ciliate. Venation is predominantly pinnate (e.g. Fig. 40) but triplinerved leaves occur in some species of *Rhodomyrtus* and all species of *Rhodamnia*, and parallel venation occurs in a few *Melaleuca* (Fig. 43) species. In pinnately veined taxa, the venation is sometimes eucamptodromus but in most cases it is weakly to strongly brochidodromus, with a well-defined intramarginal vein forming in the latter. Where the leaves are much reduced, particularly in the Chamelaucieae, the midvein is often the only vein visible.

Leaves are predominantly opposite (rarely whorled), although significant numbers of species have alternate or spirally arranged leaves. Change in phyllotaxy between juvenile and adult plants is well known in eucalypts, which have opposite juvenile leaves (Fig. 52D) that may be petiolate, subsessile or sessile (sometimes perfoliate); in the adult plant, the leaves become alternate ('disjunct opposite') and their morphology often changes. Other genera, for example, *Melaleuca*, also show the change from opposite-leaved juveniles to alternate- or spiral-leaved adults. An uncommon condition has been recorded in some opposite-leaved species of *Xanthostemon* where the juvenile leaves are spirally arranged. Weberling (1966) has recorded rudimentary stipules in some Myrtaceae and has suggested that they are present in most species. Indeed, these can usually be found but some authors have used other terms, like enations and acicular hairs, to describe them. Although most species lack obvious stipules, small, distinctly paired stipules are sometimes present (for example, in *Calytrix*). Vegetative bud scales are not a common feature of Myrtaceae, but have characteristic form in *Plinia* (Barrie 2004) and are particularly distinctive in *Metrosideros* sens. str., which has sympodial growth, and its close relative *Carpolepis*, which is monopodial. *Lophostemon* and *Syncarpia* both have scaly buds associated with terminal, false whorls of leaves.

Hairs, when present, are usually simple; Briggs and Johnson (1979) call these simple hairs 'standard' Myrtaceous hairs. They are described as acute, relatively thick-walled and are unicellular with no basal cell; other hair-types occur less commonly. Equally or unequally armed biramous hairs are found in several South American genera, particularly in *Myrcugenia*, *Calyptranthes* and *Eugenia*, although these hairs are also unicellular. Infundibular (funnel-shaped) hairs occur in a group of neotropical *Eugenia* species. Multicellular hairs occur in at least one species of the usually glabrous genus *Syzygium* and in *Hypocalymma*, a few species of *Homalocalyx* and *Balaustion microphyllum*. Amongst the eucalypts, the genera *Angophora* and *Corymbia* often have conspicuous 'bristle-glands' that are basically emergent oil glands. Some other eucalypts may show groups of blunt, thin-walled, unicellular hairs that form pseudostellate, radiating clusters over glands in the leaf (Ladiges 1985); these are particularly conspicuous in juvenile plants. Two species of *Rhodamnia* have truly stellate hairs (Snow 2007).

Vegetative Anatomy. Leaves of Myrtaceae are commonly bifacial and hypostomatous but in some cases the leaves are isobilateral (e.g. *Eucalyptus*, *Osbornia* and *Melaleuca*) and amphistomatous. In some groups, particularly eucalypts, leaves are sometimes also oriented vertically by a 90° twist of the petiole. In amphistomatous taxa, the stomatal density is lower than in the hypostomatous ones. The anatomy of these isobilateral leaves reflects the external morphology in having layers of palisade mesophyll on both sides of the leaf. The vascular tissue of lateral veins in species with relatively soft, mesic leaves is embedded in the mesophyll and enclosed by a simple bundle-sheath but, in broad-leaved species from more arid, or seasonally dry, areas or that occur on nutrient-poor soils, a number of other anatomical features may occur, such as a hypodermis (e.g. in *Metrosideros* sens. str., *Xanthostemon* [from one to five cells deep], *Lophostemon* and *Psidium*) and/or columns of fibres extending from the vascular bundle to the adaxial surface. A conspicuous feature of most myrtaceous leaves is the presence of epithelium-lined secretory cavities, more commonly known as oil glands; these usually occur adjacent to both the upper and lower epidermises. Their presence has been considered one of the fundamental features of the family. These are present in virtually all genera, including *Heteropyxis*, although their density in leaf and other tissues varies markedly. Glands begin as small groups of cells that rapidly develop into a ball of cells; the cells in the centre of this ball separate to form a secretory structure that is, thus, schizogenous in origin. *Psiloxylon* has long been known to have glands in the leaves but these do not produce any essential oils; they were for a long time considered to be of the same type of gland that occurs in Bixaceae and some Flacourtiaceae.

Transverse sections of the midrib or petiole show the arcuate to U-shaped vascular trace, occasionally with the ends incurved, approaching a flattened-cylindrical shape. The vascular bundle shows bicollateral (intraxylary) phloem, a feature of widespread occurrence in the order Myrtales. Transverse sections of the petiole may also show other features that reflect stem anatomy; the most significant of these are oil ducts and oil glands in the pith. Oil ducts are a feature of a small number of genera and are found in a few basic configurations. There are two basic types: small ducts that occur both inside and outside the vascular strand and relatively large ducts that occur only within the vascular strand. The former arrangement has been recorded only for three genera, *Lophostemon*, *Syncarpia* and *Welchiodendron*, while the latter is found in some members of the eucalypt group (most species of *Corymbia* and one species of *Angophora*) as well as in the genera *Lindsayomyrtus*, *Kjellbergiodendron* and *Whiteodendron*, which are not closely related to eucalypts. Oil glands in the pith have been noted in parts of *Eucalyptus* (the *Eudesmia* and *Symphiomyrtus* groups), in *Xanthostemon*, and in *Ristantia* and its allies *Sphaerantia* and *Mitrantia*.

The anatomy of the wood of Myrtaceae has been well summarised by van Vliet and Baas (1985) and is presented in a shortened and updated form here. The family shows the combined occurrence of intraxylary phloem and vestured pitting in common with other members of the order Myrtales. A few genera of Myrtaceae sometimes show uncommon wood anatomical characters like scalariform perforation plates and helical wall thickenings in the vessels and/or the fibres. Schmid and Baas (1985) suggest that these features have commonly been retained in taxa in cool, mesic regions, an observation that has been made in other families, too. They also suggest that the sporadic occurrence of scalariform perforation plates indicates that this is the ancestral (plesiomorphic) condition for the family and probably for the Myrtales as well.

The wood is generally diffuse-porous with solitary vessels, although radial multiples of 2–3 vessels occur in the *Acmena-Syzygium* group and some genera in the eucalypt group. Vessel diameter is variable, (14–)30–200(–260) μm, with the narrower vessels occurring in shrubby members of the family and the wider ones most commonly in eucalypts. Intervessel pitting, when present, is generally alternate but opposite, scalariform pitting has been recorded in *Neomyrtus*. Perforation plates are usually simple but are rarely scalariform, as discussed above. Vessel-ray pitting and vessel-parenchyma pitting may be similar to intervessel pitting or may be elongated to scalariform or, in some genera in the *Eucalyptus* alliance, may be large and simple. A rare feature of the wood is included phloem, which has been recorded in a few species in the related taxa *Syzygium*, *Waterhousea* and *Anetholea* as well as in *Lindsayomyrtus*; in these species, it takes the form of arcs of phloem with associated fibres and parenchyma.

Fibres in the majority of genera, including *Heteropyxis*, are in the form of fibre-tracheids with conspicuously bordered pits, while libriform fibres with simple to minutely bordered pits are found in only a few genera. More rarely, septate fibres are found, notably in *Psiloxylon* where they are also crystalliferous. Parenchyma is usually apotracheal and diffuse, rarely paratracheal, and more rarely paratracheal and confluent. Rays are generally strongly heterogeneous in the derived, fleshy-fruited taxa and weakly heterogeneous in the plesiomorphic, capsular-fruited genera. Crystals, where present, are mostly small and rhomboidal, usually occurring in strands in the axial parenchyma. In the Myrtales, silica grains in the rays are almost entirely restricted to the Vochysiaceae and part of the Myrtaceae, with rare occurrences in the Combretaceae. Within Myrtaceae, silica grains are apparently restricted to two groups of capsular genera: those traditionally allied to *Metrosideros* and the 'bottlebrush' group of genera, e.g. *Melaleuca*, *Callistemon*, and *Calothamnus*.

Bark anatomy has been studied only for some taxa of Myrtaceae, predominantly the Australian capsular genera. Bark anatomy is very variable. Fibres may be solitary, or scattered in groups or large bundles, or even in tangential bands; they are absent altogether in *Tristania*. Fibre length is a character that has been used to characterise groups of genera, e.g. to distinguish *Syzygium* and its allies from *Eugenia* sens. str. Three basic types of phellem, suberised, sclerosed and parenchymatous, have been recorded in Myrtaceae and various combinations of these appear likely to be indicative of generic relationships. For example, the combination of all three types is common to the genera of the tribe Leptospermeae and alternating layers of suberised and parenchymatous phellem is characteristic of the tribe Melaleuceae. Crystals in the phloem parenchyma are of three different forms, druses, rhomboidal or elongated-pointed. The first two may occur in concentric layers of cells or be randomly distributed, but the third type is never grouped. Silica in the bark is found in the ray parenchyma cells, as in the wood, and the recorded occurrence of silica is similar to records from wood. Oil glands or ducts have been recorded in the bark of a few genera, oil glands in *Xanthostemon*, *Ristantia* and some groups within *Eucalyptus*, and oil ducts in *Syncarpia*.

Inflorescence Structure. The extensive analysis of Myrtaceous inflorescences by Briggs and Johnson (1979) rejects the typological approach of Troll and his disciples and proposes a more descriptive one based on organography, with some new terminology. These authors view the primitive inflorescence as one with opposite and decussate phyllotaxy that shows flexibility in the development, or otherwise, of a terminal flower, thus cutting across the traditional division between determinate and indeterminate flowering shoots. They also made the distinction between the unit inflorescence and the conflorescence, a secondary aggregation of unit inflorescences, with further levels of aggregation possible. Although the family appears to be primitively flexible, the majority of inflorescences are determinate. Briggs and Johnson recognise seven main inflorescence types (listed in decreasing complexity of branching): the panicle, thyrsoid, metabotryoid, botryoid, dichasium, triad and monad. Monads and triads occasionally have an additional node on the central axis; these are designated metaxymonads and metaxytriads and presumably represent reduced forms of metaxytriads and botryoids respectively. Unit inflorescences may show further reduction of some axes to produce umbel-like inflorescences, such as those found commonly in *Eucalyptus* (derived from dichasia) or in *Backhousia* (derived from a thyrsoid), or capitate inflorescences, as in *Choricarpia* (derived from a thyrsoid). This more precise usage has advantages over traditional terminology, which would include thyrsoids, metabotryoids and true panicles under a much broader concept of the panicle (Landrum and Kawasaki 1997). Inflorescences, particularly the fewer-flowered types, are generally axillary, with terminal panicles restricted to a few genera such as *Myrcia*, *Syzygium* and *Corymbia*. Cauliflory and ramiflory are relatively uncommon, but do occur in a few genera such as *Syzygium*, *Eugenia* and *Gossia*.

Traditional descriptive language likewise has not recognised the concept of the conflorescence in this family. Conflorescences are aggregations of unit inflorescences, usually monads or triads, often with some reduction or modification of the leaves. The bottlebrush group of genera (*Callistemon* and allies) have dense spike-like conflorescences made up of monad or triad units subtended by deciduous bracts; the vegetative growth of the shoot continues after flowering has begun. In the more derived parts of the *Baeckea* group of genera, aggregations of monads may form striking conflorescences: some species of *Darwinia* produce bell-like conflorescences at the shoot apices, with the flowers in the upper axils surrounded by an involucre of petaloid leaves at the next-lower nodes, the remarkable genus *Actinodium* has conflorescences resembling a daisy, and some species of *Verticordia*, *Chamelaucium* and *Pileanthus* have flowers with long pedicels that form corymb-like conflorescences at the shoot apex. Aggregations of monads also occur in some *Eugenia* species: lateral, determinate leafy shoots have become bracteate and, depending on the degree of reduction, may give the appearance of being a raceme, fascicle or glomerule (Landrum and Kawasaki 1997, their Fig. 12).

The other phenomenon that may occur in Myrtaceous inflorescences is what Briggs and Johnson call 'accessory branching', where additional, secondary axes occur at a node, above or below the primary ones; such axes are also described as 'superimposed' by Landrum and Kawasaki (1997). Examples of this are the vertical series of monads that occur in some species of *Myrceugenia*, the accessory dichasia within the dichasia of some species of *Lophostemon* and the superimposed inflorescences found in *Choricarpia*.

Flower Structure. Flowers are generally epigynous with a well-developed hypanthium that extends above the ovary summit; they are actinomorphic, rarely somewhat zygomorphic. Slight zygomorphy is the result of either compression of the hypanthium or unequal staminal bundle development. A compressed hypanthium occurs in some species of *Micromyrtus* but this affects mostly the lower part of the flower and the open flower appears ± actinomorphic. Unequal staminal fascicles are found in some species of *Calothamnus* where the claws are broad in the upper staminal fascicles and narrow in the lower ones, and in *Lamarchea* where the claws are of similar width but the upper fascicles are longer than the lower. In both these cases, the stamens are the most conspicuous part of the flower and, although free from each other, are presented as if they were a tube. This flower form is somewhat analogous to the situation in the genus *Octamyrtus* where the proliferation of petals, which are held erect to simulate a tube, appears to be an adaptation to bird-pollination. Another floral syndrome that is apparently pollination-related is found in the relatively unrelated genera *Uromyrtus* and *Ugni*, which have flowers borne on strongly recurved pedicels and their anthers clustered tightly around the style.

The perianth is 4- or 5-merous, with a relatively small number of species that have 3 or more than 5 segments in each whorl. The genus *Octamyrtus* has 4 sepals but the number of petals has increased to 6, 8 or 12 as part of the distinctive pollination syndrome in the genus. In the mangrove genus *Osbornia*, the 8 perianth segments are only barely differentiated into calyx and corolla, and are persistent. The most frequent modification of the perianth involves the formation of various kinds of opercular structures that may be derived from one or both perianth whorls and may be completely or incompletely closed. The best known forms are those found in the tribe Eucalypteae where mono-operculate (both sepaline and petaline) and bi-operculate taxa are common, although the apparently closed perianth in the genus *Stockwellia*, and possibly in *Eucalyptopsis* as well, is formed by the elongation of the hypanthium with a much-reduced perianth borne at its apex (Carr et al. 2002). All opercula in the Eucalypteae are circumscissile as they are in some other taxa, for example, some species of *Syzygium*, and representatives of the tribe Myrteae from all three traditional subtribes. A circumscissile hypanthium can also occur in *Myrciaria* and in both calyptrate and non-calyptrate species of the genus *Siphoneugena*, and is also a feature of the *Eucalyptopsis* group and some species of *Asteromyrtus*. A common variant of the calycine calyptra is the type that opens by irregular splitting or tearing; this is uncommon amongst capsular genera, where it occurs only in *Pleurocalyptus*, but is more common in the Myrteae (for example, species of the genera *Calycorectes*, *Psidium* and *Calyptranthes*, Fig. 50C). Petaline opercular structures can consist of the imbricate petals that cohere and fall as a unit; this type of operculum is recorded for *Metrosideros* sect. *Calyptropetala* and some *Syzygium* species. Another notable feature of perianth in the family is the usually heavily dissected perianth of *Verticordia* species, in which the sepals (and sometimes also the petals) are divided into fimbriate or plumose segments giving the flowers a feathery appearance.

Flowers are usually bisexual but occasional variations in sexuality have been recorded. *Psiloxylon* is dioecious and has unisexual flowers:

the male with a pistillode, with the female lacking staminodes (Fig. 40; Schmid 1980). Flowers of *Heteropyxis* are also functionally unisexual, but appear bisexual with pistillodes and staminodes (Fernandes 1971). Plants of *Pimenta dioica* have long been known to be functionally male and female despite all flowers appearing hermaphrodite. Another species of *Pimenta*, *P. guatemalensis*, appears androdioecious but is likely to be functionally dioecious. Research on *Decaspermum parviflorum* in Malesia (Kevan and Lack 1985) and on some species of *Eugenia* in southern Africa (van Wyk and Lowrey 1988) has shown that the pollen of apparently bisexual flowers was nonviable and served as an attractant to pollinators. Other apparently dioecious species have been reported in *Myrcia* and *Calyptranthes*. Andromonoecy has arisen in a number of genera: *Leptospermum*, *Melaleuca*, *Phymatocarpus*, *Regelia*, *Beaufortia*, *Conothamnus*, *Lysicarpus* and a few eucalypts, as well as in the rainforest genera *Sphaerantia*, *Mitrantia* (Wilson and Hyland 1988) and *Psidium* (Soares-Silva and Proença 2006). In most of these genera, there are relatively high numbers of flowers in the inflorescence or conflorescence, and male flowers lack a pistillode.

The family is generally considered to be characterised by numerous distinct stamens but there is considerable variation in the androecium. Variations in core Myrtaceae include distinct staminal aggregation opposite the petals and reduction of stamen number. Stamen numbers can be as low as 4–8 in *Myrrhinium*, 4 or 5 in a few species of *Syzygium* (those species at one time referred to the genera *Aphanomyrtus* or *Tetraeugenia*) or 5 in *Ochrosperma* and some species of *Malleostemon*, *Micromyrtus* and *Thryptomene* or variably down to 2 or 3 in some taxa, while one species of *Babingtonia* sens. str. regularly has only 3 stamens per flower. Staminal aggregations or fascicles occur in most major clades of the capsular Myrtaceae where they are particularly prominent in many genera of the tribe Melaleuceae. However, staminal aggregations also occur in the Lophostemoneae, Kanieae, Tristanieae, Eucalypteae, Syzygieae (the few species at one time referred to the genus *Pareugenia*), and the Leptospermeae and Chamelaucieae. The staminal fascicles are mostly opposite the petals but exceptions occur: *Astartea* has stamens in small bundles opposite the sepals, and '*Pareugenia*' has fascicles that are numerous (8–16 in the type species) and not strictly associated with particular perianth segments.

Numerous distinct stamens is the common condition in the core Myrtaceae (subfamily Myrtoideae sens. lat.) but, in the Psiloxyloideae, the diplostemonous condition is found in *Psiloxylon* (Fig. 40) and there is a probable partial loss of the antesepalous whorl in *Heteropyxis*. In Myrtaceae sens. str., Johnson and Briggs (1985) conclude that the basic condition is obdiplostemony with frequent suppression of the antesepalous whorl. They based this conclusion on the published observation of small groups of antesepalous stamens in some specimens of *Arillastrum* and from the observation by Bunniger (1972) of antesepalous staminal primordia developing after the antepetalous primordia in a species of *Luma*. Others have suggested that the family is obhaplostemonous and that the numerous stamens commonly found is the result of proliferation from these antepetalous stamens. More recent work on floral development in the family has indicated that, in most genera in the family, staminal primordia arise mainly at, or adjacent to, the bases of the developing petals and, except in genera with very few stamens, the stamens differentiate centripetally so that the mature androecium may have two to many rows. In the genera included in these studies, stamen initiation occurred either within a pre-staminal bulge or directly on the floral apex. The timing of the appearance of stamens in the pre-staminal bulge, where it occurs, has the potential to produce staminal fascicles; but changes in spatial/temporal factors controlling flower development may have considerable effect on floral morphology. Thus, the final form of the androecium depends on whether the stamens expand radially to occupy the available space on the rim of the hypanthium or remain clustered in front of the petals, with or without fusion. Despite the evidence (from the androecium of *Psiloxylon*) that diplostemony may be the plesiomorphic condition in the family, these studies show that polyandry in the Myrtaceae cannot be explained within the classical framework of organogeny. Rather than obhaplostemony with subsequent proliferation of a single stamen, the evidence points to the pre-staminal bulge being a de novo development in core Myrtaceae. The aestivation of the stamens in the mature bud appears to be a significant character: they are erect in *Heteropyxis* and *Psiloxylon* but for the most part inflexed in the remainder of the Myrtaceae.

Stamen form is quite variable and this is frequently used in generic delimitation. The anthers are almost always uniformly tetrasporangiate in the family, but Tobe and Raven (1987, 1990) have shown that in *Heteropyxis* and *Psiloxylon* each locule of the anther opens by a separate slit, while in most other genera the pair of locules in each anther cell open by a common slit. The only exceptions to the tetrasporangiate state have been found in the Chamelaucieae: *Corynanthera* has trisporangiate anthers that open by a single slit (Green 1980) and *Malleostemon* species have anthers that appear to be bisporangiate (perhaps even unisporangiate). Typical stamens have slender, tapering filaments with anthers that are dorsifixed and usually versatile, have a prominent gland at the tip of the connective, and dehisce by longitudinal slits. Variations, however, occur that affect every element of stamen morphology. In *Kania*, *Uromyrtus* and *Ugni*, the anther connective is expanded into an elongated, glandular tip. *Uromyrtus* is also characterised by having a number of smaller glands present in the connective, a phenomenon recorded also across a range of other taxa of Myrteae, in a few species of *Syzygium*, and in the capsular Xanthostemoneae and in *Metrosideros*. In the case of the Xanthostemoneae, the connective is expanded into a broad flap of tissue that covers the point of attachment of the filament, giving the anther the false appearance of being basifixed. Anthers of *Acmena* are rather distinctive within the Syzygieae in having divergent anther sacs so that they appear reniform. Sterile anthers have been recorded in the genera *Lysicarpus* and *Ristantia*; the flowers are bisexual but the function, if any, of the sterile anthers is not known. In *Arillastrum*, the stamens are distinct and in four bundles opposite the petals, with the outer rows consisting of long staminodes. Basifixed, adnate (non-versatile) anthers occur in various genera across the family: in *Uromyrtus* and *Ugni* amongst the Myrteae; in *Calothamnus*, which dehisces by longitudinal slits, and the related genera *Beaufortia*, *Phymatocarpus*, *Regelia* and *Eremaea*, which dehisce by pores or transverse/oblique slits; and also in numerous genera in the Chamelaucieae. In the latter tribe, numerous androecial variants occur: anthers that dehisce by pores or short, divergent slits are common; staminodes that alternate with fertile stamens are found in *Darwinia*, *Homoranthus*, *Verticordia* and *Chamelaucium*; dimorphic anthers occur in *Pileanthus*, which has strongly divergent anther cells on alternate stamens; strongly flattened filaments are characteristic of *Rinzia* and some species referred to *Astartea*; strongly geniculate filaments have been recorded for *Malleostemon* and occur also in some groups within *Baeckea* sens. lat.

Ovary position varies from fully superior in the genera *Psiloxylon* (ovary sessile to very shortly stipitate) and *Heteropyxis* (ovary sessile), to adnate only at the base in the Xanthostemoneae, to half-inferior in most capsular species, to fully inferior. Ovary vascularisation has been studied in some groups and found to be indicative of fundamental divergence between the traditional myrtoid genera (e.g. *Myrtus*, *Eugenia*), which have a trans-septal supply, and members of the *Syzygium–Acmena* group, which have an axial supply. Axial vascularisation is most common in the family but a trans-septal supply has also been recorded for the related genera *Lophostemon* and *Welchiodendron*. Basic vascularisation cannot always be determined with any certainty in taxa that are unilocular or have basal placentation. Parietal placentation is uncommon but is typical of the myrtoid genera *Rhodamnia*, *Lithomyrtus* and a few species of *Rhodomyrtus*. Incompletely septate ovaries have been recorded for other genera, such as *Lophostemon*, *Myrteola*, *Acca*, *Mosiera* and *Myrtus*. Locule number is commonly between one and five but, in a few taxa, much higher numbers occur. Significant examples of this are found in the Myrteae, where *Decaspermum* and *Campomanesia* have up to 12 and 18 loculi respectively, and in the Leptospermeae, where one species of *Leptospermum* has up to 11 loculi and *Neofabricia* has up to 12.

Stigmas are dry and usually papillate but in *Angophora* and *Corymbia* have a mop-like appearance due to their unusually long papillae; something similar has been observed in a few species of *Hypocalymma*. Stigma shape varies from narrow and tapered to distinctly capitate, although the stigma is very broad and distinctly three-lobed in *Psiloxylon* and narrowly 2–4-lobed in the New Caledonian species currently known as *Eugenia myrtopsidioides*. The style base is sometimes sunken and this character-state is found in *Heteropyxis* and *Psiloxylon* (Psiloxyloideae). The style base of all members of the Tristanieae and Metrosidereae is sunken into the ovary summit and is located adjacent to the placentas, even in *Tepualia* where the placentas are basal. In taxa of the *Baeckea* group (tribe Chamelaucieae) where this is documented, the style is also inserted in a pit with its base adjacent to the lower part of the placenta and with the vascular tissue between them descending from the point of attachment of the peltate placenta. Elsewhere in the Chamelaucieae, some taxa have unilocular, indehiscent ovaries and many of them, e.g. *Pileanthus* and *Micromyrtus*, have terminal styles regardless of the placenta position but others, like some species of *Malleostemon*, retain the stylar pit. In all other genera, both capsular (e.g. *Cloezia*, *Tristaniopsis*, *Eucalyptus*) and non-capsular (e.g. *Syzygium*, *Eugenia*, *Myrtus*), the style is terminal on the ovary.

FLOWERING PHENOLOGY AND POLLINATION. A detailed study of flowering phenology of eight sympatric species (Proença and Gibbs 1994) found a range of flowering strategies that are representative of those found in the family. The first of these is the mass-flowering (or 'Big Bang') type with synchronised flowering of an individual plant, although some species exhibit 'pulse-bang' flowering with two mass-flowering events about a week apart. The second strategy has been called the 'Cornucopia' type with mass flowering sustained over a month or so. The third strategy is the 'steady-state' type where the plant produces a few flowers a day over an extended period of time (up to 90 days). Flowering commonly occurs at the transition between dry and wet seasons in seasonal climates, but post-fire flowering can also occur.

Outbreeding is probably widespread, although both self-compatible and self-incompatible species exist. Flowers are protandrous and some sophisticated outcrossing mechanisms have been recorded. In *Chamelaucium*, *Darwinia*, *Homoranthus* and some species of *Verticordia*, anthers extrude an oily mass of pollen onto a cluster of hairs on the style below the stigma, which then functions as a pollen presenter. In *Darwinia fascicularis*, for example, the pollen is released while the flower is still in bud, so that style elongation at anthesis extends the pollen presenter well beyond the flower.

Bee-pollination with pollen as the reward is a common pollination system. Bird- and mammal-pollination is also known to occur in *Syzygium*, for example, with nectar as the primary reward. Much pollination is generalised or unspecialised and may be effected by a range of animals, like beetles (observed on *Angophora*), nectar-feeding fruit bats (observed on *Syncarpia*) and even lizards (recorded on *Metrosideros*). Various floral adaptations, probably linked to bird-pollination, were mentioned above. Bird pollinators have also been suggested for those *Darwinia* species with bell-shaped inflorescences, for the brush-like conflorescences of *Callistemon*, *Calothamnus*, *Beaufortia* and *Regelia*, and for the orange-red pomegranate-like flowers of *Balaustion*. Two other species of *Darwinia*, *D. sanguinea* and *D. virescens*, have their flowers held at ground level and there is speculation that this may be a mammal-pollination syndrome.

More specialised pollination syndromes have been recorded in a few cases. Bird-pollination in the New World genera *Acca* and *Myrrhinium* has been linked to an unusual reward. There is no nectar but birds come to eat the petals, which change colour and become sweet and juicy as the anthers dehisce. Specialised bee pollination has been documented for two species of *Verticordia* where euglossine bees harvest the oily matrix in which the pollen is extruded (Houston et al. 1993).

POLLEN. Pollen form in Myrtaceae is relatively uniformly tricolporate, radially symmetrical, isopolar, mainly oblate, and triangular with straight or curved sides (Gadek and Martin 1981; Patel et al. 1985). Pike (1956) divided the pollen types into three categories: (1) longicolpate; (2) syn- or parasyncolpate; and (3) brevi- or brevissimicolpate. Within this general scheme, there is variation in size and exine surface patterning. *Octamyrtus* has remarkably large pollen grains reaching 50 μm in equatorial diameter. At the other end of the scale, with equatorial diameter in the range 7–9 μm, is the pollen of *Tristania*, which appears to lack colpi altogether (triporate). Colpi also seem to be absent in *Actinodium*.

Heteropolar pollen has been recorded in *Ristantia pachysperma* and in some species of *Syzygium*. Consistently bicolporate pollen has been recorded for *Pileanthus* in the Chamelaucieae and *Tepualia* in the Metrosidereae, although occasional bicolporate or tetracolporate grains can occur in some other species. The pollen of *Heteropyxis* and *Psiloxylon* fit the general myrtaceous pattern: in *Heteropyxis*, the polar island (apocolpium) is variably developed, so that grains may be longicolpate, syncolpate or parasyncolpate; in *Psiloxylon*, the polar islands are very large and the genus differs from all other parasyncolpate taxa in this regard. Tetrads have also been observed in a few taxa of the Myrteae, viz. *Myrtus communis* and *Psidium littorale*, although monads also occur in the same individuals. Some generalisations can be made concerning correlation of pollen characters with taxonomy: pollen in the Myrteae is very often longicolpate and in parts of the Chamelaucieae it is brevi- or brevissimicolpate. The remainder are largely syn- or parasyncolpate, with few exceptions. The ultrastructure of myrtaceous pollen is similar to that of other angiosperms except that there is a granulate infratectal layer and a granulate or alveolate layer around the pores (Gadek and Martin 1982) but this has also been observed in some other core members of the Myrtales (Patel et al. 1985).

Embryology. Embryologically, virtually all members of the family share seven basic character-states with other Myrtales (Tobe and Raven 1983). However, there is some systematically significant variation in the mode of embryo-sac formation and in the integuments. As in the vast majority of Myrtales, core Myrtaceae all have the monosporic Polygonum type of embryo-sac formation. The genera of Psiloxyloideae, *Heteropyxis* and *Psiloxylon*, on the other hand, have the bisporic Allium type. This mode of development is rare in Myrtales, being recorded elsewhere only in *Alzatea* (Alzateaceae).

Ovules in the Psiloxyloideae are hemitropous, although with some curvature of the embryo sac in *Psiloxylon* making it hemi-campylotropous. In the main part of the family (subfamily Myrtoideae) three ovule types, anatropous, hemitropous and campylotropous, all occur. Campylotropous ovules, with their curved embryo sacs, usually produce seeds with curved embryos. The ovules are almost always bitegmic with an outer 2–6-layered integument and an inner 2-layered integument and the micropyle formed from both integuments. In some eucalypts, members of the 'Symphiomyrtus' group, the ovules have two integuments but one is resorbed to give seed with the testa derived from only a single integument. In at least some species of *Thryptomene* and *Darwinia*, although the ovules are bitegmic, the outer integument does not fully enclose the ovule and the micropyle is formed by the inner integument alone. Unitegmic ovules are rare, occurring only in members of the *Syzygium* group (tribe Syzygieae), where it has been suggested to be the result of loss of the inner integument.

Apomixis has been reported for individuals of *Psidium cattleyanum* with odd ploidy levels, particularly heptaploid ($2n = 77$); this suggestion was based on genetic uniformity of progeny using isozymes. However, although apomixis in this species would not be surprising, these reports are unconfirmed. Apomixis has also been suggested for some species of *Eugenia*. On the other hand, apomixis has definitely been documented in *Syzygium*, particularly in *Syzygium jambos* where it is associated with polyembryony. In the mature embryo sac of this species, most of the nucellar tissue, with the exception of the nucellar cap, has disappeared and the cells in the nucellar cap separate with each having the potential to form an embryo. These embryos can form only after pollination and fertilisation, since their development is dependent on the production of endosperm. Polyembryony has also been recorded for a number of other *Syzygium* species, notably in *S. malaccense*, *S. jambos* and *S. cumini*. It has also been found in the Australian *S. paniculatum*, but without further examination for evidence of apomixis. In the Myrteae, polyembryony has been recorded in some cultivars of the species of *Myrciaria* known as 'Jaboticaba' (sometimes included in *Plinia*).

Fruit, Seed and Mature Embryo. Fleshy Myrtaceous fruits have usually been described as berries. These fleshy fruits are generally derived from an inferior ovary, so that the fleshy portion is predominantly derived from the hypanthium, not the ovary itself. A berry-like fruit derived from a fully superior ovary is known only in *Psiloxylon*. Drupe-like fruits occur in *Myrtella* and *Lithomyrtus*, which have a thin flesh covering a solid mass made up of the ± fused, bony seed coats. Drupes have also been recorded for *Piliocalyx* and the closely related genus *Acmena* but, in the latter at least, this is a subwoody, sclerenchymatous layer surrounding the single seed and may be derived from the integument. Drupe-like fruits have also been reported in *Stereocaryum* and *Hexachlamys*, but the exact nature of the 'endocarp' is uncertain and these reports might be referring only to a thickened testa. Many of these types of fleshy fruits are presumed to be adaptations for dispersal by birds, as well as bats and other small mammals, although riparian species of *Syzygium*, with aerenchymatous flesh, are likely to be dispersed by water.

Fleshy fruits have evolved repeatedly in the family but occur mainly in two tribes: the Myrteae and the Syzygieae, although within the latter tribe the genus *Anetholea* has a dry, indehiscent fruit. Fleshy, or somewhat fleshy fruits also occur sporadically in genera traditionally grouped with capsular Myrtaceae. These include the genus *Kjellbergiodendron* (Lophostemoneae), and the species *Kunzea pomifera* and *Leptospermum semibaccatum* (Leptospermeae). The fleshy-fruited genus *Xanthomyrtus* was long considered to belong to the tribe Myrteae but recent evidence, from morphological and molecular data, groups it with capsular genera. The mangrove *Osbornia* has a leathery fruit that is indehiscent and adapted for water dispersal. *Lindsayomyrtus* has a unique fruit type with a strongly exserted, somewhat leathery pericarp that tears irregularly to release 1–3 large seeds.

Dry, capsular fruits are common throughout the rest of the family but vary from being almost fully exserted (e.g. most Xanthostemoneae) to partly exserted (e.g. many Kanieae) to included. Dry, indehiscent fruits also occur: *Anetholea* and all members of the small tribe Backhousieae (2 genera, 11 species) have such fruits, which are not particularly woody, and also a few species of the otherwise capsular genus *Kunzea*. In the tribe Chamelaucieae, a large number of taxa have indehiscent fruits and this appears to have evolved independently in various lineages. There is even a recently described genus, *Astus*, that has fruits with one indehiscent and 2 dehiscent loculi; this phenomenon has been termed 'heterocarpidic' (Rye and Trudgen 2005) and apparently occurs in at least one other species group within *Baeckea* s.l. In the Chamelaucieae, indehiscent fruit types can be consistently multilocular, as in *Scholtzia*, but are commonly unilocular, as in *Verticordia*, *Thryptomene* and *Micromyrtus*. The small, indehiscent unilocular fruits of Chamelaucieae, sometimes described as 'nuts', act as the dispersal unit. In a few genera, perianth modifications assist in dispersal. The persistent, spreading, awned sepals of *Calytrix* and the much-divided, feathery perianth of *Verticordia* are prime examples.

Seed set is partly correlated with fruit type. Having few ovules and few seeds set is considered a derived state. This state occurs relatively often in fleshy-fruited genera, like *Syzygium* and *Eugenia* and their allies, but rarely in capsular-fruited genera. Two genera are exceptions to this generalisation: *Barongia* and *Whiteodendron* are consistently one-seeded. Most dry-fruited genera have numerous ovules, even though fertile seeds frequently make up only a small percentage of seed-like objects (seed + 'chaff') in the fruit; in these taxa there is no fixed pattern of seed set and all ovules appear to have the potential to become seeds. The occurrence of ovulodes has been relatively poorly documented until recent research focussing on the Eucalypteae (Bohte and Drinnan 2005). This study distinguishes two classes of ovulode: the congenital and the residual. Congenital ovulodes are destined never to become seeds, while residual ovulodes develop from the last initiated primordia, which fail to develop fully.

Seeds are typically exalbuminous but scanty traces of endosperm have been recorded in some genera of Myrteae (*Myrtus*, *Psidium*, *Acca* and *Gossia*); this is presumably the result of the embryo incompletely utilising the endosperm during its development, or perhaps the seeds were examined when not quite mature. Seed coats in the Myrteae take a number of forms: they may be hard and bony, with a circular operculum, and either shiny (e.g. *Myrtus*) or dull (e.g. *Psidium*); they may be crustose (e.g. *Pilidiostigma*) or membranous (e.g. *Campomanesia*). In *Campomanesia*, the seeds are enclosed by the glandular locule wall that forms a false seed coat, the actual testa being exceedingly thin and membranous. The fleshy-fruited taxa with large, single seeds, the eugenioid Myrteae, and the

Syzygieae, do not usually have strongly developed seed coats. A rare exception is found in the recently described *Eugenia lithosperma*, which has a sclerotic coat. The usual seed coat is papyraceous at best and thought to be absent altogether in acmenoid genera like *Acmena*, *Acmenosperma*, *Waterhousea* and *Piliocalyx* (Biffin et al. 2006). In these latter taxa, the embryo has darker, intrusive intercotyledonary material within the apparently fused, undifferentiated cotyledonary tissue. It has been suggested that this intrusive material may be chalazal in origin, like the tissue that makes up much of the seed coat in several southern African species of *Eugenia* s.l. (van Wyk and Botha 1985). Corner (1976) believed that pachychalazal seed coats occurred in at least some species of '*Eugenia*' currently referred to *Syzygium*.

Seeds in capsular and dry-fruited genera are quite variable. The classic 'leptospermoid' seed is narrowly linear to narrowly cuneate and is very light. It is found mainly, but not exclusively, in members of the Leptospermeae, Melaleuceae and Metrosidereae and it has been demonstrated, in *Metrosideros* at least, that it can be dispersed over long distances. Many other seed types occur amongst capsular genera, including globular, obovoid and flattened seeds, and winged seeds of various types have arisen amongst diverse genera. In *Tristaniopsis* and *Purpureostemon* the wing arises between the raphe and the embryo sac, so that the vascular trace runs around the margin of the wing. In contrast with this, wing development is distal to the raphe in *Xanthostemon crenulatus* and *X. eucalyptoides*, and in the genus *Carpolepis*. Similar wings occur in some species of *Corymbia* but not in other eucalypts, although some have narrow peripheral wings (mainly *Eucalyptus* subg. *Eudesmia*) or short, tail-like wings (*E. deglupta*). Marginal wings are also a feature of the monotypic *Homalospermum*. *Heteropyxis* has a unique wing type, with narrow wings developing at both the chalazal and micropylar ends. In *Neofabricia*, a false wing is present; one of the lower ovules develops into a seed that is fused to a wing made up of the scarious remains of the aborted ovules. Seed in the family is usually non-arillate but *Ochrosperma* and some *Rinzia* spp. have an arillate hilum. *Arillastrum* seeds bear an aril-like group of scales formed from aborted ovules that adhere to the mature seed. Indehiscent dry-fruited taxa, many of which belong to the Chamelaucieae, have seeds with a very thin testa, with the wall of the fruiting hypanthium taking its place in a functional sense. The testa in dehiscent-fruited members of Chamelaucieae is crustaceous as in other capsular genera.

While early embryology is relatively uniform across genera, final embryo morphology varies widely. Embryo form was for a long time the foundation for the three traditional subtribes in the tribe Myrteae, based on a classification of embryos as myrtoid, myrcioid or eugenioid. However, this typological scheme has now been discredited in the light of recent phylogenetic work (Lucas et al. 2007) that shows that taxa with eugenioid and myrcioid embryos are derived from ancestors with simpler, myrtoid embryos. In any case, not all embryos can be grouped readily into these three distinct categories. For example, the genus *Luma* and the recently described *Algrizea* have embryos that do not readily fit this simple classification, and the embryo of *Myrceugenia* is apparently convergent with that of the myrcioid group of genera. Myrtoid genera have embryos that are either C-shaped or circinate, with a slender hypocotyl, or variously shaped with a swollen hypocotyl and very reduced cotyledons. A group of genera with this latter embryo type has been recognised as the '*Campomanesia* complex' (including the Australian genus *Pilidiostigma* and a number of American genera including *Pimenta* and *Blepharocalyx*) but this grouping receives no support in molecular analyses. The complex ruminate cotyledons of *Acmena* and other genera of Syzygieae are not found elsewhere in the family but have arisen within a tribe where the typical cotyledons are thick and free to partly interlocking. Within Syzygieae, the genus *Anetholea* has interlocking leafy cotyledons, perhaps representing the plesiomorphic state.

In the various groups of capsular genera, cotyledons are predominantly leafy, with thick, storage cotyledons found only in *Lindsayomyrtus*. Embryos with leafy cotyledons come in diverse configurations, ranging from straight with cotyledons face-to-face (as in *Metrosideros*), individually folded (as in *Arillastrum*), partly enclosing one another (as in obvolute/convolute types found in *Tristaniopsis* and *Lophostemon*), the outer one fully enclosing the inner (as in

Kjellbergiodendron), to the somewhat myrcioid type found in *Backhousia*. The Chamelaucieae have an embryo that is rather similar to that of *Pilidiostigma* with a swollen hypocotyl, except that the embryo is mostly straight. The embryo terminates in two barely recognisable, deflexed cotyledons that are sometimes borne on a narrow neck-like extension of the hypocotyl.

Germination of embryos with storage cotyledons is, as expected, hypogeal and it is epigeal for those with leafy cotyledons. Seed viability varies considerably: in some rainforest taxa, viability can be lost within days of being shed but, in species from drier habitats, seeds can remain viable for a year or more.

KARYOLOGY. The haploid number for *Heteropyxis* and *Psiloxylon* (Psiloxyloideae) is $n = 12$. The base chromosome number in the core Myrtaceae is a remarkably regular $x = 11$, although $n = 12$ has been recorded for a few species of *Eugenia*, *Eucalyptus*, *Melaleuca* and *Metrosideros*. Chromosome numbers, however, are a potentially useful character in the tribes Chamelaucieae and Melaleuceae, where there has been dysploid reduction in some genera (Smith-White 1954; Briggs 1962; Rye 1979; Rye and James 1992). Within Chamelaucieae, virtually all reductions are found in the related genera *Actinodium*, *Darwinia*, *Homoranthus* and *Verticordia*; amongst other genera of the tribe, only two species of *Thryptomene* have been recorded with base chromosome numbers less than 11. *Darwinia* has the most extensive dysploid reduction series with base numbers in the range 5 to 7 and 9, although the majority of species are $x = 6$, including all those from eastern Australia; this same number is found in the very closely related genus *Actinodium*. The genus *Verticordia* is heterogeneous, with some species showing reductions (to 6–9) and others having the common base number $x = 11$. The latter group of species appear to have greater affinity to the genus *Chamelaucium*, which also has the base number $x = 11$. This relationship has also been demonstrated with molecular data. Interestingly, there have been frequent cases of polyploidy recorded based on these reduced numbers, so that recorded counts ($2n$) range from 24 up to 36 (tetraploid) and 42 (hexaploid). Within Melaleuceae, dysploid reduction to $x = 10$ has occurred in *Conothamnus*, *Phymatocarpus* and *Beaufortia*, with some species of the latter showing further reduction to $x = 8$ (Rye 1979).

Polyploidy on the common base number $x = 11$ is found on occasion in other parts of the family. Recorded occurrences of polyploidy are tetraploids in *Psidium*, *Eugenia*, *Syzygium*, *Baeckea* sens. lat., *Scholtzia*, *Hypocalymma*, *Calytrix*, *Leptospermum* and *Melaleuca*; hexaploids in *Eugenia* and *Syzygium*; octaploids in *Psidium*; decaploids in *Syzygium*. Triploids have been recorded in *Eugenia* and *Melaleuca*, and pentaploids and heptaploids in *Psidium*. In a few controlled crosses involving species of *Kunzea* and species of *Leptospermum* (de Lange and Murray 2004), a few of the progeny were found to have somatic chromosome numbers of $2n = 23$.

PHYTOCHEMISTRY. The presence of oil glands that produce essential oils is one of the fundamental features of the family Myrtaceae sens. lat. *Heteropyxis* also has oil-bearing glands in the leaves and these produce a rather simple oil containing over 90% monoterpenes; in *Psiloxylon* the oil-producing function of the glands appears to have been lost. Amongst Myrtaceae s. str., oils are usually dominated by monoterpenes or sesquiterpenes, but are often a mixture of both, with more complex terpenes, like triterpenes, occurring only occasionally, usually in minor amounts. Less commonly, there can be a significant presence of β-triketones, alkyl derivatives and/or aromatic compounds. Species dominated by aromatics include the Caribbean species *Pimenta racemosa*, the Bay Rum tree, although there is at least one variety whose oil has a strong monoterpene component. The Asian species *Syzygium aromaticum*, the source of cloves, and the Australian monotypic genus *Anetholea* both produce oil that contains over 90% aromatic compounds. The uncommon aromatic compound 2,4,6-trimethyoxytoluene has been recorded as the major component (80–89%) of the essential oil of *Stockwellia*. There has been some suggestion that the percentage distribution of major components of essential oils is of systematic value, although this is complicated by chemical races of species that can have markedly different oil profiles (as in *Pimenta*, mentioned above). There may be some merit in these suggestions, but the data need to be reassessed in light of the demise of the old two-subfamily paradigm.

A wide survey of the occurrence of polyhydroxyalkaloid (PHA) compounds in the family (Porter et al. 2000) found that the distribution of these compounds showed a fairly strong correlation with the traditional, narrower concept of Myrtoideae (e.g. Schmid 1980), but they were generally absent from the 'Leptospermoideae'. This result is incongruent with molecular phylogenies that do not support monophyly of either concept, so the occurrence of similar compounds may be the result of convergent evolution. On the other hand, flavonoid profiles of genera in Johnson and Briggs's *Leptospermum* alliance supported its two suballiances but not the alliance itself, in agreement with recent molecular analyses that suggest they are not closely related. Flavonoid and PHA chemistry suggest the placement of the anomalous genus *Luma* amongst eugenioid taxa but this is not supported by molecular phylogenetic analyses, which place it closer to *Myrceugenia*.

Distribution and Habitats. The family Myrtaceae has a predominantly southern hemisphere distribution, with the majority of genera occurring in Australia and South America. Despite the presence in Africa of *Heteropyxis*, one of the two genera that are sister to the core Myrtaceae (the other being the monotypic genus *Psiloxylon* from the nearby Mascarenes), only three or four genera occur naturally in subequatorial Africa and adjacent Madagascar. *Myrtus* is found in north Africa and other areas around the Mediterranean. North of the equator, the family has a significant presence in Asia and the Americas, with the distribution extending to China and India on the one hand and Mexico, the Caribbean and Florida on the other. Understandably, the numbers of genera tend to diminish towards the extremes of the distribution, both north and south. For example, in the northern hemisphere there are three genera recorded for India, four for China and four for Florida; in the southern hemisphere there are six in New Zealand, seven in Chile and eight in Tasmania.

Species of Myrtaceae are found in a wide range of habitats and are able to thrive on nutrient-poor soils and in seasonally dry regions. In Australia, there is a concentration of species along the eastern side and in the southwest of the continent. Members of the family can be found as low shrubs on coastal sands or above the tree-line (c. 1,800 m), tall shrubs in rangelands, or trees in the rainforest or gallery forest, or in dry, remote parts of central Australia. However, even in the centre of the continent, these trees (eucalypts) do not occur naturally in the driest environments but are found along dry watercourses where they are better able to take advantage of available water when it does rain. This pattern of ecological diversification is paralleled in South America and Africa where Myrtaceae have occupied a broad range of habitats. For example, in Brazil, around 50% of Myrtaceae are thought to be endemic to a region comprising the coastal tropical and subtropical moist forest (*Mata Atlântica*), the adjacent tropical dry forest, semi-deciduous forest, tropical savannas, and mangrove forest that occupies a wide band extending from the Atlantic coast of Brazil inland to include some parts of neighbouring Paraguay and Argentina. In this complex environment there are high levels of diversity, probably indicative of microhabitat diversification. A similar range of habitats is occupied by Myrtaceae in Africa. Species there have been recorded from swamp forest, lowland forest, gallery forest, montane forest, fire-prone savannas and coastal dune vegetation.

Palaeobotany. The early fossil record of Myrtaceae is relatively scanty and there is great difficulty in applying tribal, or other, concepts to the records that do exist. The earliest fossil record attributed to Myrtaceae is generally considered to be *Myrtaceidites* pollen from the upper Cretaceous of Gabon, Africa, with other late Cretaceous records from Borneo, China, Argentina and southern Australia. Myrtaceous pollen is fairly distinctive but rarely diagnostic below family level and, as a result, placement of the earliest fossil pollen is problematic. It is, therefore, not certain whether *Myrtaceidites* represents core Myrtaceae or an ancestor of a sister taxon like *Heteropyxis* or *Psiloxylon*. From this time through to the early Tertiary, fossil pollen is increasingly frequently encountered and represents identifiable lineages like Myrteae, Metrosidereae and Eucalypteae, indicating rapid expansion of the family.

Identification based on fossil leaves is often imprecise and confused by the application of names of extant genera, like *Eucalyptus* or *Eugenia*, to these fossils. Equally, the form genus *Rhodomyrtophyllum*, widely recorded in Eurasia, does not necessarily relate to the extant genus *Rhodomyrtus*. However, the leaf fossil of *Eucalyptus*, from the Miocene of New Zealand, does appear to represent that genus and provides evidence that it was at one time present there. Similarly, the leaf fossil attributed to *Xanthomyrtus*, from the Eocene of north-eastern Tasmania, does seem to show strong similarities to leaves of extant taxa from New Caledonia and Malesia. Generally speaking, however, the geographic extent of fossil leaves of Myrtaceae from the early Tertiary of Antarctica, Australia and New Zealand, Eurasia and the Americas indicates that the family was widely distributed during the Eocene, and perhaps even earlier. Fossil wood with myrtaceous affinity is also widespread during the same period and has been reported from Australia, India, Antarctica, South America and Eurasia.

Two distinctive myrtaceous macrofossils from northern America are worthy of note. The unusual fossil genus *Syzygioides*, described from the Eocene of Colorado, has some specimens showing attached leaves and fruits. It has leaves that are said to be *Syzygium*-like but the inflorescences appear to be cymose, and the semi-exserted fruits are thought to have been leathery, rather than fleshy. This combination of vegetative and reproductive structures does not closely resemble any known genus in the family, and *Syzygioides* is thus considered to represent an extinct lineage of unknown affinity. Fossil fruits reported from the late Palaeocene of North Dakota and the early Eocene of British Columbia, which have seeds similar to those of extant *Mosiera*, have been referred to a new genus, *Paleomyrtinaea*. In this case, the fossil can be placed confidently in the tribe Myrteae.

Amongst capsular groups, there are a number of distinctive macrofossils. The earliest eucalypt fruit fossils, found in south-eastern Queensland and possibly representing *Corymbia*, are dated to the Palaeocene or Eocene. Other eucalypt group macrofossils, dated to the Miocene, have been found in northern New South Wales (fruits) and Victoria (buds). Recently, leaves and fruits with likely myrtaceous affinity have been recovered from Early Eocene sediments at Laguna del Hunca, Argentina; Gandolfo et al. (2006) have put forward a strong case that these fossils represent a species of *Eucalyptus*. Basinger et al. (2007) described a new genus, *Tristaniandra*, based on late flower and fruit material of Middle Eocene age. This material does not exactly match any extant genus but the staminal and capsule characters indicate close affinity with members of the tribe Kanieae.

The overall picture from the fossil record is consistent with an origin of the family prior to the Tertiary, followed by rapid diversification and dispersal. This hypothesis is supported by molecular analyses that show that, although there is good statistical support for most of the major clades in the family, the relationships between them are unsupported. This reflects the intense period of evolutionary change that reached its peak during times of increasing aridity in the Oligocene/Miocene and led to the development of sclerophyllous vegetation in Australia.

Economic Importance. Within the family, species from a number of genera are valuable sources of timber. Of the many species of *Eucalyptus* being grown in plantations, the Tasmanian Blue Gum, *E. globulus*, is the most important and most widely cultivated. Other taxa of particular significance are the Brush Box, *Lophostemon confertus*, and numerous species of *Syzygium*. Eucalypts have also been a source of kino, a resinous substance used in medicines and tanning.

Essential oils are another economically valuable product, extracted from leaves by steam distillation. The earliest widely used essential oil was clove oil. Clove oil is extracted from the Clove tree, *Syzygium aromaticum*, and its constituents include eugenol, eugenyl acetate, caryophyllene and iocaryophyllene. The principal active ingredient of clove oil is eugenol, which is known to depress sensory receptors involved in pain perception and, for this reason, was widely used in the past for relief from toothache. An economically important oil is Tea-Tree oil, usually extracted from *Melaleuca alternifolia*, which has a high level (c. 40%) of terpinen-4-ol, an active antimicrobial and anti-inflammatory agent. Other important oils include Niaouli, from the *Melaleuca quinquenervia* group, which has nerolidol as a major component (>70%) plus significant linalool content (c. 20%); various lemon-scented oils (from *Backhousia citriodora*,

Corymbia citriodora and *Leptospermum petersonii*), which have high levels of citral and citronellal; and *Eucalyptus* oils that have prominent cineole or piperitone (peppermint) components. Among New World genera, the oil of *Pimenta racemosa* (bay rum oil) is the best known; the principal components of the commercial chemotype are eugenol (>50%) and myrcene (c. 25%). There are strong industry standards for commercial oils. Chemotypes can occur that have very different levels of major components and, even where major chemotypes do not occur, levels of undesirable components such as methyl eugenol (a potential carcinogen) need to be regulated.

Two species of eucalypts, namely *E. youmanii* and *E. macrorhyncha*, have been cultivated commercially for the production of rutin (a flavonoid glycoside), a compound often used in the treatment of capillary fragility. Harvesting of the leaves of these species begins after their fourth year in plantation, yielding more than 5% rutin after processing.

Species grown for fruit or culinary spices are predominantly taken from fleshy-fruited genera. The guava, *Psidium guajava*, is probably the most widely grown fruit and known particularly for its juice. The Strawberry Guava (*Psidium cattleyanum*), Jaboticaba (*Myrciaria cauliflora*), Feijoa or Pineapple Guava (*Acca sellowiana*), Pitanga or Surinam cherry (*Eugenia uniflora*) and the Chilean murta (*Ugni molinae*) are also regionally important foods. In Asia, various *Syzygium* species, particularly *S. cumini*, *S. aqueum*, *S. samarangense* and *S. jambos*, are also eaten. The major spices produced are clove (*Syzygium aromaticum*) and allspice (*Pimenta dioica*), although there is a growing market in Australia for lemon myrtle (*Backhousia citriodora*) and aniseed myrtle (*Anetholea anisata*) as flavourings. A number of species (mostly eucalypts) are important as nectar sources for honey production.

Besides cultivation for production of timber, essential oils or foods, many Myrtaceae are important in horticulture as hedges (e.g. *Syzygium*, *Myrtus*), flowering ornamentals (*Callistemon*, *Melaleuca*) and even for the florist industry (mainly cultivars of the Geraldton Wax, *Chamelaucium uncinatum*).

A few species are economically significant weeds. *Psidium guavaja* and *Psidium cattleyanum* are readily dispersed by birds, and have become naturalised in many places in the tropics and have proven to be quite invasive. *Melaleuca quinquenervia*, a species that prefers moist sites, has become a serious problem in Florida wetlands. The Chinese species *Rhodomyrtus tomentosa* is increasingly problematic in Florida and Hawaii.

Subdivisions and Relationships Within the Family. For around 150 years, classification of the Myrtaceae was generally based on the system of three tribes proposed by de Candolle in his *Prodromus*: Myrteae, Leptospermeae and Chamelaucieae. These tribes accommodated genera with fleshy, capsular and dry indehiscent fruits respectively. A variant of this system recognised just two groups, the fleshy-fruited and dry-fruited taxa. There were also versions of these that recognised the groups at the rank of subfamily, rather than tribe. So, in recent times it was customary (with few exceptions) to see the family as naturally divisible into two subfamilies, the fleshy-fruited Myrtoideae and the capsular-fruited Leptospermoideae. There were also two genera, *Psiloxylon* from the Mascarenes, and *Heteropyxis* from southern Africa, that were variously linked to the core Myrtaceae. *Heteropyxis* has essential oil in its leaves and a capsular fruit, and was often included within subfamily Leptospermoideae, although many classifications placed it in its own family, Heteropyxidaceae. *Psiloxylon*, on the other hand, lacks essential oils (although it does have gland-like cavities in the leaves) and has a berry-like fruit derived from a shortly stipitate, superior ovary; it has usually been treated as a distinct family, Psiloxylaceae.

Briggs and Johnson (1979) published a new view of the family that recognised an informal set of 'alliances' and 'suballiances'. These were intended to be seen as roughly equivalent to tribes and subtribes. This proposed classification was presented within a two-subfamily framework but showed significant departures from earlier classification. Principally, their *Chamelaucium* alliance diverged from traditional classifications by including the small-flowered capsular genera from the *Baeckea* group, albeit in separate suballiances. The earlier classifications had given strong weight to the small, indehiscent, nut-like fruits of the *Chamelaucium* group that are derived from unilocular ovaries, and considered them significantly different from the small, capsular fruits of the *Baeckea* group that are derived from 2- to 3-locular ovaries. In their subsequent analysis of relationships, Johnson and Briggs (1985) abandoned these suballiances but went even further and rejected the subfamilies altogether. The latter conclusion was based on their contention that *Syzygium* and its relatives constituted a group of fleshy-fruited plants that was phylogenetically separate from the other genera usually assigned to the Myrtoideae. This separation had already been suggested based on studies of flower, wood and bark anatomy but the hypothesis that these differences were indicators of significant divergence was new. Johnson and Briggs' analysis also included *Heteropyxis* and *Psiloxylon* and hypothesised that these genera were early branches from the myrtaceous phyletic line. They could see no compelling evidence for including them in a single segregate family and recognised both as unigeneric families.

Despite the strong morphological differences between *Heteropyxis* and *Psiloxylon*, the embryological studies mentioned earlier revealed some fundamental similarities between them and more recent studies of molecular data have provided very strong support for affinity between them. An analysis of sequence data from the chloroplast *trn*K region (Wilson et al. 2005) was the basis for a new classification of the family. The data produced an estimate of the phylogeny of the family with little basal resolution but a number of well-supported clades that accommodated most genera. Monophyly of Myrtaceae s. str. had moderate support but a broader circumscription, including *Psiloxylon* and *Heteropyxis*, was more strongly supported. A new infrafamilial classification was proposed that recognised two subfamilies (Psiloxyloideae and Myrtoideae) and 17 tribes. The most recent analysis, of nuclear ITS data by Biffin et al. (2007), generated phylogenetic trees with rather different topologies from those generated from chloroplast sequence data but the well-supported clades were basically the same.

Classification of Myrtaceae

I. Subfam. Psiloxyloideae Schmid (1980).
1. Psiloxyleae A.J. Scott (1980).
2. Heteropyxideae Harv. (1868).

II. Subfam. Myrtoideae Sweet (1827).
3. Xanthostemoneae Peter G. Wilson (2005).
4. Lophostemoneae Peter G. Wilson (2005).
5. Osbornieae Peter G. Wilson (2005).
6. Melaleuceae Burnett (1835).
7. Kanieae Engl. (1930).
8. Backhousieae Peter G. Wilson (2005).
9. Metrosidereae Peter G. Wilson (2005).
10. Tristanieae Peter G. Wilson (2005).
11. Syzygieae Peter G. Wilson (2005).
12. Myrteae DC. (1827).
13. Eucalypteae Peter G. Wilson (2005).
14. Syncarpieae Peter G. Wilson (2005).
15. Lindsayomyrteae Peter G. Wilson (2005).
16. Leptospermeae DC. (1827).
17. Chamelaucieae DC. (1827).

Key to the Genera

1. Fruit a leathery or fleshy berry, rarely drupe-like; leaves opposite 2
- Fruit usually dry, mostly a dehiscent capsule, occasionally leathery or fleshy (if so, then leaves always alternate); leaves opposite or alternate 64
2. Mangroves; flowers and fruit with eight, ± undifferentiated perianth segment **10. *Osbornia***
- Non-mangroves; sepals and petals distinct (occasionally calyptrate) 3
3. Cotyledons thick, plano-convex or fused into a mass; hypocotyl usually shorter than cotyledons 4
- Cotyledons submembranous, never fused into a mass; hypocotyl as long as or longer than cotyledons 22
4. Inflorescences usually terminal (sometimes rami- or cauliflorous), often paniculate; young shoots and flowers usually glabrous 5
- Inflorescences usually axillary, dichasial or raceme-like or flowers solitary; young shoots and flowers often pubescent 9
5. Anther cells globose and divergent 6
- Anther cells linear, parallel 7
6. Flowers calyptrate **42. *Piliocalyx***
- Flowers with small, distinct perianth segments **41. *Acmena***
7. Cotyledons fused with a ramified intercotyledonary inclusion 8
- Cotyledons free **44. *Syzygium***
8. Flower buds clavate; fruit cylindrical, turbinate or pyriform **40. *Acmenosperma***
- Flower buds turbinate; fruit depressed-globular **43. *Waterhousea***
9. Hypanthium extended into a tube beyond the ovary summit 10
- Hypanthium barely extended beyond the ovary summit (calyx open or closed) 13
10. Hypanthium circumscissile below the stamens after flowering 11
- Hypanthium not circumscissile 12

11. Hypanthium constricted at the ovary summit; ovules usually 3–5 per loculus; flowers pedicellate **87. *Siphoneugena***
- Hypanthium not constricted at the ovary summit; ovules 2 per loculus; flowers sessile **88. *Myrciaria***
12. Calyx free; ovules numerous **80. *Stereocaryum***
- Calyx fused, distinct only at the tips; ovules 2 per loculus **89. *Plinia***
13. Calyx fused, tips of sepals sometimes free 14
- Calyx free, 4–5(6)-lobed 17
14. Calyx fused, opening as an operculum **91. *Calyptrogenia***
- Calyx not opening as an operculum, buds opening by splits 15
15. Cotyledons free **86. *Neomitranthes***
- Cotyledons fused (at least at the margins) 16
16. Fruit with woody endocarp-like layer **80. *Stereocaryum***
- Fruit lacking endocarp-like layer **81. *Calycorectes***
17. Sepals caducous **83. *Pseudanamomis***
- Sepals persistent 17
18. Ovules apical, pendulous; prophylls exceeding the flower **47. *Meteoromyrtus***
- Ovules axile; prophylls shorter than the flower 18
19. Cotyledons fused 19
- Cotyledons free 20
20. Calyx 4-merous; ovules few per loculus (sometimes numerous) **78. *Eugenia***
- Calyx 4–6-merous, but usually 5-merous; ovules numerous **79. *Hexachlamys***
21. Hypocotyl much shorter than the cotyledons **82. *Myrcianthes***
- Hypocotyl about as long as the cotyledons **84. *Luma***
22. Cotyledons much wider than the hypocotyl, usually foliaceous and folded 23
- Cotyledons narrower than to as wide as the hypocotyl, rarely wider but, if so, then never folded 30
23. Calyx fused 24
- Calyx free 27
24. Calyx tearing irregularly **96. *Marlierea***
- Calyx calyptrate 25
25. Prophylls persistent; petals 4 **85. *Myrceugenia***
- Prophylls caducous at or before anthesis; petals rarely present 26
26. Inflorescence usually with 9 or more flowers; ovules 2 per loculus **94. *Calyptranthes***
- Inflorescence with up to 9 flowers; loculi multi-ovulate **93. *Mitranthes***
27. Flowers 4-merous **85. *Myrceugenia***
- Flowers 5-merous 28
28. Inflorescence a triad of sessile flowers; loculi >3-ovulate **92. *Algrizea***
- Inflorescence a panicle; loculi 2-ovulate 29
29. Pollen sacs of anthers at slightly different levels; hairs often appressed, silky, silvery, yellowish or reddish-brown **97. *Gomidesia***
- Pollen sacs all at same level; hairs spreading or plant ± glabrous **95. *Myrcia***
30. Stamens dark red; petals slightly fleshy 31
- Stamens white, pink or yellow; petals not fleshy 32
31. Stamens 4–8; testa hard **74. *Myrrhinium***
- Stamens 60 or more; testa subcrustose to hard **73. *Acca***
32. Testa nearly always membranous or cartilaginous; hypocotyl swollen, much wider than the small cotyledons 33
- Testa nearly always hard, often bony; hypocotyl not swollen, as wide as or only slightly wider than the cotyledons (rarely narrower) 38
33. Ovary (3)4–18-locular; seeds one per loculus, the locule wall forming a false testa **69. *Campomanesia***
- Ovary 1–3-locular; seeds more than one per loculus, the locule wall not forming a false testa 34
34. Flowers 4-merous; placenta ± peltate **76. *Chamguava***
- Flowers 4- or 5-merous; placenta not peltate 35
35. Embryo ± straight, 1–2 cm long; leaves with domatia **68. *Legrandia***
- Embryo C-shaped or spiral; leaves lacking domatia 36
36. Testa glandular-tuberculate **56. *Pilidiostigma***
- Testa not glandular-tuberculate 37
37. Placenta central in the loculus; ovules in 3 or more rows **59. *Blepharocalyx***
- Placenta ± apical; ovules clustered **77. *Pimenta***
38. Flowers yellow; cotyledons wider than the hypocotyl **38. *Xanthomyrtus***
- Flowers white, cream or pink; cotyledons narrower than to as wide as the hypocotyl 39
39. Testa dull or rough; sepals partly to fully fused **70. *Psidium***
- Testa generally lustrous and smooth; sepals distinct or partly fused 40
40. Flowers nodding; anthers elongate or sagittate 41
- Flowers erect; anthers not conspicuously elongate 42
41. Prophylls not persistent **55. *Uromyrtus***
- Prophylls persistent **61. *Ugni***
42. Seeds horizontal, separated by false horizontal septa; stigma capitate 43
- Seeds not separated by false septa; stigma capitate, slightly dilated or narrow 45
43. Branchlets glabrous **54. *Archirhodomyrtus***
- Branchlets tomentose 44
44. Petals 6, 8 or 12, long, erect, distinct but forming a tube-like structure **52. *Octamyrtus***
- Petals 4–5, spreading **51. *Rhodomyrtus***
45. Leaves distinctly triplinerved **53. *Rhodamnia***
- Leaves usually pinnately nerved 46
46. Seeds ± fused into an endocarp-like structure 47
- Seeds free, embedded in fleshy tissue 48
47. Prophylls leaf-like, persistent; embryo thick, curved **58. *Myrtella***
- Prophylls not leaf-like, caducous; embryo slender, circinate **57. *Lithomyrtus***
48. Prophyllsfoliaceous, persistent 49
- Prophyllscaducous at or about anthesis 50
49. Anthers basifixed **64. *Myrtastrum***

- Anthers dorsifixed, versatile **60. *Myrteola***
50. Flowers regularly 4-merous 51
- Flowers mostly 5-merous 56
51. Sepals partly fused, tearing into 4 at anthesis; ovary 4-locular **75. *Accara***
- Sepals free; ovary 2–4-locular 52
52. Placenta ± apical or subapical 53
- Placenta ± axile 55
53. Inflorescence a monad, leaves glabrous 54
- Inflorescence a monad or triad, leaves hairy **65. *Lenwebbia***
54. Ovules 2 per loculus; seeds few per fruit; stigma capitate **67. *Amomyrtella***
- Ovules many per loculus; seeds many per fruit; stigma narrow **62. *Lophomyrtus***
55. Seed coat smooth 56
- Seed coat papillose **72. *Curitiba***
56. Embryo circinate **50. *Gossia***
- Embryo C-shaped **71. *Mosiera***
57. Stigma capitate 58
- Stigma narrow or only slightly dilated 59
58. Ovary 4–5(–8)-locular; ovules 2 per locule; seeds oriented vertically in each locule in the fruit, held together by a membrane **48. *Decaspermum***
- Ovary 2–3-locular; ovules 10 or more per loculus, biseriate; seeds oriented horizontally in the fruit, not held together by a membrane **54. *Archirhodomyrtus***
59. Ovary mostly 3–5-locular; sepals often partly fused **46. *Calycolpus***
- Ovary 2–3-locular; sepals free 60
60. Inflorescence of monads 61
- Inflorescence of triads, metabotryoids or monads in bracteate short shoots 63
61. Mature fruit black **45. *Myrtus***
- Mature fruit not black 62
62. Fruit red or yellowish **63. *Neomyrtus***
- Fruit whitish to grey with small, dark spots **49. *Austromyrtus***
63. Ovary 2–3-locular; placenta attached to upper septum; embryo C-shaped **66. *Amomyrtus***
- Ovary 2-locular; placenta attached to lower septum; embryo circinate **50. *Gossia***
64. Plants functionally dioecious; stamens not inflexed in bud, anthers tetralocular at anthesis; base chromosome number $x = 12$ 65
- Plants usually monoecious; stamens inflexed in bud, anthers bilocular at anthesis; base chromosome number $x = 11$ 66
65. Ovary stipitate; style ± obsolete; fruit a berry **1. *Psiloxylon***
- Ovary sessile; style well-developed; fruit a capsule **2. *Heteropyxis***
66. Embryo with cotyledons equal to or longer than the hypocotyl; ovary 2–5(–11)-locular; trees or shrubs 67
- Embryo with cotyledons much smaller than the hypocotyl; ovary 1–3-locular; shrubs or subshrubs (Chamelaucieae) 123
67. Flowers and fruits with five main veins in the hypanthium (Metrosidereae) 68
- Flowers and fruits with numerous veins in the hypanthium 72
68. Vegetative buds with many broad-based protective scales 69
- Vegetative bud scales absent, narrow-based or resembling reduced leaves 70
69. Growth sympodial, apical buds usually aborted **35. *Metrosideross*.str.**
- Growth monopodial **34. *Carpolepis***
70. Placentas basal **32. *Tepualia***
- Placentas axile 71
71. Petals cohering and falling as calyptra **35. *Metrosideros*** (sect. *Calyptropetala*)
- Petals distinct **33. *Mearnsia*** s.l.
72. Ovulodes present (Eucalypteae) 73
- Ovulodes usually absent (if present, then placentation apical) 79
73. Ovary 2-locular 74
- Ovary usually 3–5-locular 77
74. Leaves opposite 75
- Leaves whorled **101. *Allosyncarpia***
75. Calyx calyptrate 76
- Calyx lobes free **98. *Arillastrum***
76. Inflorescences 3-flowered; hypanthium splitting into 4 segments at anthesis **100. *Stockwellia***
- Inflorescences 7-flowered; hypanthium not splitting at anthesis **99. *Eucalyptopsis***
77. Adult leaves opposite; perianth free **102. *Angophora***
- Adult leaves mostly alternate; flowers calyptrate
78. Inflorescences much-branched; bristle-glands present on young growth; oil ducts often present **103. *Corymbia***
- Inflorescences condensed; bristle-glands absent; oil ducts absent **104. *Eucalyptus***
79. Oil ducts present in the pith of petioles and other axes 80
- Oil ducts absent from pith (oil glands sometimes present) 85
80. Oil ducts small and numerous, occurring in both the pith and the cortex 81
- Oil ducts few and large, restricted to the pith 83
81. Infructescence syncarpou **105. *Syncarpia***
- Fruits free from each other 82
82. Fruit conical; placentas apical; seeds winged **9. *Welchiodendron***
- Fruit not conical; placentas axile; seeds linear **8. *Lophostemon***
83. Ovary inferior; fruit succulent **6. *Kjellbergiodendron***
- Ovary semi-superior; fruit exserted from hypanthium, dry or semi-succulent 84
84. Fruit a dehiscent capsule; stamens in distinct, fused bundles **7. *Whiteodendron***
- Fruit semi-succulent, indehiscent or opening by irregular splits; stamens barely connate at the base **106. *Lindsayomyrtus***
85. Seeds semi-discoid usually borne on a rod-like placenta; fruit a ± spherical capsule 86
- Seeds rarely semi-discoid, placenta not rod-like; fruit globular, ovoid, ellipsoid or conical 88

86. Calyx calyptrate, tearing irregularly at anthesis **4. *Pleurocalyptus***
- Calyx free 87

87. Fertile seeds with a conspicuous, membranous wing **5. *Purpureostemon***
- Fertile seeds lacking a wing or wing very small **3. *Xanthostemon***

88. Inflorescences pedunculate; flowers usually pedicellate 89
- Inflorescences not pedunculate; flowers sessile or pedicellate (usually shortly) 102

89. Fruit indehiscent; ovary usually only 2-locular 90
- Fruit dehiscent; ovary 2–3(4)-locular 92

90. Ovary half-inferior to almost superior; plants hairy 91
- Ovary inferior; plant glabrous **39. *Anetholea***

91. Flowers sessile in compact heads **31. *Choricarpia***
- Flowers pedicellate **30. *Backhousia***

92. Leaves opposite or irregularly ternate 93
- Leaves alternate 100

93. Seeds numerous, small and narrow 94
- Seeds few (mostly 1–3), large, ovoid to angular 98

94. Placentation basal 95
- Placentation axile 97

95. Stigma capitate; outer stamens with large, sterile anthers **27. *Lysicarpus***
- Stigma only slightly dilated; stamens all fertile 96

96. Ovules in a ± circular series on the placenta; ovules usually 4–8 per loculus **29. *Cloezia***
- Ovules scattered on the placenta; ovules 12 or more per loculus **28. *Kania***

97. Stamens fused in erect bundles opposite the petals **36. *Tristania***
- Stamens free **37. *Thaleropia***

98. Capsule exserted; seed solitary 99
- Capsule included; seeds 1–3 **22. *Sphaerantia***

99. Stamens fused in long staminal fascicles opposite the petals **21. *Barongia***
- Stamens distinct but in groups of 5–8 opposite the petals **25. *Basisperma***

100. Placentas axile to apical; stamens usually fused into short fascicles; seeds often winged **26. *Tristaniopsis***
- Placentas basal; stamens distinct but sometimes grouped; seeds ovoid, not winged 101

101. Ovary 3-locular; capsule rounded, valves strongly lignified **23. *Ristantia***
- Ovary 2-locular; capsule conical, valves lightly lignified **24. *Mitrantia***

102. Anthers dorsifixed, versatile or subversatile 103
- Anthers erect and basifixed, immobile 118

103. Stamens not grouped, distinct (rarely fused into a short tube at the base) 104
- Stamens clustered or fused into distinct bundles (rarely fused into a tube) 114

104. Flowers solitary or in loose clusters or heads 105
- Flowers in spike-like inflorescences 113

105. Flowers ± sessile in dense heads 106
- Flowers usually pedicellate, solitary or in loose clusters 108

106. Stamens fused into a ring at the base **111. *Asteromyrtus***
- Stamens free 107

107. Stamens 10 (rarely fewer), one opposite each sepal and petal **108. *Taxandria***
- Stamens in groups opposite the sepals, none opposite the petals **107. *Agonis***

108. Hypanthium shallow; stamens shorter than petals 109
- Hypanthium relatively deep; stamens rarely shorter than petals 112

109. Seeds few, usually 1–2 per loculus, compressed, often shortly winged 110
- Seeds numerous, narrowly obovoid to cuneate or linear **116. *Leptospermum***

110. Fruit 4-angled **115. *Homalospermum***
- Fruit not angular 111

111. Branching dichotomous; ovules spreading to pendulous **110. *Pericalymma***
- Branching not dichotomous; ovules erect **109. *Paragonis***

112. Stamens much shorter than the petals **112. *Angasomyrtus***
- Stamens equal to or longer than the petals **113. *Kunzea***

113. Sepals persistent **113. *Kunzea***
- Sepals caducous **11. *Callistemon***

114. Ovules two per loculus **15. *Conothamnus***
- Ovules several per loculus 115

115. Stamens in separate bundles 116
- Stamen bundles fused into a tube **14. *Lamarchea***

116. Ovules mostly four per loculus, collateral in superposed pairs **18. *Petraeomyrtus***
- Ovules usually more numerous, scattered 117

117. Flowers and fruits in a globular head; staminal bundles joined in a ring at base; hypanthium usually circumscissile **111. *Asteromyrtus***
- Flowers and fruits in spikes, heads or clusters; staminal bundles free from each other; hypanthium persistent **12. *Melaleuca***

118. Stamens shorter than petals; ovary 5–12-locular; seeds winged **114. *Neofabricia***
- Stamens exceeding petals; ovary 3-locular; seeds not winged 119

119. Stamens in distinct bundles with relatively long claws 120
- Stamens distinct and grouped, or in bundles with short, broad claws 122

120. Anthers opening by longitudinal slits **13. *Calothamnus***
- Anthers opening by ± terminal slits or pores 121

121. Ovules solitary in each loculus; anthers opening by transverse slits **16. *Beaufortia***
- Ovules 2 or more per loculus; anthers opening by pores or lateral slits **17. *Regelia***

122. Flowers in dense heads; anthers opening by transverse slits **19. *Phymatocarpus***
- Flowers solitary or in groups of 2–5; anthers opening by lateral, longitudinal slits **20. *Eremaea***

123. Ovary 2–3-locular; fruit usually dehiscent 124

- Ovary 1-locular; fruit indehiscent 139
124. Hypanthium elongated, urceolate *141. Balaustion*
- Hypanthium shallow 125
125. Seeds reniform 126
- Seeds discoid or angular 131
126. Stamens with slender filaments, distinct or slightly united at the base 127
- Stamens with broad, ± flat filaments, distinct or fused 130
127. Stamens >20; flowers usually ± sessile *127. Hypocalymma*
- Stamens fewer than 20; flowers pedicellate 128
128. Seeds arillate; ovules 2 per loculus *119. Ochrosperma*
- Seeds not arillate; ovules 4 or more per loculus 129
129. Prophylls persistent; some stamens opposite centre of petals *136. Euryomyrtus*
- Prophylls caducous; no stamens opposite centre of petals *135. Triplarina*
130. Stamens 10, ± free from each other **118. Rinzia**
- Stamens 10–20, fused into groups or a ring **126. Astartea** s.l. (*Cyathostemon*)
131. Stamens grouped into 5 bundles opposite the sepals *126. Astartea*
- Stamens separate, not grouped 132
132. Anthers adnate, dehiscing by pores or short, oblique slits 133
- Anthers basifixed, dehiscing by long slits 137
133. Inflorescence usually of solitary flowers 134
- Inflorescence usually of (2–)3 or more flowers 135
134. Sepals with a dorsal lobe ('compound'); capsule apex convex *143. Kardomia*
- Sepals simple, obtuse; capsule apex ± flat *144. Harmogia*
135. Fruits indehiscent *140. Scholtzia*
- Fruits dehiscent 136
136. Peduncles very short; anthers dehiscing by terminal pores *138. Babingtonia* s.str.
- Peduncles long; anthers dehiscing by short slits *142. Sannantha*
137. Inflorescence usually of solitary flowers 138
- Inflorescence many-flowered *137. Stenostegia*
138. Hypanthium ribbed when dry; seeds reniform *134. Astus*
- Hypanthium smooth; seeds discoid to cuboid *117. Baeckea* s.str.
139. Stamens 8; flowers 4-merous, crowded into daisy-like heads *132. Actinodium*
- Stamens 5 or more, flowers 5-merous, usually axillary, rarely in bracteate heads 140
140. Fertile stamens 10 (rarely fewer), alternating with staminodes; or stamens 20, dimorphic 141
- Stamens usually all fertile, uniform, staminodes absent 145
141. Sepals deeply bilobed; stamens dimorphic *128. Pileanthus*
- Sepals not deeply bilobed; staminodes present 142
142. Sepals entire to shortly ciliate or erose 143
- Sepals with a single slender projection from the tip or divided into plumose or linear lobes 144
143. Style much longer than the petals *131. Darwinia*
- Style not exceeding the petals *129. Chamelaucium*
144. Sepals with a single slender projection or 2–12 linear divisions; petals entire to erose; ovules in 2 longitudinal rows *130. Homoranthus*
- Sepals divided into plumose segments or numerous hair-like divisions; petals often ciliate; ovules in a single row *133. Verticordia*
145. Stamens numerous, usually >15 (if fewer, then sepal with a long awn); leaves usually alternate (very rarely opposite or whorled); stipules often present, very small 146
- Stamens mostly up to 10 (rarely to 30); leaves opposite; stipules absent 147
146. Embryo straight; sepals often awned *124. Calytrix*
- Embryo with narrow, curved apex; sepals usually obtuse, scarious *125. Homalocalyx*
147. Staminal filaments geniculate *139. Malleostemon*
- Staminal filaments not geniculate 148
148. Anthers opening by a single pore, bearing a stalked appendage *121. Corynanthera*
- Anthers opening by a 2 pores or slits 149
149. Hypanthium ± cylindrical; ovules borne apically *120. Micromyrtus*
- Hypanthium obconical or urceolate; ovules ± basal 150
150. Connective gland clavate, as large as the anther cell, directed towards the outside of the flower *122. Aluta*
- Connective gland truncate, about half the length of the anther cell, directed towards the inside of the flower *123. Thryptomene*

Genera of Myrtaceae

I. Subfam. Psiloxyloideae Schmid (1980).

Plants dioecious; leaves spirally arranged, secretory cavities present (but not containing essential oils in *Psiloxylon*); stamens not inflexed in bud, anthers tetralocular at anthesis; ovules with bisporic, Allium-type embryo sac. $x = 12$.

I.1. Tribe Psiloxyleae A.J. Scott (1980).

Male flowers with two whorls of stamens and a pistillode; female flowers with staminodes, ovary superior, shortly stipitate; style very short, inserted into the top of the ovary; stigma 3-lobed; ovules hemi-campylotropous. Fruit a berry; seed unwinged.

1. *Psiloxylon* Thouars ex Tul. Fig. 40

Psiloxylon Thouars ex Tul., Ann. Sci. Nat., Bot. IV, 6: 138 (1856).

Small trees; dioecious. Leaves simples, entire and spirally arranged; lamina gland-dotted but lacking essential oil. Inflorescences axillary or ramiflorous, consisting of fascicles of 3 to 7 pedicellate flowers. Flowers 5-merous, sepals and petals imbricate in the bud. Fruit a thin-walled, 3-locular berry. Seeds numerous with a deeply scalariform-reticulate testa; embryo straight, the cotyledons flat, as long as the hypocotyl.

One sp., *P. mauritianum* (Hook.f.) Baill., a rare endemic to the islands of Réunion and Mauritius.

Fig. 40. Myrtaceae. *Psiloxylon mauritianum*. **A** Branchlet with flower buds. **B** Fruiting branchlet. **C** Flower bud. **D** Male flower. **E** Female flower. **F** Ovary sectioned. **G** Fruit, sectioned. **H** Seed. **I** Embryo. (Hooker 1860)

I.2. Tribe Heteropyxideae Harv. (1868).

Male flowers with up to 10 stamens in two whorls and a small pistillode; female flowers with staminodes and a sessile, superior ovary; style inserted into a pit on the ovary summit; stigma capitate; ovules hemitropous. Fruit a capsule; seed winged.

2. *Heteropyxis* Harv.

Heteropyxis Harv., Thes. Cap. 2: 18 (1863); Fernandes, Mitt. Bot. Staatssamml. München 10: 207–234 (1971), key.

Shrubs and small trees. Leaves simples, entire and spirally arranged; lamina gland-dotted and containing essential oil. Inflorescences terminal or axillary panicles. Flowers 4–5-merous, functionally unisexual; sepals and petals imbricate in the bud. Fruit a 2(–3)-locular loculicidal capsule. Seeds winged at each end, few to numerous; testa scalariform-reticulate; embryo straight, the cotyledons flat, as long as the hypocotyl.

Three species, endemic to south-eastern Africa (Malawi, Mozambique, South Africa, Swaziland and Zimbabwe).

II. Subfam. Myrtoideae Sweet (1827).

Plants with bisexual flowers, rarely andromonoecious; leaves spirally arranged or opposite, secretory cavities present, containing essential oils; stamens inflexed in bud, anthers bilocular at anthesis; ovules with monosporic, Polygonum-type embryo sac. $x = 11$.

II.3. Tribe Xanthostemoneae Peter G. Wilson (2005).

Juvenile leaves spiral, adult leaves sometimes opposite. Ovary almost superior, attached on a broad base; placentas upright or projecting into the loculus with ovules at right angles to it and borne in a ring or arc. Fruit subspherical; seeds semicircular, flat; embryo accumbent.

Three genera are currently recognised. However, preliminary morphological analysis suggests that *Pleurocalyptus* is probably embedded in *Xanthostemon* and that two small groups of Australian species may be generically distinct. These hypotheses need to be tested.

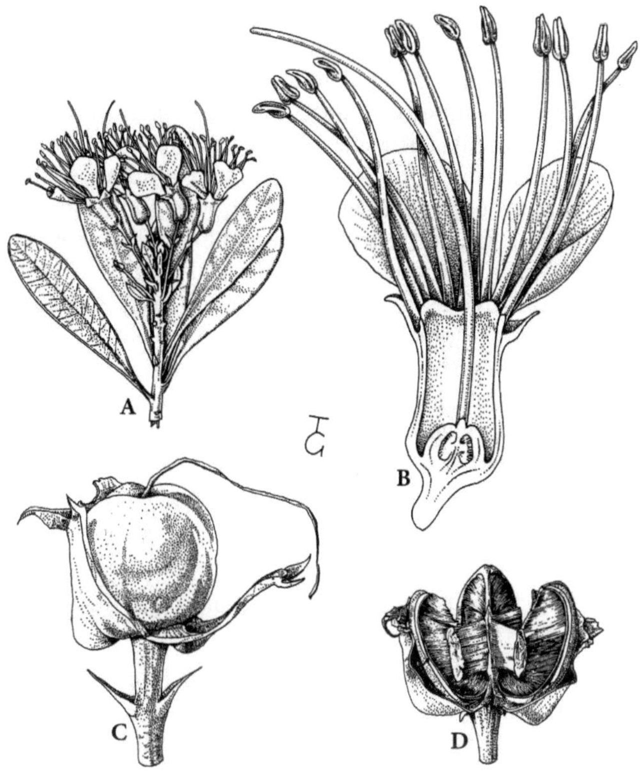

Fig. 41. Myrtaceae. *Xanthostemon auranticus*. **A** Flowering branchlet. **B** Flower, longitudinal section. **C** Unripe fruit. **D** Dehisced fruit; note the woody placentas. (Dawson 1992; drawn by T. Galloway)

3. ***Xanthostemon*** F. Muell. Fig. 41

Xanthostemon F. Muell., Hooker's J. Bot. Kew Gard. Misc. 9: 17 (1857); Wilson, Telopea 3: 451–476, rev. Aust. spp.; Dawson, Fl. Nouv. Caléd. 18: 162–216 (1992).

Shrubs to tall trees. Leaves of all species alternate in juveniles, some species becoming opposite. Inflorescences of axillary monads, triads, botryoids, metabotryoids, thyrsoids or panicles. Flowers 4- or 5-merous, yellow, white or red; stamens exceeding petals, numerous, free; anthers with a broad connective, gland-tipped, often with smaller glands elsewhere in the connective, dorsifixed near the base, the point of attachment enclosed by the connective; ovary almost superior to half-inferior, mostly 3–5-locular, surrounded by or included in a saucer- or cup-shaped hypanthium; style tapering to the small stigma; placentation axile, the ovules attached to the centre of the axis or to a rod-like placenta upright in the basal angle, or projecting from the axis horizontally or obliquely into the loculus; ovules numerous, flat and radially arranged, forming a complete or interrupted ring. Capsule globular, loculicidal, with the hypanthium partly enclosing, or flattened under, the fruit. Seeds flattened, semicircular to deltoid, occasionally winged; embryo folded with cotyledons accumbent to the hypocotyl.

Circa 45 spp., N & E Australia, New Caledonia, New Guinea, Solomon Islands, eastern Indonesia (Sulawesi, Maluku), Philippines.

4. ***Pleurocalyptus*** Brongn. & Gris

Pleurocalyptus Brongn. & Gris, Bull. Soc. Bot. France 14: 264 (1867).

Shrubs or trees. Leaves alternate; lamina bullate. Inflorescences axillary monads or triads. Flowers 5-merous; sepals fused into a calyptra that tears irregularly at anthesis; petals free, yellow; stamens numerous, much longer than petals; anthers with a broad connective, gland-tipped, dorsifixed near the base, the point of attachment enclosed by the connective; ovary almost superior, 4–6-locular included in the cup-shaped hypanthium; style slightly longer than the stamens, stigma small, narrower than the style; placentas axile, rod-like, projecting horizontally into the loculus; ovules numerous, flat and radially arranged, forming a complete ring. Capsule globular, loculicidal, with the hypanthium partly enclosing the fruit. Seeds flattened, semicircular to deltoid; embryo with broad cotyledons lying face-to-face; hypocotyl accumbent.

Two spp., New Caledonia.

5. ***Purpureostemon*** Gugerli

Purpureostemon Gugerli, Feddes Repert. Spec. Nov. Regni Veg. 46: 230 (1939).

Shrub. Leaves opposite or alternate (even on the same individual). Inflorescences a cluster of axillary monads. Flowers 5-merous, red; stamens much exceeding the petals, numerous, free; anthers with a broad connective, gland-tipped, often with smaller glands elsewhere in the connective, dorsifixed near the base, the point of attachment enclosed by the connective; ovary almost superior, 3–4-locular, surrounded by the shallow hypanthium; style as long as the stamens, stigma small, narrower than the style; placentas rod-like, oblique from the basal angles of the loculi; ovules forming a complete ring. Capsule broadly 3-lobed, loculicidal, the hypanthium flattened under the fruit. Seeds flattened, semicircular to deltoid, with a translucent wing; embryo with cotyledons lying face-to-face; hypocotyl accumbent.

One sp., *P. ciliatus* (Forst.) Gugerli, endemic to New Caledonia.

II.4. Tribe Lophostemoneae Peter G. Wilson (2005).

Adult leaves spirally arranged; oil ducts present in stems and petioles. Ovary inferior to half-inferior; placentas axile or apical, vascular supply to ovary trans-septal; ovules anatropous. Fruit a capsule or indehiscent and fleshy (*Kjellbergiodendron* only). Seeds variable, embryo with cotyledons enclosing one another.

6. *Kjellbergiodendron* Burret

Kjellbergiodendron Burret, Notizbl. Bot. Gart. Berlin-Dahlem 13: 101 (1936).

Small to medium-sized trees; stems and petioles with oil ducts. Inflorescence of panicles or thyrsoids. Flowers 5-merous, white; hypanthium shallow, margin more or less at the same level as the ovary summit; stamens very numerous, the filaments of varying lengths and united into triangular fascicles opposite the petals; ovary inferior, 2-locular; placentas axile, towards the base, ovules spreading to ascending; style inserted in a pit on the ovary summit; stigma slightly dilated. Fruit fleshy with a spongy pericarp; seed solitary, attached basally; embryo with thick, very unequal cotyledons, the outer, larger one completely enclosing the inner, smaller one.

A small genus, currently considered to comprise one species, *K. celebicum* (Koord.) Merr., from eastern Indonesia (Sulawesi and Maluku).

7. *Whiteodendron* Steenis Fig. 42

Whiteodendron Steenis, Acta Bot. Neerl. 1: 436 (1952).

Small to medium-sized trees; stems and petioles with oil ducts. Inflorescence terminal or axillary, panicles or metabotryoids. Flowers 5-merous, white; hypanthium very shallow, margin just lower than the ovary summit; stamens very numerous, the filaments of varying lengths and united into elongated fascicles opposite the petals and fused into a tube at the base; ovary half-inferior, 3-locular; placentas axile, towards the base, ovules spreading to ascending; style rather short; stigma small. Fruit almost superior, a small, slightly woody capsule; capsule dehiscing by 3 valves to reveal the solitary, basally attached seed; embryo with very unequal cotyledons, the outer, larger one completely enclosing the inner, smaller one.

One sp., *W. moultonianum* (W.W.Sm.) Steenis, from Borneo (Sarawak and Brunei).

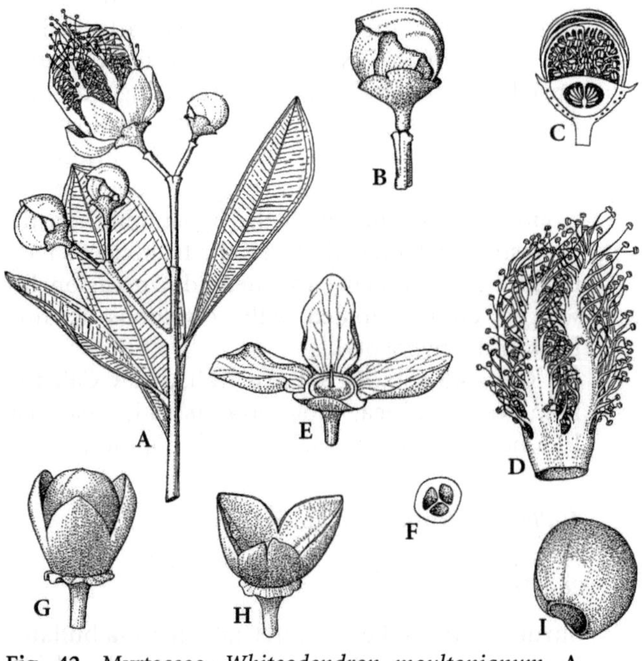

Fig. 42. Myrtaceae. *Whiteodendron moultonianum*. **A** Flowering branch. **B** Flower bud. **C** Longitudinal section of bud. **D** Androecium. **E** Flower beyond anthesis. **F** Ovary in cross section. **G** Dehisced fruit with protruding seed. **H** Fruit without seed. **I** Seed showing the hilum. (van Steenis 1953; drawn by R. van Crevel)

8. *Lophostemon* Schott

Lophostemon Schott, Wiener Z. Kunst 3: 772 (1830).

Trees or rarely shrubs; terminal buds covered with scales; stems and petioles with oil ducts that contain milky oil. Leaves alternate (opposite in juveniles), crowded in false whorls at the end of the branchlets, petiolate. Inflorescences axillary, cymose. Flowers 5-merous, white; sepals usually persistent in fruit, caducous in 1 species; stamens numerous, fused into 5 fascicles opposite the petals; ovary half-inferior, 3-locular; placentas axile, ovules numerous, anatropous; stigma capitate. Fruit a loculicidal capsule not, or scarcely, exceeding the fruiting hypanthium, 3-valved. Seeds linear; embryo straight, cotyledons obvolute or convolute.

Four spp., eastern Australia & Papua New Guinea; *L. confertus* (R.Br.) Peter G. Wilson & J.T. Waterh. is widely utilised for its timber.

9. *Welchiodendron* Peter G. Wilson & J.T. Waterh.

Welchiodendron Peter G. Wilson & J.T. Waterh., Austral. J. Bot. 30: 440 (1982).

Shrubs or trees, lacking bud scales; stems and petioles with oil ducts that contain a yellowish oil. Inflorescences axillary dichasia. Flowers 5-merous, yellow; sepals persistent; stamens numerous, fused into 5, somewhat bilobed fascicles opposite the petals; ovary half-inferior, 3(rarely 4)-locular; placentas apical, ovules borne in 2 overlapping tangential rows; style short, stigma slightly dilated. Fruit a conical capsule, strongly exserted, loculicidal, placentas falling as a unit. Fertile seeds winged; embryo straight, cotyledons obvolute or convolute.

One sp., *W. longivalve* (F.Muell.) Peter G. Wilson & J.T. Waterh., found in both Papua New Guinea and Australia (north Queensland).

II.5. Tribe Osbornieae Peter G. Wilson (2005).

Mangroves. Leaves opposite, petiole base expanded to enclose axillary bud. Flowers with a poorly differentiated perianth, appearing 8-merous; ovary 2-locular. Fruit leathery, indehiscent; seed solitary, embryo straight with cotyledons lying face-to-face.

10. *Osbornia* F. Muell.

Osbornia F. Muell., Fragm. 3: 30 (1862).

Shrubs or small trees; mangroves. Leaves emarginate, venation obscure; petiole base expanded into a pocket covering the axillary bud. Inflorescence of sessile axillary monads or triads. Flowers densely white-hairy, 8-merous; perianth broadly based, undifferentiated, persistent; stamens usually numerous, in one whorl, free; anthers dorsifixed, versatile; ovary inferior, 1–2-locular (incompletely septate); placentation axile, ovules numerous; style stout, stigma narrow. Fruit a leathery berry; seeds 1(2).

One sp., *O. octodonta* F. Muell., coastlines from northern Australia to the Philippines.

II.6. Tribe Melaleuceae Burnett (1835).

Trees or shrubs; leaves mostly spiral. Inflorescence often a spike-like (an aggregation of solitary flowers or triads commonly called 'bottlebrushes') or head-like clusters of flowers; flowers red, pink, yellow or white; stamens often fasciculate, anthers fixed or versatile, dehiscing by longitudinal slits, pores or apical slits. Fruit a very woody capsule, often with delayed dehiscence; seeds linear.

Generic limits in this tribe are under review. The transfer of all species of *Callistemon* to *Melaleuca* has already been proposed (Craven 2006) and on-going research suggests that some other genera are also in doubt. However, *Melaleuca* is likely to be shown to be polyphyletic and all relevant genera are listed here until there is fuller resolution of relationships in this group.

11. *Callistemon* R.Br.

Callistemon R.Br., Voy. Terra austral. 2 (App. 3): 547 (1814).

Shrubs or small trees; new growth villous. Inflorescences spike-like, oblong to cylindric; at first terminal, then axis usually growing into a leafy shoot. Flowers 5-merous, red, white or greenish; hypanthium urceolate to campanulate; stamens numerous, long, distinct or rarely shortly united at base; anthers versatile; ovary half-inferior, 3- or 4-locular; style with capitate stigma. Fruit a woody, loculicidal capsule, often persistent on stem and enlarging for many years; seeds numerous.

A popular genus in horticulture with c. 30 spp., endemic to Australia, all States.

12. *Melaleuca* L. Fig. 43

Melaleuca L. Mantissa Plant. 1: 14, 105 (1767); Syst. Nat. ed. 12, 2: 509 (1767); Craven & Lepschi, Austral. Syst. Bot. 12: 819–927 (1999), key.

Shrubs or trees. Leaves opposite, alternate or irregularly arranged; petiolate, sessile or peltately attached. Inflorescences terminal or axillary, spike- or head-like, the terminal bud usually growing on, occasionally aborting. Flowers 5-merous, white, yellowish, pink, red or mauve; sepals small, triangular to semicircular; petals free; stamens numerous, fused into bundles opposite the petals; anthers dorsifixed, versatile; ovary half-inferior, 3-locular, with several to many erect ovules in each loculus; style filiform, stigma simple or capitate. Fruit a loculicidal capsule, only rarely exserted from the fruiting hypanthium; seeds linear. $2n = 22, 24, 33$.

About 220 rather diverse spp., mostly endemic to Australia but with a few species extending to New Guinea and some other parts of Malesia.

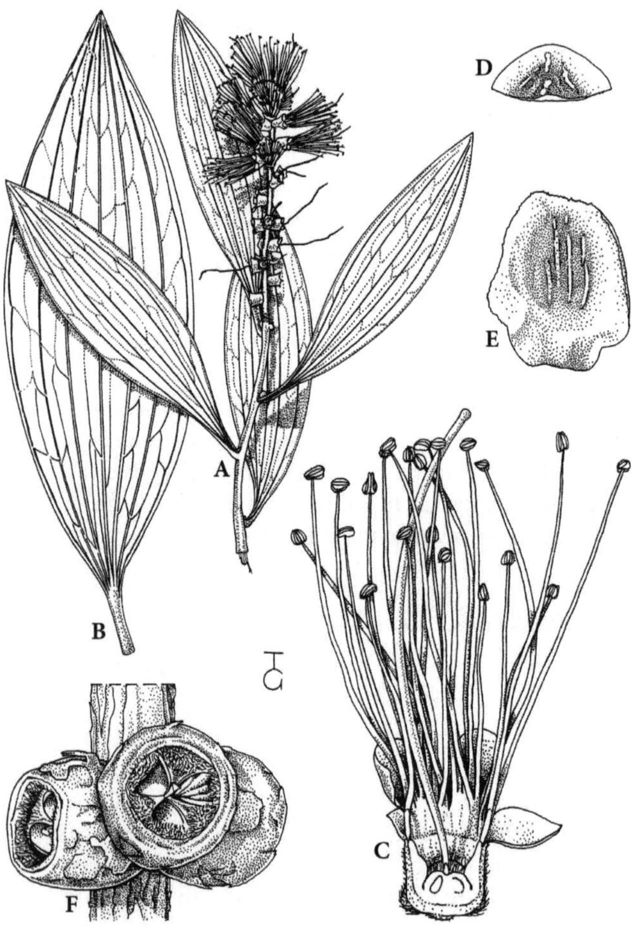

Fig. 43. Myrtaceae. *Melaleuca quinquenervia*. **A** Flowering branchlet. **B** A large leaf. **C** Flower in longitudinal section. **D** Sepal. **E** Petal. **F** Dehiscing fruits. (Dawson 1992; artwork by T. Galloway)

13. *Calothamnus* Labill.

Calothamnus Labill., Nov. Holl. Plant. Sp. 2: 25, t. 164 (1806); Hawkeswood, Nuytsia 6: 67–126 (1987), part rev.

Shrubs or small trees; branchlets glabrous or glabrescent. Inflorescences on older stems behind the apex; the flowers in axillary triads forming clusters or unilateral to cylindrical spike-like conflorescences. Flowers 4- or 5-merous, greenish, orange-red or red; sepals and petals free; stamens numerous, fused into bundles opposite the petals, the distinct filaments terminal on long claws (unequally broad in some species); anthers basifixed, opening by longitudinal slits; ovary half-inferior, 3-locular, with several ovules in each loculus; style tapering, stigma small. Fruit a woody, loculicidal capsule, the valves included in the fruiting hypanthium. Seeds linear-cuneate to oblong; embryo straight, cotyledons flat, longer than the hypocotyl.

A genus of 37 species; south-western Australia.

14. *Lamarchea* Gaudich.

Lamarchea Gaudich. in Freycinet, Voy. Uranie 483, t. 110 (1830).

Shrubs or tall shrubs; branchlets finely pubescent. Inflorescences on older stems behind the apex; the flowers scattered in axillary monads; buds covered by caducous, imbricate scales. Flowers 5-merous, green or orange-green to red; sepals and petals free; stamens numerous, in 5 groups opposite the petals, fused in the basal half into a hirsute tube; anthers dorsifixed, versatile, opening by longitudinal slits; ovary half-inferior, 3-locular, with numerous ascending ovules in each loculus; style slender, stigma dilated. Fruit a very woody, loculicidal capsule, the valves included in the fruiting hypanthium. Seeds few fertile, ± erect, oblong-cuneate; embryo not documented.

Two spp., south-western Australia.

15. *Conothamnus* Lindl.

Conothamnus Lindl., Swan River App.: 9 (1839).

Shrubs, often andromonoecious; branchlets silky-hairy. Inflorescences at the apex of the shoot; the flowers axillary, in small, head- or spike-like clusters. Flowers 5-merous, yellow to white; sepals and petals free; petals 0 in one species; stamens numerous, grouped opposite the petals, distinct or fused into bundles; anthers dorsifixed towards the top of the connective, versatile, opening by longitudinal slits; ovary half-inferior, 3-locular, with a single ovule in each loculus; style slender, stigma small. Fruit a woody, loculicidal capsule, the valves included or exserted from the fruiting hypanthium. Seeds ovoid to oblong; embryo straight, cotyledons as long as the hypocotyl, flattened plano-convex. $2n = 20$.

Three spp., south-western Australia.

16. *Beaufortia* R.Br.

Beaufortia R.Br., Hortus Kewensis ed. 2, 4: 418 (1812).

Shrubs or small trees, often andromonoecious. Leaves opposite, rarely alternate. Inflorescences at the apex of the shoot; the flowers axillary, in dense head- or spike-like clusters. Flowers 5-merous, red or deep pink; sepals and petals distinct; stamens few to numerous, grouped opposite the petals, fused into bundles, the distinct filaments terminal on a long, slender claw; anthers basifixed, opening at the apex by transverse slits; ovary inferior to half-inferior, 3-locular, with a single, peltately attached ovule in each loculus (sometimes 1 or 2 ovulodes present); style slender, stigma small. Fruit a woody, loculicidal capsule, the valves included in the fruiting hypanthium. Seeds obovoid-angular; embryo straight, cotyledons flat, longer than the hypocotyl. $2n = 16, 20$.

A genus of 17 species from south-western Australia.

17. *Regelia* Schauer

Regelia Schauer, Linnaea 17: 243 (Aug.-Oct. 1843); Dissertatio Phytographica de Regelia, Beaufortia et Calothamno 11 (Sep.-Nov. 1843).

Shrubs, often andromonoecious; branchlets glabrous or hairy. Leaves often distinctly 5-veined; peltately attached or shortly petiolate. Inflorescences at the apex of the shoot; the flowers axillary triads, in dense, head- or spike-like clusters. Flowers 5-merous, pink-purple, rarely red; sepals and petals free; stamens numerous, fused into bundles opposite the petals, the distinct filaments terminal on slender claws; anthers basifixed, opening by terminal slits or pores; ovary half-inferior, 3-locular; ovules usually 4 per loculus, collateral in 2 pairs on a peltate placenta; style slender, stigma small. Fruit a woody, loculicidal capsule, the valves included in the fruiting hypanthium. Seeds ovoid-angular; embryo straight, cotyledons flat.

A genus of 5 species from south-western Australia.

18. *Petraeomyrtus* Craven

Petraeomyrtus Craven, Austral. Syst. Bot. 12: 678 (1999).

Shrubs, andromonoecious; vegetative buds scaly. Inflorescences capitate, pseudoterminal with the axis usually growing on after anthesis, few- to several-flowered. Flowers in monads, bracteate and bracteolate, 5-merous; stamens numerous; filaments fused into fascicles opposite the petals; anthers basifixed, ± versatile, 2-celled, dehiscing by longitudinal slits; ovary half-inferior, 3-locular; ovules 4 per loculus, collateral in 2 pairs on an axile-basal peltate placenta, the proximal pair fertile and the distal pair sterile, the former much larger and lateral-basally attached; stigma not dilated. Fruit a scarcely woody, loculicidal capsule. Seeds narrowly obovoid to obovoid with a membranous testa; embryo straight, cotyledons about half the embryo length, obvolute.

A single species, *P. punicea* (Byrnes) Craven, which occurs on the sandstone plateau and outlying sandstone massifs of the Alligator River region of the Northern Territory, Australia.

19. *Phymatocarpus* F. Muell.

Phymatocarpus F. Muell., Fragmenta 3: 120 (1862).

Shrubs or tall shrubs; apical buds scaly, branchlets glabrous. Inflorescences at the apex of the shoot; the flowers axillary, in small, head-like clusters, surrounded by bud scales. Flowers 5-merous, deep pink; sepals and petals free; stamens few to numerous, grouped opposite the petals, ± distinct or fused at the base into bundles; anthers basifixed, opening at the apex by transverse slits; ovary half-inferior, 3-locular, with several to many ovules in each loculus; style slender, stigma small. Fruit a woody, loculicidal capsule, the valves included in the fruiting hypanthium. Seeds few fertile, ± erect, flattened obovoid; embryo straight, cotyledons flat. $2n = 20$.

A genus of 3 species from south-western Australia.

20. *Eremaea* Lindl. Fig. 44

Eremaea Lindl., Sketch Veg. Swan R.: 11 (1839); Hnatiuk, Nuytsia 9: 172–180 (1993), rev.

Shrubs or small trees; apical buds scaly, branchlets usually with a 2-layered indumentum. Inflorescences at the apex of the shoot; the flowers axillary, solitary or in small, head-like clusters, surrounded by bud scales. Flowers 5-merous, pink, deep pink, orange, or violet; sepals and petals free; stamens numerous, grouped opposite the petals, distinct or fused into bundles; anthers basifixed, opening by short, lateral or oblique slits; ovary half-inferior, 3-locular, with several to many ovules in each loculus; style slender, stigma small. Fruit a woody, loculicidal capsule, the valves included or exserted from the fruiting hypanthium. Seeds obovoid-angular, the fertile seeds with wings on the angles; embryo straight, cotyledons longer than the hypocotyl, obvolute.

A genus of 14 species from south-western Australia.

II.7. Tribe Kanieae Engl. (1930).

Trees or shrubs; leaves opposite or alternate. Inflorescences paniculate or cymose. Flowers yellow or white; stamens frequently in bundles; style terminal on the ovary, rarely in a shallow pit; style base not adjacent to placentas. Fruit a capsule.

Fig. 44. Myrtaceae. *Eremaea* x *codonocarpa*. **A** Flowering twig. **B** Leaves. **C** Fruiting twig. **D** Flower. **E** Stamen. **F** Style and top of ovary. **G** Transverse section of ovary. **H** Fruits. **I** Seeds and ovulode. (Hnatiuk 1993; artist Margaret Wilson)

The generic constitution of this tribe is still uncertain and further investigation is desirable. *Cloezia* is tentatively included here, although Wilson et al. (2005) considered it incertae sedis based on analysis of molecular data.

21. *Barongia* Peter G. Wilson & B. Hyland Fig. 45

Barongia Peter G. Wilson & B. Hyland, Telopea 3: 257 (1988).

Trees. Juvenile leaves spirally arranged; adult leaves opposite. Inflorescence a panicle. Flowers 5-merous, yellow; hypanthium shallow, margin equal to or slightly lower than the ovary summit; stamens very numerous, yellow, the filaments of varying lengths and united into linear, brush-like fascicles opposite the petals; ovary half-inferior, 3(4)-locular, incompletely septate at the apex; placentas basal, ovules anatropous; style short, terminal on the ovary; stigma slightly dilated. Capsule not very woody, loculicidal but with a single cavity containing one, very rarely two, relatively large seed(s). Seed angular; embryo with broad cotyledons that do not enclose one another but are sharply reflexed from the top of the hypocotyl, then folded back on themselves.

One species, *B. lophandra* Peter G. Wilson & B. Hyland, restricted to a small area in tropical Queensland, Australia.

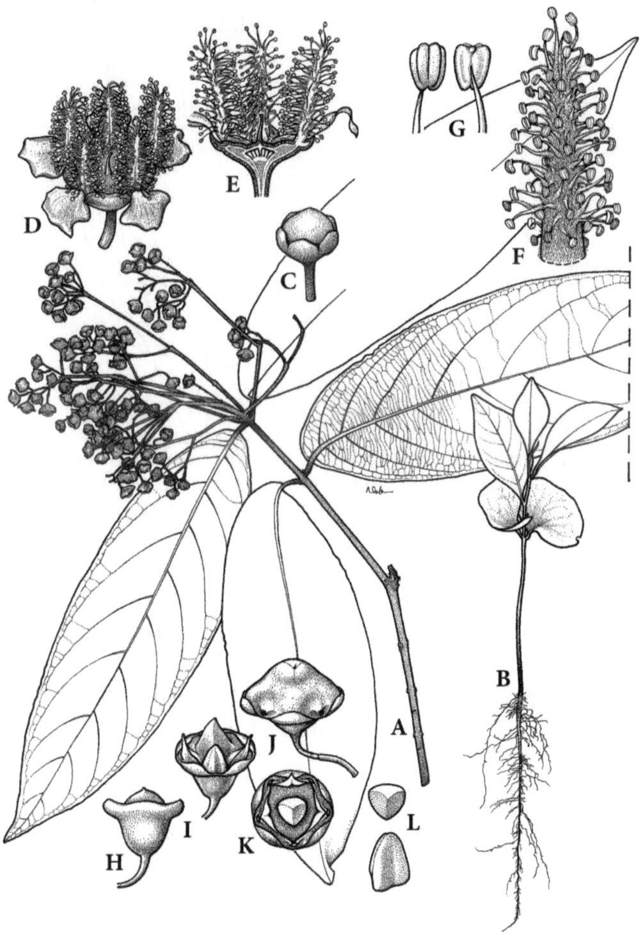

Fig. 45. Myrtaceae. *Barongia lophandra*. **A** Flowering branchlet. **B** Seedling with spiral eophylls. **C** Flower bud. **D** Flower. **E** Ditto, longitudinally sectioned. **F** Staminal bundle. **G** Stamens, front and back view. **H** Fruit. **I** Young fruit with remnants of corolla. **J** Mature capsule. **K** Loculicidally dehiscing capsule with solitary seed. **L** Seeds. (Wilson and Hyland 1988; drawn by T. Nolan; copyright: Centre for Plant Biodiversity Research)

22. *Sphaerantia* Peter G. Wilson & B. Hyland

Sphaerantia Peter G. Wilson & B. Hyland, Telopea 3: 260 (1988).

Trees, andromonoecious; oil glands present in the pith of young stems and petioles. Leaves opposite; venation brochidodromous. Inflorescences terminal or axillary thyrsoids or metabotryoids. Flowers 4- or 5-merous; hypanthium dish-shaped, exceeding the ovary summit; stamens numerous, the filaments of various lengths, aggregated into five basally connate fascicles opposite the petals; ovary half-inferior, 2–3-locular; placentas basal, ovules numerous, anatropous; style inserted in a slight depression on the ovary summit, not exceeding the stamens; stigma dilated, convex. Fruit only lightly lignified, truncate-globose; capsule included within the fruiting hypanthium, loculicidal, containing one or more seeds. Embryo with circinate cotyledons.

A genus of 2 species, from north Queensland, Australia.

23. *Ristantia* Peter G. Wilson & J.T. Waterh.

Ristantia Peter G. Wilson & J. T. Waterh., Austral. J. Bot. 30: 442 (1982).

Trees; oil glands present in the pith of young stems and petioles. Leaves alternate; venation eucamptodromous. Inflorescence a panicle or metabotryoid, axillary; flowers mostly hermaphrodite; petals 4 or 5, white to cream; sepals 4 or 5, not, or barely, apparent in the fruit; hypanthium dish-shaped, exceeding the ovary summit; stamens numerous, the filaments of various lengths, with at least some of them irregularly grouped in front of the petals; sterile anthers present, lacking an oil gland; ovary inferior, 3-locular, placentas basal, ovules 2 per loculus, anatropous; style terminal on the ovary, shorter than or barely exceeding the stamens; stigma capitate. Fruit strongly lignified; capsule globose, exserted from the fruiting hypanthium, loculicidal, containing one or more seeds. Seed obovoid; embryo with circinate cotyledons.

A genus of 3 species, from north Queensland, Australia.

24. *Mitrantia* Peter G. Wilson & B. Hyland

Mitrantia Peter G. Wilson & B. Hyland, Telopea 3: 264 (1988).

Trees, andromonoecious; oil glands frequent in the pith of young stems and petioles. Leaves alternate; venation weakly brochidodromous. Inflorescence a terminal or axillary thyrsoid. Flowers 5-merous, white; hypanthium cup-shaped, exceeding the ovary summit; stamens in a single series, aggregated into five groups opposite the petals; filaments free; ovary half-inferior, 2-locular; placentas basal, ovules few, anatropous, arranged in a semicircle on each placenta; style terminal on the ovary; stigma dilated. Capsule conical, exserted, lightly lignified, loculicidal, the base cupped by the fruiting hypanthium. Seeds 1 or 2; embryo with circinate cotyledons.

One species, *M. bilocularis* Peter G. Wilson & B. Hyland, restricted to a small area in tropical Queensland, Australia.

25. *Basisperma* C.T. White

Basisperma C.T. White, J. Arnold Arbor. 23: 84 (1942).

Small trees, oil glands present in the pith of young stems and petioles. Juvenile leaves opposite; adult leaves irregularly opposite to alternate. Inflorescence an axillary dichasial cyme. Flowers 5-merous, white; hypanthium cup-shaped, exceeding the ovary summit; stamens aggregated into five groups opposite the petals; filaments ± free; ovary almost superior, 2-locular; placentas basal, ovules few, anatropous, scattered on each placenta; style terminal on the ovary; stigma dilated. Capsule obovoid, strongly exserted, lightly lignified, loculicidal, the base cupped by the fruiting hypanthium. Seeds 1–3 with a reddish, wrinkled testa; embryo with folded cotyledons.

One species, *B. lanceolatum* C.T. White, a rare species from southern Papua New Guinea.

26. *Tristaniopsis* Brongn. & Gris Fig. 46

Tristaniopsis Brongn. & Gris, Bull. Soc. Bot. France 10: 371 (1863).

Trees or shrubs; branchlets hairy. Inflorescence axillary, cymose. Flowers 5-merous, usually yellow; stamens usually numerous and fused into short fascicles opposite the petals, the free ends inflexed; ovary half-inferior, usually 3-locular; ovules usually pendulous from the top of the axis; stigma capitate. Fruit a loculicidal capsule, strongly exserted from the hypanthium.

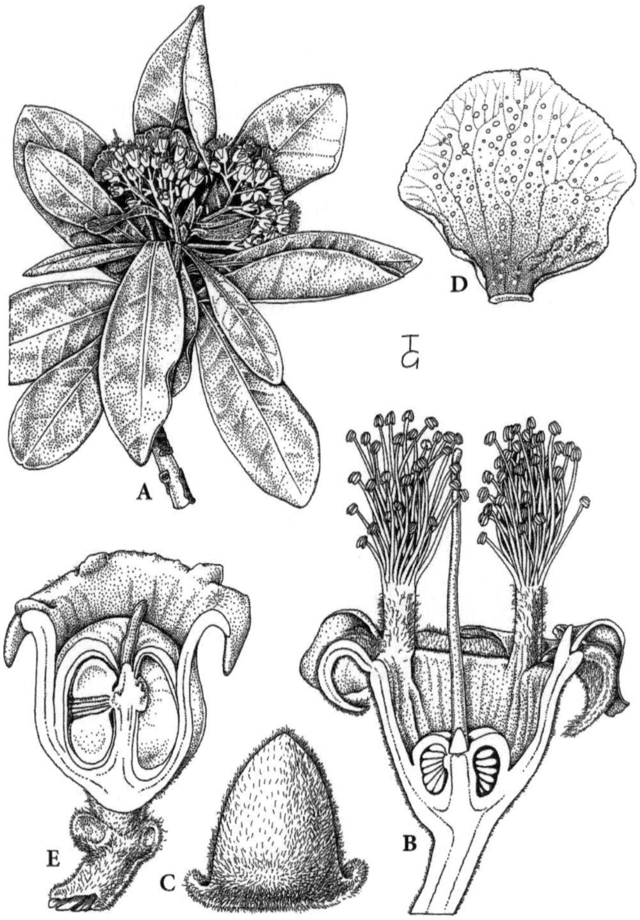

Fig. 46. Myrtaceae. *Tristaniopsis ninndoensis*. **A** Flowering branchlet. **B** Flower, longitudinal section. **C** Sepal. **D** Petal. **E** Fruit, longitudinal section. (Dawson 1992; drawn by T. Galloway)

Seeds mostly winged; embryo straight, cotyledons obvolute or convolute.

A genus of c. 40 species: Australia, New Caledonia, Malesia to Burma and Thailand.

27. *Lysicarpus* F. Muell.

Lysicarpus F. Muell., Trans. & Proc. Philos. Inst. Victoria 2: 61 (1857).

Andromonoecious trees. Leaves alternate to irregularly ternate. Inflorescences axillary triads (3-flowered cymes). Flowers 5-merous, white; hypanthium cup-shaped, exceeding the ovary summit; stamens numerous, free, with gaps opposite the sepals; outer filaments long, bearing sterile anthers; inner stamens shorter, fertile; ovary half-inferior, 3-locular; placentas oblique in the basal angles of the loculi; ovules few, anatropous, arranged in a circle on each placenta; style terminal on the ovary; stigma capitate. Fruit a thin-walled capsule, exserted from the fruiting hypanthium. Seeds linear; embryo straight with cotyledons lying face-to-face.

The single species, *L. angustifolius* (Hook.) Druce, is endemic to eastern Queensland.

28. *Kania* Schltr.

Kania Schltr., Bot. Jahrb. Syst. 52: 119 (1914).

Trees or shrubs. Inflorescence axillary, cymes or panicles. Flowers 5-merous, yellow; stamens in a single whorl, 2–5 times the number of petals, somewhat unequal in length, anthers with conspicuous, apically expanded connectives; ovary half-inferior, usually 3-locular; ovules scattered on the ± basal placenta; style terminal, remote from the placenta, stigma small. Fruit a loculicidal capsule, exserted from the hypanthium. Seeds linear; embryo straight, cotyledons lying face-to-face.

A genus of 6 species, New Guinea and the Philippines.

29. *Cloezia* Brongn. & Gris

Cloezia Brongn. & Gris, Bull. Soc. Bot. France 10: 576 (1863); Dawson, Fl. Nouv. Caléd. 18: 24–46 (1992), key.

Shrubs to small trees. Inflorescences usually axillary, cymes or monads. Flowers 5-merous, yellow or white; stamens in a single whorl, as long as the petals; anthers dorsifixed, versatile, connective sometimes expanded apically; ovary half-inferior, 3-locular; ovules few in a ± circular series on the basal placenta; style terminal, remote from the placenta, stigma small. Fruit a loculicidal capsule, exserted from the hypanthium. Seeds linear; embryo straight, cotyledons lying face-to-face.

A genus of 5 species, endemic to New Caledonia.

II.8. Tribe Backhousieae Peter G. Wilson (2005).

Trees; leaves opposite. Inflorescences cymes or panicles. Flowers white or yellowish; stamens free, numerous; style terminal on the ovary or in a shallow pit, style base not adjacent to placentas; ovary usually bilocular. Fruit dry, indehiscent. Seed with a thin testa, cotyledons incumbent to hypocotyl.

30. *Backhousia* Hook. & Harv.

Backhousia Hook. & Harv., Bot. Mag. 71, t. 4133 (1845).

Trees or shrubs. Inflorescences axillary or terminal, variable. Flowers 4- or 5-merous; sepals often prominent, often petaloid, persistent in fruit; petals usually shorter than sepals, persistent in fruit; stamens numerous, free in several whorls; anthers versatile; ovary half-inferior, 2-locular, summit domed, pubescent; ovules few, arranged in a U-shape on an axile placenta; style long, stigma not dilated. Fruit indehiscent or rarely schizocarpic, thin-walled, summit domed, surrounded by persistent sepals. Seeds depressed-ovoid, usually only 1 or 2 per fruit: embryo with an arcuate hypocotyl and incumbent cotyledons.

Around 8 species, endemic to eastern Australia.

31. *Choricarpia* Domin

Choricarpia Domin, Biblioth. Bot. 89: 472 (1928).

Trees; young stems ± 4-angled. Inflorescences dense, pedunculate, often more than one per axil, borne on a short lateral axis, with globose, many-flowered heads, but flowers ± distinct. Flowers 4- or 5-merous; sepals small; petals minute or absent; stamens numerous, free, in 1 row; ovary half-inferior, 2-locular; ovules solitary in each loculus, basal; style long, stigma slightly dilated to lobed. Fruit thin-walled, indehiscent, crowded in dense globose heads with ± persistent stamens. Seeds obovoid; embryo folded with cotyledons incumbent to the hypocotyl.

A genus of 2 species, endemic to eastern Australia.

II.9. Tribe Metrosidereae Peter G. Wilson (2005).

Trees or shrubs, growth monopodial, sometimes sympodial (in *Metrosideros* sens. str.). Leaves opposite. Inflorescences thyrsoids, metabotryoids, botryoids, cymes or flowers solitary. Flowers red or white, sometimes yellow; stamens free, numerous; ovary half-inferior; style inserted in the apex of the ovary, style base adjacent to placentas; ovary usually trilocular. Fruit a capsule. Seed linear; embryo straight; cotyledons lying face-to-face.

Although a well-defined tribe, there are still some questions regarding generic limits, particularly in relation to the species included here in the concept of *Mearnsia*.

32. *Tepualia* Griseb.

Tepualia Griseb., Abh. Königl. Ges. Wiss. Göttingen 6: 119 (1854).

Shrub or small tree; branchlets ± 4-angled when young, forming auriculate appendages at the petiole bases, the buds covered with scales. Inflorescences of axillary monads. Flowers 5-merous, each flower bearing 2 linear prophylls at the base of the hypanthium when in bud; sepals triangular, free; petals suborbicular to obovate, white; stamens numerous, uniseriate; anthers dorsifixed, versatile; ovary half-inferior, usually 3-locular; placenta basal; ovules numerous, scattered on the placenta; style gynobasic, stigma not dilated. Fruit a loculicidal capsule. Seeds linear, released through the top of the capsule, the capsule later degenerating between the 5 thickened ribs of the hypanthium; embryo straight, cotyledons lying face-to-face.

The single species, *T. stipularis* Griseb., is found in wet habitats in central and southern Chile and adjacent Argentina.

33. *Mearnsia* Merr.

Mearnsia Merr., Philipp. J. Sci., Bot. 2: 283 (1907).
Metrosideros subgen. *Mearnsia* (Merr.) J.W. Dawson (1976).

Trees, shrubs or root-climbing lianes; vegetative buds not conspicuously scaly, growth monopodial. Inflorescence terminal, axillary or borne on stems behind the leaves, varying in complexity from cymes to panicles. Flowers 5-merous, red; stamens in 1 or 2 whorls, 3–10 times the number of petals, somewhat unequal in length, anthers versatile, connective with one or more oil glands; ovary inferior, usually 3-locular; ovules numerous, scattered on the placenta; stigma not or only slightly dilated. Fruit a loculicidal capsule, included in the hypanthium. Seeds linear; embryo straight; cotyledons lying face-to-face.

A genus of c. 18 species: New Zealand, New Caledonia, New Guinea, Solomon Islands and the Philippines.

34. *Carpolepis* (J.W. Dawson) J.W. Dawson

Carpolepis (J.W. Dawson) J.W. Dawson, Bull. Mus. Natl. Hist. Nat., B, Adansonia 4: 466 [1984] (1985).
Metrosideros subgen. *Carpolepis* J.W. Dawson (1976).

Shrubs or trees, sometimes initially epiphytic; vegetative buds scaly, growth monopodial. Leaves opposite or verticillate. Inflorescences of triads borne in the axils of the scales of the expanding vegetative bud. Flowers 5-merous; sepals very unequal; petals free, yellow; stamens distinct, numerous, in 1–3 whorls, equal to or longer than the petals; anthers dorsifixed, connective with one or more oil glands; ovary half-inferior to inferior, usually 3-locular; ovules numerous, scattered on the placenta; stigma not dilated. Fruit a loculicidal capsule, partly exserted from the hypanthium. Fertile seeds with an oblique wing; embryo straight; cotyledons lying face-to-face.

A genus of 3 species, endemic to New Caledonia.

35. *Metrosideros* Gaertn. Fig. 47

Metrosideros Banks ex Gaertn., Fruct.1, t. 3, f. 10 (1788).

Shrubs or trees, sometimes initially epiphytic; aerial adventitious roots frequent; vegetative buds scaly, growth sympodial. Inflorescences of (monads or) triads borne in the axils of the scales of the expanding vegetative bud. Flowers 5-merous, usually red, rarely white or yellow; stamens numerous, in a single whorl, much exceeding the petals; anthers dorsifixed, versatile, connective with one or more oil glands; ovary half-inferior to inferior, usually 3-locular; ovules numerous, scattered on the placenta; stigma not or only slightly dilated. Fruit a loculicidal capsule, included or exserted from the hypanthium. Fertile seeds linear; embryo straight; cotyledons lying face-to-face. $2n = 22$.

Over 30 species; New Zealand, New Caledonia and Pacific islands (excluding Micronesia), extending to the Bonin Islands in the north, Hawaii in the northeast and Pitcairn Island in the southeast.

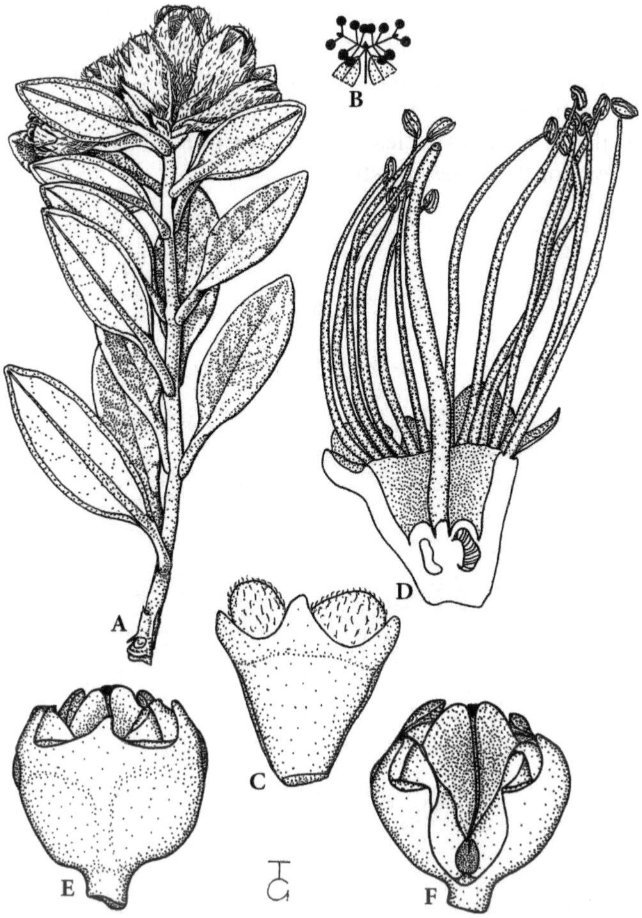

Fig. 47. Myrtaceae. *Metrosideros microphylla*. **A** Flowering branchlet. **B** Diagram of inflorescence. **C** Flower, stamens and style removed. **D** Flower, longitudinal section. **E** Dehiscing capsule. **F** Ditto, with pericarp partly removed. (Dawson 1992; drawn by T. Galloway)

II.10. Tribe Tristanieae Peter G. Wilson (2005).

Trees or shrubs. Leaves opposite, growth monopodial. Inflorescences thyrsoids, cymes or flowers solitary. Flowers yellow or orange to red; stamens free, usually fewer than 20; ovary half-inferior, style inserted in the apex of the ovary, style base adjacent to placentas; ovary usually trilocular. Fruit a capsule or berry (*Xanthomyrtus* only). Seed linear; embryo straight; cotyledons lying face-to-face or embryo with hypocotyl accumbent (*Xanthomyrtus* only). Pollen grains relatively small with a smooth exine.

Xanthomyrtus is certainly not a member of the tribe Myrteae but has only weak to moderate support as sister to *Tristania* and *Thaleropia* in analyses of molecular data. The genus has a fleshy rather than capsular fruit and differs in many other respects from these two genera; further research is required to clarify relationships amongst this group.

36. *Tristania* R.Br.

Tristania R.Br., Hort. Kew. ed. 2, 4: 417 (1812).

Shrubs or small trees. Inflorescences axillary, mostly triads or dichasia. Flowers 5-merous, yellow; calyx persistent; petals elliptic; stamens numerous, in 1 series, erect, fused into fascicles opposite the petals; anthers dorsifixed; ovary half-inferior, 3-locular; ovules covering an axile placenta; style as long as or longer than the stamens, stigma small. Fruit a loculicidal capsule, not exserted beyond the hypanthium, 3-valved. Seeds linear-cuneate; embryo straight; cotyledons about as wide as the hypocotyl, lying face-to-face.

One species, *T. neriifolia* (Sims) R.Br., endemic to the Sydney region of eastern Australia.

37. *Thaleropia* Peter G. Wilson

Thaleropia Peter G. Wilson, Austral. Syst. Bot. 6: 255 (1993).

Tall trees with scaly bark. Leaves with lateral veins numerous, at an angle of nearly 90° to the midvein, upper surface of the midvein with a narrow raised ridge. Inflorescence a many-branched thyrsoid showing regular abortion of the terminal flowers at the first to third bifurcations on the lateral dichasial units. Flowers 5-merous, yellow or red; hypanthium many-veined; sepals free, unequal; petals free; stamens 5–13(–15), anthers dorsifixed, with a small, gland-tipped connective; pollen grains very small, triporate; ovary half-inferior, usually 3-locular; style base immersed in the ovary summit to the level of the placentas; stigma small, convex; placentas axile; ovules very numerous. Fruit a lightly lignified capsule, enclosed within the fruiting hypanthium. Seeds linear to cuneate; embryo straight; cotyledons about as wide as the hypocotyl, lying face-to-face.

A genus of 3 species; NE Australia and New Guinea.

38. *Xanthomyrtus* Diels

Xanthomyrtus Diels, Bot. Jahrb. Syst. 57: 362 (1922); Scott, Kew Bull. 33: 461–484 (1979), key.

Trees or shrubs; branchlets hairy, often conspicuously glandular. Inflorescence of monads or triads. Flowers yellow, 4(–5)-merous, sessile; stamens usually numerous, 1(–2)-seriate, distinct; ovary inferior, 2–3(–4)-locular; ovules 10–20, arranged around the margin of the axile placenta; stigma small. Fruit a fleshy berry, reddish to blue-black. Seeds many, small, with a crustaceous testa; embryo with broad cotyledons lying face-to-face; hypocotyl accumbent.

A genus of 23 species, Philippines to New Caledonia.

II.11. Tribe Syzygieae Peter G. Wilson (2005).

Trees, rarely shrubs, lacking simple hairs. Leaves opposite. Inflorescences mostly terminal and paniculate; ovary inferior, mostly 2-locular; placentas axile or apical, ovules pendulous or radiating; vascular supply to ovary axial. Fruit indehiscent, mostly fleshy (except *Anetholea*). Seeds usually solitary; embryo usually with thick cotyledons, occasionally ± ruminate, rarely leafy (*Anetholea*).

Recent research (Biffin et al. 2006, 2007) has further confirmed the monophyly of this group but has concluded that all the genera should be included within a single, large genus, *Syzygium*. These analyses do support monophyly of many of the segregate genera but the lack of resolution between the clades has led to a more inclusive approach being taken. This approach is not followed here but, even if segregates are recognised, some genera do require reconsideration. For example, *Piliocalyx* should perhaps be regarded as merely a calyptrate variant of *Acmena*.

39. *Anetholea* Peter G. Wilson Fig. 48

Anetholea Peter G. Wilson, Austral. Syst. Bot. 13: 434 (2000).

Glabrous tree. Inflorescence a terminal or axillary metabotryoid. Flowers 4- or 5-merous, white; sepals ± equal, triangular, not overlapping in bud; petals free; stamens numerous in 1–2 rows opposite the sepals but in 2–3 rows opposite the petals; anthers dorsifixed towards the base, connective lacking a conspicuous gland; ovary inferior, 2-locular; style tapering to a narrow stigma; placentas borne at the apex of a strong, axile vascular strand; ovules anatropous, c. 6–7 per placenta. Fruit dry, indehiscent, with persistent sepals. Seeds usually one per fruit; embryo straight; cotyledons obvolute, germination epigeal.

One species, *A. anisata* (Vickery) Peter G. Wilson, a rare species endemic to north-eastern New South Wales, Australia.

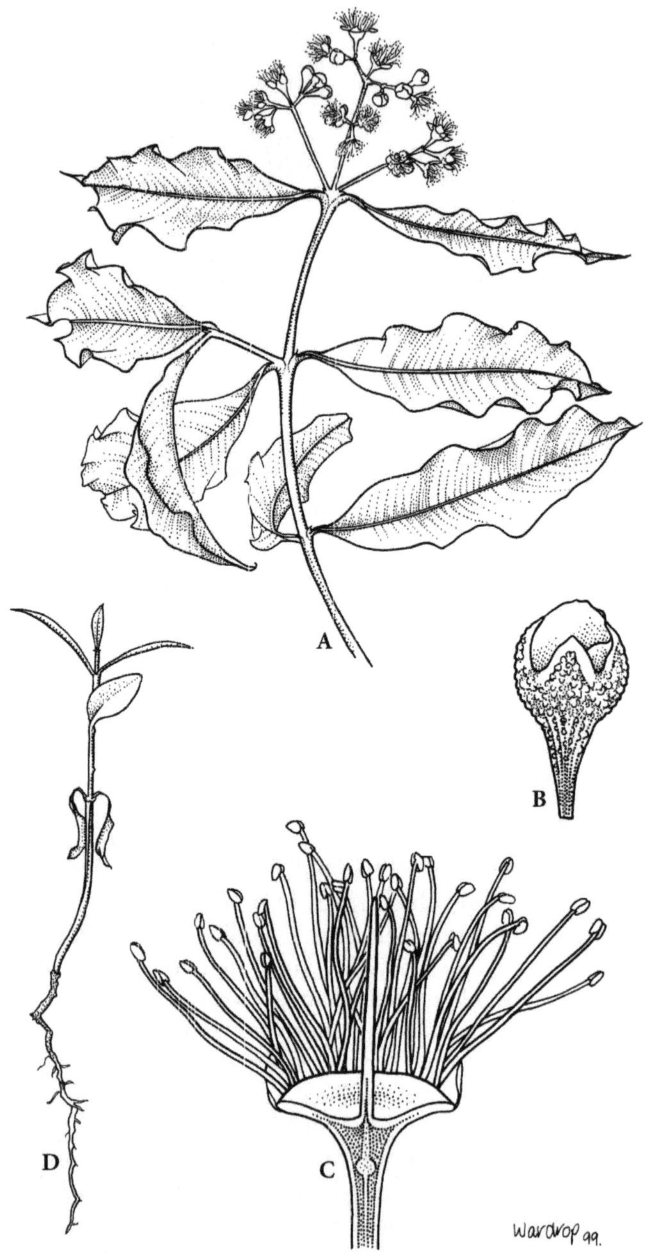

Fig. 48. Myrtaceae. *Anetholea anisata*. **A** Flowering branch. **B** Flower bud. **C** Longitudinal section of flower. **D** Seedling showing epigeal germination with leafy cotyledons. (Wilson et al. 2000; artist: C. Wardrop)

40. *Acmenosperma* Kausel

Acmenosperma Kausel, Ark. Bot. II, 3: 609 (1957).

Trees with glabrous branchlets. Inflorescence usually a panicle, terminal or in the upper axils, bracts caducous, absent at anthesis. Flowers with clavate buds; sepals 4–5, rounded, small and inconspicuous; petals 6–8, ± orbicular; stamens of medium length, anthers sacs parallel; ovary 2–3-locular, ovules pendulous in two vertical rows; style about as long as the stamens. Fruit a berry, cylindrical to pyriform, sepals not conspicuous; pericarp succulent, marked (in section) by radiating cells. Seed solitary; testa apparently absent; cotyledons ruminate, the intercotyledonary inclusion ramifying from the side or apex.

A genus of 2 species, Australia to India and S China.

41. *Acmena* DC.

Acmena DC., Prodr. 3: 262 (1828).

Shrubs or trees; branchlets glabrous. Inflorescence usually a panicle, terminal or in the upper axils, bracts often absent at anthesis. Flowers 4–5-merous, buds clavate to turbinate, usually small, sepals rounded or apiculate, petals orbicular, often irregular in outline; filaments short; anthers reniform with divaricate anther-sacs; ovary 2-locular; ovules pendulous from upper axile placentas; style usually shorter than the stamens. Fruit a succulent to leathery berry; sepals small or inconspicuous; the apex of the fruit often marked by a circular line. Seed solitary, testa apparently absent, cotyledons ruminate with an intercotyledonary inclusion ramifying from the apex.

A genus of 15 species from Australia and Malesia.

42. *Piliocalyx* Brongn. & Gris

Piliocalyx Brongn. & Gris, Bull. Soc. Bot. France 12: 185 (1865); Ann. Sci. Nat. V, 3: 225 (1865).

Shrubs or trees; branchlets glabrous. Inflorescences terminal thyrsoids, metabotryoids or botryoids. Flowers 4-merous, buds obovoid-clavate to turbinate; sepals fused into a calyptra; petals minute, fused to the operculum; stamens numerous, biseriate; anthers with divergent anther-sacs; ovary 2-locular; ovules 5–12, pendulous from upper axile placentas; style usually shorter than the stamens; stigma small. Fruit a berry, globular to depressed globular; pericarp succulent to leathery, topped by the persistent fruiting hypanthium. Seed solitary, testa apparently absent, cotyledons ruminate with an intercotyledonary inclusion ramifying from the apex.

Around 10 species; New Caledonia, New Hebrides, Fiji.

43. *Waterhousea* B. Hyland

Waterhousea B. Hyland, Austral. J. Bot., Suppl. Ser. 9: 138 (1983).

Trees or rarely shrubs, usually glabrous. Inflorescence terminal or axillary, usually a panicle. Flowers 4- or 5-merous; sepals free, small; stamens numerous; anthers dorsifixed, versatile; anther sacs parallel; ovary ± inferior, 2-locular; ovules pendulous from upper axile placentas; style about as long as the stamens; stigma small. Fruit a firm berry, globose to depressed globose with the persistent, unexpanded free portion of the hypanthium remaining at the apex. Seed solitary; testa apparently absent; cotyledons ruminate, the intercotyledonary inclusion ramifying from the base.

A genus of 4 species endemic to Australia.

44. *Syzygium* Gaertn.

Syzygium Gaertn., Fruct. 1: 166 (1788); Hyland, Austral. J. Bot., Suppl. Ser. 9: 31–138 (1983), rev. Aust. spp.; Dawson, Fl. Nouv. Caléd. 23:7–144 (1999), key.
Cleistocalyx Blume (1849).

Trees or rarely shrubs, usually glabrous. Inflorescence terminal or axillary (occasionally ramiflorous or cauliflorous), usually a panicle. Flowers 4(5)-merous, calyx occasionally calyptrate; petals distinct or occasionally forming calyptra; stamens numerous; anthers dorsifixed, versatile, anther sacs parallel; ovary ± inferior, 2(3)-locular, ovules several on an axile placenta; style about as long as the stamens, stigma small. Fruit succulent, flesh firm or spongy, usually purple to red, calyx usually persistent. Seed usually solitary, rarely 2 or more; testa soft, leathery or papery; cotyledons thick, distinct. $2n = 22, 44, 66, 110$.

Over 500 species, tropical and subtropical regions of the Old World.

II.12. Tribe Myrteae DC. in Schltdl. (1827).

Trees, sometimes shrubs; hairs simple, occasionally T-shaped. Leaves opposite. Ovary inferior, mostly 2- or 3-locular; placentas axile; ovules radiating, sometimes pendulous; vascular supply to ovary trans-septal. Fruit indehiscent, fleshy. Seeds usually numerous; embryo variable, cotyledons small to leafy or thick.

Historically, Myrteae have been divided into 3 subtribes based on generalised embryo morphology: (1) Subtribe Myrtinae with relatively small cotyledons, narrower than the hypocotyl and a hard seed coat, (2) Subtribe Myrciinae with folded, leafy cotyledons encircled by a long hypocotyl and a membranous to crustaceous seed coat, and (3) Subtribe Eugeniinae with thick, fleshy cotyledons that vary from separate to fused and a membranous seed coat. These divisions have been shown to be largely artificial and Lucas et al. (2007) recognise seven informal groups plus a couple of isolated genera, *Myrtus* and one of two species of *Blepharocalyx* in their analysis. *Myrtus* is found to be sister to all other members of the Myrteae and *Blepharocalyx* is sister to all other South American taxa.

'Ungrouped 1':

45. *Myrtus* L.

Myrtus L., Sp. pl.: 471 (1753).

Tall shrubs or small trees; hairs simple. Inflorescence an axillary monad. Flowers 5-merous; prophylls narrow, caducous; sepals and petals free; stamens very numerous, multiseriate; style longer than the stamens, narrowing to a small convex stigma; ovary 3-locular; ovules numerous in each locule on an axile placenta. Fruit crowned by the calyx; seeds few; seed coat hard and shiny; embryo C-shaped; cotyledons nearly as long as the hypocotyl.

A genus of 2 species: Mediterranean, Middle East and north Africa.

46. *Calycolpus* O. Berg

Calycolpus O. Berg, Linnaea 27: 348, 378 (Jan. 1856, '1854').

Trees and shrubs; hairs simple or two-armed. Leaves opposite, rarely subopposite. Inflorescence uniflorous or a very short bracteate shoot of 2–6 flowers. Flowers normally 5-merous; calyx open, the calyx lobes often flared below, often with an apical leafy appendage; ovary 2–6-locular, the locular wall in fruits sometimes glandular; ovules 8–30 per locule, the placenta a U-shaped pad or mound of tissue or essentially round and peltate. Fruit crowned by the calyx lobes. Seeds few to numerous, reniform; seed coat relatively hard, shiny; embryo C-shaped; hypocotyl much longer than the cotyledons.

Around 10 species, mostly in north-eastern South America.

47. *Meteoromyrtus* Gamble

Meteoromyrtus Gamble, Kew Bull. 1918: 241 (1918).

Shrubs or small trees; young branchlets villous. Inflorescences of axillary monads, frequently with accessory flowers from the same axil. Flowers 4-merous; hypanthium barely extended above the ovary; sepals and petals distinct; prophylls linear, borne at the base of the flower and exceeding it; stamens numerous, multiseriate, free, attached over the inner surface of the hypanthium; anthers dorsifixed at the base, dehiscing by longitudinal slits; ovary inferior, 2-locular; ovules c. 4 per loculus, pendulous from upper axile placentas; style exceeding the stamens, stigma dilated. Fruit, seed and embryo not documented.

The single species, *M. wynadensis* (Bedd.) Gamble, is rare and restricted to the western Ghats (India). Its phylogenetic position is uncertain.

'Australasian group':

48. *Decaspermum* J.R. & G. Forst.

Decaspermum J.R. & G. Forst., Charact. Gen.: 37 (1775); Scott, Kew Bull. 34: 59–67. (1979), key spp. Aust. Pacific; Scott, Kew Bull. 35: 403–411 (1980), key spp. Asia.
Pyrenocarpa H.T.Chang & R.H.Miao (1975).

Trees or shrubs; new growth glabrous or hairy. Inflorescence axillary, occasionally terminal; flowers solitary or in cymes, botryoids, metabotryoids or panicles. Flowers usually 5-merous; sepals persistent; petals pink or white; stamens numerous; anthers globular; ovary inferior, 3–12-locular; ovules 2–4 per loculus, collateral; stigma capitate. Fruit vertically ribbed. Seeds borne vertically and held tightly together by a membranous capsule; testa bony; embryo curved, cotyledons very small.

About 30 species, China to Australia and Pacific Islands.

49. *Austromyrtus* (Nied.) Burret s.str.

Austromyrtus (Nied.) Burret, Notizbl. Bot. Gart. Berlin, 15: 500 (1941); Snow & Guymer, Syst. Bot. Monogr. 65 (2003), rev.

Trees or shrubs. Leaves opposite, petiolate. Inflorescences in leaf axils solitary; flowers solitary. Flowers 5-merous; hypanthium not extending above the ovary summit; sepals persistent in fruit, ascending above body of fruit; petals free; stamens numerous, free; ovary inferior, 2-locular; ovules 6–10 in each loculus; placentation axile; stigma small. Fruit a soft berry, globose to subglobose. Seeds 1 to many, not arranged in regular rows, testa hard or bony; embryo curved, cotyledons small.

A genus of 3 species, endemic to Australia.

50. *Gossia* N. Snow & Guymer Fig. 49

Gossia N. Snow & Guymer, Syst. Bot. Monogr. 65: 31 (2003), Aust. spp.

Shrubs or large trees; branchlets rounded, compressed or 4-angled. Inflorescence a monad, or raceme-like or panicle-like. Flowers 4- or 5-merous; hypanthium not extending beyond the ovary summit; sepals mostly persistent in fruit and held above the body of the fruit; petals free; stamens numerous, free; ovary inferior, 2-locular; ovules few to many in each loculus; placentation axile; stigma small. Fruit a soft berry, yellow-orange or reddish, or more usually dark purple to black at maturity. Seeds usually 2–10, not obviously arranged in rows, testa bony to leathery; embryo usually helical; cotyledons small.

A genus of c. 30 species, Malesia to New Caledonia, Fiji, Australia.

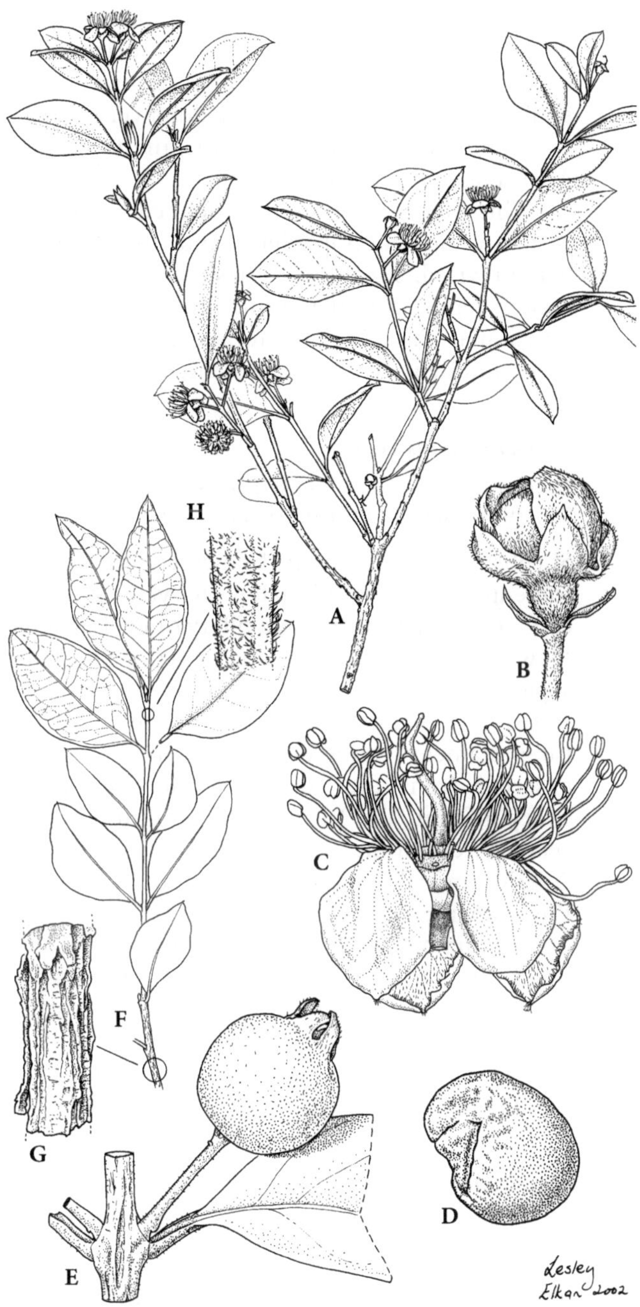

Fig. 49. Myrtaceae. *Gossia fragrantissima*. **A** Habit. **B** Bud. **C** Flower. **D** Seed. **E** Fruit in situ. **F** Branch with new growth. **G** Detail of mature stem. **H** Detail of hairs on young stem. (Drawn by L. Elkan)

51. *Rhodomyrtus* (DC.) Rchb.

Rhodomyrtus (DC.) Rchb., Deutsche Bot. Herbarienbuch Nom.: 77 (1841).

Shrubs or small trees; branchlets hairy. Leaves triplinerved or pinnately nerved. Inflorescence of axillary monads, triads or 7-flowered dichasia. Flowers 4–5-merous, white, pink or red, pedicellate; stamens numerous, free, in 3–6 series; anthers versatile; ovary inferior, usually 3–4-locular; hypanthium not, or scarcely, exceeding the ovary summit; ovules in 2 series on an axile placenta; style slender; stigma capitate. Fruit a dry or fleshy berry. Seeds borne horizontally, usually many, ± reniform, with a hard testa; embryo curved or coiled, cotyledons very small.

About 18 species, southern China to eastern Australia and New Caledonia.

This genus, as presently defined, is almost certainly not monophyletic and its status is being revised. Snow (2009) has recognised a segregate genus, *Kanakomyrtus*, endemic to New Caledonia, which is characterised by linear stylar lobes and densely glandular stamens.

52. ***Octamyrtus*** Diels

Octamyrtus Diels, Bot. Jahrb. Syst. 57: 373 (1922); Craven & Sunarti, Gard. Bull. Singapore 56:147–152 (2004), key.

Shrubs or small trees; branchlets villous or tomentose. Leaves pinnately nerved. Inflorescence of monads or triads, axillary or cauliflorous. Flowers 4-merous, white, pink or red, pedicellate; sepals 4; petals 6, 8 or 12, markedly unequal, rather long, distinct but forming a tube-like structure; stamens numerous, free, in 4–6 series, exserted from the corolla; anthers basifixed; ovary inferior, 4(–7)-locular; hypanthium not, or scarcely, exceeding the ovary summit; ovules in 2 series on axile placentas; style about as long as the stamens; stigma capitate. Fruit a globose to oblong, dry berry. Seeds borne horizontally, separated by vertical and horizontal false septa, numerous, ± reniform; embryo curved, cotyledons very small.

Six species, Malesia (Moluku, Irian Jaya, Papua New Guinea).

53. ***Rhodamnia*** Jack

Rhodamnia Jack, Malayan Miscellany 2(7): 48 (1822); Snow, Syst. Bot. Monogr. 82 (2007) rev, Australian spp.

Trees or shrubs. Leaves 3-veined from base and with numerous ± transverse veins. Inflorescences axillary, solitary or clustered. Flowers 4-merous; sepals free, persistent in fruit; petals free, obovate to ± circular; stamens very numerous; ovary ± inferior, 1-locular; hypanthium not extending beyond the ovary summit; placentas usually 2, parietal, each bearing several ovules; style slender; stigma usually capitate. Fruit a small berry. Seeds few, globose to reniform, testa bony; embryo curved, cotyledons very small.

A genus of c. 28 species, China to New Caledonia, Australia.

54. ***Archirhodomyrtus*** (Nied.) Burret

Archirhodomyrtus (Nied.) Burret, Repert. Spec. Nov. Regni Veg. 50: 59 (1941).

Trees or shrubs; new growth glabrous or minutely pubescent. Inflorescences axillary, flowers solitary or more rarely in 2- or 3-flowered cymes. Flowers 5-merous; hypanthium urceolate, not extending beyond the ovary summit; sepals distinct; petals distinct; stamens numerous; anthers ± globose; ovary ± inferior, 2- or 3-locular; ovules numerous, in 2 rows; stigma peltate. Fruit a berry, smooth. Seeds borne horizontally, numerous, not separated by false septa; testa bony; embryo curved; cotyledons very small.

A genus of 5 species, New Caledonia, Australia.

55. ***Uromyrtus*** Burret

Uromyrtus Burret, Notizbl. Bot. Gart. Berlin-Dahlem 15: 490 (1941); Snow, Syst. Bot. 26: 733–742 (2001), rev. Australian spp.

Shrubs or small trees; new growth silky or glabrous. Inflorescences axillary, flowers often solitary, or in pairs 3-flowered clusters. Flowers 5-merous, pendent; sepals rounded, persistent; petals white or pink; stamens numerous, multiseriate; anthers with an apical appendage; ovary ± inferior, 2- or mostly 3-locular; ovules in 2 vertical rows on the placenta; style slender; stigma small, not dilated. Fruit a fleshy berry, pendent. Seeds few to numerous, small; testa bony; embryo curved; cotyledons very small.

About 20 species, chiefly New Guinea to New Caledonia, Australia.

56. ***Pilidiostigma*** Burret

Pilidiostigma Burret, Notizbl. Bot. Gart. Berlin-Dahlem 15: 547 (1941); Snow, Syst. Bot. 29: 393–406 (2004), rev.

Shrubs or small trees. Inflorescences axillary, 1-flowered or raceme-like. Flowers 4- or 5-merous; sepals obtuse, small, persistent in fruit; petals free; stamens numerous, in 2 or more series; anthers sub-basifixed; ovary ± inferior, 2- or 3-locular; placentation axile, ovules in 1 row, few to several per loculus; style thick, stigma peltate. Fruit a succulent berry. Seeds few; testa tuberculate, soft; embryo folded; cotyledons small.

A genus of 5 species, endemic to Australia.

57. *Lithomyrtus* N. Snow & Guymer

Lithomyrtus F. Muell. ex N. Snow & Guymer, Austrobaileya 5: 182 (1999).

Shrubs or small trees. Inflorescence of axillary monads; prophylls scale-like, persistent or caducous in fruit. Sepals 5, imbricate in bud; stamens numerous, multiseriate; anthers dorsifixed, versatile, dehiscing by longitudinal slits; ovary 1-locular; ovules (1)2(3) per locule, placentation parietal; style flexuous or mostly straight; stigma small, rarely capitate. Fruit yellowish to olive-green, a hard berry containing a stone-like mass of bony, tightly fused seeds. Embryo circinate; cotyledons circinate, relatively thin.

A genus of 11 species, tropical Australia (Western Australia, Northern Territory, Queensland).

58. *Myrtella* F. Muell.

Myrtella F. Muell., Descr. Notes on Papuan Plants 1: 105 (1877).

Shrubs or small trees; branchlets distinctly four-winged. Inflorescence of axillary monads; prophylls foliaceous, mostly persistent in fruit. Sepals 5, valvate in bud; stamens numerous, uni- or multiseriate; anthers dorsifixed, versatile, dehiscing by longitudinal slits; ovary 2–3(4)-locular; ovules 2–4 per locule; placentation parietal; style flexuous, or mostly straight, glabrous; stigma small to scarcely capitate. Fruit red or dark blue to brown or nearly black, a hard berry containing a stone-like mass of bony, slightly fused seeds. Embryo slightly curved to crescent-shaped; cotyledons straight, relatively thick.

A genus of 2 species, Papua New Guinea, Irian Jaya, Guam; reported but not verified from the Caroline and Mariana Islands (Diels 1921).

'Ungrouped 2':

59. *Blepharocalyx* O. Berg

Blepharocalyx O. Berg, Linnaea 27: 348, 412 (1856, '1854'); Landrum, Fl. Neotrop. Monogr. 45: 115–130 (1986), key.

Trees or shrubs; hairs simple. Inflorescence uniflorous or with 3 to c. 35 flowers in a dichasium or panicle. Flowers 4-merous; calyx open with 4 lobes or closed and tearing into 4 equal, triangular lobes, the lobes caducous at anthesis; ovary normally 2-locular; ovules 4–17 per locule. Fruit crowned by a square scar. Seeds 1 to about 11; seed coat submembranous; embryo ± straight or C-shaped, the hypocotyl swollen, the cotyledons very small.

A genus of 3 species, Caribbean to southern Chile.

Blepharocalyx is an anomalous genus with a somewhat eugenioid embryo. However, in cross-section a circular vascular core is visible that is never found in that embryo type. Molecular analyses (Lucas et al. 2007) suggest that *Temu*, one of the genera currently considered a synonym of *Blepharocalyx*, may warrant recognition. The status of the other included genus, *Marlieriopsis*, is uncertain.

'Myrteola group':

60. *Myrteola* O. Berg

Myrteola O. Berg, Linnaea 27: 393 (1856).

Subshrubs or shrubs; hairs simple. Leaves opposite, decussate, often markedly 4-ranked. Inflorescence an axillary monad. Flowers 4- or 5-merous; prophylls foliaceous, persistent; stamens 9–75; ovary 2–3-locular, the septum between locules sometimes not complete; ovules 2–14, biseriate. Fruit a berry. Seeds small, hard, shiny, smooth; seed coat thin; embryo C-shaped; cotyledons nearly as long as the hypocotyl. $2n = 44$ (*M. nummularia*, Moore 1983).

Three species in South America: Colombia and Venezuela to southern Chile and Argentina, as well as the Juan Fernandez and Falkland Islands.

61. *Ugni* Turcz.

Ugni Turcz., Bull. Soc. Imp. Nat. Moscou 21: 579 (1848), Flora 31: 711 (1848).

Shrubs, often densely branched; hairs simple. Inflorescence of monads. Flowers white, 5-merous, nodding; prophylls persistent; sepals free; petals forming an ericoid, lampshade-like structure; stamens numerous, as long as the petals;

anthers sagittate; ovary (2–)3-locular; ovules few to many, 2–4-seriate. Fruit crowned by the calyx lobes. Seeds small, numerous; seed coat hard, shiny; embryo C-shaped, hypocotyl as long as the cotyledons.

A genus of four species from Mexico to Chile.

62. *Lophomyrtus* Burret

Lophomyrtus Burret, Notizbl. Bot. Gart. Berlin-Dahlem 15: 489 (1941).

Shrubs or small trees; branchlets pubescent, terete. Inflorescence of solitary, axillary monads, on long pedicels. Flowers 4-merous; sepals persistent; stamens numerous, free; anthers dorsifixed towards the base; ovary inferior, 2–3-locular; hypanthium not extending beyond the ovary summit; placentas linear, pendulous from the apex of the loculus; ovules numerous, in a single row on each placenta; style slender; stigma barely dilated. Fruit red to black; seeds reniform, small with a hard, pale testa; embryo curved; cotyledons very small. $2n = 22$.

A genus of 2 species, endemic to New Zealand.

63. *Neomyrtus* Burret

Neomyrtus Burret, Notizbl. Bot. Gart. Berlin-Dahlem 15: 493 (1941).

Shrubs or small trees; branchlets glabrous, 4-angled. Inflorescence of solitary, axillary monads, on long pedicels. Flowers 5-merous; sepals persistent; stamens numerous, free; anthers basifixed; ovary ± inferior, unilocular; hypanthium not extending beyond the ovary summit; ovules borne on 2 parietal placentas; style slender; stigma barely dilated. Fruit yellow to red; seeds ± reniform, small with a hard, smooth testa; embryo curved, cotyledons very small. $2n = 22$.

One species, *N. pedunculata* (Hook.f.) Allan, endemic to New Zealand.

64. *Myrtastrum* Burret

Myrtastrum Burret, Notizbl. Bot. Gart. Berlin-Dahlem 15: 494 (1941).

Shrubs; branchlets pubescent, terete. Leaves conspicuously gland-dotted. Inflorescence of solitary, axillary monads, on long pedicels. Flowers 5-merous; sepals almost as long as the petals, persistent; stamens numerous, free; anthers basifixed; ovary inferior, 2–3-locular; hypanthium not extending beyond the ovary summit; placentas axile; ovules in 2 rows on each placenta; style slender; stigma dilated. Fruit red to black. Seeds suborbicular to reniform, small with a hard, smooth, shiny testa; embryo curved; cotyledons very small.

One species, *M. rufo-punctatum* (Brongn. & Gris) Burret, endemic to New Caledonia.

'Pimenta group':

65. *Lenwebbia* N. Snow & Guymer

Lenwebbia N. Snow & Guymer, Syst. Bot. Monogr. 65: 25 (2003).

Trees or shrubs; branchlets not winged. Inflorescences terminal or axillary, flowers solitary or in threes. Flowers 4-merous; hypanthium not extending above the ovary summit; stamens numerous, free; ovary inferior, 3- or 4-locular; ovules many in each loculus, not in regular rows; placentation apically axile, up to 16 ovules per placenta; stigma small. Fruit a soft berry, globose to subglobose; seeds small, testa bony; embryo C-shaped; cotyledons very small.

A genus of 2 species, endemic to north-eastern Australia.

66. *Amomyrtus* (Burret) D. Legrand et Kausel

Amomyrtus (Burret) D. Legrand et Kausel, Lilloa 13: 145 (1948, '1947').

Shrubs or trees; hairs simple. Inflorescence a monad or bracteate shoot of c. 6 flowers. Flowers 5-merous; prophylls caducous; ovary 2–3-locular, placentas upper-axile with the ovules radiating from the margin. Fruit a berry. Seeds hard, few; embryo C-shaped; cotyledons short, folded back against the hypocotyl.

A genus of 2 species; Chile and Argentina.

67. *Amomyrtella* Kausel

Amomyrtella Kausel, Ark. Bot. II, 3: 514 (1956).

Tree; hairs simple. Inflorescences monads, usually borne in opposite pairs on short bracteate shoots. Flowers 4-merous; stigma dilated; stamens numerous; ovary 2- or 3-locular; ovules 2(3) per locule, pendulous on a subapical placenta. Fruit subglobose. Seeds usually 2–4, the seed coat shiny, the embryo C-shaped, with short, reflexed cotyledons.

One species, *Amomyrtella guili* (Speg.) Kausel, endemic to northern Argentina.

68. *Legrandia* Kausel

Legrandia Kausel, Revista Argent. Agron. 9: 321 (1944); Landrum, Fl. Neotrop. Monogr. 45: 130–133 (1986), descr.

Small tree; branchlets 4-angled. Leaf lamina often with domatia below. Inflorescence of monads with foliaceous prophylls borne below a short anthopodium. Flowers 4-merous; sepals and petals distinct; stamens very numerous, multiseriate; style about as long as the stamens; stigma small; ovary 2–3-locular; ovules several, borne in 2 rows on upper-axile placentas. Fruit yellowish (drying reddish brown). Seeds up to 5 per fruit, but usually solitary, with a membranous seed coat; embryo ± straight, ellipsoidal to subglobose; cotyledons very small.

A genus comprising a single species, *L. concinna* (Phil.) Kausel, endemic to Chile.

69. *Campomanesia* Ruiz & Pav.

Campomanesia Ruiz et Pav., Prodr.: 72, t. 13 (1794); Landrum, Fl. Neotrop.Monogr. 45: 7–72 (1986), key.

Trees or shrubs; hairs simple. Inflorescence a monad, dichasium or bracteate shoot. Flowers usually 5-merous; calyx open or closed and tearing at anthesis; stigma dilated; ovary (3)4–18-locular, the locular wall in fruits strongly glandular in the mature fruit and serving as a false seed coat; ovules 4–20 per locule, biseriate. Fruit crowned by the remnants of the calyx or by a circular scar. Seeds usually 1–4 (rarely more), the seed coat a thin membrane usually not detectable in the mature fruit; embryo spiral; the hypocotyl swollen, and the cotyledons very small.

About 30 species, in tropical and subtropical South America.

70. *Psidium* L.

Psidium L., Sp. pl.: 470 (1753); Landrum, Sida 20: 1449–1469 (2003), rev. *P. salutare* complex.

Trees and shrubs; branchlets glabrous or pubescent with simple hairs. Inflorescence a monad, triad or dichasium (rarely a bracteate shoot). Flowers usually 5-merous; calyx open and tearing between the lobes or not, or closed and opening by irregular tears or as a calyptra; stigma usually capitate; ovary ± inferior, (2)3–5(6)-locular; ovules few to numerous in each locule, on an axile, peltate placenta. Fruit crowned by the calyx lobes, remnants of the calyx, or by a circular scar. Seeds few to numerous; seed coat bony, with a dull or rough surface; embryo C-shaped, cotyledons very small. $2n = 22, 44, 55, 77, 88$.

A genus of 70 or more species, growing from Mexico and the Caribbean to northern Argentina. *P. guajava* L., the guava, is widely cultivated and naturalised.

71. *Mosiera* Small

Mosiera Small, Manual Southeast. Fl.: 936 (1933); Salywon, J. Bot. Res. Inst. Texas 1: 899–900 (2007).

Trees or shrubs; hairs simple. Inflorescence uniflorous or a bracteate shoot of usually no more than 2 flowers. Flowers 4-merous; calyx open; ovary 1–4-locular; ovules c. 10 per locule, borne on a linear placenta. Fruit elongate, angular, crowned by the calyx lobes. Seeds small, seed coat relatively hard or leathery, smooth; embryo C-shaped; cotyledons much shorter than the hypocotyl.

A genus of c. 20 species; Florida, Mexico, Guatemala and the Caribbean.

72. *Curitiba* Salywon & Landrum

Curitiba Salywon & Landrum, Brittonia 59: 302 (2007).

Trees or shrubs; sparsely puberulent. Inflorescence uniflorous or triflorous, borne singly or in groups of 2–4 at the nodes. Flowers 4-merous; hypanthium 4-angled to slightly winged; sepals free, persistent; ovary 2-locular; ovules numerous, radially arranged on axile placentas. Fruit crowned by the calyx lobes. Seeds small, numerous; seed coat hard, papillose; embryo C-shaped; cotyledons shorter than the hypocotyl.

One species, *C. prismatica* (D. Legrand) Salywon & Landrum, native to southern Brazil.

73. *Acca* O. Berg

Acca O. Berg, Linnaea 27: 135 (in clave), 138 (1856, '1854'); Landrum, Fl. Neotrop. Monogr. 45: 133–72 (1986), key.
Feijoa O. Berg, Linnaea 29: 258 (1858).

Trees or shrubs; hairs simple. Inflorescence uniflorous. Flowers 4-merous; calyx open; petals reddish to pink, fleshy when mature; stamens numerous, stiff and red, more or less straight, erect in bud; ovary 3–4-locular; placenta basically axile but sometimes partially parietal; ovules numerous in each locule. Fruit crowned by the calyx lobes. Seeds numerous, small, lentil-shaped; seed coat hard or subcrustaceous; embryo spiral, the cotyledons strap-like, as long as the hypocotyl.

A genus of 3 species; two in the Andes of Peru and one in eastern temperate South America, *A. sellowiana* (O. Berg) Burret, the Pineapple Guava or Feijoa, cultivated for its edible fruits.

74. *Myrrhinium* Schott

Myrrhinium Schott in Sprengel, Syst. Veg. 4(2): 404 (1827); Landrum, Fl. Neotrop. Monogr. 45: 142–148 (1986), key.

Trees or shrubs; hairs simple. Inflorescence a dichasium of 3–7 flowers, solitary or aggregated in groups of 2 to c. 8 on short bracteate shoots. Flowers 4-merous; sepals distinct, persistent; petals red, pink, or purplish, fleshy when mature; stamens 4–8, stiff, red, twice folded in the bud; ovary 2-locular; ovules 5–14 per locule, in 2 series on the placenta. Fruit succulent. Seeds few, small, cochleate; seed coat hard and shiny; embryo C-shaped; cotyledons as long as the hypocotyl.

A monotypic genus, *M. atropurpureum* Schott, with two varieties ranging from the Andes to SE and S Brazil.

75. *Accara* Landrum

Accara Landrum, Syst. Bot. 15: 221 (1990).

Shrubs; essentially glabrous; hairs when present simple. Inflorescence uniflorous. Flowers 4-merous; calyx closed or opening only as a small apical pore in the bud, tearing regularly into 4 persistent, triangular lobes at anthesis; ovary 4-locular; ovules numerous on axile placentas. Fruit crowned by the calyx lobes. Seeds small, numerous; seed coat hard, shiny; embryo C-shaped; cotyledons as long as the hypocotyl.

One species, *A. elegans* (DC.) Landrum, native to Brazil.

76. *Chamguava* Landrum

Chamguava Landrum, Syst. Bot. 16: 21 (1991).

Shrubs or trees; hairs simple; twigs tending to branch dichasially. Leaves with brochidodromous venation. Inflorescence axillary or ramiflorous, monads or clusters of up to 4 flowers. Flowers 4-merous, whitish, sessile or shortly pedunculate; prophylls small, caducous; sepals fused or free, mostly persisting; ovary 2-locular, the placenta usually subpeltate; ovules few to many, multiseriate, radiating from the placenta. Fruit a globose berry. Seeds few, ± reniform; embryo crescent-shaped; cotyledons much smaller than the thickened hypocotyl.

A genus of 3 species; southern Mexico and Central America (Guatemala to Panama).

77. *Pimenta* Lindl.

Pimenta Lindl., Collect. Bot. ad t. 19 (1821); Landrum, Fl. Neotrop. Monogr. 45: 72–115 (1986), key.

Trees and shrubs; hairs simple. Inflorescence a dichasium or a panicle of 3–15 flowers. Flowers 4-merous; calyx open; ovary 2-locular; ovules 3–6 per locule, on a subapical placenta. Fruit crowned by the calyx lobes. Seeds usually 1 or 2; seed coat membranous or hard and shiny; embryo spiral, the hypocotyl swollen; cotyledons very small.

A genus of 15 species; mostly Caribbean but with one species in Brazil.

'Eugenia group':

78. *Eugenia* L.

Eugenia L., Sp. pl.: 470 (1753); Verdcourt, Kew Bull. 54: 41–62 (1999), rev. Afr. spp.; Snow, Syst. Bot. 33: 343–348 (2008), rev. Malagasy spp.
Jossinia Commers. ex DC. (1828).
Monimiastrum J.Guého & A.J.Scott (1980).

Trees or shrubs; hairs simple or two-armed. Inflorescence uniflorous, a bracteate shoot, or rarely a dichasium, the bracteate shoots sometimes with an abbreviated axis bearing 2 to many flowers. Flowers always 4-merous; sepals free; ovary 2-locular; ovules 2–many on pad-like, axile placentas. Fruit crowned by the calyx lobes. Seeds 1–2; seed coat membranous or crusty; embryo with fused cotyledons. $2n = 22$, 33, 44, 66.

A genus of c. 550 species; growing from Mexico and the Caribbean to northern Argentina. There are also a number of species in the Pacific: New Caledonia (c. 60 species) and the Philippines (c. 11 species), Africa and neighbouring islands (c. 120 species).

The recently published phylogeny of the *Eugenia* group (van der Merwe et al. 2005) includes a monophyletic subgroup of Mascarene species, which includes the lectotype of *Jossinia* plus the single species of *Monimiastrum* included in the study. Snow (l.c.) has formally reduced *Monimiastrum* to the synonymy of *Eugenia*.

79. *Hexachlamys* O. Berg

Hexachlamys O. Berg, Linnaea 27: 137, 345 (1856, '1854').

Trees or shrubs; hairs simple or two-armed. Inflorescence uniflorous, a bracteate shoot, or rarely a dichasium; the bracteate shoots sometimes with an abbreviated axis bearing 2 to many flowers. Flowers 5(–7)-merous; sepals free, long and acute; prophylls exceeding the ovary; ovary 2–3-locular; ovules 2–few on pad-like, axile placentas; locular walls often pilose. Fruit crowned by the reflexed sepals or a circular scar, drupe-like with a crustaceous to woody inner layer. Seeds 1–2; seed coat membranous or crusty; embryo ± spherical, cotyledons fused.

A genus of 15 species: South America. Considered by some to be a synonym of *Eugenia*.

80. *Stereocaryum* Burret

Stereocaryum Burret, Notizbl. Bot. Gart. Berlin-Dahlem 15: 546 (1941).

Trees or shrubs; branchlets tomentose. Inflorescences of monads or triads. Flowers 4-merous; hypanthium extended well above the ovary, splitting at anthesis; sepals and petals distinct; stamens numerous, multiseriate, distinct, attached over the inner surface of the hypanthium; anthers dorsifixed at the base; ovary inferior, 2-locular; ovules numerous and scattered on an axile placenta; style subulate, stigma small. Fruit large, pericarp almost woody. Seed solitary, large with a thick testa; cotyledons free.

A genus of c. 8 species endemic to New Caledonia.

81. *Calycorectes* O. Berg

Calycorectes O. Berg, Linnaea 27: 136, 317 (1856, '1854').

Trees or shrubs; hairs simple or two-armed. Inflorescence uniflorous, a bracteate shoot, or rarely a dichasium, the bracteate shoots sometimes with an abbreviated axis bearing 2 to many flowers. Flowers usually 4-merous; sepals fused in bud and tearing irregularly at anthesis; ovary 2-locular; ovules 3–many on pad-like, axile placentas. Fruit crowned by the remnants of the calyx, or by a circular scar. Seeds 1–2; seed coat membranous or crusty; embryo with fused cotyledons.

A genus of c. 36 species: Mexico to Brazil. Considered by some to be a synonym of *Eugenia*.

82. *Myrcianthes* O. Berg

Myrcianthes O. Berg, Linnaea 27: 136, 315 (1856).
Reichea Kausel (1940).

Trees or shrubs; hairs simple. Leaves opposite or ternate. Inflorescence of monads or, more often, triads or 7-flowered dichasia. Flowers 4- or 5-merous; sepals free; ovary usually 2-locular; ovules few to numerous borne on central to subapical axile placentas. Fruit crowned by the sepals. Seeds 1–2; seed coat membranous; embryo subglobose; cotyledons separate, plano-convex, much longer than the short hypocotyl.

Around 30 species: mainly Andean, extending from Mexico to Chile.

83. *Pseudanamomis* Kausel

Pseudanamomis Kausel, Ark. Bot. II, 3: 511 (1956).

Trees; branches glabrous. Leaves opposite, petiolate. Inflorescence almost umbel-like, with 3 or more flowers. Flowers (4–)5-merous; hypanthium with woolly indumentum; sepals free, caducous; ovary 2–3-locular; ovules 5 or 6 per loculus, borne on a protruding placenta. Fruit yellow. Seeds 1–3 per fruit, seed coat smooth, subcarnose; embryo ellipsoid to reniform; cotyledons large and partly connate with a short hypocotyl.

One widespread species, *P. umbellulifera* (Kunth) Kausel, found in the Dominican Republic, Netherlands Antilles, Puerto Rico, Trinidad and Tobago, Guyana, Venezuela and Colombia.

84. *Luma* A. Gray

Luma A. Gray, U.S. Expl. Exped., Phan.: 535, t. 66 (1854); Landrum, Fl. Neotrop. Monogr. 45: 148–158 (1986), key.
Myrceugenella Kausel (1942).

Shrubs or small trees; hairs simple. Inflorescence of monads, triads or dichasia. Flowers 4-merous; prophylls linear, membranous, caducous; ovary 2-locular; ovules c. 6–14 per locule. Fruit a fleshy, somewhat spongy berry. Seeds lenticular; embryo with suborbicular, thinly plano-convex, distinct cotyledons; hypocotyl cylindrical, about as long as the cotyledons.

A genus of 2 species: Chile and Argentina.

85. *Myrceugenia* O. Berg

Myrceugenia O. Berg, Linnaea 27: 5 (1855); Linnaea 27: 131. (1856); Landrum, Fl. Neotrop. Monogr. 29 (1981), rev.
Nothomyrcia Kausel, Lilloa 13: 147 ('1947', publ. 1948).

Trees or shrubs; hairs usually two-armed. Inflorescence usually uniflorous, less often a dichasium or a bracteate shoot, accessory inflorescences often present. Flowers 4-merous; calyx open or in two species closed and falling as a calyptra; prophylls usually persisting until the fruit matures; ovary 2–4-locular; ovules 2–c. 20 per locule, in 2 rows on each placenta. Fruit a yellow to dark red berry, crowned by the persistent sepals or by a circular scar. Seeds 1–5; seed coat membranous; embryo myrcioid.

A genus of c. 40 species: from temperate and subtropical Brazil to southern Chile and the Juan Fernández islands.

Myrceugenia has often been grouped with the Myrcioid genera, but is anomalous there. Molecular data, supported by some morphological characters (the consistently bracteolate 4-merous flowers and ovaries with multiovulate locules), indicate that it is a member of the *Eugenia* group of genera.

'Plinia group':

86. *Neomitranthes* D. Legrand

Neomitranthes D. Legrand in Legrand & Klein, Fl. Ilustrada Catarinense 1 (Mirt.): 671 (1977).

Trees or shrubs; glabrous or with simple hairs. Inflorescence an axillary bracteate shoot often reduced to a glomerule; calyx calyptrate and tearing irregularly at anthesis, usually splitting the staminal ring; petals 0–4; ovary usually 2-locular; ovules 2–6 per locule. Fruit crowned by a circular scar. Seeds 1–2; seed coat membranous; embryo with free, plano-convex cotyledons and an inconspicuous or absent hypocotyl.

A genus of up to 17 species from north-eastern to south-eastern Brazil.

87. *Siphoneugena* O. Berg

Siphoneugena O. Berg, Linnaea 27: 136, 344 (Jan. 1856, '1854'); Proença, Edinb. J. Bot. 47: 239–271 (1990), rev.

Trees or shrubs; hairs simple. Inflorescences in axils of fallen leaves, occasionally in upper axils, usually raceme-like bracteate shoots that vary from elongated to compact. Flowers 4-merous, the hypanthium extending well beyond the ovary but contracted near its summit, circumscissile and falling with the calyx as a unit after anthesis; calyx open or closed, tearing irregularly or falling as a calyptra; petals white, pubescent on both surfaces; ovary 2-locular; ovules (2)3–5(–7) per locule, axile or almost basal. Fruit purplish-black, crowned by a circular scar. Seeds 1(–4); seed coat membranous; embryo with free, plano-convex cotyledons.

A genus of 8 species, distributed from Puerto Rico to northern Argentina.

88. *Myrciaria* O. Berg

Myrciaria O. Berg, Linnaea 27: 136, 320 (1856).
Paramyrciaria Kausel, Lilloa 32: 345 ('1966', publ. 1967).

Trees or shrubs; hairs simple. Inflorescence a bracteate shoot often reduced to a glomerule. Flowers 4-merous, the hypanthium extending well beyond the ovary, circumscissile at the summit of the ovary and mostly falling with the calyx as a unit after anthesis; calyx open; ovary 2-locular; ovules usually 2 per locule. Fruit crowned by a circular scar. Seeds 1–2; seed coat membranous; embryo a solid mass or with 2 free, plano-convex cotyledons. $2n = 22$.

A genus of c. 22 species, growing from Mexico and the Caribbean to northern Argentina.

Cultivated 'Jaboticaba' species are often polyembryonic. 'Jaboticaba' species have been referred to the genus *Plinia* but their tubular, circumscissile hypanthia indicate that they belong in *Myrciaria* (*M. cauliflora* (Mart.) O. Berg).

89. *Plinia* L.

Plinia L., Sp. pl.: 516 (1753); Barrie, Novon 14: 380–400 (2004), synopsis Mesoamerican spp.

Trees or shrubs with smooth bark that exfoliates in plates; indumentum of simple hairs; vegetative buds protected by scales. Inflorescence cauliflorous, a bracteate shoot that is often reduced to a glomerule and subtended by an involucre-like collection of bracts. Flowers 4-merous, usually silky-pubescent, subsessile to sessile; calyx fused and nearly closed in bud, tearing irregularly at anthesis into four reflexed lobes; petals caducous; ovary 2-locular; ovules 2 per locule. Fruit crowned by remnants of the calyx. Seeds 1–2; seed coat membranous; embryo of 2 large, separate, plano-convex cotyledons and a reduced hypocotyl. $2n = 22$.

A genus of c. 40 species: Central America and the Caribbean to Brazil.

90. *Hottea* Urban

Hottea Urban, Ark. Bot. 22A(10): 40 (1929).

Trees or shrubs with fissured bark; branchlets with spreading hairs, sometimes glabrescent. Inflorescences axillary or occasionally from the axils of scale leaves at the beginning of a year's growth. Flowers solitary or several; hypanthium not extended beyond the summit of the ovary; calyx closed in bud, splitting irregularly into two or more lobes at anthesis; petals unequal, 1–5; stamens 2–3-seriate; style about as long as the stamens, narrowing to a punctiform stigma; ovary 2-locular; ovules c. 6 per loculus, in two rows, borne towards the apex of the loculus. Fruit a 1(–2)-seeded berry; testa chartaceous; embryo with cotyledons plano-convex, mostly distinct but connate at the margins.

A poorly known genus of c. 9 species from the Caribbean (Cuba, Dominican Republic, Haiti).

91. *Calyptrogenia* Burret

Calyptrogenia Burret, Notizbl. Bot. Gart. Berlin-Dahlem 15: 545 (1941).

Trees or shrubs with fissured bark; branchlets glabrous. Inflorescences ramiflorous, borne below the leaves, probably from the axils of caducous scale leaves, or rarely from the upper axils. Flowers solitary or in a bracteate shoot that is reduced to a glomerule; calyx closed, opening as an operculum; petals 5, caducous; stamens numerous, multiseriate; style about as long as the stamens, narrowing to a punctiform stigma; ovary 2-locular; ovules numerous (c. 12) per locule, borne on pad-like, axile placentas. Fruit crowned by a corona. Seeds 1–2; seed coat probably membranous; embryo large, undivided.

A poorly known genus of six species from the Caribbean (Dominican Republic, Haiti, Jamaica); possibly congeneric with *Hottea*.

92. *Algrizea* Proença & NicLugh.

Algrizea Proença & NicLugh., Syst. Bot. 31, 320 (2006).

Shrubs, hairs simple. Inflorescence of solitary or paired axillary, pedunculate triads of sessile flowers. Flowers 5-merous; hypanthium shortly extended beyond the ovary summit; sepals free, well-developed; prophylls persistent; ovary 2-locular; ovules 3–6(–8) per locule. Fruit crowned by the sepals. Seeds 2–3; seed coat membranous; embryo with hypocotyl enclosed by thickened cotyledons.

One species, *A. macrochlamys* (DC.) Proença & NicLugh., endemic to the state of Bahia, Brazil.

'Myrcia group':

93. *Mitranthes* O. Berg

Mitranthes O. Berg, Linnaea 27: 316 (1856).

Small trees or shrubs; branching often dichotomous; subglabrous or tomentose. Inflorescences of 1–9 flowers, usually borne in pairs on short shoots in the apical axils. Calyx closed in the bud, calyptrate and circumscissile; petals apparently lacking; prophylls caducous at about anthesis; ovary 1-, 3- or 5-locular; ovules numerous in each locule. Fruit crowned by a circular scar. Seeds 1–2 or more; embryo myrcioid.

Seven species (all apparently rare): one in Cuba, the others in Jamaica.

94. *Calyptranthes* Sw. Fig. 50

Calyptranthes Sw., Prodr. 5, 79 (1788).

Trees or shrubs; branching often dichotomous; hairs frequently two-armed. Inflorescence usually a pair of panicles. Calyx closed and falling or partially detaching as a calyptra; petals 0–5, very small; prophylls caducous at about anthesis; ovary 2(3)-locular; ovules 2 per locule. Fruit crowned by a circular scar. Seeds 1–2; seed coat membranous; embryo myrcioid.

Over 100 species: from Mexico and the Caribbean to northern Argentina.

95. *Myrcia* DC. ex Guill. Fig. 51

Myrcia DC ex Guill., Dict. Class. Hist. Nat. 11: 378, 401, 406 (1827).
Mozartia Urb. (1923).

Trees or shrubs; hairs mostly two-armed. Inflorescence a panicle. Flowers 5(4)-merous; hypanthium thick-walled, not or barely extended beyond the ovary summit; sepals free; prophylls caducous; ovary 2–3(4)-locular; ovules 2 per locule. Fruit crowned by the sepals or remnants of the calyx. Seeds 1–2; seed coat membranous to crustaceous; embryo myrcioid.

A genus of over 500 species; Mexico and the Caribbean to northern Argentina.

The segregate genera *Gomidesia* and *Marlierea* are closely related to *Myrcia* and some have suggested they be united (Landrum and Kawasaki 1997). The *Gomidesia* anther character, however, considered the most important to distinguish it from *Myrcia*, is often difficult to observe. Nevertheless, species assigned to *Gomidesia* generally have a distinctive indumentum on the leaves and inflorescence, and a calyx with relatively short, rounded lobes. The species of the *Marlierea* complex are distinguished from *Myrcia* by a calyx that tears irregularly as the flower opens and that is sometimes partially caducous in the fruit. In molecular analyses, there is support for *Gomidesia* but *Myrcia* appears paraphyletic (Lucas et al. 2007).

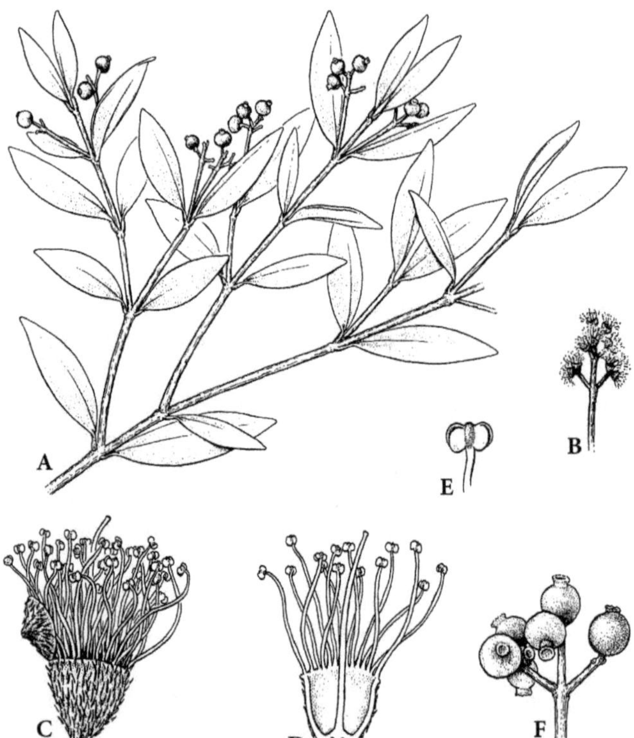

Fig. 50. Myrtaceae. *Calyptranthes thomasiana*. **A** Fruiting branchlet. **B** Inflorescence. **C** Flower with calyptra lid. **D** Longitudinal section of flower. **E** Anther. **F** Fruiting branch. (Acevedo Rodríguez 1996; reproduced with permission of the artist Bobbi Angell)

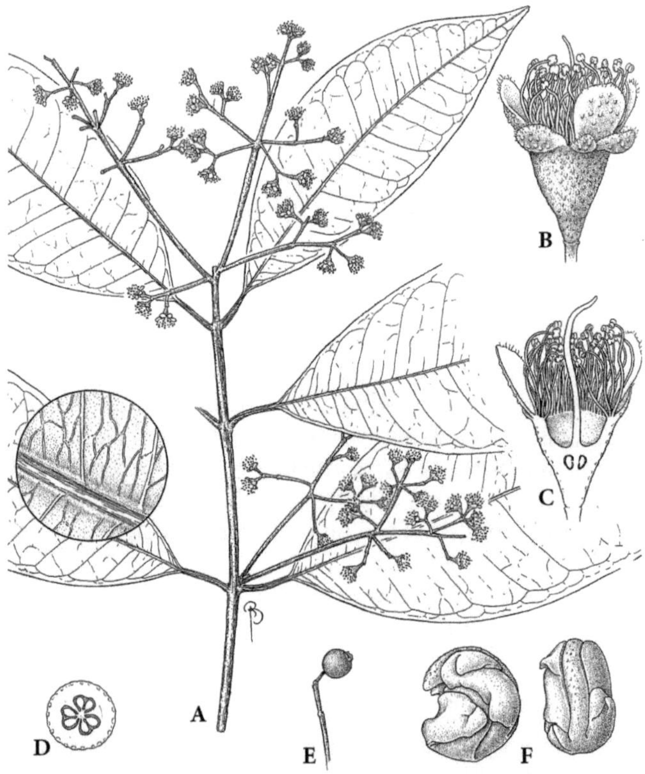

Fig. 51. Myrtaceae. *Myrcia gigas*. **A** Flowering branchlet, with detail of adaxial leaf surface (left). **B** Flower. **C** Longitudinal section of flower. **D** Transverse section of ovary. **E** Fruit. **F** Two views of myrcioid embryo showing folded cotyledons and prolonged radicula. (Acevedo Rodríguez 1996; reproduced with permission of the artist Bobbi Angell)

96. *Marlierea* Cambess.

Marlierea Cambess. in Saint-Hilaire, Fl. Brasil. Mér. 2, ed. fol.: 269; ed. qu.: 373 (1833).

Trees or shrubs; hairs simple or two-armed. Inflorescence a panicle, usually aborted above the first node and appearing paired. Flowers 5(4)-merous; hypanthium thin-walled, extended beyond the ovary summit; calyx closed in bud or the tips free, splitting irregularly at anthesis; prophylls caducous; petals small and inconspicuous or absent; ovary 2–3(4)-locular; ovules 2 per locule. Fruit crowned by the remnants of the calyx. Seeds 1–2; seed coat membranous to crustaceous; embryo myrcioid.

A genus of c. 100 species; tropical South America east of the Andes.

97. *Gomidesia* O. Berg

Gomidesia O. Berg, Linnaea 27: 5, in clave (1855, '1854').

Trees or shrubs; hairs characteristically silky, appressed, tawny to reddish-brown. Inflorescence a panicle. Flowers 5(4)-merous; hypanthium thick-walled, not or barely extended beyond the ovary summit; sepals free, rounded; prophylls caducous; petals small and inconspicuous or absent; anthers incompletely 4-locular with the pollen sacs displaced and the upper ones appearing to open extrorsely; ovary 2–3(4)-locular; ovules 2 per locule. Fruit crowned by the calyx lobes. Seeds 1–2; seed coat membranous to crustaceous; embryo myrcioid.

A genus of c. 60 species; Central and South America, and the W Indies.

II.13. Tribe Eucalypteae Peter G. Wilson (2005).

Trees sometimes shrubby ('mallees'); hairs mostly absent. Juvenile leaves opposite, adult leaves opposite or alternate; oil ducts sometimes present in stems and petioles. Inflorescences various, occasionally paniculate and terminal. Flowers often with calyptrate perianth; stamens numerous, free, borne on a stamenophore; ovary half-inferior, 2- or more locular; placentas axile, ovulodes present. Fruit a capsule. Seeds variable; embryo with cotyledons deflexed, enclosing one another.

98. *Arillastrum* Baillon

Arillastrum Panch. ex Baillon, Hist. Pl. 6: 363 (1877).

Trees; young parts bearing short, branched, multicellular hairs. Inflorescences usually sessile triads (rarely dichasia) borne on long peduncles in the axils of reduced leaves. Flowers 4-merous, white; sepals and petals distinct; stamens numerous in 4 antepetalous clusters, sometimes with small groups of short stamens opposite the sepals; outer rows of antepetalous stamens sterile, lacking anthers, the inner stamens fertile and up to half the length of the staminodes; anthers dorsifixed, versatile; ovary inferior, 2-locular; hypanthium barely exceeding the ovary summit; placentation axile, ovules numerous, ovulodes present; style short, stigma slightly dilated, convex. Fruit a very woody capsule. Fertile seeds one per loculus, hemispherical to reniform, with the scale-like unfertilised ovules adhering to them; embryo with broad cotyledons, folded back on themselves and obvolute.

One species, *A. gummiferum* (Brongn. & Gris) Baillon, endemic to New Caledonia.

99. *Eucalyptopsis* C.T. White

Eucalyptopsis C.T. White, J. Arnold Arb. 32: 139 (1951).

Trees; mostly glabrous except for a few, very short hairs on vegetative buds. Inflorescences terminal or axillary panicles, the branches of which are sessile, 3–7-flowered dichasia on elongated branches; flowers sometimes connate at the base. Flowers seemingly 4-merous; sepals fused into a calyptra, petals absent; stamens numerous, in 4 clusters or not apparently grouped, all fertile; anthers dorsifixed near the base; ovary half-inferior, 2-locular; hypanthium funnel-shaped, exceeding the ovary summit, the free part sometimes splitting but ultimately caducous; placentation axile; ovules numerous, ascending; ovulodes present; style short, stigma slightly dilated, convex. Fruit a woody capsule, included in, or exserted from, the fruiting hypanthium. Fertile seeds one, oblong, plano-convex; embryo not documented.

A genus of 2 or more species: from eastern Indonesia (Maluku and Irian Jaya) and Papua New Guinea.

100. *Stockwellia* D.J. Carr, S.G.M. Carr & B. Hyland

Stockwellia D.J. Carr, S.G.M. Carr & B. Hyland, Bot. J. Linn. Soc. 139: 416 (2002).

Trees; mostly glabrous except for the short, branched and multicellular hairs on the flowers. Inflorescences terminal or axillary metabotryoids, the branches of which are sessile triads on elongated peduncles; triads subtended by caducous prophylls. Flower buds connate at the base up to the level of the ovary summit; hypanthium extended above the ovary summit, almost closed, with 4 very small sepals and petals at the apex; hypanthium splitting at anthesis into 4 ± equal, triangular segments (and finally circumscissile); stamens numerous, in clusters at the tips of the segments, all fertile; anthers dorsifixed, versatile, connective not conspicuously gland-tipped; ovary half-inferior, 2-locular; placenta axile, peltate; ovules numerous, ascending, hemitropous; ovulodes present; style bent and tapering, stigma narrow, truncate to slightly dilated. Capsule woody with a distinct rim derived from the hypanthial margin; fruit included in, or exserted from, the fruiting hypanthium. Fertile seeds one or two, oblong, plano-convex; embryo with cotyledons deeply notched at both base and apex; cotyledons obvolute.

One species, *S. quadrifida* D.J. Carr, S.G.M. Carr & B. Hyland, restricted to a small area of north Queensland, Australia.

101. *Allosyncarpia* S.T.Blake

Allosyncarpia S.T.Blake, Austrobaileya 1: 43 (1977).

Trees; mostly glabrous except for a few, very short hairs on vegetative buds. Leaves mostly ternate. Inflorescences raceme-like, made up of ternate, pedunculate triads subtended by early-caducous, linear bracts; flowers with hypanthia connate at the base up to the level of the ovary summit. Flowers 5-merous; sepals free, valvate, petals present; stamens numerous, all fertile; anthers dorsifixed near the base; ovary inferior, 2-locular; hypanthium cup-shaped, exceeding the ovary summit, the free part circumscissile and caducous; placentation axile, ovules not numerous; ovulodes present; style not exceeding the stamens, stigma slightly dilated. Fruit a woody capsule, subglobose, strongly exserted from the fruiting hypanthium. Fertile seeds one (rarely two), depressed-ovoid; embryo with broad, contorted cotyledons.

A genus of one species, *A. ternata* S.T. Blake, from sandstone escarpments in northern Australia (Northern Territory).

102. *Angophora* Cav.

Angophora Cav., Ic. descr. Plant. 4: 21, t. 338 (1797); Leach, Telopea 2: 749–779 (1986), rev.

Shrubs or trees. Leaves dimorphic, juvenile leaves opposite, sessile, cordate and auriculate in early stages, hispid with simple unicellular hairs and raised oil glands, adult leaves opposite, usually glabrous. Inflorescences comprising unit 3–7-flowered umbellasters, aggregated into terminal conflorescences. Sepals 4 or 5, distinct, reduced to persistent projections on rim of hypanthium; petals imbricate, clawed; stamens numerous in several whorls, regularly inflexed in bud; anthers versatile, dehiscing by parallel longitudinal slits; ovary half-inferior, usually 3-locular, enclosed by hypanthium except at top; ovules numerous, 1 per loculus maturing to seed, remainder forming chaff. Capsule papery or thinly woody, ovoid or campanulate, often strongly ribbed, mostly hispid, nectary disk narrow, flat or depressed. Seeds broad-elliptic, irregularly flattened; cotyledons folded.

A genus of 15 species; endemic to eastern Australia.

103. *Corymbia* K.D. Hill & L.A.S. Johnson Fig. 52

Corymbia K.D. Hill & L.A.S. Johnson, Telopea 6: 214 (1995); Steane et al., Austral. Syst. Bot. 15: 49–62 (2002), mol. phylog.; Parra-O. et al., Taxon 55: 653–663 (2007), mol. phylog.

Trees, sometimes mallee-like. Leaves dimorphic; juvenile leaves opposite; adult leaves usually disjunct opposite. Inflorescences compound, terminal or lateral, often extensively branched, generally based on regular 7-flowered umbellasters. Calyx calyptrate, shed at anthesis; corolla of ± distinct petals, adherent to the calyx (but separable) or ± fused to the calyx (and shed with the outer calyptra), or fused into an inner calyptra and falling after the outer calyptra and sometimes with it; stamens numerous, usually in several continuous whorls; filaments regularly inflexed in bud; anthers versatile oblong, dehiscing through parallel slits; ovary half-inferior, usually 3-locular, enclosed by hypanthium except at top; ovules numerous; ovulodes present. Fruit a capsule, becoming woody in most species; disk depressed; valves enclosed.

About 115 species, all but 5 endemic to Australia (4 species native, 1 species New Guinea).

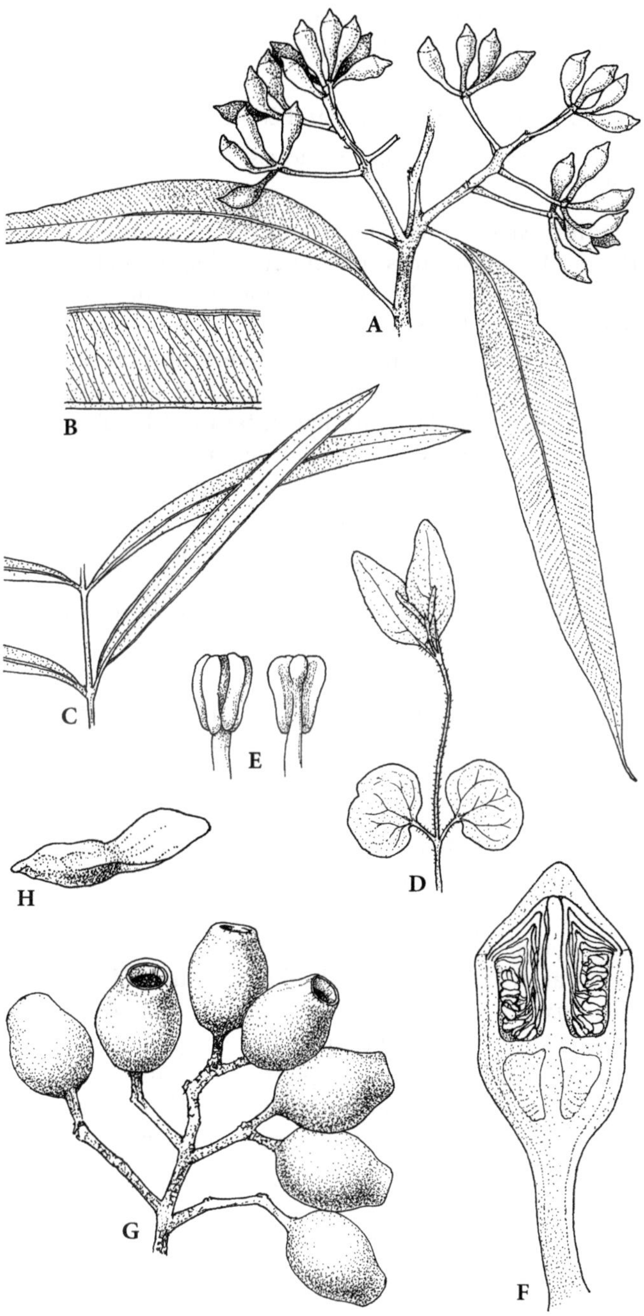

Fig. 52. Myrtaceae. *Corymbia hendersonii*. **A** Habit. **B** Leaf detail. **C** Juvenile leaves. **D** Seedling with cotyledons. **E** Stamens. **F** Longitudinal section of bud. **G** Fruit. **H** Seed. (Drawn by R. Roden)

104. *Eucalyptus* L'Hérit. Fig. 53

Eucalyptus L'Hérit., Sertum Anglicum: 18 (1789); Steane et al., Austral. Syst. Bot. 15: 49–62 (2002), mol. phylog.

Trees or multi-stemmed shrubs ('mallees'), the latter often forming an underground lignotuber. Leaves dimorphic, juvenile leaves opposite, adult leaves usually disjunct opposite, similifacial or less commonly dorsiventral; oil glands usually present in leaves. Inflorescences usually based on condensed dichasia, forming umbellasters, mostly solitary and axillary but sometimes aggregated into compound terminal, pseudoterminal or axillary structures; individual umbellasters usually 7-flowered, but ranging from 1 to >30 flowers in some species. Calyx and/or corolla fused into a calyptra which persists to anthesis; stamens numerous, usually in several continuous whorls; anthers versatile or adnate, dehiscing by separate or confluent slits or pores; ovary half-inferior, 2–7-locular, enclosed by hypanthium except at top; ovules numerous; ovulodes present. Fruit a capsule, becoming woody in most species.

Over 800 species, all but 16 species endemic to Australia (9 species New Guinea, Indonesia, Philippines).

II.14. Tribe Syncarpieae Peter G. Wilson (2005).

Trees, leaves opposite and forming false whorls at the apices. Oil ducts present; oil reddish-orange. Flowers fused into a 7-flowered unit; stamens free, numerous; ovary inferior; placentas basal in the locules. Fruit an aggregation of capsules. Seeds linear, embryo with cotyledons enclosing one another.

105. *Syncarpia* Ten.

Syncarpia Ten., Index Sem. Horto Bot. Neapol. Coll. 1839, app. 12 (1839).

Trees; terminal buds covered with scales; stems and petioles with oil ducts that contain a clear reddish oil. Leaves opposite but crowded into false whorls of 4 at the shoot apex. Inflorescences axillary, 7-flowered dichasia, flowers sessile and fused at base; peduncles relatively long. Flowers 4- to 5-merous, white; sepals small; petals free; stamens numerous, free; ovary ± inferior, 3-locular; stigma capitate. Fruit capsular, loculicidal; fruiting hypanthia from each dichasium fused to

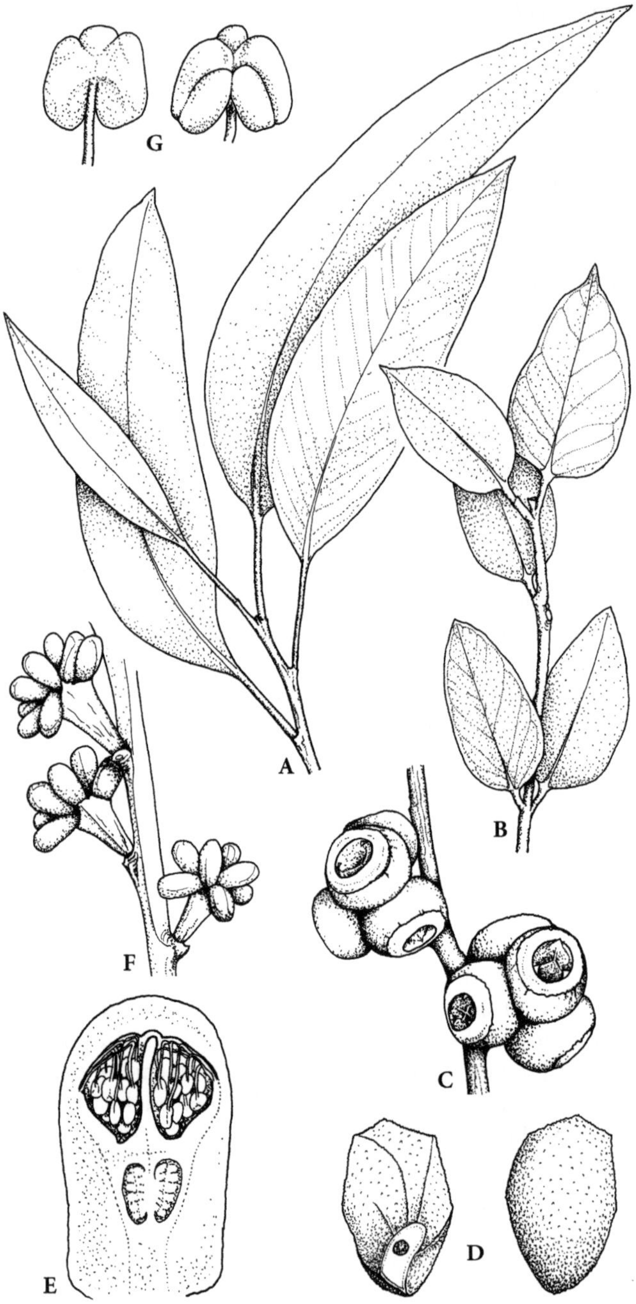

Fig. 53. Myrtaceae. *Eucalyptus bensonii*. **A** Mature leaves. **B** Juvenile leaves. **C** Fruits. **D** Seeds, front and back views. **E** Longitudinal section of bud. **F** Branch with inflorescence in bud. **G** Stamens, front and back views. (Drawn by R. Roden)

form a woody, multiple fruit. Seeds linear; embryo straight; cotyledons obvolute.

A genus of 3 species, endemic to eastern Australia.

II.15. Tribe Lindsayomyrteae Peter G. Wilson (2005).

Trees; leaves alternate, bluish-purple when young; oil ducts present in petioles and stems. Flowers yellow; sepals very short; stamens numerous, free, the inner ones shorter; ovary ± inferior. Fruit 1–3-lobed depending on seed set; embryo with thick cotyledons.

106. *Lindsayomyrtus* B. Hyland & Steenis

Lindsayomyrtus B. Hyland & Steenis, Blumea 21: 190 (1973).

Medium-sized trees; stems and petioles with oil ducts. Inflorescences in upper axils, panicles, metabotryoids or botryoids. Flowers 5-merous, white; hypanthium very shallow, margin level with the ovary summit; stamens very numerous, mostly distinct but somewhat fused at the base opposite the petals; ovary half-inferior, 3-locular; placentas axile, ovules ascending; style tapering to a small stigma. Fruit almost superior, 1–3-lobed (depending on number of seed set); pericarp thin and semi-succulent, tearing irregularly to release the basally attached seed. Embryo with very thick cotyledons, often somewhat unequal in size.

One species, *L. racemoides* (Greves) Craven, from eastern Indonesia (Maluku and Irian Jaya), Papua New Guinea and Australia (north Queensland).

II.16. Tribe Leptospermeae DC. in Schltdl. (1827).

Shrubs or trees; leaves mostly spiral. Inflorescences of solitary flowers or triads sometimes aggregated into head-like clusters of flowers, very rarely into bottlebrush-like structures (*Kunzea baxteri*). Flowers red, pink, yellow or white; stamens free, anthers versatile, dehiscing by longitudinal slits; ovules anatropous or hemitropous. Fruit a woody capsule, sometimes with delayed dehiscence. Seeds linear or ellipsoid to obovoid, occasionally winged.

107. *Agonis* (DC.) Sweet

Agonis (DC.) Sweet, Hortus Brit. ed. 2: 209 (1830); Wheeler & Marchant, Nuytsia 16: 393–433 (2007), rev.
Agonis sect. *Ataxandria* Benth. (1867).

Shrubs or trees; branchlets with simple hairs. Inflorescence axillary; flowers sessile in ± globose clusters of sessile monads, subtended by numerous, inconspicuous basal bracts. Flowers 5-merous, white; sepals and petals distinct, persistent (petals gradually tapered to the base); hypanthium shortly extended above the ovary; stamens mostly 15–30, free, 3–6 opposite each petal, none opposite sepals; anthers dorsifixed, versatile, dehiscing by longitudinal slits; ovary ± inferior, 3-locular; ovules erect on the axile-basal placenta, 3–7(–14) per locule; style short, set in a depression in the ovary summit; stigma capitate to peltate. Fruit dry, woody, loculicidally dehiscent. Seeds 1–2 per locule, narrowly ellipsoid to obovoid, compressed, very shortly winged at one end; embryo straight to slightly curved, the cotyledons plano-convex and large relative to the hypocotyl.

A genus of 4 species, endemic to the southwest of Australia.

108. *Taxandria* (Benth.) J.R. Wheeler & N.G. Marchant

Taxandria (Benth.) J.R. Wheeler & N.G. Marchant, Nuytsia 16: 406 (2007).
Agonis sect. *Taxandria* Benth., Fl. Austral. 3: 97 (1867).

Shrubs or trees; branchlets with simple hairs. Inflorescence axillary; flowers sessile in ± globose clusters of sessile monads, each flower subtended by a pair of prophylls and below these a bract. Flowers 5-merous, white, rarely pink; sepals and petals distinct, persistent (petals distinctly clawed); hypanthium shortly extended above the ovary; stamens mostly 10, distinct, one opposite each sepal and one opposite each petal; anthers dorsifixed, versatile, dehiscing by longitudinal slits; ovary ± inferior, 3-locular; ovules erect on the axile-basal placenta, 2–3 per locule; style short, set in a deep depression in the ovary summit; stigma capitate to peltate. Fruit dry, woody, loculicidally dehiscent. Seeds usually 1 per locule, narrowly ellipsoid to obovoid, compressed, very shortly winged at one end; embryo straight to slightly curved, the cotyledons plano-convex and large relative to the hypocotyl.

A genus of 11 species, endemic to the southwest of Australia.

109. *Paragonis* J.R. Wheeler & N.G. Marchant

Paragonis J.R. Wheeler & N.G. Marchant, Nuytsia 16: 430 (2007).

Shrubs; branchlets with simple hairs. Inflorescence axillary; flowers sessile, solitary or in few-flowered clusters, terminating or in the upper axils of short shoots, each flower and cluster subtended by several broad bracts and prophylls. Flowers 5-merous, white; rarely pink; sepals and petals free; petals gradually tapered to the base, not persistent; hypanthium shortly extended above the ovary; stamens distinct, numerous in a single continuous whorl; anthers dorsifixed, versatile, dehiscing by longitudinal slits; ovary ± inferior, 3-locular; ovules erect on the axile-basal placenta; style short, set in a depression in the ovary summit; stigma capitate to peltate. Fruit dry, woody, loculicidally dehiscent. Seeds usually 1 per locule, narrowly ellipsoid to obovoid, compressed, very shortly winged at one end; embryo straight to slightly curved, the cotyledons plano-convex and large relative to the hypocotyl.

Monotypic, *P. grandiflora* (Benth.) J.R. Wheeler & N.G. Marchant endemic to the southwest of Australia.

110. *Pericalymma* (Endl.) Endl.

Pericalymma (Endl.) Endl., Genera Plantarum 2: 1230 (1840); Cranfield, Nuytsia 13: 7–23 (1999), rev.

Dwarf to tall shrubs; branchlets pilose or glabrous, flexuose. Inflorescences at the ends of short branchlets, axillary monads, sessile and subtended by 3–4 bracts. Flowers 5-merous, pink to white; sepals and petals distinct; hypanthium shortly extended above the ovary; stamens 10 or more in a single whorl; anthers dorsifixed, versatile, dehiscing by longitudinal slits; ovary ± inferior, 3-locular; ovules hemitropous, several per loculus, on an axile-basal placenta; style stout, stigma capitate to peltate. Fruit a woody capsule, loculicidally dehiscent. Seeds usually one per locule, narrowly ellipsoid to obovoid, compressed, winged; embryo straight to slightly curved, the cotyledons plano-convex and large relative to the hypocotyl.

A genus of 4 species, south-western Australia.

111. *Asteromyrtus* Schauer

Asteromyrtus Schauer, Linnaea 17: 242 (1843); Craven, Austral. Syst. Bot. 1: 377–385 (1988), rev.
Sinoga S.T. Blake (1958).

Shrubs to trees; branchlets with simple hairs; bud scales present. Leaves with parallel venation.

Inflorescence terminal or ramiflorous, a ± globose head made up of sessile monads. Flowers 5-merous, white, cream, orange or red; sepals and petals free; hypanthium extended above the ovary summit, the free part circumscissile or persisting in the fruit; stamens numerous, filaments fused at their bases into a ring and usually aggregated into long fascicles opposite the petals; anthers dorsifixed, versatile, dehiscing by longitudinal slits; ovary 3-locular (very rarely 5-locular); ovules erect, several to numerous per locule, the placenta axile-basal, erect; style slender, the stigma capitate to peltate. Fruit a dry, woody, loculicidal capsule. Seeds several per locule, narrowly ellipsoid to obovoid, compressed, winged; embryo compressed, straight to slightly curved, the cotyledons plano-convex and large relative to the hypocotyl.

A genus of 13 species, endemic to northern Australia.

112. ***Angasomyrtus*** Trudgen & Keighery

Angasomyrtus Trudgen & Keighery, Nuytsia 4: 435 (1983).

Shrubs; branchlets shortly pubescent. Inflorescences of sessile axillary monads borne towards the apex of the shoot. Flowers 5-merous, white or pale pink; free part of hypanthium much exceeding the ovary summit; sepals and petals distinct; sepals persistent; stamens numerous, in two series, shorter than the petals; anthers dorsifixed, subversatile; style stout, narrowing to a dilated stigma; ovary inferior, mostly 2–3(4)-locular; placentation apical; ovules few and pendulous. Fruit a non-woody capsule, loculicidal. Seeds obovoid; embryo straight, cotyledons flat, longer than the hypocotyl.

One species, *A. salina* Trudgen & Keighery, endemic to salt lakes in the southwest of Western Australia. Unlikely to be generically distinct from *Kunzea*.

113. ***Kunzea*** Rchb.

Kunzea Rchb., Consp. Regni Veg.: 175 (1828); Toelken, J. Adelaide Bot. Gard. 17: 29–106 (1996), rev. W.A. spp.

Shrubs or small trees; branchlets hairy. Leaves usually alternate but opposite in a few species. Inflorescences usually a pseudoterminal, head-like cluster of sessile or subsessile monads or, less commonly, the flowers stalked and solitary or in clusters of 2 or 3; bracts sometimes enlarged, forming an involucre. Flowers 5-merous, white, yellow or pink to purple; free part of hypanthium usually much exceeding the ovary summit; sepals persistent; petals distinct; stamens numerous, in one or more series, exceeding the petals; anthers dorsifixed, versatile; ovary half-inferior, mostly 2- or 3-locular; placentation apical with ovules few and pendent or axile and ovules numerous. Fruit a non-woody capsule, usually loculicidal but indehiscent in a few species (one succulent). Seed oblong to obovoid; embryo straight, cotyledons flat, longer than the hypocotyl. $2n = 22$.

A genus of c. 46 species, Australia (E & W) and New Zealand.

114. ***Neofabricia*** Joy Thomps.

Neofabricia Joy Thomps., Telopea 2: 380 (1983).

Shrubs or small trees; branchlets hairy, sometimes glabrescent. Inflorescences axillary, condensed shoots composed of monads (rarely triads). Flowers 5-merous, bracteolate, sessile to very shortly pedicellate, yellow, white or cream; hypanthium shortly exceeding the ovary summit, spreading; stamens numerous, in several irregular rows, usually no longer than the petals; anthers dorsifixed at the base of the connective, not versatile, connective gland-tipped; style short, set into the ovary summit; stigma slightly dilated to capitate; ovary half-inferior, mostly 5–12-locular; placenta axile; ovules several, anatropous, spreading to pendulous. Fruit a loculicidal capsule; seeds usually 1 per loculus with a wing formed from fused, aborted ovules.

A genus of 3 species, Australia (north Queensland), endemic.

115. ***Homalospermum*** Schauer

Homalospermum Schauer, Linnaea 17: 242 (1843).

Tall shrubs; branchlets variably hairy. Inflorescences of axillary monads in the upper leaf axils. Flowers 5-merous, bracteolate, sessile or almost so, white; hypanthium shortly exceeding the ovary summit, spreading; sepals and petals distinct; stamens numerous, usually shorter than the petals, in a single whorl; anthers dorsifixed, versatile, connective gland-tipped: style slender, set into the ovary summit; stigma capitate; ovary half-inferior, mostly 4-locular; placentas axile; ovules numerous, hemitropous. Fruit a woody, loculicidal capsule. Seeds only 1–2 per loculus, somewhat flattened; embryo with cotyledons broader and longer than the hypocotyl.

One species, *H. firmum* Schauer, endemic to south-western Australia.

116. *Leptospermum* J.R. & G. Forst.

Leptospermum J.R. & G. Forst., Charact. Gen. 36 (1775); Thompson, Telopea 3: 301–348 (1989), rev.; Bean, Austrobaileya 3: 643–659 (1992), key N Australia & Malesia. *Leptospermopsis* S. Moore (1920).

Shrubs or trees. Inflorescences axillary, condensed shoots composed of monads or, rarely, flowers solitary. Flowers 5-merous, bracteolate, often pedicellate, usually white or pink, rarely red; hypanthium shortly exceeding the ovary summit; sepals persistent or caducous; stamens numerous, usually shorter than the petals; anthers dorsifixed, versatile; style short, set into the ovary summit; stigma often capitate; ovary mostly 3–5-locular; placenta axile; ovules few to numerous, anatropous. Fruit a loculicidal capsule; seeds linear, rarely winged. $2n = 22, 44$.

A genus of 79 species, mostly Australian with 2 species endemic to Malesia and 2 extending to New Zealand and New Guinea respectively.

As currently defined, this is not a monophyletic taxon. The type falls into a group of species that has persistent or relatively persistent fruits. This group occurs on the east coast of Australia, with its range extending into Malesia where it is often found at higher altitudes.

II.17. Tribe Chamelaucieae DC. in Schltdl. (1827).

Shrubs, mostly glabrous. Leaves opposite, less commonly spiral. Inflorescences of solitary flowers. Flowers red, pink, yellow or white; stamens usually few, mostly distinct but occasionally in clusters opposite the sepals; anthers rarely versatile, dehiscing by slits or pores; ovules anatropous or hemitropous. Fruit a capsule, or indehiscent; seeds reniform or ovoid to linear; embryo with very small cotyledons on a slender neck appressed to a relatively massive hypocotyl.

A number of distinct subgroups are indicated in molecular analyses; however, no subtribes are formally recognised, since resolution of all relationships is incomplete.

117. *Baeckea* L.

Baeckea L. Sp. Pl.: 358 (1753); Bean, Telopea 7: 245–268 (1997), rev.

Shrubs. Leaf margins entire, crenate or finely ciliate-toothed. Inflorescences axillary, 1-flowered or cymose. Flowers 5-merous, white to deep pink; sepals simple, persistent in fruit; petals distinct; stamens 5–12, none opposite centre of petals; filaments straight; anthers versatile, dehiscing by long parallel slits, gland-tipped; ovary half-inferior, 2(3)-locular; placentation axile, ovules 6–12 per loculus; stigma capitate. Fruit a loculicidal capsule. Seeds discoid to cuboid, not arillate.

Seventeen species, Australia, Malesia, SE Asia. Eastern Australia: 16 species (endemic). *Baeckea*, in the strict sense (Bean 1997), is restricted to eastern Australia and parts of Asia.

The generic concept, as traditionally applied, is polyphyletic and this generic name is still used in a broad sense in some parts of Australia. The genus has been under review for some years and a number of new genera have been described. Numerous other genera are likely to be recognised, or brought back from synonymy, in the near future. Current research is focussed on Western Australia where Rye (2009b) has recently revived the genus *Oxymyrrhine* for a small group of species.

118. *Rinzia* Schauer

Rinzia Schauer, Linnaea 17: 239 (1843); Trudgen, Nuytsia 5: 415–439 (1986), rev.

Small shrubs. Inflorescences of axillary monads or triads with very short peduncles. Flowers 5-merous; prophylls persistent; stamens usually 5–10, with one opposite each perianth segment or antesepalous stamens missing; filaments flattened with anthers dorsifixed, attached to the adaxial surface near the apex; style slender or stout with a small or capitate stigma; ovary 3(4)-locular; ovules 2–9 per loculus on an axile, peltate placenta. Fruit a globular loculicidal capsule. Seeds reniform, dark brown, arillate in most species; embryo with 2 small cotyledons on a hypocotylar neck folded down and back against the relatively massive hypocotyl.

A genus of 12 species, endemic to SW Western Australia. Not monophyletic.

119. *Ochrosperma* Trudgen

Ochrosperma Trudgen, Nuytsia 6: 11 (1987).

Shrubs. Inflorescences axillary, flowers solitary or in pairs on a common peduncle; prophylls caducous or persistent. Flowers 5-merous, white or faintly flushed with pink; calyx lobes simple, strongly keeled, persistent in fruit; petals distinct; stamens 5–8, none opposite the centre of petals: filaments straight; anthers versatile, dehiscing by long parallel slits, gland-tipped; ovary half-inferior, 3-locular; 2 ovules in each loculus, placentation axile; style inserted in a pit on the ovary summit; stigma capitate. Fruit a loculicidal capsule, barely exserted from the hypanthium, 3-valved, valves opening widely on dehiscence. Seeds reniform, not angular, arillate, yellowish, papillose.

A genus of 5 species, endemic to Australia.

120. *Micromyrtus* Benth.

Micromyrtus Benth., Gen. Pl. 1(2): 700 (1865); Bean, Austrobaileya 4: 455–476 (1997), rev. Qld; Green, Nuytsia 3: 195–204 (1980); Green, Nuytsia 4: 317–331 (1983); Rye, Nuytsia 15: 101–122 (2002), part rev. WA; Rye, Nuytsia 16: 117–147 (2006), part rev. WA.

Shrubs. Inflorescences axillary, consisting of 1–3 flowers on a common peduncle, often forming terminal clusters. Flowers 5- or rarely 6-merous, white, pink or yellow; sepals small; petals free; stamens equal to the number of petals and opposite them, or twice as many; ovary half-inferior, 1-locular, placentation apical, or subapical to lateral, ovules 2–10, collateral, never superposed. Fruit an indehiscent nut, scarcely enlarged from flower; seed usually solitary.

A genus of c. 22 species, endemic to Australia.

121. *Corynanthera* J.W. Green

Corynanthera J.W. Green, Nuytsia 2: 368 (1979).

Shrubs. Inflorescences a subterminal, spike-like aggregation of solitary, axillary flowers. Flowers 5-merous, yellow, somewhat zygomorphic, subtended by 2 persistent prophylls; hypanthium dorsiventrally compressed; sepals small, unequal; stamens in two whorls of 5, the antepetalous ones bearing a large stalked appendage and the antesepalous ones slightly shorter and with a smaller appendage; anthers ± globular, trisporangiate, dehiscing by a central pore; ovary inferior, 1-locular, placentation lateral; ovules 2, collateral. Fruit indehiscent, scarcely enlarged from flower; seed solitary; embryo not documented.

A monotypic genus that is very close to *Micromyrtus*; the single species, *C. flava* J.W. Green, is endemic to south-western Australia.

122. *Aluta* Rye & Trudgen

Aluta Rye & Trudgen, Nuytsia 13: 347 (2000).

Shrubs. Leaves opposite, small, oil glands prominent. Inflorescences of solitary flowers, sometimes grouped in subterminal clusters. Flowers 5-merous, white or pink; hypanthium pitted or wrinkled; sepals with membranous margins; stamens 5, c. 10 or 15–c. 20; anthers with a prominent clavate connective gland; ovary half-inferior, 1-locular, placenta sub-basal; ovules in 2 or 3 superposed pairs. Fruit an indehiscent nut with a prominently reticulate-pitted disk; seed usually solitary.

A genus of 5 species, endemic to Western Australia.

123. *Thryptomene* Endl.

Thryptomene Endl., Stirp. herb. hügel., 192 (1838); Rye & Trudgen, Nuytsia 13: 509–528 (2001), part rev. WA.

Shrubs. Leaves opposite, small; oil glands especially visible on the lower surface. Inflorescences usually of solitary flowers or pairs of flowers in the leaf axils. Flowers 5- or rarely 6-merous, white or pink; sepals and petals 6, similar to each other, petaloid, ± circular; stamens equal to the number of sepals and opposite them or up to twice as many or rarely 15–30; ovary half-inferior, 1-locular; placenta basal or lateral; ovules 2, or 2 plus 2 and superposed. Fruit an indehiscent nut; seed usually solitary. $2n = 22$, rarely 18 or 20 (Rye 1979).

A genus of c. 28 species, endemic to Australia.

124. *Calytrix* Labill.

Calytrix Labill., Nov. Holl. Plant. Sp. 2: 8, t. 146 (1806); Craven, Brunonia 10: 1–138 (1987), rev.
Calythropsis C.A.Gardner (1942).

Shrubs. Leaves alternate, rarely opposite or whorled; stipules, if present, very small. Inflorescences axillary, 1-flowered, pedunculate with a pair of persistent or caducous, distinct or connate prophylls (a 'cheiridium'). Flowers 5-merous, white, pink, yellow or purple; hypanthium usually long and tubular; calyx lobes spreading, usually long-attenuate or awned, persistent in fruit and turning dark red to purple; petals distinct, not persistent; stamens numerous in one or more series; ovary ± inferior, 1-locular; placenta ± basal; ovules 2 or rarely 3 or 4; stigma usually small; style usually persistent. Fruit dry, indehiscent, enclosed in persistent hypanthium and calyx. Seed solitary; embryo straight, the cotyledons small. $2n = 22, 44$.

About 75 species, endemic to Australia.

125. *Homalocalyx* F. Muell.

Homalocalyx F. Muell., Hooker's J. Bot. Kew Gard. Misc. 9:309 (1857); Craven, Brunonia 10: 139–158 (1987), rev. *Wehlia* F. Muell. (1876).

Shrubs; with or without bud scales; stipules present, small. Inflorescences l-flowered, one to many per branchlet, in the axils of leaves (these sometimes reduced to bracts), the flower pedunculate and subtended by a pair of persistent, distinct or connate prophylls (a 'cheiridium'). Flowers 5-merous; hypanthium short; sepals imbricate, persistent or caducous; petals imbricate, caducous; stamens indefinite, l–2-seriate, the filaments usually inflexed in bud (rarely erect), the anthers dorsifixed and versatile, dehiscing by longitudinal slits; ovary unilocular, the ovules usually 2 (rarely 4) on an axile-basal placenta; style slender, persistent or caducous, stigma small. Fruit dry, indehiscent; seed solitary; embryo with 2 small cotyledons on a hypocotylar neck folded down and back against the relatively massive hypocotyl. $2n = 22$.

A genus of 11 species, endemic to western and northern Australia.

126. *Astartea* DC.

Astartea DC., Prodr. 3: 210 (1828); Rye, Nuytsia 16: 149–156 (2006).

Shrubs, rarely small trees; branchlets glabrous. Leaves opposite or in opposite clusters. Inflorescences axillary, 1-flowered or cymose. Flowers 5-merous, white or pink; sepals simple, persistent in fruit; petals distinct; stamens 15–30 or more, clustered in fascicles opposite the sepals; filaments straight or incurved; anthers dehiscing by longitudinal slits; ovary half-inferior, 3-locular; placentas peltate, axile; ovules 2–10 per loculus; stigma capitate. Fruit a loculicidal capsule. Seeds angular, with flat sides and a rounded back, not arillate.

This genus, in the strict sense, consists of 20 species (c. 10 unnamed) and is endemic to south-western Australia.

The genus is being revised; in the traditional, broader sense, it is paraphyletic and includes species referable to *Cyathostemon* Turcz. Recently, Rye and Trudgen (2008) have recognised a generic segregate, *Seorsus*, with a strongly disjunct distribution in Borneo and Australia.

127. *Hypocalymma* (Endl.) Endl.

Hypocalymma (Endl.) Endl., Gen. Pl., 1230 (1840); Strid & Keighery, Nord. J. Bot. 22: 535–572 (2003), rev.

Shrubs; branchlets terete, angular or winged. Inflorescences axillary; flowers solitary or in pairs (reduced lateral shoots); prophylls linear or ovate, caducous or persistent. Flowers 5-merous, usually white of pink, sometimes yellow; stamens numerous, in one or two series, equalling the petals, distinct or shortly fused at the base; anthers longitudinally dehiscent; ovary half-inferior, mostly 2- or 3-locular; ovules 1–many in each loculus, borne on an axile or apical placenta; style about as long as the stamens; stigma small or capitate. Fruit a capsule. Seeds ± reniform, 1–2 per loculus; embryo with a thick hypocotyl and small cotyledons. $2n = 22, 44$.

About 14 species, south-western Australia.

128. *Pileanthus* Labill.

Pileanthus Labill., Nov. Holl. Plant. Sp. 2: 11, t. 149 (1806); Keighery, Nuytsia 15: 37–51 (2002), rev.

Shrubs; branchlets glabrous. Inflorescences of monads borne in the upper axils. Flowers 5-merous; prophylls scarious, connate and enclosing the bud, circumscissile at anthesis; hypanthium silky-pubescent; sepals 10, yellow; petals 5, white, pink or orange to red; stamens 20, filaments broadening at the base and joined in a sort tube; anthers contiguous or separated by a fork in the filament, dehiscing by slits; ovary inferior, 1-locular; placentation basal, peltate; ovules 4–10, in two rows. Fruit indehiscent; seed usually solitary; embryo not documented. $2n = 22$.

Around 6 species; endemic, south-western Australia. Closely related to the following five genera despite the lack of staminodes that are otherwise characteristic of the group.

129. *Chamelaucium* Desf.

Chamelaucium Desf., Mém. Mus. Hist. nat., Paris 5: 39, t. 3, 4 (1819).

Shrubs. Inflorescences of axillary monads, sometimes clustered towards the apex forming corymb-like conflorescence. Flowers 5-merous, white to pink or red, rarely yellowish; prophylls scarious, enclosing the bud and usually caducous; hypanthium broadly conical to cylindrical, often ribbed; sepals entire or ciliate; petals entire or fringed; stamens 10, alternating with staminodes; anthers ± oblong, opening by slits; ovary inferior, 1-locular; placenta peltate, basal; ovules 4–8; style not exceeding the petals, glabrous or bearing a ring of hairs below stigma; stigma slightly dilated. Fruit indehiscent. Seed usually 1. $2n = 22, 44, 66$.

A genus of 23 species, endemic, south-western Australia.

130. *Homoranthus* Schauer Fig. 54

Homoranthus A.Cunn. ex Schauer, Linnaea 10: 310 (1836); Craven & Jones, Austral. Syst. Bot. 4: 513–33 (1991), rev.

Shrubs. Inflorescence of monads in the upper leaf axils; sometimes flowers clustered into a corymb-like head or a 1–4-flowered axillary conflorescence; flowers enclosed by 2 prophylls in bud, caducous before or at anthesis. Flowers 5-merous, yellow or white to orange or red; hypanthium extending above the ovary, mostly 5-ribbed; sepals undivided or with 2–12 slender laciniate lobes or projections; stamens 10, alternating with 10 staminodes; anthers globose, opening by terminal pores; ovary half-inferior, 1-locular, placentation basal; ovules 2–10; style projecting from flower, bearing a ring of hairs below stigma. Fruit indehiscent. Seed solitary; embryo with 2 small cotyledons on a hypocotylar neck folded down and back against the relatively massive hypocotyl. $2n = 12, 18$.

A genus of c. 22 species, endemic to eastern and south-eastern Australia.

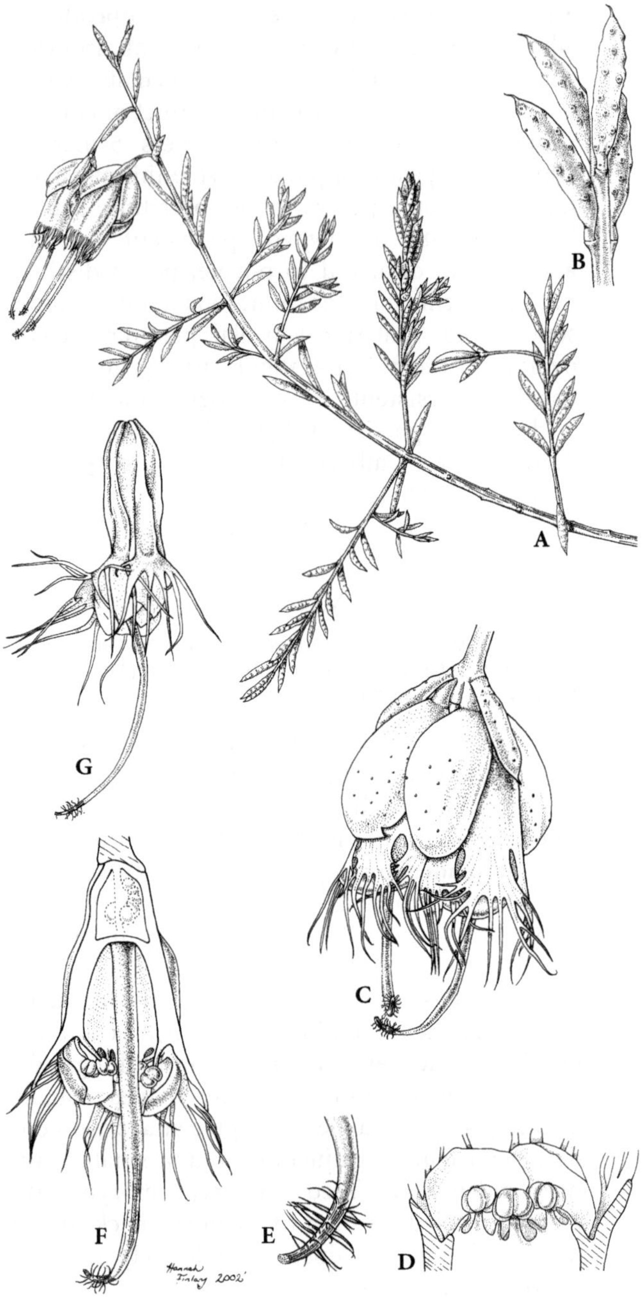

Fig. 54. Myrtaceae. *Homoranthus darwinioides*. **A** Habit. **B** Leaves. **C** Pseudo-inflorescence. **D** Flower detail showing anthers. **E** Stigma. **F** Longitudinal section of flower. **G** Fruit enclosed in hypanthium. (Drawn by H. Finlay)

131. *Darwinia* Rudge

Darwinia Rudge, Trans. Linn. Soc. London, Bot. 11: 299, t. 22 (1815); Briggs, Contr. New South Wales Natl. Herb. 3: 129–150 (1962), rev. NSW spp.
Genetyllis DC. (1828).

Shrubs. Leaves opposite or alternate. Inflorescences an aggregation of axillary flowers at apex of shoot subtended by green or coloured bracts. Flowers white, yellow or red, 5-merous, each flower enclosed by 2 prophylls when in bud; sepals usually small, ± petaloid; petals usually enclosing lower part of style; stamens 10, alternating with 10 staminodes; anthers globose, opening by terminal pores; ovary inferior, 1-locular; placenta basal; ovules 2–10; style projecting from flower, usually bearing a group of hairs below stigma. Fruit indehiscent;. Seed usually 1. $2n = 10, 12, 14, 18, 24, 28, 42$.

A genus of c. 45 species; endemic to Australia. Not a monophyletic group.

132. *Actinodium* Schauer

Actinodium Schauer, Linnaea 10: 311 (1836).

Shrubs. Inflorescences a dense, daisy-like aggregation of axillary flowers at the shoot apex, subtended by an involucre of bracts; outer flowers sterile and ray-like. Flowers white or pink, 4-merous, each flower subtended by a bract and bearing 2 linear, petaloid prophylls at the base of the pedicellate flower; sepals petaloid; petals similar to sepals; stamens 8, 4 long and 4 short, the longer ones opposite the petals; anthers globose, opening by terminal pores; ovary inferior, 1-locular; placenta basal; ovule 1; style projecting from flower, bearing a group of hairs below stigma. Fruit indehiscent, 1-seeded. $2n = 12$.

A genus of 2 species; endemic, south-western Australia. Probably not distinct from *Darwinia*.

133. *Verticordia* DC.

Verticordia DC., Prodr. 3, 208 (1828); A. George, Nuytsia 7: 231–394 (1991), key; E. George, *Verticordia*: the turner of hearts (2002), illus.

Shrubs or, rarely, small trees. Leaves opposite, rarely alternate or whorled. Inflorescences usually a dense aggregation of axillary monads forming corymb-, raceme- or spike-like conflorescences. Flowers red, white, yellow, pink or purple; prophylls scarious, caducous or persistent; hypanthium sometimes with reflexed herbaceous appendages at the apex and often with a tuft of hairs at the base; sepals variously divided into plumose, pectinate or fimbriate lobes; petals entire or variously divided; stamens 10, alternating with staminodes; anthers globose to oblong, opening by pores or slits; ovary inferior, 1-locular; placenta basal or lateral; ovules 1–13; style glabrous or bearing a group of hairs below stigma; stigma small to slightly dilated. Fruit indehiscent. Seeds 1–2. $2n = 12, 16, 18, 22, 32, 44$.

A genus of c. 97 species; endemic, western and northern Australia.

Verticordia is polyphyletic and forms a number of clades spread through the entire 'staminode' group (Barrett, Wilson and Copeland, unpubl. data). The staminode group has a major division based on anther type, and species currently referred to *Verticordia* fall into both groups.

134. *Astus* Trudgen & Rye

Astus Trudgen & Rye, Nuytsia 15: 502 (2005).

Shrubs; branchlets glabrous. Leaves opposite. Inflorescences of solitary, axillary flowers; peduncles very short. Flowers 5-merous, white to pink; hypanthium angled to ribbed opposite the sepals; sepals simple, sometimes petaloid, persistent; petals distinct, scarcely clawed; stamens usually 10, evenly distributed or opposite the sepals; filaments incurved, terete but often flattened at the base; anthers dorsifixed, opening by long parallel slits, prominently gland-tipped; ovary inferior, 3-locular but appearing 2-locular; placentas axile, with up to 7 ovules per loculus; stigma peltate or capitate. Fruit a capsule, with one indehiscent and two dehiscent loculi. Seeds reniform, exarillate.

A recent segregate from *Baeckea*, this genus consists of 4 species, endemic to south-western Australia.

135. *Triplarina* Raf.

Triplarina Raf., Sylva Tellur.: 104 (1838); Bean, Austrobaileya 4: 353–367 (1995), rev.

Shrubs. Leaves with oil dots conspicuous on lower surface, scattered or in 2 parallel rows. Inflorescences axillary, often with 2 flowers arising separately in leaf axils. Flowers 5-merous, white to deep pink; calyx lobes simple, persistent in fruit; petals free; stamens 14–18; filaments straight; anthers versatile, dehiscing by long parallel slits, gland-tipped; ovary half-inferior,

3-locular; ovules 8–13 per loculus, arranged in 2 or 3 rows on a peltate placenta; stigma capitate. Fruit a loculicidal capsule. Seeds small and reniform, not angular, not arillate.

A genus of 7 species, endemic to eastern Australia.

136. *Euryomyrtus* Schauer

Euryomyrtus Schauer, Linnaea 17: 239 (1843); Trudgen, Nuytsia 13: 543–566 (2001).

Shrubs. Stipules small, hair-like. Inflorescences of axillary, pedunculate monads (rarely pairs of flowers). Flowers 5-merous; prophylls ovate to cordate, persistent; anthopodium usually present; stamens usually 1–5 times the number of petals, if fewer than 10, the antesepalous stamens missing; filaments slender, anthers dorsifixed, versatile, connective with a prominent apical oil gland; style slender or stout with a slightly dilated stigma; ovary half-inferior, 3-locular; ovules 2–5 per loculus on an axile, peltate placenta. Fruit a globular loculicidal capsule. Seeds reniform, brown; embryo with 2 small cotyledons on a hypocotylar neck folded down and back against the relatively massive hypocotyl.

A genus of 6 species, endemic to eastern and southern Australia (New South Wales, Victoria, Tasmania, South Australia, Western Australia).

137. *Stenostegia* A.R. Bean

Stenostegia A.R. Bean, Muelleria 11: 127 (1998).

Glabrous shrubs. Inflorescences solitary, axillary, cymose, many-flowered and umbel-like. Hypanthium obconical; calyx lobes 5, compound, obtuse; petals 5, orbicular; stamens shorter than the petals, all free; anthers versatile, opening by long parallel slits; style simple, terete; stigma capitate; ovary 3-locular; ovules 8–12 per loculus, arranged in two oblique rows along placenta. Fruits capsular, hemispherical, valves enclosed. Seeds discoid, with flat sides and rounded backs.

A genus of only one species, *S. congesta* A.R. Bean, endemic to northern Australia where it is a relict found in deep ravines.

138. *Babingtonia* Lindl.

Babingtonia Lindl., Edwards's Bot. Reg. 28, t. 10 (1842); Wilson et al., Austral. Syst. Bot. 20: 302–318 (2007).

Shrubs; stems often ridged. Leaves clustered on short side shoots, leaf margins entire. Flowering shoot appearing raceme-like, individual inflorescences axillary, 1–many-flowered. Flowers 5-merous, white to pink; petals distinct, caducous; stamens usually 3–c. 12, commonly in loose groups ± opposite the sepals and not opposite centre of petals; filaments somewhat flattened, ± straight; anthers adnate, dehiscing by pores at the apices of the cells; ovary ± inferior, usually 3-locular; ovules around 7–10 per loculus; placenta furrowed, peltate; style short, set into a deep pit on the ovary summit; stigma capitate. Fruit a loculicidal capsule. Seeds angular, not arillate. This generic name has been applied widely, to include a diverse range of species that have anthers dehiscing by pores.

As defined here, it is a genus of only a few species, endemic to Western Australia.

139. *Malleostemon* J.W. Green

Malleostemon J.W. Green, Nuytsia 4: 296 (1983).

Shrubs. Inflorescences of axillary monads or triads, sometimes crowded and appearing spike-like. Flowers 5-merous, white or pink, each one subtended by 2 persistent or caducous prophylls; hypanthium smooth, rugose or 5-ribbed; stamens mostly 10, sometimes 5, the filaments geniculate towards the apex; anthers uni- or bilocular, dehiscing by one or two slits; ovary inferior, 1-locular; placentation lateral, peltate; ovules 4–8, radially arranged. Fruit indehiscent, scarcely enlarged from flower. Seed solitary; embryo with 2 small cotyledons on a hypocotylar neck curved down against the relatively massive hypocotyl.

A genus of 6 species, south-western Australia.

140. *Scholtzia* Schauer

Scholtzia Schauer, Linnaea 17: 241 (1843).

Shrubs; branchlets glabrous. Leaves opposite or in opposite clusters. Inflorescences of loose or crowded axillary umbelliform cymes, or flowers rarely solitary. Flowers 5-merous, white to pink; sepals simple, petaloid or scarious; petals free; stamens usually 5–30, evenly distributed or opposite the sepals; filaments straight or incurved; anthers opening by pores or short slits; ovary inferior, usually 2-locular; placentas axile, usually with 2 superposed ovules per loculus; stigma capitate. Fruit a capsule, indehiscent in some species. Seeds angular. $2n = 22, 44$.

Over 15 species, endemic to south-western Australia.

141. ***Balaustion*** Hook.

Balaustion Hook., Ic. Plant. 9, t. 852 (1851).

Shrubs or subshrubs. Leaves opposite, often clustered in axillary short shoots, very shortly petiolate. Inflorescences axillary, flowers solitary on a short peduncle; prophylls ovate, persistent. Flowers 5-merous, deep red; stamens numerous, in a single series, equalling the petals, distinct; anthers dorsifixed; ovary half-inferior, mostly 3-locular, included in a deep, campanulate hypanthium; ovules in 2 rows on the axile placenta; style exceeding the stamens; stigma dilated. Fruit a capsule. Seeds ± discoid, with flat sides and rounded backs.

As presently recognised, a genus of 2 species endemic to Western Australia.

Molecular data show that this is not a monophyletic group and that the distinctive flower form is the result of convergence. On the basis of molecular and morphological data, Rye (2009a) has now recognised the segregate genus *Cheyniana*, reducing *Balaustion* sens. str. to a single species.

142. ***Sannantha*** Peter G. Wilson

Sannantha Peter G. Wilson, Austral. Syst. Bot. 20: 313 (2007).

Shrubs or small trees. Inflorescences axillary triads or umbel-like dichasia, rarely of solitary flowers; the peduncles frequently 1.5–2 times longer than the anthopodia, which are mostly slender and distinct (lateral axes of dichasial inflorescences reduced, so that the prophylls are clustered at the apex of the peduncle); hypanthium more or less smooth; sepals simple or with some development of a dorsal lobe; petals entire or fringed; stamens 7–14, located towards the margins of, or opposite, the sepals, geniculate; anthers dehiscing by short slits; connectives protuberant, containing large oil glands; ovary ± inferior, 3-locular; ovules radially arranged around the placenta leaving the apex exposed. Fruit dehiscent. Seeds flattened, D-shaped.

A genus of 15 species in E Australia, New Caledonia.

143. ***Kardomia*** Peter G. Wilson

Kardomia Peter G. Wilson, Austral. Syst. Bot. 20: 316 (2007).

Shrubs or tall shrubs. Flowers mostly solitary; hypanthium ridged, rugose or papillose, rarely smooth, usually tapering into the anthopodium; sepals 'compound', the inner lobe thin-textured, obtuse, and the dorsal lobe thickened and commonly acute; stamens 4–13, located towards the margins of, or opposite, the sepals, weakly geniculate; anthers pale with rounded connectives containing oil glands; ovules radially arranged on the placenta and obscuring its apex. Fruit dehiscent. Seeds truncately D-shaped with flat sides and rounded backs.

A genus of 6 species endemic to eastern Australia.

144. ***Harmogia*** Schauer

Harmogia Schauer, Linnaea 17: 238 (1843); Wilson et al., Austral. Syst. Bot. 20: 302–318 (2007).
Baeckea sect. *Harmogia* (Schauer) Benth. & Hook.f. (1838).

Shrub. Flowers solitary; hypanthium smooth, abruptly narrowing to a short anthopodium so that the prophylls are borne close to hypanthium; sepals simple, obtuse and entire; petals rounded, entire; stamens 7–10, 1–2 located towards the margins of, or opposite, the sepals; filaments not, or only slightly, geniculate; anthers dark, with rounded connectives containing oil glands; ovary 3(4)-locular; ovules 8–12, radially arranged, obscuring the apex of the placenta. Fruit a capsule. Seeds angular, with flat sides and rounded backs.

A genus of one species, *H. densifolia* (Sm.) Schauer, endemic to eastern Australia.

Selected Bibliography

Acevedo Rodríguez, P. (1996) Flora of St. John, U.S. Virgin Islands. Mem. New York Bot. Gard. 78: 1–581.
Barrie, F.R. 2004. Synopsis of *Plinia* (Myrtaceae) in Mesoamerica. Novon 14: 380–400.
Basinger, J.F., Greenwood, D.R., Wilson, P.G., Christophel, D.C. 2007. Fossil flowers and fruits of capsular

Myrtaceae from the Eocene of South Australia. Canad. J. Bot. 85: 204–215.

Bean, A.R. 1997. A revision of *Baeckea* (Myrtaceae) in eastern Australia, Malesia and south-east Asia. Telopea 7: 245–268.

Biffin, E., Craven L.A., Crisp M.D., Gadek P.A. 2006. Molecular systematics of *Syzygium* and allied genera (Myrtaceae): evidence from the chloroplast genome. Taxon 55: 79–94.

Biffin, E., Harrington, M.G., Crisp, M.D., Craven L.A., Gadek, P.A. 2007. Structural partitioning, paired-sites models and evolution of the ITS transcript in *Syzygium* and Myrtaceae. Mol. Phylog. Evol. 43: 124–139.

Bohte, A., Drinnan, A.N. 2005. Ontogeny, anatomy and systematic significance of ovular structures in the 'eucalypt group' (Eucalypteae, Myrtaceae). Plant Syst. Evol. 255: 17–39.

Briggs, B.G. 1962. The New South Wales species of *Darwinia*. Contr. N.S.W. Natl. Herb. 3: 129–50.

Briggs, B.G., Johnson, L.A.S. 1979. Evolution in the Myrtaceae—evidence from inflorescence structure. Proc. Linn. Soc. N.S.W. 102: 157–256.

Brongniart, A., Gris, A. 1868. Déscription de quelques plantes remarquables de la Nouvelle Calédonie. Nouv. Arch. Mus. Hist. nat. Paris 4: 1–48, 15 t.

Bunniger, L. 1972. Untersuchungen über die morphologische Natur des Hypanthiums bei Myrtales- und Thymeleales-Familien. II. Myrtaceae. III Vergleich mit den Thymelaeaceae. Beitr. Biol. Pflanzen 48: 79–156.

Carr, D.J., Carr, S.G.M., Hyland, B.P.M., Wilson, P.G., Ladiges, P.Y. 2002. *Stockwellia quadrifida* (Myrtaceae), a new Australian genus and species in the eucalypt group. Bot. J. Linn. Soc. 139: 415–421.

Corner, E.J.H. 1976. The seeds of dicotyledons. New York: Cambridge University Press.

Craven, L.A. 2006. New combinations in *Melaleuca* for Australian species of *Callistemon* (Myrtaceae). Novon 16: 468–475.

Dawson, J.W. 1992. Flore de la Nouvelle Calédonie et Dépendances. 18. Myrtaceae-Leptospermoideae. Paris: Muséum national d'Histoire naturelle.

de Lange, P.J., Murray, B.G. 2004. Chromosome numbers in *Kunzea* (Myrtaceae). Austral. J. Bot. 52: 609–617.

Diels, L. 1921. Die Myrtaceen Mikronesiens. Bot. Jahrb. Syst. 56: 529–534.

Fernandes, A. 1971. Contribution à la connaissance du genre *Heteropyxis* Harv. Mitt. Bot. Staatssamml. München 10: 207–234.

Gadek, P.A., Martin, H.A. 1981. Pollen morphology in the subtribe Metrosiderinae of the Leptospermoideae (Myrtaceae) and its taxonomic significance. Austral. J. Bot. 29: 159–184.

Gadek, P.A., Martin, H.A. 1982. Exine ultrastructure of myrtaceous pollen. Austral. J. Bot. 30:75–86.

Gandolfo, M., Gonzalez, C., Zamaloa, M., Cuneo, N., Wilf, P. 2006. *Eucalyptus* (Myrtaceae) macrofossils from the early Eocene of Patagonia, Argentina. Abstract 473, Botany 2006 conference, Botanical Society of America.

Green, J.W. 1979. *Corynanthera*, a new genus of Myrtaceae (Subfamily Leptospermoideae, Tribe Chamelaucieae). Nuytsia 2: 368–372.

Hnatiuk, R.J. 1993. A revision of the genus *Eremaea* (Myrtaceae). Nuytsia 9: 137–222.

Hooker, J.D. 1860. On *Fropiera*, a new Mauritian genus of calycifloral exogens, of doubtful affinity. J. Proc. Linn. Soc. (Bot.) 5: 1–2.

Houston, T.F., Lamont, B.B., Radford, S., Errington, S.G. 1993. Apparent mutualism between *Verticordia nitens* and *V. aurea* (Myrtaceae) and their oil-ingesting bee pollinators (Hymenoptera: Colletidae). Austral. J. Bot. 41: 369–380.

Johnson, L.A.S., Briggs, B.G. 1985 ('1984'). Myrtales and Myrtaceae - a phylogenetic analysis. Ann. Missouri Bot. Gard. 71: 700–756.

Kevan, P.G., Lack, A.J. 1985. Pollination in a cryptically dioecious plant *Decaspermum parviflorum* (Lam.) A. J. Scott (Myrtaceae) by pollen-collecting bees in Sulawesi, Indonesia. Bot. J. Linn. Soc. 25: 319–330.

Ladiges, P.Y. 1984. A comparative study of trichomes in *Angophora* Cav. and *Eucalyptus* L'Hérit. - a question of homology. Austral. J. Bot. 32: 561–574.

Landrum, L.R., Kawasaki, M.L. 1997. The genera of Myrtaceae in Brazil: an illustrated synoptic treatment and identification keys. Brittonia 49: 508–536.

Lucas, E.J., Harris, S.A., Mazine, F.F., Belsham, S.R., Nic Lughadha, E.M., Telford, A., Gasson, P.E., Chase, M. W. 2007. Suprageneric phylogenetics of Myrteae, the generically richest tribe in Myrtaceae (Myrtales). Taxon 56: 1105–1128.

Moore, D.M. 1983. Flora of Tierra del Fuego. Oswestry: Nelson.

Patel, V.C., Skvarla, J.J., Raven, P.H. 1985 ('1984'). Pollen characters in relation to the delimitation of Myrtales. Ann. Missouri Bot. Gard. 71: 858–969.

Pike, K.M. 1956. Pollen morphology of Myrtaceae from the south-west Pacific area. Austral. J. Bot. 4: 13–53.

Porter, E.A., Nic Lughadha, E., Simmonds, M.S.J. 2000. Taxonomic significance of polyhydroxyalkaloids in the Myrtaceae. Kew Bull. 55: 615–632.

Proença, C.E.B., Gibbs, P.E. 1994. Reproductive biology of eight sympatric Myrtaceae from central Brazil. New Phytol. 126: 343–354.

Rye, B.L. 1979. Chromosome number variation in the Myrtaceae and its taxonomic implications. Austral. J. Bot. 27: 547–573.

Rye, B.L. 2009a. A reduced circumscription of *Balaustion* and a description of the new genus *Cheyniana* (Myrtaceae: Chamaelaucieae) with three new species. Nuytsia 19: 129–148.

Rye, B.L. 2009b. Reinstatement of the Western Australian genus *Oxymyrrhine* (Myrtaceae: Chaemaelaucieae) with three new species. Nuytsia 19: 149–165.

Rye, B.L., James, S.H. 1992. The relationship between dysploidy and reproductive capacity in Myrtaceae. Austral. J. Bot. 40: 829–848.

Rye, B.L., Trudgen, M.E. 2005. A new heterocarpidic fruit type for the Myrtaceae, with dehiscent and indehiscent loculi, in two genera from Western Australia. Nuytsia 15: 485–493.

Rye, B.L., Trudgen, M.E. 2008. *Seorsus*, a new Gondwanan genus of Myrtaceae with a disjunct distribution in Borneo and Australia. Nuytsia 18: 235–257.

Schmid, R. 1980. Comparative anatomy and morphology of *Psiloxylon* and *Heteropyxis*, and the subfamilial and tribal classification of Myrtaceae. Taxon 29: 559–595.

Schmid, R., Baas, P. 1984. The occurrence of scalarifrom perforation plates and helical vessel wall thickenings in wood of Myrtaceae. I.A.W.A. Bull. 5: 197–215.

Scott, A.J. 1980. Notes on Myrtaceae in the Mascarenes with some recombinations for taxa from Aldabra, Malaya, New Caledonia. Kew Bull. 34: 473–498.

Smith-White, S. 1954. Cytological studies in the Myrtaceae IV. The sub-tribe Euchamaelaucinae. Proc. Linn. Soc. N.S.W. 79: 21–28.

Snow, N. 2007. Systematics of the Australian species of *Rhodamnia* (Myrtaceae). Syst. Bot. Monogr. 82.

Snow, N. 2009. *Kanakomyrtus* (Myrtaceae): a new endemic genus from New Caledonia with linear stigmatic lobes and baccate fruits. Syst. Bot. 34: 330–344.

Soares-Silva, L.H., Proença, C.E.B. 2006. An old species revisited and a new combination proposed in *Psidium* Myrtaceae. Kew Bull. 61: 199–204.

Steenis, C.G.G.J. van 1952. *Kjellbergiodendron* and *Whiteodendron*, Malaysian Myrtaceae—Leptospermoideae, Metrosiderinae. Acta Bot. Neerl. 1: 435–442.

Steenis, C.G.G.J. van 1953. Misecellaneous botanical notes V. Acta Bot. Neerl. 2: 298–307.

Tobe, H., Raven, P.H. 1983. An embryological analysis of Myrtales: its definition and characteristics. Ann. Missouri Bot. Gard. 70: 71–94.

Tobe, H., Raven, P.H. 1987. Embryology and systematic position of *Heteropyxis* (Myrtales). Am. J. Bot. 74: 197–208.

Tobe, H., Raven, P.H. 1990. Embryology and systematic position of *Psiloxylon* (Myrtales). Bot. Bull. Acad. Sin. 31: 119–127.

van der Merwe, M.M., van Wyk, A.E., Botha, A.M. 2005. Molecular phylogenetic analysis of *Eugenia* L. (Myrtaceae) with emphasis on southern African taxa. Plant Syst. Evol. 251: 21–34.

van Vliet, G.J.C.M., Baas, P. 1985 ('1984'). Wood anatomy and classification of the Myrtales. Ann. Missouri Bot. Gard. 71: 783–800.

van Wyk, A.E., Botha, R. 1984. The genus *Eugenia* (Myrtaceae) in southern Africa: ontogeny and taxonomic value of the seed. S. African J. Bot. 3: 63–80.

van Wyk, A.E., Lowrey, T.K. 1988. Studies on the reproductive biology of *Eugenia* L. (Myrtaceae) in southern Africa. Monogr. Syst. Bot. Missouri Bot. Gard. 25: 279–293.

Weberling, F. 1966. Additonal notes on the Myrtaceous affinity of *Kania eugenioides* Schltr. Kew Bull. 20: 517–520.

Wilson, P.G., Hyland, B.P.M. 1988. New taxa of rainforest Myrtaceae from northern Queensland. Telopea 3: 257–271.

Wilson, P.G., O'Brien, M.M., Quinn, C.J. 2000. *Anetholea* (Myrtaceae), a new genus for *Backhousia anisata*: a cryptic member of the *Acmena* Alliance. Austr. Syst. Bot. 13: 429–435.

Wilson, P.G., O'Brien, M.M., Heslewood, M.M., Quinn, C.J. 2005. Relationships within Myrtaceae sensu lato based on a *mat*K phylogeny. Plant Syst. Evol. 251: 3–19.

Nitrariaceae

Nitrariaceae Bercht. & J. Presl (1820), nom. cons.
Nitrarioideae Engl. (1896).

M.C. Sheahan

Shrubs 0.5–2 m tall, often with spines. Leaves alternate or fasciculate, petiolate to subsessile, simple, fleshy, often with short appressed hairs, cuneate at base and acute, 3-lobed or retuse at apex; stipules minute, intrapetiolar, free, often caducous. Flowers 5-merous, bisexual (dioecious in *Nitraria billardieri*), small, pedicellate, solitary or aggregated into lax scorpioid cymes, caducous bracts sometimes present; sepals distinct, imbricate or united at base, persistent; petals concave, yellowish-green or white; disk small, fleshy; stamens (10)15, without appendages; filaments inserted at margin of disk; anthers oblong, dorsifixed, versatile; ovary superior, sessile, oblong-pyramidal, silky, 3(6)-carpellate, simple, narrowing gradually with 3(6) decurrent stigmatic lobes at apex; ovule 1 per carpel, pendulous, apotropous, bitegmic, crassinucellar. Fruit a berry-like drupe with long bony endocarp, 1-seeded, derived from single fertile loculus. Seed pendulous, ovoid-acuminate, exalbuminous but contains oil. $2n = 24, 48, 60$.

One genus and 5–8(12) spp., growing in the salt deserts, salt marshes and coastal sand dunes of the Old World: SE Europe, N Africa, SW and Central Asia and China; also one species in Australia (*N. billardieri* DC.).

Vegetative Morphology. Rigid, more or less evergreen shrubs with many erect and spreading branches which may be appressed-pubescent at first, and often have spines at the end of side branches. Leaves are simple, fleshy, often greyish-green with pubescence when young; they are alternate or grow in clusters on brachyblasts. The plants are typically drought deciduous, with few small thick leaves produced in dry periods; in severe droughts stem abscission may also take place. When rains return, new growth of larger thinner leaves emerges from the brachyblasts. The shape of the leaf is obovate or oblong-spathulate; in *N. retusa* the leaf is more or less triangular and the shape changes as the leaf develops, the new leaves having a more markedly tridentate apex when they first unfold but, at maturity, the largest leaves have a more retuse apex. There is an extensive root system.

Vegetative Anatomy (see Sheahan and Cutler 1993). The leaf surface is waxy and resistant to wetting. Venation is pinnate, reticulate and brochidodromous; the smaller veins are branched, and there are large clusters of dilated tracheids associated with higher-order veins and veinlet endings, especially at margins and apex; areoles are irregular in shape and size. Kranz anatomy is absent, and there is no evidence of C_4 activity. Trichomes are unicellular, sparse, caducous, 120–220 μm long; they are unbranched, and appressed towards the leaf apex; in shape they are straight or clavate, with a warty surface. The epidermal cells are polygonal; the stomata small and somewhat sunken, in surface view mainly anomocytic, a few paracytic or weakly actinocytic; in TS, the guard cells have an outer and sometimes an inner ledge and thickened upper and lower walls. The mesophyll is more or less dorsiventral with 3–4 layers of palisade cells and water storage tissue in the centre of the leaf. The petiole, where present, is more or less semi-circular in TS, often with a semi-circular main vascular bundle and two very small lateral bundles.

In the stem there is an unusually extensive development of the secondary phloem parenchyma; the sclerenchyma takes the form of thick-walled fibre strands in the cortex, associated with lignified parenchyma. Phloem sieve elements are usually narrow (diameter 5–8 μm), with compound sieve plates. *N. retusa* is reported to have S-type plastids, with diameter 1.0 μm and 5 more or less globular starch grains (Behnke 1988). Periderm differentiation is in the inner cortex; the phellem cells have thin suberised walls. The xylem forms a complete cylinder early. The vessels are solitary or in radial chains or clusters, grouped together in radial bands separated by fibres; sometimes weakly storied; some have spiral thickening. In *N. retusa* the vessels are in two size classes. Vessel elements are narrow and short, circular in TS, with simple perforations and minute alternate bordered pits; the ends are transverse to oblique and pit vesturing is absent. Imperforate elements are mainly libriform fibres, some with living contents; vasicentric fibres are also present. Axial parenchyma is abundant, paratracheal or aliform-confluent, with fusiform cells, forming concentric bands. Rays are 1–3(4) cells wide and of variable height, the cells square to upright with lignified walls.

In both stem and leaf there may be abundant large mucilage cells and cavities in the mesophyll of the leaf, and the cortex, phloem, rays and pith of the stem. There are some acicular crystals and small druses in epidermal cells, sometimes also in the mesophyll, and rhomboidal crystals in the stem. Tannin cells may be abundant in *N. retusa* and *N. schoberi*.

Fig. 55. Nitrariaceae. *Nitraria schoberi*. **A** Flowering twig. **B** Flower. **C** Fruit. *Nitraria sphaerocarpa*. **D** Fruiting twig. (Takhtajan 1981)

Flower Structure. The flowers are aggregated in forked scorpioid cymes, with small subtending caducous bracts (Fig. 55). They are basically pentamerous, actinomorphic and hypogynous; they are usually bisexual but *N. billardieri* is reported to be mostly dioecious (Noble and Whalley 1978). The sepals are somewhat fleshy, and may be fused, or free and imbricate. There is a free hypanthium, and there are sunken nectaries between stamens and petals. *Nitraria* usually has 15 stamens (rarely 10); this condition is interpreted by Ronse Decraene and Smets (1991) as based on haplostemony, but with antesepalous triplets of stamens derived from a complex primordium (but see Fig. 55B showing pairs of antepetalous stamens).

Three carpels fuse to form the 3-locular (sometimes 6-locular, Takhtajan 2009) superior ovary; it is ovoid, oblong or pyramidal in shape, tapering somewhat towards the apex, which has three (6) vertical stigmatic lobes. Each locule contains a single pendulous ovule.

Embryology. Information mainly from Lahham and Al-Eisawi (1986) who included *N. retusa* in their study, and Li and Tu (1991, 1994) who looked at *N. sibirica*. There is a 2-layered glandular tapetum; microspore tetrads are tetrahedral, and the pollen is 2-celled at shedding. The ovule is bitegmic and apotropous with a ventral raphe, and there is a well-developed nucellus and no endothelial tapetum. Embryo sac development is of the Polygonum type; embryogeny is of the solanad type.

Pollen Morphology. Information from Erdtman (1952), Lahham and Al-Eisawi (1986), and Xi and Zhang (1991). In *N. retusa*, pollen grains are prolate (36×27 μm), 3-colporate, with ora more or less circular. The wall is 3 μm thick; the exine is psilate to perforate, intectate, with a faint striation visible at high focus; the sexine as thick as the nexine.

Karyology. Noble and Whalley (1978) report $2n = 24$ for *N. retusa* and *N. schoberi*, and $2n = 48$ for *N. billardieri*. There is some information on *N. sibirica* in Yang et al. (1996), who report $2n = 24$; however, Ma et al. (1990) give $2n = 60$ for the same species.

Fruit and Seed. The fruit is a fleshy syncarpous drupe which develops from the trilocular ovary; it is one-seeded by abortion and may be red, purple or yellow. The fruit wall has 3 layers: a thin exocarp (with epidermis, hypodermis and 6-layered parenchyma), a pitted woody ovoid-conical mesocarp, and an endocarp of 8 or more layers of parenchyma cells (Li and Tu 1991). The seeds are straight, have a green embryo and no endosperm. Dissemination is probably by birds and small mammals. Noble and Whalley (1978) studied the dispersal of *N. billardieri* in Australia and observed that the seeds were often eaten by emus, which appeared to improve germination. The seeds can remain viable in dry sandy soils for long periods until conditions are suitable for germination.

Phytochemistry. Information mainly from Hegnauer (1973), and Saleh and El-Hadidi (1977). *Nitraria* is reported to contain the flavonoids rutin, kaempferol and isorhamnetin. According to Saleh and El-Hadidi (1977), *N. retusa* is unlike members of Zygophyllaceae in having isorhamnetin-3-rhamnogalactoside, rather than the isorhamnetin-3-rutinoside found in, for example, most *Zygophyllum* species. Alkaloids present are nitrarin and carbolin. The saponins which are found in Zygophyllaceae are absent, and there is no starch in the seeds.

Affinities. The taxonomic position of *Nitraria* has long been the subject of disagreement. In spite of dissimilarities in anatomy, embryology, biochemistry and plastid type, as well as in flower and fruit structure, it was in the past often included in Zygophyllaceae, or considered to be close to that family. More recently, *Nitraria* has consistently appeared in molecular analyses in a position widely separated from the zygophyllaceous genera, as sister to *Peganum*, *Malacocarpus* and *Tetradiclis* in a clade containing members of Sapindales (Sheahan and Chase 1996; Gadek et al. 1996; Savolainen et al. 2000).

APG III (2009) favours the inclusion of Tetradiclidaceae (=Peganaceae) in Nitrariaceae, but there are many dissimilarities between these families. For example, leaves in *Nitraria* are simple and fasciculate but in *Peganum* and *Malacocarpus*, they are pinnatifid with an unusual arrangement of epidermal cells; the fruit of *Nitraria* is drupaceous, unlike the loculicidal capsules of *Peganum* and *Tetradiclis* and the berry of *Malacocarpus*. Ronse Decraene et al. (1996) saw similarities in floral development between *Nitraria* and *Peganum* 'at first glance', but qualified this to find many more dissimilarities. Sheahan and Chase (1996) pointed out that the *rbc*L sequences of these two genera are as divergent as those of other families in Sapindales, and concluded that they should constitute separate families within Sapindales.

Distribution and Habitats. *Nitraria* is a halophyte: the English name Nitre-bush is from the Latin word for saltpetre, referring to the fact that it can thrive in saline soil. It is among the most salt-tolerant of woody plants, with up to 14% of its dry weight in the form of NaCl (Levitt 1972).

N. retusa grows in semi-arid and saline deserts and in salt marshes, in the Middle East, Arabia and northern and eastern Africa. *N. senegalensis*, sometimes regarded as a subspecies of *N. retusa*, is found in north-west Africa. *N. schoberi* has a wider distribution; it is a native of saline desert areas, stony soils and coastal sand dunes of the eastern Mediterranean and northern Africa, the near East, West and Central Asia; in China it is replaced by *N. sibirica*. A number of other species, sometimes regarded as no more than geographical variants of *N. schoberi*, have a more restricted distribution: *N. komarovii* round the Caspian, and *N. sphaerocarpa* and *N. tangutorum* in China. *N. billardieri*, which has been regarded by some as a synonym of *N. schoberi*, is found in southern temperate Australia and is the only representative of the family in the southern hemisphere. Contrasting with common observations, Pan et al. (1999) proposed as many as 12 species throughout the world, basing their classification on distribution patterns, pollen morphology, and the structure of the pericarp.

Economic Importance. The fruits of *N. retusa* have been used as an intoxicant in Arabia, and the fruits of *N. schoberi* and related species are eaten by indigenous peoples. The plant is used as a source of salt, and as firewood in some desert areas; camels and goats are reported to browse on the succulent leaves (Agnew 1980).

Furthermore, *Nitraria* bushes serve to stabilise wind-blown sand and loess soils and thus reduce erosion, and can be used as 'phytoameliorators' of saline soils because of the great salt capacity of the leaves (Noble and Whalley 1978).

A single genus:

Nitraria **L.** Fig. 55

Nitraria L., Syst. Nat.: 1044 (1759).

Description as for the family.

Selected Bibliography

Agnew, A.D.Q. 1980. Zygophyllaceae. In: Townsend, C.C., Guest, E. (eds.) Flora of Iraq, vol. 1. Baghdad: Ministry of Agriculture.

APG III. 2009. An update of the Angiosperm Phylogeny Group classification for the orders and families of flowering plants: APG III. Bot. J. Linn. Soc. 161: 105–121.

Behnke, H.-D. 1988. Sieve-element plastids - systematic relationship of Rhizophoraceae, Anisophylleaceae and allied groups. Ann. Missouri Bot. Gard. 75: 1387–1409.

Bobrov, E.G. 1974. Zygophyllaceae. In: Shishkin, B.K., Bobrov, E.G. (eds.) Flora of the U.S.S.R., vol. 14 (translation of the 1949 Russian edition). Jerusalem: Israel Program for Scientific Translations.

Chevalier, A. 1949. Les *Nitraria*, plantes utiles des déserts salés. Revue Bot. appl. Agric. Trop. 29: 595–601.

Erdtman, G. 1952. Pollen morphology and plant taxonomy. Angiosperms. Stockholm: Almqvist and Wiksell.

Gadek, P., Fernando, E.S., Quinn, C.J., Hoot, S.B., Terrazas, T., Sheahan, M.C., Chase, M.W. 1996. Sapindales: molecular delimitation and infraordinal groups. Am. J. Bot. 83: 802–811.

Goldblatt, P., Johnson, D.E. 2003. Index to Plant Chromosome Numbers 1998-2000. Monogr. Syst. Bot. Missouri Bot. Gard. 94.

Hegnauer, R. 1973. Chemotaxonomie der Pflanzen, vol. 6. Basel: Birkhäuser.

Lahham, J.N., Al-Eisawi, D. 1986. Pollen morphology of Jordanian Zygophyllaceae. Candollea 41: 325–328.

Levitt, J. 1972. Responses of plants to environmental stresses. New York: Academic Press.

Li, S., Tu, L. 1991. The studies on the embryology of *Nitraria sibirica* Pall. IV. The developmental anatomy of the fruit and seed. Acta Scient. Nat. Univ. Intramongolicae 22: 389–395.

Li, S., Tu, L. 1994. The embryology and its systematic significance of *Nitraria*. Bull. Bot. Res. 14: 255–262.

Ma, X.H., Ma, X.Q., Li, N. 1990. Chromosome observation of some drug plants in Xinjiang. Acta Bot. Boreali-Occid. Sinica 10: 203–210.

Noble, J.C., Whalley, R.D.B. 1978. The biology and autecology of *Nitraria* L. in Australia. I. Distribution, morphology and potential utilisation. Aust. J. Ecol. 3: 141–163.

Pan, X.-L., Shen, G.-M, Chen, P. 1999. A preliminary research on taxonomy and systematics of genus *Nitraria*. Acta Bot. Yunn. 21: 287–295.

Porter, D.M. 1974. Disjunct distributions in the New World: Zygophyllaceae. Taxon 23: 339–346.

Ronse Decraene, L.P., Smets, E.F. 1991. Morphological studies in Zygophyllaceae. I. The floral development and vascular anatomy of *N. retusa*. Am. J. Bot. 78: 1438–1448.

Ronse Decraene, L.P., De Laet, J., Smets, E.F. 1996. Morphological studies in Zygophyllaceae. II. The floral development and vascular anatomy of *Peganum harmala*. Am. J. Bot. 83: 201–215.

Saleh, N.A.M., El-Hadidi, M.H. 1977. An approach to the chemosystematics of the Zygophyllaceae. Biochem. Syst. Ecol. 5: 121–128.

Savolainen, V., Fay, M.F., Albach, D.C., Backlund, A., Van der Bank, M., Cameron, K.J., Johnson, S.A., Lledó, M.D., Pintaud, J.-C., Powell, M., Sheahan, M.C., Soltis, D.E., Soltis, P.S., Weston, P., Whitten, W.M., Wurdack, K.J., Chase, M.W. 2000. Phylogeny of the eudicots: a nearly complete familial analysis based on *rbc*L gene sequences. Kew Bull. 55: 257–309.

Sheahan, M.C., Chase, M.W. 1996. A phylogenetic analysis of Zygophyllaceae R.Br. based on morphological, anatomical and *rbc*L DNA sequence data. Bot. J. Linn. Soc. 122: 279–300.

Sheahan, M.C., Cutler, D.C. 1993. Contribution of vegetative anatomy to the systematics of the Zygophyllaceae R.Br. Bot. J. Linn. Soc. 113: 227–262.

Takhtajan, A.L. 1981. Plant life, vol. V (2). Moskau: Proswyeschtschenye.

Takhtajan, A. 2009. Flowering plants, 2nd edn. Dordrecht: Springer.

Xi, Y., Zhang, J. 1991. The comparative studies of pollen morphology between *Nitraria* and Meliaceae. Bot. Res. (Inst. Bot. Acad. Sinica) 5: 47–58.

Yang, D.K., Qin, Y.Q., Zhou, J.Y., Li, G., Zhu, L.H. 1996. A study on chromosomes of *Tribulus terrestris* and *Nitraria sibirica*. Guihaia 16: 161–164.

Rutaceae

Rutaceae Durande (1782), nom. cons.
Cneoraceae Vest (1818), nom. cons.
Ptaeroxylaceae J. Leroy (1960).

K. Kubitzki[1], J.A. Kallunki[2], and M. Duretto with Paul G. Wilson[3]

Trees or shrubs, sometimes scandent, rarely herbs, sometimes spiny or aculeate, usually (but not in all Cneoroideae) with schizogenous (mostly pellucid) glands containing volatile oils on leaves, young branchlets, inflorescences, flower parts, pericarp and cotyledons, and with oil cells in parenchymatous tissue; cork subepidermal. Leaves alternate, opposite, or whorled, entire or toothed, variously compound, unifoliolate, or simple, the rhachis sometimes winged; stipules 0. Inflorescences panicles, thyrses, racemes, spikes, botryoids, sciadioids, heads, umbels, or cincinni. Flowers bisexual or unisexual, usually 3–5-merous, actinomorphic or rarely somewhat zygomorphic, hypogynous or sometimes perigynous; sepals distinct to completely connate; petals distinct or sometimes coherent or connate for part of their length; stamens twice the number of petals or equal to it or sometimes more numerous (50–00 in *Clymenia*), obdiplostemonous, the antepetalous stamens often transformed into staminodia or 0; filaments distinct or sometimes coherent or connate for part of their length; anthers 4-sporangiate, longitudinally dehiscent, introrse or sometimes latrorse; disk (sometimes 0) intrastaminal, nectariferous; gynoecium often inserted on gynophore, of 1–5 (–many) carpels, ranging from more or less apocarpous with distinct or only proximally connate ovarioles and stylodia usually joined in a common style or at least with joined stigmas to completely syncarpous; each carpel with 1 locule and 1–several (rarely many) ovules; ovules bitegmic or very rarely (*Glycosmis*) unitegmic, crassinucellar, anatropous or hemitropous, more or less epitropous; placentation axile, very rarely parietal. Fruit of 1–5 distinct or proximally connate follicles, drupes (these rarely winged), samaras, samaroids, or grading to syncarpous with carpels connate into a loculicidal or rarely septicidal capsule or berry, the latter sometimes with a hard shell. Seeds rarely winged, often explosively ejected with the separating dorsilateral endocarp and sometimes furnished with the persistent ventral endocarp; endosperm + or 0; embryo relatively large, straight or curved, with flattened, plano-convex, conduplicate, or conduplicate-plicate cotyledons.

A family of 154 genera and about 2,100 species; nearly cosmopolitan, but mainly tropical and subtropical; most diverse in Australasia.

Vegetative Morphology and Anatomy. The family comprises tall, buttressed tropical rainforest trees, smaller trees or shrubs, which, particularly in the New World, often show pachycaulous tendencies, and rarely herbs. *Spathelia* is a leptocaul monocarpic tree with giant leaves crowded at the apex of the unbranched stem. In *Dictyoloma*, growth of the branches is terminated by the production of inflorescences, and continues by repeated sympodial branching. In more xeric vegetation, particularly in Australia and South Africa, the family is represented by low, scleromorphic shrubs; their shoots are often covered with glandular tubercles. Stems of some *Zanthoxylum* and of *Toddalia* are retrorse-aculeate, the spines sometimes being raised by a corky cushion that widens towards the base (knobthorns; Barber 1892). In *Zanthoxylum*, spines occur also on the twigs and/or on the leaves; they are often stout or laterally compressed, or form decurrent wings on the internodes (Fig. 56; *Z. rolandri*), or appear paired at the leaf base like pseudostipules.

In Aurantioideae, solitary or paired axillary spines are found on one or both sides of an axillary

[1]M. Groppo, T.G. Hartley, D.J. Mabberley and T.H. Trinder-Smith are thanked for advice and information.

[2]neotropical Rutaceae, mostly Angostura Alliance

[3]Boronia Alliance

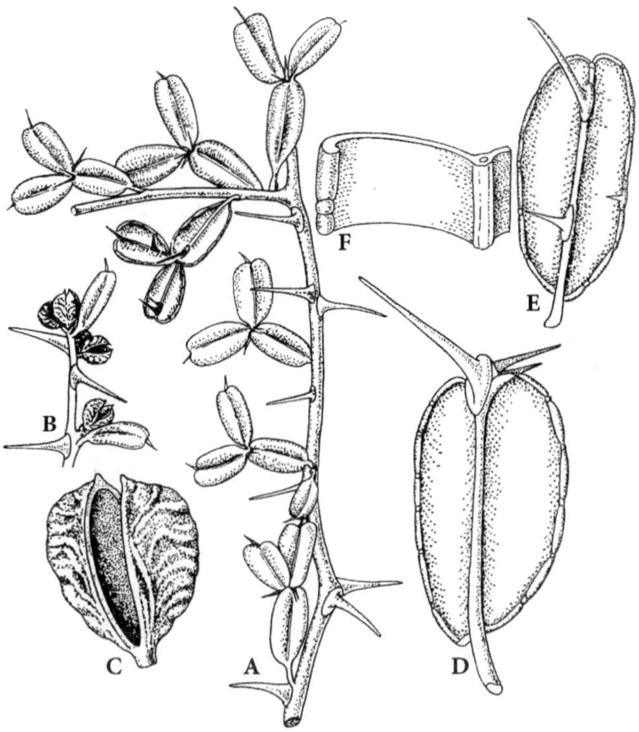

Fig. 56. Rutaceae. *Zanthoxylum rolandri.* **A** Phyllodial branch. **B** Fruiting branch. **C** Fruit with endocarp and seed ejected. **D** Phyllodium, lower side. **E** Leaflet, lower side. **F** Section of leaflet. (Beurton 1986; drawn by the author)

bud; they prevail in juvenile plants. Many authors interpreted them as homologous to the first leaf (ves) of this bud, but reports of scales or rudimentary leaves on the spines and study of their anatomical structure favour their interpretation as metamorphic axes (Uphof 1935). When the scandent, spiny *Luvunga* leans on a firm support, the spines become strongly recurved and thickened, and obviously are irritable.

Leaves are impari- or paripinnate (bipinnate in *Dictyoloma* and *Bottegoa*), pinnately trifoliolate (i.e. reduced imparipinnate with the terminal leaflet subtended by an extension of the rhachis), digitately trifoliolate (i.e. the sessile or petiolulate leaflets arising from one point), or single-bladed. The latter may be unifoliolate (i.e. reduced compound leaves, as evidenced by their petiole being distally swollen and/or articulated with the blade) or simple (petiole neither distally swollen nor articulated with blade). Genera that comprise species with unifoliolate or simple leaves are herein described as unifoliolate. Phyllodia are found in some *Zanthoxylum* (Fig. 56) and several Aurantioideae, such as *Citropsis*. *Citrus* (*Eremocitrus*) *glauca*, a tree of particularly arid and hot habitats, has isobilateral leaves, which are oriented one edge to light.

Secretory schizogenous cavities containing volatile oils, which are nearly universally present, are the most striking feature of the family. On the leaves, they appear as translucent pellucid dots but are also found on petioles, young axes, the inflorescences and flowers, particularly on the ovary hoods of many genera of the Diosma and Boronia Alliances and at the tip of the stamens of the Diosma Alliance. In *Phellodendron*, often said to lack such cavities, they occur in the sinuses of the crenation of the leaf margins. Plant anatomists have disagreed about whether the secretory glands have a schizogenous or lsyigenous origin, but the work of Turner et al. (1998) refuted the lysigenous concept on the grounds that it was based on artefactual preparations of the gland tissue. Because the epithelial cells of the glands in *Citrus* were shown to remain living long after the initial secretory phase (Turner et al. 1998), it is possible that they could replace the monoterpenes lost by evaporation or catabolism. Some genera of the Cneoroideae lack secretory cavities but have secretory cells, which contain oil or perhaps resin, in the pith and primary cortex. Secretory cells (in addition to the secretory cavities) are widespread also in Rutoideae.

Trichomes are simple or compound. When compound, they vary from fasciculate to stellate to scale-like. Very small trichomes occur in many of the species and range down to around 0.1 mm in length. Glandular hairs are 1-celled or have a stalked multicellular head. The tuberculate multicellular trichomes of *Dictamnus* contain an oil gland elevated above the surface of the axes and leaves.

The leaves are hypostomatic or more rarely amphistomatic, as in *Ruta* and some *Zanthoxylum*, *Acmadenia* and *Coleonema*. From several genera, including *Medicosma*, *Melicope*, *Brombya*, *Euodia*, *Boronia* and *Philotheca*, sclerotic modifications of the tracheal elements at the vein endings in the leaf mesophyll are known (Hartley 1985; Wilson 1998); such terminal foliar sclereids have been found to occur in about 70% of the *Boronia* species, where they vary from thin-walled and tracheid-like to thick-walled and many-armed (Rao and Bhattacharya 1981).

The nodes are trilacunar or sometimes unilacunar. The wood is generally diffuse-porous; vessels are rounded and tend to have thickened walls (the Cuban *Plethadenia* being extreme in this respect; see Heimsch 1942, their Fig. 20) and to be aggregated in multiples, chains or clusters; perforation is usually simple; intervascular and vessel-parenchyma and vessel-ray pitting is alternate; rays are mostly homogeneous and only rarely uniseriate; parenchyma is in many cases scanty paratracheal with or without terminal or non-terminal bands, generally showing transitions to abundant paratracheal types; fibres have simple or indistinctly bordered pit pairs and usually lack septations (in contrast to related families; Heimsch 1942). Oxalate is found as druses, solitary crystals, raphides, styloids or crystal sand and is often deposited in large masses in the pith and cortex.

INFLORESCENCES. Rutaceae exhibit a wide array of inflorescence forms, including panicles, thyrses, botryoids, racemes, sciadioids, heads and solitary flowers (Troll 1964, 1969). Thyrso-paniculate inflorescences appear to be the most generalised type and prevail among the Australasian and Asian rainforest genera. Highly elaborate panicles (*Zanthoxylum*, *Ptelea*, *Phellodendron*) and thyrses (*Tetradium danielli*, as *Evodia d.*) have been analysed and drawn by Troll (1969). In *Zanthoxylum*, panicles are often

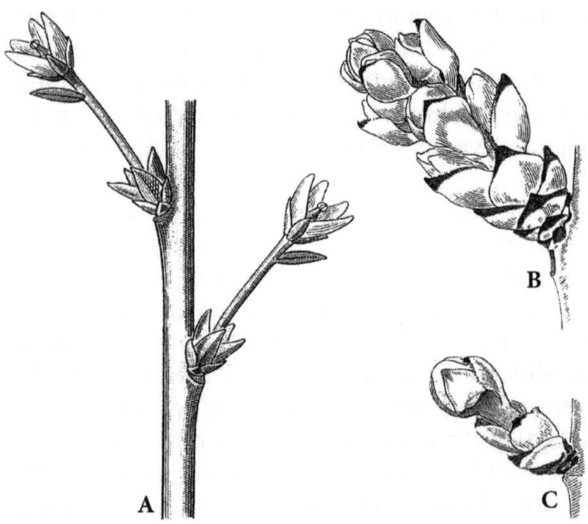

Fig. 57. Rutaceae. *Orixa japonica*. **A** Female flowers. **B, C** Male flowers in few- to 1-flowered botryoids. (Troll 1969)

reduced to botryoids, and botryoids are often accompanied by 3- to 1-flowered reductive forms, as, e.g. in *Orixa* (Fig. 57). The reduction from panicles (*Aegle*, *Limonia*) to botryoids can also be observed in Aurantioideae. The elongate, unbranched inflorescences of *Pilocarpus* start as spikes and, with the development of pedicels, become racemes.

In boronioids, panicles are widespread and take corymbose shapes, often with diminishing ramification of the lateral inflorescence branches and condensation of the main axis (i.e. sciadioids, e.g. *Phebalium*), or are reduced to botryoids or solitary flowers. In some *Eriostemon*, reduction of the pedicels leads to cephaloid inflorescences, and the bird-pollinated *Chorilaena* and *Diplolaena* stand out by their bell-shaped aggregated panicles (Fig. 58). In *Diplolaena*, these are surrounded by showy outer bracts and thus represent an "inflorescence blossom" (Classen-Bockhoff et al. 1991.) In diosmoids, the flowers are often aggregated in sciadioids at the tip of the flowering branches (e.g. *Agathosma*).

No detailed analyses of the inflorescences of neotropical Rutaceae, *Pilocarpus* excepted, are available, but there are indications of the widespread occurrence of scorpioid cymes as elements of the inflorescences of the Angostura Alliance (Figs. 72B, 73B). Within the predominantly northern hemisphere Ruta Alliance, there is clear evidence for such inflorescences. *Boenninghausenia* has thyrsoids with cincinnate cymes as the lateral branches of the inflorescence (Fig. 59; Troll 1964: 90), and *Thamnosma* also has scorpioid cymes (Troll 1969: 567 seq). The inflorescences of these two genera have often been misinterpreted as panicles and racemes respectively. There is much variation of inflorescence structure in *Ruta* where, i.a., thyrses with paired cincinni, pleiochasia and botryoids are found.

FLORAL MORPHOLOGY. Flowers in Rutaceae are usually more or less actinomorphic; clear zygomorphy is found in *Dictamnus* and in the Angostura Alliance. Basically, the flowers are hermaphrodite, pentacyclic, isomerous and obdiplostemonous (with antepetalous carpels; see, for example, the floral diagrams of *Bouchardatia* and *Ruta* in Engler 1931).

In the Angostura Alliance, the androecium is strongly affected by the increasing zygomorphy

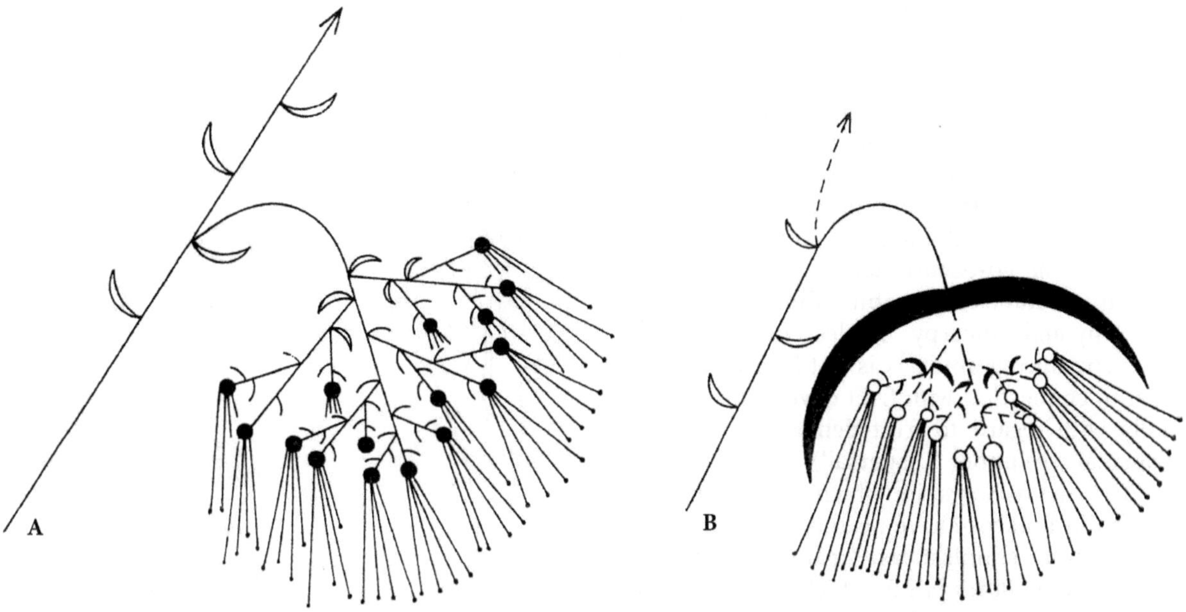

Fig. 58. Rutaceae. Inflorescences of *Diplolaena* (**A**) and *Chorilaena* (**B**). (Classen-Bockhoff et al. 1991)

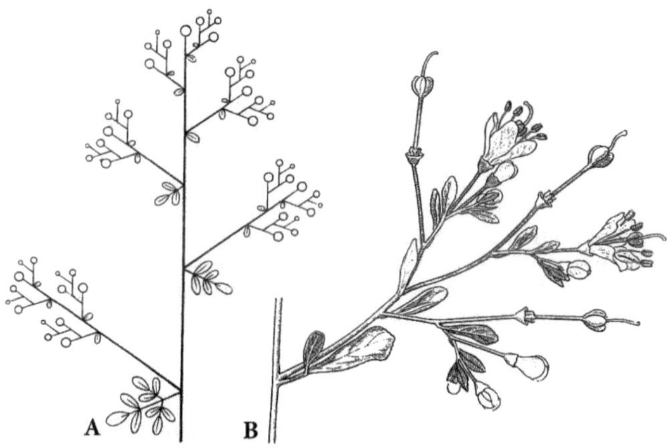

Fig. 59. Rutaceae. *Boenninghausenia albiflora*. **A** Scheme of the thyrsoid. **B** One branch of the thyrsoid. (Troll 1964)

of the corolla. In many genera, only two of the originally five stamens are fertile, whereas three are staminodial (Fig. 73D). Sometimes, more than three staminode-like objects are present (Fig. 71D); these may be small outgrowths of the two petals or of the genuine staminodes ("pseudo-staminodes"; J.R. Pirani, pers. comm. to J.A.K., Dec. 2008). The two fertile stamens are placed in the upper part of the zygomorphic corolla and flank the 'innermost' petal, i.e. the petal of which the lateral margins are interior to those of the adjacent petal. The anthers of these two stamens often have basal appendages, which in some genera are fused to each other. The functional significance of this complication is unknown.

The gynoecium in Rutaceae varies from complete apocarpy to syncarpy. The carpels show various degrees of congenital or postgenital connation (Gut 1966; Ramp 1988). In the fully syncarpous condition, congenital connation extends from the base of the ovary to the stigma. Yet, very many members of the family are completely or largely apocarpous (or rather, 'partly apocarpous'), insofar as the ovarian parts of the individual carpels, termed ovarioles[4] in our generic descriptions, are distinct or only proximally connate, whereas their stylodia typically are joined by postgenital connation or conglutination. In this case, the gynoecium is held together by the joined stylodia (the common style), which at maturity fall off so that the fully apocarpous condition is established.

Postgenital connation typically occurs among the stylodia or, more rarely, only the stigmas, and results in the formation of a compitum, a common inner space in which the pollen tube

[4]Introduction of this term (proposed by Baumann-Bodenheim 1954) was necessary because using 'ovary' to refer to the ovarian portion of both mono- or multicarpellate gynoecia as well as to the ovarian portions of the individual carpels of the latter could cause confusion.

transmitting tissues (PTTs) of all carpels are interconnected. Thus, a compitum provides the advantage of centralised distribution of and selection among pollen tubes typical of a syncarpous gynoecium, whereas in the apocarpous gynoecium that lacks a compitum, pollen tube selection occurs in each carpel/stylodium separately (Carr and Carr 1961; Endress 1982; Armbruster et al. 2002). In the co-occurrence of apocarpy and a compitum, Rutaceae combine the advantages of apocarpy and syncarpy. A selective advantage of apocarpy during anthesis is not known but may exist in connection with fruit development, especially with regard to fruit dehiscence and seed dispersal (see under 'Dispersal').

The many ways in which a compitum is formed in Rutaceae have been revealed by Ramp (1988). In the simplest case (many *Zanthoxylum*, *Euodia*, some *Boronia*), in which the stigmas are coherent but the stylodia are distinct, the pollen tubes germinate on the stigmas and grow separately through the stylodia. In *Ruta* and *Boenninghausenia*, in which the compitum is restricted to the stigmas and the stylodia are joined for their whole length, the PTT extends through the stylodia in distinct strands down to the secretory surface of the ovary locules. In others (e.g. *Choisya*, *Phellodendron*, *Pilocarpus*), a common central PTT is present in the distal part of the style from where it continues to the ovary in separate strands. A completely connate inner cylinder of PTT through the length of the common style is frequent in the family, as in the Diosma Alliance, the Boronia Alliance, the Angostura Alliance and in *Flindersia*. In the PTT of Aurantioideae and *Harrisonia*, the ventral sutures of the stylodia are open, rather than being postgenitally closed as in the majority of the family.

Stigmas of Rutaceae are either wet or dry and usually papillose (Heslop-Harrison and Shivanna 1977; Ramp 1988). Dry, papillose stigmas have been reported in one *Skimmia*, in some members of the Boronia Alliance and of the Diosma Alliance (i.e. *Coleonema* and *Agathosma*), in *Ruta* and in *Boenninghausenia*. Wet, papillose stigmas have been reported from *Dictamnus*, members of the Zanthoxylum Alliance and of the Diosma Alliance (i.e. *Calodendrum*), *Flindersia*, some *Choisya* and the Aurantioideae. Non-papillose stigmas have been reported from *Citrus*, *Erythrochiton*, *Ptelea*, some *Choisya*, *Correa* and one *Skimmia*. In some species of *Citrus* and in *Fortunella*, the pollen grains germinate on the inner surface of a tubular stigma (Schill et al. 1985).

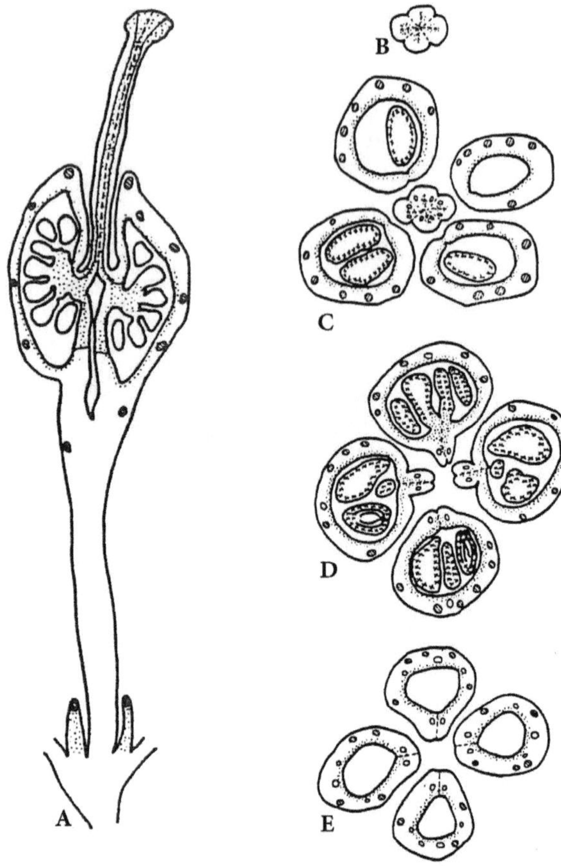

Fig. 60. Rutaceae. *Boenninghausenia albiflora*. **A** Longitudinal section of the strongly anacrostylous gynoecium with pronounced ovary hoods. **B** Transverse section of style. **C–E** Transverse sections through decreasing levels of the ovary. (Ramp 1988; drawn by author)

In most apocarpous gynoecia, the ovarioles are raised above the base of the common style, so that the latter seems to be immersed in the ovary (gynobasic styles; anacrostylous gynoecia). The raised backs of the ovarioles have been termed "ovary hoods" ("Fruchtknotenkapuzen"; Gut 1966). A real elevation of the ovary wall leading to an enlargement of the ovary cavity (*Ruta*, *Boenninghausenia*; Fig. 60) must be distinguished from a thickening of the apical wall of the ovarioles, which is seen in some boronioids and in some diosmoids (*Eriostemon*; Fig. 61) and produces the 'horns' that are characteristic of the gynoecia and fruits of the latter group.

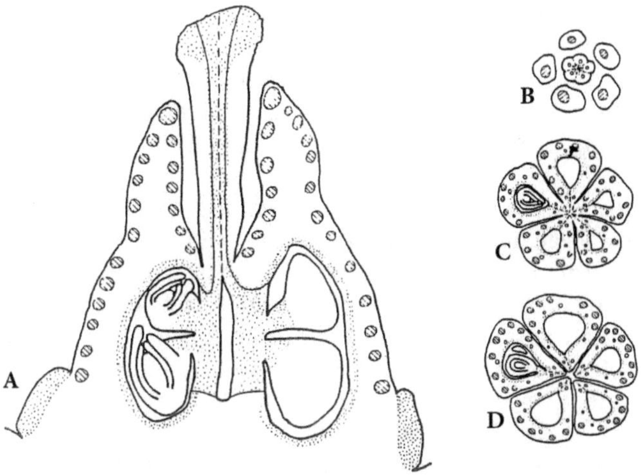

Fig. 61. Rutaceae. *Eriostemon myoporoides*. **A** Longitudinal section of the anacrostylous gynoecium; pollen tube transmitting tissue dotted; oil glands hatched. **B** Transverse section through style and ovary hoods, and **C, D** through the ovary. (Ramp 1988; drawn by author)

Disk and gynophore represent the principal nectar-secreting tissue of the flower and are intercalated between the androecium and gynoecium. The disk can appear in various shapes, such as cupular or cushion-shaped. The gynophore is an elongation of the base of the gynoecium, which sometimes may be thickened and covered by nectar-secreting tissue, as in *Dictamnus*, *Euodia* and *Phellodendron* (Fig. 64). In *Zanthoxylum*, the nectar-secreting floral base may be considered transitional between gynophore and disk. *Cneorum* has a peculiar nectariferous androgynophore (Caris et al. 2006).

Embryology. The anther wall comprises a persistent epidermis, a fibrous endothecium, two or three ephemeral middle layers, and a glandular tapetum with multinucleate cells. Pollen grains are shed in the 2-celled stage. The ovules are bitegmic (*Glycosmis* is unitegmic, Boesewinkel and Bouman 1978), anatropous and crassinucellate; the zigzag micropyle is formed by both integuments and usually directed upwards. There are usually 2 ovules per carpel, but sometimes 1 in some apocarpous and syncarpous taxa. Elevated numbers are found in several groups such as *Bosistoa*, *Bouchardatia*, the Ruta Alliance, *Dictamnus*, *Chloroxylon*, *Platydesma*, *Flindersia* and the higher Aurantioideae (Citreae). A several-layered nucellar cap is frequently present and typical of the family. Embryo sac development is of the Polygonum type, rarely the Allium type. Endosperm development is of the Nuclear type. The chalazal end of the embryo sac develops tubular projections or forms an aggressive haustorium. Nucellar polyembryony is widespread in the family but, in *Citrus*, fertilisation has been found essential for the development of these embryos. More details can be found in Mauritzon (1935), Davis (1966), Johri et al. (1992), and the literature listed therein. Polyembryony has been observed in species of *Conchocarpus* and *Almeidea* (J. Kallunki, pers. obs.).

Pollen Morphology. Pollen grains are usually 3-colporate and prolate with a reticulate tectum and lalongate endoapertures. This is the widespread and, most likely, basal condition, which usually is only moderately modified as described in the following. In Rutoideae, exine sculpture relatively often tends to be striate, less often microperforate or very coarsely reticulate, rarely echinate or baculate (*Erythrochiton*, some *Angostura*, *Nycticalanthus*, *Spiranthera*), and pollen in the Angostura Alliance is sometimes spherical and then often has 4–6, very short colpi (some *Angostura*, *Galipea*, *Sigmatanthus*, *Ticorea*; Morton and Kallunki 1993). Otherwise, the tricolpate condition is strongly fixed, and deviations such as tetracolpate grains in *Adenandra* (Victor and van Wyk 1999), *Melicope* and *Skimmia laureola* (Hartley 2001a), and pantocolporate grains such as in *Almeidea*, *Conchocarpus cuneifolius* and *Andreadoxa* are rare. The pollen of *Leptothyrsa*, with its deeply ridged striate-reticulate exine, is unlike that of any other Rutacea. In Aurantioideae, comparable progressions are found, and genera often considered to be basal (such as *Micromelum*, *Murraya* and *Merrillia*) have striate, 3-colpate grains, whereas *Bergera* is distinguished by 3-colporate, foveolate/microperforate grains, *Citrus* and related genera have reticulate or microperforate, 4/5-colporate grains, and the Bael Fruit Trees (*Balsamocitrus*, *Aeglopsis*, *Afraegle*) have microperforate, 4- or 4/5-colporate grains (Grant et al. 2000; Mou and Zhang 2009a).

Karyology. A compilation of chromosome numbers of Rutaceae was given by Fedorov (1969), to which important additions and corrections were presented by Guerra (1984) and Stace et al. (1993). There has been some uncertainty about the original base number in the family. Starting from the observations of the predominance of $n = 18$ and 36 in the probably less specialised Xanthoxyleae and Flindersieae and of $n = 9$ in Aurantioideae, Smith-White (1954) favoured $x = 9$ as the most likely base chromosome number. In contrast, Stace et al. (1993) emphasised that $n = 9$ is dominant only in Aurantioideae and otherwise is restricted to *Boronia* (where it may be due to a dysploid decrease from $n = 18$), *Haplophyllum* and *Thamnosma*, and that $n = 10$ is found only in *Boenninghausenia*, *Ruta* and *Chloroxylon*. The notable absence of $n = 9$ from the majority of the Rutoideae, they argued, would favour $n = 18$ as the original base number of the family, which through frequent dysploid decrease, less frequent dysploid increase, and occasional polyploidy may have given rise to the chromosome numbers found in the family. They also objected to the acceptance of $x = 9$ as the base number because it would imply a high degree of polyploidy in the family, for which they saw little evidence, and because polyploidy, in contrast to diploidy, appeared to them unsuited as a genetic system for macroevolutionary processes. Because in the boronioids and diosmoids much more complex patterns are found, Stace et al. (1993) invoked substantial dysploid change from the putative base number, and in boronioids such changes appeared to them compatible with morphological evolution. Under this premise, *Calodendron*, with $n = 27$, would have to be explained as the result of a dysploid rise from $n = 18$, which appears conceivable, as $n = 19$, 21, 24 and 25 are known to occur in the Diosma Alliance. Alternatively, Goldblatt and Williams (1987) postulated *Calodendron* to be an hexaploid based on $x = 9$. *Choisya* also has $n = 27$.

Stace et al. (1993) hypothesised that the base number of Aurantioideae originated by descending dysploidy; this would be compatible with the low numbers ($n = 9$, 10) in the Ruta Alliance (see above) and in *Chloroxylon* because, in several DNA sequence analyses, these taxa are resolved in close association with, or basal to, Aurantioideae.

Detailed analyses of heterochromatin banding patterns of Aurantioideae have revealed a correlation between the amount of heterochromatin and the relative advancement of taxa as expressed in apomorphic characters, often with a preferential localisation of heterochromatin on the larger chromosomes and close to the centromers in different lineages (Guerra et al. 2000).

Pollination and Reproductive Systems. Basically, Rutaceae appear entomogamous (Engler 1931; Kuhlmann and Kühn 1947; Armstrong 1979), with smaller Hymenoptera and Diptera as predominating pollinators in the humid forest biome. Phalaenophily is the likely mode of pollination in the Amazonian *Spiranthes* and *Nyticalanthus*, and butterflies and moths take part in the pollination of *Galipea jasminiflora* (Piedade and Ranga 1993) and probably many of its relatives. An assortment of butterflies visiting flowers of *Conchocarpus oppositifolius* were observed at the time the type collection was made. Flowers of *Erythrochiton gymnanthus* at the type locality opened in the evening and may be moth-pollinated. In a population of *Toxosiphon lindenii* in Costa Rica, flowers opened in the morning, presented copious nectar but no odour, and were visited several times by two species of hummingbirds (all pers. obs., J.A. Kallunki). Bird pollination is known from semi-open habitats such as the Cerrado of South America (*Hortia brasiliana*, Barbosa 1999) and the heath shrubland of Australia, where Armstrong (1979) has found it to be an important pollination mode in Australian boronioid genera such as *Correa*, *Nematolepis*, *Chorilaena* and *Diplolaena*. The same author also highlights the diversity of the pollen vectors in boronioids, among which Coleoptera feature strongly in the genera *Eriostemon*, *Phebalium* and *Asterolasia*.

Erythrochiton fallax, similarly to some Annonaceae, has perennating inflorescences that sometimes trail along the ground partly buried in leaf litter, except for the flowering tip that produces flowers at some distance from the plant itself. Their pollinators are still unknown but, given the large, bright-red calyx and zygomorphic corolla, hummingbirds would be expected.

Bat pollination seems to be unknown in the family.

Rutaceae are usually protandrous, and dichogamy seems to be frequent, with a distinct male phase of pollen release followed by a female phase devoted to pollen reception. The only comprehensive study of the breeding system with regard to the nature and distribution of self-incompatibility is that of *Zieria* by Armstrong (2002). In the monoecious *Dictyoloma vandellianum*, Pirani (1989) observed in the field that all male flowers of one individual release their pollen either before or after the period of receptivity of the female flowers and that, at the population level, flowering periods are not synchronised.

An interesting case of secondary pollen presentation is known from *Adenandra* (Marloth 1920), where the stamens and staminodes have striking glands that produce a viscid fluid. Through movements of one stamen and staminode after the other over successive days when the stigma is still undeveloped and pressed against the ovary, the pollen is glued by the anther secretion to the top of the curved style. From there, the pollen is taken away by insects in search of nectar.

The higher incidence of dioecy in the Old World Rutoideae in comparison with the New seems to us remarkable but inexplicable: of 83 Old World genera, 19 are clearly dioecious or, as in some large genera such as *Zanthoxylum*, vary from the dioecious to the monoecious condition; among the New World genera, in contrast, as far as we know only one (*Pitavia*) seems to be dioecious.

Fruit and Seed. Many Rutoideae have carpels that are largely free from each other and, at anthesis, are held together only by the common style, which is shed when the fruits mature. In most dehiscent-fruited Rutoideae that forcibly expel their seeds, the epidermis of the ovary cavities undergoes periclinal divisions and yields the thin, dry and pergamentaceous dorsilateral endocarp (Hartl 1957), whereas the remaining epicarp becomes hard and woody. In the placental region, the endocarp is thinner and separates from the epicarp and from the rest of the endocarp. This portion of the endocarp, the ventral endocarp (Figs. 62E, 69H), is membranaceous to subfleshy, has on its outer surface a torn strip of axile epicarp tissue, and either is deciduous or persists on the seed (Hartley 2001a). It remains attached to the hilum of the seeds in some rainforest genera of Australasia such as *Coatesia*, *Euodia*, *Brombya*, *Medicosma* and *Bosistoa* (Hartley 1997: 191; 2001a, b) and in some members of the New World Angostura Alliance (see Kallunki 1992: 112), and it is well developed and nearly omnipresent in the Australian Boronia Alliance and the south African Diosma Alliance (Fig. 63). The remainder of the endocarp, the dorsilateral endocarp (Figs. 62D, 69G), is elastically expelled with the seed, sometimes over considerable distances, as has been described for *Dictamnus* (Geisenheyner 1915), and *Esenbeckia* and *Metrodorea* (Kuhlmann and Kühn 1947). In some genera (*Zanthoxylum*, *Geijera*, *Tetradium*, *Melicope*, *Tetractomia* and *Perryodendron*), the endocarp is not discharged and the seed remains in the dehisced fruit until being taken by a disperser (Fig. 67E; Hartley 2001b). In the Ruta Alliance, the endocarp is instrumental in the dehiscence of the fruit but does not expel the seeds (Boesewinkel 1977a, b, c). The fruits of the drupaceous Rutoideae are more or less fleshy, and protection for the seed in the gut of the disperser is provided by a woody, cartilaginous or coriaceous endocarp, often enhanced by a thick sclerotesta, a woody mesocarp, or both.

Fruits of Aurantioideae are baccate; their edible part originates from a peculiar development of the endocarp, whereas the rest of the pericarp usually is not palatable and varies from leathery to woody. The thin, wax-covered exocarp is underlain by the outer part of the mesocarp (flavedo) that contains carotenoids and numerous oil

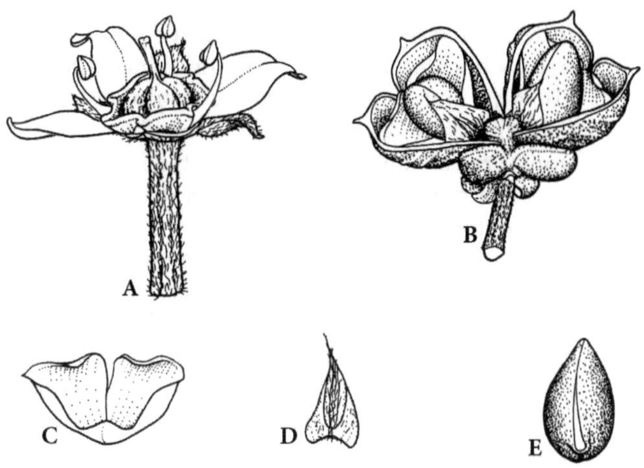

Fig. 62. Rutaceae. *Euodia hortensis*. **A** Flower. **B** Ditto, one sepal, one petal, and one stamen removed. **C** Fruit showing the position of endocarp and seeds therein. **D** Dorsilateral endocarp. **E** Ventral endocarp with strip of axile epicarp tissue on its abaxial surface. **F** Seed with raphe. Scale bar=3 mm. (Hartley 2001a)

Fig. 63. *Agathosma capensis*, seeds with well developed ventral endocarp (elaiosome), ×10. (Photo W. Stuppy, courtesy of Millenium Seed Bank, Royal Bot. Gds Kew)

glands, whereas the inner part of the mesocarp (albedo) consists of a whitish, relatively dry, spongy tissue that extends into the radial faces between the developing carpels. The endocarp is a thin layer that covers the carpels and produces either a mucilaginous pulp or distinct pulp vesicles (juice sacs) that compactly fill the locules. In the Citrus Alliance, the vesicles are large club-shaped structures that originate from the dorsal walls of the carpels; they are filled with juicy tissue derived from subepidermal layers. Between the vesicles and on the flank of the carpels, the endocarp forms glandular emergences ("Schleimzotten") that secrete mucilaginous or pulpy material (Schneider 1968). In Aurantioideae such as *Murraya* and *Triphasia* that lack pulp vesicles, a several-layered endocarp alone or the enations from it release pulpy material into the fruit locules.

Seed structure in Rutoideae is determined by the predominance of the testa as the mechanical layer, whereas the tracheidal tegmen usually remains insignificant in its protective function (Gallet 1913; Corner 1976; Boesewinkel 1977a, b, c, 1980; Boesewinkel and Boumann 1978; Wilson 1998; Hartley 2001b). The testa may have an inner layer of dense, black sclerenchyma, the sclerotesta, which varies from thin to thick and bony. When the sclerotesta is well developed, the outer layer of the seed coat may be developed as a spongy-fleshy sarcotesta. In *Zanthoxylum*, the endotesta is formed by periclinal divisions from the inner epidermis of the outer integument. Quite often, the endosperm is more or less reduced (consistently in the Diosma Alliance and in Aurantioideae), and the embryos may be variously complicated with folded cotyledons (Angostura Alliance, Fig. 72M). The seed coats of *Ruta*, *Boenninghausenia*, *Ptelea* and *Dictamnus* have an exotestal palisade, in which the cells are somewhat papillate and thickened mostly on the outer walls, but hardly lignified (Corner 1976). In *Skimmia*, where the pyrene is the dispersal unit, the tegumentary derivatives degenerate during seed development to a pellicle and the protective function is taken over by the endocarp (Boesewinkel 1977a, b, c).

A novel seed coat character was reported by Wilson (1998) for the Boronia Alliance, which consists of a circular, chalazal aperture located at the base of the seed and is observable on the inner surface of the testa. Hartley later (2003) reported this structure also for 12 genera of the Euodia Alliance. This led one of us (J.A.K.) to examine the seeds of the Angostura Alliance, where the chalazal aperture was found in seeds of all genera of this group, and also in seeds of two *Metrodorea*, in *Dictyoloma*, in five *Esenbeckia*, in two *Choisya* and in *Decatropis*. As long as the distribution of this character is not known from related families, it may be considered as peculiar to the Rutaceae.

In the seeds of many genera of the Angostura Alliance, the seed coat invaginates to a greater or lesser extent behind the hilum and forms a bump or a peg, the subhilar rumination. In *Euxylophora*, a narrow hilum extends along the adaxial edge of the seed from apex to base, at which point the testa forms a channel-like subhilar rumination that follows the abaxial edge of the seed to the chalaza near the micropyle. It seems that the seed of *Euxylophora*, unlike that of any other of the Angostura Alliance, is obcampylotropous (J.A.K., pers. obs.), though the embryo is straight.

In Aurantioideae, the seed coat is predominantly exotestal with lignified and often fibrous, longitudinally elongate elements; it is overlaid by a mucilaginous outer wall. In the exo- and endotesta, often one or several layers of crystal cells are present, which are not seen in Rutoideae. The endosperm is usually lacking or nearly completely absorbed. Sometimes, the inner integument is obliterated (*Glycosmis*) or the embryo is folded (*Micromelum*). Polyembryony is known in

Citrus and related genera and is of nucellar origin. Although as many as 13 embryos have been observed in one seed, hardly more than three reach germination.

DISPERSAL. Some dehiscent-fruited genera, such as *Flindersia* and *Chloroxylon*, have winged seeds that are passively released from the dehisced fruit and eventually wind-dispersed. Non-winged seeds that persist in the dehisced fruit (see 'Fruit and Seed' above) have a seed coat composed of a hard inner sclerotesta and a spongy-fleshy outer sarcotesta. The sclerotesta enables the seeds to maintain viability when passed through the gut of an animal, and the sarcotesta is the reward for the vector; it is covered by a shiny pellicle, which enhances the seed's visibility in the dehisced fruit. These seeds appear highly specialised for endozoochory and are presumably dispersed by arboreal, seed-eating birds (Hartley 2001b). Interestingly, seeds of the syncarpous drupe of *Acronychia* and its close allies, such as *Sarcomelicope* and *Picrella*, retain remnants of a spongy tissue that obviously is non-functional and seems to be carried over from a *Melicope*-like ancestor (Hartley 1991). Those of the apocarpous, samaroid drupes of *Pentaceras* have a sclerotesta and a rather spongy outer tissue, which also may be a leftover, in this case from a *Zanthoxylum*-like ancestor.

The seeds of the active ballists usually either lack a sclerotesta or have a thin, brittle sclerotesta and appear unsuited for endozoochory, but often possess persistent ventral endocarps (Fig. 63) that are ant-attracting elaiosomes. These appendages are nearly omnipresent in the seeds of the Boronia Alliance and the Diosma Alliance (Berg 1975), both of which are richly diversified in the heath and sclerophyll vegetation of Australia (Wilson 1970, 1998; Armstrong 2002) and the fynbos of the Cape (Trinder-Smith et al. 2007) respectively. Worldwide, these two regions are known for harbouring the greatest concentration of ant-dispersed plant species, with 1,500 in Australia and 1,200 in South Africa, against 300 in the northern hemisphere (Berg 1975; Westoby et al. 1982; Beattie and Hughes 2002). Southern hemisphere elaiosomes are firm and long-lived; they may retain their attractiveness for several years (Berg 1975; Beattie and Hughes 2002) and at least those of Rutaceae contain no lipids (T. Hartley, pers. comm. 7 Apr. 2009). Those of the northern hemisphere myrmecophytes, which usually are understorey herbs, are soft, regularly contain lipids, desiccate and become unattractive within a few days. Australian taxonomists avoid the term elaiosome in favour of ventral endocarp (or placental endocarp), in which we follow them in the taxonomic descriptions. The predominance of myrmecochory among boronioids and diosmoids appears to be related to their subxeric and nutrient-poor habitats. Dispersal by vertebrates would require fleshy fruits, and their production and attraction of vertebrate dispersers would sap valuable nutrient resources. In contrast, dispersal by ants would imply a much smaller investment in food bodies for attraction and would render possible dispersal to favourable micro-habitats through seed burial and consequent escape from predation and fire (Milewski and Bond 1982).

PHYTOCHEMISTRY. Among the flowering plants, Rutaceae exhibit one of the richest and most diverse arrays of secondary plant constituents (Price 1963). The members of the family are usually characterised by a specific combination of compounds belonging to different classes, among which the quinolones and acridones derived from anthranilic acid stand out for being highly diversified and largely restricted to the Rutaceae. There is hardly a genus within Rutoideae and Aurantioideae that does not produce compounds of this class and, to the present day, reports of novel examples of these compounds are filling the pages of journals such as *Phytochemistry* and *Journal of Natural Products*. Other groups of substances contributing to the characteristic phytochemical profile of Rutaceae include coumarins, acetophenones and limonoids. Comprehensive surveys of the phytochemistry of the family have been published by Hegnauer (1973, 1990) and Waterman and Grundon (1983), and comparative and systematic implications of the phytochemistry have been discussed particularly by Hegnauer and Waterman in their contributions to symposium volumes edited by Waterman and Grundon (1983) and by Waterman (1993).

Prenyl substitution, in which the 5-carbon hemiterpene is attached as a C- or O-bounded substituent to the alkaloid, coumarin, flavonoid, chromone or acetophenone moiety, greatly enhances the structural diversity and biological activity of the secondary metabolites of the family. Cyclisation of the hemiterpene substituent can lead to furan and pyrane systems, and furo- and pyranocoumarins and furo- and pyranoquinoline alkaloids are highly characteristic metabolites of the family.

As mentioned above, *alkaloids based on anthranilic acid* as a nitrogen source are the most abundant type of alkaloid in Rutaceae. Condensation of anthranilic acid with an acetate unit leads to simple quinolones, which abound in *Haplophyllum*, but usually simple quinolones co-occur with or are transformed into more complex alkaloids. Prenylation readily occurs at the C-3 position of the quinoline and can result in dihydrofuroquinolines, furoquinolines, linear pyranoquinolines and angular pyranoquinolines. The most widespread compounds are oxygenated at C-2 and C-4 and are variously substituted in the aromatic nucleus (Waterman 1993). Acridones are anthranilate alkaloids in which anthranilic acid is linked to a polyketide made up of three acetate units; they are known from some 17 genera distributed over the whole family and undergo all of the types of substitution that are common in Rutaceae. Systematically interesting is the loss of C-1 and C-3 oxygenation in the alkaloids of *Ruta*, *Boenninghausenia* and *Thamnosma*.

Tyrosine- or *phenylalanine-derived alkaloids* are restricted to five genera of Rutaceae, where they are known to occur in different structural types but have not served as bases for the production of novel compounds beyond those found in Annonaceae, Berberidaceae and Papaveraceae, in which they have their main representation. In Rutaceae, they are represented most commonly by benzophenanthridines, protopines, aporphines and berberines. Their systematic significance has stimulated interesting phylogenetic speculations that will be discussed in the following section.

Compounds containing a *tryptamine* nucleus include the simple carboline alkaloids and the canthinones, which both abound in Simaroubaceae and only rarely occur in Rutaceae. Indoloquinazoline alkaloids represent an interesting combination of anthranilic acid and tryptamine; they are known from *Zanthoxylum*, *Bouchardatia*, *Vepris*, *Euxylophora* and *Hortia*.

Methylcarbazoles are a striking group of alkaloids the biogenetic origin of which remains unresolved but most plausibly would seem to result from cyclisation from a 2-prenylindole (Waterman 1993). They co-occur in Aurantioideae with alkaloids that possess a 3-prenylindole skeleton; a dimer of 3-prenylindoles, yuehchukene, links *Micromelum* with *Murraya* and *Merrillia* (see Table 1, p. 297); it has received pharmacological attention because of its antifertility (anti-implantation) activity.

Imidazole alkaloids are known to occur in Rutaceae in the two genera *Pilocarpus* and *Casimiroa*.

Among cinnamate-derived compounds, *coumarins* are widely though not universally distributed in the Rutaceae. All coumarins found in the family possess 7-oxygenation, and further elaboration frequently includes oxygenation at C-6 or prenylation at C-6 or C-8. Coumarins oxygenated at C-6 and C-7 with unusual further prenylation occur in *Ptaeroxylon* and *Cneorum*. C-6 prenylated coumarins frequently give rise to linear furo- and pyranocoumarins. C-8-prenylated coumarins seem to be concentrated in few taxa such as *Murraya* and its relatives and in *Phebalium*. Coumarins based on a pattern of 5-,7-oxygenation with 6,8-prenylation are found in many Australian Boronieae and also in *Zanthoxylum* and *Hortia* (Waterman 1993).

The cinnamate-acetate-derived *flavonoids* are represented by poly-methoxylated flavanones, flavones and flavonols, the octamethoxyflavone exoticin being an extreme example. These compounds are widespread in Aurantioideae and typically occur in *Casimiroa*. The most widespread flavonoid of Aurantioideae seems to be hesperidin, a flavanone rhamnoglucoside that has no appreciable taste; in contrast, neohesperidin, differing only in the sugar moiety, is one of the intensely bitter flavanones of the *Citrus* fruits. Prenylation of flavonoids seems to be rare in Rutaceae.

Acetophenones and chromones originate through the cyclisation of four or five acetate and malonate units respectively. Both groups undergo prenylation of the aromatic nucleus and subsequent modification of the hemiterpenoid group. The distribution of *acetophenones* is limited, with exception of the widely distributed *Zanthoxylum*, to the Australasian genera of the Euodia Alliance (Quader et al. 1991). In *Zanthoxylum*, *Geijera* and *Acronychia*, modification of the aromatic nucleus is restricted to additional oxygenation, whereas in *Euodia*, *Melicope* and *Acradenia* methylation or prenylation occurs. *Chromones* are more restricted in distribution; in Rutoideae, they are known from *Angostura*, *Skimmia*, *Flindersia* and *Macluroden-dron*, and from Cneoroideae; Ptaeroxylon type chromones have been found in *Dictyloloma*, *Spathelia*, *Harrisonia*, *Cneorum*, *Ptaeroxylon* and *Cedrelopsis* (Gray 1983).

Limonoids, apart from the alkaloids derived from anthranilic acid, are the most characteristic compounds of Rutaceae. Due to their enormous structural diversity and their wide range of biological activity, they have attracted considerable interest. Limonoids originate from the tetracyclic triterpenes tirucallane and, more rarely, euphane. Structural modification is initiated through oxidation and cyclisation of the C-17 side chain, leading to protolimonoids, and proceeds by ring fission and recyclisation of the ring-A and ring-D. In Rutaceae, limonoids so far have been recorded from some 30 genera; Dreyer (1983) listed them for 23 genera. In the genera where they are present, they often appear as more than one structural type (data from Dreyer), such as in *Citrus* (17 types), *Dictamnus* (5), *Casimiroa* (5) and *Calodendrum* (4). In Cneoroideae, *Dictyoloma*, *Spathelia* and *Harrisonia* are linked by similar limonoids, and the same is true for *Cneorum* and *Cedrelopsis*, which display the most strongly modified limonoids within the family. In Meliaceae, an enhanced complexity of limonoids is brought about by the extension of ring fission to rings B and C, but limonoids in which B-ring fission has taken place have also been observed in the rutaceous *Hortia* and *Flindersia*.

Volatile oils are universally present in the family and usually occur in large amounts in taxa growing under xeric conditions. In their composition, monoterpenes, sesquiterpenes and phenylpropanes take part. In Aurantioideae, the genus *Murraya* is dominated by sesquiterpenes, whereas *Bergera*, which has been segregated from *Murraya*, contains monoterpenes (Li et al. 1988).

SIGNIFICANCE OF THE 'PROTO-RUTACEAE'. The chemical versatility of Rutaceae has attracted great interest from phytochemists (see Price 1963), and the demonstration of the chemical similarities between Rutaceae and Ranunculales, particulary Papaveraceae, but also Umbelliferae and Compositae, led Hegnauer (1963, 1973, 1983) to suggest phylogenetic relationships among these groups. The possession of 1-benzyltetrahydroisoquinoline (1-btiq) alkaloids by both Papaveraceae and Rutaceae was considered a particularly strong argument, and Waterman (1983) linked the five rutaceous genera containing these alkaloids as 'Proto-Rutaceae'. He further hypothesised that, in the chemical evolution of the family, the biogenetically "primitive" 1-btiq alkaloids of the 'Proto-Rutaceae' were superseded by the "advanced" rutaceous anthranilate-derived alkaloids, when also the synthesis of coumarins and limonoids was acquired (Waterman 1983). The alternative, as Waterman (1983) pointed out, a biogenetic origin of 1-btiq alkaloids independent from that in Ranunculales, appeared most unlikely to him. In the light of present-day available DNA-sequence data, however, a close phylogenetic link between Ranunculales and Rutaceae has to be excluded (see next section), and, although the findings of molecular systematics are expressed as sister group relationships and not as ancestor-descendant relationships, we are unable to imagine a pedigree of Rutaceae in which the capacity to produce 1-btiq alkaloids could have been carried over from its alleged or factual Papaveraceae-like ancestor and preserved until the present. Thus, the acceptance of chemical convergence as an explanation for the occurrence of the 1-btiq alkaloids in Rutaceae (and in Rhamnaceae and Combretaceae, where these compounds have also been found, albeit with limited diversity) would appear more likely. However, Waterman's hypothesis of an evolutionary replacement of 1-btiq alkaloids by anthranilate-derived alkaloids seems fully acceptable because, for biogenetic reasons, it would be very unlikely that early members of the Rutaceae, after acquiring the capacity to synthesise anthranilate-derived alkaloids, should have reverted to channelling phenylalanin/tyrosin into the synthesis of 1-btiq alkaloids. It is remarkable that all genera that produce 1-btiq alkaloids produce *at least one of the typical rutalean metabolites* and thus are "pure-bred" Rutaceae: *Zanthoxylum* and *Toddalia* synthesise anthranilate-derived alkaloids and coumarins; *Zanthoxylum* contains acetophenones; *Phellodendron* and *Fagaropis* produce limonoids; *Tetradium* exhibits the whole metabolic spectrum found in core Rutaceae (Waterman and Khalid 1981; Ng et al. 1987). The limited diversification of rutaceous compounds in these genera may be related to their small size: *Toddalia* is monotypic, and *Phellodendron* and *Fagaropsis* comprise only two and four species respectively; only in *Zanthoxylum* (>200 spp.!) is the chemical diversity considerable (Waterman 1986). That the size of a group generally tends to determine the extent of its chemical diversification is known from many large plant families; the Compositae, for instance, contain an enormous array of different groups of secondary metabolites (Zdero and Bohlmann 1990). Thus, 1-btiq alkaloids in 'Proto-Rutaceae' and anthranilate-derived alkaloids in other Rutaceae may represent alternatives in the exploitation of the alkaloid theme. It is unlikely, however, that the 'Proto-Rutaceae' will be close to the common ancestor of the family or represent a 'vital link' with the Papaverales.

Family and Subfamily Relationships. Within angiosperms, Rutaceae are recovered by DNA-based cladistic studies as part of the order Sapindales, which are placed in the malvids (Eurosids II) as part of the rosids within the eudicots (APG II 2003; Soltis et al. 2005). Such studies hypothesise that the closest relatives of Rutaceae are Simaroubaceae and Meliaceae to which Rutaceae are sister (e.g. Salvo et al. 2008; Wang et al. 2009) and that these families collectively are sister to Sapindaceae. Among the three subfamilies of Rutaceae recognised in this study, Rutoideae (incl. Toddalioideae and Flindersioideae) and Aurantioideae are closely linked through common morphological and chemical characters, such as the schizogenous oil glands and the anthranilate-derived alkaloids, and are resolved as a coherent group by all molecular analyses. Cneoroideae stand further apart morphologically but are held together by the peculiar Ptaeroxylon type chromones and are somewhat (through *Spathelia* and *Dictyoloma*) linked to Rutoideae, but also exhibit morphological and chemical similarities with Simaroubaceae. Rutoideae plus Aurantioideae are here called 'core Rutaceae', whereas 'Rutaceae s.l.' is meant to include Cneoroideae.

Secondary or Primary Apocarpy? There are good reasons for considering Rutaceae and various other eudicot lineages that combine apocarpy with a compitum as secondarily apocarpous (Endress et al. 1983). In contrast, Ramp (1988) has pointed out that, in some species of *Zanthoxylum*, the lack of fusion of the stigmas would favour the acceptance of primary apocarpy and that, in apocarpous lineages, the joining or connation of the stylodia would have made pointless the selective pressure towards syncarpy. Results of gene sequence analyses, however, militate against a close relationship between Sapindales and the apocarpous Ranunculales and basal angiosperms, from which apocarpy could therefore hardly have been directly inherited. Because of this, an evolutionarily secondary acquisition of apocarpy in Rutaceae is the most likely assumption. This view could imply that in Rutaceae the syncarpous condition would be more primitive than apocarpy, but this may apply mainly to the rutaceous stem group, whereas probable further developments in the gynoecium morphology of the crown group may make it impossible to unravel in detail the evolutionary history of the gynoecium in Rutaceae.

Generic Relationships and Subdivision of the Family. The most influential comprehensive taxonomic treatment of Rutaceae is the one proposed by Engler (1931), in which seven subfamilies and numerous tribes and subtribes were recognised. A modernised treatment of Aurantioideae was given by Swingle (1943, reprinted by Swingle and Reece 1967). In recent years, the usefulness of these taxonomic schemes and particularly the reliability of fruit structure, upon which Engler heavily relied, have been questioned by various authors (Hartley 1981, 2001b; Waterman 1983; Chase et al. 1999; Scott et al. 2000; Groppo et al. 2008; Bayer et al. 2009) and found to be untenable, as is exemplified by the grouping of several genera that originally were placed in three different subtribes into the sole genus *Vepris* (Mziray 1992). Nevertheless, the generic descriptions, keys and references to literature provided by Engler (1931) and Swingle (1943) are seminal works that will remain the rule of comparison for all future work on the family.

Here we are trying to trace relationships among the genera of subfam. Rutoideae, Aurantioideae and Cneoroideae on the basis of the evidence available to us. The main basis comprises (1) the classical and the many modern taxonomic revisions and comparative studies (Hartley 2001a as one for all), (2) the bulk of chemical evidence that has been made available by Hegnauer (1973, 1990) and Waterman (1983, 1993) and is also contained in numerous original articles usually not cited here, and (3) the gene sequence analyses of Rutaceae available at the time of writing (October 2009), mainly those by Chase et al. (1999), Scott et al. (2000), Samuel et al. (2001), Morton et al. (2003), Poon et al. (2007), Groppo et al. (2008), Salvo et al. (2008) and Bayer et al. (2009). The available information and our knowledge of the family are too inadequate for proposing a new, phylogenetically based taxonomic structure for the family; instead, we are dividing Rutaceae into a minimal number of informal groupings, the Alliances. These are largely congruent with existing tribes or subfamilies but do not represent equivalent products of the evolutionary process (sister groups), as their cladistic relationships are still not settled. We have refrained from creating monotypic Alliances for isolated genera that are assembled at the end of the taxonomic treatment of Rutoideae because, seeing the lack of even a remote relative for each of them, doing so would be uninformative. The high number of monotypic genera (43 out of a total of 154) also seems to point to the imperfect understanding of the systematics of the family.

Subfam. Rutoideae. **Bosistoa Alliance.** The five Australasian rainforest genera *Bosistoa*, *Bouchardatia*, *Acradenia*, *Dinosperma* and *Lunasia*, informally recognised by Hartley (2001b) as the *Bosistoa*-tribe, in contrast to most other Rutoideae consistently lack a sclerotesta in the seeds. Some of them have more than two ovules per carpel, and *Bosistoa* has bud scales that could be reminiscent of a former, more austral distribution under a more seasonal climate. Hartley (2001b) considered these genera to be relics and placed them at the beginning of his treatment of Australasian-Malesian Rutaceae. They have forcibly ejected seeds, and *Bosistoa* and *Bouchardatia* (the only genera for which chemical information is available) contain compounds typical of Rutaceae, e.g. indolo- and furoquinolines, acridone alkaloids, limonoids and, in *Bosistoa*, prenylated phoroglucinol derivatives and acetophenones. In the gene sequence analysis of Poon et al. (2007), *Dinosperma* and *Bosistoa* appear basal to *Euodia*, *Melicope* and *Acronychia* in a clade that is sister to a clade of the 'Proto-Rutaceae'.

Euodia Alliance. A number of opposite-leaved genera with dehiscent fruits and seeds with a thin and brittle (sclero-)testa are grouped around the genus *Euodia*. Among these genera, *Myrtopsis*, *Brombya* and *Medicosma* appear closely related, and *Euodia* itself may have descended from a *Boronella*-like ancestor. In these genera, the endocarp is forcibly expelled with the seeds when the fruit dehisces. Their seeds, which are never shiny, often retain the ventral endocarp tissue, which is presumed to act as an attractant for dispersal by ants. Hartley (2001a), who has studied the relationships among these genera, their seed structure and their possible mode of dispersal, has proposed that two other lineages also have descended from an *Euodia*-like ancestor. In both of them, the endocarp persists in the dehisced fruit. The first lineage comprises *Perryodendron* (probably bird-dispersed) and *Tetractomia* (with wind-dispersed winged seeds), while the second is represented by the large and widely distributed genus *Melicope*. Its fruits are attractively coloured; the seeds are not discharged when the fruit dehisces but remain attached to the locule and are provided with a thick sclerotesta, a well-developed sarcotesta and a shiny, black pellicle, supposedly attractive to dispersers. These three genera (like *Zanthoxylum*, *Tetradium*, *Geijera*, and the genera of the Ruta Alliance) differ significantly from the other dehiscent-fruited Rutoideae, which forcibly expel the endocarps together with the seeds. Madagascan *Ivodea* may be related to the follicular Australasian genera. Hawaiian *Platydesma* is appended here with much uncertainty; the fruit is capsular and contains shiny seeds with an osseous testa; Stone (1962) indicated *Medicosma* as possibly related to *Platydesma* (see also under the generic treatment of this genus).

The syncarpous-drupaceous genus *Acronychia* had never been considered a relative of *Melicope* until Hartley (1991) revealed subtle similarities between the two genera, including a spongy tissue in the testa of *Acronychia* that apparently is non-functional and may be carried over from a *Melicope*-like ancestor. The relationship between *Acronychia* and *Melicope* is supported by both the molecular data (Poon et al. 2007) and the common possession of prenylated acetophenones and C-2 and C-3 oxygenated acridones (Ng et al. 1987), whereas coumarins, present in some *Melicope*, are 'lost' in *Acronychia* and its other close relatives. These relatives include seven genera with drupaceous fruits in whose seeds the remnants of a sarcotesta are retained (Hartley 1982). Like *Acronychia* and related genera, the African *Vepris* s.l. (including the erstwhile genera now synonymised with it) lacks coumarins, is syncarpous, and contains typical rutaceous compounds such as furo- and pyranoquinolines, acridones, limonoids and the phenylethylamine candicine (Dagne et al. 1988).

Although morphological traits of the alternate-leaved genera *Neoschmidea* and *Halfordia* indicate a close relationship between them (Hartley 2003), no close relatives for them have been recognised; the former is nearly apocarpous and the latter completely syncarpous. In the molecular analysis of Poon et al. (2007), *Halfordia* appears close to *Bosistoa* and *Dinosperma*, whereas in the analysis of Groppo et al. (2008) *Halfordia* is basal to four boronioid genera. The alternate- and simple-leaved follicular *Coatesia*, which has forcibly expelled seeds, is appended here with hesitation.

The Chilean genus *Pitavia* does not show any clear affinity to the neotropical genera and is also geographically isolated from them, because it occurs in the warm-temperate forest region of southern Chile outside the neotropical realm. In a molecular analysis undertaken by M. Groppo, Riberão Prêto, SP, Brazil (unpubl. data, letter of 6 Oct. 2009) and employing *rps*-16 and *trn*L-F sequences, *Pitavia* appears sister to *Flindersia* + *Lunasia* embedded among Australasian and Pacific Rutoideae. Because its precise relationship remains uncertain, it is treated among the unplaced genera of Rutoideae (see also under 'Distribution and Habitats').

Unlike alkaloids derived from anthranilic acid that are diverse and widespread in Rutaceae, 1-benzyltetrahydroisoquinoline (1-btiq) alkaloids were known only from the genera *Zanthoxylum*, *Tetradium*, *Phellodendron* and *Toddalia* (the 'Proto-Rutaceae', see above), to which *Fagaropsis* later was added (Ng et al. 1987), and perhaps from *Orixa*, a report in need of verification. When revising *Tetradium*, Hartley (1981) discovered its unexpected close similarity to *Phellodendron* (Fig. 64), which is also apparent from the work of Ramp (1988). Because these genera differ radically in fruit structure, they had been placed by Engler (1931) in different subfamilies. However, sterile and staminate specimens of both appear much alike and can be distinguished only by the position of the axillary bud, which in *Tetradium* is exposed in the leaf axil and in *Phellodendron* sunk in the base of the petiole. Hartley (1981) further asserted that these genera are close to *Zanthoxylum*, and that *Zanthoxylum*, *Tetradium* and *Phellodendron* are related to one another in a linear sequence; he considered the drupaceous *Toddalia* equally related to *Zanthoxylum* but not forming part of this sequence. These phylogenetic hypotheses were confirmed by the gene sequence analyses of Poon et al. (2007), and *Fagaropsis*, which was not included in that analysis because of its production of 1-btiq alkaloids, most likely belongs into this lineage too. The geographically isolated genus *Orixa*, known to contain widespread furoquinolines and 5-oxygenated furocoumarins (which are also present in *Zanthoxylum*), reluctantly is attached to the aforementioned genera, pending the confirmation of the presence of 1-btiq alkaloids. (*Orixa* is in a weakly supported clade with *Skimmia* and *Casimiroa* in the analysis of Poon et al. 2007 but, in contrast to the statement by these authors, these genera do not contain acridones.)

Zanthoxylum, *Tetradium*, *Toddalia* and *Fagaropsis* are all specialised for endozoochory (Hartley 2001a, b). *Toddalia* and *Fagaropsis* have drupaceous fruits and seeds with a thick sclerotesta, and the dehiscent-fruited *Zanthoxylum* and *Tetradium* have seeds that, as in *Melicope*, are not discharged when the fruit deshisces and are provided with a well-developed sclerotesta, sarcotesta and shiny pellicle. *Geijera* resembles *Zanthoxylum* not only in fruit and seed structure but also in the presence of prenylated coumarins, acetophenones and, of course, quinolone alkaloids. (*Geijera*, with $n = 54$ and 81, is obviously a palaeopolyploid.) *Pentaceras* has apocarpous, samaroid drupes, and a seed coat with a thin sclerotesta and a rather spongy outer tissue, and

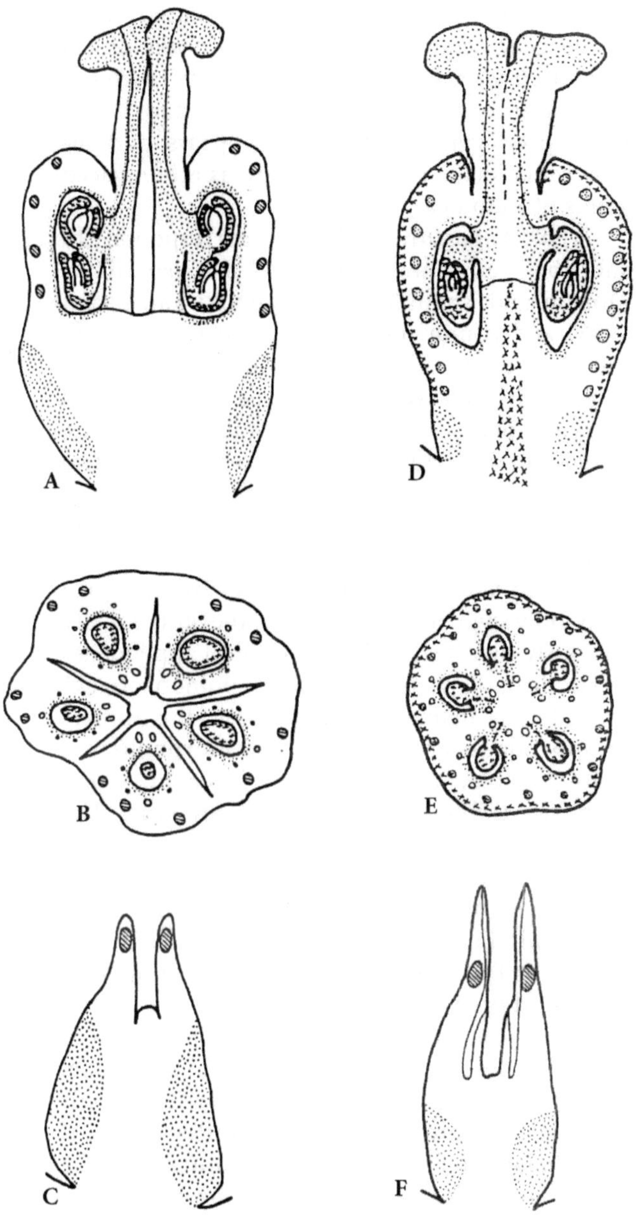

Fig. 64. Rutaceae. **A–C** *Tetradium daniellii* (as *Evodia d.*), **D–F** *Phellodendron amurense*. **A, D** Longitudinal section of the slightly anacrostylous gynoecia; pollen tube transmitting tissue (above) and glandular tissue of gynophore (below) dotted, oil glands hatched. **B, E** Transverse section of ovary. **C, D** The rudimentary distinct carpels of male flowers. (Ramp 1988; drawn by author)

shares canthinones with *Zanthoxylum*, to which it may be related. *Crossosperma*, which has the strange combination of a drupe with winged seeds and appears strongly isolated, may have its closest relative in *Phellodendron* (Hartley 1997) and here is placed close to the genera producing 1-btiq alkaloids.

The **Boronia Alliance** (Engler's tribe Boronieae) is here conceived to comprise 18 genera, with all New Caledonian genera formerly included in it now transferred to the Euodia Alliance (see Hartley 1995). The molecular analysis by Groppo et al. (2008) recovers four boronioid genera (but not *Boronia*) nested within the Euodia Alliance, with *Halfordia* sister to them; *Boronia* appears in the sister clade with *Melicope, Sarcomelicope, Medicosma* and *Acronychia*. This might imply diphyly for the boronioids, a concept that, in view of the highly diagnostic embryo structure of the group, appears less likely; it cannot be excluded that the long branch to *Boronia* (79 changes, in comparison with 25 changes to the clade with *Melicope*, etc.) might be the reason for the implied diphyly, and *Boronia* is here treated in boronioids. Some of the subtribes distinguished by Engler (1931: Eriostemoninae and Nematolepidinae) have been revealed as not natural (Mole et al. 2004), but several informal subgroups can be recognised (see taxonomic treatment). It should be noted, however, that in the future, boronioids may prove to be nested in the Euodia Alliance; in addition to the topology given by Groppo et al. (2008), morphological traits such as the distribution of circular apertures in the seed coats (see 'Fruit and Seed') and the terminal foliar sclereids (see Morphology and Anatomy) may point into this direction. The closeness of the boronioids with some members of the Euodia Alliance is also evident from the fact that *Boronella, Myrtopsis* and *Picrella* have been transferred from the former to the latter group, and that *Medicosma* and *Neoschmidea* (Euodia Alliance) according to Hartley (1985, 2003) appear to differ from *Boronia* and *Philotheca* (Boronia Alliance) in hardly anything but the morphology of the embryo.

Most genera of the strictly southern African **Diosma Alliance** are low shrubs largely confined to the fynbos vegetation. *Calodendron* is exceptional in its chromosome number (see under 'Karyology'), in synthesising various limonoids (unknown in all other members of the alliance), and in being a tall tree, but it shares important morphological traits with the shrubby members of the group, including the exalbuminous seeds, gland-tipped anthers, and horned ovaries and fruits. Pending a stronger molecular support for its sister relationship with the other diosmoid genera than that given by Trinder-Smith et al. (2007), we would consider the shrubby, chemically depauperate genera as descended from a *Calodendron*-like ancestor.

Polyaster Alliance. The genera *Polyaster*, *Plethadenia*, *Decatropis*, *Decazyx* and *Peltostigma* are largely restricted to Central America. They are alternate-leaved; their flowers are invariably diplostemonous and mostly apocarpous; fruits are follicular; and all, *Plethadenia* excepted, have two ovules per carpel. *Peltostigma* has increased stamen and carpel numbers. They are chemically little known; *Decatropis* contains oxazoles. *Choisya* and *Megastigma* have many features in common with the above five genera but *Choisya* is opposite-leaved, and *Megastigma* has drupaceous fruits with a fleshy-fibrous pericarp. Here they are doubtfully attached to the Polyaster Alliance.

Prima facie, we would see no reason why these genera could not be associated with the Australasian Euodia Alliance, as is implicit in Engler's (1931) classification; in the molecular analysis of Groppo et al. (2008), however, in which of these genera only *Choisya* was included, it is resolved in a clade that contains mainly American genera (plus two diosmoids). Therefore, and because the evolution of Old World and New World Rutaceae appears to have proceeded separately for a long period, for the time being we prefer to recognise these genera separately from the Euodia Alliance.

The **Balfourodendron Alliance**, characterised by samaroid fruits, corresponds to subtribe Pteleinae (Engler 1896), but the placement of *Ptelea* with *Balfourodendron* and *Helietta* is not supported by the gene sequence analysis of Groppo et al. (2008), who also point to differences in the floral and fruit morphology among these genera. Consequently, their samaroid fruits appear to be a result of convergence, and *Ptelea* is relegated to the unplaced genera of Rutoideae.

The capsular-fruited genera of the **Esenbeckia Alliance** (Engler's Pilocarpinae) are resolved as the strongly supported sister group of the Balfourodendron Alliance (Groppo et al. 2008), but again Engler's name-giving genus, *Pilocarpus*, differs significantly from the rest of subtribe Pilocarpinae (Kaastra 1982: 21) and likewise is not placed with it in the gene sequence analysis. *Pilocarpus* differs from *Metrodorea* and *Esenbeckia* by its racemose inflorescence, reticulate pollen exine and nearly apocarpous gynoecium, which appears particularly remarkable; it is treated here as an unplaced genus of Rutoideae.

The **Angostura Alliance** corresponds to subtribe Galipeinae Kallunki (as Cuspariinae, Morton and Kallunki 1993; Kallunki and Pirani 1998) and comprises 26 genera all restricted to the Neotropics. These genera depart from the rest of the New World Rutaceae by tendencies to zygomorphy in the flower, tubular corollas, reduction of fertile stamens from five to two with transformation of stamens into staminodes, modification of anthers, and loss of endosperm. Characters useful at the generic level are also found in pollen, seed and embryo. In most taxa with only two fertile stamens, the anthers are variously modified by basal or apical appendages or sterile basal portions above the point of attachment to the filament and, in some, the anthers or appendages are united. Pollen morphology is quite variable, and most genera (except *Conchocarpus*) exhibit a consistent pollen type. The exine can be reticulate, foveolate to perforate, or echinate; apertures are usually three and equatorial, but are more than three in many taxa and distributed over the surface of the grain in a few. Fruits are syncarpous or apocarpous, and each carpel dehisces along the ventral suture, releasing a bony endocarp that twists to expel the seed, sometimes with a thinner, axial part (the ventral endocarp) still attached. Seeds of most are campylotropous and differ in shape, superficial relief and sheen, shape and size of the hilum, and relative positions of the micropyle, hilum and chalazal area. The testa of some taxa exhibit a dark crustaceous layer of radially elongate cells, from which the tubercles on the seeds of some genera originate. In most seeds with such a layer, the testa invaginates behind the hilum to a greater or lesser extent, forming a bump, a peg, or (in *Euxylophora*) a channel (here called the hilar rumination). In taxa lacking a crustaceous layer, the testa is papery (and in dissection tends to tear, rather than break). In a few genera, the embryo is straight and the cotyledons unfolded, but the seed itself is campylotropous. In most, however, the embryo is curved and the cotyledons are conduplicate and, in most, plicate as well.

Phytochemical studies of the genera in the Angostura Alliance have demonstrated a universal presence of furoquinolines and the more or less isolated presence of acridones (*Almeidea*, *Conchocarpus*, *Ertela*, *Ravenia*), indoloquinazolines (*Conchocarpus*, *Euxylophora*), coumarins (*Angostura*, *Conchocarpus*, *Galipea*, *Rauia*, *Raulinoa*), polymethoxyflavones (*Neoraputia*) and C-glycosylflavones (*Angostura*, *Almeidea*, *Ertela*). This information confirms that the group is clearly rutaceous but this is hardly helpful in elucidating generic relationships.

Of the six genera of the Angostura Alliance that were included in the molecular study of Groppo et al. (2008), five formed a strongly supported group, from which *Adiscanthus* was excluded. The latter was paired with *Hortia*, the only New World member of Engler's Toddalioideae that, despite its drupaceous fruit, has been suggested to belong to the Angostura Alliance based on chemical constituents. The following discussion is based on our observations of morphological characters (including pollen; see Morton and Kallunki 1993), with occasional references to unpublished preliminary phylogenetic analyses of morphological data (for all genera) and of *trn*L-F and *rps*16 data (for 13 genera so far; M. Groppo, ongoing), in both of which *Hortia* was included.

Hortia, *Adiscanthus*, *Leptothyrsa* and *Euxylophora* have free petals and stamens, all of which are fertile, unappendaged anthers, and straight embryos with unfolded cotyledons. *Hortia* and *Adiscanthus* share valvate petals that are bearded (except in *H. nudipetala*) on the adaxial surface, and *Hortia* and *Euxylophora* share syncarpous ovaries, prolate pollen, pliable cotyledons, intrastaminal disks (though in the latter it is reduced to a thin tissue connecting the staminal bases) and idoloquinazoline alkaloids. In both *Adiscanthus* and *Leptothyrsa*, intrastaminal disks are absent and seeds are of similar shape with a dull thin-crustaceous testa, no hilar rumination, an expanded hilar scar, persistent ventral endocarp and very thick, stiff, unfolded cotyledons. Although the hilum of most members of the Angostura Alliance is subcircular, that of these four genera is either broad or narrow but elongated, and the seeds are 'hilar'. Each of these genera, however, is very distinct. In the unpublished analyses, these four genera appeared basal to the rest of the Angostura Alliance.

Like these four, *Nycticalanthus* and *Spiranthera* have free petals and stamens (all of which are fertile), and unappendaged anthers. Like *Adiscanthus* and *Leptothyrsa*, they have gynophores, thin disks, hilar seeds (at least in *Spiranthera*) and thick, stiff, unfolded cotyledons but, unlike them, they have intectate pollen, embryos of which the radicle is curved towards the axis (rather than erect), and unequal cotyledons of which the surfaces (rather than edges) face the axis. Although at first sight *Nycticalanthus* seems just to be an extreme of *Spiranthera*—larger leaves, flowers, and inflorescences and fruits with longer beaks—its flowers are zygomorphic, petals and stamens coherent to (and fall with) the tubular calyx, and petals quincuncial, rather than imbricate in bud like those of most genera in the Alliance. Still, we believe they are close relatives. *Spiranthera* and *Nycticalanthus* were paired in the morphological analysis.

Almeidea, too, has five free petals and five free stamens with fertile, unappendaged anthers, but it differs from the six genera discussed above by a curved embryo with conduplicate-plicate cotyledons (common elsewhere in the Angostura Alliance), by seeds with a subcircular hilum, a crustaceous testa and a hilar rumination, and by pantoporate pollen, found elsewhere so far only in *C. cuneifolius* and *Andreadoxa*. In the morphological analysis, these five genera were basal to the rest, suggesting that alternate leaves, free petals and stamens, completely fertile androecium and unappendaged anthers are basal in the subtribe.

All (but a few species in three of the) other genera are diandrous (or triandrous in the case of some *Conchocarpus*). *Conchocarpus* (Kallunki and Pirani 1998) exhibits greater variation in some characters than is the case for other genera. Unlike most diandrous taxa, its anthers are unappendaged. Within it, one group related to *C. insignis* stands out by its combination of essentially free sepals that overlap at anthesis, anthers broadly connected to the filaments, and (where known) thick, unfolded cotyledons, which are uncommon in this genus. Species related to *C. insignis* that are included in Groppo's molecular analysis form a group separate from the other species of the genus. Atypical species include *C. oppositifolius* with opposite leaves; *C. macrophyllus* with a pink corolla, filaments of the two fertile stamens connate, and a testa that is crustaceous but lacks a hilar rumination; and *C. cuneifolius* with coherent anthers, pantoporate semi-tectate densely baculate pollen, and connate carpels. The genus may be polyphyletic.

Erythrochiton was previously circumscribed to include *Toxosiphon* and *Desmotes* primarily because of their large, often white or red, valvate calyces (Kallunki 1992). The three genera, however, were widely separated in the morphological analysis, suggesting that the valvate calyx evolved more than once. *Erythrochiton* was paired with *Angostura* and, although they differ in their calyces, they both have connate petals, an apocarpus gynoecium, appendaged anthers, a crustaceous testa, and intectate pollen. In the morphological analysis, *Toxosiphon*, with *Neoraputia*, *Galipea* and *Rauia*, formed a group characterised by coherent petals, a syncarpous gynoecium and a crustacous testa, and *Toxosiphon* and *Neoraputia trifoliata* were sister to *N. magnifica*. In *Toxosiphon* and *Neoraputia*, calyx aestivation is valvate and the pollen is spheroidal and reticulate. The anther appendages of *Toxosiphon* and *N. trifoliata* (but not other congeners) are connate. *Neoraputia trifoliata* differs from other species in the genus, and from *Toxosiphon*, also by opposite leaves and anthers that are sterile at base, and *Neoraputia* may be polyphyletic. In the molecular analysis, in which *Toxosiphon* was not included, *Neoraputia* was associated instead with four apocarpous genera (incl. *Erythrochiton*).

Although both morphological and molecular analyses suggested a relationship between *Galipea* and *Rauia*, they have dissimilar anthers, cotyledons and pollen. The similarity in pollen of *Rauia*, *Euxylophora* and the *Conchocarpus* species with palmately compound leaves (e.g. *C. transitionalis*; Morton and Kallunki 1993) is likely a parallel development.

Ticorea appeared with the more basal members of the Angostura Alliance in the morphological analysis, perhaps because in four of the five species all stamens are fertile. Unlike the anthers of any other genus with a fully fertile androecium, however, those of *Ticorea* are modified, i.e. sterile above and appendaged below the attachment to the filament. Because such anthers occur elsewhere in the Alliance only in androecia with a reduced number of fertile stamens, the stamens of *Ticorea* may represent a reversal from two to five, and the genus may be more closely related to the diandrous, syncarpous genera discussed above.

Apocaulon, *Decagonocarpus*, *Ertela*, *Lubaria*, *Raputia*, *Ravenia* and *Raveniopsis* have long seemed a natural group to us and were grouped in the morphological analysis. *Ertela* and *Ravenia* were strongly associated in the molecular analysis. Although *Apocaulon* has alternate and the others opposite leaves, all exhibit free, overlapping sepals, connate petals, basally appendaged (exc. in *Ertela*) and often laterally coherent anthers, reticulate pollen grains, apocarpous gynoecia, and conduplicate, bilobed cotyledons. Such cotyledons are known elsewhere in the Angostura Alliance in *C. ramiflorus* and in *Rauia* (where, like those of *Raputia*, they are stiff, rather than pliable like those of the others). Within *Ravenia*, considerable variation exists. *Ravenia biramosa* and *R. rosea*, for example, share simple leaves, terminal inflorescences, connate and appendaged anthers, a crustaceous testa lacking a hilar rumination, and a disk taller than the ovary and with an erect margin. *Ravenia spectabilis* differs from these by trifoliolate leaves, connate but unappendaged anthers, and a disk shorter than the ovary; *R. infelix* is alone with extra-axillary inflorescences, free unappendaged anthers, and a disk with an incurved margin that cloaks the ovary. In fact, they did not form a group in the morphological analysis. As can be seen in the key, qualitative characters that will distinguish *Raveniopsis* from *Ravenia* are hard to find; species described in *Raveniopsis* are smaller plants with smaller flowers that occur primarily in the Guayana Highland, while those of *Ravenia* occur in Amazonia, eastern Brazil and the West Indies. They share an assortment of morphological characters and perhaps should be combined.

Contrary to expectation, *Desmotes* was included in this group in the morphological analysis. Although it does have in common with this group opposite leaves, appendaged anthers and apocarpous gynoecium, it differs by spinulose pollen, conduplicate-plicate cotyledons, coherent petals and the large valvate calyx.

Not much can be said with certainty about the relationships of four very distinct, monospecific, diandrous, apocarpous genera. *Naudinia* is known only by the type collection and the protologue. Its anthers are sterile and connate at base (but unappendaged), pollen reticulate, and petals connate and red. We have not seen fruit or seed. *Andreadoxa*, with free petals and gemmate, intectate pollen, is known by several collections from perhaps a single tree. *Sigmatanthus* and *Raputiarana* both have curved corollas, long-appendaged anthers clearly sterile at base, crustaceous testae and reticulate pollen, but the former differs from the latter by seeds that are smooth (vs. tuberculate) and inflorescences that are unbranched and raceme-like (vs. 1–3-times bifurcate). *Sigmanthus* is paired with *Erythrochiton* in the molecular analysis and with *Desmotes* in the morphological analysis, perhaps due, in both cases, to valvate calyx lobes.

Rutaneblina, which was described without indication of tribal affiliation, may be related to basal genera of the Angostura Alliance (i.e. *Leptothyrsa*, *Adiscanthus*, *Euxylophora*) or perhaps to *Hortia* because of its free petals, stamens and ovaries; follicular fruits; hilar seeds; flat, pliable cotyledons; and absence of a disk. Its pollen resembles that of *Adiscanthus* (Morton and Kallunki 1992). The seeds may be endospermous and, if so, would resemble those of *Esenbeckia* subg. *Oppositifolia*.

Hortia appears in the analysis of Groppo et al. (2008) in a strongly supported clade with *Adiscanthus*, with which it shares adaxially bearded, valvate petals, but has also morphological similarities with *Euxylophora*. *Hortia* stands alone, however, with regard to its pollen grains that, due to its very thick foot-layer and thick tectum with reduced columellae, are unlike all known so far in the family (Morton and Kallunki 1993). Both contain indoloquinazolines, and *Hortia* contains coumarins, which otherwise are absent from the Angostura Alliance. All this evidence may underline the isolated positions of *Rutaneblina* and *Hortia* in relation to the Angostura Alliance, to which they are doubtfully appended.

Engler's subtribe Toddaliinae, which comprised, i.a. *Casimiroa*, *Vepris*, *Toddalia*, *Acronychia*, *Halfordia*, *Hortia* and *Skimmia*, appears to us as a classical 'waste-basket' taxon held together by little more than a fleshy exocarp. While the work of Hartley has shown that, among Australasian Rutaceae, many genera with dehiscent fruits are related to genera with drupaceous fruits, the relationships of the Neotropical genera *Casimiroa*, *Amyris* and *Stauranthus* still need to be explored. The latter two genera are here treated as the **Amyris Alliance**, but the generic status of *Stauranthus* still needs confirmation. See *Casimiroa* in the section 'Unplaced Genera of Rutoideae'.

The circumscription of the **Ruta Alliance** corresponds to Engler's (1931) tribe Ruteae without *Dictamnus*, a group well characterised by an elevated number of ovules inserted in two series on capitate placentas, by a style with distinct pollen-tube-transmitting strands (Ramp 1988), and by a persistent endocarp. Chemically, the Ruta Alliance is typically rutaceous and contains many coumarins, including simple coumarins and those with an unusual substituent (the 1,1-dimethylallyl side-chain at C-3), and acridones devoid of an oxygen constituent at C-3 and, in some cases, also at C-1 (Waterman et al. 1988). A gene sequence analysis of the genera of Ruteae with exception of the rare *Psilopeganum* (Salvo et al. 2008) resolves the group as monophyletic and confirms the exclusion of *Dictamnus*. *Ruta* and *Haplophyllum*, which traditionally had been treated as subgenera of a broadly circumscribed *Ruta* (Engler 1931), are not closely related (Townsend 1986): *Ruta* is sister to a clade formed by *Boenninghausenia* and *Thamnosma*, and *Haplophyllum* is sister to *Cneoridium*. These results are in accord with the morphological findings by Townsend (1986).

Molecular analyses consistently resolve *Chloroxylon* in a clade with *Ruta*, which, however, must be considered carefully because the branch to *Ruta* is very long (81 changes vs. 25 to *Chloroxylon* in Groppo et al. 2008). Appearing unlikely at first glance, various pieces of evidence may underpin such an association: *Chloroxylon* shares with *Ruta* details of floral structure mentioned by Groppo et al. (2008), including an elevated number of ovules, with *Ruta* and *Boenninghausenia* the chromosome number $n = 10$, and with the whole Ruta Alliance the possession of a persistent endocarp and characteristic lignans of the naphthalene type (O'Sullivan 1983). Therefore, *Chloroxylon* here is doubtfully attached to the Ruta

Alliance, pending confirmation by a broader molecular analysis. If confirmed, then this relationship might point to an austral (*Chloroxylon* occurs in India and Madagascar) and woody origin of the shrubby/herbaceous Ruta Alliance.

Dictamnus, for which Engler (1931) created a distinct subtribe within his Ruteae, is set apart from this tribe, i.a. by differences in floral development (Ramp 1988), the separating endocarp, the chromosome number, the lack of naphthalene-type lignans, and the presence of highly oxidised limonoids. Certain similarity of the exotestal seed coat between *Dictamnus* and the Ruta Alliance would then be due to parallelism. Here, *Dictamnus* is treated as an unplaced genus of Rutoideae.

Unplaced Genera of Rutoideae. The placement of *Flindersia* with *Chloroxylon* in Engler's Flindersioideae has been rejected on morphological grounds (Hartley 2001b) and is not supported by molecular work either; in the topologies of Chase et al. (1999) and Groppo et al. (2008), but not of Scott et al. (2000), *Flindersia* appears in a clade with *Lunasia*, and in the unpublished *rps*-16 and *trn*L-F topologies of M. Groppo (unpubl. data), *Flindersia* + *Lunasia* appear as sister to *Pitavia*. Nevertheless, these three genera have little more in common than the lack of a sclerotesta, and, within Rutoideae, no close relatives of either *Flindersia* or *Pitavia* are obvious. Yet, as suggested earlier by Scott et al. (2000) and Groppo et al. (2008), there is hardly support for treating *Flindersia* in a subfamily distinct from the broadly construed Rutoideae.

Skimmia, *Dictamnus* and *Casimiroa*, three other isolated genera, are also structurally and chemically typically rutaceous. *Skimmia* formerly had been considered a close relative of *Halfordia* but differs from it and all other Australasian members of the Euodia Alliance in fruit and seed structure (Hartley 2001b), and it has an unusual chromosome number. In the molecular analyses of Chase et al. (1999) and Groppo et al. (2008), *Casimiroa* is resolved in a clade with *Skimmia* and *Dictamnus*, and in the analysis of Poon et al. (2007), *Casimiroa* and *Skimmia* are part of a larger clade. These associations are hardly acceptable in morphological terms (and may disappear in analyses comprising a larger number of genera), and the three genera are treated as unplaced. Reasons for treating *Ptelea* and *Pilocarpus* as unplaced genera are given above in the discussions of the Balfourodendron and Esenbeckia Alliances respectively.

Subfam. Aurantioideae. The foundations of the modern classification of Aurantioideae were laid by Engler in his treatment of the family (Engler 1896, 1931) and significantly elaborated by Tanaka (e.g. 1936, 1954, 1961) and Swingle (1943, reprinted in Swingle and Reece 1967). Many phylogenetic insights have been provided by the DNA-based phylogenetic work of Chase

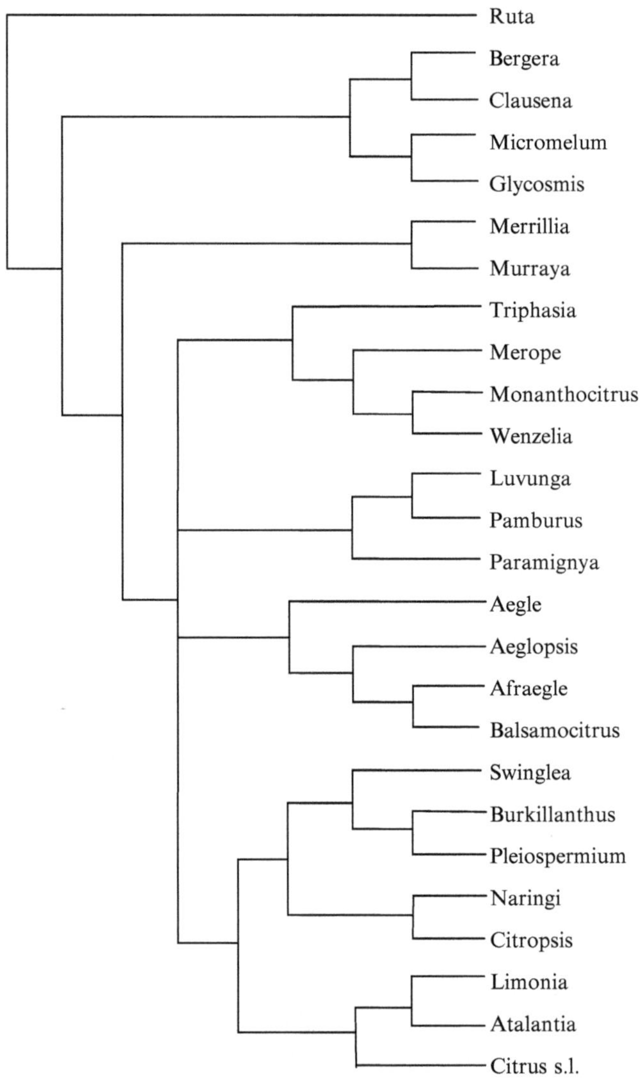

Fig. 65. Rutaceae. Phylogenetic relationships among the genera of Aurantioideae. (Simplified, with permission, from R. Bayer et al. 2009)

et al. (1999), Samuel et al. (2001) and Morton et al. (2003), among others, but above all by the analysis of Bayer et al. (2009), who present a highly resolved phylogeny of the subfamily and an expanded concept of the genus *Citrus* (Fig. 65). It is based on several carefully selected chloroplast DNA sequence regions and covers all genera of the subfamily except the little-known genus *Limnocitrus*. Further work of systematic relevance for Aurantioideae deals with palynological (Grant et al. 2000), karyological (Guerra et al. 2000) and phytochemical aspects, the mass of the latter having been made accessible by Hegnauer (vol. 6, 1973, and vol. 9, 1990), Waterman (1983), Waterman and Grundon (1983) and da Silva et al. (1987, 1988).

The monophyly of the subfamily had never been questioned and is supported both by structural and molecular evidence. Yet, the highly elaborate tribal and subtribal classifications of Engler (1931) and Swingle (1943), largely based on the structure of the fruit and particularly the pulp and pulp vesicles, cannot be upheld in the light of the new data. Instead, within Aurantioideae, two major clades can be recognised. The first, corresponding to our **Bergera Alliance**, comprises two pairs of closely related genera, *Bergera* and *Clausena* and *Micromelum* and *Glycosmis*, which were part of Swingle's tribe Clauseneae. The reinstatement of *Bergera*, which had been included in *Murraya*, was suggested by the chemical work of Chang et al. (1988) and has since been confirmed by various lines of evidence (see Samuel et al. 2001; Mou and Zhang 2009a).

The genera of the Bergera Alliance are all unarmed and bear pinnate leaves with alternate leaflets and unwinged petioles and rhachises, have two superposed ovules, and show low amounts of heterochromatin banding. The dry fruits and contortuplicate cotyledons of *Micromelum* are reminiscent of the condition in the neotropical Angostura Alliance.

All remaining genera of the subfamily are recovered in another major clade (Bayer et al. 2009), here called the **Citrus Alliance**. This is largely identical with Swingle's tribe Citreae (= Aurantieae Reichb.) but differs from it in containing the two closely related genera *Murraya* s. str. and *Merrillia*. The shift of these genera from the Bergera Alliance (Clauseneae) to the Citrus Alliance (Aurantieae) was suggested by Samuel (2001) and is supported by Bayer et al. (2009), in whose topology these two genera are recovered as a basal subclade of the Citrus Alliance. This makes sense also in morphological terms because *Murraya* s.str. and *Merrillia* appear relatively primitive in that they are unarmed and have odd-pinnate leaves with alternate leaflets. Chemically they are similar to *Micromelum* (see Table 1). Innovations emerging in the rest of the Citrus Alliance include the appearance of axillary spines, either solitary or paired and straight or curved; the reduction of the pinnate leaves to 3- or 1-foliolate leaves with articulate and winged petioles and rhachises; an increase of flower size; an occasional multiplication of stamens, carpels and/or ovules; an increase of fruit size with the elaboration of a leathery or a woody pericarp and specialised pulp vesicles; an increase of heterochromatin banding; and the diversification of coumarins, limonoids and oxygenated flavonoids, the latter reaching their peak in *Citrus* s.l. In his classification of Aurantieae, Swingle (1943) distinguished three subtribes, which were based primarily on pericarp and pulp vesicle characters. More recently, however, has it become apparent that Swingle's subtribes are not reflected in the chloroplast DNA sequence data (Fig. 65) and are not supported by the available chemical and karyological evidence. Obviously, character progressions within Citreae have not been unidirectional and may have occurred in a parallel fashion (Samuel et al. 2001), but, even so, some associations among genera of Aurantieae recognised by previous authors are expressed in the molecular trees.

Table 1. Chemical markers in the Bergera Alliance and basal genera of the Citrus Alliance

	1[a]	2	3	4	5
Micromelum	+	+	+	+	+
Glycosmis	+				
Bergera		+			
Clausena	+				
Murraya			+	+	+
Merrillia			+	+	+

[a]1 carbazole alkaloids, 2 8-prenyl coumarins, 3 6-prenyl coumarins, 4 yuehchukene, 5 polyoxygenated flavonols. (From Chang et al. 1988)

A close relationship among *Merope*, *Monanthocitrus* and *Wenzelia* had been recognised by Swingle (1943) and is supported by the analysis of Bayer et al. (2009). These genera, the Wenzelia group, have unifoliolate or simple leaves with short, non-articulated petioles and an elevated number of ovules, and all occur in Papuasia. *Triphasia* is sister to the Wenzelia group. *Oxanthera*, like the other genera in the Wenzelia group, once was thought to lack pulp vesicles and had been placed in that group by Swingle (1943), but now has been resolved as embedded in *Citrus*, a position in agreement with its elevated number of stamens and the presence of pulp vesicles that recently have been found to occur (see Bayer et al. 2009).

The obvious relationship between the woody lianoid genera *Luvunga* and *Paramygnia* has been confirmed by the study of Bayer et al. (2009), in which these genera appear in a weakly supported group together with the unispecific, very thorny and much branched *Pamburus*. These genera are native to South Asia, and in flowers and fruits superficially resemble those of *Citrus* but are diplostemonous and have fruit locules filled with mucilaginous pulp, rather than pulp vesicles. The four genera of the Bael Fruit group of Swingle (1943) form a strongly supported group that is notable for its hard-shelled fruits in which the seeds are embedded in mucilaginous fluid. The south Asian *Aegle* differs from the three African genera *Aeglopsis*, *Afraegle* and *Balsamocitrus* in the elevated number of stamens and locules and woolly seeds. It is noteworthy that Swingle's other Hard-shelled Citrus Fruit trees (of his subtribe Balsamocitrinae), especially *Limonia* and *Feroniella*, do not go together with the Bael Fruit trees but come out in different positions on the molecular tree (Bayer et al. 2009). Although Swingle (1943) had been quite certain about the close relationship among the hard-shelled citrus trees, the multiplicity of their origin (see below) surpasses all expectations. On the other hand, Swingle himself (1943) had doubts about whether *Swinglea* was correctly included among the Hard-shelled Citrus Fruit trees and, indeed, it is not (Bayer et al. 2009). The disjunction between *Limonia* (a member of the Wood Apple Group of Swingle) and *Aegle* (belonging to his Bael-Fruit Group) in the molecular trees is less surprising, despite Swingle's (1943) certainty that the two hard-shelled groups were related.

The great similarity between *Pleiospermium* and *Burkillanthus* in leaf, spine and flower structure had been noted by Swingle, and a subclade comprising these genera together with *Swinglea* is strongly supported in the topology of Bayer et al. (2009). These genera may represent different evolutionary stages in the origin of pulp vesicles. In *Swinglea*, the seeds are immersed in a glutinous fluid apparently secreted by ellipsoid mucilage glands 5–10 mm long that open into the locule. *Burkillanthus* has rather thick and semi-translucent pulp vesicles that line the walls of the fruit locules and, as the fruit ripens, apparently collapse and set free a mucilaginous fluid. Within *Pleiospermium*, there is a gradation from short compact pulp vesicles 1–2 mm long in one species to long pointed ones in others. Moreover, the pulp vesicles in this group differ from those in *Citrus* in being corticated and releasing mucilage into the locule.

Naringi and *Citropsis*, which share great similarities in leaf and fruit structure, had been considered as closely related by Swingle (1943), a view that is confirmed by the analyses of Morton et al. (2003) and Bayer et al. (2009). The pulp vesicles of *Naringi* are rudimentary, and those of *Citropsis* taper gradually from a broad base and appear relatively unspecialised in comparison with those of *Citrus* (Bayer et al. 2009).

A clade consisting of the unispecific *Limonia* and the genus *Atalantia* forms the sister group to *Citrus* s.l. It came as a surprise that the two genera of Swingle's 'Wood Apples', *Limonia* and *Feroniella*, are not sister taxa (Bayer et al. 2009), as long had been supposed. In hindsight this makes sense, however, because *Limonia* shares with the non-*Citrus* genera the diplostemony, whereas *Feroniella*, now embedded in *Citrus*, has four times as many stamens as petals, as is the rule in *Citrus* s.l. *Atalantia*, which is sister to *Limonia*, comprises two sections, one with well-developed, juicy pulp vesicles, and another with reduced, almost vestigial ones. *Severinia* may belong into the latter section of *Atalantia* but more study is needed before its generic status can be clarified.

The expanded generic concept for *Citrus* now comprises all genera referred to by Swingle (1943) as "True Citrus Fruit Trees", plus *Clymenia*, *Oxanthera* and *Feroniella*. Two major groups, a 'largely southern clade' and a 'northern clade', can be distinguished (Bayer et al. 2009). The first group comprises mainly wild species from New Guinea and Australia ('*Microcitrus*' and '*Eremocitrus*'), New Caledonia ('*Oxanthera*'), the Bismarck archipelago ('*Clymenia*') and India (*C. medica* and *C. indica*). *Citrus medica* is the only cultivated species of this group and, because of its spread by man, its original range is not known. In the northern clade, *C.* ('*Poncirus*') *trifoliata* and '*Feroniella*' are the most aberrant members of *Citrus* in the summer-green 3-foliolate leaves of the former and the paracarpous ovary and odd-pinnate leaves of the latter. Other species within the northern clade, which exist as wild plants or little altered land races, include *C. maxima*, *C. reticulata*, *C. halimii*, *C. cavalerei* and *C. hystrix*. These wild species and *C. medica* from the southern clade and possibly other now extinct species may have given rise to the great diversity of *Citrus* cultivars known today (see also under 'Economic Importance').

Subfam. Cneoroideae (Spathelioideae), originally restricted to *Spathelia*, is here extended on the basis of molecular studies that have recovered *Spathelia*, *Dictyoloma*, *Harrisonia*, *Cneorum*, *Bottegoa* and *Ptaeroxylon* (*Cedrelopsis* was not included) as a clade sister to Rutaceae-Rutoideae + Aurantioideae (Chase et al. 1999; Savolainen et al. 2000; Scott et al. 2000). It is a quite diverse assemblage of genera of highly disjunct distribution, which differ from the core Rutaceae in morphological and chemical characters. In Takhtajan's (1997) classification, these genera were treated as two families, Ptaeroxylaceae and the monogeneric Cneoraceae. *Dictyoloma* and *Spathelia* (but not *Harrisonia*, as erroneously indicated by Takhtajan 1997) have multicellular oil glands but these, in contrast to the core Rutaceae, are restricted to the margin of the leaflets, and *Dictyoloma* may be unique in that the cell walls of the multicellular glands are not or hardly resorbed (Blenk 1884), a report in need of verification. In the remaining genera, multicellular oil glands are lacking but oil cells are present in their parenchymatous tissues. *Dictyoloma*, *Spathelia* and *Harrisonia* have appendages on the stamen filaments (also known in *Citrus* '*Feroniella*' and frequent in Simaroubaceae). The group deserves further study.

All Cneoroideae genera are held together by the possession of chromones (otherwise occurring in Rutaceae very erratically), but typical rutaceous compounds are also present. They include prenylated coumarins in *Cneorum*, *Ptaeroxylon* and *Cedrelopsis* and limonoid derivatives in *Spathelia*, *Cneorum*, *Harrisonia* and *Cedrelopsis* (Waterman 1983; Mulholland et al. 2000); anthranilate-derived or other alkaloids are known only from *Spathelia*. *Cneorum* stands out in having highly modified limonoids, the pentanortriterpenoids. Waterman (1983: 395 seq.) has combined these data into a quite convincing phylogenetic hypothesis, according to which this subfamily branched off from ancestral Rutaceae/Rutales at a time when the capacity to synthesise coumarins, limonoids and some simple anthranilate alkaloids was already available (later more or less lost in the different lineages), but prior to the inception of the quassinoid pathway as exists in Simaroubaceae.

Distribution and Habitats. Rutaceae have a wide distribution in the tropics and the warm-temperate regions of the southern hemisphere; only a limited number of genera are found in the north temperate regions. Both in structural diversity and numbers of taxa, the family is best developed in Australasia including Malesia and is highly diverse in the Cape region of South Africa; it is also represented in SE Asia, Africa/Madagascar, and North, Central and South America.

Subfam. Rutoideae. The **Bosistoa Alliance** is almost completely eastern Australian; only one genus (*Lunasia*) extends from northeast Australia widely into Malesia; all are rainforest taxa.

The **Euodia Alliance**, too, consists primarily of rainforest species but some, for instance, several *Zanthoxylum*, occur in (sub)xeric vegetation such as that of the Greater Antilles, and the deciduous *Phellodendron* and *Orixa* are found in the temperate region of the northern hemisphere. Among the New Caledonian taxa, quite a few grow in heathland ('maquis') on schistose and ultramafic soil.

The distribution of the majority of the genera is clearly Australasian: four of the 37 genera are restricted to Australia, and another eight are endemic to New Caledonia (where 14 of the 37 are represented). *Geijera*, *Medicosma* and *Halfordia* occur in Australia, New Caledonia and New Guinea; *Halfordia* also is found in Melanesia, whereas its close relative *Neoschmidea* is strictly New Caledonian. *Euodia* is found in New Caledonia and NE Australia, and its close allies occur in Papuasia, Melanesia, Malesia and mainland Asia. *Melicope*, a speciose lineage probably descended also from an Euodia-like ancestor, has its most primitive stock in Australasia and is distributed from Malesia eastwards to the Hawaiian and Marquesas Islands and in the west to the Indo-Himalayan and the Malagasy regions. *Acronychia* and eight closely related genera, which may have descended from a *Melicope*-like ancestor, are Australasian, with *Maclurodendron* and one *Acronychia* reaching mainland Asia via Malesia. *Vepris*, which may be close to *Acronychia*, is mainly African and Madagascan and, like the little known Madagascan *Ivodea*, may have its closest relatives among the Australasian genera. *Pitavia*, an unplaced genus of Rutoideae with Australasian affinities, is endemic to the Coastal Cordillera of southern Chile and possibly is a remnant of the Palaeogene circumantarctic flora, or may have arrived there by long-distance dispersal. The latter view may find support in the fact that *Pitavia* co-occurs in the Chilean Coastal Cordillera with other taxa that have austral affinities, such as *Nothofagus alessandri*, *Berberidopsis corallina* and *Gomortega keule*.

The link of *Zanthoxylum* and its allies with the Australasian region is not as obvious as that of the previous genera. *Zanthoxylum* has a pantropical distribution including the warm-temperate region in North America and Asia. The distribution of its close relatives is strongly disjunct: *Tetradium* occurs from Himalayan Asia east to Japan and Malesia, whereas *Phellodendron* is found from south-western China to Japan and south to Taiwan. Among the other close relatives of *Phellodendron*, *Fagaropsis* occurs in Africa and Madagascar, and *Crossosperma* is endemic to New Caledonia. *Toddalia* is distributed in southern and tropical Africa, Asia and Malesia.

The **Boronia Alliance** is strictly Australian; two species found outside Australia (one *Zieria* in New Caledonia, one *Leionema* in New Zealand) are clearly nested within the respective genera. The members of the Boronia Alliance are bound to heathland, woodland and eucalypt forest; rarely do they grow along rainforest margins (some *Nematolepis*) or in wet eucalypt forest and coastal communities (*Chorilaena*). Nine mostly small genera occur in south-western West Australia, and five of them, including *Geleznowia*, *Muirantha*, *Rhadinothamnus*, *Chorilaena* and *Diplolaena*, are endemic to this region. Eastern New South Wales is another area with a great concentration of boronoid Rutaceae. The large genera *Philotheca*, *Asterolasia* and *Phebalium* grow in southern Australia and, except for some more widely distributed species of *Phebalium*, are exclusive to this region. Most boronioids have scleromorphic foliage, which is thought to be a consequence mainly of the low fertility of soils abounding in south-western and south-eastern Australia (see Beadle 1981).

The **Diosma Alliance** is restricted to southern Africa and, with exception of *Calodendron*, is found preferentially in fynbos vegetation. Most genera occur in the Western Cape Province of South Africa, particularly its south-western part, where *Phyllosma*, *Coleonema*, *Macrostylis*, *Sheilanthera*, *Adenandra* and *Euchaetis* are centred. The other genera have a wider distribution that includes the Eastern and Western Cape Provinces, and *Agathosma* ranges to Lesotho and Kwazulu-Natal. The aberrant *Calodendron* is distributed from the Western Cape Province to Tanzania.

The **Polyaster Alliance** is mostly found in Mexico and Central America, with *Choisya* extending to the SW U.S.A.; *Plethadenia* occurs in Cuba and Hispaniola, and *Peltostigma* in Jamaica and from Mexico to Peru.

The **Esenbeckia Alliance** comprises three genera of which *Metrodorea* and *Raulinoa* are strictly South American. *Esenbeckia* ranges from Mexico and the West Indies to northern Argentina and has two distribution centres, one in Mexico, and the other in south-eastern Brazil. It is largely absent from Amazonia (although collections of *E. pilocarpoides* and *E. grandiflora* more recent than Kaastra's (1982) monograph have filled that gap somewhat) and grows in a variety of habitats from thorn scrub and dry woodland to moist primary forest.

Of the two genera of the **Balfourodendron Alliance**, *Helietta*, growing in mesic to xeric habitats, is distributed among several disjunct areas from Mexico and adjacent Texas and Cuba south to Argentina and Paraguay. *Balfourodendron* occurs in (semi)deciduous forests from north-eastern Brazil to Argentina and Paraguay.

The **Angostura Alliance** ranges from Mexico and the West Indies to tropical South America. The six apparently less-derived pentandrous or tetrandrous genera with hardly zygomorphic flowers, unappendaged anthers and plano-convex cotyledons (*Adiscanthus*, *Leptothyrsa*, *Euxylophora*, *Spiranthera*, *Nycticalanthus* and *Almeidea*, the latter with conduplicate-plicate cotyledons) are all South American and tropical lowland forest dwellers. The more-derived 21 genera, characterised by diandrous flowers with tendencies towards appendaged, glandular-punctate anthers and conduplicate-plicate cotyledons, are also South American but eight of them are also represented in Central America and/or the West Indies. Most genera of this Alliance grow in humid lowland forest, and a few in montane forest (e.g. *Raveniopsis* in the Guayana Highland). There is little evidence for deciding whether the origin of this group was in North or South America, and its presence in the Greater Antilles may speak for its old age.

Of the **Amyris Alliance**, the large genus *Amyris* is distributed mainly in Mexico and Central and South America; one species extends to Texas and Florida, U.S.A.; *Stauranthus* is Central American.

The genera of the **Ruta Alliance** occur mainly in the northern hemisphere, where they grow preferably in open mesic or xeric habitats. *Boenninghausenia* is Asian, *Psilopeganum* endemic to China, and *Cneoridium* restricted to south-western North America. *Ruta* ranges from southwest Asia to the Mediterranean and Macaronesia, and *Haplophyllum* is distributed from East Asia to the Mediterranean and to Arabia and North and Northeast Africa. *Thamnosma* occurs in the south-western U.S.A. and adjacent Mexico, and in Somalia and southern Africa.

Subfam. Aurantioideae is distributed in Africa, South and SE Asia, Malesia and Australasia, with some *Wenzelia* extending to Fiji and Hawaii. Most genera are mainly Indomalayan, with some of their species reaching Australia/New Caledonia (*Micromelum*, *Bergera*, *Clausena*, *Murraya*, *Luvunga*, *Paramignya* and *Citrus*) or tropical Africa (three spp. of *Clausena* and all spp. of three of the four Bael Fruit tree genera and of *Citropsis*). *Citrus* is exceptional because the majority of its wild species are Australian (Bayer et al. 2009). Most Aurantioideae occur in humid evergreen forest from the lowland to the montane belt; *Clausena anisata* grows in eastern Africa up to an altitude of 3,000 m. The two *Naringi* spp. have a notable drought resistance, which probably is surpassed only by that of the eastern Australian desert lime, *C.* '*Eremocitrus*' *glauca* that withstands severe drought and hot dry winds.

The distribution of Subfam. Cneoroideae is highly disjunct. *Spathelia* and *Dictyoloma* are neotropical, the former in lowland forests of the Greater Antilles and cis-Amazonian South America, and the latter in semidry woodland in trans-Amazonian South America. The sole species of *Bottegoa* grows in woodland and savannas of north-eastern and eastern Africa, and *Cedrelopsis* is restricted to humid and subxerophytic forest in Madagascar. *Harrisonia* is distributed from tropical Africa through Southeast Asia and Malesia to northern Australia, usually under seasonal conditions. *Cneorum* occurs in the western Mediterranean, the Canary Islands and Cuba.

Fossils, Age, and Distributional History. A critical synopsis of the distribution of Rutaceae in space and time, primarily based on fossil seeds and fruits, but also on leaf and wood remains, was given by Gregor (1989). In the Cretaceous and Palaeocene, the family was represented by form genera such as *Rutaspermum*, *Toddaliospermum* and *Pteleaecarpum*. Various 'species' accommodated in the fossil taxon *Rutaspermum* may represent *Zanthoxylum* (Tiffney 1980). Tertiary fossils have been assigned to about ten modern genera, among which *Zanthoxylum*, *Toddalia* and *Tetradium* make their earliest appearance in the Palaeocene. *Zanthoxylum* is on record from the Palaeocene of Europe, the Eocene of Peru, and the early Oligocene of eastern North America (Tiffney 1980). *Toddalia* appears in Europe from the Upper Palaeocene to the Pliocene and in the north-eastern U.S.A. in the Oligocene. Fossil seeds attributed to *Euodia costata* (Chandler) Tiffney are known from the Upper Palaeocene to the Upper Eocene of southern England (Tiffney 1980), and those attributed to *E. lignitica*

Tiffney are known from the Oligocene of the north-eastern U.S.A. (Tiffney 1981). Both, however, have been recognised by Hartley (2001a, b) as belonging in *Tetradium*, where he found them closely similar to *T. glabrifolium*, the most primitive extant species of the genus, which is widespread in Asia from the Himalayas to Japan and Malesia. *Phellodendron* seeds are known from the Early Oligocene of eastern North America, from the Mid Oligocene to the Pliocene of Europe and from the Pliocene of Asia and Japan. It is remarkable that the oldest known fossils of the family (*Zanthoxylum*, *Toddalia* and *Tetradium*) all are north-Tethyan, belong to the group of the five genera that produce 1-btiq alkaloids, and are specialised for bird dispersal. *Toddalia* has drupaceous fruits, and *Zanthoxylum* and *Tetradium* have seeds with a strong sclerotesta that are not forcibly expelled from the follicular fruit but taken by avian dispersers. The avian fossil record clearly shows that, by the Palaeocene and the Late Cretaceous, wide-ranging birds already existed (see Mayr 2005). It is obvious, therefore, that the appearance of these highly evolved genera in the Palaeocene sets only a minimum age of the crown group of Rutaceae (s.str.).

To rationalise distributional patterns in Rutaceae, vicariance events have been invoked as the important factor (Hartley 2001a, b; Ladiges and Cantrill 2007; Beattie et al. 2008), and age estimates for disjunctions have been calibrated against the timing of plate movements suggested by geologists. Thus, an age estimate for the Australian-Madagascan disjunction of *Melicope* of at least 71 Ma was suggested (Hartley 2001a), which implies a land connection via India, and the disjunction in *Euodia* between Australia and New Caledonia was determined at 66 Ma or before (Hartley 2001a). The latter age estimate may also apply to the genera of Rutoideae and perhaps of Aurantioideae that occur in New Caledonia, north-eastern Australia and adjacent regions of the south-eastern Pacific. It has been hypothesised (Beattie et al. 2008) that the Australasian "southern" *Citrus* lineage over the past 37 Ma may have dispersed from Australia and Papuasia north to Malesia and SE Asia—perhaps endozoochorously by bats and birds but substantially on terranes that shifted from (what was to become) New Guinea to form present Halmahera and most of the Philippines, bringing forth the "northern" *Citrus* clade.

Because most age estimates for Rutaceae based on molecular dating conflict with time frames that would be congruent with a vicariance explanation for the observed distribution patterns[5], some authors have explained these distributions by long-distance dispersal. Pfeil and Crisp (2008), who present an astonishingly recent date for the Aurantioideae clade, also append geological evidence according to which much of New Caledonia was probably submerged under the sea until 35 Ma and only since then could have been colonized from other land masses by long-distance dispersal. However, because of the tectonic activity at the eastern margin of the Australian plate in the Australian-New Caledonian region in the Middle Eocene, terrestrial habitats there may have been widespread and may have existed as recently as the Late Eocene (see review by Ladiges and Cantrill 2007). Even if present-day New Caledonia had been submerged for part of the Tertiary, continuity of land for the biota of a 'greater New Caledonia' may have extended back to the Palaeocene/Eocene, which would provide a time frame congruent with a vicariance explanation of the floristic connections of the New Caledonian flora with the southwest Pacific region.

Incidentally, fossil time-markers are sometimes misinterpreted in molecular dating as indicating the absolute age of a lineage, when older fossils are not available. This pratice is criticised by Olmstead and Tank (2008), who politely equate it to "overprecision in molecular dating". It is revealing that fossils used to fix the first appearance of a lineage often closely resemble extant species, but in the absence of fossil precursors, these seem to appear 'out of nowhere', implying gaps of an indefinite extent in the prior record (see also Heads 2009). This may apply to the Eocene material of *Ailanthus* (Simaroubaceae)

[5]Rutaceae s.str.: Muellner et al. 2007: c. 76 Ma; Pfeil and Crisp 2008: 36.4–56.8 Ma, mean = 47.6; Rutaceae s.l.: Wikström et al. 2001: 37–39 Ma; Muellner et al. 2007: 93.3–72.9 Ma; Pfeil and Crisp 2008: 53.3–72.7 Ma, mean = 62.7) Aurantioideae: Muellner et al. 2007: c. 49 Ma; Pfeil and Crisp 2008: 12.1–28.2 Ma, mean = 19.8).

used by Muellner et al. (2007) and Pfeil and Crisp (2008) for the calibration of their trees and to the fossil seeds of *Zanthoxylum* and *Tetradium* mentioned above, which hardly represent basal taxa within their respective lineages.

The geographical origin of Rutaceae remains in the dark. Cneoroideae, the basal clade, is pantropical with a trans-Atlantic and/or trans-Pacific disjunction (*Spathelia* + *Dictyoloma* vs. the rest). Among Rutoideae, Australasia is the centre of diversity for the Bosistoa Alliance, the Euodia Alliance and the Boronia Alliance. The Diosma Alliance, though morphologically close to Australasian taxa, remains ambiguous in regard to its closest relatives. The *trn*L-*trn*F sequences (Scott et al. 2000; Groppo et al. 2008) join it with Australasian rutoids, whereas *rps*16 groups it with American taxa (Groppo et al. 2008). The Ruta Alliance is mainly north-Tethyan but through *Chloroxylon* (India and Madagascar) it may have an austral link. New World Rutoideae are a relatively coherent group, in which many character progressions observable in the Old World are repeated. The Polyaster Alliance is particularly similar to, and previously has been associated with, the Australasian Euodia Alliance (Engler 1931), and the present geographical cleft between the two is easily bridged by the Tertiary fossils. Thus, for Rutaceae, as for so many other tropical forest plant families, a north-Tethyan origin may be conceivable (Kubitzki and Krutzsch 1998), whereas the Neotropics, Australasia and South Africa could be secondary centres of diversification.

Economic Importance. Citrus fruit trees are the most important crop plants of Rutaceae, whose cultivation has a very long tradition. In an ancient Chinese document from the third millenium b.c., reference is made to baskets filled with "small oranges and pummeloes" (Webber 1967). On the basis of written documents, works of plastic art, and other kinds of testimony, Tolkowsky (1938) reconstructed the spread of various species and horticultural groups of *Citrus* from the Far East to India, the Middle East, the Mediterranean and North Africa, and Europe. The first citrus to come to Europe was the citron, *Citrus medica*, which Alexander the Great is said to have taken from India and Persia to the Mediterranean, where initially it was cultivated for its fragrance. After the introduction of the citron, oranges and lemons, kumquats, leech limes and others were brought to Europe by the Muslims via the Iberian Peninsula and Sicily (Ramón-Lacca 2003). Sweet orange, mandarin and grapefruit came much later, between the fifteenth and nineteenth centuries, in connection with the trade with Portugal's and Britain's Asian colonies.

Citrus species attracted great horticultural interest and, in Renaissance gardens, they were grown in tubs and wintered under cover, until the construction of an Orangérie at the Palace of the Louvre in 1617 inspired imitations, which in the 1840s were superseded by modern greenhouses. Many Australian (esp. *Correa*) and some South African (*Diosma*) genera are widely used in horticulture; some *Boronia* are popular as cut flowers.

Systematic citrus breeding was initiated in 1893 by the US Department of Agriculture in Florida. During the last 100 years, the cultivation and processing of citrus fruits has developed into a gigantic industry. Pertinent commercial, genetic and horticultural aspects are treated in 'The Citrus Industry' (Reuther, Webber and Batchelor, eds., 1967–1973). Citrus is cultivated in a belt approximately 40° north and south of the equator, but the finest-quality oranges and most acidic lemons are grown under irrigation in seasonal climate in subtropical regions. In the equatorial areas with equable climate, citrus is grown preferentially for local consumption. Apart from the fresh fruit, citrus juices are a product of ever growing importance, with the essential oil of the peel as the most important by-product. Many cleaning products now use rutaceous (particularly *Citrus*) ingredients for their antimicrobial and fungicidal properties, whose economic value can be expected to increase in a market inclined to 'go green' by using natural products that do not harm the environment. Polymethoxylated flavones from tangerine and orange peel are a major ingredient in a dietary supplement that exhibits dramatic effects in lowering cholesterol. The leaves of Australian boronoid Rutaceae are distilled to obtain their diverse monoterpenes, and those of *Boronia megastigma* is distilled as perfume.

Among other rutaceous fruit trees, *Casimiroa edulis* (white sapote, zapote blanco, and matasano or matasán in Guatemala) is widely distributed in Central America and much cultivated. Its fruit resembles an early apple in shape, size and colouring, and also in texture and the pleasant smell of its flesh. The popular belief that eating considerable amounts of the fruit will induce sleep has perhaps a sound base; the glucoside casimirosine extracted from it is said to have a hypnotic and sedative effect.

Formerly, the alkaloid pilocarpine extracted from *Pilocarpus* was used for the treatment of glaucoma, in which it has been superseded by medicaments that have fewer secondary effects.

The bark of *Angostura trifoliata* (Willd.) T.S. Elias (= *Galipea officinalis* J. Hancock) is the base of the Angostura bitter, an additive produced since 1824 in Angostura (today Ciudad Bolívar, Venezuela) and later manufactured in Trinidad.

The timbers of few Rutaceae have gained attention in the world market and are available only intermittently and in low quantities. *Amyris balsamifera* and other species of the same genus have relatively hard but easily workable and scented timber. *Chloroxylon swietenia*, Ceylon sandlewood, is used for furniture and cabinet-making, as is the West Indian sandlewood, *Zanthoxylum flavum*, whereas the African *Z. macrophyllum* and related species are of more regional use. The timber of Amazonian *Euxylophora* is much esteemed but also of regional use. Most of the species of *Flindersia* are regionally important as commercial timbers; their uses range from railway sleepers and fencing to general construction (*F. australis* and *F. ifflaiana*) and cabinet-making (*F. brayleyana* and *F. pimenteliana*).

Key to the Subfamilies

1. Fruit dehiscent or indehiscent, follicular, capsular, or drupaceous, without pulp produced by pulp vesicles; seeds albuminous or exalbuminous, with only 1 embryo **Keys to the Genera of Rutoideae and Cneoroideae of the Old World** (this page) **and the New World** (p. 307)
– Fruits baccate, non-winged, sometimes with leathery or woody pericarp containing pulp often produced by pulp vesicles of the carpel walls; seeds with mucilaginous coat, exalbuminous, sometimes with more than 1 embryo **Key to the Genera of Aurantioideae** (p. 309)

Key to the Genera of Rutoideae and Cneoroideae of the Old World, including Australia and Oceania

1. Cotyledons linear, about the same width as the hypocotyl; endosperm copious; mostly small-leaved shrubs or undershrubs of xeromorphic vegetation, most in Australia. **Boronia Alliance** 3
– Cotyledons orbicular to ovate, elliptic, elliptic-oblong, or obovate, considerably wider than the hypocotyl; endosperm present or not 2
2. Leaves simple; anthers with apical gland; fruit capsular or dehiscing into horned mericarps; seeds mostly explosively released; endosperm 0; mostly xeromorphic shrubs or small trees, southern Africa. **Diosma Alliance** 21
– Leaves various; anthers usually without apical gland; fruit various; seeds explosively released or not; endosperm present or not; mostly comparatively large-leaved shrubs or trees of rainforests 31
3. Leaves opposite, simple or pinnate or bipinnate; flowers 4(5)-merous 4
– Leaves alternate, simple; flowers usually 5-merous, rarely 4- or 6–8-merous 7
4. Leaves simple; calyx hemispherical, sometimes lobed; petals connate and forming a tube, or lately separating and then often still connate at base and stamens spreading **40. *Correa***
– Leaves simple or pinnate; sepals distinct; petals distinct, usually spreading, sometimes urceolate; stamens erect or inwardly curved 5
5. Stamens 8 (rarely 4 caducous) **37. *Boronia***
– Stamens 4 6
6. Disk distinctly 4-lobed; leaves simple or ternate **38. *Zieria***
– Disk entire; leaves simple **39. *Neobyrnesia***
7. Leaves lepidote on abaxial surface, sometimes also so on branches, perianth, stamens and ovary 8
– Leaves and stems glabrous or variously hairy but not lepidote, rarely (*Asterolasia, Eriostemon*) stellate-lepidote on petals 12
8. Flowers sessile, in dense or compact heads or rarely solitary 9
– Flowers pedicellate, in an open inflorescence or solitary 10
9. Carpels 2–4 **53. *Microcybe***
– Carpels 5 **52. *Phebalium***
10. Inflorescence umbellate or rarely flowers solitary; anthers basifixed, with a prominent apical gland **52. *Phebalium***
– Inflorescence cymose or solitary; anthers versatile, without an apical gland 11
11. Sepals connate; anthers obtusely apiculate **50. *Rhadinothamnus***
– Sepals distinct; anthers retuse at apex **49. *Nematolepis***
12. Leaf margin deeply sinuate or lobed **51. *Chorilaena***
– Leaf margin entire 13

13. Inflorescence heads surrounded by large, imbricate involucral bracts in 3 or 4 rows **54. *Diplolaena***
- Inflorescence not surrounded by large involucral bracts 14
14. Corolla cylindrical 15
- Corolla spreading 16
15. Stamens 5, alternating with 5 staminodes, densely hairy on abaxial surface **46. *Drummondita***
- Stamens 10, all fertile, pilose at swollen base **48. *Muiriantha***
16. Sepals larger than petals, coloured and similar to inflorescence bracts **44. *Geleznowia***
- Sepals much smaller than petals, not showy 17
17. Staminal filaments glabrous, divergent 18
- Staminal filaments variously hairy, rarely glabrous, erect or pyramidally arranged 19
18. Sepals minute or 0; carpels 1–5 **45. *Asterolasia***
- Sepals obvious; carpels 5 **47. *Leionema***
19. Petals 5-veined, stellate-lepidote; staminal filaments with a subapical verrucosity **41. *Eriostemon***
- Petals 1-veined, glabrous or with simple hairs; staminal filaments smooth 20
20. Anther and apiculum pilose **42. *Crowea***
- Anther and apiculum glabrous **43. *Philotheca***
21. Trees with broad leaves; ovary on long gynophore; fruit with blunt spines **55. *Calodendrum***
- Shrubs or small willowy trees with narrow leaves; gynophore 0 or very short; fruit usually with a single horn per carpel 22
22. Flowers incomplete; petals 0 **65. *Empleurum***
- Flowers complete; petals 5 23
23. Style about same length as petals, exceeding the claw if any; flowers terminal, in clusters and/or in the axils of the leaves 24
- Style much shorter than the petals, not exceeding the claw if any; flowers solitary, or few, or terminal, in reduced racemes, never in dense heads or clusters 26
24. Staminodes present in a variety of shapes **56. *Agathosma***
- Staminodes 0 or vestigial 25
25. Ovary (2)3(4)-carpelled **62. *Macrostylis***
- Ovary 5-carpelled **57. *Phyllosma***
26. Petals transversely bearded at the throat **63. *Euchaetis***
- Petals pubescent or glabrous at the throat 27
27. Staminodes situated in a groove in the petal, connate below (except *C. nubigenum*) **58. *Coleonema***
- Staminodes free from the petals or 0 28
28. Anther with a stipitate apical gland **60. *Adenandra***
- Anther with a sessile or immersed apical gland 29
29. Anthers and ovary hairy **61. *Sheilanthera***
- Anthers glabrous and ovary glabrous or with a few spiky hairs or with stalked glands 30
30. Disk spreading-crenulate; staminodes 0 or vestigial **64. *Diosma***
- Disk circumvallate; staminodes of simple shapes present **59. *Acmadenia***
31. Leaves and cortex with solitary oil cells; pellucid lysigenous oil glands 0 32
- Leaves with pellucid lysigenous oil glands 36
32. Leaves simple, alternate; [stamen filaments inserted in pits of the androgynophorous disk; fruit disintegrating from central column into 3–4 drupelets]. W Mediterranean **154. *Cneorum***
- Leaves pinnate or 3-foliolate 33
33. Leaves opposite; [fruit a 2-valved capsule]. Southern Africa **150. *Ptaeroxylum***
- Leaves alternate 34
34. Stamens 8–10; filaments with basal scale; [ovary 4–5-celled; fruit a drupe]. Africa, S Asia, Malesia and Australia **153. *Harrisonia***
- Stamens 4 or 5; stamen filaments without basal scale 35
35. Ovary 2-celled; ovule 1 per locule; fruit indehiscent, broadly winged. Africa **151. *Bottegoa***
- Ovary 3–5-celled; ovules (1)2(3) per cell; fruit capsular. Madagascar **152. *Cedrelopsis***
36. Herbs and halfshrubs; ovules usually more than 2 per carpel (except some*Haplophyllum*); endosperm copious; [flowers hermaphrodite, sometimes slightly zygomorphic] 37
- Woody; ovules usually 1 or 2 per carpel, more rarely up to 8; seeds with or without endosperm 42
37. Flowers slightly zygomorphic; endocarp elastically discharged with the seeds when the fruit dehisces; embryo straight; [seeds 3(2) per follicle] **116. *Dictamnus***
- Flowers regular; endocarp persistent in dehisced fruit; embryo curved 38
38. Carpels (3)4–5 39
- Carpels 2(3) 41
39. Carpels on elongate gynophore, apocarpous except for the common style; fruit follicular. E Asia, Malesia **109. *Boenninghausenia***
- Carpels ± connate; fruit capsular 40
40. Petals denticulate or fimbriate; stamens filaments glabrous; leaves all compoundly or simply pinnate. Macaronesia to SW Asia **108. *Ruta***
- Petals entire; filaments usually bearded within; leaves simple or trisect. From the Mediterranean to E Asia **112. *Haplophyllum***
41. (Half)shrubs; leaves simple, sometimes deeply divided; flowers paniculate to racemose. Africa, N America **111. *Thamnosma***
- Herbaceous; leaves 3-foliolate; flowers in leafy cincinnate cymes. C China **110. *Psilopeganum***
42. Fruit dehiscent 43
- Fruit indehiscent 65
43. Stylodia extremely short, joined in a very short common style; stigma broad, discoid, deeply lobed; disk 0; [fruit follicular; leaves opposite, whorled or alternate, simple or 1-foliolate]. Madagascar **14. *Ivodea***
- Stylodia evident, joined in a common style; stigma not broadly discoid, slightly lobed if at all; disk usually present 44
44. Leaves alternate 45

– Leaves opposite 52
45. Fruit follicular; seeds not winged 46
– Fruit capsular; seeds apically winged 51
46. Plants often armed; leaves mostly impari- or paripinnate (1–15-jugate) or (1)3-foliolate; [flowers unisexual or rarely bisexual; perianth of 4–9 undifferentiated segments or biseriate with 3–6 sepals and 3–8 petals; gynoecium ± apocarpous]. Tropics and warm-temperate regions of the northern hemisphere **28. *Zanthoxylum***
– Plants unarmed; leaves simple or 1-foliolate 47
47. Sepals and petals each 3. Malesia, New Guinea, NE Australia **5. *Lunasia***
– Sepals and petals each 4 or 5 48
48. Inflorescences cauligerous, subtended by caducous bud scales. E Asia **34. *Orixa***
– Inflorescences axillary or terminal, in *Geijera* rarely ramigerous 49
49. Leaves sessile; stamens 10. New Caledonia **26. *Neoschmidea***
– Leaves petiolate; stamens 5 50
50. Petals valvate in bud, hooked adaxially at apex; endocarp and seed persistent in fruit. New Guinea, New Caledonia, Australia **33. *Geijera***
– Petals imbricate in bud, not hooked at apex; endocarp and seed forcibly discharged. E Australia **25. *Coatesia***
51. Stamens 5, alternating with 5 filament-like staminodes; gynoecium 5-carpellate; capsule septicidal. Australia, New Guinea **117. *Flindersia***
– Stamens 10; gynoecium 3-carpellate; capsule loculicidal. Madagascar, S India **114. *Chloroxylon***
52. Endocarp and seed persistent in dehisced fruit 53
– Endocarp forcibly discharged with seed when fruit dehisces 56
53. Testa with thick inner layer of sclerenchyma, an outer layer of spongy tissue (sometimes 0 in *Tetradium*), and a shiny, black pellicle 54
– Testa ± thin and brittle, with neither spongy tissue nor a shiny pellicle 55
54. Leaves imparipinnate, 1–9-jugate; inflorescences terminal and/or upper-axillary; flowers mostly functionally unisexual; stamens 4 or 5; fruit of distinct or basally connate follicles. S and E Asia, Malesia **29. *Tetradium***
– Leaves digitately 3- or 1-foliolate or simple; inflorescences axillary to cauligerous; flowers bisexual or functionally unisexual; stamens 4 or 8 or rarely 4–8; fruit of basally connate follicles or grading to completely syncarpous (the carpels united in a 4-loculed capsule). Most diverse in New Guinea and Hawaii, but also in Australia, the western Pacific, Malesia, S and SE Asia, and Madagascar **13. *Melicope***
55. Stamens 8; seeds exalate. Moluccas, New Guinea and New Britain **11. *Perryodendron***
– Stamens 4, alternating with 4 staminodes; seeds alate. Malesia to Solomon Islands **12. *Tetractomia***
56. Inflorescences terminal or terminal and upper-axillary 57
– Inflorescences axillary; [flowers 4-merous] 62
57. Young branches with stellate to lepidote trichomes; [leaves simple; flowers 4–5-merous; stamens 8 or 10; fruit of basally connate follicles or grading to almost completely syncarpous (the carpels united in a 5-loculed loculicidal capsule)]. New Caledonia **8. *Myrtopsis***
– Young branches with simple trichomes 58
58. Flowers 5(6)-merous 59
– Flowers 4-merous 60
59. Buds with 2 pairs of scales enclosing leaf and floral primordia; ovules 4–6 per carpel. E Australia **1. *Bosistoa***
– Buds naked; ovules 2 per carpel **3. *Acradenia***
60. Leaflets mostly with pocket-domatia in axils of secondary veins; ovules 6–8 per locule. E Australia **2. *Bouchardatia***
– Pocket domatia 0; ovules 1 or 2 per locule 61
61. Leaves (im)paripinnate, digitately 3- or 1-foliolate; cotyledons flattened, convolute and folded in seed. E Australia **4. *Dinosperma***
– Leaves simple; cotyledons plano-convex, not folded. New Caledonia **6. *Boronella***
62. Flowers functionally unisexual; basal portion of first-formed inflorescences usually perennial, becoming ± woody and producing additional lower-axillary or ramuligerous inflorescences. New Guinea, New Caledonia, E Australia **10. *Medicosma***
– Flowers functionally bisexual; inflorescences annual 63
63. Fruit a capsule; stamen filaments largely fused into a wide tube bearing the distinct anthers; ovules 5–8 per carpel. Hawaiian Islands **15. *Platydesma***
– Fruit follicular; stamen filaments distinct; ovules 2 per carpel 64
64. Petals valvate; staminal filaments sublinear; follicles with short stylar beak; leaves digitately 3-foliolate or simple. New Guinea, NE Australia and eastwards to Samoa and Niue **7. *Euodia***
– Petals narrowly imbricate; staminal filaments dilated, ± petaloid; follicles not beaked; leaves 1-foliolate. NE Australia **9. *Brombya***
65. Leaves alternate 66
– Leaves opposite 70
66. Leaves pinnately or digitately compound 67
– Leaves simple or 1-foliolate 69
67. Climber with retrorse-aculeate stems; [fruit a 1–7-loculed drupe]. Africa, S and SE Asia **31. *Toddalia***
– Erect woody plants, not aculeate 68
68. Leaves digitately 1-, 3–9(–12)-foliolate; stamens 4–8 (10); fruit a 1–4-seeded drupe. Africa to SW India **24. *Vepris***
– Leaves imparipinnate, 3–7-jugate; stamens 10; fruit of 1–5 distinct, winged, 1-loculed drupes. E Australia **35. *Pentaceras***
69. Flowers 5-merous; stamens 10; ovules 2 per locule. New Guinea, Australia **27. *Halfordia***
– Flowers 4–5(–7)-merous; stamens 4 or 5; ovule 1 per locule. S and E Asia to Philippines **121. *Skimmia***

70. Flowers 5-merous, stamens 5; [disk 0; buds small, sunk in base of petioles]. E and SE Asia **30. *Phellodendron***
- Flowers 4-merous (but see variable numbers of petals and stamens in *Fagaropsis*) 71
71. Flowers bisexual 72
- Flowers functionally unisexual 75
72. Stamens 8; [fruit of 4 basally connate drupes or grading to completely syncarpous (a single 4- or 4–8-locular drupe)]. Australia, New Guinea, Malesia and mainland Asia **16. *Acronychia***
- Stamens 4 (alternating with 4 staminodes in *Dutailliopsis* and *Dutaillyea*) 73
73. Fruit a 1-loculed drupe; ovarioles distinct (3 of the 4 abortive and caducous), joined apically in a common style; leaves 1-foliolate. E Australia **17. *Pitaviaster***
- Fruit a 4-loculed drupe, like the gynoecium completely syncarpous; leaves digitately 3-foliolate 74
74. Branches with simple trichomes; staminodes distinct; outer surface of endocarp manifestly serrate-winged and sharply tuberculate. New Caledonia **22. *Dutailliopsis***
- Branches with stellate to lepidote trichomes; the staminodes adnate to petals; outer surface of endocarp smooth. New Caledonia **20. *Dutaillyea***
75. Leaves imparipinnate (2–4-jugate); [fruit a drupe; sepals 4; petals 4(–6); stamens 4–8]. C, E and NE Africa, Madagascar **32. *Fagaropsis***
- Leaves digitately 1–5-foliolate 76
76. Stamens 8 77
- Stamens 4 (sometimes alternating with 4 staminodes in *Picrella*, and rarely so in *Comptonella*) 79
77. Trichomes simple; leaves digitately 3–5-foliolate; style practically 0; [fruit 4- or 4–8-loculed; seeds laterally flattened, asymmetrically obovate-triangular, fimbriate-winged at dorsal margin]. New Caledonia **36. *Crossosperma***
- Trichomes simple and fasciculate; leaves 1-foliolate; style well developed 78
78. Staminal filaments ciliate or ciliolate, those in functionally carpellate flowers, like the petals, persistent or subpersistent in fruit; fruit of 1–4 distinct drupes or a 4-loculed drupe. E Australia east to Fiji **18. *Sarcomelicope***
- Stamen filaments glabrous, sublinear, those in functionally carpellate flowers, like the petals, caducous in fruit; fruit a 4-loculed drupe. Malesia, SE Asia **23. *Maclurodendron***
79. Trichomes simple; floral disk lobed or undulate; ovarioles distinct; style deeply immersed; fruit of 1–4 distinct drupes. New Caledonia **19. *Picrella***
- Trichomes (rarely 0) stellate to lepidote; floral disk 0; ovarioles grading from basally to almost completely connate, joined subapically or apically in a common style; fruit of 1–4 basally connate drupes or grading to almost completely syncarpous and then forming a 4-loculed drupe. New Caledonia **21. *Comptonella***

Key to the Genera of Rutoideae and Cneoroideae of the New World

1. Leaves bipinnate [fruit separating into five 3–4-seeded follicles]. Brazil, Peru, Bolivia **149. *Dictyoloma***
- Leaves pinnate, digitately 1–5(–12)-foliolate, or simple 2
2. Leaves pinnate 3
- Leaves 1- or digitately 3–5(–12)-foliolate or simple 12
3. Leaves opposite 4
- Leaves alternate 5
4. Fruit 1–5 follicles; flowers haplostemonous [xerophytic, glandular-verrucose shrubs]. Cuba, Hispaniola **67. *Plethadenia***
- Fruit a 1-carpelled drupe; flowers diplostemonous. Southern U.S.A., West Indies, Mexico, Central and South America **106. *Amyris*** p.p.
5. Single-stemmed, hapaxanthic trees [leaves often gigantic, 10–100-jugate; fruit indehiscent, 2–3-alate]. West Indies and northern South America **148. *Spathelia***
- Branched, pollakanthic trees or shrubs 6
6. Fruit or mericarps indehiscent 7
- Fruit dehiscent 9
7. Fruit 1–5 dry, narrowly winged mericarps. Southern Mexico to Honduras **69. *Decatropis***
- Fruit a drupe 8
8. Carpels 2; drupe sometimes accompanied by a second, often not fully developed drupe; embryo curved. Mexico to Nicaragua **72. *Megastigma***
- Carpel 1; drupe always solitary; embryo straight. S U.S.A., West Indies, Mexico, Central and South America **106. *Amyris*** p.p.
9. Plants often aculeate; flowers in New World species usually unisexual; seeds remain attached to dehisced fruit. Pantropical and warm temperate E Asia and E North America **28. *Zanthoxylum***
- Plants not aculeate; flowers invariably bisexual; seeds released from dehisced fruit 10
10. Flowers haplostemonous; inflorescence an elongate raceme. Mexico to Argentina, Antilles **118. *Pilocarpus*** p.p.
- Flowers diplostemonous; inflorescence with cymose branches 11
11. Leaves imparipinnate, 5–12-jugate. Mexico **66. *Polyaster***
- Leaves paripinnate, 2–5-jugate. Cuba and Hispaniola **67. *Plethadenia***
12. Leaves opposite (in some *Ravenia* appearing alternate when one leaf of a pair is very reduced) 13
- Leaves alternate 30
13. Flowers diplostemonous 14
- Flowers haplostemonous 17
14. Fruit dehiscent, 1–5 follicles. Mexico, south-western U.S.A. **71. *Choisya***
- Fruit indehiscent (drupe or berry) 15
15. Fruit a berry, embryo curved. SW North America **113. *Cneoridium***
- Fruit a drupe; embryo straight 16

16. Carpels 5, free, 1–5 of them maturing as separate drupelets. Southern Chile **119. *Pitavia***
- Carpel 1. S U.S.A., West Indies, Mexico, Central and South America **106. *Amyris*** p.p.
17. Erect herbs, sometimes suffruticose; outermost sepal very large, mimicking a bract and concealing the corolla; [leaves 3-foliolate, opposite below and alternate above]. SW Mexico to Peru, Brazil and Bolivia **103. *Ertela***
- Shrubs or trees; sepals sometimes unequal but none of them large enough to conceal the corolla 18
18. Fruit indehiscent 19
- Fruit dehiscent 20
19. Fruit a compound samara with a single (3)4-seeded body completely surrounded by (3)4 broad wings; domatia present near midrib on abaxial surface of leaf. Brazil, Paraguay, Argentina **76. *Balfourodendron***
- Fruit splitting into (3)4–5 dorsally winged mericarps; domatia 0. North, Central and South America, Cuba **77. *Helietta***
20. Fruit a (4)5-parted capsule (carpels at least partly connate after dehiscence) 21
- Fruit 1–5 follicles (separate after dehiscence) 24
21. Petals more than 1 cm long; anthers appendaged at base; [leaves 3-foliolate] **93. *Neoraputia*** (*trifoliata*)
- Petals less than 0.5 cm long; anthers not appendaged at base 2
22. Flowers 4-merous; twigs armed with opposite branch-spines; leaves simple. S Brazil **75. *Raulinoa***
- Flowers 5-merous; twigs unarmed; leaves simple or 1- or 3-foliolate 23
23. Petioles furnished with basal hood-shaped sheath enveloping the terminal bud. South America **74. *Metrodorea***
- Petiolar sheath 0. West Indies, Texas, Central and South America **73. *Esenbeckia***
24. Calyx red, valvate, more than 1.5 cm long; corolla red, lobes unguiculate. Panama **91. *Desmotes***
- Calyx green (rarely red), imbricate; corolla red, green, or white, lobes not unguiculate 25
25. Inflorescence cauligerous, 1–4 monochasia at a node, elongating as successive flowers develop. Northern South America **100. *Raputia***
- Inflorescence terminal or axillary on leafy stems 26
26. Calyx cupular, with short equal lobes not overlapping at anthesis 27
- Calyx of ± free, ± unequal sepals overlapping (sometimes broadly) at anthesis, the outer two usually more than twice as wide (or in *R. urbanii* more than twice as long) as the inner ones 28
27. Inflorescence a terminal thyrse with sessile partial inflorescences; petals white, coherent, imbricate in bud. Brazil **84. *Conchocarpus oppositifolius***
- Inflorescence a terminal monochasium; petals red or orange, connate, induplicate-valvate in bud. Northern South America **97. *Decagonocarpus***
28. Upper petal free, four lower petals connate into a lobed lip; seeds smooth. Venezuela, Costa Rica **99. *Lubaria***
- All petals connate or coherent; seeds tuberculate 29
29. Sepals broad, the outer two broadly overlapping the others in flower, all as long as the fruit and concealing it initially. Greater Antilles and from Honduras to Brazil and Peru **101. *Ravenia***
- Sepals obviously unequal, the outer two overlapping somewhat but not concealing the others, all or at least the inner three usually shorter than, and not concealing, the fruit. Northern South America **98. *Raveniopsis*** and *Ravenia urbanii*
30. Decumbent, succulent herbs; leaves 3-foliolate, often appearing basal. Venezuela **102. *Apocaulon***
- Woody; leaves various, borne on branches 31
31. Fruit indehiscent 32
- Fruit dehiscent, with separating endocarp 37
32. Fruit a samara with a broad circular wing surrounding the 2-seeded body. North America **120. *Ptelea***
- Fruit a drupe or berry 33
33. Carpel 1 34
- Carpels 2–5(8) 35
34. Flowers haplostemonous; disk 0. Central America **107. *Stauranthus***
- Flowers diplostemonous; disk gynophorous. S U.S.A., West Indies, Mexico, Central and South America **106. *Amyris*** p.p.
35. Carpels 3–4, united ventrally, maturing as separate drupelets; [stamen filaments inserted in the disk]. Cuba (also Canary Islands and western Mediterranean) **154. *Cneorum***
- Carpels (2–)4–5(–8), connate, maturing as a single drupe or berry 36
36. Leaves simple, up to 1.25 m long; petals usually barbate in lower half; carpels 5. Panama, South America **104. *Hortia***
- Leaves digitately (1)3–5(7)-foliolate; petals not barbate; carpels (2–)4–5(–8). SE U.S.A. to Costa Rica **115. *Casimiroa***
37. Stamens 10–45; carpels 6–10. Jamaica, Mexico to Peru **70. *Peltostigma***
- Stamens up to 10; carpels up to 5 38
38. Carpels 2(3) 39
- Carpels 5 40
39. Flowers diplostemonous, bisexual; fruit a bilobed, gynophorous capsule; 4–8 seeds per carpel. Southern U.S.A. and Mexico (also Somalia and southern Africa) **111. *Thamnosma***
- Flowers haplostemonous, unisexual; fruit subcircular, tardily dehiscent in 2 valves; 1 seed per carpel. Mexico (Baja California) **120. *Ptelea*** p.p. (*Taravalia*)
40. Flowers diplostemonous; stamens 10; filaments connate below. Mexico to Costa Rica **68. *Decazyx***
- Flowers haplostemonous; stamens 4 or 5 (and in some taxa, 2 or 3 sterile) 41
41. All (4)5 stamens fertile 42
- Only 2(3) stamens fertile 51
42. Leaf blades with circular pits (domatia) in axils of secondary veins on abaxial surface; [petals forming a tubular corolla; anthers sterile at base above

point of attachment to the filament and appendaged at base]. Northern South America **96. *Ticorea*** p.p.
- Leaf blades not domatiiferous 43
43. Petals less than 0.5 cm long 44
- Petals more than 1 cm long 46
44. Leaves subsessile (petiole virtually 0), simple; inflorescence a condensed dichasium; intrastaminal disk 0. Venezuela **105. *Rutaneblina***
- Leaves petiolate, simple or digitately foliolate; inflorescence a raceme or panicle; intrastaminal disk present 45
45. Inflorescence an elongate raceme; fruit of usually fewer than 5 follicles. Mexico to Argentina, Antilles **118.*Pilocarpus*** p.p.
- Inflorescence a panicle; fruit a capsule. West Indies, Texas, Central and South America **73. *Esenbeckia*** p.p.
46. Leaves simple 47
- Leaves 3-foliolate 50
47. Flowers 4-merous; inflorescence an extra-axillary, racemiform thyrse. Northern South America **80. *Leptothyrsa***
- Flowers 5-merous; inflorescence a terminal (sometimes corymbiform) thyrse 48
48. Inflorescence a corymbiform thyrse; petals valvate; disk reduced to a band of tissue adherent to a short, broad gynophore; seed tear-shaped. Northern South America **79. *Adiscanthus***
- Inflorescence a thyrse; petals imbricate; disk cupular or 0; seed flat or reniform 49
49. Disk 0; anthers narrow; large tree; fruiting follicles 5, apically truncate; seed flat, glossy. Brazil and Peru **78. *Euxylophora***
- Disk cupular; anthers ovate or oblong, blunt; small trees; fruiting follicles 1–3; seed ellipsoid-reniform, dull. E Brazil **83. *Almeidea***
50. Flowers actinomorphic; calyx cupular; disk cupular or cylindric, embracing short gynophore; follicles short beaked. Tropical South America **82. *Spiranthera***
- Flowers zygomorphic; calyx tubular; disk far exceeded by the gynophore; follicles long-beaked. Tropical South America **81. *Nycticalanthus***
51. Leaf blades with circular pits (domatia) in axils of secondary veins on abaxial surface; [petals forming a tubular corolla; fertile stamens 2]. Northern South America **96. *Ticorea diandra***
- Leaf blades not domatiiferous 52
52. Trichomes echinoid or stellate. Cuba, Nicaragua south to Bolivia and Brazil **89. *Angostura***
- Trichomes when present simple 53
53. Calyx more than 1 cm long, valvate 54
- Calyx less than 0.5 cm long, usually imbricate, rarely valvate 55
54. Calyx white or green; corolla densely woolly without; petals coherent; follicles with a narrow dorsal wing; seeds glabrous. Central and South America **92. *Toxosiphon***
- Calyx red or white; corolla glabrous without; petals connate; follicles without dorsal wing; seeds pubescent. Costa Rica to tropical South America **90. *Erythrochiton***
55. Carpels connate axially and at least at base laterally in flower and in fruit at maturity 56
- Carpels free in flower and in mature fruit 59
56. Anthers lacking basal appendages 57
- Anthers appendaged at base 58
57. Petals linear, acute; anthers acute at apex, free; cotyledons conduplicate, thick, stiff. South America **94. *Rauia***
- Petals oblanceolate, rounded; anthers rounded at apex, coherent; cotyledons conduplicate-plicate, flexible. Eastern Brazil **84. *Conchocarpus cuneifolius***
58. Inflorescence a double cincinnus; fruit densely tomentulose. Tropical South America **93. *Neoraputia*** p.p.
- Inflorescence a thyrse; fruit glabrous. Costa Rica to Brazil and Bolivia **95. *Galipea***
59. Anthers sterile at base above point of attachment to filament and appendaged below this point; appendages free; usually 5 carpels maturing 60
- Anthers not sterile at base, appendaged or not; appendages where present connate; usually fewer than 5 carpels (often only 1) maturing 61
60. Leaves 5–7-foliolate; inflorescence a long-pedunculate dichasium branched 1–several times, with monochasial branches. Colombia, Peru, Brazil **87. *Raputiarana***
- Leaves 3-foliolate; inflorescence terminal, raceme-like. Northeast Brazil **88. *Sigmatanthus***
61. Petals forming a tubular corolla, red. Colombia **85. *Naudinia***
- Petals free, usually white, rarely pink or yellow 62
62. Anthers connate by the appendages. Eastern Brazil **86. *Andreadoxa***
- Anthers free. From Nicaragua to Bolivia and Brazil **84. *Conchocarpus*** p.p.

Key to the Genera of Aurantioideae

1. Fruit a juicy or more rarely a dry berry with a thin-walled or leathery pericarp always made up of thin-walled cells (*Burkillanthus* with thin woody endocarp) 2
- Fruit with hard pericarp made up of lignified cells 26
2. Ovary locules each containing very many (<15) ovules; [large-fruited trees with spiny twigs and 1–3-foliolate leaves] 3
- Ovary locules each containing 1–12 ovules 4
3. Ovary 5-locular; fruit ovoid, with a thin peel dotted with numerous oil-glands covering a thin, hard, woody endocarp; seeds glabrous **140. *Burkillanthus***

- Ovary 8–10-locular; fruit longitudinally ribbed, with very thick, leathery rind and soft endocarp; seeds hairy **139. *Swinglea***
4. Stamens twice as many as petals 5
- Stamens 3–4 (or sometimes many more) times as many as petals; [leaves 1-foliolate or simple] ***Citrus*** s.l. 22
5. Ovary locules each containing 4–8(–10) ovules 6
- Ovary locules each containing 1–2 ovules 9
6. Leaves imparipinnate; fruit up to 10 cm long; seed coat densely scaly **127. *Merrillia***
- Leaves simple; fruit up to 4.5 cm long; seeds glabrous 7
7. Ovules 4 per locule; fruit strongly angled; seeds very large, to 3.3 cm long, elongate, reniform, pointed at attached end **129. *Merope***
- Ovules 6–8 per locule; fruit not angular; seeds less than 2 cm long, not pointed at attached end 8
8. Petals about 0.5 cm long; seeds spotted **130. *Monanthocitrus***
- Petals 2.0–3.5 cm long; seeds not spotted **131. *Wenzelia***
9. Twigs unarmed; leaves imparipinnate with alternate leaflets, or rarely 1- or 3-foliolate 10
- Twigs armed with axillary solitary or paired spines (rarely so in *Atalantia* '*Severinia*'); leaves simple, 1- or 3-foliolate, or imparipinnate but then the leaflets opposite 14
10. Petals valvate in bud; ovary locules often twisted; cotyledons thin, convolute and/or folded **122. *Micromelum***
- Petals imbricate; ovary locules not twisted; cotyledons thick, plano-convex, not convolute 11
11. Style persistent in fruit; ovules 1(2) per locule; buds and new growth covered with dense, fine-woolly, rust-coloured pubescence; [leaves 1–3-foliolate or impari-4-jugate] **123. *Glycosmis***
- Style caducous; ovules 2(1) per locule; new growth without dense fine-woolly pubescence 12
12. Flower buds globose or broad-ovoid; ovary usually borne on glabrous gynophorous disk; style short and thick; stigma flattened, scarcely broader than style **125. *Clausena***
- Flower buds subcylindric or ellipsoid to obovoid; ovary surrounded by disk; style slender, longer than ovary; stigma broader than style, usually capitate 13
13. Inflorescences few(–10)-flowered; stamen filaments filiform; seeds villous or glabrous; stem and root bark straw-coloured or whitish **126. *Murraya***
- Inflorescences many-flowered; stamen filaments dilated; seeds glabrous; stem and root bark dark brown **124. *Bergera***
14. Leaves imparipinnate 5–9-foliolate; petioles and rhachis segments broadly winged 15
- Leaves 1–3-foliolate or simple; petioles not or only narrowly winged 16
15. Ovules 1 or 2 per locule; fruit 0.6–1.5 cm in diam. Asia **143. *Naringi***
- Ovule 1 per locule; fruit 2–3 cm in diam. Africa **144. *Citropsis***
16. Petioles without pulvinoid structure, not articulated with blade 17
- Petioles articulated with leaf blade 19
17. Leaves simple or 3-foliolate; spines paired; flowers solitary or in groups of 2 or 3 **128. *Triphasia***
- Leaves simple; spines solitary or 0; flowers in racemes, corymbs, panicles, or fascicles 18
18. Spines abundant; leaves thick, drying to velvety grey-green; flowers in racemes; fruit 2.5 cm diam.; pulp vesicles 0 **132. *Pamburus***
- Spines rare; leaves not velvety grey-green when dry; flowers mostly in corymbs; fruit 0.6–1.2 cm diam.; pulp vesicles sessile **146. *Atalantia*** p.p.
19. Clambering lianas with recurved spines 20
- Erect shrubs or trees; spines straight 21
20. Leaves palmately 3-foliolate; petioles very long **133. *Luvunga***
- Leaves 1-foliolate; petioles short **134. *Paramignya***
21. Leaves 3-, 2- or 1-foliolate, articulated with winged petioles **141. *Pleiospermium***
- Leaves 1-foliolate or simple, articulated with wingless petioles **146. *Atalantia*** p.p.
22. Unarmed trees or shrubs 23
- Armed with solitary or paired axillary spines 24
23. Leaves blunt, usually emarginate at tip, glaucous; stamens 3–4 times as many as petals; ovary 2–7-loculate **147. *Citrus*** ('*Oxanthera*')
- Leaves acuminate at tip, not glaucous; stamens 10–20 times as many as petals; ovary 14–16-loculate **147. *Citrus*** ('*Clymenia*')
24. Leaves very thick and leathery, with appressed, greyish indumentum, isobilateral, one edge to light; stamens (3)4 times as many as petals; ovules 2 per locule **147. *Citrus*** (*C.* '*Eremocitrus*' *glauca*)
- Leaves not particularly thick or leathery, dorsiventral; stamens 4 or more times as many as petals; ovules (2)4–8 per locule 25
25. Deciduous; leaves 3-foliolate; [flowers produced from scaly buds on last year's twigs] **147. *Citrus*** ('*Poncirus*')
- Mostly evergreen; leaves 1-foliolate **147. *Citrus***
26. Ovary with axile placentas, 6–15-locular; leaves 3-foliolate or simple 27
- Ovary paracarpous with numerous parietal placentas, 1-locular; leaves pinnate 30
27. Leaves simple; [stamens 2–3 times as many as petals; ovary 5–8-loculate]. Africa **137. *Aeglopsis***
- Leaves trifoliolate 28
28. Stamens twice as many as petals; disk small; [ovary 5–8-loculate]. Africa **135. *Balsamocitrus***
- Stamens more numerous; disk large 29
29. Stamens more than 4 times as many as petals; ovary 8–20-loculate; seeds woolly. Indomalayan region **136. *Aegle***
- Stamens 3 or 4 times as many as petals; ovary 8 (–10)-loculate; seeds glabrous. Africa **138. *Afraegle***

30. Stamens about twice as many as petals; anthers twice as long as the filaments; filaments unappendaged; seeds pilose. Asia **145. *Limonia***
- Stamens about 4 times as many as petals; anthers much shorter than filaments; filaments with basal appendages; seeds glabrous. Asia **147. *Citrus*** ('*Feroniella*')

Genera of Rutaceae

In the generic descriptions, the plants are assumed to be evergreen, flowers to be actinomorphic, and flowers bisexual unless stated otherwise.

I. Subfam. Rutoideae Arn. (1832).

Toddalioideae K. Koch (1869).
Flindersioideae Luerss. (1881).

A. Bosistoa Alliance (Bosistoa tribe, informally proposed by Hartley 2001b).

Leaves usually alternate; flowers usually diplostemonous; ovules sometimes 8–4(2–1); fruit follicular; dorsilateral endocarp discharged with seed; ventral endocarp persistent; seed coat pergamentaceous, lacking a sclerotesta.

Five genera, Australasia, Malesia.

1. *Bosistoa* F. Muell. ex Benth.

Bosistoa F. Muell. ex Benth., Fl. Austral. 1: 359 (1863); Hartley, J. Arnold Arb. 58: 416–436 (1977), rev.
Pagetia F. Muell. (1866).

Trees; trichomes simple; buds enclosed by two pairs of scales. Leaves opposite, imparipinnate (1–6-jugate), or 1-foliolate. Inflorescences terminal or terminal and upper-axillary panicles. Flowers 5-merous, diplostemonous; petals valvate; disk pulvinate to columnar; ovarioles proximally connate and joined laterally in a common style; ovules 4–6 per locule. Fruit of 1 or 2 basally connate follicles. Seed solitary; endosperm 0; cotyledons plano-convex, suborbicular or broadly elliptic.

Four spp., eastern Australia, lowland rainforests.

2. *Bouchardatia* Baill.

Bouchardatia Baill., Adansonia 7: 350 (1867).

Shrub or small tree; trichomes simple. Leaves opposite, imparipinnate (1–2-jugate) or 1- 2- or 3-foliolate, leaflets with pocket-domatia. Inflorescences terminal or terminal and upper-axillary panicles. Flowers 4-merous, diplostemonous; petals imbricate; disk gynophoroid; ovarioles proximally connate, joined subapically in a common style; ovules 6–8 per locule. Fruit of 1–4 basally connate follicles. Seeds solitary; endosperm scant or 0; cotyledons plano-convex, elliptic.

One sp., *B. australis* Baill., eastern Australia.

3. *Acradenia* Kippist

Acradenia Kippist, Proc. Linn. Soc. Lond. 2: 201 (1853); Hartley, J. Arnold Arb. 58: 171–181 (1977), rev.

Shrubs or trees; trichomes simple; buds naked. Leaves opposite, digitately 3-foliolate. Inflorescences terminal and/or axillary, corymbiform panicles. Flowers 5(–6)-merous, diplostemonous; petals imbricate; disk pulvinate to columnar; ovarioles (4)5, each with a prominent upper abaxial gland, proximally connate, joined laterally in a common style; ovules 2 per carpel. Fruit of 1–5 basally connate follicles. Seed solitary; endosperm scant; cotyledons plano-convex, ovate. $n = 19$.

Two spp., eastern Australia and Tasmania.

4. *Dinosperma* T.G. Hartley Fig. 66

Dinosperma T.G. Hartley, Adansonia III, 19: 190 (1997).

Shrubs or trees; trichomes simple. Leaves (sub) opposite or whorled, occasionally alternate, imparipinnate or paripinnate (1–2-jugate), digitately 3-foliolate, or 1-foliolate. Inflorescences terminal and/or axillary panicles. Flowers 4-merous, diplostemonous; petals valvate or imbricate; disk annular, pulvinate, or columnar; ovarioles proximally connate, joined subapically in a common style; ovules 2 per locule. Fruit of 1–4 basally connate follicles; endocarp separate and discharged with seed, or rarely adnate at base. Seeds 1(2) per follicle; testa pergamentaceous to coriaceous but lacking a sclerotesta; endosperm scant or 0; cotyledons flattened, convolute and folded in seed, transversely elliptic in seedling.

Four spp., eastern Australia.

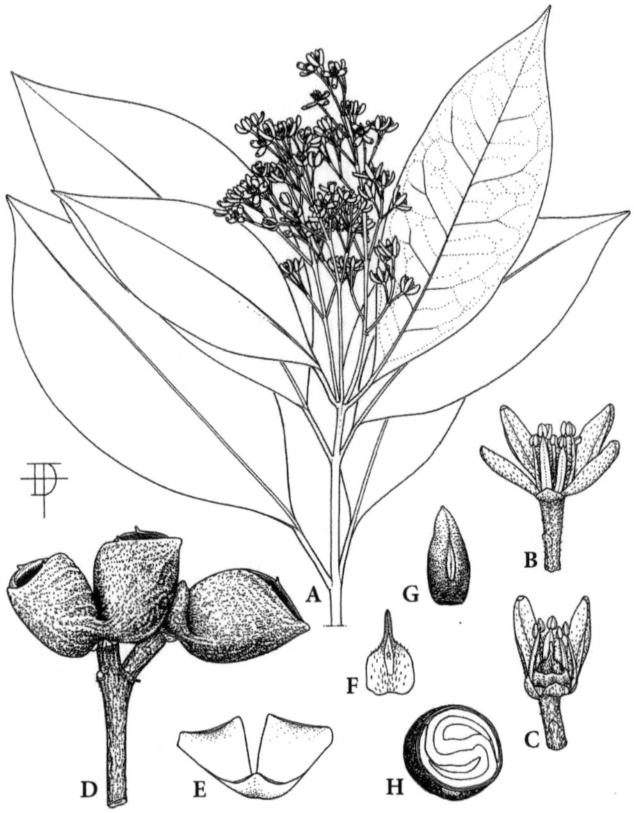

Fig. 66. Rutaceae. *Dinosperma melanophloia*. **A** Flowering twig. **B** Functionally male flower. **C** Functionally female flower. **D** Fruits. **E** Dorsilateral endocarp. **F** Ventral endocarp. **G** Seed. **H** Transverse section of seed. (Hartley 1997; artwork by D. Fortescue)

5. *Lunasia* Blanco

Lunasia Blanco, Fl. Filip. ed. 1: 783 (1837); Hartley, J. Arnold Arb. 48: 460–475 (1967), rev.

Dioecious shrubs or small trees; trichomes fasciculate to lepidote. Leaves alternate, simple. Inflorescences axillary, panicle- or raceme-like, with flowers in capitate clusters. Flowers functionally unisexual, 3-merous, haplostemonous; petals valvate; disk 0 (?); ovarioles (rudimentary in male flowers) proximally connate, joined apically by 3 basally coherent stylodia; ovules 1 per locule. Fruits 1–3 basally connate follicles. Seeds solitary; endosperm scant or 0; cotyledons plano-convex, elliptic.

One sp., *L. amara* Blanco, Philippines south to Java, New Guinea and north-eastern Australia.

B. Euodia Alliance (Zanthoxyleae Hook. 1862 p. p., Toddalioideae Engl. 1893 p.p.).

Mostly rainforest trees or shrubs; flowers mostly 5(4)-merous, diplo-(haplo-)stemonous; ovarioles distinct or basally connate and stylodia joined in a common style, or gynoecium syncarpous; ovules 2, 1, or rarely more than 2; fruits dehiscing or drupaceous; dorsilateral endocarp in dehiscent fruits usually discharged with seed; ventral endocarp sometimes well developed; seeds with or rarely without sclerotesta, sometimes with sarcotesta, albuminous or exalbuminous. Thirty two genera, nine of which monotypic, mostly in the Old World, one genus in South America.

6. *Boronella* Baill.

Boronella Baill., Adansonia 10: 302 (1872); Hartley, Bull. Mus. Natl. Hist. Nat. Paris B IV, Adansonia 17: 107–111 (1995), key.

Shrubs; branchlets with cortex articulated at nodes; trichomes, if present, simple. Leaves opposite or whorled, simple. Inflorescences terminal or terminal and upper-axillary, condensed panicles. Flowers 4-merous, diplostemonous; disk present; ovarioles proximally connate, joined apically in a common style; ovules 1 or 2 per locule. Fruit of 1–4 basally connate follicles; endocarp elastically discharged with the seed. Seed solitary; testa thin and brittle; endosperm copious; embryo straight; cotyledons plano-convex, suborbicular, wider than hypocotyl.

About six spp., New Caledonia.

7. *Euodia* J.R. Forst. & G. Forst. Fig. 62

Euodia J.R. Forst. & G. Forst., Char. Gen. Pl.: 7 (1775), ed. 2: 13 (1776); Hartley, Allertonia 8, 1: 1–65 (2000), rev.

Shrubs or trees; trichomes simple. Leaves opposite, digitately 3-foliolate or 1-foliolate. Inflorescence axillary (subterminal) panicles or racemes, or flowers solitary. Flowers 4-merous; petals valvate, hooked at apex; stamens 8 or 4; staminal filaments sublinear; anthers apiculate; disk annular or cupular; ovarioles proximally connate, subpically with contiguous or coherent stylodia; ovules 2 per carpel. Fruit of 1–4 basally connate follicles each with a short stylar beak; endocarp elastically discharged with seed at dehiscence. Seeds solitary or 2; testa thin and brittle, dull to lustrous, with sclerotesta; some spp. with ventral endocarp attached to hilum; endosperm copious; cotyledons flattened, elliptic.

Seven spp., New Guinea and north-eastern Australia eastwards to Samoa and Niue.

Traditionally (Engler 1931), *Euodia* comprised about 120 spp. but Hartley (1981) recognised the heterogeneity of this assemblage and reinstated *Tetradium* (which had been treated as a section of *Euodia* by Engler) and transferred the bulk of the *Euodia* to *Melicope*.

8. ***Myrtopsis*** Engler

Myrtopsis Engler in Engler & Prantl, Nat. Pflanzenfam. III, 4: 137 (1896).

Shrubs; young branches with stellate to lepidote trichomes. Leaves opposite, simple. Inflorescences paniculate, terminal or terminal and upper-axillary. Flowers 4- or 5-merous, diplostemonous; disk 10-lobate; ovarioles proximally more or less connate and joined laterally or towards base in a common style; ovules 2 or 1 per carpel. Fruit of 1–5 basally connate follicles or grading to almost completely syncarpous and then forming a 5-loculed loculicidal capsule; endocarp elastically discharged with the seed. Seed solitary; testa dull to lustrous.

About 9 spp., New Caledonia.

9. ***Brombya*** F. Muell.

Brombya F. Muell., Fragm. 5: 4 (1865).

Shrubs or trees; trichomes simple. Leaves opposite, 1-foliolate. Inflorescences axillary, paniculate. Flowers 4-merous, diplostemonous; sepals adaxially glabrous, persistent in fruit; petals valvate or narrowly imbricate; stamen filaments flattened, broad and ± petaloid; antepetalous anthers usually sterile; ovarioles proximally connate, joined (sub) apically in a common style; ovules 2 per carpel. Fruit of 1–4 basally connate follicles; endocarp discharged with seed when fruit dehisces. Seeds solitary or 2; testa thin and brittle, lustrous, with sclerotesta; ventral endocarp attached to hilum; endosperm copious; cotyledons flattened, elliptic.

Two spp., north-eastern Australia.

10. ***Medicosma*** Hook.f

Medicosma Hook.f. in Benth. & Hook., Gen. Pl. 1: 296 (1861); Hartley, Aust. J. Bot. 33: 27–64 (1985).

Monoclinous or dioecious shrubs or trees; trichomes simple to stellate or scale-like; buds naked. Leaves (subalternate) opposite to (sub) verticillate, digitately 1- or 3(4, 5)-foliolate. First-formed inflorescences upper-axillary, the basal portion of these usually perennial, becoming ± woody and producing additional lower-axillary or ramuligerous inflorescences. Flowers bisexual or functionally unisexual, 4-merous; stamens (rudimentary in female flowers) 8 (4); disk pulvinate to cupular; ovarioles (rudimentary in male flowers) proximally connate, joined (sub) apically in a common style; ovules 2 (1) per carpel. Fruit of 1–4 1- or 2-seeded follicles; endocarp discharged with seed when fruit dehisces. Seeds solitary or 2; testa thin and brittle, dull to rather lustrous, with sclerotesta; some spp. with ventral endocarp attached to hilum; endosperm copious; cotyledons flattened, elliptic.

Twenty five spp., southern New Guinea, eastern Australia, and New Caledonia. Rainforests, littoral forests, maquis, and lowland savannahs.

11. ***Perryodendron*** T.G. Hartley

Perryodendron T.G. Hartley, Adansonia III, 19: 198 (1997).

Tree, sometimes vast; trichomes simple. Leaves opposite, 1-foliolate. Inflorescences axillary panicles. Flowers 4-merous, diplostemonous; petals valvate; disk 8-crenulate; ovarioles proximally connate, joined (sub)apically in a common style; ovules 2 per locule. Fruit of 1–4 follicles; endocarp persistent in dehisced fruit, adnate at least dorsally; seeds 1or 2 per follicle, exalate; testa thin, brittle, with inner layer of black sclerenchyma; endosperm copious; embryo straight.

One sp., *P. parviflora* (C.T. White) T.G. Hartley, Moluccas, New Guinea and New Britain.

12. ***Tetractomia*** Hook.f.

Tetractomia Hook.f., Fl. Brit. Ind. 1: 490 (1875); Hartley, J. Arnold Arb. 60: 127–153 (1979), rev.
Terminthodia Ridley (1915).

Shrubs or trees; trichomes simple or 0; buds naked. Leaves opposite, 1-foliolate. Inflorescences upper-axillary panicles, sometimes reduced to 1 or few flowers. Flowers 4-merous; petals valvate or slightly imbricate; stamens 4, alternating with 4 staminodes; disk flattened or somewhat pulvinate; ovarioles proximally connate, stylodia twisted together (sub)apically in a style; ovules 2 per carpel. Fruit of 1–4 basally connate, boat-shaped follicles; endocarp persistent in dehisced fruit, adnate. Seeds 1 or 2 per follicle, winged; testa pergamentaceous, dull to lustrous, extended towards the apex of the follicle as a hyaline wing; endosperm copious, fleshy; embryo straight, elliptic.

Six spp., Sumatra and Malay Peninsula eastwards to Solomon Islands. From near sea level to 3,300 m, in lowland, heath and montane forest and subalpine habitats.

13. *Melicope* J.R. Forst. & G. Forst. Fig. 67

Melicope J.R. Forst. & G. Forst., Char. Gen. Pl.: 28 (1775) and ed. 2: 55 (1776); Hartley, Allertonia 8, 1: 1–328 (2001), rev.
Euodia J.R. Forst. & G. Forst., Char. Gen. Pl.: 7 (1775), ed. 2: 13 (1776), pro maj. parte.
Pelea A. Gray (1854); see Hartley & Stone, Taxon 38: 119–123 (1989).
Boninia Planch. (1872).
Tractocopevodia Raizada & V. Naray. (1946).
Evodiella B.L. Linden (1959).

Often dioecious shrubs or trees; trichomes simple, or fasciculate, stellate, or sublepidote. Leaves opposite or whorled, digitately 3-foliolate or 1-foliolate. Inflorescences axillary to cauligerous panicles, often reduced to few or solitary flowers. Flowers bisexual or functionally unisexual, 4-merous; petals apically hooked; stamens 4 or 8 or rarely 4–8; anthers rounded, obtuse, or mucronate at apex; stamen filaments sometimes connate into a wide tube bearing the apically free anthers; disk pulvinate to annular, cupular, or plane; ovarioles grading from proximally to completely connate, joined (sub)apically in a common style; stylodia sometimes separating as the fruit matures; ovules 2 or 1, rarely 5–8 per carpel. Fruit of 1–4 basally connate follicles or grading to completely syncarpous (the carpels united into a 4-loculed, loculicidal capsule); endocarp cartilaginous, adnate or separate, but neither it nor seed discharged when fruit dehisces. Seeds solitary or 2; testa with thick sclerotesta, sarcotesta, and shiny, black pellicle; endosperm copious; cotyledons ± flattened, elliptic. $2n = 12, 18, 19, 36$.

About 234 spp., Malagasy and Indo-Himalayan regions east to Hawaii and Marquesas Islands and south the New Zealand.

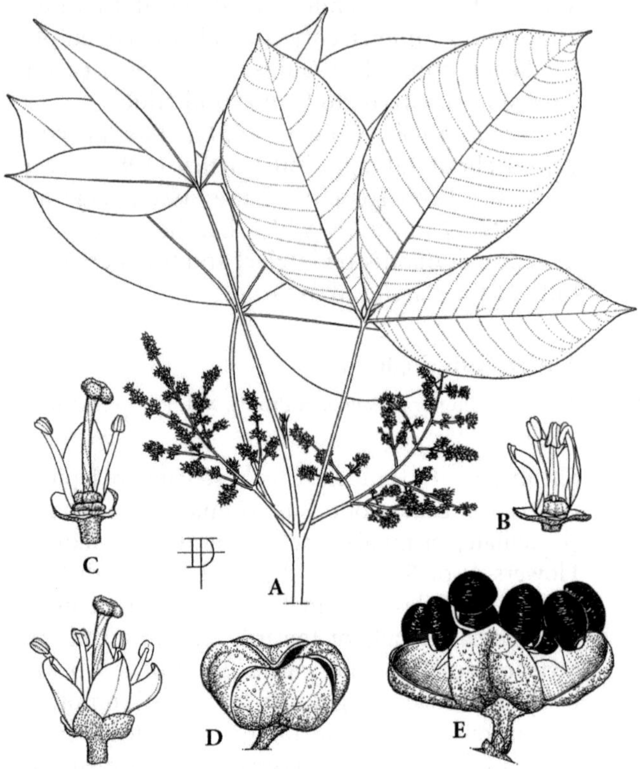

Fig. 67. Rutaceae. *Melicope latifolia*. **A** Flowering branchlet. **B** Functionally male flower, two petals and one stamen removed to show the reduced gynoecium surrounded by disk. **C** Two functionally female flowers, one with two sepals, three petals, and two stamens removed. **D** Partially dehisced fruit. **E** Fully dehisced fruit. (Hartley 2001a; artwork by D. Fortescue)

14. *Ivodea* Capuron Fig. 68

Ivodea Capuron, Adansonia II, 1: 73 (1961), and 7: 487–500 (1967), rev.

Dioecious shrubs. Leaves opposite, whorled, or alternate, 1-foliolate. Inflorescences axillary and terminal, paniculate, the females reduced and raceme-like. Flowers functionally unisexual, 4 (5)-merous, haplostemonous or rarely diplostemonous (female flowers with 0, 4 or 8 staminodes); calyx small; petals valvate-induplicate; disk 0; ovarioles proximally connate; the stylodia joined in a very short common style with a broad discoid, deeply lobed stigma (stylodia and stigmatic lobes distinct in *I. trichocarpa* Capuron); ovules 2 per carpel, collateral. Fruit of 4–5 (3–1) follicles. Seeds solitary; testa membranaceous, brittle; endosperm 0; cotyledons thick, at base narrowly emarginate.

Ten spp., Madagascar, humid evergreen and dry deciduous forests.

Capuron (1961) compared this genus with *Melicope* which, however, differs in having a disk and endosperm and, above all, in having a thick sclerotesta and sarcotesta both of which are not well developed (if at all) in *Ivodea*. He (Capuron 1967) also reported that several species of *Ivodea* from the East Coast of Madagascar have densely set leaves at the ends of the branches, among which vegetable mould and other debris accumulates and decomposes, providing a habitat for myriapodes, insects and even soil-inhabiting animals.

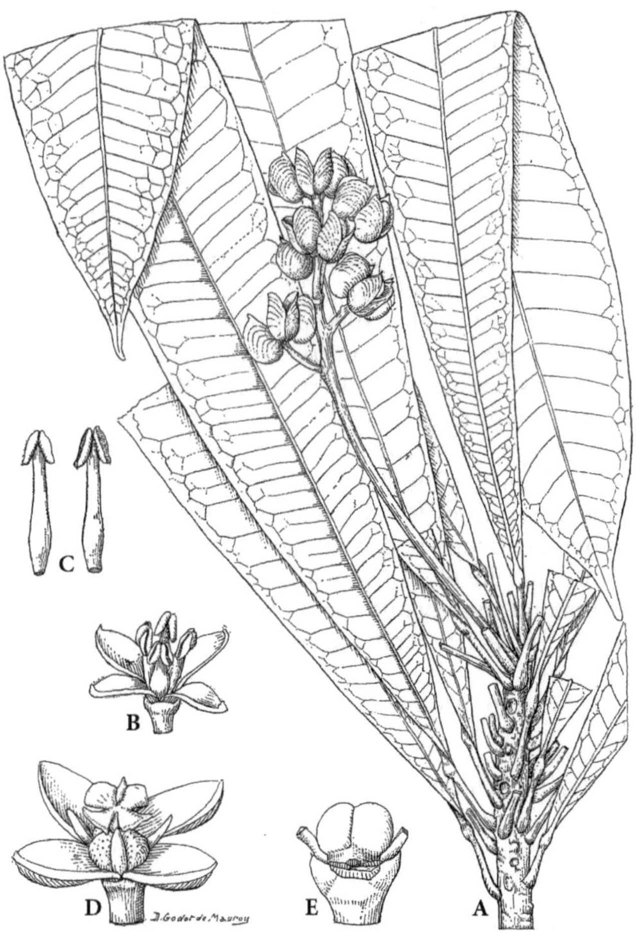

Fig. 68. Rutaceae. *Ivodea nana*. **A** Fruiting branch. **B** Male flower. **C** Stamens, ventral and dorsal view. **D** Female flower with petaloid stigmas and staminodes. **E** Ovary, stigma fallen off. (Capuron 1967; artwork by D. Godot de Mauroy)

15. *Platydesma* H. Mann

Platydesma H. Mann in Proc. Boston Soc. Nat. Hist. 10: 317 (1866); Stone, J. Arnold Arb. 43: 410–427 (1962).

Shrubs or small trees. Leaves (sub)opposite, simple. Inflorescences short axillary cymes or racemes. Flowers 4-merous, diplostemonous; stamen filaments largely connate into a wide tube bearing the free anthers; disk plane, 4- or 8-lobed; gynoecium syncarpous; ovary 4-lobed; style columnar, with linear stigmas, sometimes the stylodia separating as the fruit matures; ovules 5–8 per carpel. Fruit a dry, 4-lobed capsule with the carpels remaining connate, loculicidally dehiscent, the thin glabrous endocarps finally separable. Seeds black, shining; testa osseous; embryo thin, broad, rounded, enclosed by endosperm. $n = 18$.

Four spp. with several varieties, Hawaiian Islands.

Harbough et al. (2009) suggested on the basis of molecular findings that this genus is nested within Hawaiian *Melicope*. The large, fleshy, hermaphrodite flowers, the largely connate stamens, and copious nectar production qualify its species as a bird-pollinated offshoot of *Melicope* (W. Wagner, pers. comm., 25 March 2010), which the authors plan to integrate into an expanded concept of *Melicope*.

16. *Acronychia* J.R. Forst. & G. Forst.

Acronychia J.R. Forst. & G. Forst., Char. Gen. Pl.: 27 (1775); ed. 2: 53 (1776), nom. cons.; Hartley, J. Arnold Arb. 55: 469–523, 525–567 (1974), rev.
Pleiococca F. Muell. (1875).

Trees, shrubs or rarely climbers; trichomes simple. Leaves opposite, digitately 3- or 1-foliolate. Inflorescences axillary and/or ramuligerous, paniculate, subcorymbose, or reduced to few or one flower. Flowers 4-merous, diplostemonous; petals valvate; disk pulvinate, 8-lobed; ovarioles 4(–8), grading from proximally to completely connate, joined apically in a common style; ovules 2 per carpel. Fruit indehiscent, of 4(–8) basally connate, 1-locular drupelets or grading to completely syncarpous to form a single 4- or 4–8-locular drupe. Seeds solitary or 2; testa thin and brittle to thick, with sarcotesta and sclerotesta; endoperm copious; cotyledons flattened, ovate to elliptic. $n = 17, 18$.

Fourty-eight spp., Australasia (most speciose in Australia and New Guinea), with 1 sp. ranging to Java and another to mainland Asia.

17. *Pitaviaster* T.G. Hartley

Pitaviaster T.G. Hartley, Adansonia III, 19: 200 (1997).

Shrub or tree; trichomes simple. Leaves opposite, 1-foliolate. Inflorescences axillary panicles. Flowers 4-merous, haplostemonous; petals valvate, hooked at apex; anthers rounded or emarginate at apex; disk annular; ovarioles distinct (three of them abortive and caducous), joined apically in a common style; ovules 2 per locule. Fruit a 1-loculate drupe (3 of the 4 carpels abortive and caducous) with woody mesocarp and cartilaginous endocarp. Seed solitary, testa very thin and fragile, with sclerotesta; endosperm copious; cotyledons flattened, ovate.

One sp., *P. haplophyllus* (F. Muell.) T.G. Hartley, eastern Australia.

18. *Sarcomelicope* Engl.

Sarcomelicope Engl. in Engler & Prantl, Nat. Pflanzenfam. III, 4: 122 (1896); Hartley, Aust. J. Bot. 30: 359–372 (1982), rev., & in Bull. Mus. Natl. Hist. Nat., Paris IV, B Adansonia 8: 183–189 (1986), key.
Bauerella Borzi (1897).

Dioecious shrubs or trees; trichomes simple or substellate. Leaves opposite to verticillate, 1-foliolate. Inflorescences axillary, paniculate to racemose, or flowers solitary. Flowers functionally unisexual, 4-merous, diplostemonous; staminal filaments ciliolate, those in carpellate flowers, like the petals, (sub)persistent in fruit; disk pulvinate, crenulate; ovarioles grading from distinct to completely connate, joined subapically in a common style; ovules 2 per carpel. Fruit of 1–4 distinct drupes or grading to completely syncarpous to form a 4-loculed drupe. Seeds solitary or 2; testa thin and brittle to thick, with sclerotesta; endosperm copious; cotyledons flattened, elliptic.

Nine spp., eastern Australia east to Fiji; rainforest, littoral forest, maquis, and dry scrub.

19. *Picrella* Baill.

Picrella Baill., Adansonia 10: 150 (1871); Hartley & Mabberley, Adansonia III, 25: 251–259 (2003), rev.
Zieridium Baill. (1872); Hartley, Bull. Mus. Natl. Hist. Nat. B, IV, Adansonia 5: 392 (1983).

Dioecious shrubs; trichomes simple. Leaves opposite, digitately 3- or 1-foliolate, or simple. Inflorescences axillary or axillary and infrafoliar, thyrsiform or reduced to compound or simple racemes or solitary flowers. Flowers usually functionally unisexual, 4-merous; sepals connate at base; petals distinct, valvate in bud, adaxially hooked at apex; stamens 4, rudimentary in female flowers; staminodes 4, sometimes adherent to petals; disk 4–8-lobed or undulate; ovarioles distinct or proximally connate, joined subapically in a common style; stigma 4-branched or ± peltate and 4-lobed; ovules 2 or 1 per locule. Fruit of 1–4 distinct drupes. Seeds 1 (2); testa black, shiny, with thick bony sclerotesta; endosperm copious; embryo straight; cotyledons flattened, elliptic.

Three spp., New Caledonia.

20. *Dutaillyea* Baill.

Dutaillyea Baill., Adansonia 10: 327 (1872), Hartley, Bull. Mus. Natl. Hist. Nat. B, IV, Adansonia 6: 29–35 (1984).

Shrubs or small trees; trichomes stellate to lepidote. Leaves opposite, (1)3-foliolate, with at least the terminal leaflet petiolulate. Inflorescences axillary, few-flowered panicles. Flowers 4-merous; stamens 4, alternating with 4 epipetalous staminodes; disk pulvinate, 8-lobed; gynoecium completely syncarpous; ovules 2 per locule, superposed; stigma shallowly 4-lobed, hardly differentiated from style. Fruit a 4-loculate drupe with a single stylar scar. Seeds 2 or 1 per locule; testa with thick, black sclerotesta and spongy outer layer bounded by shiny, black pellicle; endosperm fleshy; embryo straight.

Two spp., New Caledonia.

21. *Comptonella* Baker f.

Comptonella Baker f., J. Linn. Soc., Bot. 45: 281 (1921); Hartley, Bull. Mus. Natl. Hist. Nat. B, IV, Adansonia 5: 391–413 (1983), rev.

Dioecious or rarely monoecious or monoclinous shrubs or trees; trichomes stellate, lepidote or 0. Leaves opposite, 1- or 3-foliolate. Inflorescences paniculate or reduced to few-flowered racemes, fascicles, or solitary flowers, axillary, ramuligerous, or cauline. Flowers usually unisexual, tetramerous; 4 staminodes rarely present; disk 0; ovarioles grading from nearly free to almost completely connate, joined (sub)apically in a common style; ovules 2 per carpel. Fruit developing of 1–4 carpels (the undeveloped usually persistent), drupaceous, grading from nearly apocarpous to almost completely syncarpous, each carpel with a separate stylar scar. Seeds 1 or 2 per locule; testa with spongy to rather fleshy outer and bony inner layer; endosperm copious, fleshy; embryo straight. $n = 18$.

Eight spp., New Caledonia.

22. *Dutailliopsis* T.G. Hartley

Dutailliopsis T.G. Hartley, Adansonia III, 19: 210 (1997).

Tree; trichomes simple. Leaves (sub)opposite or whorled, digitately 3-foliolate. Inflorescences paniculate, axillary. Flowers 4-merous; stamens 4, alternating with 4 staminodes; disk pulvinate, 8-lobed; gynoecium completely syncarpous; stigma punctiform; ovules 2 per locule. Fruit a 4-loculate drupe; endocarp with tuberculate and serrate-winged outer surface. Seed solitary, with yellowish, persistent funiculus; testa with inner layer of black sclerenchyma; endosperm copious; embryo straight; cotyledons flattened.

One sp., *D. gordonii* T.G. Hartley, southern New Caledonia, on ultrabasic soil.

23. *Maclurodendron* T.G. Hartley

Maclurodendron T.G. Hartley, Gard. Bull. Singapore 35: 4 (1982), and ibid. pp. 1–19, rev.

Dioecious small trees; trichomes simple and fasciculate. Leaves opposite, 1-foliolate. Inflorescences axillary panicles or racemes. Flowers functionally unisexual, 4-merous, diplostemonous (pollen lacking in pistillate flowers); disk applanate or pulvinate, 8-lobate; gynoecium completely syncarpous, rudimentary in staminate flowers; style straight; stigma capitate, 4-lobate; ovules 2 per carpel. Fruit a 4-loculed drupe. Seeds 2(1) per locule; outer seed coat spongy, inner bony; endosperm copious; embryo straight or slightly bent.

Six spp., Sumatra and Malay Peninsula east to the Philippines and north to Vietnam and Hainan, from near sea level to 1,500 m.

24. *Vepris* Comm. ex A.Juss.

Vepris Comm. ex A.Juss., Mém. Mus. Paris 12: 509 (1825); Mziray, Symb. bot. Ups. 30 (1): 1–95 (1992).
Teclea Delile (1843).
Toddaliopsis Engler (1895).
Araliopsis Engler (1896).
Diphasia Pierre (1898).
Oricia Pierre (1898).
Humblotiodendron Engl. (1917).
Oriciopsis Engl. (1931).
Tecleopsis Hoyle & Leakey (1932).
Diphasiopsis F.A. Mendonça (1961).

Shrubs or trees, dioecious or polygamous, mostly evergreen. Leaves alternate or rarely subopposite, digitately 1-, 3–9(–12)-foliolate. Inflorescences terminal, axillary or cauliflorous panicles, racemes, glomerules or few-flowered fascicles. Flowers bisexual or functionally unisexual, small, relatively inconspicuous, usually 4-merous but petals occasionally 3 or 5; male flowers: stamens 4–8(–10), inserted at base of disk; pistil vestigial; female flowers: staminodes 0–8; gynoecium 1–4-carpellate, syncarpous; stigma sessile; ovules 1 or 2 per carpel. Fruit drupaceous, generally semi-fleshy, entire or lobed, with 1–4(5) usually 1-seeded cells. Seed coat thin; endosperm copious, fleshy, or more rarely sparse or 0. $n = 16$.

About 80 spp., throughout tropical Africa, including Zanzibar, Madagascar and the Mascarenes, and in tropical Arabia and southwest India where it is represented by one species each.

In the circumscription of this genus, we are following the suggestion of Mziray (l.c.). Until recently, the eight genera treated as synonyms of *Vepris* had been treated as distinct genera, although their close affinity to *Vepris* s.str. and their doubtful generic status have been noted by several workers. There is overlap in the variation of virtually all morphological traits among the nine genera, including leaf morphology, floral merism and fruit structure. The absence of an endosperm in some members of the group shows the lability of this trait in the group; for instance, an endosperm is present in some but not all of the 10 species of *Vepris* s.str. of the Flora Zambesiaca area (see Mendonça in Fl. Zambesiaca 2, 1).

25. *Coatesia* F. Muell.

Coatesia F. Muell., Fragm. 3: 26 (1862).

Trees; trichomes simple. Leaves alternate, simple. Inflorescences terminal or terminal and upper-axillary panicles. Flowers 5-merous, haplostemonous; petals imbricate, not hooked at apex; ovarioles proximally connate, laterally joined in a common style; ovules 2 per locule. Fruit of 1–5 basally connate follicles; endocarp elastically discharged with seed. Seeds solitary; testa thin and brittle, lustrous, with sclerotesta; endosperm 0; cotyledons plano-convex, transversally elliptic.

One sp., *C. paniculata* F. Muell., eastern Australia.

26. ***Neoschmidea*** T.G. Hartley

Neoschmidea T.G. Hartley, Adansonia III, 25: 7–12 (2003).

Shrubs; trichomes simple or compound. Leaves alternate, simple, sessile or subsessile. Inflorescences few-flowered axillary cymes, or flowers solitary. Flowers 5(very rarely 4)-merous, diplostemonous; sepals valvate in bud, persistent in fruit; petals distinct, imbricate or valvate in bud, hooked adaxially at apex; disk pulvinate; ovarioles proximally connate and laterally joined in the common style; stigma capitate; ovules 2 per locule, subcollateral. Fruit of 1–5 basally connate follicles, often with stylar beak and/or extra-stylar spur-like appendage; endocarp at dehiscence discharged with seed. Seeds solitary; testa thin and brittle with dense, black sclerotesta; endosperm copious; embryo straight; cotyledons plane, flattened, elliptic-oblong, about twice as wide as the hypocotyl.

Two spp., New Caledonia, in maquis on ultramafic and schistose grounds.

27. ***Halfordia*** F. Muell.

Halfordia F. Muell., Fragm. 5: 43 (1865).

Shrubs or trees; trichomes simple and fasciculate. Leaves alternate, simple. Inflorescences loose terminal or terminal and upper-axillary panicles. Flowers 5-merous, diplostemonous; petals valvate or imbricate; disk 10-costate; gynoecium 3–5-carpellate, completely syncarpous or with apical septicidal fissures; style short; ovules 1 per carpel. Fruit a 3–5-loculed drupe, sometimes with apical septicidal fissures; locules 1-seeded. Seeds solitary; testa thin and brittle, with sclerotesta; endosperm copious; cotyledons flattened, elliptic-oblong.

One sp., *H. kendack* (Montrouz.) Guillaumin, New Guinea and eastern Australia east to New Caledonia and New Hebrides.

28. ***Zanthoxylum*** L.

(*Xanthoxylum*, *Xanthoxylon* and *Zanthoxylon* are variant spellings).
Zanthoxylum L., Sp. Pl. 1: 270 (1753); Hartley, J. Arnold Arb. 47: 171–221 (1966), rev. Males. spp.; Beurton, Plant Syst. Evol. 189: 165–191 (morph.); Reynel, C. (1995). Systematics of neotropical *Zanthoxylum* (Rutaceae) with an account on the wood anatomy of the genus. Ph.D. Thesis, University of Missouri, St. Louis.
Fagara L., Syst. ed. 10: 897 (1759), nom. cons.

Dioecious, monoecious or polygamous, deciduous or evergreen trees or rarely scandent shrubs, often aculeate or spinose, sometimes with knobs tipped with spines (knobthorns); trichomes simple, bifid or stellate. Leaves alternate, (im)paripinnate to 1-foliolate, sometimes phyllodic. Inflorescences terminal panicles, axillary or infrafoliar botryoids, or cymose fascicles. Flowers functionally unisexual or bisexual; perianth of 4–9 irregularly uniseriate, undifferentiated segments, or biseriate with 3–6 distinct or connate sepals and imbricate or valvate petals; stamens 3–8 (0 in female flowers); disk small, often pulvinate, or obscure; gynoecium (1)2–5-carpelled (0 in male flowers); carpels acrostylous or more rarely anacro- to basistylous; ovarioles proximally coherent or distinct and joined apically or subapically in a common style, or stylodia distinct; the stigmas capitate, distinct or coherent; ovules 2 per carpel, collateral. Fruit of 1–5 distinct or basally connate follicles, very rarely a capsule; endocarp often separate, but neither it nor seed discharged when fruit dehisces. Seeds 1(2); testa with thick sclerotesta, sarcotesta, and shiny, black or reddish pellicle, at maturity often hanging from the carpels on slender funicles; endosperm fleshy, copious, or rarely scant; cotyledons flattened or plano-convex, orbicular to elliptic. $n = 16, 18, 32, 33, 34, 35, 36, 68$.

About 225 spp., pantropical and extending to warm-temperate regions in eastern Asia and eastern North America.

29. ***Tetradium*** Lour.

Tetradium Lour., Fl. Cochinchin.: 91 (1791); Hartley, Gard. Bull. Sing. 34: 91–131 (1981), rev.

Dioecious or monoecious, evergreen or deciduous, trees or shrubs; trichomes simple. Leaves opposite, imparipinnate, 1–9-jugate. Inflorescences terminal and/or upper-axillary corymbose panicles. Flowers mostly unisexual, 5- or 4-merous, haplostemonous; petals apically hooked; disk in male flowers conic or cylindric, in female flowers pulvinate or barrel-shaped, persistent in fruit as a short stipe; ovarioles distinct or proximally connate, joined subapically in a common style; stigma peltate, lobed; ovules 2 or 1 per carpel; rudimentary carpels in male flowers divergent, finger-like, free to base or connate towards the base. Fruit of 1–5 1- or 2-seeded follicles; endocarp and seed persistent in dehisced follicle. Seeds with sclerotesta, sarcotesta, and shiny pellicle; endosperm copious; embryo straight; cotyledons broad-elliptic, plano-convex. $n = 18$, 36, 38, 39, 40.

Nine spp., Himalayas eastwards to Japan and south to Java and Sumbawa.

30. *Phellodendron* Rupr.

Phellodendron Rupr., Bull. Cl. Phys.-Math. Acad. Imp. Sci. Saint-Pétersbourg II, 15: 353 (1857); Ma, Cao, Liu, Yu & Han, Edinburgh J. Bot. 63: 131–151 (2006), rev.

Dioecious, deciduous trees; bark corky; buds solitary, small, sunken in the base of the petioles. Leaves opposite, imparipinnate, 3–5-jugate; foliar glands restricted to the notches of the leaf margin. Inflorescences corymbose panicles, terminal and axillary. Flowers functionally unisexual, 5-merous, haplostemonous; male flowers with small disk around the rudimentary distinct carpels; female flowers with 5 staminodes; disk 0; gynoecium completely syncarpous; gynophore short; style very short or 0; stigma broadly peltate, 5-lobed, persistent; ovule 1 per carpel. Fruit a 5-angled drupe, with 5 compressed, cartilaginous pyrenes; seeds with black, crustaceous testa, endosperm scant; cotyledons flat. $n = 38$, 39, 40.

Only two of the numerous described spp. are recognised by Ma et al. (*Ph. amurense* Rupr. and *P. sinense* C.K. Schneider), south-western China northeast to south-eastern Russia (Far East) and Japan and south, discontinuously, to Taiwan, and widely cultivated in (warm-) temperate regions of the northern hemisphere.

31. *Toddalia* Juss.

Toddalia Juss., Gen. Pl.: 371 (1789), nom. cons.; Mziray, Symb. bot. Ups. 30, 1: 1–95 (1992).

Dioecious or polygamous woody scramblers or climbers; stems usually retrorse-aculeate. Leaves alternate, (1, 2)3-foliolate. Inflorescences terminal and/or axillary panicles or racemes. Flowers functionally unisexual, 4–5(6)-merous, haplostemonous; male flowers with vestigial gynoecium; female flowers: staminodes 5, gynoecium 4–7-carpellate, completely syncarpous, on short, stout, disk-like gynophore; style 0; stigma 5-lobed; ovules 2 per carpel. Fruit a 4–7-loculed drupe. Seed usually 1 per locule, glossy; testa with black sclerenchyma surrounded by compact parenchymatous tissue; endosperm copious; embryo curved. $n = 18$, 36.

One sp., *T. asiatica* (L.) Lam., from Africa east to the Philippines, often in secondary vegetation.

32. *Fagaropsis* Mildbr. ex Siebenlist

Fagaropsis Mildbr. ex Siebenlist, Forstwirtsch. Deutsch-Ostafr.: 90 (1914); Mziray, Symb. bot. Ups. 30 (1): 1–95 (1952).
Clausenopsis (Engler) Engler (1931).

Dioecious, usually deciduous shrubs or trees, the latter often buttressed. Leaves opposite, imparipinnate, 2–4-jugate. Inflorescences terminal panicles. Flowers unisexual; sepals 4; petals 4(–6); stamens 4–8 (0 in female flowers); gynoecium 3–5-carpellate, completely syncarpous; ovary subglobose, 3–5-lobed and -locular (vestigial in male flowers), at base surrounded by annular disk; style short; stigma 3–5-lobed; ovule 1 per carpel. Fruit a subglobose drupe, with usually 2–4 thin-walled pyrenes. Seeds albuminous; embryo curved.

Four spp., tropical central and eastern Africa to Somalia (2) and Madagascar (2), in upland semideciduous forests. The wood is green-yellow.

33. *Geijera* Schott

Geijera Schott, Rutac.: 7 (1834).

Trees or shrubs; trichomes simple. Leaves alternate, simple. Inflorescences terminal and/or axillary or ramigerous panicles, or flowers solitary. Flowers (3)4- or 5-merous, haplostemonous; petals valvate in bud, hooked adaxially at apex; disk thick, fleshy; ovarioles proximally connate, joined apically in a common style; ovules 2 per carpel. Fruit of 1–5 basally connate follicles, sometimes tardily dehiscent; endocarp adnate or separate, but neither it nor seed discharged when fruit dehisces. Seeds solitary; testa with thick sclerotesta, sarcotesta, and shiny black pellicle; endosperm rather scant; cotyledons somewhat plano-convex, suborbicular or broadly elliptic. $n = 54, 81$.

About six spp., New Guinea, northern, eastern and southern Australia, and New Caledonia.

34. *Orixa* Thunb. Fig. 57

Orixa Thunb., Nov. Gen. Pl. 3: 56 (1783).

Dioecious, deciduous shrub. Leaves alternate, simple, entire. Inflorescences ramigerous, subtended by caducous bud scales; male flowers in patent botryoids, females solitary on thin peduncles provided with 2–3 sterile bracts. Flowers functionally unisexual, 4-merous, haplostemonous; petals shortly clawed, imbricate, patent; disk 4-lobed, flat; male flowers: anthers ovoid, gynoecium 0; female flowers: staminodes 4; ovarioles proximally connate, the apical stylodia joined apically in a short common style; stigma capitate, 4-lobed; ovule 1 per carpel. Fruit of 2–4 basally connate follicles; endocarp cartilaginous, elastically discharged with seed when fruit dehisces. Seed solitary, ovoid-globose; testa black, thin and brittle; endosperm scant; embryo slightly curved. $n = 20$.

One sp., *O. japonica* Thunb., mountain forests of southern and middle Japan, S Korea, China.

35. *Pentaceras* Hook.f.

Pentaceras Hook.f. in Bentham & Hooker, Gen. Pl. 1: 298 (1862), nom. cons.

Trees; trichomes simple. Leaves alternate, imparipinnate, 3–7-jugate. Inflorescences terminal panicles compound of racemes. Flowers 5-merous, diplostemonous; petals valvate; ovarioles on gynophorous disk, distinct, joined laterally in a common style; ovules 2 per carpel. Fruit of 1–5 distinct, winged, 1-loculed drupes (samaras). Seeds solitary or 2; testa with thin sclerotesta and rather spongy outer tissue; endosperm scant; cotyledons plano-convex, elliptic.

One sp., *P. australe* (F. Muell.) Hook.f. ex Benth., eastern Australia.

36. *Crossosperma* T.G. Hartley

Crossosperma T.G. Hartley, Adansonia III, 19: 204 (1997).

Tree, apparently dioecious; trichomes simple. Leaves opposite, digitately 3- or 5-foliolate. Inflorescences axillary, ramigerous, or cauligerous panicles. Flowers functionally unisexual, 4-merous, diplostemonous; disk narrowly ovoid (male flowers) or barrel-shaped (female flowers); gynoecium completely syncarpous; style practically 0; stigma very broad, with deltoid, apically incised lobes; ovules 1(2) per locule. Fruit a 4- or 4–8-loculed drupe; pericarp fleshy; pyrenes laterally flattened, with cartilaginous endocarp. Seeds solitary, laterally flattened, asymmetrically obovate-triangular; testa thin, fleshy, fimbriate-winged at dorsal margin; endosperm scant.

Two spp., New Caledonia.

C. Boronia Alliance (Tribe Boronieae Bartl. 1830).

Perennial subshrubs to small trees; ovarioles proximally distinct or basally connate, rarely connate to middle or nearly to apex (*Asterolasia*); stylodia joined to form a common style; ovules 1 or 2. Follicles usually smooth, rarely transversely ridged (*Diplolaena*, some *Correa*), dehiscing explosively ventrally and somewhat dorsally, with separating dorsilateral endocarp. Seeds with copious fleshy endosperm; ventral endocarp persistent or caducous; embryo straight, narrow, the cotyledons not wider than the hypocotyl. Australian, with outliers in New Caledonia and New Zealand, eighteen genera with about 390 spp.

Boronia Group (Genera 37–39). Leaves opposite, simple or compound. Inflorescence usually cymose, or flowers solitary. Flowers pedicellate, 4-merous (*Boronia scabra* subsp. *attenuata* Paul G. Wilson often 5-merous); sepals distinct; petals distinct, spreading, rarely urceolate; stamens distinct, usually erect or inwardly curved.

37. *Boronia* Sm.

Boronia Sm., Tracts Nat. Hist., 288 (1798); Wilson, Nuytsia 1: 197–207 (1971); Weston et al., Austr. J. Bot. 32:187–203 (1984); Wilson, Nuytsia 12: 119–154 (1998), W Austr. spp.; Duretto & Ladiges, Austr. Syst. Bot. 11: 636–665 (1999); Duretto, Muelleria 12: 1–131 (1999), keys to sections *Alatae*, *Algidae* & *Valvatae*; Duretto, Muelleria 17: 19–135 (2003), key to sections *Cyanothamnus* & *Boronia* (E Australian spp.).

Subshrubs or shrubs, rarely small trees; glabrous or hairs simple or stellate; leaves simple, ternate, imparipinnate or bipinnate. Flowers diplostemonous; stamens 8 (rarely 4 caducous), all or only 4 fertile; disk prominent, usually entire. Seed: hilum elliptical or linear, raphe usually basal, sometimes forming a pulpy mass, ventral endocarp thick and persistent or membranous and caducous. $n = 7, 8, 9, 10, 11, 16, 18, 36$.

148 spp., non-arid Australia, including Tasmania, in heath, woodland, forest. Six sections are recognised: sect. *Cyanothamus* (Lindl.) F. Muell. (23 spp.); sect. *Imbricatae* Engler (1 sp.); sect. *Boronia* (58 spp. in 2 series); sect. *Alatae* Duretto (1 sp.); sect. *Algidae* Duretto (3 sp.); sect. *Valvatae* (Benth.) Engl. (62 spp. in 5 subsections, 9 series, 8 subseries).

38. *Zieria* Sm.

Zieria Sm., Trans. Linn. Soc. Lond. 4: 216 (1798); Armstrong, Austr. Syst. Bot. 15: 277–463 (2002), rev. & biol.; Duretto & Forster, Austrobaileya 7: 473–544 (2007), rev. Queensland spp.

Shrubs to small trees; hairs simple or stellate; leaves simple or ternate. Flowers haplostemonous; stamens antesepalous, all fertile; disk with 4 distinct gland-like lobes opposite stamens. Seed: hilum thin, raphe basal, ventral endocarp thick, persistent. $n = 18, 36$.

Circa 60 spp., eastern Australia, including north Queensland and Tasmania, in heath, woodland, eucalypt forest and rainforest, 1 sp., *Z. chevalieri* Virot in serpentine heath (maquis) in New Caledonia.

39. *Neobyrnesia* J.A.Armstr.

Neobyrnesia J.A.Armstr., Telopea 1: 399 (1980).

Shrubs; hairs simple; leaves simple. Flowers haplostemonous; stamens antesepalous, all fertile; disk entire, narrow. Seed: hilum linear and in a shallow groove, raphe basal, ventral endocarp thick, persistent. $n = 12$.

Only one sp., *N. suberosa* J.A.Armstr., dissected sandstone country, the Northern Territory, Australia.

Correa Group (Genus 40). Leaves opposite, simple. Inflorescence cymose, or flowers solitary. Flowers pedicellate, 4-merous; calyx hemispherical, sometimes lobed; petals connate and tubular, sometimes becoming free, and then sometimes spreading.

40. *Correa* Andr.

Correa Andr., Bot. Repos. 1: t. 18 (1798); Wilson, Trans. R. Soc. South Australia 85: 21–53 (1961), rev.; Wilson, Nuytsia 12: 89–105 (1998), rev.

Shrubs or small trees; hairs stellate. Flowers diplostemonous; calyx connate; stamens all fertile; filaments distinct; disk 8-lobed. Seed: hilum central and elliptical, raphe basal, ventral endocarp thick, persistent. $n = 16$.

Eleven spp., south-eastern Australia, including Tasmania and south-eastern Western Australia, in forest and heath.

Eriostemon Group (Genera 41–54). Leaves alternate, simple. Inflorescence axillary or terminal, cymose or umbellate, or flowers solitary. Flowers sessile or pedicellate, 5-merous (*Philotheca virgata* (Hook.f.) Paul G. Wilson 4-merous); sepals distinct or connate, rarely 0; petals distinct or rarely connate (*Leionema sympetalum* (Paul G. Wilson) Paul G. Wilson, *Nematolepis phebalioides* Turcz., *Philotheca tubiflora* A.S. George), spreading or rarely tubular. Stamens distinct or connate, erect or spreading.

41. *Eriostemon* Sm. Fig. 69

Eriostemon Sm., Trans. Linn. Soc. Lond. 4: 221 (1798); Wilson, Nuytsia 12: 241–242 (1998); Bayly et al., Austr. Syst. Bot. 11: 13–22 (1998).
Eriostemon p.p., quoad sect. *Eriostemon*, Wilson, Nuytsia 1: 19–25 (1970), rev.

Shrubs or small trees; hairs simple or stellate, sometimes lepidote on petals; leaves entire. Flowers solitary, axillary, pedicellate, diplostemonous; prophylls minute; sepals distinct; petals spreading, imbricate, mostly 5-veined from base, stellate lepidote, caducous; stamens all fertile, erect or pyramidally arranged, included; filaments distinct, woolly-ciliate; anthers glabrous, apiculate or not, non-glandular; disk thick, entire. Seed: hilum elliptic, raphe in centre of adaxial face, ventral endocarp thick, persistent. $n = 17$.

Two spp., eastern Queensland and New South Wales, Australia, in heathland, woodland and eucalypt forest.

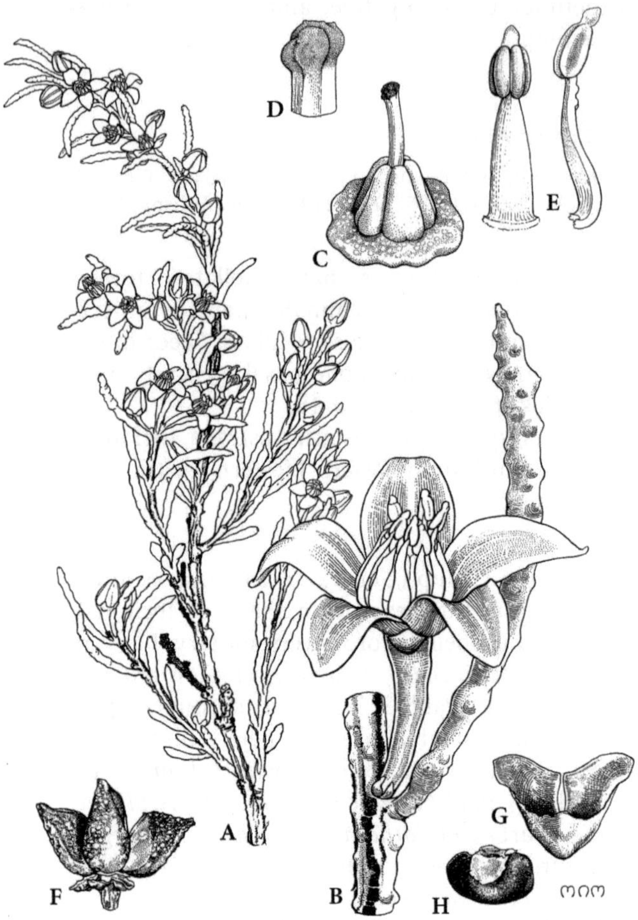

Fig. 69. Rutaceae. *Eriostemon wonganensis*. **A** Flowering branch. **B** Flower. **C** Gynoecium and disk. **D** Stigma. **E** Stamens. **F** Fruit. **G** Dorsilateral endocarp. **H** Seed with attached ventral endocarp. (Paul G. Wilson 1982; artwork by Margaret Wilson)

42. *Crowea* Sm.

Crowea Sm., Trans. Linn. Soc. Lond. 4: 222 (1798); Wilson, Nuytsia 1: 14–19 (1970), rev., and ibid. 11: 429–430 (1997), rev.

Shrubs; vegetative parts glabrous or branches puberulous; leaves entire or toothed. Flowers solitary, axillary or terminal to a short branch, pedicellate; prophylls minute, diplostemonous; sepals distinct, imbricate; petals spreading, imbricate, 1-veined, glabrous, persistent; stamens all fertile, pyramidally arranged, included; filament distinct, linear, margins pilose; anther dorsifixed, sparsely pilose, apiculum large, bearded, non-glandular; disk narrow. Seed: hilum a shallow groove, raphe basal, fleshy, ventral endocarp thick, persistent. $n = 19$.

Three spp., one in south-western Western Australia, two in south-eastern Australia; in heathland, woodland and eucalypt forest.

43. *Philotheca* Rudge

Philotheca Rudge, Trans. Linn. Soc. Lond. 11: 298 (1816); Wilson, Nuytsia 12: 239–265 (1998), key & synopsis.
Eriostemon Sm. p.p. [except sect. *Eriostemon* (= *Eriostemon* s.str.)], including the sections *Nigrostipulae*, *Erionema*, *Osmanthos*, *Corynonema*, *Cyanochlamys* and *Gymnanthos*, as revised by Wilson, Nuytsia 1: 19–60 (1970).

Shrubs or small trees, glabrous or hairs simple or stellate; leaves entire. Inflorescence axillary or terminal, a few- to many-flowered cyme, rarely a cluster or raceme, or flowers solitary; bracts small. Flowers pedicellate, diplostemonous; sepals distinct; corolla spreading, rarely tubular; petals 1-veined, imbricate or rarely valvate; stamens all fertile, erect or pyramidally arranged, included; filaments distinct or connate in lower half, linear, usually pilose; anthers glabrous, apiculum minute, terminal glands 0, 2–many; disk usually narrow. Seed: hilum elliptic to linear, raphe variable; ventral endocarp thick and persistent or membranous and caducous. $n = 14$, 28.

53 spp. in southern Australia, including Tasmania; in heathland, woodland and eucalypt forest. Four sections recognised: sect. *Philotheca* (33 spp.); sect. *Corynonema* (Paul G. Wilson) Paul G. Wilson (3 spp.); sect. *Cyanochlamys* (F. Muell.) Paul G. Wilson (2 spp.); sect. *Erionema* (F. Muell.) Paul G. Wilson (15 spp.).

44. *Geleznowia* Turcz.

Geleznowia Turcz., Bull. Soc. Imp. Naturalistes Moscou 22: 12, t. 1 (1849).

Glabrous shrubs; leaves entire. Flowers terminal, solitary or in small clusters, surrounded by petaloid bracts often exceeding flower. Flowers diplostemonous; sepals distinct, larger than petals; petals spreading, imbricate; stamens all fertile, bent to one side of ovary, included or equal to petals, glabrous; filaments distinct, filiform; anthers versatile, apiculum 0, terminal gland 0; disk 0. Seed: hilum circular, raphe large and filling the adaxial margin, ventral endocarp thick, persistent. $n = 14$.

One sp., *G. verrucosa* Turcz., south-western Western Australia, in heathland.

45. *Asterolasia* F. Muell.

Asterolasia F. Muell., Trans. Phil. Soc. Victoria 1: 9 (1855); Wilson, Nuytsia 6: 7–8 (1987).
Urocarpus J.Drumm. ex Harv. (1855); Wilson, Nuytsia 1: 206–207 (1971), synopsis.
Asterolasia sect. *Urocarpus* (J.Drumm. ex Harv.) Benth. (1863).

Shrubs; hairs stellate; leaves entire. Inflorescences axillary or terminal, umbel-like cymes, subtended by leafy or petaloid bracts. Flower pedicellate; sepals minute or 0, distinct; petals spreading, induplicate-valvate; stamens 10(–25), all fertile, divergent, included; filaments filiform, distinct, glabrous or rarely stellate hairy; anthers basifixed, glabrous, minutely apiculate or not; terminal gland 0–1; disk inconspicuous; carpels 1–5. Seed: hilum short and linear, raphe minute, ventral endocarp caducous. $n = 13, 14$.

18 spp., southern Australia, in heath and *Eucalyptus* woodland or forest.

46. *Drummondita* Harv.

Drummondita Harv., Hooker's J. Bot. Kew Gard. Misc. 7: 53 (1855); Wilson, Nuytsia 12: 86–87 (1998).
Philotheca sect. *Drummondita* (Harv.) F. Muell. (1883).

Shrubs, glabrous, or hairs simple; leaves entire. Flowers terminal, usually solitary, ebracteolate, sessile or shortly pedicellate, diplostemonous; sepals distinct; corolla tubular; petals imbricate, glabrous; stamens 5, antepetalous, erect, exserted, alternating with 5 slightly longer staminodes; filaments 3/4-connate, densely sericeous outside, variably pilose within; anthers dorsifixed, usually with some hairs and with minute, non-glandular apiculum; disk narrow. Seed: hilum a large cavity, ventral endocarp persistent. $n = 14$.

Nine spp., one in The Northern Territory and Queensland, eight in south-western Western Australia. Found mainly in heath and woodland, often in rocky situations.

47. *Leionema* Rudge

Leionema (F. Muell.) Paul G. Wilson, Nuytsia 12: 270–277 (1998).
Eriostemon sect. *Leionema* F. Muell. (1862).
Eriostemon sect. *Chorilaenopsis* F. Muell. (1862).
Phebalium sect. *Leionema* (F. Muell.) Benth. (1863), Wilson, Nuytsia 1: 99–118 (1970), rev.

Shrubs or small trees, glabrous or hairs simple or stellate; leaves entire or toothed. Inflorescence axillary or terminal, cymose or flowers solitary, prophylls minute. Flowers pedicellate, diplostemonous; sepals distinct; corolla spreading or tubular; petals valvate; stamens all fertile, divergent, subequal to petals or exserted; filaments distinct, glabrous; anthers versatile, apiculum usually 0, non-glandular; disk entire. Seed: hilum linear to narrowly elliptic; raphe fleshy, basal or sub-basal; ventral endocarp thick, persistent. $n = 16$.

26 or 27 spp., 25 or 26 in eastern Australia, including north Queensland and Tasmania, where found in heathland, woodland and eucalypt forest, and one, *L. nudum* (Hook.) Paul G. Wilson, confined to forests on the North Island of New Zealand.

48. *Muiriantha* C.A.Gardner

Muiriantha C.A.Gardner, J. & Proc. R. Soc. Western Australia 27: 181 (1942).

Shrub; hairs stellate and simple; leaves entire. Inflorescence terminal or axillary, 2–4-flowered cyme or flower solitary; prophylls foliaceous. Flowers pedicellate, diplostemonous; sepals distinct; corolla tubular; petals imbricate; stamens all fertile, erect, equal to petals; filaments distinct, slender with a small densely pilose scale just above base on inner side; anthers versatile, apiculate minute, terminal gland present; disk small, crenate. Seed: hilum superficial, linear, raphe small. $n = 14$.

One sp., *M. hassellii* (F. Muell.) C.A.Gardner, south-western Western Australia, in eucalypt woodland over heath.

49. *Nematolepis* Turcz.

Nematolepis Turcz., Bull. Soc. Imp. Naturalistes Moscou 25: 158 (1852); Wilson, Nuytsia 12: 267–288 (1998), synopsis.
Phebalium sect. *Eriostemoides* Endl. (1840), Wilson, Nuytsia 1: 92–96 (1970).

Lepidote shrubs or small trees; leaves entire. Inflorescences axillary, cymose or flowers solitary, prophylls small. Flowers pedicellate, diplostemonous; sepals distinct; corolla spreading, rarely tubular; petals imbricate; stamens all fertile, divergent, c. equal to petals; filaments distinct, glabrous or stellate-hairy; anthers versatile, glabrous, non-glandular, retuse at apex, apiculum 0;

disk prominent. Seed: hilum superficial, narrowly elliptic, raphe small, thin, ventral endocarp membranous, caducous. $n = 16$.

Seven spp., one in south-western Western Australia, six in south-eastern Australia including Tasmania, in heath, woodland, forest and sometimes on rainforest margins.

50. *Rhadinothamnus* Paul G.Wilson

Rhadinothamnus Paul G.Wilson, Nuytsia 1: 197 (1971); Wilson, Nuytsia 12: 267–288 (1998), morph. & synopsis. *Phebalium* sect. *Gonioclados* Paul G.Wilson (1970); Wilson, Nuytsia 1: 96–99 (1970), rev.

Lepidote shrubs; leaves entire or bilobed. Inflorescence axillary, cymose or flower solitary; prophylls small. Flowers pedicellate, diplostemonous; sepals connate; corolla spreading or tubular; petals valvate; stamens all fertile, erect, included, glabrous; filaments distinct; anthers versatile, obtusely apiculate, non-glandular; disk short. Seed: hilum narrowly elliptic surrounded by hilar strands, raphe sub-basal, shrunken.

Three spp., south-western Western Australia, found in heath, swamps and rocky areas.

51. *Chorilaena* Endl.

Chorilaena Endl., Enum. Pl.: 17 (1837); Keighery, J. W. Austr. Nat. 13: 156–158 (1977).

Shrubs; hairs stellate; leaves deeply sinuate. Inflorescence an axillary, 6-flowered, reflexed umbel; bracts ± equal to flowers. Flowers pedicellate, diplostemonous; calyx connate; petals erect, valvate; disk small, lobed; stamens all fertile, erect, exserted; filaments distinct, thickened and bearded adaxially towards base, otherwise glabrous, slender above; anthers glabrous or pilose, apiculum minute, non-glandular. Seed: hilum narrow-elliptic and surrounded by hilar strands, raphe short, shrivelled. $n = 14$.

One sp., *C. quercifolia* Endl., south-western Western Australia, in wet eucalypt forest and coastal communities.

52. *Phebalium* Vent.

Phebalium Vent., Jard. Malmaison 2: 102, t. 102 (1805). *Phebalium* sect. *Phebalium*, Wilson, Nuytsia 1: 60–92 (1970), rev.

Lepidote shrubs or small trees; leaves often glandular-punctate. Inflorescence terminal, umbellate or rarely flowers solitary; prophylls minute. Flowers pedicellate, diplostemonous; sepals basally connate; petals spreading, imbricate; stamens all fertile, divergent, exserted, glabrous; filaments distinct; anthers basifixed, apiculum 0, terminal gland present; disk obscure. Seed: hilum linear, raphe fleshy and shrivelled, situated in lower half of the adaxial face, ventral endocarp thin, caducous. $n = 16, 32$.

28 spp., southern and eastern Australia, including north Queensland and Tasmania, in heath, woodland and forest.

53. *Microcybe* Turcz.

Microcybe Turcz., Bull. Soc. Imp. Naturalistes Moscou 25: 166 (1852); Wilson, Nuytsia 12: 83–88 (1998).

Shrubs; hairs stellate or lepidote; leaves entire or glandular-punctate. Inflorescence terminal, a compact head subtended by leaf-like foliar bracts, or flowers solitary. Flowers sessile, usually 5-merous, diplostemonous; sepals distinct or connate; petals spreading, imbricate; stamens all fertile, divergent, exserted, glabrous; filaments distinct, stellate-lepidote or pilose towards base; anthers ± basifixed, apiculum 0, terminal gland present; disk 0; carpels 2–4. Seed: hilum linear, raphe fleshy and shrivelled, situated in lower half of the adaxial face, ventral endocarp thin, caducous.

Four spp., southern Australia, in heath and woodland.

54. *Diplolaena* R.Br.

Diplolaena R.Br. in M.Flinders, Voy. Terra Austr. 2: 546 (1814); Keighery, J. W. Austr. Nat. 13: 156–158 (1977); Wilson et al., Nuytsia 12: 107–118 (1998), key.

Shrubs; hairs stellate; leaves entire. Inflorescence terminal, dense, campanulate, heads, surrounded large, often showy, appressed imbricate bracts. Flowers sessile, usually 5-merous, usually diplostemonous; sepals 0; petals linear, membranous; stamens all fertile, equal to perianth or exserted; filaments erect, distinct, stellate-pilose adaxially towards base; anthers versatile, minutely glandular-apiculate; disk lobed; carpels 5(–8). Seed: hilum linear, raphe basal, ventral endocarp caducous. $n = 13, 14$.

15 spp., south-western Western Australia, in heath and woodland.

D. Diosma Alliance (Tribe Diosmeae DC. 1822).

Shrubs or halfshrubs, rarely trees accumulating coumarins but very rarely anthranilate-derived alkaloids; leaves simple. Flowers usually 5-merous, mostly perfect, usually diplostemonous (one stamen whorl staminodial); anthers with a usually sessile terminal gland; ovarioles basally connate and subapically joined in a common style; ovules (1)2. Fruits follicular or capsular, mostly horned. Seeds mostly explosively released, exalbuminous, usually (*Calodendrum* excepted) furnished with persistent ventral endocarp; embryo straight; cotyledons fleshy. South Africa to (*Calodendrum*) East Africa, eleven genera with about 276 spp. Found almost exclusively in Cape fynbos vegetation.

The study of plastid DNA sequences by Trinder-Smith et al. (2007) has revealed that several of the genera are polyphyletic, whereas the tribe appears monophyletic. Pending a more complete analysis, generic boundaries are in need of a profound revision, and genera are here given in the circumscription by Williams (1981–1984, see under the genera).

55. *Calodendrum* Thunb.

Calodendrum Thunb., Nov. gen. 2: 41 (1782), nom. cons.

Tree. Leaves opposite or ternate. Inflorescences thyrses. Disk small, cupuliform; gynophore long, filiform. Capsule shortly stalked, thick-woody, pentagonous, with blunt spines, 5-locular, septifragous; seeds horizontal, of the size of hazelnuts. $n = 27$.

Two spp., E Africa to Western Cape Province; *C. capense* (L.f.)Thunb., Cape chestnut, cultivated and ornamental.

56. *Agathosma* Willd.

Agathosma Willd., Enum. hort. berol.: 259 (1809); Pillans, J. S. Afr. Bot. 16: 55–183 (1950), rev.
Barosma Willd. (1809).

Shrubs or undershrubs sometimes multi-stemmed from the base. Leaves opposite, whorled, or alternate. Inflorescences in terminal or axillary clusters, or flowers solitary. Flowers bisexual or unisexual; petals clawed or sessile at base, rarely up to 8; staminodes often petaloid, usually with a gland behind the apex, rarely adnate to base of petals; disk cup-shaped, mostly crenulate. Fruit capsular; seeds "with peripheric endosperm" (Pillans 1950).

150+ spp., South Africa: Western and Eastern Cape Provinces, Kwazulu-Natal and Lesotho, with the greatest concentration in the Western Cape Province.

57. *Phyllosma* H. Bolus

Phyllosma H. Bolus, Bot. Jahrb. Syst. 24: 547 (1898); Williams, J. S. Afr. Bot. 47: 755–764 (1981), rev.

Shrubs. Leaves alternate, petiolate. Inflorescences 2–4-flowered. Petals clawed, glabrous or minutely pubescent; staminodes vestigial or 0; disk closing over the gynoecium at first. Fruit with scattered stalked glands, horned; ventral endocarp black.

Two spp., South Africa, south western and western parts of Western Cape Province.

58. *Coleonema* Bartl. & Wendl.

Coleonema Bartl. & Wendl., Diosmeae in Beitr. Bot. 1: 55 (1824); Williams, J. S. Afr. Bot. 47: 63–102 (1981); rev.

Shrubs mostly arising fom a single stem. Leaves alternate, gland-dotted in 2 rows. Flowers solitary, terminal or axillar, subtended by prophylls and 3–8 bracts; petals with the claw usually connate with staminode; disk circumvallate, partly free; style and stamens filaments short. Fruit hornless or with horns short. $n = 17, 18$.

Eight spp., South Africa, south-western and southern parts of Western and Eastern Cape Provinces.

59. *Acmadenia* Bartl. & Wendl.

Acmadenia Bartl. & Wendl., Diosmeae in Beitr. Bot. 1: 59 (1824); Williams, J. S. Afr. Bot. 48: 169–240 (1982), rev.

Shrubs; branches short, glabrescent. Leaves sessile or short-petiolate. Inflorescence terminal; prophylls much reduced. Calyx deeply divided; petals mostly clawed, bearded and bent inwards at the throat; staminodes short or vestigial; pollen triporate; disk circumvallate, exceeding ovary. Fruit normally 5-carpellate; seeds usually black and shining, with white or partly black ventral endocarp.

32 spp., Western and Eastern Cape Provinces of South Africa, mostly in southern and south western parts.

60. *Adenandra* Willd.

Adenandra Willd., Enum. pl. hort. berol. 1809: 256 (1809), nom. cons.; Strid, Op. Bot. 32: 1–112 (1972), rev.

Sympodially branched shrubs. Leaves alternate or more rarely subopposite, subsessile, usually 3–12 mm long, petiole sometimes with two basal appendages. Inflorescences terminal, corymbose, subcapitate or subumbellate. Anthers tipped by a stipitate, finally reflexed gland, alternating with 5 clavate, gland-tipped staminodes; ovary usually covered with stipitate glands; style usually recurved in young flowers; stigma discoid to capitate. Fruiting carpels almost free, dehiscing along adaxial suture. Seeds black, smooth; endocarp light yellow, cartilaginous splitting into 2 shells at maturity. $2n = 28, 38, 42, 48, 50$ (Strid 1972).

18 spp., south-western parts of Western Cape Province.

61. *Sheilanthera* Williams

Sheilanthera Williams, J. S. Afr. Bot. 47: 761 (1981).

Shrubs. Leaves alternate. Inflorescences terminal, 1–3-flowered. Petals narrowly clawed, thinly pubescent at throat; staminodes 0; stamen filamens and style short and glabrous; disk cirvumvallate, largely exceeding ovary; stigma capitellate; ovary densely pubescent.

One sp., *S. pubens* Williams, South Africa, south-western parts of Western Cape Province.

62. *Macrostylis* Bartl. & Wendl.

Macrostylis Bartl. & Wendl., Diosmeae in Beitr. Bot. 1: 191 (1824); Williams, J. S. Afr. Bot. 47: 373–403 (1981), rev.

Shrubs. Leaves alternate, rarely opposite, usually sessile. Inflorescence terminal, 1–14-flowered. Flowers very small; petals crisp-bearded about the middle; staminodes usually vestigial; disk envelops ovary, nectariferous; stigma simple or capitellate; style usually largely exceeding petals; ovary (2)3(4)-carpellate. Fruit with horns up to 5 mm long. Seeds with white ventral endocarp.

Ten spp., South Africa, south-western parts of Western Cape Province.

63. *Euchaetis* Bartl. & Wendl.

Euchaetis Bartl. & Wendl., Diosmeae in Beitr. Bot 1: 15 (1824), Williams, J. S. Afr. Bot. 47: 157–193 (1981), rev.

Shrubs, usually arising from a single stem. Leaves alternate or opposite. Flowers solitary, twin or aggregated into heads of up to 60 florets, subtended by 1(2) bracts and the prophylls; petals clawed, the claw transversely bearded above, ciliate at margin; staminodes 5, rarely 0; disk circumvallate or spreading, exceeds ovary; stigma capitellate or capitate; style short, up to 1.6 mm long; ovary (3)5-carpellate. Fruit horned. $n = 14$.

23 spp., South Africa, south and south-western parts of Western Cape Province.

64. *Diosma* L.

Diosma L., Sp. Pl.: 198 (1753); Williams, J. South Afr. Bot. 48: 329–407 (1982), rev.

Shrubs normally arising from a single stem, glabrescent. Leaves alternate or less usually opposite. Inflorescences terminal, few-flowered compound racemes, or flowers solitary, short-pedicellate or sessile; prophylls usually much reduced. Petals connate with the lobes of the disk; disk conspicuous, sinuate-crenulate, overtopping the ovary; staminodes 0 or vestigial; style and stamen filaments very short. Fruit with horns erect or spreading. $n = 15$.

In the revised circumscription of Williams 28 spp., South Africa, western extremities of Northern and Eastern Cape Provinces and Western Cape Province.

65. *Empleurum* Aiton

Empleurum Aiton, Hort. Kew. 3: 513 (1789); Williams, J. S. Afr. Bot. 50: 425–434 (1984), rev.

Polygamous shrubs or small trees, glabrous. Leaves alternate, glandular-serrate. Flowers pseudoaxillary, pedicellate, often in pairs. Calyx 4-lobed; petals 0; stamens 4; staminodes and disk 0; ovary 1(2)-carpellate; style 0; stigma very large. Fruit long beaked.

Two spp., South Africa, mountains of Western and Eastern Cape Provinces.

This is the only anenophilous lineage of the Diosma Alliance, which is reflected in the reduced floral structure and the comparatively large anthers and small pollen grains.

E. Polyaster Alliance (Rutaceae-Xanthoxyleae Hook.f. 1862 p.p.).

Leaves alternate; flowers actinomorphic, 4–5-merous, diplostemonous (*Peltostigma* with inconsistent merosity); petals distinct; gynoecium syncarpous or apocarpous but stylodia usually joined into a common style (distinct in *Plethadenia* and *Decatropis*); ovules 2 per carpel (1 in *Plethadenia*); fruit follicular with seeds forcibly expelled or rarely (*Decatropis*) of indehiscent mericarps.

Five small genera, mostly North and C American; *Choisya* and *Megastigma*, which differ in some characters from the other three genera, are doubtfully attached.

66. *Polyaster* Hook.f.

Polyaster Hook.f. in Benth. & Hook., Gen. Pl. 1: 299 (1862).

Shrub. Leaves alternate, imparipinnate, 5–12-jugate, leaflets opposite; rachis narrowly alate. Inflorescence an axillary, once- or twice-forked dichasium with monochasial branches. Flowers 5-merous; sepals valvate; petals imbricate; filaments pubescent adaxially at expanded base; anthers dorsifixed in a notch, orbicular, shorter than filaments, not glandular-punctate; disk cushion-like, gynophorous, obscurely 10-lobed; ovarioles distinct; style simple; stigma 5-lobed; ovules collateral. Fruit of 2–5, 2-valved mericarps with thin endocarp. Seeds 2 per mericarp, markedly reniform.

One sp., *P. boronoides* Hook.f., Mexico.

67. *Plethadenia* Urb.

Plethadenia Urb., Symb. Antill.7: 237 (1912) & 9: 177 (1924); Beurton, Willdenowia 30: 115–123 (2000), rev.

Small, glandular-verrucose shrubs. Leaves alternate or (sub)opposite, paripinnate, (1)2–5-jugate; leaflets (sub) opposite; rachis narrowly alate. Inflorescence an axillary, few-flowered cyme. Flowers 4-merous; petals white, valvate; filaments dilated and pubescent towards base; anthers nearly basifixed, suborbicular, more or less equal to filament, not glandular-punctate; disk flat or short-cupular, undulate; ovarioles connate ventrally; stylodia distinct, short; stigmas connate at anthesis; ovule 1 per carpel. Fruit of 1–4 basally subconnate, shortly apiculate follicles, with separating, elastic endocarp. Seed 1 per follicle, reniform.

Two spp., E Cuba, Hispaniola.

68. *Decazyx* Pittier & S.F. Blake

Decazyx Pittier & S.F. Blake, Contr. U.S. Natl. Herb. 24 (1): 9 (1922); Chiang, Bol. Soc. Bot. Mex. 43: 1–3 (1982).

Trees. Leaves alternate or when crowded subopposite, 1-foliolate; petioles basally or at both ends thickened. Inflorescence a cluster of terminal and axillary, slender, racemiform thyrses with flowers in glomerules. Flowers 5-merous; petals white, imbricate; filaments connate below for half their length and caducous as a ring; anthers (sub)basifixed, ovate-apiculate, shorter than filaments, not glandular-punctate; disk cushion-like, gynophorous; ovarioles distinct; style simple; stigma capitate. Fruit of 1–5 follicles cohering only at the base, with separating, elastic endocarp. Seeds 1 or 2 per follicle, suborbicular; endosperm scant or 0; testa coriaceous; embryo curved; cotyledons conduplicate (*D. esparzae*) or incumbent (*D. macrophylla*).

Two species, Mexico, Guatemala, Honduras and Costa Rica.

69. *Decatropis* Hook.f.

Decatropis Hook.f. in Benth. & Hook., Gen. Pl. 1: 298 (1862).

Small trees or shrubs. Leaves alternate, imparipinnate, 2–4-jugate; leaflets opposite. Inflorescence a terminal or upper-axillary panicle. Flowers 5-merous; petals white, valvate; filaments subterete; anthers small, elliptic, shorter than filament, not glandular-punctate; disk cushion-like, gynophorous; ovarioles distinct; stylodia distinct but apically joined in a 5-lobed stigma. Fruit of 2–5, distinct, reniform or lunate, indehiscent mericarps, each furnished with two narrow, longitudinal wings. Seed 1 per mericarp, reniform, smooth, dull, albuminous; testa papery; embryo curved; cotyledons not conduplicate (probably incumbent), elliptic.

Two spp., southern Mexico, Guatemala, Honduras.

70. *Peltostigma* Walp.

Peltostigma Walp., Rep. 5: 387 (1846).
Pachystigma Hook., Icon. Pl. t. 698 (1844).

Shrubs and small trees. Leaves alternate, digitately 3–5- or 1-foliolate. Inflorescences terminal, few-flowered racemes or axillary, 2–5-flowered glomerules, or solitary flowers. Flowers with inconsistent merosity; sepals 3–4, caducous, the inner ones larger than the outer; petals 3–8, white, imbricate; stamens 10–45, filaments subulate; anthers sagittate, not glandular-punctate; disk annular, fleshy, nectariferous; carpels 6–10, connate, the style very short; stigma broad, lobed; ovules superposed or collateral. Fruit dehiscing into 6–10 follicles united only at the base, with separating endocarp. Seeds 1 or 2 per follicle, with elongate hilum and fleshy funicle; endosperm 0; testa coriaceous; embryo straight; cotyledons plano-convex, thick.

Two spp., Jamaica, Central America, Ecuador and Peru.

Doubtfully attached to Polyaster Alliance:

71. *Choisya* H.B. & K.

Choisya H.B. & K., Nov. gen. sp. 6: 4, t. 513 (1823); Muller, Am. Midl. Nat. 24: 729–742 (1940), rev.

Shrubs. Leaves opposite, digitately 3- or 5–12-foliolate. Inflorescences terminal or axillary diffuse panicles, or flowers solitary. Flowers (4)5-merous, diplostemonous; petals white, imbricate; filaments flat below, subulate above; anthers ovate, dorsifixed, shorter than filaments, not glandular-punctate; disk cushion-like, gynophorous, pubescent like the distinct, gland-tipped ovarioles; stylodia conglutinate into a common style; stigma capitate, 5-lobed; ovules 2 per carpel. Fruit dehiscing into 2–5 follicles each with a dorsal horn and persistent stylodium base and dehiscent ventrally to the dorsal horn, with separating endocarp. Seeds 1 or 2 per carpel, elliptic-reniform, albuminous; testa coriaceous or crustaceous; embryo slightly curved; cotyledons incumbent, elliptic. $n = 27$.

Six spp., Mexico and south-western U.S.A. (Arizona), in xeric to mesic vegetation.

72. *Megastigma* Hook.f.

Megastigma Hook.f. in Benth. & Hook., Gen. Pl. 1: 299 (1862); Standley & Steyermark, Fl. Guatem., Fieldiana, Bot. 24: 414–415 (1946).

Pubescent, deciduous shrubs or small trees. Leaves alternate, imparipinnate, 5–8-jugate. Inflorescences terminal few-flowered racemes. Flowers 4-merous but carpels 2, diplostemonous; petals white, imbricate; filaments terete, slender; anthers basifixed, ovate, shorter than filaments, not glandular-punctate; disk conical, gynophorous; ovarioles connate only ventrally; style short and thick; stigma large, capitate, 2-lobed; ovules 2 per locule, collateral. Fruit drupaceous, usually only one carpel maturing with second one persistent at its base; exocarp fleshy, endocarp fibrous, curved in conformation with the seed. Seed 1, markedly reniform; testa thin, fragile; embryo curved; cotyledons incumbent, elliptic, wider than radicle.

Two spp., southern Mexico, Guatemala, Honduras, Nicaragua.

F. Esenbeckia Alliance (Pilocarpinae Engler 1896 without *Pilocarpus*).

Leaves opposite or alternate; flowers actinomorphic, haplostemonous; anther tips mucronate; gynoecium syncarpous; ovules 2 per carpel; fruit capsular, with separating dorsilateral endocarp; seeds with elongate hilum; endosperm usually very thin or 0; cotyledons plano-convex, thick, stiff.

Three genera, mostly neotropical.

73. *Esenbeckia* Kunth in HBK. Fig. 70

Esenbeckia Kunth in HBK., Nov. Gen. sp. 7: 191 (246 quarto edn) 1825; Kaastra, Fl. Neotropica 25: 24–116 (1982), rev.

Shrubs or small trees, rarely subshrubs. Leaves alternate or (sub)opposite, simple or digitately 1–5-foliolate. Inflorescences terminal and axillary, thyrso-paniculate. Flowers 5(7)-merous; petals yellowish, sometime pink-tinged, valvate or imbricate; filaments subulate, usually flattened at base, swollen abaxially at base in subg. *Esenbeckia*; anthers usually mucronate; disk annular or cupular, 5- or 10-lobed, adnate to base of ovaries; ovarioles connate, sometimes only at base; style simple; stigma capitate or clavate. Fruit a (4)5-locular capsule, smooth, muricate, or tuberculate, with or without apophyses, dehiscing septicidally and loculicidally, with separating endocarp. Seeds 1(2) per carpel, carinate abaxially, shiny or dull, with elongate hilum, albuminous (in sect. *Oppositifolia*) or not; testa crustaceous; embryo straight.

Thirty spp., West Indies, and from U.S.A. (Texas) and Mexico to north Argentina, rarely represented in Amazonia, ranging from dry woodland to moist forest. Three subgenera distinguished by Kaastra.

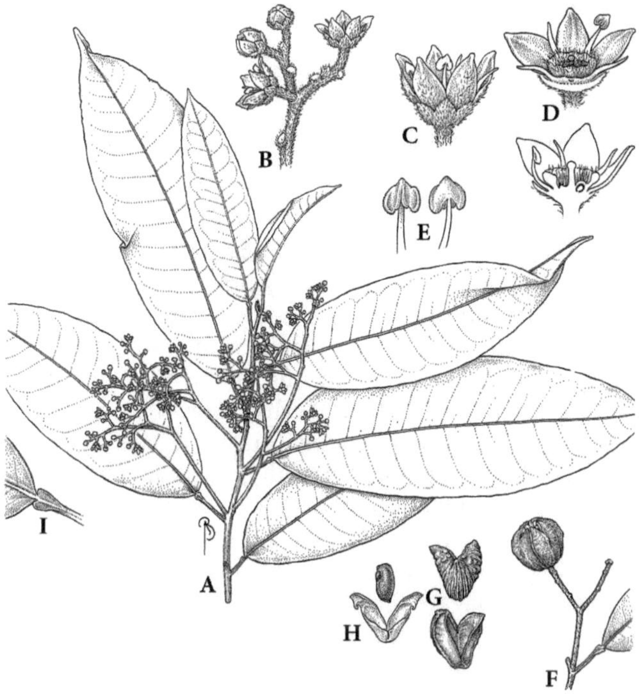

Fig. 70. Rutaceae. *Esenbeckia cowanii.* **A** Flowering branchlet. **B** Detail of inflorescence. **C** Flower. **D** Flower with two petals removed (above) and medial section of flower (below). **E** Stamen, adaxial (left) and abaxial (right) views. **F** Fruit. **G** Mericarp, dehisced, ab- and adaxial views. **H** Seed and endocarp. **I.** Leaf base. (Reproduced with permission of the artist Bobbi Angell)

74. *Metrodorea* St.-Hil.

Metrodorea St.-Hil., Fl. Br. mér. 1: 81 (1825); Kaastra, Fl. Neotropica 33: 116–130 (1982), rev.

Shrubs or trees. Leaves opposite, 1-3-foliolate; terminal buds of upper leaf pair completely enveloped by the petiolar sheaths which fall with the leaves. Inflorescences terminal or axillary panicles. Flowers 5-merous; petals valvate; filaments subulate, accumbent between lobes of disk; disk annular, more or less tuberculate, lobed, adnate to base of ovary; ovarioles connate; style simple; stigma subcapitate, 5-lobed. Fruit a capsule, with or without apophyses, dehiscing septicidally and loculicidally. Seed 1 or 2, oblong, keeled abaxially, reticulate-colliculate, dull; testa coriaceous; embryo straight.

Five spp., from Surinam through Brazil to Bolivia, moist and dry forests.

75. *Raulinoa* Cowan

Raulinoa Cowan, Sellowia 12: 90 (1960); Kaastra, Fl. Neotropica 33: 130–132 (1982).

Shrub. Some branchlets short and spine-tipped. Leaves opposite, simple. Inflorescences axillary on branchlets and spines, greatly reduced, cymose, racemiform, or subfasciculate. Flowers 4-merous; petals dark red, imbricate; filaments subulate, surrounded by disk at base; disk annular or cupular, smooth or slightly tuberculate, adnate to base of gynoecium; ovarioles each with a tuberculate apophysis; style simple, persistent; stigma 4-lobed. Fruit a (3)4-locular capsule, each carpel with reflexed dorsal apophysis, dehiscent septicidally and loculicidally. Seed 1 per carpel, ovoid, reticulate-colliculate; cotyledons unequal, eared.

One sp., *R. echinata* Cowan, Brazil, Sta. Catarina, on inundated river banks.

G. Balfourodendron Alliance (Pteleinae Engler 1896, without *Ptelea*).

Flowers actinomorphic, haplostemonous, 4–5-merous; gynoecium syncarpous; ovules 2 per carpel; fruit samaroid, not dehiscent; seeds enclosed in non-dehiscent cartilaginous endocarp; endosperm thin; embryo slightly curved or straight; cotyledons plano-convex.

Two genera, North, Central and South America.

76. *Balfourodendron* Mello ex Oliv.

Balfourodendron Mello ex Oliv., Hooker's Icon. Pl. 13: t. 1203–1204 (1877); Pirani, Brittonia 50: 357–360 (1998), rev.

Trees or treelets. Leaves opposite, digitately 3-foliolate, domatiiferous. Inflorescence a terminal compound thyrse, usually longer than leaves. Flowers tetramerous; petals whitish, distinct, imbricate; filaments subulate; anthers dorsifixed, versatile, ovoid, shorter than the filaments; disk cupular, forming an erect, undulate rim, surrounding the ovary, embracing bases of filaments; ovary 4-lobed; style terete; stigma capitate; ovules collateral. Fruit a (3)4-locular samara with (3)4 expanded, rounded wings from base to apex of each carpel. Seed 1 per carpel; testa thin; embryo straight, linear; cotyledons plano-convex, oblong.

Two spp., north-eastern to southern Brazil and adjacent Paraguay and Argentina, in (semi) deciduous forests.

77. *Helietta* Tul.

Helietta Tul., Ann. Sci. Nat. III, 7: 280 (1847); Pirani, Brittonia 50: 360–380 (1998), rev.

Shrubs or trees. Leaves (sub)opposite, digitately 3-foliolate. Inflorescence a terminal, simple or compound thyrse. Flowers 4–5-merous, haplostemonous; sepals persistent in fruit; petals distinct, imbricate; filaments complanate; anthers dorsifixed, versatile, ovoid, subacute or apiculate, shorter than filaments; disk cupular, forming an erect, undulate rim embracing bases of filaments; ovary lobed; style 1; stigma discoid to capitate, lobed; ovules 2, collateral. Fruit a samarium splitting from base to apex into (3)4–5 indehiscent mericarps, each with an ascending wing. Seeds 1 per mericarp, albuminous, oblong, with elongate hilum; testa thin, fragile; embryo straight, linear; cotyledons plano-convex, oblong.

Eight spp. with disjunct distributions in Mexico and adjacent U.S.A. (Texas), in Cuba, in northern South America, in Peru, and in south-eastern Brazil and adjacent Paraguay and Argentina, in mesic to dry habitats.

H. Angostura Alliance (subtr. Galipeinae Kallunki 1998, Cuspariinae Engl. 1893).

Flowers often slightly zygomorphic; petals basally joined by capillinection or connation; stamens 5(4), all fertile, or 2(3) fertile and flanking the petal with two inner margins and 3(2) transformed into staminodes, sometimes with appendages that look like the staminodes (pseudostaminodes) adnate to the two petals that alternate with the staminodes; anthers often basally appendaged and dorsally glandular-puncate; gynoecium syncarpous or apocarpous but distally usually joined in a common style; ovules 2 per carpel; dorsilateral endocarp discharged with the seed; ventral endocarp rarely persistent (but hardly acting as an elaiosome); chalazal area usually visible on outer surface of seed; seed coat sometimes with subhilar rumination; endosperm rarely well developed, often very thin in mature seed; cotyledons often conduplicate-plicate, sometimes bilobed at apex. 27 genera, 11 of which monotypic, all Neotropical.

78. *Euxylophora* Huber

Euxylophora Huber, Bol. Mus. Para. Emilio Goeldi 6: 84 (1909).

Tall tree; wood yellow. Leaves alternate, simple. Inflorescence a terminal, corymbiform thyrse. Flowers actinomorphic, 5-merous; calyx cupular, 5-denticulate, persistent; petals white, distinct, linear, patent, imbricate in bud; stamens all fertile, distinct, inserted on fleshy ring; filaments terete; anthers dorsifixed well below the middle, not glandular-punctate abaxially, cuspidate and ultimately bifid front to back due to separation of abaxial and adaxial surfaces at apex, lacking basal appendages; disk 0; ovarioles connate; style simple; stigma clavate. Fruit a schizocarp dehiscing into 5 apically truncate follicles. Seeds 1 or 2 per follicle, lenticular, black, shiny, with slit-like hilum from apex to base along adaxial edge of the seemingly obcampylotropous seed; testa crustaceous, subhilar rumination from base to near apex of seed along abaxial edge; embryo straight; cotyledons orbicular, plano-convex, fleshy, enclosed in endosperm.

One sp., *E. paraensis* Huber, Brazil (eastern Amazonia) and Peru (acc. to Pennington et al. 2004), furnishing valuable timber (Pao amarelo).

79. *Adiscanthus* Ducke

Adiscanthus Ducke, Arch. Jard. Bot. Rio de Janeiro 3: 186 (1922); Gereau, Candollea 45: 368 (1990).

Shrubs or small unbranched trees. Leaves alternate, simple; blade very long. Inflorescence a terminal corymbiform thyrse. Flowers actinomorphic, 5-merous; calyx cupular, sinuate, persistent; petals purplish abaxially, white and white-bearded adaxially, distinct at anthesis, valvate in bud; stamens all fertile; filaments distinct, flattened at base; anthers dorsifixed, not glandular-punctate abaxially, unappendaged; disk obsolete, reduced to a band of tissue adherent to a short, broad gynophore; ovarioles laterally connate; style simple; stigma capitate, slightly 5-lobed. Fruit of 1–5 follicles. Seed 1 per follicle, tear-shaped, mottled, glabrous, with elongate hilum and persistent ventral endocarp;

testa thin-crustaceous; subhilar rumination 0; embryo straight; cotyledons plano-convex, thick, stiff.

One sp., *A. fusciflorus* Ducke, lowland forests of N South America (Venezuela, Brazil, Peru).

80. *Leptothyrsa* Hook.f.

Leptothyrsa Hook.f., Gen. Pl. 1: 284 (1862); Gereau, Candollea 45: 370 (1990).

Shrubs or small trees, not ramified. Leaves alternate, simple, very long, crowded at tip of stems. Inflorescence an extra-axillary, racemiform thyrse. Flowers actinomorphic, 4-merous; calyx cupular, with rounded lobes; petals white, coherent in lower half, the lobes imbricate in bud; stamens all fertile; filaments terete, distinct; anthers dorsifixed, linear, longer than filaments, not glandular-punctate abaxially, lacking appendages; disk a very low cushion subtending the ovary; ovarioles connate ventrally; style simple; stigma clavate, 4-sulcate. Fruit of 1–4 follicles. Seed 1 per follicle, tear-shaped, smooth, with broad elliptic hilum and persistent ventral endocarp; testa thin-crustaceous; subhilar rumination 0; endosperm 0; embryo straight; cotyledons plano-convex, thick, stiff.

One sp., *L. sprucei* Hook.f., lowland forests of Brazil, Colombia and Peru.

81. *Nycticalanthus* Ducke

Nycticalanthus Ducke, Notizbl. Bot. Gart. Mus. Berlin-Dahlen 11: 341 (1932); Pirani, J. R., Flora da Reserva Ducke: Rutaceae, in Rodriguésia 56(86): 189–204 (2005).

Tree. Leaves alternate, 3-foliolate. Inflorescence a terminal thyrse of dichasial partial inflorescences with monochasial branches. Flowers slightly zygomorphic, 5-merous, nocturnal; calyx tubular, 5-dentate, caducous with attached petals and stamens; petals white, distinct, narrow-oblanceolate, unequal, imbricate in bud; stamens all fertile, distinct; filaments subterete, filiform; anthers basifixed, not glandular-punctate, lacking appendages; disk a short undulate rim encircling the base of a conspicuous gynophore; ovarioles connate ventrally, lobed at apex; style simple, filiform; stigma capitellate. Fruit a schizocarp dehiscing into 1–5 slightly flattened, dorsally carinate, long-beaked follicles elevated on gynophore up to 1.8 cm long. Seed 1 per follicle, oblong, smooth, glabrous; testa thin-coriaceous, subhilar rumination 0; cotyledons plano-convex, thick, unequal.

One sp., *N. speciosus* Ducke, Brazil, Amazon lowland.

82. *Spiranthera* A. St.-Hil.

Spiranthera A. St.-Hil., Bull. Sci. Soc. Philom. Paris 1823: 130 (1823); Bamps & Robyns, Bull. Jard. Bot. Nat. Belg. 54: 483–485 (1984), key.

Shrubs or trees. Leaves alternate, 3-foliolate. Inflorescence a terminal thyrse. Flowers ± actinomorphic, 5-merous; calyx cupular, 5-lobed, caducous; petals white, distinct, linear, usually quincuncial (infrequently imbricate) in bud; stamens all fertile, distinct; filaments subterete, elongate; anthers basifixed, slightly bifid at base, not glandular-punctate abaxially, lacking appendages, coiled after anthesis; disk cupular or cylindric, thin, embracing the gynophore, entire or unevenly laciniate, the lobes often pubescent; ovarioles ventrally connate, apically lobed, borne on short gynophore; style simple; stigma clavate to slightly capitate. Fruit a schizocarp dehiscing into 1 or 2(–5) beaked follicles. Seed 1 per follicle, ellipsoid, smooth, glabrous; testa crustaceous; subhilar rumination 0; embryo curved; cotyledons plano-convex, thick, stiff, often unequal.

Four spp., from northern South America to Peru and Bolivia.

83. *Almeidea* St.-Hil.

Almeidea St.-Hil., Bull. Soc. Philom. 1823: 129 (1823); I.M. Silva, Bradea 4 (46): 362–367 (1987) and Revisão taxon. do gênero *Almeidea* St.-Hil., M.Sc. Thesis, Univ. Fed. Rio de Janeiro (1988).

Small trees. Leaves alternate, 1-foliolate or simple. Inflorescence a (sub)terminal thyrse. Flowers actinomorphic, 5-merous; calyx cupular, 5-dentate, persistent; petals pink or lilac, distinct, imbricate in bud; stamens all fertile (rarely some sterile in some flowers); filaments distinct, flattened at base, subulate at apex, free from petals; anthers dorsifixed below middle, not glandular-punctate abaxially, lacking basal appendages; disk cupular, shorter than ovary; ovarioles distinct; style simple; stigma capitate, 5-lobed. Fruit of 1–3 follicles. Seed 1 per follicle, ellipsoid-reniform, sometimes beaked, sometimes with more than one embryo; testa coriaceous;

subhilar rumination 0; embryo curved; cotyledons conduplicate-plicate, fleshy, cordate.

Five spp., Brazil, Atlantic coastal range from Bahia to Paraná.

84. *Conchocarpus* J.C. Mikan

Conchocarpus J.C. Mikan, Del. Faun. Fl. Bras. t. 2 (1820); Kallunki & Pirani, Kew Bull. 53: 266–334 (1998), rev.

Shrubs or small trees, often unbranched; trichomes simple. Leaves alternate (exc. in *C. oppositifolius* Kallunki), 1- or digitately 3–7-foliolate, or simple. Inflorescence a terminal or extra-axillary thyrse or dichasium. Flowers actinomorphic or slightly zygomorphic; calyx (4)5-lobed, sometimes deciduous as a dentate ring; petals (4)5, usually white, rarely pink, coherent by pubescence, forming a tube, imbricate in bud; fertile stamens 2(3 or 5); staminodes 3(2 or 0); filaments and staminodes free from each other, flattened and adherent to petals at base; anthers basifixed, glandular-punctate or not abaxially, usually distinct, lacking basal appendages; disk cupular, sometimes inflexed and concealing the ovary; carpels 5; ovarioles usually distinct; stylodia and stigmas usually coherent [distinct in *C. longifolius* (A. St.-Hil.) Kallunki & Pirani)]. Fruit of 1–5 rarely basally connate follicles. Seed 1 per follicle, ellipsoid-reniform, smooth, rarely tuberculate, glabrous, sometimes with more than one embryo; testa coriaceous (rarely crustaceous); subhilar rumination 0; embryo curved (rarely straight); cotyledons conduplicate-plicate and usually pliable, conduplicate and sometimes unequal and stiff, or plano-convex, thick and stiff.

About 45 spp., from Nicaragua to northern Bolivia and south-eastern Brazil.

Highly variable but held together by the combination of the woody habit, alternate leaves, small calyx, coherent petals, flattened staminal filaments, completely fertile, basally unappendaged anthers, cupular disk, and free carpels.

85. *Naudinia* Planch. & Linden

Naudinia Planch. & Linden, Ann. Sci. Nat. III, 19: 79 (1853).

Shrub. Leaves alternate, 1-foliolate. Inflorescence an extra-axillary cyme. Flowers zygomorphic; calyx cupular, 5-dentate; corolla red, with long gradually expanded tube and 5 lanceolate lobes induplicate-valvate in bud; fertile stamens 2; staminodes 3; filaments and staminodes flattened and connate at base and adherent to corolla tube; anthers basifixed, not glandular-punctate abaxially, sterile at base above point of attachment to filament and connate there, lacking basal appendages; disk cupular; carpels 5; ovarioles distinct; style simple; stigma capitate, 5-lobed. Fruit of 3–5 follicles laterally compressed and dorsally keeled. Seed 1 per follicle, reniform; testa brown, shining; cotyledons plicate.

One sp., *N. amabilis* Planch. & Linden, Colombia.

86. *Andreadoxa* Kallunki

Andreadoxa Kallunki, Brittonia 50: 59 (1998).

Trees. Leaves alternate, 1-foliolate. Inflorescence an axillary thyrse, several borne near apex of branch. Flowers slightly zygomorphic; calyx cupular, subentire, persistent; petals 5, yellow, free, imbricate in bud; fertile stamens 2; staminodes 5 (the 2 adnate to petals probably pseudostaminodes); filaments narrow-trullate; anthers basifixed, conspicuously glandular-punctate abaxially, with flat apiculum, connate by basal appendages; disk cupular, entire; carpels 5; ovarioles distinct; style simple; stigma subcapitate. Fruit of 1(2) suborbicular follicles. Seed 1 per follicle, smooth, glabrous; testa papery, subhilar rumination 0; embryo curved; cotyledons conduplicate-plicate, rounded at apex, cordate at base.

One sp., *A. flava* Kallunki, Atlantic coastal range of Bahia, Brazil.

87. *Raputiarana* Emmerich

Raputiarana Emmerich, Rodriguésia 30 (45): 285 (1978); Ducke, Arch. Jard. Bot. Rio de Janeiro 5: 143–144 (1930) (sub *Raputia*).

Shrub. Leaves alternate, digitately 5–7-foliolate. Inflorescence a long-pedunculate, once- or several-forked dichasium with monochasial branches. Flowers zygomorphic; calyx cupular, sinuate; petals 5, creamy white or pinkish, coherent at base in a short tube, the corolla geniculate near base in bud, 2-lipped, the lobe of innermost petal erect, the other 4 recurved, imbricate in bud; fertile stamens 2; staminodes 3; filaments of fertile stamens connate with the 2 adjacent staminodes, the 3rd staminode free, all adherent to the corolla tube; anthers not glandular punctate abaxially, sterilized at base above the point of attachment to the filament, with elongate basal appendages; disk cupular, denticulate, exceeding the ovary; carpels 5; ovarioles distinct; style simple; stigma capitate. Fruit of usually 5, apically truncate follicles. Seeds 2 per follicle, ovoid, sometimes tuberculate on angles, glabrous; testa crustaceous, a subhilar rumination present; embryo curved; cotyledons conduplicate-plicate.

One sp., *R. subsigmoidea* (Ducke) Emmerich, Amazonian Brazil, Peru, Colombia.

Raputia heptaphylla Pittier, Costa Rica, Venezuela (D.F.) and perhaps Colombia (Darien), probably belongs to this genus as well. Although similar in most characters, the few specimens lack basal appendages.

88. *Sigmatanthus* Huber ex Emmerich

Sigmatanthus Huber ex Emmerich, Rodriguésia 30 (45): 267 (1978).

Small tree or shrub. Leaves alternate, trifoliolate. Inflorescence terminal, raceme-like. Flowers zygomorphic; calyx subcampanulate, 5-lobed, persistent, the lobes valvate in bud; petals 5, white or pale pink, connate at base in a short tube, the corolla geniculate near base in bud, 2-lipped, the lobe of innermost petal erect, the other 4 recurved, lobes imbricate in bud; fertile stamens 2; staminodes 5 (2 probably pseudostaminodes); filaments and staminodes distinct, flat at base, adherent to corolla tube; anthers not glandular-punctate abaxially, basally sterilized above the point of attachment to the filament, with ligulate appendages below this point; disk cupular, slightly 5-dentate, inflexed and covering the ovary; carpels 5; ovarioles distinct; style simple; stigma capitate. Fruit of usually 5 follicles. Seeds 2 per follicle, subglobose, smooth, glabrous; testa crustaceous, a subhilar rumination present; embryo curved; cotyledons conduplicate-plicate.

One sp., *S. trifoliatus* Huber ex Emmerich, northeast Brazil (Pará to Rio Grande do Norte).

89. *Angostura* Roemer & Schultes Fig. 71

Angostura Roemer & Schultes, Syst. Veg. 4: 188 (1819); Kallunki & Pirani, Kew Bull. 53: 257–266 (1998), rev.; Beurton, Willdenowia 34: 277–289 (2004), 8th species.
Bonplandia Willd. (1802) non Cav. (1800).
Cusparia Humb. (1807), nom. provis.

Trees or shrubs; trichomes echinoid or stellate. Leaves alternate, (1)3-foliolate or sometimes digitately 4–7-foliolate. Inflorescences terminal, long- to short-pedunculate thyrses. Flowers subactinomorphic; calyx cupular, 5-lobed, persistent; petals 5, white, connate into a straight tube with spreading lobes imbricate in bud; fertile stamens 2; staminodes 5 (2 probably pseudostaminodes); filaments of stamens and staminodes connate into a tube adnate to corolla tube for most of its length; anthers basifixed, not glandular-punctate abaxially, with denticulate or bifid basal

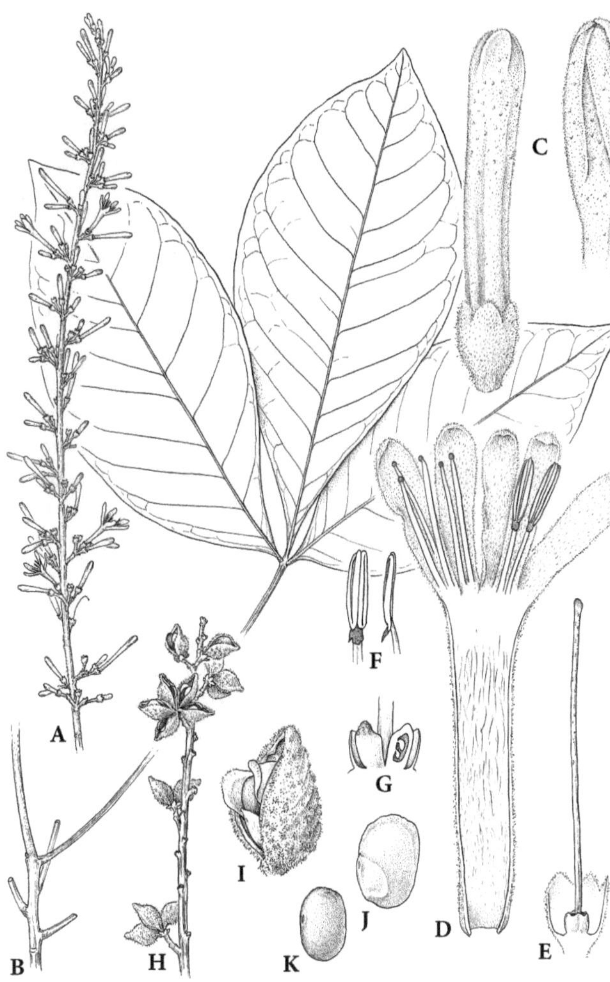

Fig. 71. Rutaceae. *Angostura granulosa*. **A** Inflorescence. **B** Leaf. **C** Bud. **D** Corolla, opened lengthwise. **E** Pistil and disk. **F** Anther with appendage, adaxial and lateral views. **G** Carpels, one sectioned and showing two superposed ovules. **H** Infructescence. **I** Fruiting mericarp. **J** Seed enclosed by endocarp, showing thinner ventral portion. **K** Seed, lateral view. (Reproduced with permission of the artist Bobbi Angell)

appendages; disk cupular; carpels 5(6); ovarioles distinct but tightly pressed against each other; style simple; stigma subcapitate. Fruit of 1–5 follicles. Seed 1 per follicle, globose or oblong, smooth or rugulose, glabrous; testa crustaceous, a subhilar rumination present; embryo curved; cotyledons conduplicate-plicate.

Eight spp., Cuba, Nicaragua to Venezuela and south to northern Bolivia and southern Brazil.

90. *Erythrochiton* Nees & Mart.

Erythrochiton Nees & Mart., Nova Acta Leop.-Carol. 11: 151, 165, t. 18 (1823); Kallunki, Brittonia 44: 123–139 (1992), rev.
Erythrochiton sect. *Erythrochiton* Engler (1895).

Shrubs or small trees, often un(few)branched. Leaves alternate, 1-foliolate, often clustered at apex of stem. Inflorescence an extra-axillary, pedunculate dichasium, sometimes perennating, or, in *E. hypophyllanthus* Planchon & Linden, epiphyllous on lower leaf surface. Flowers actinomorphic or zygomorphic; sepals 5, white, green, or red, valvate in bud, forming a large, campanulate to tubular calyx; petals 5, white, connate in a straight tube, the lobes imbricate in bud; fertile stamens (4)5 or only 2 and then staminodes (1)3 or 5 (then 2 probably pseudostaminodes); filaments adnate to corolla tube; anthers basifixed, not glandular-punctate abaxially, when only 2 fertile appendaged at base and often cohering by appendages; disk tubular, taller than ovary; carpels 5; ovarioles distinct; style simple; stigma capitate and shallowly 5-lobed. Fruit of (4)5 follicles rarely connate at base. Seeds 1 or 2 per follicle, ovoid or when 2 cuboidal, tuberculate, tomentulose; testa crustaceous, a subhilar rumination present; cotyledons conduplicate-plicate.

Seven spp. ranging from Costa Rica across northern South America and south to Bolivia and Brazil; understorey of tropical forests from lowland to 1,200 m.

91. *Desmotes* Kallunki

Desmotes Kallunki, Brittonia 44: 115 (1992).

Shrub. Leaves opposite, simple. Inflorescences terminal, subsessile, 1–2-flowered. Flowers zygomorphic; sepals 5, red, valvate, connate for half their length, up to 2 cm long; petals 5, red, adherent to filaments and staminodes and forming a straight tube, the lobes unguiculate, imbricate in bud; fertile stamens 2; staminodes 5 (2 probably pseudostaminodes); filaments and staminodes adherent to corolla; anthers basifixed, not glandular-punctate, with flattened bilobed basal appendage; disk cupular, taller than ovary; carpels 5; ovarioles distinct; style simple; stigma clavate. Fruit of 5 follicles, enclosed in persistent calyx. Seeds 1 or 2 per follicle, oblong- or ovoid-reniform, unevenly tuberculate, glabrous; testa crustaceous, a subhilar rumination present; embryo curved; cotyledons conduplicate-plicate, deeply bilobed at apex.

One sp., *D. incomparabilis* (Riley) Kallunki (= *Erythrochiton i.*), known only from Panama, Veraguas: Coiba Island.

92. *Toxosiphon* Baill.

Toxosiphon Baill., Adansonia 10: 311 (1872); Kallunki, Brittonia 44: 117–123 (1992), rev.
Erythrochiton sect. *Toxosiphon* (Baillon) Engl. (1895).

Shrubs or small trees, often unbranched. Leaves alternate, 1- or 3-foliolate, often clustered at apex of stem. Inflorescence a terminal long-pedunculate dichasium. Flowers slightly zygomorphic; sepals 5, often white, valvate in bud, free at anthesis; petals 5, white, coherent in a straight or curved tube, pubescent, the lobes imbricate in bud; fertile stamens 2; staminodes 3–5 (then 2 probably pseudostaminodes); filaments distinct, flat, adherent to corolla by dense pubescence; anthers basifixed, not glandular-punctate abaxially, coherent by basal appendages; disk cupular, shorter than ovary; carpels 5; ovarioles ventrally connate; style simple; stigma capitate. Fruit a schizocarp of (4)5 follicles dorsally and apically winged at maturity. Seeds 1 or 2 per follicle, ovoid or (when 2) cuboidal, tuberculate, glabrous; testa crustaceous, a subhilar rumination present; embryo curved; cotyledons conduplicate-plicate.

Four spp., ranging from southern Mexico through Panama and from Ecuador to northern Bolivia and adjacent Brazil, in the understorey of humid tropical forests, 0–1,200 m.

93. *Neoraputia* Emmerich ex Kallunki Fig. 72

Neoraputia Emmerich, nom. inval., Rodriguésia 30 (45): 234 (1978); Kallunki, Brittonia 61: 28–34 (2009), validation.

Trees; trichomes simple. Leaves alternate (in *N. trifoliata* sometimes opposite), digitately 1–5(7)-foliolate. Inflorescence a terminal, long-pedunculate, double cincinnus or compound dichasium with monochasial branches. Flowers subzygomorphic; calyx variously 5-dentate or lobed, sometimes bilabiate, persistent in fruit; petals 5, white, salmon, or pale green, coherent at base and forming a ± straight tube shorter than or equal to lobes, the lobes imbricate in bud; fertile stamens 2; staminodes 3 (plus 2 pseudostaminodes in *N. trifoliata*), awl-shaped at apex; filaments and staminodes distinct, flattened at base, adherent to corolla tube; anthers basifixed, oblong, not glandular-punctate abaxially, with a small rim-like basal appendage (or in *N. trifoliata*, sterilized above point of attachment to filament and with a broader basal appendage and connate by these sterile portions); disk cupular, taller than ovary; carpels 5; ovarioles connate ventrally and laterally; style simple; stigma capitate, 5-sulcate. Fruit a densely tomentellous septicidal capsule, each segment dehiscing ventrally and partly dorsally. Seeds (1)2 per follicle, subglobose to ellipsoid, smooth, glabrous; testa crustaceous, a subhilar rumination present; embryo curved; cotyledons conduplicate-plicate.

Six spp., Venezuela, French Guiana, Peru, Brazil, in tropical lowland.

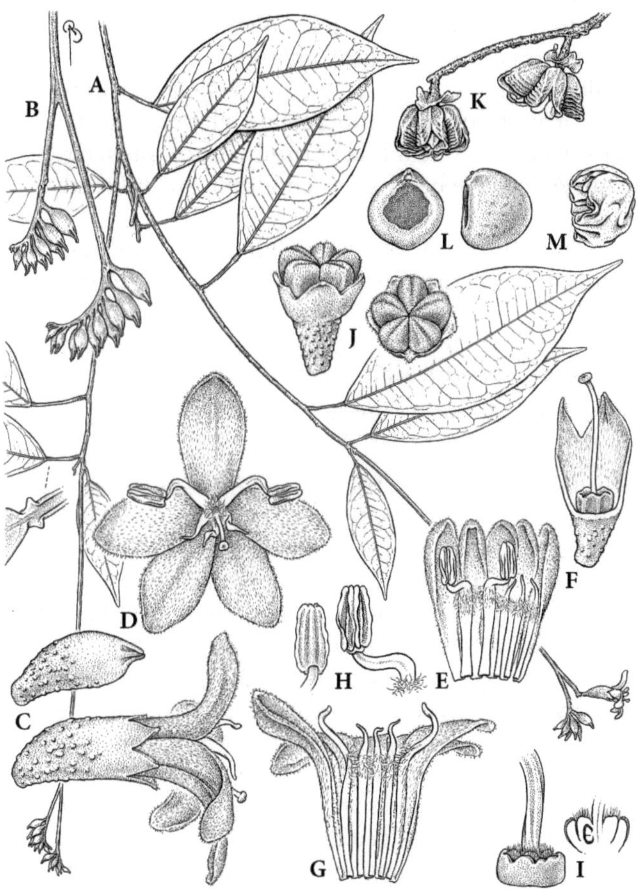

Fig. 72. Rutaceae. *Neoraputia calliantha*. **A** Flowering branch, with detail showing stipels. **B** Apex of inflorescence. **C** Flower bud and flower. **D** Flower. **E** Bud opened to reveal stamens and staminodes. **F** Calyx cut to reveal disk and style. **G** Corolla and androecium, anthers fallen. **H** Anther, abaxial and adaxial views. **I** Disk, ovary, and base of style and longitudinal section of same. **J** Young fruit. **K** Mature fruits. **L** Seed, adaxial and lateral views; note small hilum and broad chalazal area. **M** Embryo. (Reproduced with permission of the artist Bobbi Angell)

94. *Rauia* Nees & Mart.

Rauia Nees & Mart., Nova Acta Leop.-Carol. 11: 151, 167 (1823).

Shrubs or small trees. Leaves alternate, 1- or 3-foliolate. Inflorescence a terminal, pedunculate, corymboid or racemoid thyrse. Flowers actinomorphic; calyx cupular, 5-dentate, persistent in fruit; petals 5, white, narrow, free at very base but above coherent into a straight tube, narrowly imbricate in bud; fertile stamens 2; staminodes 3 or 5 (2 probably pseudostaminodes); filaments and staminodes distinct, flat, adherent by pubescence to the petals above the base; anthers basifixed, not glandular-punctate abaxially, apiculate, lacking basal appendages; disk cupular, equal to or taller than ovary; carpels 5; ovarioles ventrally ± connate; style simple; stigma clavate. Fruit a schizocarp dehiscing into (4)5 follicles. Seeds 1 or 2 per follicle, ellipsoid-reniform, sometimes with a curved micropylar beak, smooth, glabrous or pubescent; testa crustaceous, subhilar rumination present; embryo curved; cotyledons conduplicate, thick, stiff, unequal.

About 10 spp., northern South America to Rio de Janeiro.

95. *Galipea* Aubl.

Galipea Aubl., Hist. pl. Gui. fr. 2: 662, t. 269 (1775).

Trees or shrubs; trichomes simple. Leaves alternate, (1)3-foliolate. Inflorescence a terminal or extra-axillary pedunculate thyrse. Flowers slightly zygomorphic, often showy; calyx short-campanulate, 5-dentate; petals 5, white, connate in a narrow tube, the lobes imbricate in bud; fertile stamens 2; staminodes 3 or 5(6) and 2(3) probably pseudostaminodes; filaments and staminodes adnate to corolla tube at base, coherent with each other and adherent to corolla above; anthers basifixed, not glandular-punctate abaxially, sterile at base above point of attachment to filament (except in *G. dasysperma*), appendaged at base, usually (with two exceptions) connate by their sterile bases and appendages; disk cupular, shorter than or equal to ovary; carpels (4)5; ovarioles connate; style simple; stigma subcapitate, (3–)5-lobed. Fruit a 5-locular capsule, the individual segments connate (distinct in *G. dasysperma*), each dehiscing ventrally and to middle dorsally. Seed 1 per follicle, elliptic or subglobose, smooth, glabrous or (in *G. dasysperma*) pubescent; testa crustaceous, subhilar rumination 0; embryo curved; cotyledons conduplicate-plicate.

Fifteen spp., Costa Rica to Brazil and Bolivia.

96. *Ticorea* Aubl.

Ticorea Aubl., Hist. pl. Gui. fr. 2: 689, t. 277 (1775); Kallunki, Brittonia 50: 500–513 (1998), rev.

Trees or shrubs. Leaves alternate, 3(4)-foliolate, domatiiferous. Inflorescence an extra-axillary, usually long-pedunculate thyrse. Flowers actinomorphic; calyx cupular, subentire or dentate; petals 5, white, coherent in a long narrow tube, the lobes imbricate in bud; fertile stamens (4)5 or (in *T. diandra* Kallunki) 2 fertile and 3 staminodes; filaments adherent to corolla tube by woolly pubescence; anthers basifixed, glandular-punctate abaxially, sterile at base above the point of attachment to the filament and (exc. in *T. diandra*) appendaged below this point; disk cupular, pubescent on rim, persistent; carpels (4)5; ovarioles ventrally and basally connate; style simple; stigma capitate. Fruit a schizocarp dehiscing into 1–5 follicles free except at base, each with a dorsal ridge and often with two lateral ridges where abutting adjacent carpels. Seed 1 per follicle, oblong, smooth, glabrous; testa papery, subhilar rumination 0; embryo curved; cotyledons conduplicate-plicate, very stiff.

Five spp., lowland forests of the eastern slopes of the Andes in Ecuador, Peru and Bolivia, and in Brazilian Amazonia and the Guianas.

97. *Decagonocarpus* Engl.

Decagonocarpus Engl. in Mart., Fl. Bras. 12 (2): 105, t. 21 (1874).

Shrubs or small trees. Leaves opposite, simple. Inflorescence a terminal, few-flowered monochasium. Flowers slightly zygomorphic; calyx cupular, 5-lobed, persistent; petals 5, orange or red, connate into a long, slightly curved tube with recurved to spreading lobes, induplicate-valvate in bud; fertile stamens 2; staminodes 3; filaments and staminodes connate in a tube adherent in upper half to corolla; anthers basifixed, laterally coherent, not glandular-punctate abaxially, appendaged at base; disk cupular, enclosing the ovary; carpels 5; ovarioles distinct; style simple; stigma only slightly enlarged. Fruit of 5, beaked follicles. Seeds 1 or 2 per follicle, elliptic-reniform or cuboidal, tuberculate, glabrous; testa crustaceous, a subhilar rumination present; embryo curved; cotyledons conduplicate, bilobed at apex.

Two spp., on Guayana Shield in southern Venezuela, northern Brazil and eastern Colombia.

98. *Raveniopsis* Gleason

Raveniopsis Gleason, Brittonia 3: 166 (1939); Kallunki, Fl. Venez. Guay. 9: 26–30 (2005).

Shrubs or small trees; trichomes simple or stellate-lepidote. Leaves opposite (in *R. steyermarkii* some alternate), 1–3-foliolate. Inflorescence axillary or terminal, a double cincinnus, a racemiform or spiciform monochasium, a congested capitulate thyrse, or a solitary flower. Flowers slightly to markedly zygomorphic; sepals 5, free or very shortly connate, usually unequal, persistent in fruit; petals 5, white, pink, orange, or red, connate, the corolla tube cylindric, slightly curved, longer than or equal to the lobes, the lobes imbricate in bud; fertile stamens 2; staminodes 3; filaments and all (or 2 adjacent) staminodes connate into a tube, tightly adherent or adnate to corolla; anthers basifixed, glandular-punctate or not abaxially, sometimes laterally coherent, with basal appendages; disk cupular, truncate to ± lobate, shorter than or equal to ovary; carpels 5; ovarioles distinct or sometimes ventrally connate at base; style simple; stigma ± 5-lobed. Fruit of 1–5, beaked follicles. Seed 1 per follicle, ellipsoid-reniform, tuberculate, glabrous; testa crustaceous, a subhilar rumination present; embryo curved; cotyledons conduplicate, bilobed at apex (or rarely incumbent and rounded at apex).

Nineteen spp., endemic to Guayana Shield of southern Venezuela and northern Brazil, (300–) 1,000–2,600 m above sea level.

99. *Lubaria* Pittier

Lubaria Pittier, Trab. Mus. com. Venezuela 5: 279 (1929).

Tree. Leaves simple, opposite. Inflorescence a terminal, once- or twice-forked dichasium with monochasial branches. Flowers zygomorphic; sepals 5, distinct, the 2 outer larger; petals 5, white, the adaxial, innermost one distinct, the others connate into a 4-lobed lip, the lobes imbricate in bud; fertile stamens 2; staminodes 3 (4); filaments of fertile stamens connate at base, staminodes distinct, all flat and adherent to corolla at base; anthers basifixed, not glandular-punctate abaxially, laterally coherent, with basal appendages; disk cupular, inflexed and enclosing ovary; carpels 5; ovarioles distinct; style simple; stigma subcapitate. Fruit of 1–5 follicles. Seed 1 per follicle, ellipsoid-reniform, smooth, glabrous; testa crustaceous, a subhilar rumination present; embryo curved; cotyledons conduplicate, bilobed at apex.

One sp., *L. aroensis* Pittier, Costa Rica, Venezuela (Coastal Cordillera).

100. *Raputia* Aubl. Fig. 73

Raputia Aubl., Hist. pl. Gui. fr. 2: 670, t. 272 (1775); Kallunki, Brittonia 46: 279–295 (1994), rev.; Pirani, J. R., Flora da Reserva Ducke: Rutaceae, Rodriguésia 56(86): 189–204 (2005), 11th species.
Myllanthus R.S. Cowan (1960).
Achuaria Gereau (1990).

Shrubs or trees. Leaves opposite, 1- or 3-foliolate. Inflorescence cauline (rarely axillary), pedunculate, 1–4 monochasia at a node, elongating as successive flowers develop. Flowers zygomorphic; sepals 5, connate at very base, unequal, broadly overlapping; petals 5, white or green, connate at base forming a curved, 2-lipped corolla, the innermost petal in upper position, the others forming a 4-lobed, recurved lower lip, the lobes imbricate in bud; fertile stamens 2; staminodes (2)3; filaments and staminodes distinct, flattened and adherent to corolla; anthers basifixed, glandular-punctate abaxially, laterally connate, with basal appendages; disk cupular, shorter or taller than ovary; carpels 5; ovarioles distinct; style simple; stigma obliquely capitate. Fruit of 1–5 follicles. Seed 1 per carpel, oblong, smooth, glabrous; testa coriaceous, subhilar rumination 0; embryo curved; cotyledons conduplicate, thick, stiff, bilobed at apex.

Eleven spp., northern South America, mostly in lowland forest.

101. *Ravenia* Vell.

Ravenia Vell., Fl. Flumin.: 20 (1825), i, t. 49 (1827); Gereau, Candollea 45: 370 (1990).

Shrubs or small trees. Leaves opposite or appearing alternate due to reduction of one of

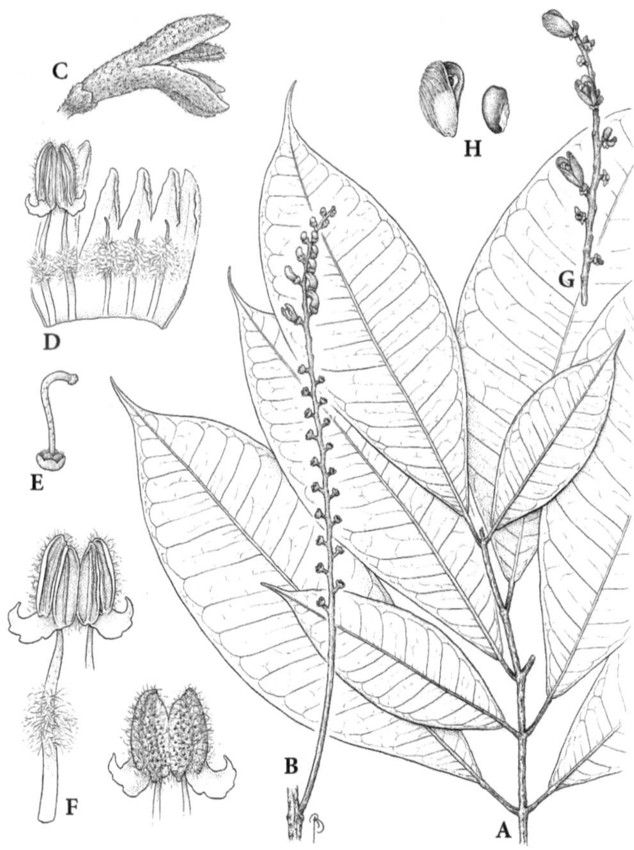

Fig. 73. Rutaceae. *Raputia simulans.* **A** Leafy branch. **B** Inflorescence. **C** Flower. **D** Corolla, opened to show two fertile stamens and three staminodes. **E** Gynoecium and cupular disk. **F** Fertile stamens with connate anthers and appendages, adaxial view (left) and abaxial view (right). **G** Infructescence. **H** Fruiting mericarp, dehisced (left); seed with ventral endocarp (right). (Reproduced with permisson of the artist Bobbi Angell)

two opposite leaves, simple or 3-foliolate. Inflorescence a terminal dichasium with two monochasial branches, a monochasium, or 1 or 2 flowers. Flowers zygomorphic; sepals 5, the 2 outer larger than inner, often investing the mature fruit; petals 5, pink or green, connate to the middle or more, the lobes imbricate in bud; fertile stamens 2; staminodes (2)3(5, of which 2 are probably pseudostaminodes); filaments and staminodes flat at base, adnate or adherent to the corolla tube; anthers basifixed, glandular-punctate abaxially, laterally coherent or not, with or without basal appendages; disk cupular, enclosing or shorter than ovary; carpels 5; ovarioles distinct; style simple; stigma capitate or with 5 short branches. Fruit of 1–5 follicles. Seeds (1) 2 per follicle, ovoid-ellipsoid, usually tuberculate, glabrous; testa crustaceous, a subhilar rumination present or not; embryo curved; cotyledons conduplicate, fleshy, bilobed at apex.

About 11 spp. on Greater Antilles and Trinidad and from Honduras to Brazil and Peru.

102. *Apocaulon* Cowan

Apocaulon Cowan, Mem. New York Bot. Gard. 8: 119 (1953).

Decumbent, pubescent, rhizomatous herbs. Leaves alternate, congested, often appearing basal, 3-foliolate. Inflorescence a long-pedunculate, double cincinnus. Flowers zygomorphic; sepals 5, strongly unequal, shortly coherent, persistent; petals 5, connate, forming a corolla with curved tube and (4)5 imbricate lobes, the 2 lobes opposite the lobe flanked by the 2 stamens forming a bilobed lip; fertile stamens 2; staminodes 3; filaments and staminodes distinct, adherent at throat of corolla tube; anthers basifixed, glandular-punctate abaxially, coherent by their adaxial surfaces, with basal appendages; disk cupular, minutely puberulent at margin; carpels (4)5; ovarioles distinct; style simple; stigma capitate, 5-lobed. Fruit of 1–5, beaked follicles. Seeds 1 per follicle, elliptic-reniform, ± tuberculate, glabrous; testa crustaceous, a subhilar rumination present; embryo curved; cotyledons conduplicate, emarginate at apex.

One sp., *A. carnosum* R.S. Cowan, montane forests on Guayana Shield in southern Venezuela.

103. *Ertela* Adans.

Ertela Adans., Fam. Pl. 2: 358 (1763); Pirani, J. R., Flora da Reserva Ducke: Rutaceae, Rodriguésia 56(86): 189–204 (2005).
Monniera Loefl. (1758) non B. Juss. ex P. Br. (1756).

Erect herbs, sometimes suffruticose, stems dichotomously or trichotomously forked above. Leaves usually opposite on lower part of stem, sometimes alternate or subopposite on upper part, 3-foliolate. Inflorescences terminal but sometimes appearing axillary, a pedunculate double cincinnus. Flowers zygomorphic; sepals 5, distinct, strongly unequal, the 2 outermost much larger and concealing the corolla or the whole inflorescence; petals 5, white, connate, forming a bilabiate corolla with a short tube, one lip formed by the innermost petal, the other lip 4-lobed, the lobes imbricate in bud; fertile stamens 2; staminodes 3; filaments connate, the staminodes distinct, all adherent to corolla; anthers basifixed, not glandular-punctate abaxially, coherent laterally in lower half, lacking basal appendages; disk cupular and shorter than ovary except opposite the 4-lobed lip where a scale-like extension exceeds the ovary; carpels 5; ovarioles distinct; style simple; stigma capitate. Fruit of 1–5 follicles. Seed 1 per follicle, ellipsoid-reniform, tuberculate, glabrous; testa crustaceous, a subhilar rumination present; embryo curved; cotyledons conduplicate, bilobed at apex.

Two spp., one widespread from SW Mexico through N South America to Peru, Bolivia and Rio de Janeiro; the other in eastern Brazil.

Doubtfully attached to Angostura Alliance:

104. *Hortia* Vand

Hortia Vand., Fl. lus. bras. spec.: 14 (1788).

Shrubs or trees; trichomes simple. Leaves alternate, simple, often clustered at branch ends, up to 1.25 m long. Inflorescence a terminal, dichasially branched, often large and showy, corymbose panicle. Flowers small, 5-merous, haplostemonous; calyx cupuliform; petals valvate, fleshy, usually barbate in lower half, pink; all stamens fertile; disk cupular or annular, 5-lobed, more or less adnate to base of gynoecium; carpels connate; ovary ellipsoid, tapering into a 5-grooved, short style; stigma small; ovules 2 per carpel, superposed. Fruit when immature a 5-locular drupe, when mature a 4–7-seeded berry lacking evident locules; epicarp coriaceous, glandular; meso- and endocarps in immature fruits fibrous or spongy and in mature fruits fleshy and succulent. Seeds conical, with circular hilum, albuminous; testa crustaceous; embryo straight; cotyledons flat, oval, fleshy.

Ten spp., Panama and tropical South America; *H. regia* Sandw. with yellow, juicy, edible fruits; *H. excelsa* Ducke (Pará) with yellow timber ('pau amarelo').

105. *Rutaneblina* Steyerm. & Luteyn

Rutaneblina Steyerm. & Luteyn, Ann. Missouri Bot. Gard. 71: 314 (1984).

Low shrubs. Leaves alternate, simple, subsessile. Inflorescence an almost sessile, condensed or corymbiform dichasium. Flowers regular, 5-merous; sepals distinct, triangular; petals creamy white or pink, distinct, contorted in bud; all stamens fertile; filaments distinct, subterete, with glandular patches at base adaxially; anthers basifixed, deltoid-ovoid; disk obsolete; ovarioles distinct; style simple; stigma capitate; ovules 2 per carpel. Fruit of 1–5, bluntly beaked follicles, with separating dorsilateral endocarp. Seed 1 per follicle, flattened-tear-shaped, smooth, shiny, with elongate hilum; testa crustaceous, a subhilar rumination absent; endosperm present?; embryo straight; cotyledons plano-convex, fleshy.

One sp., *R. pusilla* Steyerm. & Luteyn, Guayana Shield in southern Venezuela.

This genus was described without tribal affiliation but doubtfully is placed here. In seed structure, it may approach *Esenbeckia*: it may have endosperm, and the embryo is enclosed in two layers, an outer thin one attached to the chalazal area, and a thicker inner one that may be endosperm.

I. Amyris Alliance (Amyridinae Engler 1896, without *Teclea*).

Carpel 1; drupe 1-seeded; seed exalbuminous; cotyledons plano-convex, thick, stiff.

Two genera, New World.

106. *Amyris* P. Browne

Amyris P. Browne, Hist. Jamaica: 208 (1756); Gereau, Candollea 46: 227–235 (1991), rev. S. Am. spp.
Kodalyodendron Borhidi & Acuña (1973).

Glabrous shrubs or trees, sometimes armed with short, axillary spines. Leaves opposite or alternate, imparipinnate and 1–11-foliolate. Inflorescence a terminal or axillary corymbiform panicle, pseudoraceme, or flowers solitary. Flowers hermaphrodite, rarely unisexual, (3)4 (5)-merous but 1-carpellate, diplostemonous; calyx cupular; petals white, imbricate; filaments slender; anthers ovate or oblong, shorter than filaments; disk cushion-like and gynophorous or 0; carpel 1; stylodium very short and stout or 0; stigma capitate to discoid-subcapitate; ovules 2, suspended from top of the locule, collateral. Fruit a globular to ellipsoid drupe; pericarp fleshy; endocarp pergamentaceous. Seed globose or ellipsoid; testa thin, papery; embryo straight; cotyledons plano-convex, thick, stiff.

About 40 spp., West Indies, Mexico, C America, north-western South America from Venezuela to Peru, with one or two spp. in North America (Texas and Florida).

107. *Stauranthus* Liebm.

Stauranthus Liebm., Kjoeb. Vid. Meddel.: 91 (1853); Pool, Fl. Nicaragua 3: 2299 (2001).

Evergreen shrubs or small trees, functionally dioecious. Leaves alternate, 1-foliolate; petiole alate, distally thickened. Inflorescences axillary or terminal, short racemes. Flowers unisexual, 4- or 5-merous but carpel 1, haplostemonous; calyx cupular; petals white, valvate, inflexed at apex; filaments subulate; anthers orbicular, notched at base, shorter than filaments; disk 0; carpel sessile; style 0; stigma large, disk-shaped or capitate; ovule 1, apical, pendulous. Fruit a drupe; exocarp fleshy, oily; endocarp coriaceous. Seed oblong; testa thin, coriaceous; embryo straight; cotyledons plano-convex, thick, stiff.

One sp., *S. perforatus* Liebman, SE Mexico to Panamá. A second sp., *S. conzattii* Rose & Standley from Oaxaca, may belong to *Amyris*; a flowering specimen that seems to be a good match with the fruiting type specimen has diplostemonous flowers (J.A.K.).

J. Ruta Alliance (Tribe Ruteae DC. 1822 subtr. Rutinae Engl. 1893).

Halfshrubs or perennials, or rarely shrubs; flowers actinomorphic, diplostemonous; petals imbricate (valvate in the questionable *Chloroxylon*); ovules mostly 4–8, inserted in two series on a capitate placenta; dorsilateral endocarp adnate to mesocarp and persistent in dehisced fruit; synthesising rutaceous metabolites including special acridones and naphthalene coumarins. $n = 9, 10$.

Seven genera, temperate and subtropical regions of northern hemisphere (*Thamnosma* also in Africa).

108. *Ruta* L.

Ruta L., Sp. Pl.: 383 (1753).
Ruta subg. *Euruta* Engl. (1895).

Perennial herbs and halfshrubs. Leaves alternate, compoundly or simply pinnate. Inflorescences terminal and axillary thyrsoids. Lateral flowers 4-merous, terminal flowers usually 5-merous; petals denticulate or fimbriate, usually clawed; stamen filaments glabrous, narrow, attenuate; disk large, cushion-shaped; gynoecium ± syncarpous, anacrostylous; ovary lobed; style with unthickened stigma; ovules several per carpel. Fruit a capsule, dehiscing from the inner side of apex. Seeds several per locule, dorsally angled; endosperm copious; embryo straight. $n = 10, 20, 38, 39, 40$.

Seven spp., from Macaronesia and the Mediterranean region to southwest Asia.

109. *Boenninghausenia* Reichb. Fig. 59

Boenninghausenia Reichb., Consp.: 197 (1828), nom. cons.

Slender herbaceous perennials. Leaves alternate, pinnately to ternately decompound (up to 27-foliolate). Inflorescences terminal and axillary thyrses. Flowers 4-merous; petals obovate; stamens (5–)8; anther dehiscence latrorse; filaments filiform; disk broadly cupular, denticulate; carpels on elongate, thin gynophore; ovarioles distinct, joined in a strongly anacrogynous style with simple stigma; ovules 4–8 per locule. Fruit of 4 mericarps dehiscing ventrally-apically. Seeds few in each fruitlet, reniform, black, granular; endosperm fleshy; embryo curved. $n = 10$.

One (–3?) spp., Himalayas eastwards to Japan and south to Java and Lesser Sunda Islands.

110. *Psilopeganum* Hemsl. ex Forb. & Hemsl.

Psilopeganum Hemsl. ex Forb. & Hemsl., J. Linn. Soc. 23: 103, t. 3 (1886).

Herbaceous perennial. Leaves alternate, 3-foliolate. Inflorescence terminal and axillary few-flowered leafy cincinni, or flowers solitary. Flowers 4(5)-merous but carpels only 2(3); disk columnar; ovarioles proximally connate, otherwise contiguous, joined in an anacrogynous style; stigma capitate; ovules 5–6 per locule. Fruit a 2-lobed capsule with ventricidal, 4–6-seeded locules. Seeds reniform; seed coat tuberculate; endosperm carnose; embryo curved.

One spp., *P. sinense* Hemsl., China.

111. *Thamnosma* Torr. ex Frém.

Thamnosma Torr. ex Frém. in Frém. [2nd] Rep.: 313 (1845); Thulin, Nord. J. Bot. 19: 5–11 (1998), rev. Old World spp.

Halfshrubs or shrubs. Leaves alternate, narrow, simple and sometimes deeply divided. Inflorescences terminal or axillary scorpioid cymes. Flowers 4-merous but gynoecium 2(3)-carpellate; anther dehiscence latrorse; disk crenate or cupular; ovary on gynophore, syncarpous, 2(3)-lobed, anacrostylous; style thin; stigma capitate; ovules 4–8 per locule. Fruit a 2(3)-locular capsule. Seeds curved, with carnose endosperm; testa usually muricate. $n = 9$.

Eight spp., Somalia, southern Africa, southern U.S.A. and northern Mexico.

The subgeneric distinction proposed by Engler (1931) based on seed sculpture is untenable; *T. montana* Torr. & Frém. is the only species that has almost smooth seeds (Thulin 1998).

112. *Haplophyllum* A.Juss.

Haplophyllum A.Juss., Mém. Mus. Paris 12: 464 (1825), nom. cons.; Townsend, Hook. Icon. Pl. 40: 1–336 (1986), rev.
Ruta subg. *Haplophyllum* (A.Juss.) Engl. (1895).

Perennial herbs or halfshrubs. Leaves alternate, simple or more rarely 3- or 5-parted with lanceolate or linear segments. Inflorescence terminal, thyrsoid. Flowers 5-merous; petals entire, shortly clawed or gradually tapering; stamen filaments usually bearded within and glandular, distinct or monadelphous; disk pulvinate; gynoecium 3-5-carpellate, almost anacrostylous, on short gynophore; ovarioles basally connate, otherwise contiguous, distally joined in a common style with unthickened stigma; ovules (1)2–4(–8) per locule. Fruit a capsule, dehiscing from inner side of apex, or indehiscent. Seeds few per locule, reniform and dorsally rounded; endosperm copious, carnose. $n = 9$.

About 66 spp., from the western Mediterranean and northern and north-eastern Africa and Arabia through C Asia to China (Prov. Heilungkiang).

A subdivision into three sections has been proposed by Townsend (1986).

113. *Cneoridium* Hook.f.

Cneoridium Hook.f. in Benth. & Hook., Gen. Pl. 1: 312 (1862).

Low glabrous shrub. Leaves opposite, simple. Inflorescences axillary, usually 3-flowered cymes. Flowers 4-merous, diplostemonous, but gynoecium 1-carpellate; disk ring-shaped; ovary sessile with style arising from near base of carpel on one side; stigma capitate; ovules 2–4. Fruit a globose, fleshy, punctate, thin-skinned berry. Seeds 1–2, globose, with fleshy endosperm; embryo curved. $n = 18$.

One sp., *C. dumosum* (Nutt.) Hook.f., south-western North America (California and Baja California).

Doubtfully attached to Ruta Alliance:

114. *Chloroxylon* DC.

Chloroxylon DC., Prodr. 1: 625 (1824); Capuron, Adansonia II, 1: 65–82 (1961), and 7: 479–500 (1967).

Trees with green-yellow wood. Leaves alternate, (im)paripinnate. Inflorescences terminal, paniculate. Flowers 5-merous, diplostemonous, but carpels 3; petals valvate; disk 10-crenate; gynoecium fully syncarpous; style very short; stigma small; ovules 6–8 per locule in 2 rows. Fruit a loculicidal capsule with a fragile central axis. Seeds apically winged, imbricately inserted on inner margin of capsule; endosperm 0; cotyledons plano-convex, ovate. $n = 10$.

Three spp., two in Madagascar, and *C. swietenia* DC. in southern India and Sri Lanka.

K. Genera of Rutoideae lacking close relatives

115. *Casimiroa* La Llave

Casimiroa La Llave in La Llave et Lex., Nov. veg. descr. fasc. 2: 2 (1825); Martinez, An. Inst. Biol. Univ. Nac. México 22: 25–181 (1951), C Am. spp.
Sargentia S. Watson (1890); see Chiang, Taxon 38: 116–119 (1989).

Large shrubs or trees, rarely dioecious. Leaves alternate, deciduous or persistent, digitately (1)3–5(7)-foliolate. Inflorescences terminal and axillary panicles or rarely solitary flowers. Flowers hermaphrodite or unisexual by abortion of anthers or ovary, haplostemonous; sepals 4–6, united at base; petals (4)5, white, valvate; stamens 5; filaments terete; anthers cordate, shorter than filaments, not glandular-punctate; disk annular; carpels (2–)4–5(–8), connate; ovary lobed, immersed in disk; style usually short; stigma lobed or entire; ovule 1 per carpel. Fruit a berry or a drupe with an often well developed sarcocarp, resembling an apple, 2–5(–8)-locular, developing 2–5(–8) large 1-seeded pyrenes or (*C. sapota* Oerst.) seeds lacking the hard endocarp (J.A.K., pers. obs.). Seed ellipsoid, with elongate hilum, without endosperm; testa coriaceous; cotyledons plano-convex, fleshy.

About ten spp., from highlands of Mexico extending to Texas and Costa Rica.

C. edulis LaLlave ex Lex. (white sapote, zapote blanco, and matasano or matasán in Guatemala) is widely distributed in Central America and much cultivated for its fruit that resembles an early apple in shape, size and colouring and also in texture and smell of its pleasantly sweet flesh. It is believed popularly that if eaten in considerable amounts, the fruit induces sleep. The glucoside casimirosine extracted from it is said to have a hypnotic and sedative effect.

116. *Dictamnus* L.

Dictamnus L., Sp. Pl.: 383 (1753).

Perennial suffrutescent herbs, the stems covered with short, patent hairs and dark, stipitate glands which release a volatile inflammable oil. Leaves alternate, imparipinnate, 3–6-jugate. Inflorescences terminal, simple or compound racemes. Flowers showy, slightly zygomorphic, 5-merous, diplostemonous; petals narrow, the 4 upper bent upwards, the lowest declinate; stamen filaments declinate and upwardly curving; disk thick, annular; gynoecium syncarpous, on short gynophore, anacrostylous; ovary 5-lobed; style declinate; stigma small, simple; ovules 3(4) per carpel. Fruit capsular, after decomposition of the style disintegrating into 5 mericarps coherent only at base; endocarp discharged explosively with seeds at dehiscence. Seeds 3 per mericarp; testa black, thin, shiny; endosperm fleshy; embryo with thick cotyledons. $n = 15(?)$, 18.

Probably a single polymorphic sp., *D. albus* L., from warm-temperate Europe through temperate Asia to N China.

117. *Flindersia* R.Br.

Flindersia R.Br.in Flinders, Voy. Terra Austr. 2: 595, t. 1 (1814); Hartley, J. Arnold Arb. 50: 481–526 (1969), rev.; Whiffin, Aust. J. Bot. 30: 635–645 (1982), phylog.

Trees; trichomes simple or fasciculate to lepidote. Leaves alternate to opposite, impari- or paripinnate, 1–8-jugate, or 1-foliolate. Inflorescences terminal and/or axillary, paniculate. Flowers bisexual or functionally unisexual, 5-merous; stamens 5, alternating with 5 filament-like staminodes; disk cupular, deeply to shallowly folded-crenulate to folded-lobed; carpels completely connate, usually with 5 glands around base of the style; style short, slightly anacrogynous; stigma capitate, 5-ridged; ovules 2–6 per locule (1–3 on each side of a protruding placenta). Fruit a 5-valved 5-loculed septicidal capsule. Seeds 2–6 per locule, 1–3 attached on each side of each of the placentas which are much enlarged in fruit, winged at both ends or at apical end only; testa pergamentaceous, without sclerotesta; endosperm 0; embryo elliptic to oblong, cotyledons flattened, elliptic or transversely so. $n = 18$, 54.

Seventeen spp., Australia (Queensland, NS Wales, 15 spp., 11 endemic), Moluccas (Ceram) to Irian and Papua, rainforest and dry scrub.

118. *Pilocarpus* Vahl.

Pilocarpus Vahl, Eclog. 1: 29 (1797); Kaastra, Fl. Neotropica 33: 132–181 (1982), rev.; Skorupa, Revisião taxonômica de *Pilocarpus*. Doctoral Thesis, Univ. of São Paulo, Brazil (1996).

Shrubs or trees. Leaves alternate, often crowded at tips of branchlets, 1-, 2- or 3-foliolate or imparipinnate, 2–6-jugate. Inflorescences terminal or lateral on older stems, elongate racemes or spikes. Flowers 5(4)-merous, haplostemonous; calyx persistent; petals green or reddish brown, distinct, induplicate-valvate, inflexed at tip; filaments subulate, flat near base, accumbent in grooves of disk; anthers with subapical dorsal gland; disk annular to cupular, completely adnate to ovaries; carpels connate, with apophyses, sometimes on a globose gynophore; style short; stigma 5-lobed; ovules 1 or 2 per carpel. Fruit of 1–5 follicles connate only at base, with separating, elastic endocarp. Seeds 1(2) per follicle, reniform-ovate, slightly carinate abaxially, smooth, shiny, without endosperm; testa papery; embryo slightly curved; cotyledons plano-convex, thick, sometimes unequal.

About 17 spp., Mexico to North Argentina, and Greater and Lesser Antilles, in moist and dry forest and woodland; absent from the Amazon basin.

119. *Pitavia* Molina

Pitavia Molina, Saggio Chile ed. 2: 287 (1810).

Dioecious (?) small trees or shrubs; trichomes simple. Leaves simple, crenate, opposite, or in trimerous whorls. Inflorescences upper-axillary, paniculate, their lateral branches opposite. Flowers functionally unisexual, 4-merous, diplostemonous; sepals imbricate, persistent; petals ovate, imbricate; anthers dorsifixed, introrse; disk columnar, gynophorous; ovarioles distinct, gland-tipped, subapically joined in a common style; stigma narrow; ovules 2 per carpel, collateral. Fruit of (4–)1 fleshy, punctate, 1-seeded drupes with a thick fleshy mesocarp and a thin ligneous endocarp lining the seed cavity in the fruit, splitting in the sagittal plane at apex and allowing emergence of the radicle. Seed not liberated, elongate-ellipsoid, with broad and long hilum; testa leathery; endosperm scant; embryo straight; cotyledons plano-convex, fleshy.

One sp., *P. punctata* Mol., coastal Cordillera of southern Chile at about 35–37°S.

This genus from the warm-temperate forest region of southern Chile is an unplaced genus of Rutoideae and has its closest relatives in the mainly Australasian Euodia Alliance (see under 'Generic Relationships' and 'Distribution and Habitats').

Hartley (1997: 204) describes the testa as thick and fleshy, and J.A.K. has observed the testa composed of a leathery outer layer and fleshy inner layer separable from each other only with difficulty. At germination, the pericarp is perforated by the radicula (J. SanMartin, pers. comm. Dec. 2008).

120. *Ptelea* L.

Ptelea L., Sp. Pl. 1: 118 (1753); Bailey, Brittonia 14: 1–45 (1962), rev.
Taravalia Greene (1906).

Shrubs or small trees, subdioecious. Leaves alternate, 3(4–5)-foliolate. Inflorescences terminal, corymbiform panicles, sometimes few- or 1-flowered. Flowers unisexual, haplostemonous, 4–5(6)-merous but carpels 2(3); sepals deciduous, petals greenish-white, imbricate; filaments pubescent in lower half or glabrous; anthers oval, shorter than filaments; disk gynophorous, rudimentary in male flowers; carpels connate; ovary compressed perpendicular to septum, usually narrowly 2-winged; style short and slender; stigma capitate, usually 2-lobed. Fruit a subcircular, 2(3)-locular samara with a broad dorsal wing on each carpel or (*P. aptera* Perry) rarely a wingless 2-valved capsule. Seed 1 per carpel, narrow-oblong, colliculate, with elongate hilum; testa thin, fragile; embryo straight; cotyledons flat, oblong to elliptic.

A North American genus for which over 60 species have been proposed in the past, whereas three, with the monotypic *Taravalia* included, are recognised by Bailey (1962); *P. trifoliata* L., the common hop tree or stinking ash, is a widely known ornamental.

121. *Skimmia* Thunb.

Skimmia Thunb., Nov. Gen. Pl. 3: 57 (1783), nom. cons.; Huang, Acta Phytotax. Sin. 7: 344–355, t. 66–70 (1958); Taylor, Kew Mag. 4: 168–194 (1987).

Glabrous shrubs with green twigs, dioecious or flowers hermaphrodite. Leaves alternate, simple, persisting for 1-several years. Inflorescence a many-flowered terminal panicle (females sometimes reduced to 1–5 flowers). Flowers bisexual or functionally unisexual, 4–5(–7)-merous and haplostemonous but gynoecium 2–5-carpelled, completely syncarpous, inserted on disk; style as long as ovary; stigma capitate to 2–5-lobed; ovule 1 per locule. Fruit a red or black fleshy 1–5-loculed drupe with 1–4(5) cartilaginous pyrenes. Seeds ovoid; testa membranous; embryo straight, surrounded by white endosperm. $n = 15, 16, 30$.

Four spp., much cultivated, from eastern Afghanistan and the Himalayas eastwards to Japan and south to Vietnam and the Philippines.

II. Subfam. Aurantioideae Eaton (1836).

Woody; schizogenous oil glands copious in vegetative tissues, flowers and fruits; gynoecia syncarpous; fruits baccate, usually with leathery or less often woody pericarp strongly embedded with oil glands, containing mucilaginous pulp secreted into the fruit locules or contained in juicy pulp vesicles originating from the endocarp; seeds exalbuminous, sometimes with more than 1 embryo. $n = 9$, most wild spp. diploid, but in *Glycosmis* hexa- and octoploids known (Guerra et al. 2000; Mou and Zhang 2009b).

A. Bergera Alliance (= Tribe Clauseneae Wight & Arn. 1834)

Axillary spines 0; leaves impari(3–1)-pinnate; leaflets alternate; rhachis nonarticulated; ovary 2–5-loculate; ovules 1 or 2 per locule; carbazole alkaloids present; limonoids usually 0.

122. *Micromelum* Blume

Micromelum Blume, Bijdr.: 137 (1825), nom. cons.; Swingle, Bot. Citrus: 139–152 (1943).

Unarmed small trees. Leaves imparipinnate, 3–11-jugate, rarely 3- or 1-foliolate. Inflorescences terminal, paniculate corymbs, often large and many-flowered. Flowers small, 5-merous, diplostemonous; petals valvate in bud; ovary 2–6-loculate, with radial walls often twisted; style slender, caducous; stigma flattened to subglobose; ovules 2 per carpel. Fruit a subglobose or oblong dry berry; pericarp thin, gland-dotted; pulp 0. Seeds 1–3, cotyledons thin, convolute and/or folded. $n = 9$.

About 10 spp., from India, Sri Lanka and Bangladesh to south China, and south through Indochina and Malesia to Australia, New Caledonia and the south Pacific; *M. minutum* (Forst.f.) Wight & Arn. a common and widespread coastal tree of the Indopacific.

All other Aurantioideae: Petals imbricate; ovary locules not twisted; cotyledons thick, plano-convex, not folded.

123. *Glycosmis* Corr.-Serr.

Glycosmis Corr.-Serr., Ann. Mus. Hist. Nat. Paris 6: 384 (1805), nom. cons.; Stone, Proc. Acad. Nat. Sci. Philadelphia 137: 1–27 (1985), rev.; Stone, Gard. Bull. 46: 113–126 (1994).
Thoreldora Pierre (1896).

Unarmed small trees or shrubs; new growth covered with dense fine-woolly rust-coloured pubescence. Leaves imparipinnate, 1- or 3-foliolate, or simple. Inflorescences terminal and/or axillary, often densely racemose. Flowers small, perianth (4) 5-merous; sepals united to the middle; stamens (4) 10; disk annular or cylindric; ovary 2–5-loculate; ovules 1(2) per locule; style very short and thick, persistent; stigma broad, persistent. Fruit a berry, with mucilaginous pulp or dry, 1–3-seeded; pulp vesicles 0. Seeds oval, with membranaceous seed coat. $2n = 18, 54, 72$.

About 50 spp. in the Indomalayan region, *G. parviflora* (Sims) Little naturalised in tropical Africa and America.

124. *Bergera* Koenig ex L.

Bergera Koenig ex L., Mant. 2: 563 (1771).
Murraya (L.) Spreng. (1825) p.p.; Swingle, Bot. Citrus: 200–203 (1943).
Chalcas sect. *Bergera* (Koenig ex L.) Tanaka, J. Soc. Trop. Agric. (Taiwan) 1: 23–44 (1929).

Shrubs; stem and root bark dark brown. Leaves imparipinnate, 2–15-jugate, rarely (‘*B. stenocarpa*’) 1-foliolate; rhachis wingless. Inflorescences usually many-flowered. Flowers (4)5-merous, diplostemonous, small; stamen filaments dilated; disk glabrous; pistil shorter than stamens; ovary 2-loculate; style elongate, caducous; stigma capitate. Fruit small, 8–10 mm long, ellipsoid, purple-black. Seeds few; testa glabrous.

Five spp. from India to SE Asia, Malesia, Formosa and New Caledonia.

The leaves of *B. koenigii* L. are an essential ingredient of all Ceylonese curries.

Morphological (Tanaka 1929; Chang et al. 1988), phytochemical (Kong et al. 1986, 1988; Chang et al. 1988) and karyosystematic evidence (Guerra et al. 2000) points to the heterogeneity of *Murraya* in the circumscription of Swingle (1943). The reinstated genus *Bergera* differs from *Murraya* s.str. chemically (see Table 1, p. xxx) and morphologically by its wingless rhachis, many-flowered inflorescences, much smaller flowers, dilated stamen filaments, dark-brown stem and root bark, glabrous seeds, and a foveolate pollen exine.

125. *Clausena* Burm.f. Fig. 74

Clausena Burm.f., Fl. Ind.: 87, 243 (1768); Swingle, Bot. Citrus: 158–192 (1943); Molino, Bull. Mus. Natl. Hist. Nat. Paris IV, 16 B, Adansonia: 105–153 (1994), rev.

Unarmed, sometimes deciduous trees or shrubs. Leaves imparipinnate; rhachis usually not winged. Inflorescences terminal and axillary, thyrso-paniculate. Flower buds globose to

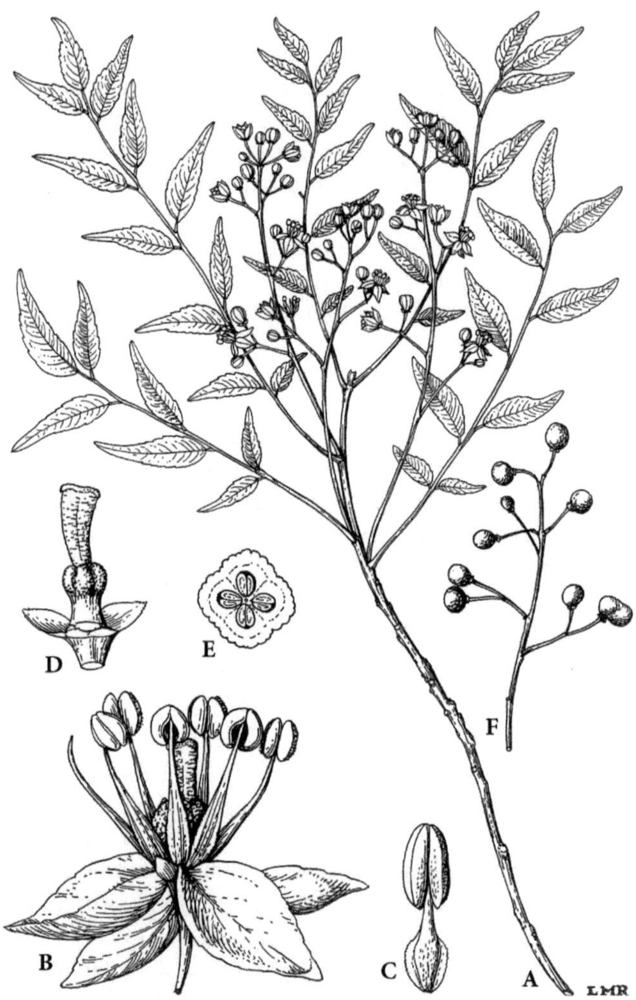

Fig. 74. Rutaceae. *Clausena anisata*. **A** Flowering branchlet. **B** Flower. **C** Stamen. **D** Gynoecium. **E** Transverse section of ovary. **F** Infructescence. (Mendonça 1963)

broad-ovoid; flowers 4–5-merous, diplostemonous; sepals fused into a cup below; nectary disk annular; gynophore well-developed, glabrous; ovary (2–)4–5-loculate, often pubescent or glandular; ovules 2 per locule; style caducous, short; stigma narrow. Fruits small, subglobose or ovate, with 2–5 segments. Seeds 1–3(–5), glabrous. $n = 9, 18$.

Fifteen spp., SW Asia, Malesia, New Guinea, Australia, one of them, *C. anisata* (Willd.) Hook.f. ex Benth. distributed from East Asia to tropical and South Africa.

B. Citrus Alliance (= Tribe Aurantieae Rchb. 1832).

Solitary or paired axillary spines usually present; leaflets of pinnate leaves shifting from alternate to opposite; leaves often reduced to become 3- and 1-foliolate with articulated and winged rhachis; flowers sometimes increased in size and tending to multiplication of stamen, carpel and ovule numbers; fruits increasing in size with elaboration of leathery or woody pericarp and fleshy pulp vesicles; carbazole alkaloids 0; coumarins, limonoids and polyoxygenated flavonols strongly diversified.

126. *Murraya* Koenig ex L.

Murraya Koenig ex L., Mant. 2: 554 (1771); Swingle, Bot. Citrus: 192–206 (1943) (excluding the spp. now in *Bergera*).
Chalcas sect. *Murraya* (Koenig ex L.)Tanaka, J. Soc. Trop. Agric. (Taiwan) 1: 23–44 (1929).

Unarmed shrubs or small trees; stem and root bark straw-coloured or whitish. Leaves imparipinnate; leaflets alternate; rhachis wingless or winged. Inflorescences few-(up to 10-)flowered, or flowers solitary. Flower buds cylindrical or long-ovoid, large. Flowers (4)5-merous, diplostemonous; stamen filaments filiform; disk annular, cushion-shaped, cylindric, or 0; ovary ovoid, 2–5-loculate; ovules 2(1) per locule, superperposed or almost collateral; style elongate, slender, caducous; stigma capitate. Fruit ovoid or subglobose, red, with mucilaginous pulp. Seeds 1–several; testa villous. $n = 9$.

Five species from India through SE Asia and Malesia to Australia (Queensland) and New Caledonia.

127. *Merrillia* Swingle

Merrillia Swingle, Phil. J. Sci. C, Bot. 13: 337 (1918); Swingle, Bot. Citrus: 206–209 (1943).

Unarmed tree; stem and root bark straw-coloured or whitish. Leaves imparipinnate; leaflets alternate, towards leaf tip increasingly larger, the terminal largest; rhachis narrowly winged. Inflorescence axillary, 2–1-flowered. Flowers large (up to 60 mm long), trumpet-shaped, slightly zygomorphic, 5-merous, diplostemonous; petals distinct; ovary 5 (6)-loculate, on well-developed gynophore, merging into the elongate style; stigma capitate; ovules 8–10 per locule. Fruit subglobose to oblong, large (10×7.6 cm), yellow-green; pericarp very thick, coriaceous, irregularly lacunose with cartilaginous walls; segments filled with clear mucilage. Seeds 8–10 per segment, lenticular; testa densely scaly.

One sp., *M. caloxylon* (Ridl.) Swingle, Thailand, Malay Peninsula, NE Sumatra, and Borneo (Sabah).

Contains a number of metabolites in common with *Murraya* s. str. (Kong et al., Biochem. Syst. Evol. 16: 47 (1988)).

128. *Triphasia* Lour.

Triphasia Lour., Fl. Cochinch.: 152 (1790); Swingle, Bot. Citrus: 236–239 (1943).
Echinocitrus Tanaka (1943).

Shrubs or small trees; spines paired or solitary. Leaves simple or 3-foliolate; petioles wingless. Flowers solitary or in groups of 2 or 3 in leaf axils, 3- or 5-merous, diplostemonous; disk annular; ovary 3–5-loculate; style slender, caducous; stigma capitate, 3- or 5-lobed; ovules 1 or 2 per locule. Fruit ovoid to subglobose, about 1.5 cm long. Seeds 1–several, embedded in mucilaginous pulp. $n = 9, 18$.

Three spp. in SE Asia, the Philippines and New Guinea; *T. trifolia* (Burm.f.) P. Wilson (limeberry, myrtle lime) with edible fruit and widely grown as an ornamental.

129. *Merope* Roem.

Merope Roem., Syn. monogr. 1: 144 (1846); Swingle, Bot. Citrus: 234–236 (1943).

Small tree; spines stout, paired or solitary. Leaves unifoliolate; petioles short, wingless. Inflorescences 1(2)-, rarely few-flowered, axillary. Flowers 5-merous, diplostemonous; stamens distinct; ovary 3(4)-loculate, on tall gynophorous disk; style short; stigma flat; ovules 2–4 per locule. Fruit an ovoid, angular berry with 3 or 4 flattened sides. Seeds few to several per fruit, oblong, flattened, overlapping.

One species, *M. angulata* (Willd.) Swingle, in tidal swamps in Myanmar, Indo-China, Malesia and New Guinea; the buoyant fruit is dispersed by the tides.

130. *Monanthocitrus* Tanaka

Monanthocitrus Tanaka, J. Arnold Arb. 9: 138 (1928); Swingle, Bot. Citrus: 225–229 (1943); Stone, Proc. Acad. Nat. Sci. Philadelphia 140: 272–274 (1988), key.

Shrubs; spines paired or solitary, subequal. Leaves simple; petioles short, wingless. Flowers small, solitary, 5-merous, diplostemonous, the buds ovoid or ellipsoid; disk annular; ovary 5-loculate; style stout; stigma 3–5-lobed; ovules 2–10 per locule. Fruit a globose to ovoid or pyriform berry; pericarp somewhat leathery, thin, glandular; locules lined with smooth, cartilaginous walls, without pulp vesicles. Seeds 6–10 per fruit, packed tightly in the locules and embedded in mucilage or scant pulp, flattened; testa somewhat spotted.

Four spp., Borneo, New Guinea (Irian Yaya, Papua New Guinea). Very close to *Wenzelia*, which differs, if at all (Stone 1988), by smaller leaves and solitary rather than paired spines.

131. *Wenzelia* Merr.

Wenzelia Merr., Phil. J. Sci. C, Bot. 10: 272 (1915); Swingle, Bot. Citrus: 214–224 (1943).

Shrubs, unarmed or with solitary spines. Leaves simple; petioles short, wingless. Flowers solitary in leaf axils, 5-merous, diplostemonous; stamens distinct; disk annular or shallowly cupular; ovary 4–5-loculate; ovules 6(8) per locule. Fruit obovoid or ellipsoid, 3–5 cm long, with coriaceous, glandular pericarp, segments containing mucilage without pulp vesicles. Seeds large, to 24 mm long, thick or flattened (subg. *Wenzelia*), or smaller, to 17 mm long, flat, the free edges with thin membranous margins (subg. *Papualimo*).

Nine spp., from southern Philippines to New Guinea (6), Bougainville Island, Solomon Islands, Fiji and Hawaii.

132. *Pamburus* Swingle

Pamburus Swingle, J. Wash. Acad. Sci. 6: 336 (1916); Swingle, Bot. Citrus: 240–243 (1943).
Atalantia sect. IV. *Pamburus* (Swingle) Engler (1931).

Much branched shrubs or small trees; spines straight, solitary (or in pairs?). Leaves simple; petioles short, neither winged nor articulated. Inflorescences short axillary racemes. Flowers 4- or 5-merous, diplostemonous; filaments distinct; disk cylindric; ovary 5- or 4-loculate; ovules 2 per locule. Fruits globose, 2.5 cm diam.; pericarp with many oil glands, orange; seeds surrounded by glutinous mucilage; pulp vesicles 0. Seeds large, thick, globose or ovoid, usually 1 or 2 per locule.

One sp., *P. missionis* (Wight) Swingle, South India, Sri Lanka.

133. *Luvunga* Wight & Arn.

Luvunga Wight & Arn., Prodr. Fl. Ind. Or.: 90 (1834); Swingle, Bot. Citrus: 244–253 (1943); Stone, Proc. Acad. Nat. Sci. Philadelphia 137: 221–223 (1985), key.

Woody lianas; spines solitary, recurved, irritable. Leaves 3(–1)-foliolate; petioles long, wingless, articulated at both ends. Inflorescences dense axillary racemes or panicles. Flowers diplostemonous, 3–5-merous; disk annular, pulvinate, or columnar; ovary 2–4-loculate; ovules 1–2 per locule, collateral or superposed; style gradually merging with ovary, caducous. Fruits ellipsoid or globose, 10–30 mm in diam.; rind thick, with oil glands on the surface and larger (oil or mucilage?) glands in deeper pericarp tissue. Seeds 1–3, embedded in mucilaginous pulp; pulp vesicles 0.

About 10 spp., from India through SE Asia and Malesia to NE Australia and New Guinea.

134. *Paramignya* Wight

Paramignya Wight, Ill. Ind. Bot.: 108, pl. 42 (1838); Swingle, Bot. Citrus: 253–269 (1943).

Woody lianas or erect shrubs; spines solitary, recurved or straight. Leaves 1-foliolate or simple; petioles short, articulated at both ends. Flowers axillary, solitary or in clusters, white, fragrant, large, 4–5-merous, diplostemonous; disk cup-shaped, conic, or columnar; ovary 3–5-loculate; ovules 1 or 2 per locule; style usually elongate, glabrous or pubescent; stigma globose to peltate. Fruits globose or obovoid, often narrowed towards base, 10–20 mm long, otherwise as in *Luvunga*. Seeds rather flat.

About 12 spp. (many more described), from India through SE Asia and Malesia to NE Australia. Perhaps not distinct from *Luvunga* (Mabberley 1998).

135. *Balsamocitrus* Stapf

Balsamocitrus Stapf, J. Linn. Soc. 37: 504 (1906); Swingle, Bot. Citrus: 463–465 (1943).

Deciduous trees; spines solitary or paired. Leaves 3(1)-foliolate. Inflorescences short axillary panicles. Flowers 5-merous, diplostemonous; disk annular; ovary 5–8-loculate; ovules many and in 2 series per locule; style short, caducous; stigma inconspicuous. Fruit large (up to 13×12 cm), ovoid-globose, with a hard, woody mesocarp, 5–8-loculate, each locule with many seeds in a liquid jelly. Seeds semi-ellipsoid, compressed; seed coat coriaceous, glabrous.

Two spp., tropical Africa.

136. *Aegle* Corr. Serr.

Aegle Corr. Serr., Trans. Linn. Soc. 5: 222 (1800); Swingle, Bot. Citrus: 453–456 (1943).

Deciduous trees; long-shoots with axillary solitary or paired spines; short-shoots with very consended internodes, spineless. Leaves (1)3(5)-foliolate. Inflorescences terminal or axillary, loosely branched. Flowers large; perianth 4- or 5-merous; calyx caducous; stamens 30–50+; disk small, annular; ovary 8–20-loculate; ovules numerous in 2 rows in each locule; style short; stigma thicker than style, longitudinally furrowed. Fruits (sub)globose, with thin, parenchymatous exocarp and woody, very hard mesocarp; segments 8–16(–20), narrow, filled with 6–10+ seeds embedded in glutinous gum that becomes hard when dry. Seeds oblong, slightly flattened; testa woolly. $n = 9, 18$.

Three spp., Indomalayan region; *A. marmelos* (L.) Corr. Serr., Indian bael tree, India, Myanmar, a Hindu sacred tree.

137. *Aeglopsis* Swingle

Aeglopsis Swingle, Bull. Soc. Bot. France 58, Mém. 8d: 237 (1911); Swingle, Bot. Citrus: 460–463 (1943).

Deciduous, much branched, very spiny small trees or shrubs; spines solitary, rarely paired. Leaves 1(2–3)-foliolate. Inflorescences axillary few- to many-flowered panicles. Perianth 3–5-merous; stamens 2 or 3 times as many as petals; disk large, lobed, enclosing the ovary; ovary 5–8-loculate; ovules 12–18 (?) per locule, in 2 series; style short; stigma cylindrical to capitate. Fruits subglobose or pyriform, up to 11×8 cm, with hard, woody mesocarp, 5–8-loculate, filled with mucilaginous juice; each locule many-seeded. Seeds ovate-compressed; seed coat coriaceous, glabrous. $n = 9$.

Five spp., W Africa, Uganda, Sudan.

138. *Afraegle* (Swingle) Engl.

Afraegle (Swingle) Engl., Pflanzenwelt Afrikas 3: 761 (1915); Swingle, Bot. Citrus: 456–460 (1943).
Balsamocitrus sect. *Afraegle* Swingle (1912).

Trees; spines solitary, rarely paired. Leaves 3-foliolate. Inflorescences small, 6–10-flowered (compound) racemes. Perianth (3–)5-merous; calyx persistent; stamens 3–4 times as many as petals; disk broad, lobed; ovary 8(–10)-loculate; ovules many per locule; style short; stigma long-ovate. Fruits large, globose or pear-shaped, with hardish mesocarp, many-seeded. Seeds up to 17 mm long and 10 mm thick, ovate or wedge-shaped at base.

Four spp., W Africa.

139. *Swinglea* Merr.

Swinglea Merr., J. Arnold Arb. 8: 131 (1927); Swingle, Bot. Citrus: 448–452 (1943).

Small tree; spines slender, straight, sharp, in pairs at the axils or else one of the spines replaced by a branch. Leaves 3-foliolate. Inflorescences axillary, few-flowered. Flowers 5-merous, diplostemonous; ovary 8–10-loculate; style well developed; stigma rounded; ovules numerous per cell. Fruit oblong, longitudinally ribbed, with very thick, leathery pericarp embedded with ellipsoid oil-glands, 4–7 mm long; the locules lined with mucilage glands, filled with gum, and surrounded with watery tissue; pulp vesicles 0. Seeds flattened, ovate, hairy, numerous in the long narrow cells.

One sp., *S. glutinosa* (Blanco) Merrill, Philippines: Luzon.

Formerly included among Hard Shelled Citroid Fruit Trees (Swingle 1943), but clearly not belonging there (Guerra et al. 2000; Samuel et al. 2001; Bayer et al. 2009).

140. *Burkillanthus* Swingle

Burkillanthus Swingle, J. Arnold Arb. 20: 255 (1939); Swingle, Bot. Citrus: 291–295 (1943).

Tree; some branches with straight, usually paired spines, others spineless. Leaves 1-, 2-, (3-) foliolate; petioles narrowly winged or merely margined, articulated. Inflorescences few-flowered axillary fascicles. Flowers 5-merous, diplostemonous; disk short, cylindrical; ovary 5-loculate; ovules 22–26 in 2 rows in each locule; style slender; stigma capitate. Fruits large, to 11×9 cm, ovoid, with a thin, glandular peel covering a thin, hard, woody endocarp; pulp vesicles sessile on dorsal locule walls, very long, swollen near the base, then cylindric, then tapering to an acute apex. Seeds very numerous, 22–27 mm long, with thin, slightly wrinkled seed coat, immersed in mucilaginous gum mingled with collapsed remnants of the pulp vesicles; embryo single, with light buff (not green) cotyledons.

A single sp., *B. malaccensis* (Ridl.) Swingle, Malay Peninsula, Sumatra.

141. *Pleiospermium* (Engl.) Swingle

Pleiospermium (Engl.) Swingle, J. Wash. Acad. Sci. 6: 426 (1916); Swingle, Bot. Citrus: 285–291 (1943).

Small trees; twigs with 1 or 2 axillary spines or unarmed. Leaves 3-, 2- or 1-foliolate; petioles articulated with leaf blades, winged or nearly wingless. Inflorescences few-flowered axillary fascicles or terminal panicles. Flowers 4- or 5-merous, diplostemonous; disk cupulate; ovary subsessile or short-stalked, 4- or 5-loculate; ovules 2 per locule; style not sharply delimited from ovary, ending in capitate stigma. Fruit globose or oblong, 2–3 cm diam., with slender pulp vesicles 1–2 to 10 mm long, disintegrating into an oily or resinous mass as the fruits ripen. Seeds about 1 cm long, flattened.

Five spp., SE Asia, W Malesia.

142. *Limnocitrus* Swingle

Limnocitrus Swingle, J. Arnold Arb. 21: 2 (1940); Swingle, Bot. Citrus: 297–299 (1943).

Shrub; branches with stout single spines. Leaves simple; petioles short, not winged, not articulated with leaf blade. Inflorescence terminal or terminal and subapical-axillary, short, corymbose. Flowers 4- or 5-merous, diplostemonous; stamen filaments distinct; anthers linear; disk annular; ovary oblong-ovate, with 15–20 narrow hirsute ridges, with a single large oil gland over each of the 4 or 5 locules; style slender; stigma slightly capitate, distended with 2 or 1 large oil glands between the stylar canals; ovules 2 per locule. Fruit (sub)globose, 3.5–4 cm diam.; pericarp gland-dotted, orange; locules filled with numerous slender, fusiform pulp vesicles. Seeds large, flattened-oval, with a creamy-white testa; embryo green.

A single sp., *L. littoralis* (Miq.) Swingle, S Vietnam, Java, Bali, in tidal swamps.

The species is little known and seems to share some characters with *Pleiospermium* but may be generically distinct.

143. *Naringi* Adans.

Naringi Adans., Fam. Pl. 2: 341 (1763).
Hesperethusa Roem. (1846); Swingle, Bot. Citrus: 299–302 (1943).

Slender trees or shrubs; spines solitary or paired. Leaves imparipinnate and 3–4-jugate to 3-, 2- or 1-foliolate; leaflets opposite; petiole and rhachis broadly winged. Inflorescences axillary, few-flowered racemes. Flowers small, 4- or 5-merous, diplostemonous; disk annular or cupular; ovary 4- or 5-loculate; ovules 1 or 2 per locule; style short. Fruit 0.6–2.5 cm across, one to all segments with 1 or 2 seeds embedded in sessile pulp vesicles of irregular size and shape. Seeds with hard, smooth seed coat. $n = 9$.

Two spp., *N. crenulata* (Roxb.) D.H. Nicolson [= *Hesperethusa crenulata* (Roxb.) Roem.]) and *N. alata* (Wall. ex Wt. & Arn.) J.L. Ellis [= *Pleiospermium alatum* (Wt. & Arn.) Swingle], in hot dry habitats from India eastwards through all continental SE Asia. The former species has 4-merous flowers and 1 ovule per ovary locule, whereas the latter is 4- or 5-merous and has 2 ovules per cell.

144. *Citropsis* (Engl.) Swingle & M. Kellerm.

Citropsis (Engl.) Swingle & M. Kellerm., J. Agric. Res. 1: 421 (1914); Swingle, Bot. Citrus: 302–322 (1943).

Shrubs or small trees; spines solitary or paired, rarely 0. Leaves imparipinnate, (1–3)5–9-foliolate; leaflets opposite; petiole and rhachis usually winged. Inflorescences short axillary racemes or cymose clusters. Flowers 4- or 5-merous, diplostemonous; disk annular; anther filaments in some spp. cohering laterally at the base in groups; anthers 2-lobed at base; ovary 4(5)-loculate; ovule 1 per locule; style long; stigma subglobose. Fruit globose, 2–3 cm diam., with pulp vesicles broad at the base and tapering to an acute tip. Seeds large (10×6×4 mm); testa hard, smooth. $n = 9$.

Eight spp. in tropical Africa.

Two subgenera distinguished by Swingle: subgen. *Citropsis*, petioles and rachis segments broadly winged; ovary not expanded at top by oil cells, and subgen. *Afrocitrus*, petiole and rachis segments narrowly winged or wingless; ovary distended at top by large oil glands.

145. *Limonia* L.

Limonia L., Sp. Pl. ed. 2: 554 (1763); Airy Shaw, Kew Bull. 1939: 293 (1939).
Feronia Corr. Serr. (1800), Swingle, Bot. Citrus: 465–468 (1943).

Polygamous deciduous or semi-evergreen tree; spines solitary, straight. Leaves imparipinnate; leaflets opposite; petiole and rhachis winged. Inflorescences compound racemes. Flowers bisexual or male, (4)5(6)-merous, diplostemonous; calyx caducous; anthers longer than filaments; disk short; ovary at first incompletely 4–6-loculate, then becoming 1-loculate with numerous ovules on 4–6 parietal placentae; style thick, attenuate; stigma oblong, fusiform. Fruit large, globose, with a hard, woody shell; the single cavity bearing numerous, densely packed seeds embedded in gum-like pulp; pulp vesicles 0. Seeds oblong, compressed; testa minutely pilose. $n = 9$.

One sp., *L. acidissima* L. (= *Feronia elephantum* Corr. Serr.), elephant apple, or wood apple, India, Sri Lanka, Java (?).

146. *Atalantia* Corr. Serr.

Atalantia Corr. Serr. in Ann. Mus. Hist. Nat. Paris 6: 383 (1805); Swingle, Bot. Citrus: 322–339 (1943).
? *Severinia* Tenore (1840).

Small trees or shrubs with solitary, stout, sharp spines, or spineless, especially the fruiting branches of old trees. Leaves 1-foliolate or simple; petioles wingless, articulated with blade. Inflorescences of axillary (terminal) fascicles, corymbs, or panicles. Flowers 3–5-merous, diplostemonous; calyx 3–5-lobed or forming a continuous envelope over the other floral organs and splitting ± irregularly into 2 or 3 parts; petals imbricate; stamen filaments distinct or ± connate into a staminal tube; disk annular or cupular; ovary (1–)3–5-loculate; ovules 1 or 2 per locule; style caducous; stigma capitate. Fruits small, subglobose; pericarp greenish-yellow when ripe and dotted with oil glands; pulp vesicles broad-based, sessile, peripheral, or filling the space in the segments left by the seeds. Seeds oblong, green within, sometimes polyembyonic.

Seventeen spp., from India and SE Asia through Malesia to New Guinea. A rootstock for *Citrus*.

Swingle (1943) distinguished two subgenera: subgen. *Atalantia*, stamen filaments more or less connate; pulp vesicles abundant; subgen. *Risso*, stamen filaments distinct; pulp vesicles few or none(?). In the circumscription of this genus, we are following Burkill (1931) (see Engler 1931: 459) and Swingle (1943) rather than Engler (1931) in including only species with pulp vesicles.

Bayer et al. (2009) showed that *Severinia* is closely allied to and "almost certainly" congeneric with *Atalantia* but a re-evaluation including a broader sampling of *Atalantia* would be desirable.

147. *Citrus* L.

Citrus L., Sp. Pl. 1: 401 (1753); Swingle, Bot. Citrus: 386–445 (1943); Bayer et. al., Am. J. Bot. 96: 668–685 (2009).
Poncirus Raf. (1838); Swingle, Bot. Citrus: 366–373 (1943).
Oxanthera Montr. (1860); Swingle, Bot. Citrus: 229–234 (1943).
Feroniella Swingle (1912); Swingle, Bot. Citrus: 468–471 (1943); see also Mabberley, Blumea 55: 73–77 (2010).
Eremocitrus Swingle (1914); Swingle, Bot. Citrus: 361–366 (1943).
Microcitrus Swingle (1915); Swingle, Bot. Citrus: 376–386 (1943).
Fortunella Swingle (1915); Swingle, Bot. Citrus: 342–361 (1943).
Pleurocitrus Tanaka (1929); Swingle, Bot. Citrus: 376–386 (1943).
Clymenia Swingle (1939); Swingle, Bot. Citrus: 373–376 (1943).

Glandular, aromatic shrubs or trees, evergreen, *C. trifoliata* deciduous and then developing from dormant buds with bud scales, usually with solitary, rarely paired spines, the older branches often spineless. Leaves 1-foliolate, rarely simple, in *C. trifoliata* 3-foliolate, in *C.* ['*Feroniella*'] *lucida* imparipinnate with opposite leaflets, bilateral, in *C.* ['*Eremocitrus*'] *glauca* isobilateral and one edge to light; petioles and rhachis segments winged or cylindrical and clearly jointed with blade (except in *C. medica* and *C. polyandra*). Inflorescences short, axillary, corymbiform racemes, or flowers in pairs or solitary in leaf axils. Flowers bisexual or sometimes bisexual and unisexual by abortion of the gynoecium; calyx (4)5(–7)-lobed; petals (4)5(–8); stamens 16–20(–60, in *C. polyandra* up to 100), polyadelphous or distinct; in *C. lucida* the filaments on inner side with a pilose linear basal appendage; disk annular to cushion-shaped, supporting the pistil; ovary (2–)5–15(–18)-loculate; ovules (2) 4–8+; in *C. lucida* the ovary locules confluent to become one lobed cavity with numerous ovules in parietal placentation; style cylindric, caducous; stigma capitate, sometimes slightly lobed. Fruit a sarcocarp (hesperidium), usually large to very large, ellipsoidal to cylindric and often mamillate at apex or pyriform to subglobular, sometimes depressed at apex; pericarp differentiated into thin, wax-covered exocarp; the outer yellowish-green to orange, leathery part of the mesocarp (flavedo) dotted with numerous oil glands; a middle, thick and spongy, white inner part of the mesocarp (albedo); the inner, membranaceous endocarp with juicy, slender-stalked, fusiform, pulp vesicles inserted mostly on dorsal walls, in '*Clymenia*' and 'subg. *Papeda*' mostly on radial walls; in '*Feroniella*' fruit with hard mesocarp composed of radially arranged woody prisms and a single cavity containing the seeds immersed in resinous gum. Seeds ellipsoidal to obovoid, plump or flattened, sometimes beaked at apex, usually a few in each locule; seed coat leathery, hard, or membranous; embryos 1–several. $n = 9$.

About perhaps 30 wild spp. indigenous to S, E and SE Asia, Malesia, Australia and the SW Pacific islands.

In the expanded generic concept that here is followed (Mabberley 2004; Bayer et al. 2009), all segregate genera that once had been split off from *Citrus* are reunited and combined with *Oxanthera*, *Clymenia* and *Feroniella*. Species numbers are particularly contentious in edible *Citrus*, because interspecific hybrids readily form between species and, as apomixis is common, can be fixed as cultivars. The numerous ancient apomictic hybrids and cultivars selected from these have been recognised as 'species' of edible *Citrus* by some authors; Tanaka (1954) proposed 145 species and, later (1961), 157. Several of them are widely cultivated and have often escaped from cultivation in warm regions of the world, and new combinations are still being synthesised for new fruit crops and rootstocks for grafting, the latter often including *C.* '*Poncirus*' *trifoliata*. Mabberley (1997) related the bulk of commercial edible *Citrus* to the following species and hybrid groups:

C. x *aurantium* L. (oranges and grapefruits, the most important group commercially)

C. x *limon* (L.) Osb. (lemons)

C. maxima (Burm.) Merr. (pomelo)

C. medica L. (citron)

C. x *taitensis* Risso (rough lemons)

C. x *aurantiifolia* (Christm.) Swingle (lime)

C. reticulata Blanco (tangerine, mandarine, satsuma)

III. Subfam. Cneoroideae Webb (1842).

Spathelioideae Engl. (1896).
Cneoraceae Vest (1818).
Ptaeroxylaceae J. Leroy (1960).

Multicellular oil glands restricted to the margin of the leaflets only in *Spathelia* and *Dictyoloma*; solitary oil cells in parenchymatous tissues generally present; flowers hermaphrodite or functionally unisexual, (3)4–5-merous, haplostemonous or rarely (*Harrisonia*) diplostemonous; stamen filaments appendaged only in *Spathelia*, *Dictyoloma* and *Harrisonia*; ovules 1, 2 or rarely (*Dictyoloma*) 4–5; fruits capsular, drupaceous, or rarely (*Dictyoloma*) follicular; endocarp crustaceous, bony or woody; chromones generally present; prenylated coumarins and limonoids in several genera; alkaloids (quinolone type) known only from *Dictyoloma*.

Seven genera, Old and New World.

148. *Spathelia* L.

Spathelia L., Sp. Pl. ed. 2: 386 (1763), nom. cons.; Cowan & Brizicky, Mem. New York Bot. Gard. 10 (2): 58–64 (1960).
Sohnreyia Krause (1914).
Diomma Engler ex Harms (1931).

Monocarpic polygamous trees with simple unbranched trunks conspicuously marked with leaf scars; multicellular oil glands only on margins of leaflets; oil cells in leaves and axes. Leaves alternate, imparipinnate, 2–100-jugate. Inflorescences terminal thyrses, up to 3 m long. Flowers 5-merous, haplostemonous; petals quincuncial; stamen filaments usually broadened or winged at base, the wing shortly free at apex; anthers dorsifixed, ellipsoid; carpels connate; ovary 2–3-locular (rudimentary in male flowers), in subg. *Spathelia* each carpel with a large, dorsal secretory cavity, in subg. *Diomma* the ovary borne on a cushion-like, nectariferous disk; style short or 0; stigma lobed; ovule 1(2) per carpel, pendulous, apotropous. Fruit 2- or 3-alate, indehiscent; endosperm present or not; embryo straight; cotyledons plano-convex.

About 18 spp., from the West Indies (Bahamas, Cuba, Jamaica) to northern South America (Guayana Highland, Venezuela, Brazil, Peru, Colombia).

The validity of the three subgenera (*Spathelia*, *Diomma* and *Sohnreyia*) distinguished by Cowan and Brizicky (1960) needs to be confirmed, as it is questionable whether the characters on which these are based (presence of a gynophore and a secretory cavity in the carpels, or of an endosperm) are correlated in the way assumed by these authors. Species with leaves with very many (up to 200) leaflets (subg. *Spathelia*) seem to be restricted to the West Indies.

149. *Dictyoloma* A.Juss.

Dictyoloma A.Juss., Mém. Mus. Paris 12: 499, t. 24 (1825), nom. cons.

Monoecious small trees; multicellular oil glands only on margins of the leaflets, their cell walls not resorbed. Leaves alternate, bipinnate. Inflorescences branched corymbose panicles; prophylls 0. Flowers functionally unisexual, 5-merous, haplostemonous; stamen filaments at the base with a woolly appendage; anthers dorsifixed, versatile, notched at base; disk thick, pubescent, 5-lobed; ovarioles distinct but the stylodia connate into a short, 5-furrowed style; stigma large, 5-lobed; ovules 4–5 in 2 series per carpel. Fruit separating into 5 follicles, each 4–5-seeded, dehiscing ventrally from base to apex; endocarp thin, crustaceous. Seeds compressed, dorsally conspicuously winged, strongly curved (micropyle, hilum, and chalazal area adjacent to each other); endosperm sparse; embryo curved around the intruded testa; cotyledons narrow-oblong.

One sp., *D. vandellianum* A. Juss. Bolivia, Peru, Brazil; the ground leaves ('barbasco negro') are used for stunning fish.

150. *Ptaeroxylon* Eckl. & Zeyh.

Ptaeroxylon Eckl. & Zeyh., Enum. 1: 54 (1835); White & Styles in Fl. Zamb. 2 (2): 547–550 (1966).

Dioecious shrub or small to medium-sized tree. Leaves opposite, paripinnate. Inflorescences contracted thyrses. Flowers tetramerous but carpels 2, haplostemonous; sepals distinct almost to the base, with open aestivation; petals imbricate; disk cushion-like, lobed; carpels connate; ovary laterally compressed, 2-locular; style with 2 spreading stigmatic lobes; ovule 1 per locule, with adaxial raphe. Fruit a capsule, dehiscing with 2 lobed valves from central column. Seeds with long terminal wing, exalbuminous; embryo curved.

One sp., *P. obliquum* (Thunb.) Radlk., from north-eastern Tanzania to the Cape Province of South Africa.

151. *Bottegoa* Chiov.

Bottegoa Chiov., Result. Sc. Miss. Stefan.-Paoli Somal. Ital. 1: 58 (1916); van der Ham et al., Kew Bull. 50: 243–265 (1995).

Shrub or small tree; oil cells in leaves and flowers. Leaves alternate, biparipinnate; leaflets isobilateral, upper side with small, opaque glands. Inflorescences depauperate axillary panicles or thyrses with up to 10 flowers. Flowers hermaphrodite(?), 4(5)-merous but carpels 2, haplostemonous; sepals spreading, slightly connate; petals valvate; stamen filaments slightly winged near base, without scale; disk gynophorous; carpels connate; ovary on short stipe, laterally compressed, winged and incompletely 2-locular; style with 2 recurved stigmatic lobes, ovule 1 per locule. Fruit indehiscent, subcircular, 3.5–5 cm wide, broadly 2-winged. Seeds 1 or 2, reniform, compressed; embryo curved; endosperm thin, rich in oil.

One sp., *B. insignis* Chiov., south-eastern Ethiopia, southern Somalia and north-eastern Kenya in *Acacia-Commiphora* woodland, bushland and wooded grassland, at 250–1,200 m altitude.

152. *Cedrelopsis* Baill.

Cedrelopsis Baill. in Grandidier, Hist. pl. Madag., Atlas: t. 257 (1893); Leroy et al., Bull. Mus. Natl. Hist. Nat. Paris IV, 12: 43–57 (1990); Leroy & Lescot, Fl. Madagas. et des Comores, Fam. 107bis (1991).

Dioecious trees or shubs. Leaves alternate, (im) paripinnate, 4–14-jugate. Inflorescences short thyrses. Flowers 5-merous, haplostemonous; disk gynophorous in female flowers, annular or cushion-like in male flowers; carpels connate; ovary oblong, 3–5-locular; style short; stigma broadened; ovules 2(3) per locule, apotropous. Fruit a 3–5-valved capsule, the individual carpels separating from a central column and then dehiscent along an adaxial suture. Seeds solitary, apically winged; endosperm usually 0; cotyledons plano-convex, elliptic.

Eight spp., Madagascar, humid and subxerophytic forest.

The descriptions and the illustrations in Leroy and Lescot indicate that the radicle is at the base of the embryo, rather than at the apex as in all other Rutaceae that J.A.K. has dissected.

153. *Harrisonia* R.Br. ex A.Juss. Fig. 75

Harrisonia R.Br. ex A.Juss. in Mém. Mus. Paris 12: 517, t. 28 (1825); nom. cons.

Shrubs, sometimes scandent, or small trees; trunk and larger branches often bearing thorns on wart-like outgrowths. Leaves alternate, imparipinnate or 3-foliolate; petiole often with a pair of stipule-like, recurved spines at base; petiole and rachis sometimes winged. Inflorescences terminal and axillary thyrses. Flowers hermaphrodite, 4–5-merous, diplostemonous; stamen filaments slender, with wooly scale attached to lower half; anthers ellipsoid or cordate; disk ring-, cushion-, or cup-shaped, or inconspicuous; carpels connate but style at base sometimes with distinct stylodia; stigma undistinguished; ovule 1 per carpel, pendulous. Fruit drupaceous, depressed-globose, ± lobed; exocarp fleshy; pyrenes up to 5, each with a perforation of the endocarp at the base of the stylar canal. Seed with thin testa; endosperm sparse; embryo curved, with horseshoe-shaped cotyledons conforming to intruded part of testa (funicle and chalaza more or less coincide and are adjacent to micropyle).

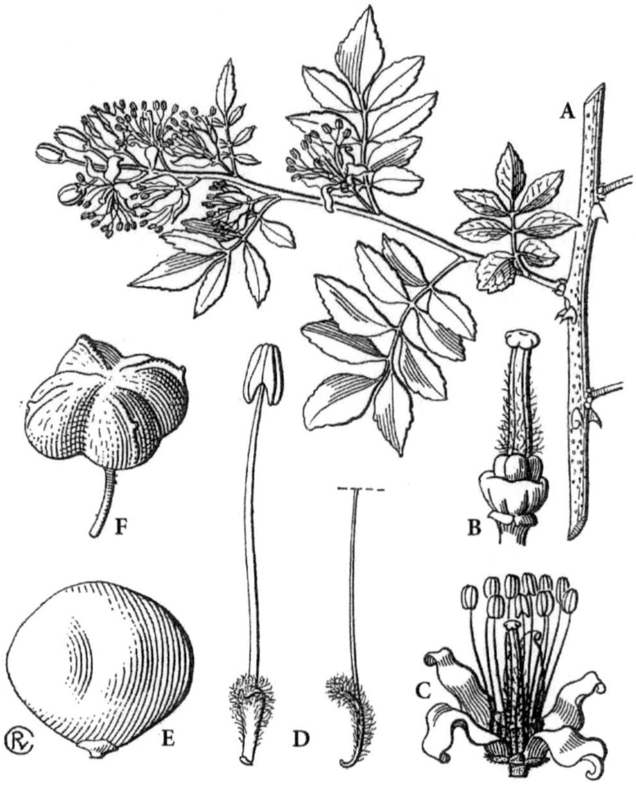

Fig. 75. Rutaceae. *Harrisonia perforata*. **A** Flowering twig. **B** Flower. **C** Pistil with disk. **D** Stamens with appendaged filaments. **E** Fruit. *Harrisonia brownii*. **F** Fruit. (Nooteboom 1962; artist C. van Crevel)

About four spp., tropical Africa, SE Asia, Indomalaysia and N Australia, under distinctly seasonal conditions.

154. *Cneorum* L.

Cneorum L., Sp. Pl. 1: 34 (1753); Lobreau-Callen & Jérémie, Grana 25: 155–158 (1986). *Neochamaelea* (Engl.) Erdtman (1952); Caris et al., Plant Syst. Evol. 257: 223–232 (2006), fl. ontog; Appelhans et al., Taxon 59: 1126–1134 (2010).

Shrubs; solitary oil cells in leaves and cortex. Leaves alternate, simple. Inflorescences axillary, few-flowered cymes or solitary flowers. Flowers 3- or 4-merous, haplostemonous; sepals and petals distinct; stamen filaments inserted in pits of the androgynophorous disk; anthers oval, shorter than filaments; carpels connate; ovary 3–4-locular and -lobed; style elongate, straight, with 3–4 patent stigmatic branches; ovules 2 per carpel. Fruit of 3–4 drupelets separating from the central column, each with a thick mesocarp and 1 pyrene;

pyrenes 2-seeded. Seeds with fleshy endosperm; embryo horseshoe-shaped.

Two subgenera, subg. *Cneorum*, glabrous, flowers 3-merous, with 'septal nectaries', one sp., *C. tricoccum* L., western Mediterranean, introduced to Cuba (see Lobreau-Callen & Jérémie l.c. and Appelhans et al. l.c.); and subg. *Neochamaelea* Engl., with T- and Y-shaped hairs, 4-merous flowers, and no septal nectaries, inflorescence adnate to petioles of subtending leaves, *C. pulverulentum* Vent., Canary Islands.

Excluded:

Hunsteinia papuana Lauterbach = *Rapanea leucantha* K. Schum. (Myrsinaceae), fide H. Sleumer, Blumea 31: 288 (1986).

Selected Bibliography

APG 2003. The Angiosperm Phylogeny Group. An update of the Angiosperm Phylogeny Group classification for the orders and families of flowering plants: APG II. Bot. J. Linn. Soc. 141: 399–436.

Armbruster, W.S., Debevec, E.M., Willson, M.F. 2002. Evolution of syncarpy in angiosperms: theoretical and phylogenetic analyses of the effects of carpel fusion on offspring quantity and quality. J. Evol. Biol. 18: 657–672.

Armstrong, J.A. 1979. Biotic pollination mechanisms in the Australian flora - a review. New Zealand J. Bot. 17: 467–508.

Armstrong, J.A. 2002. *Zieria* (Rutaceae): a systematic and evolutionary study. Austr. Syst. Bot. 15: 277–463.

Barber, C.A. 1892. On the nature and development of the corky excrescences on stems of *Zanthoxylum*. Ann. Bot. 6: 155–166, pl. VII, VIII.

Barbosa, A.A.A. 1999. *Hortia brasiliana* Vand. (Rutaceaeae): polinização por aves Passeriformes no cerrado do sudeste brasileiro. Rev. bras. Bot. 22: 99–105.

Baumann-Bodenheim, M.G. 1954. Prinzipien eines Fruchtsystems der Angiospermen. Ber. Schweiz. Bot. Ges. 64: 94–112.

Bayer, R.J., Mabberley, D.J., Morton, C., Miller, C.H., Sharma, I.K., Pfeil, B.E., Rich, S., Hitchcock, R., Sykes, S. 2009. A molecular phylogeny of the orange subfamily (Rutaceae-Aurantioideae). Am. J. Bot. 96: 668–685.

Beadle, N.C.W. 1981. The vegetation of Australia. Stuttgart: G. Fischer.

Beattie, A.J., Hughes, L. 2002. Ant-plant interactions. In: Herrera, C.M., Pellmyr, O. (eds.) Plant-animal interactions: an evolutionary approach. Oxford: Blackwell, pp. 211–235.

Beattie, G.A.C., Holford, P., Mabberley, D.J., Haigh, A.M., Bayer, R., Broadbent, P. 2008. On the origins of *Citrus*, Huanglongbing, *Diaphorina citri* and *Trioza erytreae*. In: Proc. Int. Res. Conf. on Huanglongbing, Orlando, FL, 2-5 Dec. 2008, pp. 23-56.

Berg, R. 1975. Myrmecochorous plants in Australia and their dispersal by ants. Austr. J. Bot. 23: 475–508.

Beurton, C. 1986. Phyllodienbildende *Xanthoxylum*-Sippen in Cuba I. Feddes Repert. 97: 29–41.

Beurton, C. 1994. Gynoecium and perianth in *Xanthoxylum* s.l. (Rutaceae). Plant Syst. Evol. 189: 165–191.

Blenk, P. 1884. Ueber die durchsichtigen Punkte in den Blättern. Simarubaceae. Flora 67/1884: 291–299.

Boesewinkel, F.D. 1977a. Development of ovule and testa in Rutaceae I. *Ruta*, *Xanthoxylum*, and *Skimmia*. Acta Bot. Neerl. 26: 193–211.

Boesewinkel, F.D. 1977b. Development of ovule and testa in Rutaceae II. The unitegmic and pachychalazal seed of *Glycosmis* cf. *arborea* (Roxb.) D.C. Acta Bot. Neerl. 27: 69–78.

Boesewinkel, F.D. 1977c. Development of ovule and testa in Rutaceae III. Some representatives of the Aurantioideae. Acta Bot. Neerl. 27: 341–354.

Boesewinkel, F.D. 1980. Development of ovule and seedcoat in the Rutales-Geraniales assembly. Amsterdam: Academic Proefschrift, University of Amsterdam.

Boesewinkel, F.D., Bouman, F. 1978. Development of testa and ovule in Rutaceae. II: The unitegmic and pachychalazal seed of *Glycosmis arborea* (Roxb.) DC. Acta Bot. Neerl. 27: 69–78.

Burkill, J.H. 1931. An enumeration of the species of *Paramignya*, *Atalantia* and *Citrus*, found in Malaya. Gard. Bull. 5: 212–223.

Capuron, R. 1967. Nouvelles observations sur les Rutacées de Madagascar. Adansonia II, 7: 479–500.

Caris, P., Smets, E., De Koster, K., Ronse Decraene, L.P. 2006. Floral ontogeny of *Cneorum tricoccum* L. (Rutaceae). Plant Syst. Evol. 257: 223-232.

Carr, S.G.M., Carr, D.J. 1961. The functional significance of syncarpy. Phytomorphology 11: 249–256.

Chang, K.L., Wong, K.M., Gray, A.I., Waterman, P.G. 1988. Chemotaxonomic relationship between *Murraya* and *Merrillia* (Rutaceae) (in Chinese, with English summary). Acta Phytotax. Sin. 26: 205–210.

Chase, M.W., Morton, C.M., Kallunki, J.A. 1999. Phylogenetic relationships of Rutaceae: a cladistic analysis of the subfamilies using evidence from *rbc*L and *atp*B sequence variation. Am. J. Bot. 86: 1191–1199.

Classen-Bockhoff, R., Armstrong, J.A., Ohligschläger, M. 1991. The inflorescence of the Australian genera *Diplolaena* R.Br. and *Chorilaena* Endl. (Rutaceae). Austr. J. Bot. 39: 31–42.

Corner, E.J.H. 1976. The seeds of the dicotyledons. Vols. 1, 2. Cambridge: Cambridge University Press.

Dagne, E., Yenesew, A., Waterman, P.G., Gray, A.I. 1988. The chemical systematics of the Rutaceae, subfamily Toddaloideae, in Africa. Biochem. Syst. Ecol. 16: 179–188.

da Silva, M.F. das G.F., Gottlieb, O.R. 1987. Plant chemosystematics and phylogeny. XXXI. Evolution of quassinoids and limonoids in the Rutales. Biochem. Syst. Ecol. 15: 85–103.

da Silva, M.F. das G.F., Gottlieb, O.R., Ehrendorfer, F. 1988. Chemosystematics of the Rutaceae: suggestions for a more natural taxonomy and evolutionary interpretation of the family. Plant Syst. Evol. 161: 97–134.

Davis, G.L. 1966. Systematic embryology of the angiosperms. New York: Wiley.

Dreyer, D.L. 1983. Limonoids of the Rutaceae. In: Waterman, P.G., Grundon, M.F. (eds.) Chemistry and chemical taxonomy of the Rutales. London: Academic Press, pp. 215–245.

Dugo, G., Di Giacomo, A. (eds.) 2002. Citrus. London: Taylor & Francis.

Endress, P.K. 1982. Syncarpy and alternative modes of escaping disadvantages of apocarpy in primitive angiosperms. Taxon 31: 48–52.

Endress, P.K., Jenny, M., Fallen, M.E. 1983. Convergent elaboration of apocarpous gynoecia in higher advanced dicotyledons (Sapindales, Malvales, Gentianales). Nord. J. Bot. 3: 293–300.

Engler, A. 1931. Cneoraceae, Rutaceae. In: Engler, A., Prantl, K. (eds.) Die natürlichen Pflanzenfamilien, 2nd edn, vol. 19a, pp. 184–359. Leipzig: W. Engelmann.

Federici, C.T., Fang, D.Q., Scora, R.W., Roose, M.L. 1998. Phylogenetic relationships within the genus *Citrus* (Rtutaceae) and related genera as revealed by RFPL and RAPD analysis. Theor. Appl. Genet. 96: 812–822.

Fedorov, Al.A. (ed.) 1969. Chromosome numbers of flowering plants. Leningrad: Nauka.

Fischer, T.C., Butzmann, R. 1998. *Citrus meletensis* (Rutaceae), a new species from the Pliocene of Valdarno (Italy). Plant Syst. Evol. 210: 51–55.

Gallet, F. 1913. Développement et structure anatomique du tégument séminal des Rutacées. Thèse, Université de Paris.

Geisenheyner, L. 1915. Der Schleuderapparat von *Dictamnus fraxinella* Pers. Ber. deutsch. bot. Ges. 33: 442–446.

Goldblatt, P., Williams, I. 1987. Notes on chromosome cytology of Rutaceae-Diosmeae. Ann. Missouri Bot. Gard. 74: 443–444.

Goldblatt, P., Tobe, H., Carlquist, S., Patel, V.C. 1985. Familial position of the Cape genus *Empleuridium*. Ann. Missouri Bot. Gard. 72: 167–183.

Grant, M., Blackmore, S., Morton, C. 2000. Pollen morphology of the subfamily Aurantioideae (Rutaceae). Grana 39: 8–20.

Gray, A.I. 1983. Structural diversity and distribution of coumarins and chromones in the Rutales. In: Waterman, P.G., Grundon, M.F. (eds.) Chemistry and chemical taxonomy of the Rutales. London: Academic Press, pp. 97–146.

Gregor, H.J. 1989. Aspects of the fossil record and phylogeny of the family Rutaceae (Zanthoxyleae, Toddalioideae). Plant Syst. Evol. 162: 251–265.

Groppo, M., Pirani, J.R., Salatino, M.L.F., Blanco, S.R., Kallunki, J.A. 2008. Phylogeny of Rutaceae based on two noncoding regions from cpDNA. Am. J. Bot. 95: 985–1005.

Guerra, M. dos S. 1984. New chromosome numbers in Rutaceae. Plant Syst. Evol. 146: 13–30.

Guerra, M., Galvão Bezerra dos Santos, K., Barros e Silva, A.E., Ehrendorfer, F. 2000. Heterochromatin banding patterns in Rutaceae-Aurantioideae – a case of parallel chromosomal evolution. Am. J. Bot. 87: 735–747.

Harbourgh, D.T., Wagner, W.L., Allan, G.J., Zimmer, E.A. 2009. The Hawaiian arquipelago is a stepping stone for dispersal in the Pacific: an example from the plant genus *Melicope* (Rutaceae). J. Biogeogr. 36: 230–241.

Hartl, D. 1957. Struktur und Herkunft des Endocarps der Rutaceen. Beitr. Biol. Pfl. 34: 35–49.

Hartley, T.G. 1981. A revision of the genus *Tetradium* (Rutaceae). Gard. Bull. Sing. 34: 91–131.

Hartley, T.G. 1985. A revision of the genus *Medicosma* (Rutaceae). Austr. J. Bot. 33: 27–64.

Hartley, T.G. 1991. A new combination in Australian *Acronychia* (Rutaceae). Austr. Syst. Bot. 4: 445–448.

Hartley, T.G. 1995. A new combination in *Boronella* (Rutaceae) and a view on relationships of the genus. Bull. Mus. Natl. Hist. Nat., Paris IV, Adansonia 17: 107–111.

Hartley, T.G. 1997. Five new rain forest genera of Australasian Rutaceae. Adansonia III, 19: 189–212.

Hartley, T.G. 2001a. On the taxonomy and biogeography of *Euodia* and *Melicope* (Rutaceae). Allertonia 8 (1): 1–319.

Hartley, T.G. 2001b. Morphology and biogeography in Australasian-Malesian Rutaceae. Malays. Nature J. 55: 197–219.

Hartley, T.G. 2003. *Neoschmidia*, a new genus of Rutaceae from New Caledonia. Adansonia III, 25: 7–12.

Heads, M. 2009. Inferring biogeographic history from molecular phylogenies. Biol. J. Linn. Soc. 98: 257–774.

Hegnauer, R. 1963. In: Swain, T. (ed.) Chemical plant taxonomy. London: Academic Press, pp. 389–427.

Hegnauer, R. 1973. Chemotaxonomie der Pflanzen. Bd. 6. Basel: Birkhäuser.

Hegnauer, R. 1983. Chemical characters and the classification of the Rutales. In: Waterman, P.G., Grundon, M.F. (eds.) Chemistry and chemical taxonomy of the Rutales. London: Academic Press, pp. 401–440.

Hegnauer, R. 1990. Chemotaxonomie der Pflanzen. Bd. 9. Basel: Birkhäuser.

Heimsch, C. Jr. 1942. Comparative anatomy of the secondary xylem in the "Gruinales" and "Terebinthales", of Wettstein with reference to taxonomic grouping. Lilloa 7: 83–198, pl. I–XVII.

Heinrich, G. 1969. Elektronenmikroskopische Beobachtungen zur Entstehungsweise der Exkretbehälter von *Ruta graveolens*, *Citrus limon* und *Poncirus trifoliata*. Österr. Bot. Z. 117: 397–403.

Heinrich, G. 1970. Elektronenmikroskopische Beobachtungen an den Drüsenzellen von *Poncirus trifoliata*; zugleich ein Beitrag zur Wirkung aetherischer Öle auf Pflanzenzellen und eine Methode zur Unterscheidung flüchtiger von nichtflüchtigen lipophilen Elementen. Protoplasma 69: 15–36.

Heslop-Harrison, Y., Shivanna, K.R. 1977. The receptive surface of the angiosperm stigma. Ann. Bot. 41: 1233–1258.

Hodgson, R.W. 1961. Taxonomy and nomenclature in *Citrus*. In: Price, J.R. (ed.) Proc. 2nd Conf. Int. Org. *Citrus* Virologists, Gainesville, FL, pp. 1–7.

Johri, B.M., Ambegoakar, K.B., Srivastava, P.S. 1992. Comparative embryology of angiosperms. 2 vols. Berlin Heidelberg New York: Springer.

Kallunki, J.A. 1994. Revision of *Raputia* (Cuspariinae, Rutaceae). Brittonia 46: 279–295.

Kong, Y.C., Cheng, K.F., Ng, K.H., But, P.P.H., Li, Q., Yu, S.X., Chang, H.T., Cambie, R.C., Kinoshita, T., Kan, W.S., Waterman, P.G. 1986. A chemotaxonomic division of *Murraya* based on the distribution of the alkaloids

yuehchukene and girinimbine. Biochem. Syst. Ecol. 14: 491–497.
Kubitzki, K., Krutzsch, W. 1998. Origins of east and south-east Asian plant diversity. In: Zhang Aoluo, Wu Sugong (eds.) Floristic characteristics and diversity of east Asian plants. Beijing: China Higher Education Press, and Berlin: Springer, pp. 56–70.
Kuhlmann, M., Kühn, E. 1947. A flora do distrito de Ibiti. São Paulo: Instituto de Botânica.
Ladiges, P. Y., Cantrill, D. 2007. New Caledonia-Australian connections: biogeographic patterns and geology. Austr. Syst. Bot. 20: 383–389.
Li, Q., Zhu, L.F., But, P.P.H., Kong, Y.C., Chang, H.T., Waterman, P.G. 1988. Monoterpene and sequiterpene rich oils from the leaves of *Murraya* species: chemotaxonomic significance. Biochem. Syst. Ecol. 16: 491–494.
Lobreau-Callen, D., Jérémie, J. 1986. L'espèce *Cneorum tricoccum* (Cneoraceae, Rutales) représentée à Cuba. Grana 25: 155–158.
Mabberley, D.J. 1997. A classification for edible *Citrus* (Rutaceae). Telopea 7: 167–172.
Mabberley, D.J. 1998. Australian Citreae with notes on other Aurantioideae (Rutaceae). Telopea 7: 333–344.
Mabberley, D.J. 2004. Citrus (Rutaceae): a review of recent advances in etymology, systematics, and medical applications. Blumea 49: 481–498.
Mabberley, D.J. 2010. The species of *Citrus* (Rutaceae) with pinnate leaves. Blumea 55: 73–74.
Marloth, R. 1920. Notes on the function of the staminal and staminodial gland in the flowers of *Adenandra*. Ann. Bol. Herb. 3: 38–39.
Mauritzon, J. 1935. Über die Embryologie der Familie Rutaceae. Svensk Bot. Tidskr. 29: 319–347.
Mayr, G. 2005. The Palaeogene fossil record of birds in Europe. Biol. Rev. 80: 515–542.
Mendonça, F.A. 1963. Rutaceae. In: Flora Zambesiaca 2(1), pp. 180–210. London: Crown Agents.
Milewski, A.V., Bond, W.J. 1982. Convergence of myrmecochory in Mediterranean Australia and South Africa. In: Buckley, R.C. (ed.) Ant-plant interactions in Australia. The Hague: W. Junk, pp. 89–98.
Mole, B., Udovicic, F., Ladiges, P.Y., Duretto, M.F. 2004. Molecular phylogeny of *Phebalium* (Rutaceae: Boronieae) and related genera based on ITS 1+2 regions of rDNA. Plant Syst. Evol. 249: 197–212.
Morton, C.M., Kallunki, J.A. 1993. Pollen morphology of the subtribe Cuspariinae (Rutaceae). Brittonia 45: 286–314.
Morton, C.M., Grant, M., Blackmore, S. 2003. Phylogenetic relationships of the Aurantioideae inferred from chloroplast DNA sequence data. Am. J. Bot. 90: 1463–1469.
Mou, F.-J., Zhang, D.-X. 2009a. Pollen morphology supports the reinstatement of *Bergera* (Rutaceae). Nord. J. Bot. 27: 298–304.
Mou, F.-J., Zhang, D.-X. 2009b. *Glycosmis longipetala* F.J. Mou & D.X. Zhang, a new species of Rutaceae from China. J. Syst. Evol. 47: 162–167.
Muellner, A.N., Vassiliades, D.D., Renner, S.S. 2007. Placing Biebersteiniaceae, a herbaceous clade of Sapindales, in a temporal and geographic context. Plant Syst. Evol. 266: 233–252.
Mulholland, D.L., Parel, B., Coombes, P.H. 2000. The chemistry of the Meliaceae and Ptaeroxylaceae of Southern and Eastern Africa and Madagascar. Curr. Org. Chem. 4: 1011–1054.
Ng, K.M., But, P.P.-H., Gray, A.I., Hartley, T.G., Kong, Y.-C., Waterman, P.G. 1987. The biochemical systematics of *Tetradium*, *Euodia* and *Melicope* and their significance in the Rutaceae. Biochem. Syst. Ecol. 15: 587–593.
Nooteboom, P. 1962. Simaroubaceae. In: Flora Malesiana I, 6: 193–226. Leyden: Northoff.
Olmstead, R., Tank, D.C. 2008. Over-precision in molecular dating. Abstract ID 433, Botany 2008, Bot. Soc. America.
O'Sullivan, J. 1983. Structural diversity and distribution of lignans in the Rutales. In: Waterman, P.G., Grundon, M.F. (eds.) Chemistry and chemical taxonomy of the Rutales. London: Academic Press, pp. 267–279.
Pennington, T.D., Reynel, C., Daza, A. 2004. Illustrated guide to the trees of Peru. Sherborne: David Hunt, 848 pp.
Pfeil, B.E., Crisp, M.D. 2008. The age and biogeography of *Citrus* and the orange subfamily (Rutaceae: Aurantioideae) in Australasia and New Caledonia. Am. J. Bot. 95: 1621–1631.
Piedade, L.H., Ranga, N.T. 1993. Ecologia da polinização de *Galipea jasminiflora* Engler (Rutaceae). Rev. bras. Bot. 16: 151–157.
Pirani, J.R.. 1989. Flora da Serra do Cipó, Minas Gerais: Rutaceae. Bol. Botánica, Univ. São Paulo 11: 109–119.
Poon, W.-S., Shaw, P.-C., Simmons, M.P., But, P.P.-H. 2007. Congruence of molecular, morphological, and biochemical profiles in Rutaceae: a cladistic analysis of the subfamilies Rutoideae and Toddalioideae. Syst. Bot. 32: 837–846.
Price, J.R. 1963. In: Swain, T. (ed.) Chemical plant taxonomy. London: Academic Press, pp. 428–452.
Quader, A., Armstrong, J.A., Gray, A.I., Hartley, T.G., Waterman, P.G. 1991. Chemosystematics of *Acradenia* and general significance of acetophenones in the Rutaceae. Biochem. Syst. Ecol. 19: 171–176.
Ramón-Lacca, L. 2003. The introduction of cultivated Citrus to Europe via northern Africa and the Iberian Peninsula. Econ. Bot. 57: 502–514.
Ramp, E. 1988. Struktur, Funktion und systematische Bedeutung des Gynoeciums bei den Rutaceae und Simaroubaceae. Inaug.-Diss., Phil. Fak. II, Univ. Zürich: ADAG.
Rao, T.A., Bhattacharya, J.B. 1981. Comparative morphology of foliar sclereids in *Boronia* Sm. (Rutaceae). Proc. Ind. Acad. Sci. (Plant Sci.) 90: 9–29.
Reuther, W. Webber, H.J., Batchelor, L.D. (eds.) 1967–1989. The Citrus Industry, 2nd edn. Vols. 1-5. Riverside, CA.: University of California, Div. of Agric. Sci.
Ridley, H.N. 1930. The dispersal of plants throughout the world. Ashford: L. Reeve.
Salvo, G., Bacchetta, G., Ghahremaninejad, F., Conti, E. 2008. Phylogenetic relationships of Ruteae (Rutaceae): new evidence from the chloroplast genome and comparisons with non-molecular data. Mol. Phylog. Evol. 49: 736–748.
Samuel, R., Ehrendorfer, F., Chase, M.W., Greger, H. 2001. Phylogenetic analyses of Aurantioideae (Rutaceae) based on non-coding plastid DNA sequences and phytochemical features. Plant Biol. 3: 77–87.

Savolainen, V., Fay, M.F., Albach, D.C., Backlund, A., van der Bank, M., Cameron, K.M., and 17 other authors. 2000. Phylogeny of the eudicots: a nearly complete familial analysis based on *rbcL* gene sequences. Kew Bull. 55: 257–309.

Schill, R., Baumm, A., Wolter, M. 1985. Vergleichende Mikromorphologie der Narbenoberflächen bei den Angiospermen; Zusammenhänge mit Pollenoberflächen bei heterostylen Sippen. Plant Syst. Evol. 148: 185–214.

Schneider, H. 1968. The anatomy of *Citrus*. In: Reuther, W., Batchelor, L.D., Webber, H.J. (eds.) The Citrus Industry, vol. 2, 2nd edn. Riverside, CA.: University of California, pp. 1–85.

Scott, K.D., McIntyre, C.L., Playford, J. 2000. Molecular analyses suggest a need for a significant rearrangement of Rutaceae subfamilies and a minor reassessment of species relationships within *Flindersia*. Plant Syst. Evol. 223: 15–27.

Smith-White, S. 1954. Chromosome numbers in the Boronieae (Rutaceae) and their bearing on the evolutionary development of the tribe in the Australian flora. Austr. J. Bot. 2: 287–303.

Soltis, D.E., Soltis, P.S., Endress, P.K., Chase, M.W. 2005. Phylogeny and evolution of angiosperms. Sunderland, MA: Sinauer.

Specht, R.L. 1981. Ecophysiological principles determining the biogeography of major vegetation formations in Australia. In: Keast, A. (ed.) Ecological biogeography of Australia. The Hague: W. Junk, pp. 299–333.

Stace, H.M., Armstrong, J.A., James, S.H. 1993. Cytoevolutionary patterns in Rutaceae. Plant Syst. Evol. 187: 1–28.

Stone, B.C. 1962. A monograph of the genus *Platydesma* (Rutaceae). J. Arnold Arb. 43: 410–427.

Swingle, W.T., Reece, C.P. 1967. The botany of *Citrus* and its wild relatives. In: Reuther, W., Webber, H.J., Batchelor, L.D. (eds.) The Citrus Industry, rev. edn., vol. 1. Riverside, CA: University of California, pp. 190–430.

Takhtajan, A. 1997. Diversity and classification of flowering plants. New York: Colombia University Press.

Tanaka, T. 1936. The taxonomy and nomenclature of Rutaceae-Aurantioideae. Blumea 2: 101–110.

Tanaka, T. 1954. Species problem in *Citrus*. Ueno, Tokyo: Jap. Soc. Promot. Sci.

Tanaka, T. 1961. Citrologia. Semi-centennial commemoration papers on *Citrus*. Osaka: Citrologia Supporting Foundation.

Tiffney, B.H. 1980. Fruits and seeds of the Brandon Lignite, V. Rutaceae. J. Arnold Arbor. 61: 1–40.

Tiffney, B.H. 1981. *Euodia costata*, new combination (Rutaceae) from the Eocene of southern. England. Paläontol. Z. 55: 185–190.

Tolkowsky, S. 1938. Hesperides: a history of the culture and use of Citrus fruits. London: John Bales, Sons & Curnow Ltd.

Townsend, C.C. 1986. Taxonomic revision of the genus *Haplophyllum* (Rutaceae). Hook. Icon. Plant. XL, Pt. I, II, and III. Royal Bot. Gards. Kew: Bentham-Moxon Trustees.

Trinder-Smith, T.H., Linder, H.P., van der Niet, T., Verboom, G.A., Nowell, T.L. 2007. Plastid DNA sequences reveal generic paraphyly within Diosmeae (Rutoideae, Rutaceae). Syst. Bot. 32: 847–855.

Troll, W. 1964. Die Infloreszenzen. Vol. 1. Stuttgart: G. Fischer.

Troll, W. 1969. Die Infloreszenzen. Vol. 2. Stuttgart: G. Fischer.

Turner, G.W., Berry, A.M., Gifford, E.M. 1998. Schizogenous secretory cavities of *Citrus limon* (L.) Burm. f. and a reevaluation of the lysigenous gland concept. Int. J. Plant Sci. 151: 75–88.

Uphof, J.C.Th. 1935. Wissenschaftliche Beobachtungen und Versuche an Agrumen. VII. Die Morphologie der Dornen. Gartenbauwissenschaft 9: 219–230.

van der Ham, R.W.J.M., Baas, P., Bakker, M.E., Boesewinkel, F.E., Bouman, F., van Heuven, B.J., Klaasen, R.K.W.M. 1995. *Bottegoa* Chiov. transferred to the Ptaeroxylaceae. Kew Bull. 50: 243–265.

Victor, J.E., van Wyk, A.E. 1999. Pollen morphology of *Adenandra* (Rutaceae: Diosminae) and its taxonomic implications. Grana 38: 1–11.

Wang, H., Moore, M.J., Soltis, P.S., Bell, C.D., Brockington, S. F., Alexandre, R., Davis, C.C., Latvis, M., Manchester, S. R., Soltis, D.E. 2009. Rosid radiation and the rapid rise of angiosperm-radiated forests. Proc. Natl. Acad. Sci. U.S. 106: 3853–3858.

Waterman, P.G. 1983. Phylogenetic implications of the distribution of secondary metabolites within the Rutales. In: Waterman, P.G., Grundon, M.F. (eds.) Chemistry and chemical taxonomy of the Rutales. London: Academic Press, pp. 377–400.

Waterman, P.G. 1986. A phytochemist in the African rain forest. Phytochemistry 25: 3–17.

Waterman, P.G. 1993. Phytochemical diversity in the order Rutales. In: Downum, K.R., Romeo, J.T., Stafford, H.A. (eds.) Phytochemical potential of tropical plants. New York: Plenum Press, pp. 203–233.

Waterman, P.G., Grundon, M.F. (eds.) 1983. Chemistry and chemical taxonomy of the Rutales. London: Academic Press.

Waterman, P.G., Khalid, S.A. 1981. The biochemical systematics of *Fagaropsis angolensis* and its significance in the Rutales. Biochem. Syst. Ecol. 9: 45–51.

Webber, H.J. 1967. History and development of the Citrus industry. In: Reuther, W., Batchelor, L.D., Webber, H.J. (eds.) The Citrus Industry, vol. 1, 2nd edn. Riverside, CA: University of California, pp. 1–39.

Weberling, F., Leenhouts, P.W. 1966. Systematisch-morphologische Studien an Terebinthales-Familien. Akad. Wiss. Lit., Abhandl. Math.-nat. Kl. Jg. 1965, 10.

Westoby, M., Rice, B., Shelley, J.M., Haig, D., Kohen, J.L. 1982. Plants' use of ants for dispersal at West Head, New South Wales. In: Buckley, R.C. (ed.) Ant-plant interactions in Australia. The Hague: W. Junk, pp. 75–87.

Wikström, N., Savolainen, V., Chase, M.W. 2001. Evolution of the angiosperms: calibrating the family tree. Proc. R. Soc. Lond. B 268: 2211–2220.

Wilson, P.G. 1970. A taxonomic revision of the genera *Crowea*, *Eriostemon* and *Phebalium* (Rutaceae). Nuytsia 1: 3–155.

Wilson, P.G. 1982. *Eriostemon woganensis* (Rutaceae), a new species from the Wogan Hills, Western Australia. Nuytsia 41: 47–50.

Wilson, P.G. 1998. A taxonomic review of the genera *Eriostemon* and *Philotheca* (Rutaceae: Boronieae). Nuytsia 12: 239–265.

Zdero, C., Bohlmann, F. 1990. Systematics and evolution within the Compositae, seen with the eyes of a chemist. Plant Syst. Evol. 171: 1–14.

Sapindaceae

Sapindaceae Juss., Gen.: 246 (1789), nom. cons., "Sapindi".
Aceraceae Juss. (1789), nom. cons.
Hippocastanaceae A. Rich. (1823), nom. cons.

P. Acevedo-Rodríguez, P.C. van Welzen, F. Adema, and R.W.J.M. van der Ham

Trees, treelets, shrubs, lianas or herbaceous climbers; cork superficial; stems of climbing species (i.e., *Serjania*, *Paullinia*, *Urvillea*, *Houssayanthus*, and *Thinouia*) usually with multiple vascular cylinders. Leaves pinnately or ternately compound or palmate, or rarely simple, alternate, rarely opposite; proximal leaflets seldom reduced, reflexed, and covering the stem to resemble a pair of stipules (pseudostipules), distal leaflet in most arboreal and shrubby species rudimentary; stipules present only in climbing species, minute to large. Inflorescences axillary, terminal, pseudoterminal, cauliflorous or ramiflorous, thyrso-paniculate, racemose, spicate, or fasciculate, or flowers solitary. Flowers 5-merous, regular, or less often 4-merous and obliquely zygomorphic, bisexual or more often functionally unisexual by reduction (plants monoecious or rarely dioecious); sepals distinct or connate at base; petals usually white or light yellow, rarely 0, usually ornamented by an adaxial appendage; appendages variously shaped, mostly petaloid, simple, bifurcate, or hood-shaped, basally adnate to the petal or just a prolongation of petal margins, concealing the nectary; disk extrastaminal, annular or unilateral, often lobed, cup-shaped or dish-shaped, very rarely on both sides of the stamens or intrastaminal; stamens (3–)5–8(–30); filaments distinct or connate at base, equal or unequal in length; anthers dorsifixed or basifixed, introrse, opening by longitudinal slits; sterile stamens present in pistillate flowers; gynoecia syncarpous, (1–)3(–8)-carpellate; carpels with 1, 2, or exceptionally many (7–8 in *Xanthoceras*, 8 in *Magonia*) ovules; style terminal or exceptionally gynobasic (*Deinbollia*), 2–3-branched, or with simple, 2–3-lobed, capitate stigma, sometimes (*Acer*) the style branches elongate and the style nearly 0; pistil usually rudimentary in staminate flowers. Fruit a septifragal or loculicidal capsule, a schizocarp with winged or non-winged mericarps, baccate or rarely a drupe. Seeds sessile or exceptionally (*Distichostemon*) subtended by a funiculus, variously shaped, exalate or rarely winged, naked, with a partial to complete sarcotesta, or an arillode (arising from the integuments); embryo oily or starchy, lacking endosperm, notorhizal or lomatorhizal with straight, curved or plicate, fleshy cotyledons, the radicle often separated by a deep fold in the testa that forms a radicular pocket.

Mostly tropical or subtropical, with a few genera extending to sub-temperate zones; 141 genera and about 1,900 species.

Vegetative Morphology. Most genera of Sapindaceae are predominantly medium-sized to large emergent trees or erect shrubs, less often they are tendrilled lianas or understory palm-like treelets, exceptionally sub-shrubs or scandent shrubs. The arboreal and fruticose habits are widespread throughout the distributional range of Sapindaceae, while treelets, lianas, and sub-shrubs are restricted to the tropics. There are about 500 species of lianas, all of which are exclusive to the Neotropics (with the exception of several species found in the Paleotropics), particularly to tribe Paullinieae (*Cardiospermum*, *Houssayanthus*, *Lophostigma*, *Paullinia*, *Serjania*, *Urvillea*, and *Thinouia*), accounting for ca. 60% of the Sapindaceae species in the Neotropics. Sub-shrubs are rare and are known to occur in dry vegetation subject to periodic fires. In the Paleotropics, a few species of *Allophylus*, *Laccodiscus*, and *Lepisanthes* are known to be scandent shrubs, not showing any active mechanism for ascending the host plants.

Exudates for the most part are odorless and colorless. However, exudates are red in *Pometia*, or white in numerous species of *Paullinia* and *Serjania*.

Stems are for the most part terete and smooth. However, many of the climbing species have stems that are deeply furrowed, or sharply to obtusely angled, and a few have thorny stems.

Bud scales are present in a few taxa, being predominant in subfamily Hippocastanoideae. These are simple and accrescent with the growing buds. Bud scales have also been reported for a few tropical genera, including *Exothea*, *Melicoccus*, *Talisia*, and *Sapindus* that bear minute scales, or *Talisia* (Fig. 84B) and *Pseudopteris* that bear large leaf-like cataphylls. Cataphylls are clustered at the ends of branches, on axillary buds, and at the base of inflorescences. Their size and form vary considerably. In some species they can be up to 25 cm long, while in others they are pinnatifid and a few cm in length.

Leaves are predominantly spirally arranged, and variously compound, less often opposite, simple or digitate. Compound leaves include the following types: palmate, pinnate, bipinnate, tripinnate, trifoliolate, biternate, triternate, or a combination of these. A peculiar character of numerous Sapindaceae with pinnately compound leaves is the presence of a single terminal rudimentary leaflet or process (Fig. 85A). At first glance, this structure may be confused with an undifferentiated leaflet primordium present in other families such as the Meliaceae. However, in Meliaceae, this structure is produced in pairs, and slowly developed into new leaflets. Leaflets are predominantly entire, but there is variation, with some genera having entire, crenate or serrate leaflets. Venation is variable and represented by brochidodromous, cladodromous, craspedodromous, mixed-craspedodromous, semi-craspedodromous, and palinactinodromous types (Hickey 1979). Stipules are largely absent in Sapindaceae, and restricted to the genera of the Paullinieae. Pseudostipules are present in a few species of the paleotropical genera *Alectryon*, *Blighiopsis*, *Chouxia*, *Cupaniopsis*, *Eriocoelum*, *Glenniea*, *Haplocoelopsis*, *Laccodiscus*, *Lepisanthes*, *Macphersonia*, *Otonephelium*, *Placodiscus*, and *Pometia* (Weberling 1976). Although superficially similar to stipules, these are in reality proximal leaflets reduced in size, which often clasp the stem. Petioles and leaf rhachises are terete, angled, carinate, sulcate, and sometimes narrowly to broadly winged. The base of petioles and the petiolules are more often enlarged, with the adaxial portion very often depressed or furrowed, or only very rarely nearly cylindrical.

Tendrils are found only in genera of the Paullinieae. They are opposite and coiled, and seem to be homologous to the proximal pair of cincinni (or drepania) of their thyrsoid inflorescences. Very often, the portion of the inflorescence above the tendrils is aborted, resulting in a short axillary branch that produces a pair of opposite tendrils in its distal portion.

The indumentum in the family is quite variable. Plants are either glabrous or exhibit several different kinds of indumentum, which are predominantly composed of simple, erect or curly, non-glandular trichomes. Less frequently, the indumentum may include multicellular-glandular, papilliform, fasciculate, stellate trichomes, or peltate scales. Trichomes occur as a pure stand or as a mixture of different types, and are often classified as puberulent, appressed-pubescent, tomentose, tomentulose, sericeous, velutinous, setiferous, hirsute, woolly, pilose, pilosulous, or furfuraceous. Stiff, irritating hairs occur on the outer surface of fruits of *Cnesmocarpon* and *Jagera* species.

VEGETATIVE ANATOMY. A detailed anatomical survey of Dodonaeoideae and Sapindoideae, which covers both vegetative and reproductive organs, was presented by Radlkofer (1890) and used in explaining his classification. Among other things, Radlkofer described carefully the location and structure of secretory cells that contain saponins and mucilaginous cells, which occur in the leaf epidermis. An important trait discovered by Radlkofer is the constant presence of a cylinder of sclerenchyma in the pericycle, which consists of phloem fibers and stone cells. This structure was observed in all genera of Sapindaceae as circumscribed by Radlkofer (1890) (=Dodonaeoideae + Sapindoideae), with the notable exception of *Xanthoceras* and *Valenzuelia* (the latter renamed as *Guindilia*), in which the ring of sclerenchyma is not continuous; incidentally, *Guindilia* has opposite leaves. The sclerenchyma cylinder is also constantly present in tribe Hippocastaneae, whereas in tribe Acereae sometimes (e.g., *Acer negundo*) it is well developed but sometimes (*A. pseudoplatanus*,

Dipteronia) it is represented only by isolated strands of fibers.

Nodes of Sapindaceae are trilacunar, rarely also 5-lacunar in Hippocastaneae. The vessel segments usually have simple perforation; scalariform perforation is rarely seen in Sapindoideae, and more often in Hippocastaneae. Rays are mostly 1-seriate, but in Acereae mixed 1- and pluriseriate rays are found. A comprehensive study of wood anatomy of Dodonaeoideae and Sapindoideae was published by Klaassen (1999).

Numerous woody vines of the Sapindoideae present anomalous secondary thickening of their stems. The most salient feature of this anomalous thickening can be described as multistelate, where the stem has a single central stele surrounded by three, five, or up to ten peripheral steles. Anomalous secondary thickening is predominant in *Serjania* (Fig. 76A, D) and *Paullinia* (Fig. 76B), where 58% and 12%, respectively of the species show some kind of anomaly (Acevedo-Rodríguez 1993).

INFLORESCENCES. The inflorescences in Sapindaceae are variously shaped thyrses with lateral dichasia, cincinni, drepania, or reductive forms thereof (for details, see Radlkofer 1890: 178 seq.). They are axillary, distal, cauliflorous, or

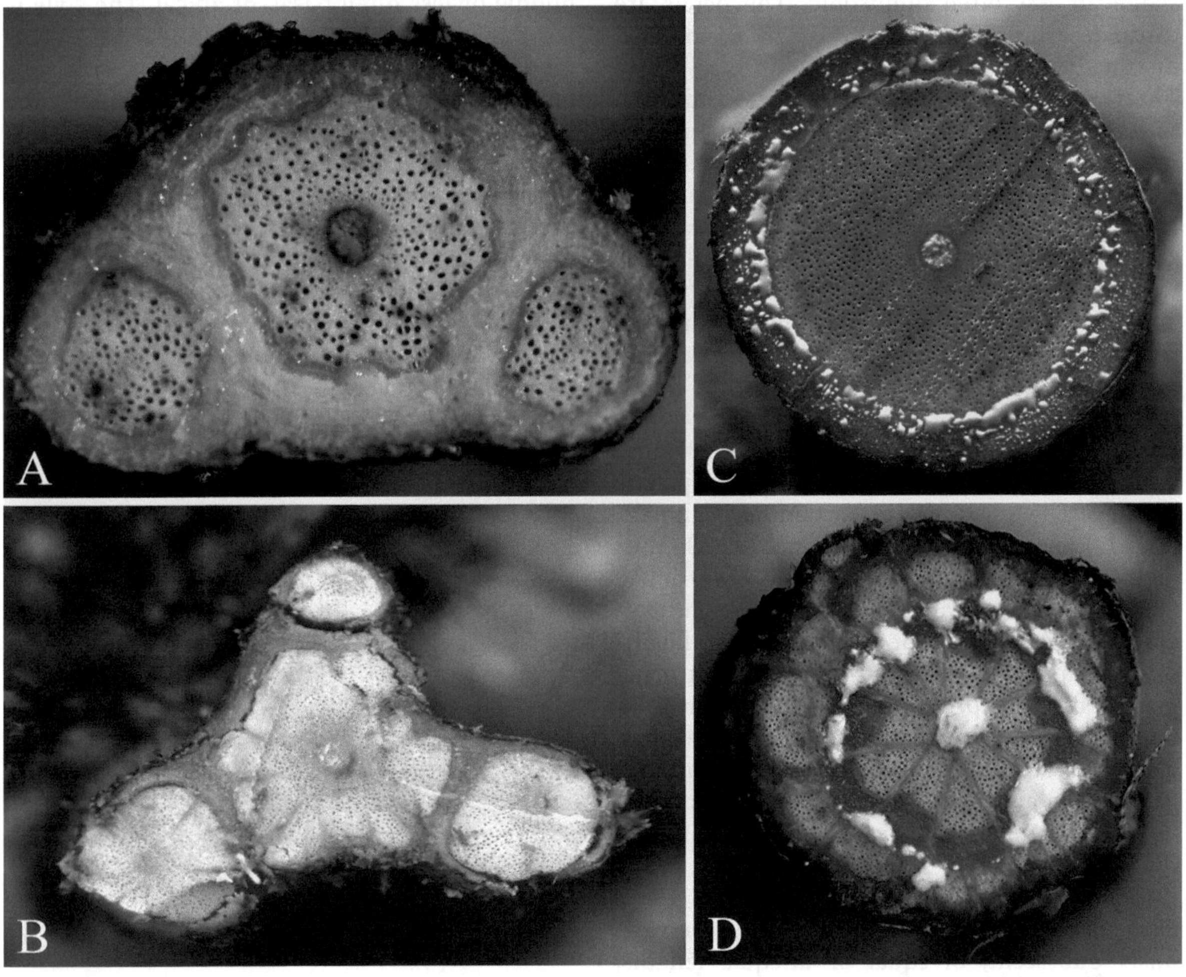

Fig. 76. Sapindaceae. Stem cross sections of Sapindaceae vines. **A** *Serjania* sp., stem with a central stele and two smaller peripheral steles. **B** *Paullinia alata*, stem with a central stele and three smaller peripheral steles. **C** *Paullinia ingifolia*, stems with single stele and white latex. **D** *Serjania grandifolia* Radlk., vascular tissue produced in alternate concentric layers alternating with connective tissue; note the white latex. (orig.)

supra-axillary, solitary or fasciculate. Some genera have racemes that seem to be derived from thyrsoid ancestors. The flowers within the inflorescences are predominantly unisexual or bisexual, but with a strong tendency for one sex to predominate.

FLOWERS. Pedicels are usually conspicuous, and have an abscission zone or articulation anywhere from base to near the apex, less often they are inconspicuous or non-articulated. The perianth is more often 5-merous, but there is considerable variation in the number of parts. Sepals are distinct (Figs. 77, 78) to completely connate (Fig. 86A, D), and may be as few as three or as many as ten. They are usually of similar size and shape, or less often dimorphic. The petals are distinct, with imbricate aestivation, inserted on the base of an extrastaminal nectary disk. Their number varies from four to six, in addition to the 5-merous corolla. However, there are numerous genera or species that completely lack a corolla. Petals are erect (Fig. 83C) or reflexed (Fig. 78), those of Sapindoideae with an adnate adaxial petaloid appendage (Figs. 83, 84), or with extended inrolled basal margins lacking appendages, or lacking appendages and inrolled margins altogether (Fig. 86E). The petaloid appendages are simple (Fig. 84D, G), bifid, hood-shaped (Fig. 83D), or corniform, and very often sericeous or tomentose. Leinfellner (1958) analyzed these appendages, emphasizing their peltate nature. The disk is extrastaminal, annular (Fig. 77), more rarely amphistaminal (Fig. 80A, B) or intrastaminal, cup-shaped, 2-, 4-, 5(–8)-lobed (Fig. 83C), of elongated corniform lobes, unilateral and semi-annular, or rudimentary, with various indumenta, or glabrous altogether. In many genera, nectar is produced as a reward for pollinators. The number of stamens is more often eight but there is considerable variation, with genera containing from five to eight or from eight to ten. Stamens are sometimes as few as four in *Cupania* or *Dictyoneura*, or as many as 20 in *Hornea*, 30 in *Deinbollia*, and 74 in *Distichostemon*. Filaments are glabrous or variously pubescent, of equal or unequal lengths, erect, spreading, or sigmoid. Anthers are basifixed or dorsifixed, oblong to linear or elliptic to ovate with an obtuse, apiculate or retuse apex, opening along longitudinal slits. The ovary is syncarpous and usually 3-carpellate. However, there are numerous genera possessing 2-carpellate ovaries in addition to the 3-carpellate ones. *Blomia* and sometimes *Alectryon* and *Nephelium* have unicarpellate gynoecia. *Chytranthus* and *Radlkofera* have gynoecia with up to eight carpels. The septae are complete, with the ovary containing the same number of locules as carpels, except in *Melicoccus* (Fig. 77) and *Zollingeria*, where the septae are partially developed, resulting in a unilocular ovary. Ovules are anatropous, hemitropous, or campylotropous, one per carpel in subfam. Sapindoideae, or two per carpel in subfam. Dodonaeoideae, and 7–8 in *Magonia* (Dodonaeoideae) and *Xanthoceras* (Xanthoceroideae); placentation is axile, commonly in the middle or less often basal, or apical. The style is terminal and simple in all Sapindaceae except for *Deinbollia*, which has a gynobasic or a subterminal style; in *Dipteronia* and some *Acer*, the style can be very short, or lacking, and has two prominent stigmatic style branches (Fig. 80C, H). The stigmatic surface is variable, very often represented by a line along the stigmatic branches, or by as many lines as carpels along the distal portion of a simple style. Stigmas are sometimes capitate, elongated-cylindrical, or lobed.

For the aspect of oblique floral monosymmetry, see the ontogenetic studies of *Koelreuteria* by Ronse Decraene et al. (2000).

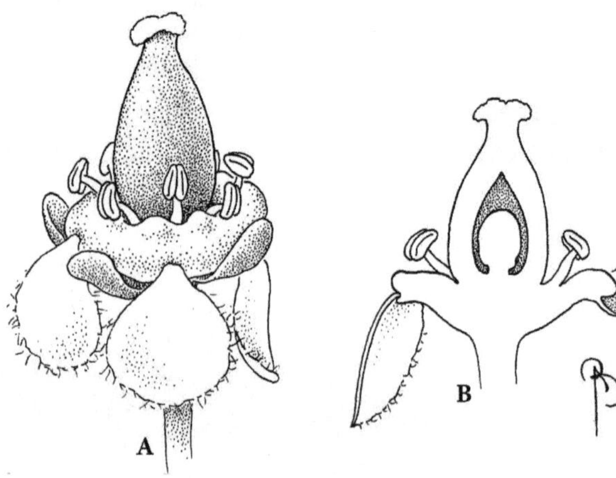

Fig. 77. Sapindaceae. *Melicoccus bijugatus* Jacq. **A** Pistillate flower. **B** Ditto, longitudinal section. (Acevedo-Rodríguez 1996; reproduced with permission of the artist Bobbi Angell)

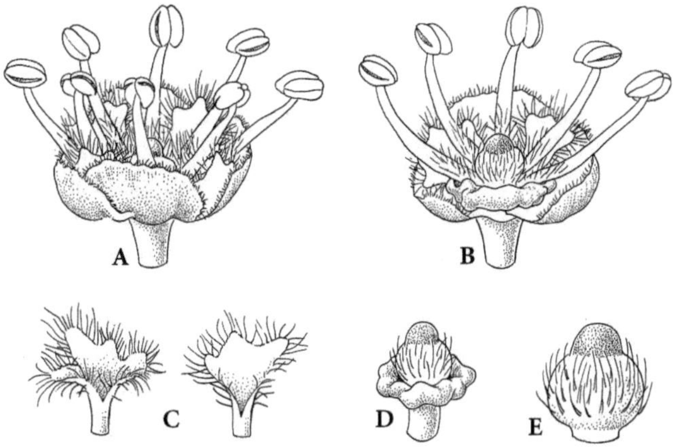

Fig. 78. Sapindaceae. *Vouarana anomala*. **A, B** Male flowers. **C** Petals, ventral view. **D** Pistillode with subtaining annular disk. **E** Pistillode. (Acevedo-Rodríguez 1997; drawn by Alice Tangerini)

Sex Distribution. Functionally bisexual flowers are rare among Sapindaceae; they are known from *Acer*, *Aesculus*, *Bizonula*, *Dodonaea*, *Exothea*, and *Handeliodendron*. Most Sapindaceae are usually labeled as polygamous, and there is evidence for a wide distribution within the family of a particular sexual system, duodichogamy, which has been unveiled in such distantly related taxa as *Cupania* (Bawa 1976), *Acer*, *Dipteronia*, *Hippocastanum*, *Deinbollia*, and *Koelreuteria* (de Jong 1976), *Sapindus* (Subba Reddi et al. 1983), *Serjania* (Acevedo-Rodríguez 1993), *Talisia* (Acevedo-Rodríguez 2003), and *Paullinia* (Somner, unpubl. data).

Duodichogamy implies a sequence of three distinct phases of flowering in which all flowers of a given individual are in the same phase. During the first phase, male flowers release functional pollen; they have a reduced pistil. When they are dropped, female flowers appear on the same inflorescence, which have a well-developed pistil and short stamens with indehiscent anthers. These phases may overlap for a few days. After fertilization of the female flowers (from flowers of an individual that is in a different phase), in a third phase hermaphrodite flowers with well-developed stamens and pistils appear but usually are effectively only male. Sometimes, as in *Acer*, the third phase may be female when the first female phase is repressed for some reason. In many Sapindaceae, the first or usually the last phase may be missing, so that the reproductive system is dichogamous. Since self-incompatibility seems to be frequent in Sapindaceae, (duo) dichogamy acts as a barrier against selfing. Most genera and species of the family share the possession of male and apparently hermaphrodite but functionally female flowers, and it is likely that they are dichogamous; in the absence of experimental data, however, this remains unproven. In the generic descriptions, these cases are labeled "falsely polygamous". True dioecy is more rarely documented in the family, as for instance for some *Acer*, *Xerospermum noronhianum*, and some *Nephelia* (van Welzen 1989). It is likely that duodichogamy is basal in the family (de Jong 1976), and van Welzen (1989) supposes that it is symplesiomorphic.

Pollination. The open, white, fragrant, and nectar-secreting flowers of tribe Hippocastaneae appear primarily bee-pollinated, such as several *Aesculus*, among which the colored spot on the petals functions as a nectar guide. Nevertheless, such flowers can also be attractive to other kinds of pollinators including Lepidoptera, and species of *Aesculus* sect. *Pavia*, particularly the red-flowered *Ae. pavia*, are pollinated by humming birds. Similarly, in *Billia* the white-flowered *B. columbiana* is probably bee-pollinated, whereas the red flowers of *B. hippocastanum* suggest bird pollination (Forest et al. 2001). In *Acer*, the transition from entomogamy to anemophily is accompanied by one from dichogamy to dioecy (de Jong 1976; Hesse 1979). In Sapindoideae, apart from nectar also pollen appears to be an important reward, because male flowers are not only much more abundant than functionally female flowers, but they also have a longer duration and are much more visited than the females (van Welzen 1989), which implies that the anthers in the female flowers act as mimics.

Pollen Morphology. The literature on pollen morphology of Sapindaceae is quite extensive. The most complete work at the family level is that of Muller and Leenhouts (1976; including 78% of genera), where pollen types were evaluated in regard to their systematic significance. More recent progress has been made by several workers (van der Ham 1990; Acevedo-Rodríguez 1993; Ferrucci and Anzótegui 1993), but their analyses apply only to some of the genera. The present overview is based on pollen data for

137 of the 141 genera currently recognized in Sapindaceae.

Sapindaceae pollen grains are usually isopolar or subisopolar monads. Tetrads occur only in *Magonia* (Fig. 79F). Several genera of Paullinieae (see below) have distinctly heteropolar grains (Fig. 79G, I). Pollen grain size is usually between 20 and 30 μm, and the grains are oblate to prolate in shape. Colporate pollen is usually suboblate to prolate (Fig. 79A–E), whereas pollen with small apertures (porate, brevicolporate) or with connected apertures (syncolporate, parasyncolporate) has a more oblate shape (Fig. 79G–L). The equatorial outline is almost circular (Fig. 79C, D) to bluntly triangular (Fig. 79G–L); the meridional outline is almost circular to more or less elliptic (Fig. 79B, E, H).

Generally, Sapindaceae pollen is 3-aperturate (Fig. 79A–L), but often small percentages of 2-and 4-aperturate grains co-occur. Colporate pollen (Fig. 79A–C, E, F) is the commonest, being known from many genera in all four subfamilies. It is a relatively basic type found in many other angiosperm families. Several other types are more restricted. Syncolporate (Fig. 79J) and parasyncolporate pollen (Fig. 79K), with and without apocolpial fields, respectively, are known only in subfamily Sapindoideae, being present in most Cupanieae, *Alectryon*, *Schleichera*, *Castanospora*, *Tristira*, and *Tristiropsis*. Parasyncolporate and syncolporate are not clear-cut character states. Moreover, several genera (e.g., *Alectryon*, *Arytera*, and *Cupania*) possess both colporate and (para)syncolporate pollen, and often intermediates as well (van der Ham 1990; van Bergen et al. 1995). Small ectoapertures occur in brevicolporate and porate pollen found in a few species of *Allophylus* (Fig. 79H), *Lepisanthes*, *Pometia* (Fig. 79L), *Talisia*, and tribe Paullinieae. Pollen of several Paullinieae is heteropolar with a demisyncolporate aperture system. *Cardiospermum* pollen has short demicolpi on the proximal side of the pollen, whereas pollen of several related genera has a demisyncolporate aperture system, and a few species of *Serjania* and *Urvillea* (Fig. 79G, I) have a syncolporate aperture system proximally with short demicolpi distally. *Distichostemon* pollen has indistinct ectoapertures, and a few species of *Harpullia* have pollen without recognizable ectoapertures (cryptoaperturate; Fig. 79D). Sapindaceae pollen grains nearly always have lalongate, elliptic to subcircular endoapertures, though usually hidden by the ectoaperture margins. *Acer* pollen may have apertures without or with indistinct endoapertures.

The exine is usually clearly stratified, showing a tectum, a columellate infratectal layer, and a nexine. In *Diplopeltis*, *Distichostemon*, and *Dodonaea*, the infratectal layer is granular/columellate, which might relate to wind-pollination. Mostly, the nexine consists of a distinctly delimited foot layer and endexine. The endexine is thin in nonapertural parts, thickens toward the apertures, and is maximal along and under the apertures.

The ornamentation of the exine shows much variation. Striate (Fig. 79A, E), rugulate (Fig. 79H, J), perforate (Fig. 79G), and intermediate types are most common; verrucate (Fig. 79C), scabrate (Fig. 79B), and reticulate (Fig. 79D, L) types are less frequent. In subfamilies Xanthoceroideae, Hippocastanoideae, and Dodonaeoideae, rugulate pollen is rare and psilate pollen entirely absent. In subfamilies Hippocastanoideae and Sapindoideae, verrucate and scabrate ornamentation are absent

Fig. 79. Sapindaceae. Pollen micrographs. **A, G–L** Subfam. Sapindioideae. **B–D, F** Subfam. Dodonaeoideae. **E** Subfam. Hippocastonoideae. All micrographs in polar view, except for **B, E,** and **H**, which are in equatorial view. **A** *Erythrophysopsis aesculina*, colporate, striate pollen grain, ×2,100. **B** *Dodonaea truncatialis*, colporate, scabrate pollen grain (prolate equatorial view) with mesocolpial scabrae in a coarser pattern, ×1,500. **C** *Exothea copalillo*, colporate, loosely scabrate-verrucate pollen grain, ×3,150. **D** *Harpullia crustacea*, cryptoaperturate, reticulate (muri with scabrae) pollen grain, ×2,500. **E** *Handeliodendron bodinieri*, colporate, striate pollen grain (prolate equatorial view) showing verrucate colpus membrane, ×2,000. **F** *Magonia pubescens*, pollen tetrad (upper pollen grain in polar view) with striate-gemmate ornamentation, ×900. **G, I** *Urvillea ulmacea*, heteropolar, demisyncolporate, perforate-reticulate pollen grains, with proximal view showing long, fused demicolpi (**G**), and distal view showing short demicolpi (**I**), ×1,800. **H** *Allophylus quercifolius*, porate rugulate pollen grain, equatorial view showing one of the three pores, ×2,000. **J** *Cupania belizensis*, syncolporate, rugulate pollen grain, ×1,900. **K** *Matayba apetala*, parasyncolporate, perforate-indistinctly rugulate pollen grain, ×2,250. **L** *Pometia pinnata*, brevicolporate, reticulate pollen grain, ×1,500. (Photos R. van der Ham)

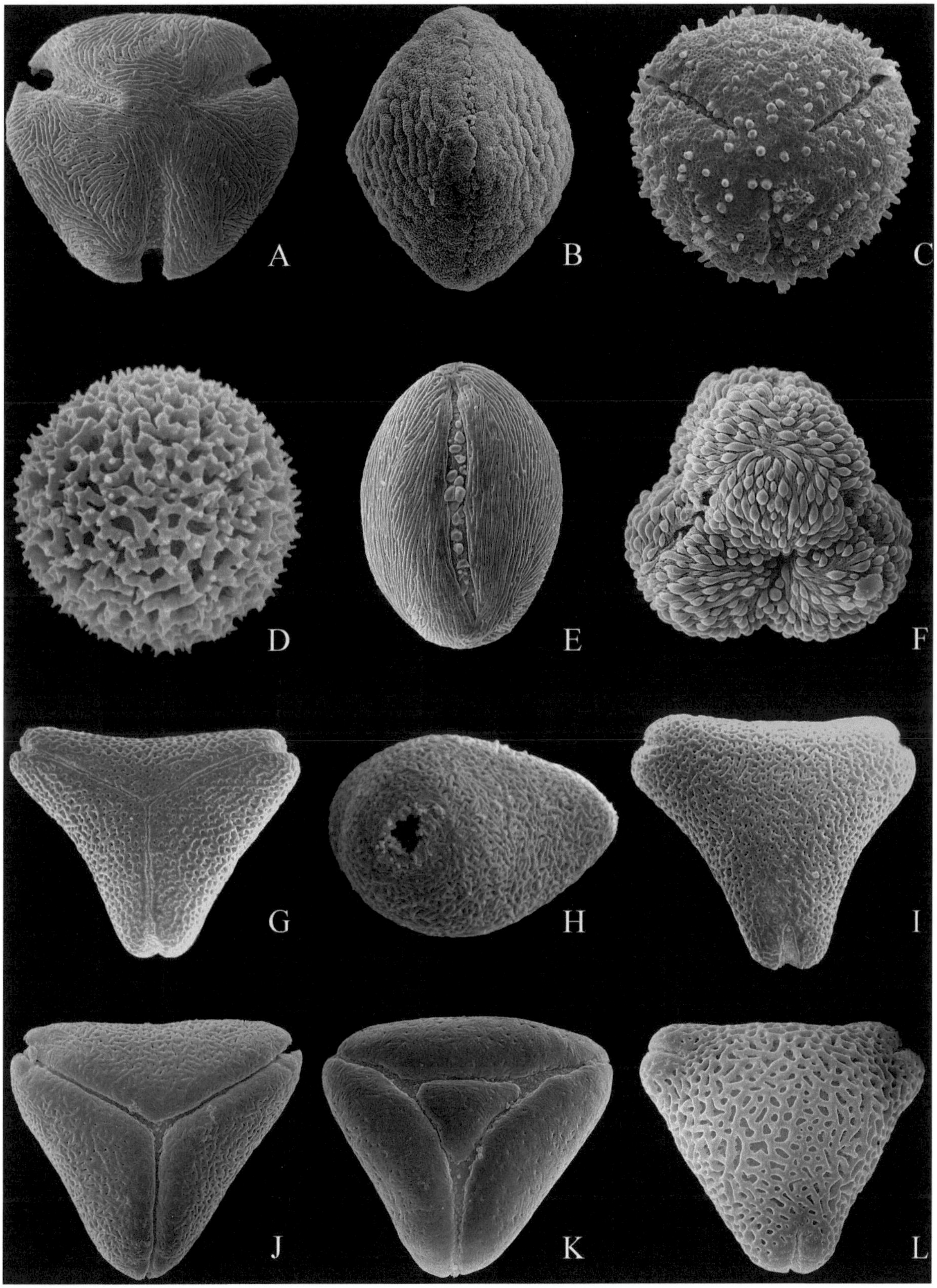
A
B
C
D
E
F
G
H
I
J
K
L

or rare (present only in some *Acer* species, and in the Malesian genera *Cubilia, Dimocarpus, Jagera,* and *Trigonachras*), but it is common in *Xanthoceras* and Dodonaeoideae (e.g., *Dodonaea, Exothea, Filicium, Ganophyllum,* and *Zanha*). The tectum is usually perforate, but less densely and less distinctly so toward the apertures. Perforate types are linked with more or less reticulate types. Simple reticulate ornamentation is rare (e.g., *Pometia*; Fig. 79L). Striate ornamentation is often associated with a colporate aperture system (Fig. 79A, E), and rugulate with a (para)syncolporate aperture system (Fig. 79J, K), which probably reflects a functional, harmomegathic relation. Ornamentation often varies within genera, while, e.g., *Dimocarpus longan* shows remarkable infraspecific variation: striate, perforate, scabrate, and intermediates thereof (van der Ham 1990).

Muller and Leenhouts (1976) recognized 12 pollen types, which were based largely on ectoaperture features and pollen grain shape. The colporate spheroidal type A (Fig. 79A–C, E), found in most genera, was considered basic. Type B1 and type B2 (Fig. 79J, K; many genera) include spheroidal to oblate pollen with a (para)syncolporate aperture system. These types intergrade and are combined here (type B1/2). This applies also to the brevicolporate oblate types A1 and C1 (Fig. 79L; 8 genera; type A1/C1). Type C2 (Fig. 79G, I; 4 genera) includes oblate pollen with a heteropolar aperture system, and type C3 (Fig. 79H; 2 genera) oblate pollen with porate apertures. Type D was described as 2-porate. However, of the two genera mentioned, *Lophostigma* has 4-aperturate pollen, while the very rare 2-porate *Allophylus* pollen is probably a variant of the normal 3-porate type (Fig. 79H). Type E (Fig. 79F) represents the tetrad pollen of *Magonia.* Type F includes colporate (per)prolate, equatorially constricted pollen, and is found only in *Diplopeltis huegelii.* Type G was described as spheroidal pollen with protruding endoapertures without ectoapertures, occurring in *Distichostemon.* However, indistinct ectoapertures appeared to be present; therefore, this pollen type is considered here as a variant of the basic colporate type. Type H (Fig. 79D) includes the spheroidal cryptoaperturate pollen occurring in a few *Harpullia* species. As far as possible, brief pollen descriptions are indicated in the generic descriptions below.

Comparison with the molecular phylogenetic trees of 64 genera by Harrington et al. (2005) suggests that the colporate spheroidal type A is indeed basal in the Sapindaceae, subfamilies Xanthoceroideae, Hippocastanoideae, and Dodonaeoideae being characterized or heavily dominated by this pollen type. Pollen types E, F, and H are each restricted to one or a few species of Dodonaeoideae. In subfamily Sapindoideae, the commonest pollen types A and B1/2 occur largely clustered—for example, type A in clade A (Lepisantheae-Sapindeae) and clade B (Nephelieae), and type B1/2 in clade C (Cupanieae), whereas type A1/C1 is found in a few isolated taxa. Pollen types C2 and C3 are restricted to clade D (Paullinieae-Thouinieae).

Embryology. Pollen grains are 2-celled when shed. The ovule is anatropous to hemitropous or campylotropous (at least after fertilization), bitegmic, crassinucellate, and the micropyle is formed by both integuments or rarely only by 7the inner one. The chalazal megaspore develops into a Polygonum type embryo sac. The endosperm is nuclear, and the tissue is digested before it becomes cellular (Davis 1966; Johri et al. 1992).

Fruit and Seed. The vast majority of Sapindaceae genera have trilocular fruits, although many also have bilocular, unilocular, or sometimes quadrilocular fruits. In addition, fruits of several genera are predominantly unilocular due to the abortion of carpels, but sometimes they may develop more than one locule. Partly trilocular fruits are produced in *Koelreuteria* and *Euphorianthus* because of their distally incomplete septa. Fruits of *Chytranthus* and *Radlkofera* contain up to 8 locules. Fruits are either indehiscent or dehiscent. Indehiscent fruits are baccate with a leathery to crustose pericarp, or exceptionally a drupe with a stony (*Tristiropsis*) or woody (*Hypelate*) endocarp. Indehiscent fruits are either unlobed, deeply lobed, or made of monocarps, these of the same number as the carpels or less, due to abortion. Dehiscent fruits are capsular or schizocarpic. The former are predominantly loculicidal and early dehiscent, or less often septicidal, septifragal, calyptrate, tardily dehiscent, or incompletely dehiscent. Their texture varies from woody, coriaceous to papery or membranous. Schizocarpic fruits are predominantly made of samaroid mericarps, and less often of sub-globose, unwinged mericarps.

Seeds are ellipsoid to sub-globose or less often lenticular or flattened, exalate or exceptionally winged (*Diplokeleba* and *Magonia*), glabrous or less often pubescent, the testa totally dry (scleroidal), with fleshy portions, or completely fleshy (sarcotestal), the scleroidal testa totally naked or partially to completely covered by an arillode (of integumental origin), which sometimes bears a funiculus-like basal projection (e.g., *Guioa*, *Mischocarpus*, and *Sarcopteryx*). Scleroidal seeds sometimes bear a small to large hilum. The embryo is usually fleshy, oily or starchy, lacking endosperm, notorhizal or lomatorhizal with straight, curved or plicate, fleshy cotyledons, the radicle often separated by a deep fold in the testa that forms a radicular pocket.

Phytochemistry. (based on Hegnauer 1973, 1990) A large amount of quebrachitol, a monomethyl ether of 1-inositol, occurs in leaves, barks, flowers, and fruits of the whole family. Cyanogenic glucosides are widely distributed in the family (but not in Hippocastanoideae); they are distributed throughout the plant body, including ripe seeds. There they are often replaced by cyanolipids, which in part also are very toxic. Polyphenolics are found in Sapindaceae mainly as coumarins, flavonoids, and proanthocyanidins. Condensed tannins, based on proanthocyanidins, can occur in large quantities, amounting to 15–20% of the dry bark in Sapindoideae. Gallotannins and their derivatives (bergenin) are also present. The family is also rich in saponins, which are contained in idioblasts within the vegetative tissues and very often in the seeds. These compounds include mono- and bidesmosidic saponins with triterpene sapogenins, and some have prominent ichthyotoxic and detergent properties. The resinous exudates of Dodonoideae are based on diterpenes and triterpenes. Non-proteinogenic amino acids of the hypoglycin type with branched carbon chains of 6 or 7 C-atoms occur free or as glutamyl peptides in the seeds, and are known from the whole family. Unspecific alkaloid tests have been positive for a large number of species of the family, but specific compounds rarely have been isolated and characterized. An exception are the purine bases (caffeine, theobromine, theophylline), which are accumulated in large amounts in the seeds and other plant parts of *Paullinia cupana* and related species, from which in Amazonia a stimulating drink ("guaraná") is prepared since time immemorial.

Dispersal. Wind-dispersed fruits consist of samaroid mericarps (Figs. 82, 83F) that are straw-colored at maturity. The genera *Diplokeleba* and *Magonia* have capsules with winged seeds that may also be dispersed by wind. Capsular fruits have green, red, orange, or yellow pericarps, with seeds that are either sarcotestal or arillate, and presumably dispersed by birds. Indehiscent baccate fruits have green, yellow, or red pericarps, and usually contain sarcotestal seeds very likely dispersed by mammals such as monkeys or bats. Several genera with inflated capsules seem to lack dispersal mechanisms; in these, fruit walls usually disintegrate, and the seeds are ultimately released by the effect of gravity.

Fossils. *Acer* fruits first occur in the late Paleocene of North America, and in the Eocene and later Tertiary the genus is well represented there but also in Europe and Asia (Manchester 1999). The early occurrence of *Acer* in North America with the coeval appearance of *Dipteronia* (McClain and Manchester 2001) and the related extinct Eocene fruit, *Deviacer*, may point to a North American origin of the group. Leaves and fruits of *Aesculus* are on record from the Maastrichtian to early Eocene of North America (Manchester 2001), and also from the Paleogene of Spitsbergen and Kamtchatka (Budantsev 1983), whereas in Europe the genus does not appear before the Miocene (see also Harris et al. 2009). Fossil seeds attributable to Sapindoideae (from genus *Sapindospermum* Reid & Chandler) are known from the Cretaceous of North America, Greenland, Europe, and Siberia, and fossil wood (*Sapindoxylum* Kräusel) has been recovered from Cretaceous beds in Egypt (Knobloch and Mai 1986). Sapindaceous flowers are reported from the middle Eocene of British Columbia (Erwin and Stockey 1990). *Koelreuteria* has an ample fossil record in the northern hemisphere that extends back to the middle Eocene (Manchester 1999).

The fossil pollen record of the family begins in the Cretaceous, represented mainly by the pollen genus *Cupanieidites*, which corresponds to the extant genus *Cupaniopsis* and several related genera (see Cookson and Pike 1954). *Cupanieidites* appeared first in the Coniacian and Santonian of Gabon, and in the Senonian of India and Brazil, became more frequent during the Maastrichtian in North America, and appeared during the Paleocene in Australia and was abundant there in the Eocene (see Muller 1981).

Family Classification. Sapindaceae were first proposed by Jussieu in 1789 in his *Genera Plantarum* as a family distinct from Aceraceae (including *Aesculus*). This concept of Sapindaceae has been followed by numerous workers such as de Candolle (1824, who proposed Hippocastanaceae as a separate family), Cambessèdes (1829), and Don (1831). Later workers, such as Reichenbach (1834), Lindley (1862), Bentham and Hooker (1862), Baillon (1874), and Blume (1878), included Aceraceae, Hippocastanaceae, and various genera currently recognized in other families such as Staphyleaceae, Sabiaceae, and Melianthaceae within Sapindaceae.

The concept of Sapindaceae held by the end of the 19th century and through most of the 20th century as a family distinct from Aceraceae and Hippocastanaceae is the result of the monumental work of Radlkofer (1888, 1890, 1931). He provided the first worldwide system of classification for the family, recognizing two subfamilies and 14 tribes, and although outdated, his work is still a useful framework for the identification of taxa and the understanding of phylogenetic relationships within Sapindaceae. His system considered the uniovulate Sapindaceae to be the most basic members within the family. In 1964, Scholz provided modern names for the two recognized subfamilies (i.e., Dodonaeoideae and Sapindoideae), and inverted their sequence, i.e., considering the multiovulate Dodoneaiodeae as basal within the family. This rearrangement was followed by Capuron (1969), who in addition, modified some of the generic concepts. Muller and Leenhouts (1976) independently proposed a rearrangement to Radlkofer's system that agreed with Scholz's general rearrangement but proposed more substantial changes. They reduced tribe Aphanieae into Lepisantheae, and suggested the inclusion of families Aceraceae and Hippocastanaceae into subfamily Dodonaeoideae, Aceraceae in its own tribe, and Hippocastanaceae as part of tribe Harpullieae.

The inclusion of Aceraceae and Hippocastanaceae by Muller and Leenhouts (1976) was slowly to be adopted. Thorne (1976), Cronquist (1981), and Takhtajan (1997) continued recognizing Aceraceae and Hippocastanaceae as distinct families. At the same time, floristic works were not confronted with this dilemma, because Sapindaceae s.s. are predominately tropical, while Aceraceae and Hippocastanaceae are largely temperate. More recently, researchers are following the suggestions of Muller and Leenhouts to include Aceraceae and Hippocastanaceae within Sapindaceae. Judd et al. (1994, 1999), who did a cladistic analysis based on morphological characters, supported the inclusion of both families within Sapindaceae. Savolainen et al. (2000), Soltis et al. (2000), and Johnson and Chase (in Klaassen 1999), from results of studies based on DNA sequence data for a small number of taxa, found Hippocastanaceae and Aceraceae to be sister groups and to be nested within Sapindaceae.

In a later study, Harrington et al. (2005), using sequencing data from two plastid genes (*mat*K and *rbc*L) for 64 genera of Sapindaceae, Aceraceae, and Hippocastanaceae, found strong support for the recognition of four major clades. Their analyses show weak support for a Hippocastaniodeae clade (Aceraceae and Hippocastanaceae) being sister or basal to the remaining Sapindaceae, which could be interpreted as support for the recognition of Aceraceae and Hippocastanaceae as distinct families. They, however, opted for the recognition of four subfamilies within Sapindaceae (i.e., Xanthoceroideae, Hippocastanoideae, Dodonaeoideae, and Sapindoideae), instead of four or five closely related families. The acceptance of four subfamilies within Sapindaceae seems to be gaining support even in the absence of strong cladistic evidence. Thorne and Reveal (2007) modified their concept of Sapindaceae to recognize four subfamilies as suggested by Harrington et al. In a more recent study, Buerki et al. (2009), using sequence data from eight plastid and nuclear genes for 104 genera of Sapindaceae s.l., found ample support for the recognition of four clades within Sapindaceae, as proposed by Harrington et al., but like these authors, they could not place the Xanthoceroid clade with certainty among the remaining clades, but left room for the interpretation of Aceraceae and Hippocastanaceae as a clade closely related to Sapindaceae s.str. The current treatment follows the approach of Harrington et al. (2005) in recognizing four subfamilies within Sapindaceae, instead of multiple families.

Tribal Classification. Radlkofer (1931) recognized 14 tribes within Sapindaceae based mostly on flower symmetry and fruit type. His system remained largely unchanged until Muller and Leenhouts (1976) cast doubts on the monophyly of the Melicocceae, and the validity of the Cupanieae as distinct from the Schleichereae and Nephelieae. They ultimately suggested the possibility of reducing the latter three tribes into one. In a more recent cladistic analysis based on morphological data, Acevedo-Rodríguez (2002) cast further doubts on the validity of the tribe Melicocceae, suggesting the genera *Castanospora*, *Tristira*, and *Tristiropsis* not to belong there. Leenhouts (1978) evaluated the tribe Nephelieae, but did not address the limits of the closely related Cupanieae and Schleichereae. These studies were followed by those of Johnson and Chase (in Klaassen 1999) who analyzed many more genera (40) than did previous authors. Their studies, although including few genera, suggest clades that conflict with the current tribal classification, except for the Paullinieae-Thouinieae clade. Their analysis also suggests Paullinieae and Thouinieae to be sister groups of a larger clade. These results contrast with those of Acevedo-Rodríguez (1993), Harrington et al. (2005), and Buerki et al. (2009) where the genera of Paullinieae are nested within the Thouinieae, therefore calling for the merging of the Paullinieae and Thouinieae tribes.

The work of Harrington et al. (2005) found support for various lineages within their proposed four subfamilies. Some of these lineages are strongly supported, while others are only weakly so. In summary, they found support for the Acereae and Hippocastaneae clades within subfamily Hippocastanoideae; a dehiscent fruit clade and an indehiscent fruit clade within subfamily Dodonaeoidea; clade Lepiantheae-Sapindeae, clade Nephelieae, clade Cupanieae, and clade Paullinieae-Thouinieae within subfamily Sapindoideae. Buerki et al. (2009) found strong support for the topologies obtained by Harrington et al. (2005) for the first three subfamilies, but differ by proposing 10 groups within subfamily Sapindoideae. Buerki et al.'s results provide strong support for the *Paullinia*, *Melicoccus*, *Tristiropsis*, *Macphersonia*, *Sapindus*, *Schleichera*, and *Delavaya* groups, and only weak support for the *Blomia*, *Cupania*, and *Koelreuteria* groups. However, the relationship among these clades is not well supported in their analyses. Some of the proposed topologies seem inconclusive, since some of the morphologically well-defined genera are suggested to be either paraphyletic or polyphyletic, and many topologies are only weakly supported. With exception of Paullinieae (expanded to include Thouinieae), which has support from various analyses, the tribal classification in Sapindaceae is still largely unresolved.

In the present treatment, we follow Harrington et al. (2005) in recognizing four subfamilies and four tribes within subfamilies Hippocastanoideae and Dodonaeoideae. With regards to subfamily Sapindoideae, we will not make an attempt to divide it into tribes, as most relationships suggested by previous workers are either not supported by molecular data or are inconclusive (Harrington et al. 2005; Buerki et al. 2009). Instead, we will recognize only the tribes Paullinieae (including Thouinieae) and Melicocceae (excluding *Tristira*, *Tristiropsis*, *Castanospora*, and *Dilodendron*) for which there seems to be strong support from either morphology or DNA sequencing data; the remaining genera will be treated alphabetically as incertae sedis within subfamily Sapindoideae. We also follow Harrington et al. (2005) in including *Koeleuteria* and *Ungnadia* within subfamily Sapindoideae.

Affinities. An early, very extensive study on the relationship of Sapindaceae with other families, involving molecular data, was that of Gadek et al. (1996). They analyzed *rbc*L sequence data for representatives of all putative sapindalean families. Their results identified a sapindalean clade sister to representatives of Malvales. This premise supports a broader concept of Sapindales similar to that of Cronquist (1981) but with the exclusion of some families. More recent work has fully confirmed the work of Gadek et al. (1996); see Introduction to Sapindales in this volume.

Economic Importance. Sapindaceae are the source of numerous products, some of which are economically important, either globally or locally. Among the most important ones, fruit crops are high on the list. These include edible fruits such as litchi (*Litchi sinensis*), longan (*Dimocarpus longan*), rambutan (*Nephelium lappaceum*), and pulasan (*N. ramboutan-ake*) at a global level, and the mamoncillo or keneep (*Melicoccus bijugatus*), pitomba (*Talisia esculenta*), and cutuplií o guaya (*T. olivaeformis*) at a local level. The arillodes of *Blighia sapida* are the source of the nutritious ackee, widely consumed in Jamaica, but highly toxic when eaten unripe (Rashford 2001). Numerous species of *Paullinia* have been reported to be useful in the preparation of medicines, caffeine-rich beverages, binding and weaving material, and for fish, human and arrow poisoning (Beck 1990). The seeds of *Paullinia cupana* are the source of the important Brazilian crop guaraná, a source of caffeine and flavoring of soft drinks. Almost all Sapindaceae are used around the tropics for fish poisoning (Acevedo-Rodríguez 1990). The wood of some species of *Euphorianthus, Harpullia*, and *Schleichera* are used in the construction of houses. Numerous genera are grown as ornamentals—e.g., *Acer, Aesculus, Arfeuillea, Allophylus, Cardiospermum, Filicium, Harpullia, Koelreuteria, Sapindus*, and *Xanthoceras*. Minor products include oils from *Pappea* and *Schleichera*, and arrow poison from *Paullinia pinnata*.

Key to the Genera

1. Carpels multiovulate 2
– Carpels uniovulate 41
2. Leaves simple or unifoliolate 3
– Leaves compound or palmately lobed 8
3. Leaves opposite; carpels 2; fruits of 2 indehiscent, distally winged mericarps **2. *Acer*** (in part)
– Leaves alternate; carpels 3 or more numerous, sometimes 2 but then fruits capsular 4
4. Petals absent 5
– Petals present 7
5. Floral disk unilateral **17. *Llagunoa*** (in part)
– Floral disk obsolete or absent 6
6. Plant often viscid; sepals 3–7; stamens 5–15, in one row **13. *Dodonaea*** (in part)
– Plant not viscid; sepals 5–8; stamens 8–74, usually in two or more rows **12. *Distichostemon***
7. Plant with glandular and simple trichomes; inflorescence a thyrse, with numerous flowers; sepals of similar size; fruit a schizocarp of 3 indehiscent obovoid cocci or a crustose capsule **11. *Diplopeltis***
– Plants without glandular trichomes; inflorescence racemose, 1–3-flowered; outer two sepals smaller; fruit an inflated capsule **124. *Stocksia***
8. Leaves opposite or subopposite 9
– Leaves alternate 16
9. Leaves digitate or trifoliolate 10
– Leaves pinnate 14
10. Leaves trifoliolate 11
– Leaves digitate 12
11. Corolla zygomorphic, usually of only 4 petals; petals clawed, crested above the claw; carpels (2)3(4); stigma unlobed; fruit a loculicidal, unwinged capsule **5. *Billia***
– Corolla actinomorphic, of 5 petals; petals not clawed nor crested; carpels 2; stigmas two; fruit of two indehiscent, distally or dorsally winged mericarps **2. *Acer*** (in part)
12. Fruit of two indehiscent, distally or dorsally winged mericarps **2. *Acer*** (in part)
– Fruit a loculicidal, unwinged capsule 13
13. Sepals connate half or more of their length **4. *Aesculus***
– Sepals distinct to the base **6. *Handeliodendron***
14. Ovary 3(5)-carpellate; fruit capsular **13. *Dodonaea*** (in part)
– Ovary 2(3)-carpellate; fruit schizocarpic, of 2 winged mericarps 15
15. Mericarps completely surrounded by a wing **3. *Dipteronia***
– Mericarps with a distal wing **2. *Acer*** (in part)
16. Leaves trifoliolate 17
– Leaves pinnate 20
17. Plant bearing stellate hairs **9. *Cossinia*** (in part)
– Plant glabrous or with simple hairs 18
18. Petals 0; floral disk unilateral **17. *Llagunoa*** (in part)
– Petals 5; floral disk annular 19
19. Fruit indehiscent, 1-locular, baccate; petals lacking appendages **27. *Hypelate***
– Fruit a dehiscent, 2–3-coccate, loculicidal capsule; petals with marginal appendages **64. *Delavaya***
20. Plant bearing stellate hairs 21
– Plant glabrous or with simple hairs, lacking stellate hairs 25
21. Floral disk with 5 elongated, horn-like lobes; leaflet margins serrate; fruit thick woody, with corky endocarp **1. *Xanthoceras***
– Floral disk unilateral, semi-annular or annular, without horn-like lobes; leaflets entire or crenate; fruit crustose or papery to woody but not corky 22
22. Seeds with white or red sarcotesta on lower half or base **16. *Harpullia*** (in part)
– Seeds without sarcotesta (seed coat black or dark brown) 23
23. Abaxial surface of leaflets densely stellate pubescent; seeds glabrous **9. *Cossinia*** (in part)
– Abaxial surface of leaflets sparsely stellate pubescent; seeds hispidulose 24

24. Disk semi-annular, or nearly complete, formed by two semi-annular halves; fruits papery **7. *Arfeuillea***
- Disk annular or semi-annular, dish-shaped to pentagonous; capsules crustose **20. *Majidea***
25. Corolla 0 or rudimentary 26
- Corolla present 29
26. Fruit a dehiscent, loculicidal capsule **8. *Averrhoidium***
- Fruit indehiscent, baccate 27
27. Fruit 2-locular **25. *Ganophyllum***
- Fruit 1-locular 28
28. Leaf rachis winged; inflorescences of axillary, glomerate cymes; stamens not coiled in bud **21. *Doratoxylon***
- Leaf rachis naked; inflorescences of thyrses; stamens coiled in bud **28. *Zanha***
29. Petals lacking appendages 30
- Petals bearing appendages 36
30. Pollen dispersed as tetrads; ovules 7–8 per locule; seeds winged **19. *Magonia***
- Pollen dispersed as monads; ovules 2(3) per locule; seeds unwinged 31
31. Leaf rachis winged **26. *Hippobromus***
- Leaf rachis unwinged 32
32. Petals suborbiculate, sessile **22. *Euchorium***
- Petals ovate or clawed 33
33. Carpels 2; fruit baccate, indehiscent **23. *Exothea***
- Carpels 3(4); fruit capsular, loculicidal 34
34. Floral disk unilateral, double, the inner lobe concave, 4-dentate; sepals fimbriate-glandular; ovary with stipitate glands **18. *Loxodiscus***
- Floral disk annular, lobed; sepals not fimbriate nor glandular; ovary lacking stipitate glands 35
35. Floral disk simple; capsule fleshy, 1(3)-coccate **15. *Eurycorymbus***
- Floral disk double, the central rim tubular sometimes elongated into a gynophore; capsule membranous, with more or less compressed locules **59. *Conchopetalum***
36. Fruits indehiscent **72. *Erythrophysopsis***
- Fruits dehiscent 37
37. Fruits coriaceous or woody 38
- Fruits membranaceous, or chartaceous 39
38. Seeds pubescent **71. *Erythrophysa*** (in part)
- Seeds glabrous **138. *Ungnadia***
39. Sepals imbricate; disk semi-annular; seeds pubescent **71. *Erythrophysa*** (in part)
- Sepals valvate; disk annular; seeds glabrous 40
40. Petals yellow; appendages formed by the fimbriate margins of the sagitate blade base; floral disk elevated on a short androgynophore; stamens straight in bud; distal leaflet fully developed; capsules with incomplete septa **82. *Koelreuteria***
- Petals white; appendages formed by a narrow marginal flap; floral disk sessile; stamens geniculate in bud; distal leaflet rudimentary; capsule with complete septa **120. *Sinoradlkofera***
41. Leaves unifoliolate or simple 42
- Leaves compound 50
42. Calyx and corolla actinomorphic 43
- Calyx and corolla zygomorphic 46
43. Fruit indehiscent 44
- Fruit dehiscent 45
44. Corolla of 4 or 5 petals, anthers dorsifixed; sarcotesta 0 **115. *Sapindus*** (in part)
- Corolla 0; anthers basifixed, sarcotesta present **73. *Glenniea*** (in part)
45. Fruit a 1-locular (by abortion), loculicidal capsule; seed with lobed arillode **100. *Pappea***
- Fruit 1–3-coccate, septicidal or loculicidal-calyptrate; seed with granular sarcotesta **44. *Alectryon*** (in part)
46. Fruit indehiscent 47
- Fruits schizocarpic 48
47. Calyx 4-merous; fruit of 1–2 rounded cocci **29. *Allophylus*** (in part)
- Calyx 5-merous; fruit of 3, ellipsoid, apically keeled cocci **95. *Namataea***
48. Mericarps sub-globose, exalate **34. *Guindilia***
- Mericarps winged 49
49. Calyx 5-merous; mericarps with a dorsal wing **31. *Bridgesia***
- Calyx 4-merous; mericarps with a distal wing **40. *Thouinia*** (in part)
50. Distal leaflet well-developed 51
- Distal leaflet rudimentary 61
51. Trees or shrubs, exceptionally climbing shrubs; stipules absent 52
- Vines, lianas or climbing shrubs, sometimes not climbing in early stages; stipules minute, or exceptionally large and early deciduous 55
52. Leaves pinnate 53
- Leaves trifoliolate 54
53. Fruit schizocarpic; mericarps winged; leaves once pinnate **30. *Athyana***
- Fruit a loculicidal capsule; leaves bipinnate **66. *Dilodendron*** (in part)
54. Fruit schizocarpic, of 3 winged mericarps **40. *Thouinia*** (in part)
- Fruits indehiscent, of 1(2) unwinged monocarps **29. *Allophylus*** (in part)
55. Fruits schizocarpic, splitting into 3 1-winged mericarps 56
- Fruit capsular, winged or not winged 59
56. Flowers actinomorphic **39. *Thinouia***
- Flowers zygomorphic 57
57. Mericarps with a dorsal wing surrounding the coccus **35. *Houssayanthus***
- Mericarps with a proximal wing 58
58. Stamens with filaments of similar length; pollen cylindrical **36. *Lophostigma***
- Stamens with filaments of unequal lengths; pollen triangular **38. *Serjania***
59. Capsules woody, coriaceous or crustaceous, not inflated; pollen 3-porate **37. *Paullinia***
- Capsules papery, inflated or nearly so; pollen 3-(demi)(syn)colporate 60
60. Capsules not winged or only narrowly so **32. *Cardiospermum***
- Capsule dorsally winged **41. *Urvillea***
61. Leaves bipinnate or tripinnate 62

- Leaves once pinnate 66
62. Leaves tripinnate **133. *Tripterodendron***
- Leaves bipinnate 63
63. Sepals valvate **50. *Bizonula***
- Sepals imbricate 64
64. Anthers basifixed; placentation axile; fruit a loculicidal capsule **66. *Dilodendron*** (in part)
- Anthers dorsifixed; placentation basal; fruit indehiscent, baccate or nearly so 65
65. Stigma capitate, sessile; aril and sarcotesta present **90. *Macphersonia*** (in part)
- Stigma of 3 elongated branches; aril or sarcotesta absent **135. *Tristiropsis***
66. Calyx zygomorphic 67
- Calyx actinomorphic (slightly zygomorphic in *Tinopsis*) 96
67. Sepals connate to form a tubular or urceolate calyx 68
- Sepals distinct 81
68. Calyx bilabiate **111. *Pseudopancovia***
- Calyx of (4)5(–7) lobes, not bilabiate 69
69. Calyx urceolate 70
- Calyx tubular 71
70. Fruit 5–8-locular, not ribbed; inflorescence of densely packed flowers; bracts longer than the flowers **113. *Radlkofera***
- Fruit 3–8-locular, longitudinally ribbed; inflorescence with loosely spaced flowers; bracts shorter than the flowers **57. *Chytranthus***
71. Petals 4; disk unilateral; stamens (6)8; style terminal **99. *Pancovia***
- Petals 5; disk annular; stamens (8)12–30; style gynobasic or sub-terminal **63. *Deinbollia***
72. Fruit indehiscent 73
- Fruit dehiscent 76
73. Fruit deeply lobed or lobed-coccate **87. *Lepisanthes*** (in part)
- Fruit ovoid, obovoid or ellipsoid, not lobed 74
74. Fruit locules unwinged **87. *Lepisanthes*** (in part)
- Fruit with 3 dorsal wings 75
75. Corolla 0; disk annular; fruit trilocular **134. *Tristira***
- Corolla of 4–5 petals; disk unilateral, flattened; fruit unilocular **140. *Zollingeria***
76. Fruit schizocarpic, splitting into indehiscent mericarps 77
- Fruit capsular 83
77. Mericarps unwinged 78
- Mericarps winged 80
78. Seeds covered with sticky, saponiferous pulp **115. *Sapindus*** (in part)
- Seeds not surrounded by a sticky or saponiferous pulp 79
79. Mericarp membranaceous, bladdery **109. *Porocystis***
- Mericarps woody, not bladdery **119. *Scyphonychium***
80. Mericarps with a proximal wing **131. *Toulicia***
- Mericarps with a distal or dorsal wing 81
81. Calyx 4-merous; corolla zygomorphic **33. *Diatenopteryx***
- Calyx 5-merous; corolla actinomorphic 82
82. Leaves with 10–12 leaflets; disk annular or semi-annular; stamens 8; ovary 3-carpellate; fruits of 3 mericarps **48. *Atalaya***
- Leaves with 2–4 leaflets; disk 5-lobed; stamens 18–24; ovary bicarpellate; fruits of 2 mericarps **80. *Hornea***
83. Corolla absent **65. *Dictyoneura*** (in part)
- Corolla present 84
84. Corolla zygomorphic; capsule echinate or muricate 85
- Corolla actinomorphic; capsules smooth 86
85. Capsules echinate; petal appendages marginal **121. *Sisyrolepis***
- Capsules muricate; petal appendage basal, simple **104. *Phyllotrichum***
86. Capsules 1-locular **96. *Neotina***
- Capsules 2- or 3-locular 87
87. Seeds winged, non-arillate **10. *Diplokeleba***
- Seeds, ellipsoid, sub-globose, unwinged, arillate 88
88. Gynoecium 2-carpellate 89
- Gynoecium (2)3(4)-carpellate 90
89. Leaflets entire; anthers retuse at apex **138. *Vouarana*** (in part)
- Leaflets serrate or crenate-serrate; anthers apiculate at apex **128. *Tina*** (in part)
90. Arillode basal or lateral 91
- Arillode partially to nearly completely covering the seed 92
91. Arillode basal, surrounding the hilum; disk annular **114. *Rhysotoechia***
- Arillode at base of seed but not covering the hilum; disk pentagonal **110. *Pseudima***
92. Petal appendages absent 93
- Petal appendages present or sometimes rudimentary 94
93. Petals more or less cuneate at base; inflorescence axillary; arillode fimbriate at apex **94. *Molinaea*** (in part)
- Petals clawed at base; inflorescence cauliflorous; arillode open but not fimbriate at apex **74. *Gloeocarpus***
94. Arillode with a basal funiculus-like appendage (rudimentary in one species) **77. *Guioa***
- Arillode lacking a basal funiculus-like appendage 95
95. Petals with one or two basal appendages **62. *Cupaniopsis***
- Petals with marginal appendages or without appendages **94. *Molinaea*** (in part)
96. Petals absent 97
- Petals present 118
97. Fruits indehiscent or tardily dehiscent (pseudodehiscent) 98
- Fruits dehiscent 107
98. Sepals connate into a cupular or urceolate calyx 99
- Sepals distinct or connate only at base 101
99. Fruit tardily dehiscent (pseudodehiscent), usually 1-coccate, the pericarp smooth **123. *Stadmania*** (in part)

- Fruit indehiscent, 3-locular, or if 1-coccate, then with muricate pericarp 100
100. Gynoecium 3-carpellate; fruit usually 3-locular, 3-sulcate or 3-lobed **105. *Placodiscus***
- Gynoecium bicarpellate; fruit usually 1-coccate (1 coccus rudimentary), indehiscent, baccate; muricate **88. *Litchi***
101. Seeds exarillate **73. *Glenniea*** (in part)
- Seeds arillate 102
102. Gynoecium 2-carpellate; arillode basal 103
- Gynoecium 3(4)-carpellate; arillode nearly covering the entire seed 104
103. Pseudostipules present; floral disk glabrous **98. *Otonephelium***
- Pseudostipules 0; floral disk pubescent **67. *Dimocarpus*** (in part)
104. Arillode with a dorsal split **79. *Haplocoelum***
- Arillode covering the seed completely (no dorsal split) 105
105. Sepals connate at base (1/4 of their length); floral disk unlobed; anthers dorsifixed **118. *Schleichera***
- Sepals distinct to base; floral disk lobed; anthers basifixed 106
106. Sepals 5; stamens 8; cotyledons smooth **84. *Lecaniodiscus***
- Sepals (5)6 or 7(8); stamens 6–8(10); cotyledons with brain-like appearance (cerebriform) **49. *Beguea***
107. Calyx with imbricate aestivation 108
- Calyx with valvate aestivation 111
108. Pseudostipules present; anthers basifixed; fruits 1-locular **52. *Blighiopsis***
- Pseudostipules 0; anthers dorsifixed; fruits (1)3(4)-locular 109
109. Leaflets serrate **65. *Dictyoneura*** (in part)
- Leaflets entire 110
110. Aril with a basal funiculus-like appendage; inflorescences of thyrses or panicles **93. *Mischocarpus*** (in part)
- Aril lacking a basal funiculus-like appendage; inflorescences of racemes **51. *Blighia***
111. Fruit with circumscissile dehiscence 112
- Fruit with loculicidal dehiscence 113
112. Seed with a dorsal, white arillode **107. *Podonephelium***
- Seeds with red sarcotesta on lower half **44. *Alectryon*** (in part)
113. Seeds naked (without arillode or sarcotesta); endotesta ruminately grown together with embryo **76. *Gongrospermum***
- Seeds arillate or sarcotestal 114
114. Seeds arillate **136. *Tsingya***
- Seeds sarcotestal 115
115. Sepals distinct to the base 116
- Sepals connate at least half way to form cupular or acetabuliform calyx 117
116. Gynoecium unicarpellate; stigma capitate **53. *Blomia*** (in part)
- Gynoecium (1)2(–4)-carpellate; stigmata 2(3), elongated, and usually coiled **97. *Nephelium*** (in part)
117. Fruit warty, spiny or echinate **97. *Nephelium*** (in part)
- Fruit smooth **44. *Alectryon*** (in part)
118. Fruits indehiscent 119
- Fruits dehiscent 133
119. Seeds laterally flattened, with a longitudinal ventral hilum **106. *Plagioscyphus***
- Seeds variously shaped, hilum if present restricted to the basal-ventral part 120
120. Gynoecium 2(3)-carpellate 121
- Gynoecium 3-carpellate 128
121. Stamens 5(–7) 122
- Stamens (6–)8(–10) 125
122. Seeds exarillate **24. *Filicium***
- Seeds completely covered by an arillode 123
123. Sepals connate half of their length **108. *Pometia***
- Sepals distinct 124
124. Cataphylls present; petals lacking appendages **112. *Pseudopteris***
- Cataphylls absent; petals bearing basal or marginal appendages **129. *Tinopsis***
125. Testa fleshy (sarcotestal) 126
- Testa not fleshy 127
126. Fruit ellipsoid or globose, smooth; carpels with incomplete septa **42. *Melicoccus***
- Fruit 1–2-coccate, warty or spiny; carpels with complete septa **139. *Xerospermum***
127. Seed exarillate; calyx aestivation valvate; fruit smooth **55. *Castanospora***
- Seed arillate; calyx aestivation imbricate; fruit smooth to echinate **67. *Dimocarpus*** (in part)
128. Sepals connate into a cup-shaped calyx **123. *Stadmania*** (in part)
- Sepals distinct, or if connate, then the calyx tubular 129
129. Filaments coiled in bud 130
- Filaments short, not coiled in bud 131
130. Petal appendages marginal **90. *Macphersonia*** (in part)
- Petal appendages basal **56. *Chouxia***
131. Seeds naked (not arillate nor sarcotestal) **115. *Sapindus*** (in part)
- Seeds arillate or sarcotestal 132
132. Seeds sarcotestal **43. *Talisia***
- Seeds completely covered by a translucent arillode **54. *Camptolepis***
133. Fruit schizocarpic, splitting into winged mericarps **127. *Thouinidium*** (in part)
- Fruit capsular 134
134. Fruit echinate, warty or with hispid or setaceous hairs 135
- Fruit smooth or scrobiculate, glabrous or variously pubescent 138
135. Fruit a 1- or 2-coccate, warty or echinate capsule; petals lacking appendages **60. *Cubilia***
- Fruit a 3-locular, trigonous or trilobed, hispid capsule; petals with appendages 136
136. Disk cupular; sepals valvate; seeds arillate at base; fruits without irritating hairs **70. *Eriocoelum***

- Disk annular; sepals imbricate; seeds non-arillate, with hilar sarcotesta; fruit with irritating hairs 137
137. Fruit locules not winged **81.** ***Jagera***
- Fruit locules winged on lower dorsal portion **58.** ***Cnesmocarpon***
138. Capsules 1-locular 139
- Capsules 2- or 3-locular 143
139. Seed naked (not arillate nor sarcotestal), with large white hilum **101.** ***Paranephelium***
- Seed arillate or sarcotestal, with small hilum 140
140. Seed arillate, the arillode with dorsal split **75.** ***Gongrodiscus***
- Seed sarcotestal 141
141. Fruit smooth 142
- Fruits tuberculate to echinate **97.** ***Nephelium*** (in part)
142. Seed black with partial red sarcotesta **44.** ***Alectryon*** (in part)
- Seed completely covered by a brownish sarcotesta **53.** ***Blomia*** (in part)
143. Petals lacking appendages 144
- Petals bearing appendages 148
144. Sepals connate; seeds exarillate **122.** ***Smelophyllum***
- Sepals distinct; seeds arillate 145
145. Capsule obovoid or trigonous 146
- Capsule deeply 2- or 3-lobed 146
146. Seeds sarcotestal at its base or lower half, not fimbriate at apex **16.** ***Harpullia*** (in part)
- Seeds arillate, arillode covering at least lower half, fimbriate at apex **94.** ***Molinaea*** (in part)
147. Arillode ventrally split and fimbriate **125.** ***Storthocalyx***
- Arillode ventrally attached, covering whole seed, not fimbriate **85.** ***Lepiderema***
148. Calyx with valvate aestivation 149
- Calyx with imbricate aestivation 155
149. Sepals distinct 150
- Sepals connate at least at base 153
150. Petals with a single ventral appendage 151
- Petals with marginal appendages 152
151. Gynoecium 3-carpellate; capsule (1–2)3-coccate, wider than long **45.** ***Amesiodendron***
- Gynoecium 2-carpellate; capsule 2-lobed or elliptic, longer than wide **86.** ***Lepidopetalum***
152. Seeds sarcotestal; gynoecium bicarpellate **46.** ***Aporrhiza***
- Seed arillate; gynoecium tricarpellate **91.** ***Matayba***
153. Sepals connate at base 154
- Sepals connate at least half of their length **44.** ***Alectryon*** (in part)
154. Petals with marginal appendages; fruit locules not winged dorsally **92.** ***Mischarytera***
- Petals with ventral, bifid appendage; fruit locules with a narrow dorsal wing **116.** ***Sarcopteryx***
155. Petals with marginal appendages 156
- Petals with ventral appendages 166
156. Seeds sarcotestal 157
- Seeds arillate 159
157. Pubescence of stellate hairs **16.** ***Harpullia*** (in part)
- Pubescence of simple hairs 158
158. Pseudostipules 0; sarcotesta cupular **117.** ***Sarcotoechia***
- Pseudostipules present; sarcotesta completely covering the seed **83.** ***Laccodiscus***
159. Arillode basal 160
- Arillode covering at least lower half of the seed 163
160. Petals longer than the sepals, with crested appendages **130.** ***Toechima***
- Petals shorter than the sepals, with non-crested appendages 161
161. Petals clawed at base **69.** ***Elattostachys***
- Petals not clawed at base 162
162. Pericarp woody; seeds arillate at base, lacking a funiculus-like appendage **138.** ***Vouarana*** (in part)
- Pericarp coriaceous; seeds completely covered by an arillode, with a funiculus-like appendage **93.** ***Mischocarpus*** (in part)
163. Capsule dehiscent by a loculicidal calyptra or septicidal **47.** ***Arytera***
- Capsules loculicidal 164
164. Seeds lenticular **68.** ***Diploglottis***
- Seeds obovoid or ellipsoid 165
165. Gynoecium bicarpellate; stigma an invaginate prolongation of the style **128.** ***Tina*** (in part)
- Gynoecium tricarpellate; stigma simple, with three stigmatic lines **61.** ***Cupania***
166. Petal bearing a single appendage 167
- Petals bearing 2 appendages 171
167. Sepals connate at base 168
- Sepals distinct to base 169
168. Seed exarillate **102.** ***Pavieasia***
- Seed sarcotestal **89.** ***Lychnodiscus***
169. Petals clawed at base **132.** ***Trigonachras*** (in part)
- Petals cuneate at base 170
170. Gynoecium 3-carpellate; disk 7- or 8-lobed; ovary hirsute **103.** ***Pentascyphus***
- Gynoecium 2-carpellate; disk annular; ovary glabrous **78.** ***Haplocoelopsis***
171. Petals not clawed; appendage crested **126.** ***Synima***
- Petals clawed, appendage not crested 172
172. Seed with a 2- or 3-lobed sarcotesta around the hilum **14.** ***Euphorianthus***
- Seed not sarcotestal (naked) **132.** ***Trigonachras*** (in part)

Subfamilies, Tribes, and Genera of Sapindaceae

I. Subfam. Xanthoceroideae Thorne & Reveal (2007).

Leaves alternate; petals without appendages; ovules 7–8 per locule; disk with orange horn-like appendages.

1. ***Xanthoceras*** Bunge

Xanthoceras Bunge, Mem. Sav. Etr. Acad. Petersb. 2: 85 (1834).

Falsely polygamous trees. Leaves alternate, imparipinnate, with stellate pubescence; leaflets serrate; distal leaflet rudimentary; stipules 0. Inflorescences terminal racemose thyrses. Flowers actinomorphic, bisexual or functionally unisexual; sepals 5, distinct; petals 5, with darker coloration at base, clawed; disk with 5 erect corniform lobes, alternating with the petals; stamens 8; pollen colporate, loosely verrucate; ovary 3-carpellate, with 7–8 ovules per carpel; style filiform with capitate, 3-sulcate stigma. Fruit a 3-locular, woody loculicidal capsule, with corky endocarp. Seeds exarillate. $2n=30$.

A single species, *X. sorbifolia* Bunge, endemic to China, and commonly cultivated as an ornamental tree.

II. Subfam. Hippocastanoideae Burnett (1835).

Leaves opposite; ovules 2 per locule.

1. **Tribe Acereae** (Durande) Dumort. (1827).

Flowers actinomorphic; petals without appendages; disk annular.

2. *Acer* L. Fig. 80A–I

Acer L., Sp. Pl. 1054 (1753); de Jong, Meded. Landbou. Wageningen Nederl. 72: 1–201 (1976).

Duodichogamous or dioecious trees or shrubs. Leaves opposite, simple, unlobed, palmately lobed to deeply dissected, or trifoliolate, palmaticompound or imparipinnate, margins entire, dentate, serrate or lobed; petioles usually long; stipules 0 or rarely present. Inflorescences terminal or axillary, corymbose-thyrsoid, racemose or fasciculate. Flowers unisexual or bisexual, actinomorphic; sepals (4)5, distinct or less often connate, imbricate; petals 4–5 or 0, white or greenish, not clawed; disk extrastaminal or less often intrastaminal, amphistaminal or 0, annular or annular-lobed, glabrous or pubescent; stamens (5)8 (10, 12), the filaments equal or unequal; pollen colpate or colpor(oid)ate, striate, rugulate or reticulate, sometimes scabrate; ovary 2(3, 5, 8)-carpellate, with 2 ovules per carpel; style branches stigmatic, 2; style sometimes very short. Fruits schizocarpic, splitting into 2 samaroid, 1-seeded, mericarps, each with a long, dorsal or distal wing. Seed laterally compressed, with papery testa.

Sixteen sections and about 126 species, northern Asia, Japan, Europe, and North America.

Fig. 80. Sapindaceae. A–C *Acer platanoides*. **A** Leaf. **B** Male flower. **C** Female flower. **D** *Acer laurinum*, part of infructescence. **E** *Acer carpinifolium*, part of infructescence. **F–J** *Acer negundo*. **F** Leaf. **G** Male flower. **H** Female flower. **I** Fruit. **J** *Dipteronia sinensis*, fruit. (Takhtajan 1981; drawn by A. Schischtkin)

3. *Dipteronia* Oliver Fig. 80J

Dipteronia Oliver, Hooker's Icon. Pl. t. 1898. (Oct 1889).

Duodichogamous trees. Leaves opposite, imparipinnate; leaflets 9–17, with serrate margins; terminal leaflet fully developed; stipules 0. Inflorescences terminal, thyrsoidate. Flowers functionally unisexual, actinomorphic; sepals 5, distinct; petals 5, cream or greenish, clawed; disk extrastaminal, annular-lobed, stamens 6–8, the filaments of equal length; pollen colporate, striate; ovary 2-carpellate, with 2 ovules per carpel; stigmata 2, connate at base. Fruits seemingly schizocarpic, of 2 suborbicular, samaroid mericarps, each with a wing completely surrounding the seed locule. Seeds exarillate.

A single species, *D. sinensis* Oliver, endemic to China.

2. Tribe Hippocastaneae (DC.) Dumort. (1827).

Flowers zygomorphic; petals usually with appendages; disk unilateral.

4. *Aesculus* L.

Aesculus L., Sp. Pl.: 344. (1753).
Hippocastanum Miller (1754).

Trees or shrubs. Leaves opposite, palmati-compound; petioles usually long; stipules 0. Inflorescences terminal thyrses or racemes. Flowers bisexual or functionally staminate, zygomorphic; sepals 5, imbricate, connate to form a tubular or campanulate calyx; petals 4(5), equal or unequal, clawed; appendages 0 or minute and placed above the claw; disk unilateral, 4-lobed; stamens (5)6–8, the filaments of unequal length; pollen colporate, finely striate, sometimes scabrate, usually with heavily verrucate colpus membranes; ovary (2)3(4)-carpellate, with 2 ovules per carpel; style with punctiform or obscurely 3-lobed stigma. Fruit usually a 1-seeded loculicidal capsule, smooth or echinate. Seeds exarillate, with a large, pale hilum.

Thirteen species, southeastern Europe, Asia, Japan, and North America.

5. *Billia* Peyritsch

Billia Peyritsch, Bot. Zeit. 16: 153 (1858).

Trees. Leaves opposite, trifoliolate; margins entire; petioles long; stipules 0. Inflorescences of terminal panicles bearing bisexual and unisexual flowers. Flowers zygomorphic; sepals 5, imbricate, distinct, unequal; petals 4(5), red or white, unequal, clawed; appendages minute, two or a single bifid or dissected, ventral on the claw; disk unilateral, 4-lobed; stamens 6–8, the filaments of unequal length; pollen colporate, finely striate, with heavily verrucate colpus membranes; ovary (2)3(4)-carpellate, with 2 ovules per carpel; style with punctiform stigma. Fruit usually a 1-seeded, loculicidal capsule, smooth. Seeds exarillate, with a small, pale hilum.

Two species, from southern Mexico to northern South America.

6. *Handeliodendron* Rehder

Handeliodendron Rehder, J. Arnold Arb. 16: 65 (1935).

Trees. Leaves opposite, digitate. Inflorescence of terminal panicles. Flowers zygomorphic, bisexual; sepals 5, distinct, imbricate; petals 4–5, clawed, with 2 marginal appendages above the claw; disk unilateral, lobed; stamens (7)8, unequal; pollen colporate, striate, with heavily verrucate colpus membranes (Fig. 79E); ovary 3-carpellate, stipitate, with 2 ovules per carpel; style short, stigma capitate. Fruit a (1–)3-locular, stipitate, coriaceous, loculicidal capsule. Seeds with a double arillode at base.

A single species, *H. bodinieri* (Lév.) Rehder, endemic to China.

III. Subfam. Dodonaeoideae Burnett (1835).

Leaves alternate; petals usually without appendages.

3. Tribe Dodonaeeae Kunth ex DC. (1824).

Disk (semi)annular; ovules (1)2(3 or 8) per locule; fruits dehiscent.

7. *Arfeuillea* Pierre ex Radlk.

Arfeuillea Pierre ex Radlk. in Engl. & Prantl., Nat. Pflanzenfam. III, 5: 362 (1895); Radlk. in Engl., Pflanzenr. 98: 1467 (1933); Welzen in Santisuk & Larsen, Fl. Thailand 7: 185, 186 (1999).

Falsely polygamous trees with simple and stellate hairs. Leaves alternate, paripinnate; leaflets crenate or entire; distal leaflet rudimentary.

Inflorescences axillary or terminal thyrses, with foliaceous bracts. Flowers zygomorphic, pistillate or staminate on same inflorescence; sepals 5, petal-like, distinct, imbricate; petals 4, without appendages; disk semi-annular, or nearly complete, formed by two semi-annular halves; stamens (6) 7–8(9); pollen colporate, striate; ovary 3-carpellate, with 2 ovules per carpel. Fruit a 3-locular, loculicidal, papery capsule. Seed one per carpel, obovoid, exarillate.

A single species, *A. arborescens* Pierre ex Radlk., from Thailand and Laos, cultivated throughout the tropics as an ornamental. Doubtfully distinct from *Majidea*.

8. ***Averrhoidium*** Baill. Fig. 81

Averrhoidium Baill., Adansonia 11: 244 (1874).

Dioecious trees. Leaves alternate, paripinnate; leaflets serrate or entire; distal leaflet rudimentary; stipules 0. Inflorescences axillary thyrses. Flowers actinomorphic, functionally unisexual; sepals 5, distinct, imbricate; petals 0 or rudimentary, 1–4, without appendages; disk annular-lobed; stamens (7)8; pollen colporate, striate; ovary 3-carpellate, with 2 ovules per carpel; style filiform, with 3 stigmatic groves. Fruit a 1-locular, loculicidal, crustose capsule. Seeds 1–2 per fruit, with sarcotesta.

Four species, one in Mexico, the remaining from tropical South America.

Fig. 81. Sapindaceae. *Averrhoidium dalyi*. **A** Fruiting branch. **B** Fruits showing apical (left) and basal (right) dehiscence. **C** Seed, lateral view (left) and ventral view (right). **D** Embryo, lateral view. (Drawn by A. Tangerini)

9. ***Cossinia*** Commers. ex Lam.

Cossinia Commers. ex Lam., Encycl. 2: 132 (1786). [also spelled *Cossignia*]

Falsely polygamous trees or shrubs, with stellate pubescence. Leaves alternate, trifoliolate or imparipinnate, terminal leaflet well-developed. Inflorescences terminal paniculate or corymbose thyrses. Flowers actinomorphic or zygomorphic, bisexual or functionally unisexual; sepals 5, imbricate; petals 4–5, without appendages; disk annular or semi-annular; stamens 5–6(–8); pollen colporate, with clearly scabrate mesocolpia and indistinctly scabrate to perforate apocolpia; ovary 3-carpellate, with 2 ovules per carpel; stigma subcapitate. Fruit a septicidal, 3-locular, crustose capsule. Seeds globose, exarillate.

Three species, two from Mauritius and one from New Caledonia.

10. ***Diplokeleba*** N.E. Brown

Diplokeleba N.E. Brown, Trans. & Proc. Bot. Soc. Edinburgh 20: 50 (1894).

Falsely polygamous trees. Leaves paripinnate; leaflets entire or undulate; distal leaflet rudimentary. Inflorescences terminal, corymbose thyrses. Flowers functionally unisexual; calyx zygomorphic, the sepals 5, imbricate, outer sepals smaller; petals 5, without appendages; disk annular, cup-shaped, crenulate; stamens 8; pollen colporate, irregularly striate to rugulate; ovary 3-carpellate, with a single ovule per carpel; style filiform, with 3 stigmatic lobes. Fruit a 3-locular, woody loculicidal capsule. Seeds winged.

Two species, southern South America.

11. ***Diplopeltis*** Endl.

Diplopeltis Endl. in Endl. et al., Enum. Pl. Hugel.: 13 (1837).

Monoecious sub-shrubs. Leaves alternate, simple; stipules 0. Inflorescences terminal thyrses. Flowers zygomorphic; sepals 5, imbricate; petals 4(5), clawed, without appendages; disk unilateral, lobed, erect; stamens 8; pollen colporate, loosely to densely scabrate to finely striate; ovary 3-carpellate, with 2 ovules per carpel; style filiform. Fruit a 3-locular, loculicidal, crustose capsule or a schizocarp of 3 indehiscent cocci. Seeds arillate, ovoid, 1 or 2 per locule.

Five species, Australia.

12. ***Distichostemon*** F. Muell.

Distichostemon F. Muell., Hooker's J. Bot. Kew Gard. Misc. 9: 306 (1857).

Falsely polygamous trees. Leaves alternate, simple, entire, serrate or crenate; stipules 0. Inflorescences axillary or terminal. Flowers actinomorphic, functionally unisexual; sepals 5–8, imbricate; petals 0; disk rudimentary; stamens 8–74; pollen colporate with indistinct colpi, finely scabrate; ovary 3(–6)-carpellate, with 2 ovules per carpel; style filiform, with 3 stigmatic lobes. Fruit a 3(–6)-locular, septifragal, marginicidal, capsule, the locules dorsally or distally winged, inflated or flattened. Seeds exarillate, subtended by a conspicuous funiculus, with a conspicuous annular ring around the hilum, 2 per locule.

Six species, Australia.

13. ***Dodonaea*** Miller Fig. 82

Dodonaea Miller, Gard. Dict. Abr. ed.: 4 (1754); Leenhouts, Blumea 28: 271 (1983), reg. rev.; West, Brunonia 7: 18 (1984), reg. rev.

Dioecious or falsely polygamous-dioecious shrubs or trees, with viscous glandular hairs. Leaves alternate or rarely opposite, simple or paripinnate; distal leaflet fully developed or rudimentary. Inflorescences axillary or terminal racemes or thyrses. Flowers actinomorphic, unisexual or bisexual; sepals (3–)5(–7), imbricate or valvate; petals 0; disk 0 or rudimentary in pistillate flowers; stamens 5–15; pollen colporate, finely to coarsely scabrate, with scabrae often in coarse patterns (Fig. 79B); ovary (2)3–5(6)-carpellate, with 2 ovules per carpel; style filiform; stigma grooved or divided. Fruit a 2–6-locular, septifragal or septicidal capsule, the locules dorsally winged. Seed exarillate, on enlarged placenta. $2n = 28, 30, 32$.

Primarily Australian with 59 endemic species and 8 pantropical species.

Fig. 82. Sapindaceae. *Dodonaea viscosa* Jacq. **A** Fruiting branch. **B** Capsules. **C** *D. angustifolia*, fruit. **D** *D. polyandra*, fruit. (Adema et al. 1994; drawn by J. Wessendorp)

14. ***Euphorianthus*** Radlk.

Euphorianthus Radlk., Sitzungsber. Math.-Phys. Cl. Königl. Bayer. Akad. Wiss. München 9: 673 (1879); Leenhouts, Blumea 33: 198 (1988), rev.

Falsely polygamous trees. Leaves alternate, paripinnate; distal leaflet rudimentary. Inflorescences axillary to pseudo-terminal thyrses. Flowers actinomorphic, bisexual or functionally unisexual; sepals 5, distinct, imbricate; petals 5, clawed, with a pair or woolly appendages above the claw; disk annular; stamens (6)7–8; pollen parasyncolporate, rugulate; ovary 3-carpellate, with a single ovule per carpel; style filiform; stigma grooved to slightly lobed. Fruit an incompletely 3-locular, sub-globose, velvety, fleshy, loculicidal capsule, the endocarp sericeous. Seeds with a 2(3)-lobed sarcotesta around the hilum.

A single species, *E. euneurus* (Miq.) Leenh., found in eastern Malesia, from the Philippines and Celebes to Vanuatu. The wood is used in the construction of houses.

15. *Eurycorymbus* Hand.-Mazz.

Eurycorymbus Hand.-Mazz., Akad. Wiss. Wien. Math. Naturwiss. Kl. Anz. 59: 104 (1922).

Dioecious trees. Leaves alternate, paripinnate; leaflets serrate; distal leaflet rudimentary. Inflorescences axillary, corymbose thyrses. Flowers actinomorphic, unisexual; sepals 5; petals 5, spatulate, without appendages; disk annular, dish-shaped-crenate; stamens 8; pollen colporate, striate; ovary 3(4)-carpellate, with 2 ovules per carpel; style filiform, with 3 stigmatic branches. Fruit a 1(3)-coccate, fleshy, loculicidal capsule. Seeds exarillate, sub-globose, puberulent, one per locule, persistent on receptacle.

A single species, *E. austrosinensis* Hand.-Mazz., from China.

16. *Harpullia* Roxb.

Harpullia Roxb., Fl. Ind. 2: 441 (1824); Leenhouts & Vente, Blumea 28: 1 (1982), rev.

Falsely polygamous shrubs or trees. Indumentum of stellate hairs. Leaves alternate, paripinnate; distal leaflet rudimentary; petiole and rachis winged or unwinged. Inflorescences axillary, terminal, ramiflorous, or cauliflorous thyrses. Flowers actinomorphic, bisexual or functionally unisexual; sepals 5, distinct, imbricate; petals 5, clawed, with 2 marginal appendages above the claw or the petals sessile without appendages; disk annular; stamens 5–8; pollen colporate, sometimes cryptoaperturate, striate, striate-rugulate, rugulate, reticulate with scabrae or coarsely reticulate with finely striate to scabrate muri (Fig. 79D); ovary 2–3(4)-carpellate, with 1 or 2 ovules per carpel; style with 2 or 3 stigmatic lines. Fruit a 2–3-locular, chartaceous to woody loculicidal capsule. Seeds with an orange sarcotestal ring around the hilum or completely covered by an aril that is basally adnate to the testa. $2n=30$.

About 26 species from India, Sri Lanka, southeastern China, Malesia, Australia to New Caledonia and Tonga. Several species are cultivated for their wood or as ornamentals.

17. *Llagunoa* Ruiz & Pavón

Llagunoa Ruiz & Pavón, Prodr.: 126 (1794).

Falsely polygamous shrubs or trees. Leaves alternate, simple or trifoliolate. Flowers solitary or in axillary cymes. Flowers zygomorphic, bisexual or functionally unisexual; sepals 5, imbricate; petals 0; disk unilateral, semi-annular; stamens 8; pollen colporate, striate; ovary 3-carpellate, with 2 ovules per carpel; style subulate; stigma subcapitate. Fruit a 3-locular, 3-lobed, crustose, loculicidal capsule. Seeds exarillate. $2n=20$.

Three or four species from the Andean highlands of South America.

18. *Loxodiscus* Hook. f.

Loxodiscus Hook. f., Hooker's J. Bot. Kew Gard. Misc. 9: 200 (1857).

Falsely polygamous shrubs or trees. Leaves alternate, imparipinnate; leaflets serrate; distal leaflet rudimentary. Inflorescences terminal thyrses. Flowers zygomorphic, bisexual or functionally unisexual; sepals 5, imbricate, with fimbriate-glandular margins; petals 4(5), clawed, without appendages; disk semi-annular, double, the inner lobe concave, 4-dentate; stamens 7–8; pollen colporate, striate; ovary 3-carpellate, with stipitate glands and 2 ovules per carpel; style subulate; stigma obtuse. Fruit a 3-locular, loculicidal, membranous capsule. Seeds exarillate.

A single species, *L. coriaceus* Hook. f., from New Caledonia.

19. *Magonia* St. Hil.

Magonia St. Hil., Bull. Sci. Soc. Philom. Paris 1824: 78 (1824); Joly et al., Brittonia 32: 380–386 (1980).

Falsely polygamous trees. Leaves alternate, paripinnate; leaflets entire; distal leaflet rudimentary. Inflorescences axillary or terminal thyrses. Flowers zygomorphic, bisexual or functionally unisexual; sepals 5, distinct, imbricate; petals 5, without appendages; disk complete, half of one side consisting of 4 short glands, the other half of two concentric erect, fleshy laminae, the outer one taller; stamens 8; pollen grains in tetrads, colporate (6 pairs of colpi per tetrad), striate-gemmate (Fig. 79F); ovary 3-carpellate, with 8 ovules per carpel; style short, stigma capitate, sub-3-lobed. Fruit a large, 3-locular, trigonous, woody loculicidal capsule. Seeds flattened, surrounded by a wing. $2n=30$.

One species, *M. pubescens* St.-Hil., from Brazil, Bolivia, and Paraguay.

20. *Majidea* Kirk ex D. Oliver

Majidea Kirk ex D. Oliver, Hooker's Icon. Pl. tab. 78, 1097 (1871).

Falsely polygamous trees. Indument of fasciculate stellate hairs. Leaves alternate, paripinnate; distal leaflet rudimentary. Inflorescences axillary or terminal thyrses. Flowers actinomorphic or zygomorphic, bisexual or functionally unisexual; sepals 5, imbricate; petals 0 or only 1–2 or 4; disk annular or semi-annular, dish-shaped to pentagonous; stamens 7–8; pollen colporate, striate; ovary 3-carpellate; with 2 ovules per carpel; style elongated with punctiform stigma. Fruit a 3-locular, deeply lobed, loculicidal capsule, with crustose pericarp. Seeds exarillate. $2n = 24$.

Four or five species from Africa and Madagascar.

4. Tribe Doratoxyleae Radlk. (1890).

Disk annular; ovules (1)2(3) per locule; fruits indehiscent.

21. *Doratoxylon* Thouars ex Hook. f.

Doratoxylon Thouars ex Hook. f. in Benth. & Hook. f., Gen. 1: 408 (1862).

Falsely polygamous-dioecious shrubs or treelets. Leaves alternate, paripinnate or imparipinnate; leaflets entire or crenate; distal leaflet rudimentary; rachis narrowly winged; stipules 0. Inflorescences axillary, glomerate cymes. Flowers actinomorphic, functionally unisexual or bisexual; sepals 5, imbricate; petals 0; disk annular-5-lobed; stamens 5(6–7), the anthers dorsifixed; pollen colporate, scabrate to coarsely verrucate with striate to rugulate verrucae; ovary 2-carpellate, with 2 ovules per carpel; style short; stigma capitate. Fruits baccate, indehiscent, 1-locular. Seed ellipsoid, exarillate.

Six species, one from Mauritius and five from Madagascar.

22. *Euchorium* Ekm. & Radlk.

Euchorium Ekm. & Radlk., Repert. Spec. Nov. Regni Veg. 21: 230 (1925).

Dioecious trees. Leaves alternate, paripinnate; distal leaflet rudimentary; stipules 0. Inflorescences of axillary racemes. Flowers unisexual; calyx zygomorphic, sepals 5, imbricate, the outer two smaller; petals 5, without appendages; disk annular; stamens 8; pollen unknown; ovary 3-carpellate, with 2 ovules per carpel; style filiform, obtuse. Fruits unknown.

A single species, *E. cubense* Ekm. & Radlk., endemic to western Cuba.

In the absence of information on fruits, this genus is included in tribe Dodonaeaeae because of its resemblance to *Exothea*.

23. *Exothea* Macfad.

Exothea Macfad., Fl. Jamaica 1: 232 (1837).

Falsely polygamous dioecious trees. Leaves alternate, paripinnate; distal leaflet rudimentary. Inflorescences axillary or sub-terminal corymbose thyrses. Flowers bisexual or functionally unisexual; calyx zygomorphic, sepals 5, imbricate, the outer two smaller; petals 5, without appendages; disk annular, dish-shaped-crenate; stamens (7)8 (–10); pollen colporate, loosely scabrate to verrucate (Fig. 79C); ovary 2-carpellate, with 2 ovules per carpel; style elongated; stigma capitate. Fruits 1(2)-locular, globose, baccate, indehiscent. Seeds with coriaceous-fleshy testa.

Three species, distributed throughout the West Indies, Mexico, Guatemala, Colombia, and Ecuador.

24. *Filicium* Thwaites ex Hook. f.

Filicium Thwaites ex Hook. f. in Benth. & Hook. f., Gen. Pl. 1: 325. (1862); Thwaites, Enum. Pl. Zeyl.: 58, 408 (1864); Welzen in Fl. Males. I, 11: 754 (1994); Welzen in Fl. Thailand 7: 198, 199 (1999).

Falsely polygamous trees. Leaves alternate, paripinnate; distal leaflet rudimentary; rachis (broadly) winged. Inflorescences axillary or pseudo-terminal thyrses. Flowers actinomorphic, functionally unisexual; sepals 5, distinct, valvate; petals 5, without appendages; disk annular, 5-lobed, woolly; stamens 5; pollen colporate, loosely scabrate; ovary 2-carpellate, with a single pendant ovule per carpel; style short; stigma lobed. Fruits 1–2-locular, baccate, indehiscent, smooth, glabrous, the pericarp fleshy. Seeds exarillate. $2n = 32$.

Three or four species, mainly in east Africa and Madagascar, one extending to India and Sri Lanka, and widely cultivated as an ornamental throughout the tropics.

The taxonomic placement of *Filicium* has been much debated over decades. It was originally described as an Anacardiacea, and later transferred into Sapindaceae. In 1862, it was transferred into Burseraceae by Hooker f., followed by a transfer back into Sapindaceae by Baillon in 1874. In 1890, *Filicium* was placed by Radlkofer into the tribe Doratoxyleae of the subfamily Dodonaeoideae, in spite of its uniovular carpels; this position is supported by Harrington et al. (2005).

25. *Ganophyllum* Blume

Ganophyllum Blume, Mus. Bot. Lugd.-Bat. 1: 230 (1850) [1851]; Radlk. in Engl., Pflanzenreich 98: 1423 (1933); Leenhouts in Fl. Males. I, 11: 538 (1994).

Dioecious trees. Indument of glandular scales. Leaves alternate, paripinnate; distal leaflet rudimentary. Inflorescences axillary thyrses. Flowers actinomorphic; calyx cup-shaped, (4)5(–7)-lobed, valvate; petals 0; disk annular-lobed; stamens 5(–7); pollen colporate, irregularly verrucate; ovary 2(3)-carpellate, with 2 ovules per carpel; style short; stigma indistinctly lobed. Fruits indehiscent, 2-locular, ovoid, baccate, with papery endocarp and leathery mesocarp. Seeds exarillate.

One or two species, from western and central Africa, Andaman and Nicobar Islands, throughout Malesia to northeastern Australia, New Guinea, and the Solomons.

26. *Hippobromus* Ecklon & Zeyher

Hippobromus Ecklon & Zeyher, Enum.: 151 (1836).

Falsely polygamous dioecious trees. Leaves alternate, paripinnate; leaflets serrate; distal leaflet rudimentary; rachis winged. Inflorescences axillary thyrses. Flowers zygomorphic, bisexual or functionally unisexual; sepals 5, connate at base, imbricate; petals 5(6), shorter than the sepals, without appendages; disk annular, pentagonous; stamens 8; pollen colporate, rugulate; ovary 3-carpellate, with 2 or less often 3 ovules per carpel; style short; stigma decurrent. Fruits 1-locular, indehiscent, baccate. Seeds exarillate.

A single species, *H. alatus* Ecklon & Zeyher, Africa.

27. *Hypelate* P. Browne

Hypelate P. Browne, Civ. Nat. Hist. Jamaica: 208 (1756).

Falsely polygamous shrubs or trees. Leaves trifoliolate; stipules 0. Inflorescences axillary or terminal thyrses. Flowers bisexual or functionally unisexual; calyx zygomorphic, sepals (4)5, imbricate, the outer two smaller; petals (4)5, without appendages; disk annular, obsolete 5-lobed; stamens (7)8; pollen colporate, loosely verrucate; ovary 3-carpellate, with 2 ovules per carpel; style-short; stigma capitate. Fruit a 1-locular, indehiscent drupe with woody endocarp and fleshy mesocarp and exocarp. Seeds exarillate, with papery testa.

A single species, *H. trifoliata* Sw., endemic to the West Indies and the southeastern United States (Florida).

28. *Zanha* Hiern

Zanha Hiern, Cat. African Pl. 1: 128 (1896).
Dialiopsis Radlk. (1902) [nomen nudum] and in Engl. & Prantl., Nat. Pflanzenfam., Nachtr. 3: 207 (1907).

Dioecious trees. Leaves alternate, paripinnate; leaflets 3–7 pairs, entire or serrulate; distal leaflet rudimentary; stipules 0. Inflorescences distal, short thyrses. Flowers actinomorphic, functionally unisexual; sepals 4–5(6), imbricate, connate on lower 1/3–1/2; petals 0; disk annular or funnel-shaped; stamens (3)4–5, coiled in bud; pollen colporate, loosely scabrate; ovary 2-carpellate, with 2 ovules per carpel; style short, filiform, stigmas 2-lobed. Fruit 1-locular, indehiscent, with fleshy mesocarp and coriaceous endocarp. Seed 1 per fruit, with coriaceous or slightly fleshy testa.

Three species from southern Africa and Madagascar. Fruits of *Z. suaveolens* Capuron are said to be edible.

IV. Subfam. Sapindoideae

Leaves alternate; petal appendages usually present; disk annular or unilateral; carpels uniovulate.

5. Tribe Paullinieae Kunth ex DC. (1824).

Tribe Thouinieae Blume (1847).

Herbaceous vines, shrubs or small trees, often stipulate; leaves imparipinnate; flowers zygomorphic.

29. *Allophylus* L

Allophylus L., Sp. Pl. 348 (1753); Leenhouts, Blumea 15: 301–358 (1967); Fritsch, Kulturpfl. 18: 194 (1970); Ferrucci, Bol. Soc. Argentina Bot. 24: 200–202 (1985).

Duodichogamous or dioecious shrubs or trees, or less often scandent shrubs. Leaves alternate, trifoliolate or less often unifoliolate or digitate (in a few Asian collections); leaflets serrate, dentate or entire; petioles elongated; stipules 0. Inflorescences axillary racemes, panicles or thyrses. Flowers zygomorphic, 4-merous, functionally pistillate or staminate; sepals distinct, imbricate, the outer distinctly smaller than the inner; petals with a single, 2-lobed appendage or 2 marginal appendages; disk unilateral; stamens 8; pollen brevicolporate to porate, rugulate, sometimes striate, reticulate or psilate; ovary 2(3)-carpellate, with a single ovule per carpel; style with 2–3 stigmatic branches. Fruit an indehiscent drupe, 1–2-coccate, with crustose endocarp and fleshy exocarp. Seeds exarillate, with papery testa. $2n = 28$.

Depending on circumscription, a single species or about 250 species with circumtropical distribution. Several species are planted as ornamentals.

The notorious difficulties in species delimitation in *Allophylus* led Leenhouts (1967) to revise the status of the genus, which he found to be dominated by an enormous degree of clinal variation and intergradation among regional populations. As a consequence, he suggested to treat *Allophylus* as a single polymorphic species with the possibility of adding existing species names for informally characterizing local races. Since then, however, no progress in an understanding of the systematic structure of *Allophylus* has been made, and most botanists still continue recognizing regional populations as species.

30. *Athyana* (Griseb.) Radlk.

Athyana (Griseb.) Radlk. in T. Durand, Index Gen. Phan.: 73 (1887) [1888].

Monoecious trees. Leaves alternate, imparipinnate; leaflets serrate; distal leaflet fully developed; rachis winged; stipules 0. Inflorescences axillary or pseudo-terminal thyrses. Flowers zygomorphic, functionally unisexual; sepals 5, of similar size, valvate; petals 4, of same length, with a single petaloid hood-shaped appendage; disk semi-annular-lobed; stamens (7)8; pollen colporate, finely perforate; ovary 3-carpellate, with a single ovule per carpel; stigma capitate. Fruit schizocarpic, splitting into 3 samaroid mericarps with a distal wing. Seed exarillate.

A single species, *A. weinmannifolia* Radlk., southern South America (Peru, Bolivia, Argentina).

31. *Bridgesia* Bertero ex Cambess., nom. cons.

Bridgesia Bertero ex Cambess., Nouv. Ann. Mus. Hist. Nat. 3: 234 (1834), non Hook. (1831).

Falsely polygamous shrubs. Leaves alternate, simple, lobed or serrate; stipules 0. Inflorescence a short, axillary cyme or flowers solitary. Flowers zygomorphic, unisexual or bisexual; sepals 5, imbricate; petals 4, with a single cucullate, crested appendage; disk semi-annular, 4-lobed; stamens 8; pollen colporate, striate; ovary 3-carpellate, with a single ovule per carpel; style filiform with 3 stigmatic branches. Fruit schizocarpic, splitting into 3 samaroid, papery mericarps with a short dorsal wing. Seed exarillate.

A single species, *B. incisifolia* Bertero & Cambess., from Andean Chile.

32. *Cardiospermum* L.

Cardiospermum L., Sp. Pl.: 366 (1753).

Monoecious herbaceous vines. Leaves alternate, ternately compound or biternate; stipules minute. Inflorescences axillary thyrses bearing tendrils at base of rachis. Flowers zygomorphic, functionally staminate or pistillate; sepals 4–5, unequal, imbricate; petals 4, with a basal, hood-shaped appendage; disk unilateral, 2-lobed; stamens 8; pollen heteropolar, demi(syn)colporate proximally, perforate to reticulate; ovary 3-carpellate, with a single ovule per carpel; style with 3 stigmatic branches. Fruit a septifragal, marginicidal, inflated, capsule with thin-membranous walls. Seeds black, with a small, heart-shaped arillode surrounding the micropyle. $x=10, 11$.

About 15 species native to the Neotropics, 3 of which are widely distributed throughout the tropics. Two species are cultivated as ornamentals.

33. *Diatenopteryx* Radlk.

Diatenopteryx Radlk., Sitzungsber. Math.-Phys. Cl. Königl. Bayer. Akad. Wiss. Münch. 8: 284 (1878); Ferrucci, Sapindaceae in Spichiger & Ramella, Flora del Paraguay (1991).

Falsely polygamous shrubs or trees. Indumentum of simple hairs and scales. Leaves alternate, imparipinnate; leaflets dentate or serrate; distal leaflet rudimentary or fully developed; stipules 0. Inflorescences axillary thyrses. Flowers zygomorphic, functionally unisexual; sepals 4, distinct, imbricate; petals 4, with a basal hood-shaped, crested appendage; disk semi-annular, lobed; stamens (6–)8; pollen colporate, rugulate; ovary 2-carpellate, with a single ovule per carpel; style filiform; stigma subcapitate. Fruits schizocarpic, splitting into 2 samaroid mericarps with a long distal wing. Seed exarillate.

Two species from southern South America (Brazil, Bolivia, Paraguay, and Argentina).

34. *Guindilia* Gilles ex Hook. & Arn.

Guindilia Gilles ex Hook. & Arn., Hooker's Bot. Misc. 3: 170 (1833).
Valenzuelia Bert. ex Cambess. (1834), non Mutis (1810).

Falsely polygamous trees. Leaves alternate, simple, opposite, entire or tridentate at apex; stipules 0. Inflorescences axillary cymes. Flowers zygomorphic, bisexual or functionally unisexual; sepals 5, imbricate; petals 4(5), with a hood-shaped, crested, ventral appendage; disk unilateral, modified into a 2-lobed mount; stamens 8; pollen colporate, striate; ovary 3-carpellate, with a single ovule per carpel; style filiform; stigma 3-lobed. Fruit schizocarpic, splitting into (1–)3 sub-globose, crustose mericarps. Seed exarillate.

Three species from southern South America.

35. *Houssayanthus* Hunziker

Houssayanthus Hunziker, Kurtziana 11: 17 (1978); Ferrucci, Bonplandia 5 (19): 164–174 (1981); Ferrucci, Candollea 41: 218 (1986) and ibid. 42: 805–807 (1987); Rzedowski & Calderón de Rzedowski, Acta Bot. Mexicana 76: 89–98 (2006).

Woody vines. Cross section of stem simple or with a central and three peripheral vascular cylinders. Leaves trifoliolate, 5-pinnately compound, or biternate; stipules minute. Inflorescences axillary thyrses with a pair of tendrils at base of rachis. Flowers zygomorphic, functionally unisexual; sepals 4 or 5, unequal, imbricate; petals 4, with a basally adnate hood-shaped appendage; disk unilateral, 4-lobed; stamens 8; pollen heteropolar, demisyncolporate, perforate; ovary 3-carpellate, with a single ovule per carpel; style with 3 stigmatic branches. Fruit a schizocarp splitting into 3 mericarps with a short dorsal wing. Seeds lenticular, exarillate. $2n = 24$.

Five species, two from Mexico, one from Venezuela, and two from Brazil, Bolivia, Paraguay, and northeastern Argentina.

36. *Lophostigma* Radlk.

Lophostigma Radlk. in Engl. & Prantl, Nat. Pflanzenfam., Nachtr. 1: 228 (1897); Acevedo-Rodríguez, Syst. Bot. 18: 379–388 (1993), rev.

Woody vines. Cross section of stem with a single vascular cylinder. Leaves alternate, trifoliolate; stipules minute. Inflorescences axillary thyrses with a pair of tendrils at base of rachis. Flowers zygomorphic, functionally unisexual; sepals 5, unequal, imbricate; petals 4, much shorter than the sepals, with 2 basally adnate digitate appendages; disk unilateral, 4-lobed; stamens 8, with short filaments; pollen cylindric-ellipsoid, 4-aperturate, possibly a lengthy variant of heteropolar 3-demisyncolporate Paullinieae pollen (e.g., *Serjania*); ovary 3-carpellate, with a single ovule per carpel; style with 3 stigmatic branches. Fruit a schizocarp splitting into 3 mericarps with an elongated proximal wing. Seeds lenticular, exarillate.

Two species, one from Ecuador and Peru, the other from Bolivia.

37. *Paullinia* L.

Paullinia L., Sp. Pl.: 365 (1753); Radlkofer, Monogr. Paullinia (1895); Simpson, Fieldeana Bot. 36: 125–164 (1976), reg. rev.

Woody vines, usually producing milky sap. Cross section of stem with a central and 3–5 peripheral vascular cylinders, or with a single vascular cylinder. Leaves alternate, trifoliolate, pinnate (5-pinnate), bipinnate, biternate, or variously dissected; stipules minute to large. Inflorescences axillary thyrses with a pair of tendrils at base of rachis, or cauliflorous thyrses without tendrils. Flowers zygomorphic, functionally unisexual; sepals 4–5, unequal, imbricate; petals 4, with a basally adnate hood-shaped appendage; disk unilateral, 4-lobed; stamens 8; pollen porate, perforate; ovary 3-carpellate, with a single ovule per carpel; style with 3 stigmatic branches. Fruit a 1–3-locular, septifragal-marginicidal capsule, the locules dorsally winged or exalate, exceptionally spiny. Seeds globose, oblong or ellipsoid, with a partial to nearly complete sarcotesta. $2n=24$.

About 190 species native to the Neotropics, one species extending into Africa. Numerous species are used as fish poisons; *P. cupana* Kunth is the source of the economically important "guaraná" used in the confection of a stimulating drink or soft drinks in Brazil.

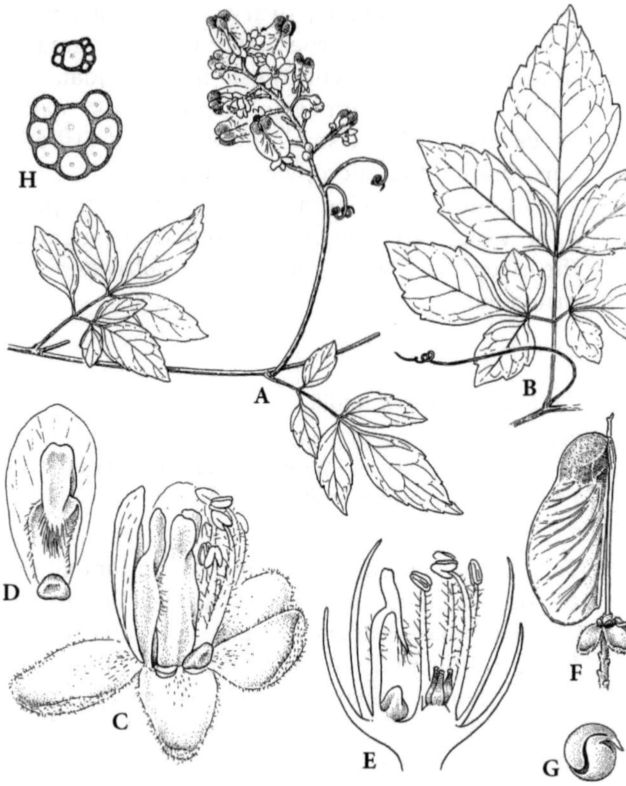

Fig. 83. Sapindaceae. *Serjania polyphylla*. **A** Fertile branch. **B** Biternate leaf with tendril. **C** Male flower. **D** Petal with adnate hood-shaped appendage. **E** Longitudinal section of male flower. **F** Winged mericarp attached to floral axis. **G** Embryo. **H** Cross section of stem showing numerous vascular cylinders. (Acevedo-Rodríguez 1996; reproduced with permission of the artist Bobbi Angell)

38. *Serjania* Plum. ex Miller Fig. 83

Serjania Plum. ex Miller, Gard. Dict. Abr. ed. 4 (1754); Acevedo-Rodríguez, Mem. New York Bot. Gard. 67: 1–93 (1993).
Chimborazoa H.T. Beck (1994); Acevedo-Rodríguez, Novon 8: 106 (1998).

Duodichogamous woody or herbaceous vines, often producing milky sap. Cross section of stem with a single or multiple vascular cylinders, usually 3, but also 8 or 10. Leaves alternate, ternately compound (ternate, biternate, or triternate) or 5-pinnately compound; stipules small and early deciduous. Inflorescences axillary or terminal thyrses. Flowers zygomorphic, functionally unisexual; sepals 4–5, unequal, imbricate; petals 4, with a basally adnate hood-shaped appendage; disk unilateral, 2- or 4-lobed; stamens 8; pollen heteropolar, demi(syn)colporate proximally, sometimes with short distal demicolpi as well, perforate, sometimes reticulate, psilate or faintly rugulate; ovary 3-carpellate, with a single ovule per carpel; style with 3 stigmatic branches. Fruit schizocarpic, splitting into three samaroid mericarps, each with a proximal wing. Seeds lenticular to globose, exarillate. $2n=24$.

About 230 species native to tropical and subtropical America.

39. *Thinouia* Triana & Planchon

Thinouia Triana & Planchon, Ann. Sci. Nat. Bot. IV, 18: 368 (1862).
Allosanthus Radlk. (1933).

Lianas. Secondary growth of stems with numerous cortical steles. Leaves alternate, trifoliolate; stipules minute. Inflorescences umbelliform thyrses, seldom bearing tendrils, axillary or aggregate into terminal thyrsoid inflorescences. Flowers actinomorphic, bisexual, 5-merous; calyx cup-shaped, the sepals valvate; petals obovate to spatulate, with a pair of short, marginal appendages; disk annular; stamens 6–8; pollen colporate, striate; ovary 3-carpellate, with a single ovule per carpel; style elongated with 3 stigmatic branches. Fruit schizocarpic, splitting into 3 mericarps, each with a distal wing. Seeds nearly spherical, exarillate.

About 12 species from Central and South America.

40. *Thouinia* Poit., nom. cons.

Thouinia Poit., Ann. Mus. Natl. Hist. Nat. 3: 70 (1804), non L. f. (1782), nom. rej.; Votava, Taxonomic revision of the genus *Thouinia* (Sapindaceae). Ph.D. dissertation, Columbia University, New York, 235 pp. (1973), rev.

Falsely polygamous shrubs or trees. Leaves alternate, trifoliolate or unifoliolate; stipules 0. Inflorescences axillary, racemose thyrses. Flowers zygomorphic, functionally unisexual; sepals 4, imbricate; petals 4, with a ventral, bifid appendage; disk unilateral, lobed; stamens 8; pollen brevicolporate, rugulate; ovary 3-carpellate, with a single ovule per carpel; style filiform, with 3 stigmatic branches. Fruit schizocarpic, splitting into 3 samaroid mericarps, each bearing a distal wing. Seeds exarillate.

About 30 species from Central America and the West Indies.

41. *Urvillea* Kunth

Urvillea Kunth in H.B.K., Nov. Gen. Spec. 5: 105 (1821).

Herbaceous to woody vines. Stems terete and lenticellate, becoming trilobate at age, producing milky sap. Leaves alternate, trifoliolate or biternate; stipules minute, deciduous. Inflorescences axillary thyrses. Flowers zygomorphic, bisexual; sepals 5, unequal, imbricate; petals 4, spatulate, with an adnate hood-shaped appendage on adaxial surface; disk unilateral, 4-lobed, receptacle enlarged into a short androgynophore; stamens 8; pollen heteropolar, demisyncolporate proximally, often with short distal demicolpi as well, perforate, indistinctly rugulate or reticulate (Fig. 79G–I); ovary 3-carpellate, with a single ovule per carpel; style with 3 stigmatic branches. Fruit a 3-locular, thin, papery, semi-inflated capsule, with a narrow, marginal (septal) wing. Seeds sub-globose, with a heart-shaped or reniform, white arillode around the hilum. $2n=22$.

About 15 species from Central and South America.

6. Tribe Melicocceae Blume (1847).

Flowers actinomorphic; fruits indehiscent, usually single seeded; seeds with sarcotesta.

42. *Melicoccus* P. Browne Fig. 77

Melicoccus P. Browne, Hist. Jamaica: 210 (1756); Acevedo-Rodríguez, Moscosoa 9: 58–61 (1997); Acevedo-Rodríguez, Fl. Neotrop. 87 (2002), rev.
Melicocca L. (1762), nom. illeg.

Dioecious or monoecious trees. Leaves alternate, paripinnate; leaflets 1–2 pairs; distal leaflet rudimentary, 0 or exceptionally present; rachis usually winged; stipules 0. Inflorescences terminal panicles or racemes. Flowers actinomorphic, functionally pistillate or staminate. Calyx cup-shaped, sepals 4(5), equal, imbricate; petals 4(5) erect or reflexed; appendages 0 or rudimentary; disk annular, slightly lobed; stamens 8; pollen colporate, striate; ovary 2-carpellate, unilocular, with a single ovule per carpel; style obsolete; stigmatic surface lobed, subcapitate. Fruit indehiscent, with leathery pericarp, sub-globose or ellipsoid. Seeds 1(2) with edible sarcotesta. $2n = 32$.

Ten species, native to South America and Dominican Republic. *Melicoccus bijugatus* Jacq. is widely cultivated in the Neotropics for its edible, tasty fruits.

43. *Talisia* Aublet Fig. 84

Talisia Aubl., Hist. Pl. Guiane 1: 349 (1775); Mennega, Acta Bot. Neerl. 21: 578–586 (1972); Acevedo-Rodríguez, Fl. Neotrop. 87: 1–179 (2003), rev.
Tapirocarpus Sagot (1882).

Duodichogamous unbranched shrubs or small to large trees. Leaves alternate, paripinnate or imparipinnate; leaflets entire; distal leaflet rudimentary. Inflorescences axillary, terminal or cauliflorous thyrses. Flowers 5-merous, actinomorphic, staminate or pistillate; calyx usually cup-shaped, sepals imbricate or valvate, equal or unequal in size; petals reflexed or ascending, with marginal appendages or with a basal, petaloid, sericeous appendage; disk annular, 5–8-lobed; stamens 5–8; pollen either colporate, striate, or brevicolporate, psilate; ovary 3-carpellate, with a single ovule per carpel; style subulate, crowned by a capitate to cylindrical stigma.

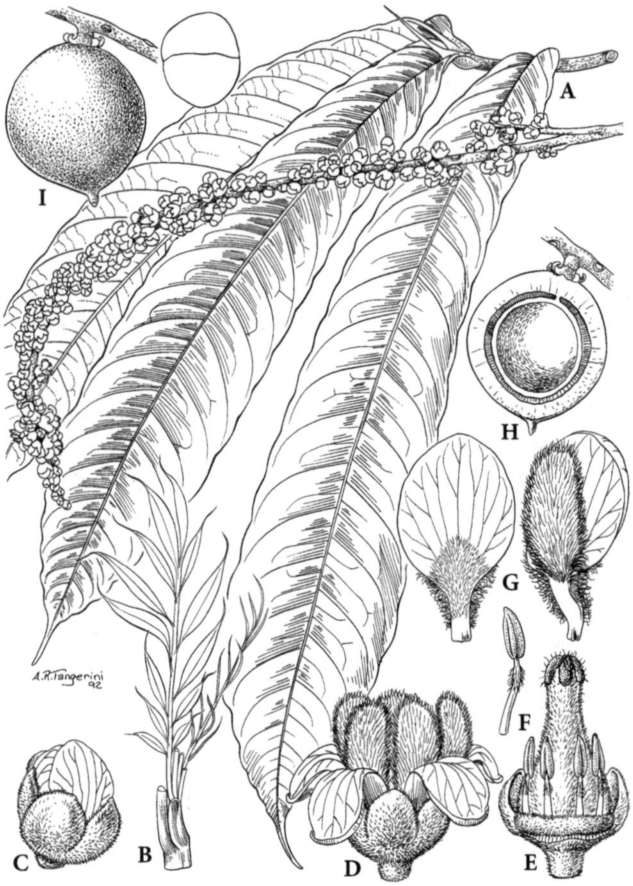

Fig. 84. Sapindaceae. *Talisia princeps* Oliver. **A** Distal portion of pinnate leaf and inflorescence. **B** Cataphylls. **C** Flower bud. **D** Pistillate flower. **E** Ditto, perianth removed to show disk, sterile stamens and pistil. **F** Stamen. **G** Abaxial and adaxial views of petal with adnate appendage. **H** Longitudinal section of fruit showing mesocarp and seed. **I** Fruit and ventral view of embryo. (Acevedo-Rodríguez 2003; drawn by A. Tangerini)

Fruit 1(–3)-seeded, indehiscent, with leathery or less often woody pericarp. Seeds entirely covered with a sarcotesta.

Fifty two species predominantly from South America, some occurring in Central America and southern Mexico.

Genera 44–140: Incertae Sedis

44. *Alectryon* Gaertn

Alectryon Gaertn., Fruct. Sem. Pl. 1: 216, pl. 46 (1788); Reynolds, Austrobaileya 2: 332–338 (1987); Leenhouts, Blumea 33: 313–327 (1988), reg. rev.

Falsely polygamous shrubs or trees. Leaves alternate, paripinnate or less often unifoliolate; leaflets serrate or entire; distal leaflet rudimentary; lower pair of leaflets (pseudostipules) sometimes clasping the stem. Inflorescences axillary thyrses or panicles. Flowers actinomorphic; calyx acetabuliform, sepals 4–5(6), valvate or sub-imbricate; petals 0 or 4–5, shortly clawed, the appendages apparently of marginal origin, connate to form a funnel-shaped structure with the petal; disk annular-lobed, glabrous; stamens (5–)8, the filaments subequal, inserted on the disk; pollen colporate to parasyncolporate, striate to rugulate; ovary (1)2–4(5)-carpellate, with a single ovule per carpel; stigma grooved or lobed. Fruits capsular, 1–3-coccate, lobed or unlobed, circumscissile or dehiscent septifragally along septum. Seed with a partial granular, red sarcotesta. $2n = 32$.

About 25 species in eastern Malesia, Australia, New Zealand, New Caledonia, and extending into the Pacific to Samoa and the Sandwich Islands. Two subgenera: *Synalectryon* and *Alectryon*.

45. *Amesiodendron* Hu

Amesiodendron Hu, Bull. Fan. Mem. Inst. Biol. 7, Bot.: 207 (1936); Lo, Acta Phytotax. Sin. 17: 36, f. 3 (1979); Leenhouts in Fl. Males. I, 11: 465 (1994).

Polygamous monoecious trees. Leaves alternate, once pinnate, paripinnate; leaflets serrate; distal leaflet rudimentary. Inflorescences axillary or terminal panicles. Flowers actinomorphic; sepals 5, distinct, valvate; petals 5, with a single basal appendage; disk annular, bowl-shaped, forming a ring on upper portion; stamens (6)7–8(9); pollen syncolporate, rugulate; ovary 3-carpellate, with a single ovule per carpel; style with two stigmatic lines. Fruits loculicidal capsules, 3-coccate, sometimes 1 or 2 cocci rudimentary. Seeds with sarcotestal ring around hilum.

One to three species, from Indochina, southern China, peninsular Malaysia, and Sumatra.

46. *Aporrhiza* Radlk.

Aporrhiza Radlk., Sitzungsber. Math.-Phys. Cl. Königl. Bayer Akad. Wiss. München 8: 338 (1878).

Falsely polygamous trees or shrubs. Leaves alternate, once pinnate, paripinnate; terminal leaflet rudimentary. Inflorescences terminal, axillary or cauliflorous thyrses. Flowers actinomorphic, functionally unisexual; sepals 5, distinct, valvate; petals 5, clawed, with a pair of appendages formed by the inflexed margin above the claw; disk annular, sub-lobed; stamens 6–8; pollen colporate, perforate; ovary 2-carpellate, with a single ovule per carpel; style short; stigma 2-lobed. Fruit a 2-locular, loculicidal capsule. Seeds with a sarcotesta on lower half. $2n = 28$.

Four to six species from tropical Africa.

47. *Arytera* Blume Fig. 85

Arytera Blume, Rumphia 3: 169 (1847); van der Ham, Blumea 23: 289–300 (1977); Turner, Blumea 38: 137–144 (1993); van Bergen et al., Blumea 40: 195–209 (1995).

Trees or shrubs. Leaves alternate, once pinnate, paripinnate; leaflets usually with domatia; distal leaflet rudimentary. Inflorescences axillary or pseudo-terminal thyrses. Flowers actinomorphic, functionally unisexual; sepals 5, connate to nearly distinct, valvate to imbricate; petals (2–)5(6), with 2 marginal appendages, sometimes clawed; disk annular, lobed; stamens (5–)7–8(–10); pollen colporate to parasyncolporate, rarely syncolporate, rugulate to striate-rugulate; ovary (1)2–3-carpellate, with a single ovule per carpel; style filiform, with 2–3 stigmatic lines or 2–3-lobed. Fruit a 1–3-coccate, loculicidal capsule. Seeds with basal arillode, apically open, covering half to nearly the entire seed.

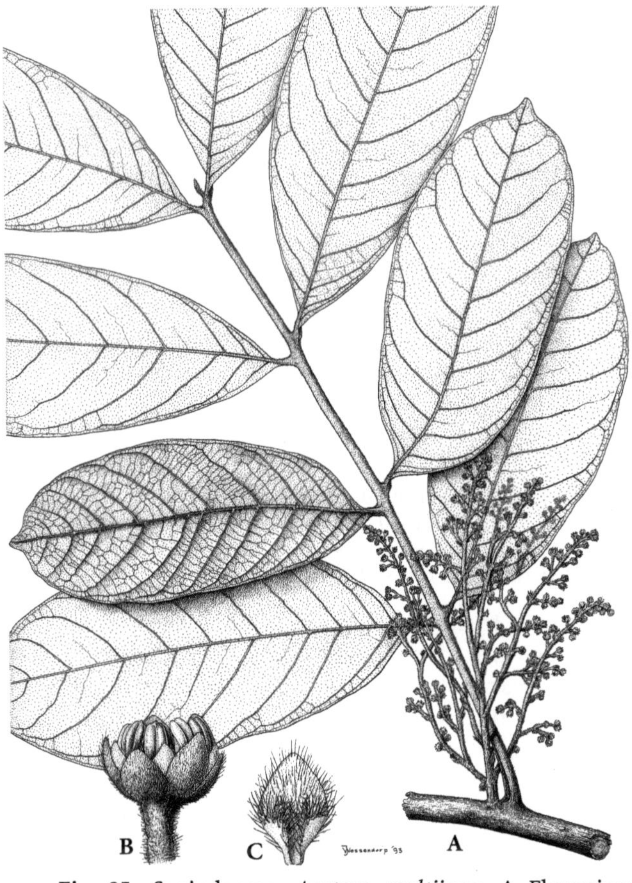

Fig. 85. Sapindaceae. *Arytera multijuga*. **A** Flowering branch. **B** Male flower. **C** Petal. (Adema et al. 1994, drawn by J. Wessendorp)

About 28 species in northeast India and southeast Asia, throughout Malesia, and Australia, Solomon Islands and Pacific Islands.

Section *Arytera* has simple hairs, sepals externally pubescent, and 2–3-lobed ovaries; Section *Azarytera* has glandular-scaly indument, glabrous sepals, and 2-lobed ovaries.

48. *Atalaya* Span. ex Blume

Atalaya Span. ex Blume, Rumphia 3: 186 (1847); Radlk. in Pflanzenreich 98: 605 (1932); Reynolds, Austrobaileya 1: 398–406 (1981), reg. rev.; Leenhouts in Fl. Males. I, 11: 479 (1994), reg. rev.

Falsely polygamous trees or shrubs. Leaves alternate, once pinnate, paripinnate or less often imparipinnate, up to 6-jugate; distal leaflet rudimentary; rachis sometimes winged. Inflorescences terminal or axillary thyrses. Flowers unisexual or bisexual; calyx slightly zygomorphic, sepals 5, distinct, imbricate, the outer 2 smaller; petals (4)5, clawed, with a pair of appendages formed by the inflexed margins above the claw; disk annular or semi-annular; stamens 8; pollen colporate, rugulate; ovary 3-carpellate, with a single ovule per carpel; placentation axile; style conical with 3 stigmatic lines. Fruits schizocarpic, splitting into (1–)3 divaricate, samaroid mericarps, each with a long, dorsal wing. Seed exarillate, laterally compressed, with papery testa.

Twelve species, mostly from northern and eastern Australia, also in the Lesser Sunda Islands, southeastern New Guinea, and South Africa.

49. *Beguea* Capuron

Beguea Capuron, Mem. Mus. Natl. Hist. Nat., B, Bot. II, 19: 105 (1969).

Dioecious trees. Leaves alternate, paripinnate; leaflets 3–7 pairs, opposite or alternate, entire;

distal leaflet rudimentary; stipules 0. Inflorescences axillary racemes or thyrses. Flowers actinomorphic, unisexual; sepals (5)6–7(8), distinct, valvate; petals 0; disk annular-lobed; stamens 6–8 (–10); pollen colporate, striate; ovary 3-carpellate, with a single ovule per carpel; style filiform with 3 stigmatic branches. Fruit indehiscent, crustose, 1-seeded. Seeds arillate, cotyledons cerebriform (brain-like).

A single species, *B. apetala* Capuron, endemic to Madagascar.

50. ***Bizonula*** Pellegrin

Bizonula Pellegrin, Bull. Soc. Bot. France 71: 299 (1924).

Hermaphroditic trees. Leaves bipinnate; distal leaflet rudimentary; stipules 0. Inflorescences terminal thyrses. Flowers actinomorphic, bisexual; sepals 5, distinct, valvate; petals 5, with a basal appendage forming a pocket; disk annular, double; stamens 12–13; pollen colporate, striate; ovary 3-carpellate, with a single ovule per carpel; style elongated, weakly lobed. Fruit not known.

A single species, *B. le-testui* Pellegrin, from Gabon, Africa.

51. ***Blighia*** Koenig

Blighia Koenig, Ann. Bot. 2: 571 (1806).

Falsely polygamous-dioecious, usually trees or shrubs. Leaves alternate, paripinnate; leaflets 1–5 pairs, entire; distal leaflet rudimentary; stipules 0. Inflorescences axillary racemose thyrses. Flowers actinomorphic, functionally unisexual; sepals 5, valvate; petals 5, connate along appendage margins to form a pouch; disk annular, 8-lobed; stamens 8–10; pollen colporate, striate; ovary 3(4)-carpellate, with a single ovule per carpel; style elongated-conical. Fruit a 3(4)-locular, loculicidal, capsule. Seeds with a basal arillode. $2n = 32$.

Three species from tropical Africa. *Blighia sapida* Koenig is widely cultivated for its edible arillodes in Jamaica and as an ornamental in Africa, the West Indies, and areas of the Neotropics.

52. ***Blighiopsis*** van der Veken

Blighiopsis van der Veken, Bull. Jard. Bot. État. 30: 413 (1960).

Dioecious trees. Leaves alternate, paripinnate; leaflets opposite or alternate, entire; distal leaflet rudimentary; pseudostipules small. Inflorescences thyrsoid. Flowers actinomorphic, unisexual; sepals (4)5(7), slightly imbricate, distinct to base; petals 0; disk annular; stamens 5(7), anthers basifixed; pollen colporate, striate; ovary 3-carpellate, with a single ovule per carpel; style short, crowned by a trigonous stigma. Fruit a tardily loculicidal, 1-locular, coriaceous capsule. Seed solitary, with basal arillode.

A single species, *B. pseudostipularis* van der Veken, Central Africa.

53. ***Blomia*** Miranda

Blomia Miranda, Annales Inst. Biol. Univ. Nac. Mexico 24: 82 (1953).
Tikalia Lundell (1961).

Falsely polygamous-dioecious trees. Leaves alternate, paripinnate; leaflets crenate, 1-4 pairs; distal leaflet rudimentary; stipules 0. Inflorescences axillary thyrses. Flowers actinomorphic, unisexual or bisexual; sepals 5, distinct, valvate; petals 0 or vestigial, with a pair of minute marginal appendages; disk annular-lobed; stamens 5–6; pollen colporate, striate; ovary 1-carpellate, with a single ovule per carpel; style short; stigma capitate. Fruit a 1-locular, tardily loculicidally dehiscent, coriaceous, red capsule. Seeds with thin sarcotesta.

A single species, *B. cupanioides* Miranda, in Mexico, Guatemala, and Belize.

54. ***Camptolepis*** Radlk.

Camptolepis Radlk. in Engl. & Prantl., Nat. Pflanzenfam., Nachtr. 2–4, 3: 207 (1907); Capuron, Mem. Mus. Natl. Hist. Nat. B, Bot. II, 19: 1–189 (1969).
Hypseloderma Radlk. (1932) [1933].

Dioecious trees. Leaves alternate, paripinnate; leaflets entire; distal leaflet rudimentary; stipules 0. Inflorescences ramiflorous, short thyrses. Flowers actinomorphic, unisexual; sepals 5, distinct, imbricate; petals 5, with a single basal, short appendage; disk annular-lobed; stamens (10–)12; pollen colporate or brevicolporate, perforate; ovary 3-carpellate, with a single ovule per carpel; style elongated; stigma 3, ellipsoid. Fruit indehiscent, (1–)3-locular. Seeds completely covered by a translucent arillode.

Four species from tropical east Africa and Madagascar.

55. *Castanospora* F. Muell.

Castanospora F. Muell., Fragm. 9: 92 (1875); Reynolds, Austrobaileya 2: 34–35 (1984).

Falsely polygamous trees. Leaves alternate, paripinnate; leaflets entire, alternate or subopposite; distal leaflet rudimentary; stipules 0. Inflorescences axillary racemes or thyrses, with lateral dichasia. Flowers actinomorphic, functionally staminate or pistillate; calyx cup-shaped, sepals 5, ovate, sub-valvate; petals 5, with 2 minute basal appendages; disk annular-lobed; stamens 8; pollen parasyncolporate, striate-rugulate; ovary 2-carpellate, with a single ovule per carpel; style short, with 2 reflexed stigmatic branches. Fruit indehiscent, crustose-fleshy, bicoccate, sometimes with a thin septa. Seeds large, chestnut-like, brown, exarillate.

A single species, *C. alphandi* F. Muell., eastern Australia.

56. *Chouxia* Capuron

Chouxia Capuron, Mem. Mus. Natl. Hist. Nat., B, Bot. II, 19: 130 (1969); Schatz et al., Adansonia III, 21: 51–62 (1999), rev.

Dioecious or monoecious trees. Leaves alternate, paripinnate; leaflets opposite or subopposite, entire; distal leaflet rudimentary; stipules 0. Inflorescences cauliflorous racemes or thyrses. Flowers actinomorphic, staminate or pistillate; sepals (4)5(6), distinct, imbricate; petals 5, with a simple basal appendage; disk annular; stamens (7)8(–10); pollen colporate, striate-reticulate; ovary (2)3-carpellate, with a single ovule per carpel; style elongated; stigma 3. Fruit 1–3-locular, indehiscent, baccate. Seeds arillate.

Six species endemic to Madagascar.

57. *Chytranthus* Hook. f.

Chytranthus Hook. f. in Benth. & Hook. f., Gen. Pl. 1: 403 (1862).
Glossolepis Gilg (1897).

Falsely polygamous dioecious treelets. Leaves alternate, paripinnate, terminal leaflet rudimentary; stipules 0. Inflorescences cauliflorous, fasciculate, racemose thyrses. Flowers zygomorphic, functionally staminate or pistillate; calyx urceolate, sepals 5, connate at base, imbricate; petals 4(5–7), with a basal, simple or hood-shaped and sometimes crested appendage; disk semi-annular or reniform; stamens 7–8(9–11); pollen colporate of brevicolporate, striate; ovary 3–8-carpellate, with a single ovule per carpel; style subulate. Fruit indehiscent, 3–8-locular, deeply ribbed, fleshy. Seeds with sarcotesta. $2n = 32$.

Twenty-six to 29 species from west tropical Africa.

58. *Cnesmocarpon* Adema

Cnesmocarpon Adema, Blumea 38: 195–201 (1993).

Trees. Leaves alternate, paripinnate; leaflets papillate beneath; distal leaflet rudimentary. Inflorescences axillary or ramiflorous thyrses. Flowers actinomorphic; sepals 5, imbricate, slightly unequal; petals 5, with 2 marginal appendages, or less often lacking appendages; disk annular or semi-annular; stamens 8; pollen syncolporate or parasyncolporate, psilate or indistinctly rugulate; ovary 3-carpellate, with a grains single ovule per carpel, style with 3 stigmatic lines. Fruit a 3-locular, fleshy, loculicidal capsule, the locules basally winged, covered with irritating hairs. Seeds obovoid, testa shiny black, with carunculoid sarcotesta at the base.

Four species in Australia and Papua New Guinea, occurring in primary lowland to montane forest.

59. *Conchopetalum* Radlk.

Conchopetalum Radlk. in T. Durand, Index Gen. Phan.: 81 (1887) [1888].

Shrubs or trees. Leaves alternate, paripinnate; leaflets entire; distal leaflet rudimentary; stipules 0. Inflorescences axillary or ramiflorous, fasciculate, pseudo-umbelliform cymes. Flowers actinomorphic, functionally unisexual; sepals 5, connate at base, imbricate; petals 5, red, without appendages; disk annular, double, the central rim sometimes tubular resembling an androgynophore; stamens 8, exserted; pollen colporate, striate-rugulate; ovary 3-carpellate, with 2 ovules per carpel; style elongated, the stigma punctiform. Fruit a 3-locular, chartaceous capsule. Seeds with brown sarcotestal ring around the hilum.

Two species endemic to Madagascar.

60. *Cubilia* Blume Fig. 86

Cubilia Blume, Rumphia 3: 100 (1847); Leenhouts, Blumea 24: 297 (1978).

Falsely polygamous trees. Leaves alternate, paripinnate; terminal leaflet rudimentary. Inflorescences terminal or pseudo-terminal, thyrsoid or corymbiform. Flowers actinomorphic; calyx urceolate, with a narrow opening and 5 minute lobes; petals 5, shorter than the calyx, without appendages; disk annular, fleshy; stamens 5 (or 6); pollen grains small (12×13 μm), colporate, scabrate; ovary 2-carpellate, with a single ovule per carpel; stigma short, with 2 stigmatic lobes. Fruit a 1–2-coccate, loculicidal, echinate, woody capsule. Seeds with a large round hilum, arillate on lower half or third.

A single species, *C. cubili* (Blanco) Adelb., from eastern Borneo, the Philippines, Celebes, and the western Moluccas Islands.

61. *Cupania* L.

Cupania L., Sp. Pl.: 200 (1753).

Duodichogamous or dioecious trees. Leaves alternate, paripinnate or imparipinnate; leaflets mostly serrate; stipules 0. Inflorescences axillary or terminal thyrses or racemes. Flowers actinomorphic, functionally unisexual; sepals 5, short, imbricate; petals 5, with a pair of marginal tomentose appendages; disk annular-lobed; stamens (4, 6) 8; pollen syncolporate (Fig. 79J) or parasyncolporate, rugulate; ovary 3-carpellate, with a single ovule per carpel; style elongated; stigma conical. Fruit a 2- or 3-locular, woody or leathery, loculicidal capsule. Seeds with a cupular arillode at base. $2n = 32$.

About 50 species from tropical and subtropical America.

62. *Cupaniopsis* Radlk.

Cupaniopsis Radlk., Sitzungsber. Math.-Phys. Cl. Königl. Bayer. Akad. Wiss. Münch. 9: 483, 498, 584 (1879); Adema, Leiden Bot. Series 15: 75 (1991), rev.

Shrubs or treelets, pubescence of simple or glandular trichomes, or scaly. Leaves alternate, paripinnate; leaflets entire or less often dentate or serrate; distal leaflet rudimentary; stipules 0. Inflorescences axillary or cauliflorous thyrses.

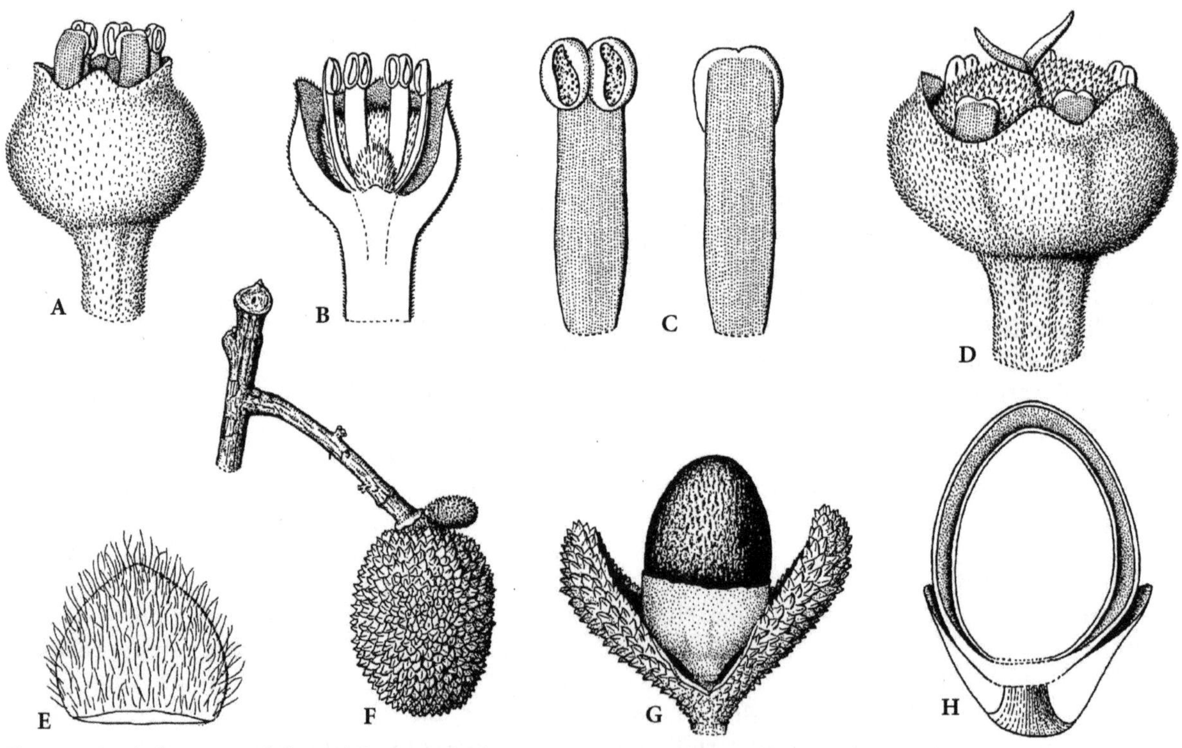

Fig. 86. Sapindaceae. *Cubilia cubili*. **A** Male flower. **B** Longitudinal section of male flower. **C** Anthers, ventral and dorsal views. **D** Female flower. **E** Petal. **F** Capsule. **G** Dehisced capsule showing seed with arillate base. **H** Longitudinal section of seed. (Adema et al. 1994, drawn by J. van der Os)

Flowers unisexual; calyx zygomorphic, sepals (4)5 (-7), distinct, imbricate, the outer two distinctly smaller; petals (4)5, with 1 or 2 basal appendages; disk annular-lobed; stamens (5–)8–14; pollen syncolporate or parasyncolporate, sometimes colporate, rugulate, striate-reticulate, reticulate or perforate, sometimes verrucate; ovary (2)3 (4)-carpellate, with a single ovule per carpel. Fruit a (1)2–3-locular, dehiscent, fleshy capsule. Seeds with basal arillode, covering half to nearly the entire seed, or exceptionally naked and sarcotestal.

Sixty species from eastern Malesia, Caroline Islands, northern and eastern Australia, and from Solomon Islands to Samoa including New Caledonia.

63. ***Deinbollia*** Schumach.

Deinbollia Schumach., Beskr. Guin. Pl.: 242 (1827).

Falsely polygamous or dioecious, shrubs or trees. Leaves alternate, paripinnate; leaflets entire; distal leaflet rudimentary; stipules 0. Inflorescences axillary or terminal thyrses. Flowers functionally unisexual; calyx zygomorphic, sepals 5, imbricate, the outer two smaller; petals 5, with a basal, deeply bilobed appendage; disk annular, cup-shaped or vase-shaped; stamens (8–)12–30; pollen colporate, perforate; ovary (2)3(5)-carpellate, with a single ovule per carpel; style gynobasic or sub-terminal, filiform. Fruits of (1)2–3(–5) indehiscent, fleshy cocci. Seeds exarillate. $2n=30$.

About 38 species from southern Africa, Madagascar, and Mascarene Islands.

64. ***Delavaya*** Franch.

Delavaya Franch., Bull. Soc. Bot. France 33: 462 (1886).

Falsely polygamous shrubs or trees. Leaves alternate, trifoliolate; leaflets serrate-denticulate; stipules 0. Inflorescences terminal thyrses. Flowers functionally unisexual; calyx zygomorphic, sepals 5, imbricate, the outer two smaller; petals 5, with a pair of marginal appendages; disk pulvinate, nearly shortly tubular; stamens 8; pollen colporate, striate; ovary 2–3-carpellate, with 2 ovules per carpel; style subulate. Fruits 2–3-coccate, loculicidal capsules. Seeds exarillate.

A single species, *D. yunnanensis* Franch., China.

65. ***Dictyoneura*** Blume

Dictyoneura Blume, Rumphia 3: 163 (1847); van Dijk, Blumea 31: 437–449 (1986).

Falsely polygamous shrubs or trees. Leaves alternate, paripinnate; leaflets serrate to lobed; distal leaflet rudimentary; stipules 0. Inflorescences axillary, spicate or racemose thyrses. Flowers unisexual; sepals (5)6, imbricate, of same length in actinomorphic calyx or the inner and/or the outermost sepals smaller in zygomorphic calyx; petals 0; disk annular; stamens (4)5(6); pollen colporate, rugulate; ovary 2(3)-carpellate, with a single ovule per carpel; style short, with 2(3) stigmatic grooves. Fruit a 2(3)-locular, subfleshy capsule, endocarp granular. Seeds with a partial cupular, ventral or nearly complete sarcotesta.

Two or three species from eastern Borneo, the Philippines, Celebes, Moluccas, and New Guinea.

66. ***Dilodendron*** Radlk.

Dilodendron Radlk., Sitzungsber. Math.-Phys. Cl. Königl. Bayer. Akad. Wiss. München 8: 355 (1878); Gentry & Steyermark, Ann. Missouri Bot. Gard. 74: 533–538 (1987), rev.

Falsely polygamous-dioecious trees. Leaves alternate, bipinnate or sub-tripinnately compound; distal leaflets on rachises fully developed or rudimentary; secondary rachises marginate, alternate; stipules 0. Inflorescences terminal thyrses. Flowers 5-merous, functionally staminate or pistillate; calyx slightly zygomorphic with one sepal larger than the others, aestivation imbricate; petals much shorter than sepals, with narrow, marginal appendages; disk annular and flattened; stamens 6–8(9); pollen colporate, striate; ovary 3-carpellate, with a single ovule per carpel; style short with 3 stigmatic lobes. Capsule loculicidal, 2–3-locular, woody. Seeds exarillate.

Three species from tropical continental America.

67. ***Dimocarpus*** Lour.

Dimocarpus Lour., Fl. Coch.: 233 (1790); Leenhouts, Blumea 19: 113–131 (1971), rev.; van der Ham, Palynosciences 2: 239–254 (1993).

Falsely polygamous trees or shrubs; indumentum of simple and stellate hairs. Leaves alternate, paripinnate, rarely unifoliolate; leaflets serrate or entire; distal leaflet rudimentary. Inflorescences terminal or less often axillary thyrses. Flowers actinomorphic, functionally unisexual; sepals 5–6, imbricate, connate at base or distinct; petals 5(6) or 0, without appendages; disk annular, 5-lobed, pubescent; stamens (6–)8(–10); pollen colporate, striate to perforate, sometimes scabrate or rugulate; ovary 2(3)-carpellate, with a single ovule per carpel; style filiform, with 2–3 stigmatic spreading lobes. Fruits of 1(2) indehiscent, or tardily dehiscent, warty, smooth or spiny mericarps. Seeds with a thin, translucent-white, fleshy arillode around the hilum. $2n = 30$.

Six species in southern and southeastern Asia from Sri Lanka and India to eastern Malesia and Australia. *Dimocarpus longan* Lour. is widely cultivated as the source of the tropical fruit Longan.

68. *Diploglottis* Hook. f.

Diploglottis Hook. f. in Benth. & Hook. f., Gen. Pl. 1: 395 (1862); Reynolds, Austrobaileya 1: 390 (1981); Reynolds, Austrobaileya 2: 328 (1987), reg. rev.; Leenhouts in Fl. Males. I, 11: 520 (1994), reg. rev.

Falsely polygamous trees. Leaves alternate, paripinnate; distal leaflet rudimentary. Inflorescences axillary or pseudo-terminal thyrses. Flowers actinomorphic, functionally unisexual; sepals 5, imbricate; petals 4–5, clawed, with a pair of appendages formed by the inflexed margin above the claw, theses sometimes crested; disk annular or semi-annular, lobed; stamens 6–9; pollen parasyncolporate, rugulate or psilate; ovary 2–3-carpellate, with a single ovule per carpel; style filiform, with 3 stigmatic grooves. Fruit a 2–3-locular, loculicidal capsule, coccate, deeply lobed or ovoid. Seeds with 2-lobed arillode covering most of the seed.

Twelve species from northeastern Australia and Papua New Guinea.

69. *Elattostachys* (Blume) Radlk.

Elattostachys (Blume) Radlk., Actes Congr. Bot. Amsterdam 1877: 101 (1879); Adema in Fl. Males. I, 11: 527 (1994).

Falsely polygamous trees or shrubs. Leaves alternate, paripinnate; distal leaflet rudimentary. Inflorescences (supra)axillary thyrses. Flowers actinomorphic, bisexual or functionally unisexual; sepals 5, valvate to slightly imbricate; petals 5, clawed with 2 marginal appendages; disk annular, dish- or cup-shaped; stamens 8; pollen parasyncolporate or colporate, striate to rugulate; ovary 3-carpellate, with a single ovule per carpel; style with 3 stigmatic lines. Fruit a 3-locular, woody, loculicidal capsule. Seeds completely arillate or only so at base.

About 20 species, from Malesia, Australia, Solomon Islands, New Hebrides, New Caledonia, Fiji, Samoa, and Tonga.

70. *Eriocoelum* Hook. f.

Eriocoelum Hook. f. in Benth. & Hook. f., Gen. Pl. 1: 400 (1862).

Monoecious or falsely polygamous trees or shrubs. Leaves alternate, paripinnate; leaflets 2–5 pairs; distal leaflet rudimentary; pseudostipules usually present. Inflorescences axillary, spicate, racemose or thyrsoid. Flowers actinomorphic, functionally unisexual; sepals 5, distinct, valvate; petals 5, with a short, pubescent ventral appendage; disk annular, cupular, 8–10-lobed-crenate; stamens 8–10; pollen colporate, striate; ovary 3-carpellate, with a single ovule per carpel; style filiform, with 3 stigmatic lobes. Fruit a 3-locular, woody, loculicidal capsule, sometimes with setaceous-hispid pubescence. Seeds with an arillode at base.

About 10 species from tropical Africa.

71. *Erythrophysa* E. Meyer ex Arnold

Erythrophysa E. Meyer ex Arnold, J. Bot. (Hooker) 3: 258 (1841) (as *Erythrophila*, corr. Sonder in Harvey & Sonder, Fl. Cap. 1: 237 (1860).

Falsely polygamous shrubs or trees. Leaves alternate, imparipinnate; distal leaflet fully developed. Inflorescences axillary or terminal thyrses. Flowers zygomorphic, bisexual or functionally unisexual; sepals 5, imbricate; petals 4(5), with a pair or digitiform, simple or dissected appendages above the claw; disk semi-annular; stamens 8; pollen colporate, striate; ovary 3-carpellate, with 2 ovules per carpel; style elongated, filiform; stigma punctiform. Fruit a 1–3-locular, membranous, inflated, loculicidal capsule. Seeds globose, exarillate, sparsely pubescent, the testa delineating the contour of cotyledons.

Five species from South Africa and Madagascar.

72. ***Erythrophysopsis*** Verdc.

Erythrophysopsis Verdc., J. Linn. Soc. Bot. 58: 202 (1962).

Polygamous shrubs or trees. Leaves alternate, imparipinnate; distal leaflet fully developed. Inflorescences axillary or terminal thyrses. Flowers zygomorphic, functionally unisexual; sepals 5, distinct, imbricate; petals 4, with a pair or digitiform, simple or dissected appendages above the claw; disk unilateral, 4-lobed; stamens 8; pollen colporate, striate (Fig. 79A); ovary 3-carpellate, with 2 ovules per carpel; style elongated, filiform; stigma punctiform. Fruits indehiscent, 1–3-locular, woody or crustose. Seeds 1–3 per fruit, sub-globose, exarillate, woolly pubescent, the testa delineating the contour of cotyledons.

A single species, *E. aesculina* (Baill.) Verdc., Madagascar.

73. ***Glenniea*** Hook. f.

Glenniea Hook. f. in Benth. & Hook. f., Gen. Pl. 1: 404 (1862); Leenhouts, Blumea 21: 91–103 (1973) and 22: 411–414 (1975).
Melanodiscus Radlk. (1887) [1888].
Hedyachras Radlk. (1920).

Falsely polygamous-dioecious or monoecious trees or shrubs; indumentum of simple or stellate hairs. Leaves alternate, unifoliolate or paripinnate; leaflets entire; distal leaflet rudimentary; pseudostipules sometimes present. Inflorescences terminal, thyrsoid. Flowers actinomorphic, functionally unisexual; sepals (3)4–5, imbricate or valvate, distinct; petals 0; disk patelliform or annular-lobed; stamens 4–8; pollen colporate, striate to striate-reticulate; ovary 2(3)-carpellate, with a single ovule per carpel; stigma lobed. Fruits indehiscent, crustose or baccate. Seed exarillate.

Eight species, 3 in tropical Africa, 1 in Madagascar, 1 in Sri Lanka, and 3 in Indochina and Malesia.

74. ***Gloeocarpus*** Radlk.

Gloeocarpus Radlk., Philipp. J. Sci. 8, Bot.: 464 (1914); Welzen, Blumea 35: 389 (1991), rev.

Trees; indument of glandular hairs. Leaves alternate, paripinnate; leaflets serrate; distal leaflet rudimentary. Inflorescences ramiflorous thyrses. Flowers functionally unisexual; calyx zygomorphic, sepals 5, distinct, imbricate, outer 2 smaller; petals 5, with 2 marginal appendages; disk annular; stamens 7; pollen syncolporate or parasyncolporate, rugulate; ovary 3-carpellate, with a single ovule per carpel; style short, filiform; stigma not lobed. Fruit a 3-locular, woody, deeply lobed, loculicidal capsule. Seeds completely covered with a thin arillode.

A single species, *G. patentivalvis* (Radlk.) Radlk., endemic to the Philippines.

75. ***Gongrodiscus*** Radlk.

Gongrodiscus Radlk., Sitzungsber. Math.-Phys. Cl. Königl. Bayer. Akad. Wiss. München 9: 503, 607 (1879); Turner & van der Ham, Bull. Mus. Natl. Hist. Nat. Paris IV, 18: 339–349 (1996).

Falsely polygamous shrubs or trees. Leaves alternate, paripinnate; distal leaflet rudimentary. Inflorescences axillary, thyrsoid. Flowers actinomorphic, bisexual or functionally unisexual; sepals 5, distinct, valvate; petals 5, with 2 marginal appendages; disk annular, 5-lobed; stamens (7) 8; pollen parasyncolporate, rugulate; ovary 3-carpellate, with a single ovule per carpel; style short; stigma sub-trilobed. Fruit a 1-locular (incomplete septa), fleshy, tardily dehiscent capsule. Seeds arillate along ventral portion.

Three species endemic to New Caledonia.

76. ***Gongrospermum*** Radlk.

Gongrospermum Radlk., Philipp. J. Sci. 8, Bot.: 469 (1914); Welzen, Rheedea 1: 60 (1991), rev.

Trees. Leaves alternate, paripinnate; leaflets papillate on lower surface; distal leaflet rudimentary. Inflorescences axillary, simple or thyrses. Flowers actinomorphic, functionally unisexual; sepals 5, valvate, distinct; petals 0; disk sub-cupular, 5-lobed; stamens 8; pollen syncolporate, rugulate; ovary 3-carpellate, with a single ovule per carpel; style pyramidal. Fruit a 1-locular, leathery, loculicidal capsule. Seeds exarillate, endotesta ruminately grown together with embryo.

A single species, *G. philippinense* Radlk., Philippines.

77. ***Guioa*** Cav.

Guioa Cav., Icon. 4: 49, t. 373 (1798); Welzen, Leiden Bot. Ser. 12: 146 (1989), rev.

Shrubs, trees, or treelets. Leaves alternate, paripinnate; leaflets entire or less often crenate or serrate; distal leaflet rudimentary; rachis terete or winged. Inflorescences axillary or terminal thyrses. Flowers functionally unisexual; calyx zygomorphic, sepals 5(6), imbricate, petaloid, outer 2 smaller; petals 5(6), usually clawed, with two marginal or ventral appendages that are usually crested; disk annular or semi-annular; stamens (7)8; pollen syncolporate or parasyncolporate, rarely colporate, rugulate to perforate or psilate; ovary 3-carpellate, with a single ovule per carpel; style apical, pyramidal, with 3 stigmatic lines. Fruit a 3-locular, obcordate, deeply 3-lobed to coccate, leathery, loculicidal capsule. Seeds almost entirely covered by an arillode that has a basal projection.

About 64 species ranging from southeastern Asia, throughout Malesia to Australia, New Caledonia, and Samoa.

78. *Haplocoelopsis* Davis

Haplocoelopsis Davis, Kew Bull. 52: 231 (1997).

Monoecious shrubs or small trees. Leaves alternate, paripinnate or imparipinnate; leaflets entire; distal leaflet rudimentary; pseudostipules present. Inflorescences axillary, racemose thyrses. Flowers actinomorphic, unisexual; sepals 5, distinct, imbricate; petals 5, with a short, bilobed ventral appendage; disk annular; stamens 8–9; pollen unknown; ovary bilobed, 2-carpellate, with a single ovule per carpel; style with 2 stigmatic lobes. Fruit a 2-locular, circular, laterally compressed, loculicidal capsule. Seeds not known.

A single species, *H. africana* Davies, from central and east Africa.

79. *Haplocoelum* Radlk.

Haplocoelum Radlk., Sitzungsber. Math.-Phys. Cl. Königl. Bayer. Akad. Wiss. München 8: 336 (1878).

Falsely polygamous-dioecious trees. Leaves alternate, paripinnate; distal leaflet rudimentary. Inflorescence a congested axillary polychasium. Flowers actinomorphic, functionally unisexual; sepals 5–6, distinct, narrowly imbricate; petals 0 or less often present; disk annular; stamens 5–6, inserted on the disk; pollen colporate, rugulate; ovary (2)3-carpellate, with a single ovule percarpel; style with 3 stigmatic lobes. Fruit 1–2-locular, indehiscent, baccate. Seeds ellipsoid or laterally compressed, with a dorsally or distally split arillode.

About 7 species from tropical Africa and Madagascar.

80. *Hornea* Baker

Hornea Baker, Fl. Mauritius: 59 (1877).

Falsely polygamous shrubs or trees. Leaves alternate, paripinnate; leaflets 2 or 4, entire; distal leaflet rudimentary. Inflorescences terminal, corymbose or thyrsoid. Flowers bisexual or functionally unisexual; calyx zygomorphic, sepals 5, orbicular, concave, distinct, imbricate, the 2 outer sepals smaller; petals 5, clawed, with 2 ventral appendages above the claw; disk 5-lobed; stamens 18–24; pollen colporate, rugulate; ovary 2-carpellate, with a single ovule per carpel; style short; stigma punctiform. Fruits of 2 mericarps with a dorsal wing. Seeds exarillate.

A single species, *H. mauritiana* Baker, endemic to Mauritius.

81. *Jagera* Blume

Jagera Blume, Rumphia 3: 155 (1847); Leenhouts in Fl. Males. I, 11: 614 (1994).

Falsely polygamous, often pachycaulous trees or shrubs. Leaves verticillate, sometimes opposite or spirally arranged, paripinnate; leaflets serrate; distal leaflet rudimentary. Inflorescences axillary thyrses. Flowers actinomorphic, bisexual or functionally unisexual; sepals 5, slightly connate at base, imbricate; petals 5, with marginal or ventral appendages; disk annular; stamens (7)8(–10); pollen syncolporate or parasyncolporate, perforate or reticulate with often finely tapering scabrae; ovary 3-carpellate, with a single ovule per carpel; style filiform with 3 stigmatic lines. Fruit a 3-locular, woody, loculicidal capsule, with stiff, irritating hairs. Seeds with a small sarcotesta around the hilum.

Two species in the Moluccas, New Guinea, and eastern Australia.

82. *Koelreuteria* Laxm.

Koelreuteria Laxm., Nov. Comm. Acad. Petrop. 16: 562, t. 18 (1772); Meyer, J. Arnold Arb. 57: 129–166 (1976); Adema in Fl. Males. I, 11: 755 (1994).

Duodichogamous trees. Leaves alternate, once or twice pinnate, imparipinnate; leaflets entire, serrate or crenate; terminal leaflet well-developed. Inflorescences terminal thyrses. Flowers zygomorphic; sepals 5, valvate; petals 4, yellow, clawed, with fimbriate involute base of lamina forming an appendage; disk annular, undulate, on a short stipe; stamens (5–)8; pollen colporate, striate; ovary 3-carpellate, with 2 ovules per carpel, the septa incomplete on distal portion; stigma entire or trifid. Fruit a 3-locular, papery, inflated, loculicidal capsule with incomplete septa; seeds 2 per locule, exarillate. $2n = 22, 30, 32$.

About four species native to Japan, southern China, Taiwan, and perhaps indigenous to Fiji. *Koelreuteria paniculata* Laxm. and *K. bipinnata* Franchet are cultivated worldwide in temperate areas as ornamentals.

83. *Laccodiscus* Radlk.

Laccodiscus Radlk., Sitzungsber. Math-Phys. Cl. Königl. Bayer. Akad. Wiss. München 9: 496 (1879).

Falsely polygamous trees or scandent shrubs. Leaves alternate, paripinnate; distal leaflet rudimentary; pseudostipules present. Inflorescences axillary or terminal thyrses. Flowers actinomorphic, bisexual or functionally unisexual; sepals 5, connate at base, imbricate; petals 5, with short marginal appendages; disk annular-lobed or 5-lobed; stamens 8–10; pollen colporate, perforate; ovary 3-carpellate, with a single ovule per carpel; style curved; stigma papillose. Fruit a 3-locular, woody, 3-lobed, loculicidal capsule. Seeds arillate (fide Fouilloy and Hallé 1973a).

About six species native to West Africa.

84. *Lecaniodiscus* Planch. ex Benth.

Lecaniodiscus Planch. ex Benth. in Hook., Niger Fl.: 250 (1849).
Chiarinia Chiov. (1932).

Falsely polygamous-dioecious shrubs or trees. Leaves alternate, paripinnate; leaflets 3–7 pairs; distal leaflet rudimentary; stipules 0. Inflorescences axillary, racemose or thyrsoid. Flowers actinomorphic, functionally unisexual; sepals 5, imbricate, distinct; petals 0 or 5, with 2 minute basal appendages; disk annular, lobed; stamens 8; pollen colporate, striate; ovary (2)3-carpellate, with a single ovule per carpel; style short, stigma subsessile and 3-lobed or with 2 reflexed stigmatic branches. Fruits crustose, 1-locular or bicoccate, indehiscent or tardily splitting from the base. Seeds nearly entirely covered by an arillode.

Two species native to tropical Africa.

85. *Lepiderema* Radlk.

Lepiderema Radlk., Actes Congr. Bot. Amsterdam 1877: 250 (1879); Reynolds, Austrobaileya 1: 488 (1982), reg. rev.; Schot, Blumea 36: 235 (1991), reg. rev.

Falsely polygamous shrubs or trees; indument of lepidote scales. Leaves alternate, paripinnate; distal leaflet rudimentary. Inflorescences axillary, ramiflorous, or pseudo-terminal racemose thyrses. Flowers actinomorphic, bisexual or functionally unisexual; sepals 5, petaloid, distinct, imbricate; petals 5, shortly clawed, without appendages; disk annular; stamens (6–)8; pollen syncolporate, parasyncolporate or colporate, rugulate-reticulate; ovary 3-carpellate, with a single ovule per carpel; style spirally twisted; stigma lobed. Fruit a 3-locular, trigonous, stipitate, woody, loculicidal capsule. Seed basally arillate.

Eight species, 6 in Australia and 2 in New Guinea.

86. *Lepidopetalum* Blume

Lepidopetalum Blume, Rumphia 3: 171 (1847); Welzen, Piskaut & Windadri, Blumea 36: 452 (1992), rev.

Falsely polygamous trees. Leaves alternate, paripinnate; distal leaflet rudimentary. Inflorescences ramiflorous, axillary or pseudo-terminal thyrses. Flowers actinomorphic, functionally unisexual; calyx shortly cupular, sepals 5(6), distinct, valvate; petals 5(–7), shorter than the sepals, with a single basal appendage bigger than the blade; disk annular or semi-annular; stamens (7)8(–10), the filaments nearly equal, the anthers dorsifixed; pollen colporate, striate to striate-reticulate; ovary 2-carpellate, with a single ovule per carpel; stigma sessile, fleshy, of two elongate lobes adnate to the distal portion of ovary. Fruit a 2-locular, leathery or woody, glabrous, loculicidal capsule. Seed with a basal to complete, orange sarcotesta.

Six species throughout Malesia and northeastern Australia.

87. *Lepisanthes* Blume

Lepisanthes Blume, Bijdr. Fl. Nederl. Ind. 5: 237 (1825); Leenhouts, Blumea 17: 33–91 (1969).
Aphania Blume (1825).
Erioglossum Blume (1825).
Otophora Blume (1847).
Hebecoccus Radlk. (Radlk. (1878).
Thraulococcus Radlk. (1878).
Aphanococcus Radlk. (1887) [1888].
Manongarivea Choux. (1927).
Sapindopsis How & Ho (1955), non Fontaine (1889).
Howethoa Rauschert (1982).

Monoecious trees, shrubs, or climbing shrubs. Leaves alternate, (im)paripinnate, sometimes unifoliolate; terminal leaflet rudimentary; pseudostipules sometimes present. Inflorescences terminal, axillary, ramiflorous or cauliflorous thyrses. Flowers bisexual or functionally unisexual; calyx zygomorphic, the sepals 5, imbricate, the outer 2 distinctly smaller; petals (2–)4(5), with marginal or ventral appendages; disk semi-annular or annular, 5-lobed or crenate; stamens (4–)8 (–18); pollen colporate or brevicolporate, rarely syncolporate, rugulate to reticulate, rarely psilate; ovary 2–3(4)-carpellate, sessile or stipitate, with a single ovule per carpel; style short, apical; stigma capitate, sometimes sessile. Fruits indehiscent, 2–3-lobed, sub-fleshy, sometimes deeply lobed or coccate with distinct monocarps. Seeds ellipsoid, obovoid to sub-globose, exarillate. $2n = 26$, 28, 30.

About 24 species in tropical Africa, Madagascar, southern and southeastern Asia from Sri Lanka to Hainan, Malesia, and northwestern Australia.

Four subgenera: subgen. *Lepisanthes*: Leaves paripinnate, without pseudostipules; petiole and rachis not winged; outer sepals sericeous outside; petals longer than sepals; fruits usually only slightly lobed, septa continuous; subgen. *Otophora*: leaves pari- or imparipinnate, without pseudostipules; petiole winged or not; outer sepals glabrous or hairy outside; petals shorter than the sepals; fruits not or slightly lobed, septa often ± interrupted; subgen. *Erioglossum*: Leaves paripinnate, without pseudostipules; petiole and rachis not winged; outer sepals glabrous outside; petals longer than the sepals; fruits lobed, septa complete; subgen. *Aphania*: Leaves paripinnate, sometimes simple, sometimes with pseudostipules (petiole and rachis winged); outer sepals glabrous; petals as long as sepals; fruits lobed, septa complete.

88. *Litchi* Sonn.

Litchi Sonn., Voy. Ind. Or. Chine 2: 230, t. 129 (1782); Leenhouts, Blumea 24: 398 (1978).
Euphoria Commers. ex Juss. (1789).

Duodichogamous trees; indument of 2-branched hairs. Leaves alternate, paripinnate; distal leaflet rudimentary. Inflorescences terminal and axillary thyrses. Flowers actinomorphic bisexual or functionally unisexual; calyx cup-shaped, with 4–5 equal, minute lobes; petals 0; disk annular; stamens (6)7(–11); pollen colporate, striate; ovary 2-carpellate, with a single ovule per carpel; stigma of 2 elongated, spreading or coiled lobes. Fruit 1-coccate (1 coccus rudimentary), indehiscent, with leathery, muricate pericarp. Seeds partly or completely covered by a translucent, fleshy, convolute arillode. $2n = 28$, 30.

A single species, *L. chinensis* Sonn., from southeastern China, Indochina, Malay Peninsula, Java, Borneo, and the Philippines. Widely cultivated in subtropical regions for its edible fruits, commonly known as *Litchi*.

89. *Lychnodiscus* Radlk.

Lychnodiscus Radlk., Sitzungsber. Math-Phys. Cl. Königl. Bayer. Akad. Wiss. München 8: 332 (1878).

Falsely polygamous shrubs or trees. Leaves alternate, paripinnate; distal leaflet rudimentary. Inflorescences axillary or terminal thyrses or panicles. Flowers actinomorphic, bisexual or functionally unis exual; sepals 5, connate at base, imbricate; petals 5, with a single ventral appendage; disk annular; stamens 10–12; pollen colporate, striate; ovary 3-carpellate, with a single ovule per carpel; stigma clavate. Fruit a (1–)3-locular, 3-lobed, loculicidal capsule; seeds sarcotestal.

About seven species from tropical Africa.

90. *Macphersonia* Blume

Macphersonia Blume, Rumphia 3: 156 (1847).

Dioecious trees. Leaves alternate, bipinnate, or once pinnate; distal leaflet rudimentary. Inflorescences axillary, racemose or spicate, or rarely thyrsoid. Flowers actinomorphic, bisexual or functionally unisexual; sepals 5, distinct, imbricate; petals 5, clawed, with appendages formed by the inflexed margins above the claw; disk annular, dish-shaped; stamens 8; pollen colporate, striate; ovary (2)3-carpellate, with a single ovule per carpel; stigma sessile, capitate or trigonous. Fruit 1–2-locular, indehiscent or tardily dehiscent, subfleshy. Seeds completely covered by a translucent arillode.

About eight species from Aldabra, Madagascar, and west tropical Africa.

91. *Matayba* Aublet Fig. 79K

Matayba Aublet, Hist. Pl. Guiane 1: 331 (1775).

Falsely polygamous-dioecious, large or small trees. Leaves alternate, paripinnate or imparipinnate; leaflets entire; distal leaflet rudimentary. Inflorescences axillary or terminal thyrses. Flowers 5-merous, actinomorphic, bisexual or unisexual; sepals short (less than 2 mm long), distinct, valvate; petals as long as the sepals or longer, with a pair of marginal tomentose appendages; disk annular, usually lobed; stamens (4–6)8; pollen syncolporate or parasyncolporate, sometimes colporate, rugulate; ovary 3-carpellate, with a single ovule per carpel; stigma trilobed or trifid. Fruit a 2- or 3-locular, trigonous or lobed, woody or leathery, loculicidal capsule. Seeds nearly globose or ellipsoid, arillate at base or seldom nearly entire.

About 50 species from tropical and subtropical America.

92. *Mischarytera* H. Turner

Mischarytera H. Turner, Blumea Suppl. 9: 210 (1995), rev.

Trees. Leaves paripinnate; distal leaflet rudimentary. Inflorescences axillary to pseudo-terminal thyrses. Flowers actinomorphic, functionally unisexual; sepals 5, connate at base, valvate; petals 5, clawed, with appendages formed by the inflexed margins above the claw, or these 0; disk annular, unlobed or lobed; stamens 7–8; pollen parasyncolporate, sometimes colporate, rugulate; ovary 3-carpellate, with a single ovule per carpel; stigma shortly 3-lobed. Fruit a 3-locular, loculicidal or loculifragal capsule. Seeds arillate.

Three species in eastern Australia and Papua New Guinea.

93. *Mischocarpus* Blume, nom. cons.

Mischocarpus Blume, Bijdr. 1825: 238 (1825); van der Ham, Blumea 23: 251 (1977).
Pedicellia Loureiro (1790), nom. rej.
Mischocodon Radlk. (1913).

Falsely polygamous shrubs or trees. Leaves alternate, paripinnate; distal leaflet rudimentary. Inflorescences pseudo-terminal, axillary or ramiflorous thyrsoids. Flowers actinomorphic, bisexual or functionally unisexual; sepals imbricate, distinct or connate into a crateriform calyx, crowned by 5 subequal lobes; petals 0, reduced or slightly longer than the sepals, with marginal appendages; disk annular or semi-annular; stamens (5–)8(9); pollen syncolporate or parasyncolporate, sometimes colporate, rugulate to rugulate-reticulate; ovary (2)3(4)-carpellate, with a single ovule per carpel; stigma 3-lobed. Fruit a (1)2-locular, coriaceous, loculicidal capsule. Seeds completely covered by an arillode with a basal, funiculus-like appendage.

About 15 species from southeastern Asia, throughout Malesia to Australia.

94. *Molinaea* Commers. ex Juss.

Molinaea Commers. ex Juss., Gen. Pl.: 248 (1789).

Falsely polygamous shrubs or trees. Leaves alternate, paripinnate; leaflets entire or serrate, distal leaflet rudimentary. Inflorescences axillary, simple or paniculate thyrses. Flowers bisexual or functionally unisexual; calyx zygomorphic, sepals 5, distinct, imbricate, the outer two smaller; petals 5, with marginal appendages or without appendages; disk annular or semi-annular; stamens 8; pollen parasyncolporate, rarely syncolporate or colporate, rugulate; ovary 3-carpellate, with a single ovule per carpel; style short; stigma 3-gonous, grooved. Fruit a 3-locular, loculicidal capsule. Seeds arillate.

About ten species from Madagascar, Mauritius, and Mascarene Islands.

95. *Namataea* D.W. Thomas & D.J. Harris

Namataea D.W. Thomas & D.J. Harris, Kew Bull. 54: 951 (1999).

Seemingly dioecious shrub. Leaves alternate, simple; stipules 0. Inflorescences cauliflorous, ramiflorous or less often axillary, racemes. Flowers unisexual; calyx zygomorphic, urceolate, the sepals 5, connate at lower quarter or third, slightly imbricate, third and fifth sepals slightly longer than remaining sepals; petals 4, a basal hood-shaped, crested appendage (the crest infundibuliform); disk reniform; stamens 7; pollen not known; ovary 3-carpellate, with a single ovule per carpel; style stout. Fruit fleshy, indehiscent, of 3 ellipsoid, cocci, each with an apical beak.

A single species, *N. simplicifolia* D.W. Thomas & D.J. Harris, Cameroon.

96. *Neotina* Capuron

Neotina Capuron, Mem. Mus. Natl. Hist. Nat. B, Bot., II, 19: 174 (1969).

Monoecious or dioecious trees. Leaves alternate or subopposite, paripinnate; distal leaflet rudimentary. Inflorescences axillary thyrses. Flowers unisexual; calyx zygomorphic, the sepals 5, imbricate, outer two sepals smaller; petals 4–5, with 2 marginal or ventral appendages; disk annular; stamens 5(6–8); pollen parasyncolporate or colporate, rugulate to rugulate-reticulate; ovary 2-carpellate, with a single ovule per carpel; style elongated, with 2 stigmatic lines. Fruit a 1-locular, sub-fleshy, loculicidal capsule. Seeds with red or orange arillode for 2/3 of their length.

Two species from Madagascar.

97. *Nephelium* L.

Nephelium L., Syst. Nat. ed. 12, 2: 623 (1767); L., Mantissa Pl.: 18 (1767); Leenhouts, Blumea 31: 373–436 (1986).

Falsely polygamous or dioecious trees or less often shrubs. Leaves alternate, paripinnate; leaflets distinctly glaucous beneath; distal leaflet rudimentary. Inflorescences axillary, pseudoterminal or terminal (in *N. cuspidatum* Blume also rami- and cauliflorous) thyrses. Flowers actinomorphic; calyx cup-shaped, crowned by 4–6 subequal, valvate lobes; petals 0 or 4–6, clawed, with a bilobed appendage; disk annular; stamens 4–10; pollen colporate, striate; ovary (1)2(–4)-carpellate, with a single ovule per carpel; style elongated; stigma usually 2-lobed. Fruits 1(2)-coccate, tardily loculicidally dehiscent, the pericarp warty to spiny, coriaceous or less often woody or corky. Seeds completely covered by edible sarcotesta. $2n=22$.

About 16 species from southeastern Asia in Yunnan and Assam to Hainan and Malesia. *Nephelium lappaceum* L. (the Rambutan) and *N. ramboutan-ake* (Labill.) Leenh. (the Pulasan) are widely cultivated for their edible fruits.

98. *Otonephelium* Radlk.

Otonephelium Radlk, Sitzungsber. Math-Phys. Cl. Königl. Bayer. Akad. Wiss. München 20: 253, 288 (1890).

Falsely polygamous trees. Leaves alternate, paripinnate; distal leaflet rudimentary; pseudostipules present. Inflorescences terminal or axillary thyrses. Flowers actinomorphic bisexual or functionally unisexual; sepals 5, imbricate, distinct; petals 0; disk annular, glabrous; stamens 5–9; pollen colporate, striate, sometimes irregularly striate or rugulate; ovary 2-carpellate, with a single ovule per carpel; stigma bifid. Fruit 1-coccate, with rudimentary cocci at base, indehiscent, baccate-crustose, with soft spines. Seeds arillate.

A single species, *O. stipulaceum* Radlk., western India. Doubtfully different from *Dimocarpus* by its glabrous disk and the presence of pseudostipules.

99. *Pancovia* Willd., nom. cons.

Pancovia Willd., Sp. Pl. 2: 285 (1799); non Fabricius (1759), nom. rej.

Falsely polygamous or dioecious, trees or shrubs. Leaves alternate, paripinnate; leaflets 2–12 pairs; distal leaflet rudimentary. Inflorescences axillary or cauliflorous, fasciculate, racemose or thyrsoid. Flowers zygomorphic, functionally unisexual; sepals 4–5, imbricate or sub-valvate; petals 3–4, clawed, with 2 inflexed or dissected appendages above the claw; disk semi-annular, unilateral; stamens (6–)8; pollen colporate, rugulate; ovary 3-carpellate, with a single ovule per carpel; style subulate; stigma sub-clavate. Fruit indehiscent, 3-locular, with fleshy to woody pericarp. Seeds exarillate, laterally compressed. $2n=32$.

Ten to 12 species native to west tropical and subtropical Africa.

100. *Pappea* Eckl. & Zeyh.

Pappea Eckl. & Zeyh., Enum. Pl. Afr. austr. extratrop. 1: 53 (1834–1835).

Dioecious trees or shrubs. Leaves alternate, simple. Inflorescences axillary, racemose or thyrsoid. Flowers actinomorphic, functionally unisexual; sepals 5, valvate; petals (4)5(6), with a pair of marginal, hairy appendages; disk annular; stamens 8(–10); pollen colporate, striate; ovary 3-carpellate, with a single ovule per carpel; style short; stigma sub-lobed. Fruit a 1-locular, fleshy, loculicidal capsule. Seeds with a lobed arillode.

One to four species native to southern Africa. The fruit of *P. capensis* (Spreng.) Eckl. & Zeyh. is said to be edible, a bland oil is expressed from the seeds (Harvey and Sonder 1894).

101. *Paranephelium* Miq.

Paranephelium Miq., Fl. Ind. Bat. Suppl. 509 (1861) [1860]; Davids, Blumea 29: 425 (1984), rev.
Mildea Miquel (1867), non Griseb. (1866).
Scyphopetalum Hiern (1875).

Falsely polygamous trees. Leaves alternate, mainly imparipinnate; distal leaflets rudimentary or well-developed. Inflorescences ramiflorous or terminal thyrses. Flowers actinomorphic; calyx shortly cupular, the sepals (4)5(–7), mainly equal, distinct or connate at base, valvate; petals (4)5(–7), often clawed, with a single appendage; disk annular, 5-lobed, cup-shaped; stamens 5–9; pollen syncolporate or parasyncolporate, rugulate; ovary mainly 3-carpellate; with a single ovule per carpel; stigma usually lobed. Fruits 1(–3)-locular, globular, smooth to densely spiny, woody, loculicidal capsules or dehiscing randomly. Seeds exarillate, with an enlarged white, round hilum.

Four species in southeast Asia from Yunnan, Myanmar, and Indochina to Sumatra, Borneo, and the Philippines.

102. *Pavieasia* Pierre

Pavieasia Pierre, Fl. Forest. Cochinch.: t. 317 (1894).

Falsely polygamous trees or shrubs. Leaves alternate, paripinnate; distal leaflet rudimentary. Inflorescences terminal thyrses. Flowers actinomorphic, bisexual or functionally unisexual; calyx shortly cupular, sepals 5, connate at base, imbricate; petals 5, with a single basal appendage; disk annular; stamens 8; pollen syncolporate, striate; ovary 3-carpellate, pilose, with a single ovule per carpel; style filiform; stigma obscurely 3-lobed at apex. Fruit a 3-locular, loculicidal capsule. Seeds exarillate.

One to three species endemic to China.

103. *Pentascyphus* Radlk.

Pentascyphus Radlk., Sitzungsber. Math.-Phys. Cl. Königl. Bayer. Akad. Wiss. München 9: 472, 495, 539 (1879).

Falsely polygamous trees or shrubs. Leaves alternate, paripinnate; leaflets alternate, entire; distal leaflet rudimentary; stipules 0. Inflorescences terminal or axillary thyrses. Flowers actinomorphic, bisexual or functionally unisexual; sepals 5, distinct, imbricate; petals 5, obovate to cuneate, with a single short, basal appendage connate to petals' margins; disk annular, 7–8-lobed; stamens 8; pollen syncolporate or parasyncolporate, rugulate; ovary 3-carpellate, hirsute, with a single ovule per carpel. Fruit unknown.

A single species, *P. thyrsiflorus* Radlk., in French Guiana, Surinam, and Brazil (Amazonas).

104. *Phyllotrichum* Thorel ex Lecomte

Phyllotrichum Thorel ex Lecomte, Notul. Syst. (Paris) 2: 8 (1911).

Falsely polygamous trees. Leaves alternate, paripinnate, with 5–6 pairs of leaflets; distal leaflets rudimentary. Inflorescences ramiflorous, racemose thyrses. Flowers zygomorphic; sepals 5, distinct, imbricate; petals 4, with a single basal appendage; disk unilateral, semi-annular; stamens 8(9); pollen syncolporate or parasyncolporate, striate; ovary 3-carpellate, densely pubescent, with a single ovule per carpel; style nearly trigonous; stigma nearly 3-lobed. Fruit a 3-locular, loculicidal, muricate capsule. Seeds ovoid, exarillate.

A single species, *P. mekongense* Thorel ex Lecomte, endemic to Laos.

105. *Placodiscus* Radlk.

Placodiscus Radlk., Sitzungsber. Math.-Phys. Cl. Königl. Bayer. Akad. Wiss. München 8: 332 (1878).

Falsely polygamous trees or shrubs. Leaves alternate, paripinnate, with 2–9 pairs of leaflets; distal leaflets rudimentary; pseudostipules sometimes present. Inflorescences cauliflorous, spicate thyrses. Flowers actinomorphic; calyx urceolate or turbinate, crowned by 5 valvate or narrowly imbricate sepals; petals 0; disk annular, dish-shaped; stamens 8; pollen colporate, striate; ovary 3-carpellate, tomentose, with a single ovule per carpel; style short; stigma minute.

Fruits baccate, 1–3-locular, 3-sulcate or 3-lobate, indehiscent. Seeds exarillate.

About ten species native to tropical Africa.

106. *Plagioscyphus* Radlk.

Plagioscyphus Radlk., Sitzungsber. Math.-Phys. Cl. Königl. Bayer. Akad. Wiss. München 8: 335 (1878).
Cotylodiscus Radlk. ibid. 8: 334 (1878).
Strophiodiscus Choux (1926).
Poculodiscus Danguy & Choux (1927).

Falsely polygamous shrubs. Leaves alternate, paripinnate; leaflets 1–5 pairs, serrate; distal leaflets rudimentary. Inflorescences axillary, racemose thyrses. Flowers actinomorphic; sepals 5, distinct or connate at base, imbricate; petals 4–5, the appendages 2 and marginal or single and basal; disk annular, vase-shaped, or unilateral, semi-annular or semi-vase-shaped; stamens (7)8; pollen colporate or brevicolporate, perforate to reticulate; ovary 2–3-carpellate, tomentose, with a single ovule per carpel; stigma apiculate. Fruits baccate, 1–3-locular, indehiscent. Seeds arillate, with a longitudinal ventral hilum.

About ten species native to Madagascar.

107. *Podonephelium* Baill.

Podonephelium Baill., Adansonia 11: 245 (1874).

Falsely polygamous-dioecious shrubs. Leaves alternate, paripinnate, with 3–6 pairs of leaflets; distal leaflets rudimentary. Inflorescences axillary thyrses. Flowers actinomorphic; calyx crateriform, with 4–7, valvate lobes; petals 0; disk cup-shaped, 8-crenate; stamens (5–)8; pollen colporate, striate; ovary 3-carpellate, tomentose, with a single ovule per carpel; stigma 3-lobed. Fruit a 1(2)-coccate, crustose, circumscissile dehiscent capsule. Seeds with white arillode mostly along dorsal portion.

Four species native to New Caledonia.

108. *Pometia* Forst. & Forst.

Pometia Forst. & Forst., Char. Gen. Pl.: 55, t. 55 (1775); Jacobs, Reinwardtia 6: 109–144 (1962).

Falsely polygamous trees, producing red exudates. Leaves alternate, paripinnate; leaflets entire or serrate, often with large orbicular glands beneath; distal leaflet rudimentary; pseudostipules present. Inflorescences terminal or axillary thyrses. Flowers actinomorphic, functionally unisexual; sepals 5, connate at base to half of their length, valvate; petals 5 without appendages; disk annular or semi-annular; stamens 5(6); pollen brevicolporate, reticulate (Fig. 79L); ovary 2(3)-carpellate, with a single ovule per carpel; style filiform; stigma obtuse, emarginate. Fruits 1(2)-locular, indehiscent, with fleshy mesocarp. Seeds fully covered by an arillode.

Two species found in Sri Lanka, Andaman and Nicobar Islands, Indochina, Taiwan, Malesia, Fiji, Samoa, and Tonga. The wood of *P. pinnata* Forst. is used as firewood in the Pacific Islands.

109. *Porocystis* Radlk.

Porocystis Radlk., Sitzungsber. Math.-Phys. Cl. Königl. Bayer. Akad. Wiss. München 8: 353 (1878).

Falsely polygamous trees. Leaves alternate, paripinnate or imparipinnate; leaflets entire; distal leaflet rudimentary or exceptionally fully developed. Inflorescences terminal or axillary thyrses. Flowers zygomorphic, bisexual or functionally unisexual; sepals 5, distinct, imbricate; petals 4, clawed, with a bifid appendage above the claw; disk unilateral, semi-annular to 4-lobed; stamens 8; pollen colporate, rugulate-reticulate; ovary 3-carpellate, with a single ovule per carpel; style short, with 2 recurved, stigmatic branches. Fruit a 3-locular schizocarp, splitting into 3 membranous, inflated, wrinkled mericarps. Seeds exarillate.

Three species, 2 from Brazil (Amazonas) and 1 from French Guiana.

110. *Pseudima* Radlk.

Pseudima Radlk., Nouv. Giornale Bot. Ital. 10: 108 (1878).

Falsely polygamous-dioecious, small to large trees. Leaves alternate, pinnately compound; distal leaflet rudimentary; stipules 0. Inflorescences axillary or terminal thyrses. Flowers bisexual or unisexual; calyx zygomorphic, sepals 5, imbricate, concave, the outer 2 sepals shorter; petals 5, longer than the sepals, lacking appendages; disk annular, 5-lobed; stamens 8 or 10, shorter than the petals; pollen colporate, perforate; ovary 2(3)-carpellate, with a single ovule per carpel; style filiform; stigma obtuse. Fruit a 2(3)-coccate, loculicidal, leathery capsule, the cocci equally developed or one of them rudimentary. Seeds large, arillate.

One species from the lowlands of tropical continental America.

111. ***Pseudopancovia*** Pellegr.

Pseudopancovia Pellegr., Bull. Soc. Bot. France 102: 228 (1955).

Falsely polygamous-dioecious, shrubs. Leaves alternate, paripinnate; distal leaflet rudimentary; stipules 0. Inflorescences axillary, spicate thyrses. Flowers zygomorphic, bisexual or unisexual; calyx 2-lipped, 4–5-lobed; petals 3–4, clawed, with a basal appendage forming a pocket; disk unilateral; stamens 7; pollen colporate, rugulate; ovary 3-carpellate, with a single ovule per carpel. Fruit unknown.

A single species, *P. heterophylla* Pellegr., endemic to west equatorial Africa.

112. ***Pseudopteris*** Baill.

Pseudopteris Baill., Adansonia 11: 243 (1874).

Falsely polygamous trees. Leaves alternate, paripinnate; leaflets opposite or alternate, entire or crenate-serrate; distal leaflet rudimentary; stipules 0. Inflorescences of axillary racemose thyrses. Flowers actinomorphic, bisexual or unisexual; sepals 5, distinct, imbricate; petals 5, smaller than the sepals, cucullate, without appendages; disk 5-lobed ; stamens 5; pollen colporate, striate; ovary 2–3-carpellate, with a single ovule per carpel. Fruit baccate, 1–3-locular, indehiscent. Seeds sub-globose, completely covered by an arillode.

Three species endemic to Madagascar.

113. ***Radlkofera*** Gilg

Radlkofera Gilg, Bot. Jahrb. Syst. 24: 300 (1897).

Falsely polygamous unbranched trees. Leaves alternate, paripinnate, with 13–20 pairs of leaflets; distal leaflet rudimentary; stipules 0. Inflorescences axillary, racemose thyrses, with elongated bracteoles. Flowers zygomorphic, bisexual or unisexual; calyx urceolate, sepals 5, connate at base, imbricate; petals 4, with a ventral appendage; disk unilateral; stamens 7–8; pollen colporate, indistinctly rugulate; ovary 5–7(8)-carpellate, tomentose, with a single ovule per carpel; style elongated. Fruit indehiscent, 5–8-locular, fusiform, with fleshy mesocarp. Seeds exarillate, orange.

A single species, *R. calodendron* Gilg, western Africa.

114. ***Rhysotoechia*** Radlk.

Rhysotoechia Radlk., Actes Congr. Bot. Amsterdam 1877: 131 (1879); Etman, Blumea 39: 41 (1994).

Falsely polygamous trees or shrubs. Leaves alternate, paripinnate. Flowers seemingly bisexual; calyx zygomorphic, the sepals 5, distinct, imbricate, 2 outer sepals smaller; petals 5, clawed, with marginal appendages or these 0; disk annular; stamens (7)8; pollen syncolporate or parasyncolporate, rugulate to perforate, sometimes psilate; ovary (2)3-carpellate, with a single ovule per carpel; style with 3 stigmatic lines. Fruit a (2)3-locular, sub-globose, obovoid, or subcordate, loculicidal capsule. Seeds with a cup-shaped arillode at base.

About 14 species from Australia, Borneo, Philippines, Sulawesi, Moluccas, and New Guinea.

115. ***Sapindus*** Plum. ex L.

Sapindus Plum. ex L., Sp. Pl.: 367 (1753); Radlkofer in Pflanzenreich 98: 630 (1932); Leenhouts in Fl. Males. I, 11: 713 (1994).

Duodichogamous trees. Leaves alternate, paripinnate or unifoliolate; leaflets 2–8 pairs, often falcate; distal leaflet rudimentary. Inflorescences terminal, thyrses. Flowers actinomorphic or partly zygomorphic, functionally pistillate or staminate; sepals 5, distinct, imbricate, the outer 2 smaller; petals 4 or 5, with a single large appendage, a transverse ridge, or a pair reduced marginal appendages; disk annular or cup-shaped; stamens 8; pollen colporate, rugulate; ovary 3-carpellate, with a single ovule per carpel; style short, stigma capitate or with 3 conivent stigmatic branches. Fruit schizocarpic, 1(2)-coccate, with 2(1) rudimentary cocci, separating into indehiscent globose mericarps, with fleshy pericarp containing much saponin. Seeds globose, exarillate. $n = 11, 15, 18$.

About ten species with tropical to subtemperate distribution. Several species used as ornamentals.

116. *Sarcopteryx* Radlk.

Sarcopteryx Radlk. Actes Congr. Bot. Amsterdam 1877: 127 (1879); Reynolds, Austrobaileya 2: 53 (1984), reg. rev.; Welzen, Blumea 36: 91 (1991), reg. rev.

Falsely polygamous shrubs or trees. Indumentum of simple hairs and red glandular hairs. Leaves alternate, paripinnate; leaflets 1–5 pairs, entire; distal leaflet rudimentary. Inflorescences axillary or terminal, simple or thyrses. Flowers actinomorphic, functionally unisexual; sepals 5, connate at base, valvate; petals 5, clawed, with 2 dissected or crenate appendages above the claw, these sometimes crested; disk annular; stamens 8; pollen syncolporate or parasyncolporate, rugulate to rugulate-reticulate; ovary 3-carpellate, with a single ovule per carpel; style filiform. Fruit a 3-locular, loculicidal, leathery-woody, capsule, each locule with a narrow, dorsal wing. Seeds completely covered by dorsally opened arillode with a basal funiculus-like appendage, the testa papery.

Twelve to 13 species, from eastern Australia, Moluccas, and New Guinea.

117. *Sarcotoechia* Radlk.

Sarcotoechia Radlk., Sitzungsber. Math.-Phys. Cl. Königl. Bayer. Akad. Wiss. München 9: 501 (1879); Leenhouts, Blumea 33: 198 (1988).

Falsely polygamous trees. Leaves alternate, paripinnate or unifoliolate; leaflets serrate or entire; distal leaflets rudimentary. Inflorescences axillary, or ramiflorous thyrses. Flowers actinomorphic; sepals 5, distinct, slightly imbricate; petals 5, shorter than the sepals, with a pair of seemingly marginal appendages; stamens (5–)7(8); pollen parasyncolporate, rugulate; ovary 2–3-carpellate, with a single ovule per carpel; style apical; stigma slightly lobed. Fruit a 2–3-locular, 2–3-lobed, fleshy, loculicidal capsule. Seeds with a cupular or reduced sarcotesta.

About 11 species from Australia (northern Queensland), Papua New Guinea, and Moluccas.

118. *Schleichera* Willd., nom. cons.

Schleichera Willd., Sp. Pl. 4, 2: 1096 (1806); Leenhouts in Fl. Males. I, 11: 727 (1994).

Falsely polygamous trees, with glandular indument. Leaves alternate, paripinnate; leaflets entire; distal leaflet rudimentary. Inflorescences axillary, racemose or thyrsoid. Flowers actinomorphic, functionally unisexual; sepals 4–6, equal, valvate, connate at base; petals 0; disk annular, dish-shaped; stamens 5–9; pollen parasyncolporate, striate; ovary 2–4-carpellate, with a single ovule per carpel; stigma lobed. Fruits 1-locular, indehiscent, coriaceous, usually with a spiny pericarp. Seed completely covered by an arillode. $2n=30$, 32.

A single species, *S. oleosa* (Lour.) Oken, distributed from Sri Lanka and India to Indochina, Malesia east to the Moluccas and Lesser Sunda Islands. The seeds are the source of macassar oil widely used as hair ointment; the wood is used as construction material.

119. *Scyphonychium* Radlk.

Scyphonychium Radlk., Sitzungsber. Math.-Phys. Cl. Königl. Bayer. Akad. Wiss. München 9: 473, 495, 519 (1879); Ferrucci, Bonplandia 6: 117–124 (1989).

Falsely polygamous trees. Leaves alternate, paripinnate; leaflets entire; distal leaflet rudimentary. Inflorescences terminal thyrsoids. Flowers functionally unisexual; calyx zygomorphic, sepals 5, distinct, imbricate, the outer 2 smaller; petals 5, clawed, twice as long as the sepals, with a bilobed minute appendage forming a pocket above the claw; disk annular, cup-shaped, 5-lobed; stamens 8; pollen colporate, perforate; ovary 2-carpellate, with a single ovule per carpel; stigma shortly bifid. Fruit, (1)2-coccate, woody, schizocarpic. Seeds exarillate.

A single species, *S. multiflorum* Radlk., native to northern and eastern Brazil, and French Guiana.

120. *Sinoradlkofera* F. Meyer

Sinoradlkofera F. Meyer, J. Arnold Arb. 58: 183 (1977).
Boniodendron Gagnep., nom. inval.; Leenhouts, Blumea 28: 45 (1982).

Falsely polygamous trees. Leaves paripinnate; leaflets serrate; distal leaflet rudimentary. Inflorescences terminal thyrses. Flowers sub-actinomorphic; sepals 5, valvate; petals 5, white, clawed, appendages marginal or 0; disk annular; stamens 8, geniculate in bud; pollen colporate, striate; ovary 3-carpellate, with 2 ovules per carpel; style subulate. Fruits 3-locular, inflated, loculicidal capsules. Seeds exarillate.

A single species, *S. minor* (Hemsley) F. Meyer, native to China.

121. ***Sisyrolepis*** Radlk.

Sisyrolepis Radlk. in F. N. Williams, Bull. Herb. Boiss. II, 5: 222 (1905); Leenhouts, Blumea 23: 336 (1977); Welzen in Santisuk & Larsen, Fl. Thailand 7: 243 (1999).
Delpya Pierre ex Radlk. (1910).

Falsely polygamous shrubs or trees. Leaves alternate, paripinnate; leaflets crenate; distal leaflet rudimentary. Inflorescences axillary thyrses. Flowers zygomorphic; sepals 5, distinct, imbricate; petals 4(5), with a pair of marginal appendages; disk semi-annular, lobulate; stamens 8(9); pollen syncolporate, finely striate; ovary 3-carpellate, with a single ovule per carpel; stigma not lobed. Fruit a 3-locular, echinate, loculicidal capsule. Seeds exarillate.

A single species, *S. muricata* (Pierre) Leenh., Thailand and Cambodia.

122. ***Smelophyllum*** Radlk.

Smelophyllum Radlk., Sitzungsber. Math.-Phys. Cl. Königl. Bayer. Akad. Wiss. München 8: 330 (1878).

Falsely polygamous shrubs. Leaves alternate, paripinnate; leaflets 3–4 pairs; distal leaflet rudimentary. Inflorescences axillary thyrses. Flowers actinomorphic; calyx cupular, sepals 5, connate, imbricate; petals 5, without appendages; disk annular, 5-lobed; stamens 8; pollen colporate, striate; ovary (2)3-carpellate, with a single ovule per carpel; style thickened. Fruit a (1)2(3)-coccate, chartaceous capsule. Seeds exarillate.

A single species endemic to South Africa.

123. ***Stadmania*** Lam.

Stadmania Lam., Tabl. Encycl. 2: 443 (1793); reg. rev.; Capuron, Mem. Mus. Natl. Hist. Nat. B, Bot. II, 19: 151–160 (1969).
Pseudolitchi Dand. & Choux (1926).

Dioecious or monoecious trees. Leaves alternate, paripinnate; leaflets (1)3–5 pairs, serrate; distal leaflet rudimentary. Inflorescences terminal or axillary, racemose thyrses. Flowers actinomorphic, functionally unisexual; calyx cupular, sepals 5, valvate; petals 5, sometimes clawed, 0 in one species; appendages simple, crested or dissected, basal or above the claw; disk annular to 5-crenate; stamens (6–)8; pollen colporate, striate or perforate, sometimes irregularly striate or rugulate; ovary 3-carpellate, with a single ovule per carpel; style short. Fruit a 1(3)-coccate, indehiscent. Seeds covered with a translucent arillode.

Six species from Mauritius, Madagascar, and eastern tropical Africa.

124. ***Stocksia*** Benth.

Stocksia Benth., Hooker's J. Bot. Kew Gard. Misc. 5: 304 (1853).

Polygamous spiny shrubs. Leaves simple, fasciculate on short branches axillary to the spines; stipules 0. Inflorescences axillary, racemose thyrses. Flowers zygomorphic, functionally unisexual; sepals 5, imbricate, distinct, the outer smaller; petals 5, clawed, without appendages; disk annular-lobed; stamens 7–8; pollen colporate, striate; ovary 3-carpellate, with 2 ovules per carpel; style distal; stigma minute. Fruit a 3-locular, membranous, inflated, loculicidal capsule. Seeds exarillate.

A single species, *S. brahuica* Benth., from Persia and Afghanistan.

125. ***Storthocalyx*** Radlk.

Storthocalyx Radlk., Sitzungsber. Math.-Phys. Cl. Königl. Bayer. Akad. Wiss. München 9: 499, 660 (1879).

Falsely polygamous shrubs or trees. Leaves alternate, paripinnate, with 2–10 leaflets; distal leaflet rudimentary; stipules 0. Inflorescences axillary, spicate or thyrsoid. Flowers actinomorphic, functionally unisexual; sepals 5, distinct, narrowly imbricate or valvate; petals 5, without appendages; disk annular; stamens 8; pollen syncolporate, perforate; ovary 3-carpellate, with a single ovule per carpel; style subulate; stigma minute. Fruit a 3-locular, trigonous or pyriform, corticose-woody loculicidal capsule. Seeds with a ventrally split, fimbriate arillode.

Four species from New Caledonia.

126. ***Synima*** Radlk.

Synima Radlk., Sitzungsber. Math.-Phys. Cl. Königl. Bayer. Akad. Wiss. Münch. 9: 501 (1879); Leenhouts & Adema in Fl. Males. I, 11: 730 (1994).

Monoecious trees. Leaves alternate, paripinnate; leaflets crenate-denticulate, without domatia or glands; distal leaflet rudimentary. Inflorescences axillary, together sometimes pseudo-terminal. Flowers actinomorphic; calyx shortly cupular, sepals 5, short, distinct, narrowly imbricate; petals 5, rhomboidal, with a pair of recurved, woolly or ciliate appendages, usually distinctly crested; disk annular; stamens 8; pollen parasyncolporate, rugulate-reticulate; ovary 3-carpellate, with a single ovule per carpel; style apical; stigma slightly lobed. Fruit a 3-locular, trigonous, loculicidal capsule. Seeds with basal or dorsal sarcotesta.

Two species; Australia (N Queensland) and southeast New Guinea. Lowland and montane rain forest, mossy oak forest, up to 900 m.

127. *Thouinidium* Radlk.

Thouinidium Radlk., Sitzungsber. Math.-Phys. Cl. Königl. Bayer. Akad. Wiss. München 8: 284 (1878).

Falsely polygamous trees. Leaves alternate, paripinnate; leaflets 2–8 pairs, serrate or entire; distal leaflet rudimentary; stipules 0. Inflorescences terminal thyrses. Flowers functionally unisexual; calyx actinomorphic or slightly zygomorphic, sepals 5, distinct, imbricate; petals (4)5, clawed, with an appendage above the claw forming a pocket; disk cup-shaped; stamens 6–8(–10); pollen colporate, perforate; ovary 3-carpellate, with a single ovule per carpel; style short, subulate. Fruit schizocarpic, splitting into 3, distally winged, samaroid mericarps. Seeds exarillate.

Six species from Mexico, Central America, and some islands in the Greater Antilles.

128. *Tina* Schult., nom. cons. prop.

Tina Schult. in Roem. & Schult., Syst. Veg. 5: XXXII, 414 (1819–1820) ('1819').
Gelonium Gaertn., Fruct. Sem. Pl. 2: 271, fig. 139 n. 8 (1791), nom. rej. prop.

Falsely polygamous shrubs or trees. Leaves alternate, paripinnate, with 2–6 pairs of leaflets; leaflets serrate or crenate-serrate; distal leaflet rudimentary; stipules 0. Inflorescences axillary thyrses. Flowers functionally unisexual; calyx actinomorphic or zygomorphic, sepals (3–)5, distinct, imbricate, in two series; petals 5, with 2 marginal appendages; disk annular; stamens (6–)8; pollen syncolporate or parasyncolporate, rugulate; ovary 2(3)-carpellate, with a single ovule per carpel; style subulate; stigma a invaginate prolongation of the style. Fruit a 2-locular, loculicidal capsule. Seeds arillate.

Six species from Madagascar.

129. *Tinopsis* Radlk.

Tinopsis Radlk. in T. Durand, Index Gen. Phan.: 78 (1887) [1888].
Bemarivea Choux (1927).

Falsely polygamous-dioecious trees. Leaves alternate, paripinnate, with 2–4 pairs of leaflets; distal leaflet rudimentary; stipules 0. Inflorescences axillary or terminal thyrses. Flowers actinomorphic, functionally unisexual; calyx slightly zygomorphic, the sepals 5, imbricate, distinct; petals 5, with a single appendage forming a pocket or with 2 marginal appendages; disk annular, slightly sulcate; stamens 5(7); pollen colporate, rugulate to rugulate-reticulate; ovary 2-carpellate, with a single ovule per carpel. Fruit a 2-locular, indehiscent or less often tardily and incompletely dehiscent capsule, usually 1-seeded. Seed arillate.

Eleven species from Madagascar.

130. *Toechima* Radlk.

Toechima Radlk., Actes Congr. Bot. Amsterdam 1877: 130 (1879); Reynolds, Austrobaileya 2: 176 (1985); Leenhouts, Blumea 33: 203 (1988); Leenhouts in Fl. Males. I, 11: 732 (1994).

Falsely polygamous trees. Leaves alternate, paripinnate; leaflets entire to serrate. Inflorescences axillary thyrses. Flowers actinomorphic; sepals 5, equal, valvate to narrowly imbricate; petals 5, clawed, with a single crested appendage; disk annular; stamens 8; pollen parasyncolporate, rugulate to rugulate-reticulate; ovary 2–3-carpellate, with a single ovule per carpel; stigma lobed. Fruit a 2–3-locular, fleshy, loculicidal capsule. Seed with a basal placental arillode.

About eight species distributed in Australia and New Guinea.

131. *Toulicia* Aublet

Toulicia Aubl., Hist. Pl. Guiane 1: 359 (1775).

Falsely polygamous-dioecious, small, unbranched trees. Leaves imparipinnate; leaflets usually falcate, opposite or alternate; distal leaflet rudimentary.

Inflorescences terminal or axillary thyrses. Flowers zygomorphic, staminate or pistillate; sepals 5, unequal, imbricate; petals 4, with a petaloid, bifid ventral appendage or with 2 marginal appendages; disk unilateral, semi-annular; stamens 8; pollen colporate, perforate; ovary 3-carpellate, with a single ovule per carpel; style with 3 stigmatic branches. Fruits schizocarpic, splitting into three, samaroid mericarps, each with a proximal wing and a papery, inflated locule. Seeds exarillate.

About 12 species from the lowlands of South America.

132. ***Trigonachras*** Radlk.

Trigonachras Radlk., Actes Congr. Bot. Amsterdam 1877: 116 (1879); Leenhouts, Blumea 33: 204 (1988).

Falsely polygamous trees. Leaves alternate, paripinnate; leaflets entire, often with glands below; distal leaflet rudimentary. Inflorescences axillary or terminal thyrses or panicles. Flowers actinomorphic, bisexual or unisexual; sepals 5, distinct, narrowly imbricate (sub)equal; petals 5, clawed, with (1)2 appendages above the claw; disk annular; stamens (7)8(9); pollen syncolporate or parasyncolporate, perforate to scabrate with often finely tapering scabrae; ovary 3-carpellate, with a single ovule per carpel; style with 3 stigmatic lines. Fruit a 3-locular, fleshy, loculicidal capsule. Seeds exarillate.

About eight species, occurring throughout Malesia, but absent from Java and the Lesser Sunda Islands.

133. ***Tripterodendron*** Radlk.

Tripterodendron Radlk., Sitzungsber. Math.-Phys. Cl. Königl. Bayer. Akad. Wiss. München 20: 290 (1891).

Falsely polygamous-dioecious trees. Leaves alternate, tripinnate; leaflets denticulate or serrate; distal leaflet rudimentary; stipules 0. Inflorescences axillary thyrses. Flowers actinomorphic, bisexual or functionally unisexual; sepals 5–6, valvate; petals clawed, 5(6–8), with a pair of appendages above the claw; disk annular; stamens 8; pollen colporate, striate; ovary 2-carpellate, with a single ovule per carpel; style short, with a marginal stigmatic line. Fruit a 2-locular, fleshy, loculicidal capsule. Seeds arillate.

A single species endemic to east-central Brazil.

134. ***Tristira*** Radlk.

Tristira Radlk., Actes Congr. Bot. Amsterdam 1877: 133 (1879); Leenhouts in Fl. Males. I, 11: 740 (1994).

Falsely polygamous trees. Leaves alternate, paripinnate; leaflets opposite or alternate, entire or dentate; distal leaflets rudimentary. Inflorescences terminal or axillary thyrses. Flowers bisexual or functionally unisexual; calyx zygomorphic, sepals 5, distinct, imbricate, the outer two smaller; petals 0; disk annular, lobed; stamens 8(9); pollen parasyncolporate, rugulate; ovary 3-carpellate, with a single ovule per carpel; style subulate, with 3 stigmatic lines. Fruits 3-locular, indehiscent drupes, with slightly fleshy exocarp and stony endocarp, each locule with a dorsal wing. Seeds exarillate.

A single species, *T. triptera* (Blanco) Radlk., eastern Philippines, Celebes, and Moluccas.

135. ***Tristiropsis*** Radlk.

Tristiropsis Radlk. in T. Durand, Index Gen. Phan.: 76 (1887)[1888]; Leenhouts in Fl. Males. I, 11: 742 (1994), rev. *Palaoea* Kanehira (1935).

Falsely polygamous trees. Leaves alternate, bipinnate; leaflets entire; distal leaflet rudimentary. Inflorescences axillary thyrses. Flowers bisexual or functionally unisexual; calyx zygomorphic, sepals 5, imbricate, the outer 2 smaller; petals 0 or 5, the appendages either forming a pocket, or a folded marginal outgrowth; disk annular; stamens 8(–13); pollen parasyncolporate, rugulate; ovary 3(–5)-carpellate, with a single ovule per carpel; stigma not lobed, grooved. Fruit a (2)3-locular, indehiscent drupe, with slightly fleshy exocarp and stony endocarp; seeds exarillate.

Three species from Borneo, the Philippines and throughout eastern Malesia to northeastern Australia, the Solomons, Marianas, and Christmas Islands in the Pacific.

136. ***Tsingya*** Capuron

Tsingya Capuron, Mem. Mus. Natl. Hist. Nat., B, Bot. II, 19: 104 (1969).

Monoecious trees. Leaves alternate, paripinnate; leaflets entire; distal leaflet rudimentary; stipules 0. Inflorescences axillary racemose thyrses. Flowers actinomorphic, functionally unisexual; sepals 5, valvate; petals 0; disk annular, pulvinate; stamens

8–10; pollen colporate, striate; ovary 3-carpellate, with a single ovule per carpel; style with 3 stigmatic lines. Fruit (immature) unilocular by abortion. Seed one per locule, arillate, with long ventral hilum.

A single species, *T. bemarana* Capuron, endemic to Madagascar.

137. *Ungnadia* Endl.

Ungnadia Endl., Atakta Bot. t. 36 (1835) [1833].

Falsely polygamous shrubs or trees. Leaves alternate, paripinnate; leaflets serrate; distal leaflet fully developed; stipules 0. Inflorescences axillary or ramiflorous, pseudo-umbelliform. Flowers zygomorphic, bisexual or functionally unisexual; sepals 5, imbricate, 3 distinct, 2 connate; petals 4–5, clawed, with a tuft of filiform appendages above the claw; disk unilateral, undulate, with androgynophore; stamens (7)8(–10); pollen colporate, finely striate; ovary 3-carpellate, stipitate, with 2 ovules per carpel; style filiform with punctiform stigma. Fruit a 3-locular, 1–2-seeded, loculicidal, coriaceous capsule. Seeds exarillate, with large white hilum. $2n=32$.

A single species, *U. speciosa* Endl., Mexico and southern United States (Texas).

138. *Vouarana* Aubl. Fig. 78

Vouarana Aubl., Pl. Guiane 2: (Suppl.) 12, fig. 374 (1775).

Falsely polygamous medium-sized trees. Leaves alternate, paripinnate; leaflets entire; distal leaflet rudimentary. Inflorescences axillary or terminal thyrses. Flowers 4–5-merous, actinomorphic or zygomorphic, bisexual or unisexual; sepals distinct, unequal, concave, imbricate; petals rhombic, shorter than the sepals, with two marginal appendages; disk annular and lobed; stamens 6–8; pollen parasyncolporate, rugulate; ovary 2-carpellate, with a single, basal ovule per carpel; stigma subulate and papillose. Fruit a 1–2-seeded, woody, loculicidal capsule. Seeds ellipsoid with a large basal arillode.

Two species from Costa Rica to northern Brazil.

139. *Xerospermum* Blume

Xerospermum Blume, Rumphia 3: 99 (1847); Leenhouts, Blumea 28: 389 (1983), rev.

Falsely polygamous or dioecious trees. Leaves alternate, paripinnate; leaflets entire, abaxially with flat, orbicular glands toward proximal portion of blade; distal leaflet rudimentary. Inflorescences axillary thyrses. Flowers actinomorphic, bisexual or functionally unisexual; sepals 4–5, ± equal, imbricate; petals 4–5, sessile to clawed, without appendages; disk annular or semi-annular; stamens 7–9; pollen colporate, striate to psilate, sometimes irregularly striate or rugulate; ovary 2 (3)-carpellate, with a single ovule per carpel; stigma lobed. Fruit indehiscent, 1–2-coccate, coriaceous to woody, granular to shortly spiny. Seed with complete, thin, sarcotesta. $2n = 32$.

Two species from Bangladesh, Indochina and western Malesia. *X. noronhianum* Blume is dioecious.

140. *Zollingeria* Kurz, nom. cons.

Zollingeria Kurz, J. Asiat. Soc. Bengal, 41, 2: 303 (1872), non Schultz.-Bip. (1854), nom. rej.; Adema, Blumea 37: 73 (1992); Welzen in Fl. Thailand 7: 248 (1999).

Falsely polygamous trees. Leaves alternate, paripinnate; leaflets entire; distal leaflet rudimentary. Inflorescences axillary thyrses. Flowers actinomorphic or zygomorphic; sepals 5, distinct, slightly to distinctly unequal, usually imbricate; petals 4–5, with or without appendages; disk annular or semi-annular; stamens 8; pollen colporate, irregularly striate; ovary 3-carpellate, unilocular, with a single ovule per carpel; stigma 3-lobed, or 3 stigmatic lines. Fruit 1-locular, 3-winged, coriaceous, indehiscent. Seed exarillate, flattened, elongated.

Three or four species from Myanmar, Laos, Thailand, and Borneo.

Insufficiently Known Genus:

141. *Chonopetalum* Radlk.

Chonopetalum Radlk., Bot. Jahrb. Syst. 56: 258 (1920).

Falsely polygamous trees. Leaves alternate, paripinnate; distal leaflet rudimentary; stipules 0. Inflorescences axillary thyrses. Flowers actinomorphic, functionally staminate or pistillate; calyx cup-shaped, sepals 5, distinct, imbricate; petals 5, with a single basal appendage as long as the petal; disk annular; stamens 8; pollen unknown; ovary 3-carpellate (as inferred from the pistillodes; pistillate flower not known). Fruit not known.

One species, *Ch. stenodictyum* Radlk., known only from the type collection from Equatorial Guinea, Africa.

Doubtful Genus:

Hirania Thulin

Hirania Thulin, Nord. J. Bot. 24(5): 510 (2007).

Dioecious ? shrubs. Leaves simple, alternate on long shoots, or congested in short, lateral shoots; stipules 0. Inflorescences terminal thyrses. Flowers zygomorphic; sepals 5, partially imbricate; petals 4, clawed, without appendages, pink; disk unilateral, of two erect segments; stamens 8.

A single species recently described from Somalia based on a specimen bearing only pistillate flowers.

We have not seen any material of this putative Sapindaceous genus, and since the original description portrays the floral disk as intrastaminal, we wonder if it really belongs in Sapindaceae or not. Additional material would be necessary to place this genus in the right family.

Selected Bibliography

Acevedo-Rodríguez, P. 1990. The occurrence of piscicides and stupefactants in the plant kingdom. Adv. Econ. Bot. 8: 1–23.

Acevedo-Rodríguez, P. 1993. Systematics of *Serjania* (Sapindaceae). Part I: A revision of *Serjania* Sect. Platycoccus. Mem. New York Bot. Gard. 67: 1–93.

Acevedo-Rodríguez, P. 1996. Flora of St. John. Mem. New York Bot. Gard. 78.

Acevedo-Rodríguez, P. 1997. Novelties of Neotropical Sapindaceae I. BioLlania 6: 143–151.

Acevedo-Rodríguez, P. 2003. Melicocceae (Sapindaceae): *Melicoccus* and *Talisia*. Flora Neotropica 87: 1–179.

Adema, F., Leenhouts, P.W., van Welzen, P.C. 1994. Sapindaceae. In: Flora Malesiana I, 11(3): 419–768.

Ahuja, M.R., Natarajan, A.T. 1957. Chromosome number of some common plants. Curr. Sci. 26: 117.

Baillon, H. 1874. Histoire des plantes. Paris.

Bawa, K.S. 1976. The reproductive biology of *Cupania guatemalensis* Radlkofer (Sapindaceae). Evolution 31: 52–63.

Beck, H.T. 1990. A survey of the useful species of *Paullinia* L. (Sapindaceae). Adv. Econ. Bot. 8: 41–56.

Beck, H.T. 1991. The taxonomy and economic botany of the cultivated guarana and its wild relatives and the generic limits within the Paullinieae (Sapindaceae). Ph.D. Dissertation, City University of New York, 510 pp.

Bentham, G., Hooker, J.D. 1862. Genera plantarum, vol. 1. London: Reeve.

Bergen, M.A. van, van der Ham, R.W.J.M., Turner, H. 1995. Morphology and evolution of *Arytera* pollen (Sapindaceae-Cupanieae). Blumea 40: 195–209.

Blume, C.L. 1878. Rumphia. Leiden.

Bowden, W.M. 1945. A list of chromosome numbers in higher plants. II. Menispermaceae to Verbenaceae. Am. J. Bot. 32: 191–200.

Budantsev, L.Y. 1983. History of the arctic flora of the early Cenophytic epoch (in Russian). Leningrad: Nauka.

Buerki, S., Forest, F., Acevedo-Rodríguez, P., Callmander, M.W., Nylander, J.A.A., Harrington, M., Sanmartin, I., Küpfer, P., Alvarez, N. 2009. Plastid and nuclear DNA markers reveal intricate relationships at subfamilial and tribal levels in the soapberry family (Sapindaceae). Mol. Phylogen. Evol. 51: 238–258.

Burnett, G.T. 1835. Outlines of botany. 2 vols. London.

Cambessèdes, J. 1829. Mémoire sur la famille des Sapindacées. Mém. Mus. Hist. Nat. 18: 1–50.

Candolle, A.P. de, 1824. Sapindaceae. In: Prodromus Systematis Naturalis Regni Vegetabilis 1. Paris: Treuttel & Würtz, pp. 601–618.

Capuron, R. 1969. Révision des Sapindacées de Madagascar et des Comores. Mém. Mus. Natl. Hist. Nat. B, Bot. II, 19: 1–189.

Chaudhuri, K. 1940. A note on the morphology and chromosome number of *Litchi chinensis* Sonner. Curr. Sci. 9: 416.

Cookson, I.C., Pike, K.M. 1954. Some dicotyledonous pollen types from Cainozoic deposits in the Australian region. Austr. J. Bot. 7: 197–219.

Cronquist, A. 1981. An integrated system of classification of flowering plants. New York: Columbia University Press.

Davis, G.L. 1966. Systematic embryology of the angiosperms. New York: Wiley.

Don, G. 1831. A general system of gardening and botany. London.

Erwin, D.M., Stockey, R.A. 1990. Sapindaceous flowers from the Middle Eocene (Allenby Fm.) of British Columbia, Canada. Can. J. Bot. 68: 2025–2034.

Ferrucci, M.S. 1981. Recuentos cromosomicos en Sapindáceas. Bonplandia 11: 73–81.

Ferrucci, M.S. 1985. Recuentos cromosomicos en *Allophylus* y *Serjania* (Sapindaceae). Bol. Soc. Bot. Argent. 24: 200–202.

Ferrucci, M.S., Anzótegui, L.M. 1993. El polen de Paullinieae (Sapindaceae). Bonplandia 6: 211–243.

Forest, F., Drouin, J.N., Charest, R., Brouillet, L., Bruneau, A. 2001. A morphological phylogenetic analysis of *Aesculus* L. and *Billia* Peyr. (Sapindaceae). Can. J. Bot. 79: 154–169.

Fouilloy, R., Hallé, N. 1973a. Sapindaceae. In: Aubreville, A., Leroy, J.F. (eds.) Flore du Cameroun. Paris: Muséum National d'Histoire Naturelle.

Fouilloy, R., Hallé, N. 1973b. Sapindaceae. In: Aubreville, A., Leroy, J.F. (eds.) Flore du Gabon. Paris: Muséum National d'Histoire Naturelle.

Fritsch, R. 1970. Chromosomenzahlen von Pflanzen der Insel Kuba I. Kulturpfl. 18: 194.

Gadek, P.A., Fernando, E.S., Quinn, C.J., Hoot, S.B., Terrazas, T., Sheahan, M.C., Chase, M.W. 1996. Sapindales: molecular delimitations and infraordinal groups. Am. J. Bot. 83: 802–811.
Guervin, C. 1961a. Contribution à l'étude cyto-taxinomique des Sapindacées et caryologiques des Melianthacées et des Didiereacées. Rev. Cyt. Biol. Veg. 23: 4–87.
Ham, R.W.J.M. van der 1990. Nephelieae pollen (Sapindaceae): form, function, and evolution. Leiden: Rijksherbarium/Hortus Botanicus.
Hardin, J.W. 1957. A revision of the American Hippocastanaceae 1. Brittonia 9: 145–171.
Hardin, J.W. 1957. A revision of the American Hippocastanaceae 2. Brittonia 9: 173–195.
Harrington, M.G., Edwards, K.J., Johnson, S.A., Chase, M.W., Gadek, P.A., 2005. Phylogenetic inference in Sapindaceae sensu lato using plastid *mat*K and *rbc*L DNA sequences. Syst. Bot. 30: 366–382.
Harris, A.J., Xiang, Qiu-Jun, Thomas, D.T. 2009. Phylogeny, origin, and biogeographic history of *Aesculus* L. (Sapindales), an update from combined analysis of DNA sequences, morphology, and fossils. Taxon 58: 108–126.
Harvey, W.H., Sonder, O.W. 1894. Flora Capensis. Vol. 1. London: Reeve.
Hegnauer, R. 1973, 1990. Chemotaxonomie der Pflanzen, Bd. 6, Bd. 9. Basel: Birkhäuser.
Hesse, M. 1979. Ultrastruktur und Verteilung des Pollenkitts in der insekten- und windblütigen Gattung *Acer* (Aceraceae). Plant Syst. Evol. 131: 277–289.
Hickey, L.J. 1979. A revised classification of the architecture of dicotyledonous leaves. In: Metcalfe, C., Chalk, L. (eds.) Anatomy of the Dicotyledons, vol. 1, 2nd ed. Oxford: Clarendon Press, pp. 25–39.
Johnson, S., Chase, M.W. 1999. Cladrogram based on unpublished data on the chloroplast rbcL gene; pag. 189. In: Klaassen, R. (ed.) Wood anatomy of Sapindaceae. IAWA J. Suppl. 2: 1–214.
Johri, B.M., Ambegoakar, K.B., Srivastava, P.S. 1992. Comparative embryology of angiosperms. 2 vols. Berlin, Heidelberg, New York: Springer.
Jong, P.C. de 1976. Flowering and sex expression in *Acer* L. (a biosystematic study). Meded. Landbouwhogesch. Wageningen 72-2: 1–201.
Judd, W.S., Sanders, R.W., Donoghue, M.J. 1994. Angiosperm family pairs: preliminary analyses. Harvard Pap. Bot. 5: 1–51.
Judd, W.S., Campbell, C., Stevens; P. 1999. Plant systematics: a phylogenetic approach. Sunderland, MA: Sinauer.
Kadry, A. 1951. Chromosome behaviour in *Cardiospermum halicacabum* L. Svensk Bot. Tidskr. 45: 414–416.
Klaassen, R.K.W.M. 1999. Wood anatomy of Sapindaceae. IAWA J. Suppl. 2: 1–214.
Knobloch, E., Mai, D.H. 1986. Monographie der Früchte und Samen in der Kreide von Mitteleuropa. Prag: Vidal Ústredni ústav geologický.
Leenhouts, P.W. 1978. Systematic notes on the Sapindaceae-Nephelieae. Blumea 24: 395–403.
Leinfellner, W. 1958. Über die peltaten Kronblätter der Sapindaceen. Österr. bot. Z. 105: 443–514.
Lindley, J. 1862. The vegetable kingdom. London.
Manchester, S.R. 1999. Biogeographical relationships of North American Tertiary floras. Ann. Missouri Bot. Gard. 86: 472–522.
Manchester, S.R. 2001. Leaves and fruits of *Aesculus* (Sapindales) from the Paleocene of North America. Int. J. Plant Sci. 162: 985–998.
Mangenot, S., Mangenot, G. 1957. Nombres chromosomiques nouveaux chez diverses dicotylédones et monocotylédones d'Afrique occidentale. Bull. Jard. Bot. État Brux. 27: 639–654.
Mangenot, S., Mangenot, G. 1958. Deuxième liste de nombres chromosomiques nouveaux chez diverses dicotylédones et monocotylédones d'Afrique occidentale. Bull. Jard. Bot. État Brux. 28: 315–329.
McClain, A.M., Manchester, S.R. 2001. *Dipteronia* (Sapindaceae) from the Tertiary of North America and implications for the phytogeographic history of the Aceroideae. Am. J. Bot. 88: 1316–1325.
Mennega, A.M. 1972. Wood structure of the genus *Talisia* (Sapindaceae). Acta Bot. Neerl. 21: 578–586.
Metcalfe, C.R., Chalk, L. 1957. Anatomy of Dicotyledons. Oxford: Clarendon Press.
Muller, J. 1981. Fossil pollen records of extant angiosperms. Bot. Rev. 47: 1–142.
Muller, J., Leenhouts, P.W. 1976. A general survey of pollen types in Sapindaceae in relation to taxonomy. In: Ferguson, I.K., Muller, J. (eds.) The evolutionary significance of the exine. Linnean Soc. Symp. Ser. 1, pp. 407–445. London: Academic Press.
Pax, F. 1902. Aceraceae. In: Engler, A. (ed.) Das Pflanzenreich IV, 163, pp. 1–89. Leipzig: W. Engelmann.
Radlkofer, L. 1888. Sapindaceae. In: Durand, T., Index generum phanerogamorum, pp. 71–83. Bruxell.
Radlkofer, L. 1890. Über die Gliederung der Familie der Sapindaceen. Sitzungsber. Math.-Phys. Cl. königl. bayer. Akad. Wiss. München 20: 105–379.
Radlkofer, L. 1931–1934. Sapindaceae. In: Engler, A. (ed.) Das Pflanzenreich IV, 165, Heft 98a-h, pp. 1–1539. Leipzig: W. Engelmann.
Ramirez, D.A. 1961. Cytology of Philippine plants. VII. *Nephelium lappaceum* Linn. Philipp. Agric. 45: 340–342.
Rashford, J. 2001. Those that do not smile will kill me: the ethnobotany of the ackee in Jamaica. Econ. Bot. 55: 190–211.
Reichenbach, C.L. 1834. Das Pflanzenreich. Leipzig.
Reynolds, S.T., West, J.G.. 1985. Sapindaceae. In: George, A.S. (ed.) Flora of Australia 25: 5–164.
Ronse Decraene, L.P., Smets, E., Clinckemaillie, D. 2000. Floral ontogeny and anatomy in *Koelreuteria* with special emphasis on monosymmetry and septal cavities. Plant Syst. Evol. 223: 91–107.
Savolainen, V., Fay, M.F., Albach, D.C., Backlund, A., van der Bank, M., Cameron, K.M., Johnson, S.A., Lledo, M.D., Pintaud, J.-C., Powell, M., Sheahan, M.C., Soltis, D.E., Soltis, P.S., Weston, P., Whitten, W.M., Wurdak, K.J., Chase, M.W. 2000. Phylogeny of the eudicots: a nearly complete familial analysis based on rbcL gene sequence. Kew Bull. 55: 257–309.
Scholz, H. 1964. Sapindaceae. In: Melchior, H. A. Engler's Syllabus der Pflanzenfamilien, vol. 2, pp. 282–284. Berlin: Gebrüder Borntraeger.
Soltis, D.E., Soltis, P.S., Chase, M.W., Mort, M.E., Albach, D.C., Zanis, M., Savolainen, V., Hahn, W.H.,

Hoot, S.B., Fay, M.F., Axtell, M., Swensen, S.M., Prince, L.M., Kress, W.J., Nixon, K.C., Farris, J.S. 2000. Angiosperm phylogeny inferred from 18S rDNA, *rbc*L, and *atp*B sequences. Bot. J. Linn. Soc. 133: 381–461.

Subba Reddi, C., Reddi, E.U.B., Reddi, N.S., Reddi, P.S. 1983. Reproductive ecology of *Sapindus emarginatus* Vahl (Sapindaceae). Proc. Indian Natl. Sci. Acad. 49B (1): 57–72.

Takhtajan, A.L. (ed.) 1981. Flowering Plants, vol. 5 (2). Moscow: Proswyeschtschenye.

Takhtajan, A. 1997. Diversity and classification of flowering plants. New York: Columbia University Press.

Thorne, R.F., 1976. A phylogenetic classification of the Angiospermae. In: Hecht, M.K., Steere, W.C., Wallace, B. (eds.) Evolutionary Biology, vol. 9. New York: Plenum Press, pp. 35–106.

Thorne, R.F., Reveal, J.L. 2007. An updated classification of the class Magnoliosida ("Angiospermae"). Bot. Rev. 73: 67–182.

Turner, H., van der Ham, R.W.J.M. 1996. A taxonomic and pollen morphological revision of the genus *Gongrodiscus* (Sapindaceae). Bull. Mus. Natl. Hist. Nat. Paris IV, 18: 339–349.

Weberling, F. 1976. Die Pseudostipeln der Sapindaceae. Akad. Wiss. Lit. Mainz, Abh. Math.-Naturwiss. Kl. 2: 5–27.

Weckerle, C.S., Rutishauser, R. 2003. Comparative morphology and systematic position of *Averrhoidium* within Sapindaceae. Int. J. Plant Sci. 164: 775–792.

Weckerle, C.S., Rutishauser, R. 2005. Gynoecium, fruit and seed structure of Paullinieae (Sapindaceae). Bot. J. Linn. Soc. 147: 159–189.

Welzen, P.C. van 1989. *Guioa* Cav. (Sapindaceae): taxonomy, phylogeny, and historical biogeography. Leiden Bot. Ser. 12, 315 pp.

Welzen, P.C. van 1999. Euphorbiaceae. In: Santisuk, C., Larsen, K. (eds.) Flora of Thailand 7 (1): 169–250.

West, J.G. 1984. A revision of *Dodonaea* Miller (Sapindaceae) in Australia. Brunonia 7: 1–194.

Simaroubaceae

Simaroubaceae DC. (1811), nom. cons.
Leitneriaceae Benth. & Hook.f. (1880).

J.W. Clayton

Trees and shrubs, occasionally with thorns; pith conspicuous; triterpenoid compounds of the quassinoid type present throughout vegetative tissues. Leaves alternate, spirally arranged, estipulate (stipules found in *Picrasma*), pinnately compound or unifoliolate (rarely trifoliolate); leaflets entire, coarsely toothed, serrate or basally lobed, sometimes with conspicuous pitted or flattened glands beneath or above; venation pinnate, brochidodromus or occasionally reticulate. Hairs mostly simple, unicellular or multicellular, sometimes glandular-capitate. Inflorescences terminal or axillary determinate thyrses, sometimes appearing raceme-like, pseudo-umbellate, catkin-like or flowers clustered in leaf axils. Flowers perfect, polygamous or unisexual, actinomorphic, bracteate (bracts large and surrounding flowers in *Leitneria*); pedicels bracteolate, occasionally jointed; sepals 4–5 (0 in *Leitneria*), connate below, calyx sometimes splitting unevenly, occasionally bearing glands; petals 4–5(–8) (0 in *Leitneria*), distinct; stamens 4–10(–18), distinct; filaments often with hairy appendage; anthers dorsifixed, basifixed or versatile, dehiscing by 2 longitudinal slits, introrse (occasionally extrorse to latrorse); ovary superior, of (1)2–5 carpels, distinct or connate basally, occasionally connate axially and deeply lobed; placentation axile; ovule one per locule, anatropous; stylodia distinct or connate into a common style, occasionally absent; stigmas on stellately spreading stigmatic branches, or a single slightly lobed or capitate stigma; fruit with 1–5 samaroid or drupaceous mericarps; exocarp thin, fleshy, occasionally dry, nut-like, often carinate; endocarp reticulate or crustaceous; testa membranaceous, cotyledons planoconvex; endosperm mostly lacking.

A family of 22 genera and about 109 species, mainly tropical and subtropical but some temperate species.

Vegetative Morphology. The family is woody, composed of large trees up to 50 m high, shrubs, subshrubs, and occasionally suffructescent plants with all the leaves basal (*Simaba*). The wood is pithy or fistulous (Cronquist 1944d), making it lightweight, and the bark and twigs are often striated. The family is typified by a bitter taste to the bark and twigs, on account of quassinoid compounds in scattered secretory cells throughout the vegetative structures (Cronquist 1981). Thorns are present in *Castela* and in *Holacantha*, where they occur at the tips of all branches (Cronquist 1944d).

Leaves are predominantly once-pinnately compound, arranged spirally around cylindrical stems. Unifoliolate leaves have evolved multiple times (based on studies of character evolution; Clayton, unpubl. data), and are characteristic of *Castela*, *Leitneria*, *Amaroria* and *Samadera*, and are found in six species of *Soulamea* (Jaffré and Fambart 2002) and two species of *Simaba* (*S. monophylla* and *S. obovata*). The leaves of *Holacantha* are reduced to scales or absent entirely, except in the seedlings (Cronquist 1944d). Leaflets are alternate, subopposite or opposite, but always opposite in *Quassia*, which has a distinctive winged and jointed rachis. Leaflet shape is diverse, but strongly asymmetrical leaf bases are common in compound leaves. Leaf margins are predominantly entire, but are serrate or coarsely toothed in temperate species of *Ailanthus*, *Picrasma* and *Brucea*. Stipules are reported from *Picrasma* (Nooteboom 1962) and *Soulamea* (Stevens 2006), but have been shown to be pseudostipules, probably derived from the basal leaflets of a pinnate leaf (Weberling and Leenhouts 1966).

Vegetative Anatomy. Wood anatomy is described in detail by Webber (1936) and Record and Hess (1943; New World genera). Growth rings are present but indistinct, and diffuse porous or ring-porous (*Ailanthus*, *Leitneria*). Wood is dominated by fibre-tracheids, except in *Holacantha* and *Castela*, in which wood fibres are libriform (Webber 1936). Vessels have spiral thickenings in *Castela*, *Holacantha* and *Leitneria*, but these are rare or absent in the rest of the family. Normal wood parenchyma cells are sparse to moderately abundant, the cells often septate and crystalliferous (Webber 1936; Record and Hess 1943). Vertical secretory canals are common in stems (Spiekerkoetter 1924), and in *Leitneria* resin ducts are described as present in the margin of the pith (Record and Hess 1943). Nodes are tri- or multilacunar, and calcium oxalate crystals are present in parenchymatous tissues (Cronquist 1981). Flattish or concave glands are common on leaf surfaces, typically towards the margin, and often associated with teeth if present. Multicellular secretory glands are found on the abaxial surface of the sepals of *Samadera* (Nair and Joseph 1957). Primarily unicellular, but also multicellular and glandular hairs are common on the inflorescence axes and floral organs (Nair and Joseph 1957; Nair and Joshi 1958; Nair and Sukumaran 1960; Nooteboom 1962).

Inflorescence Structure. Inflorescences can be axillary or terminal, and are determinate thyrses, with the dichasia often appearing fasciculate or reduced to a single flower, giving the appearance of a panicle (sensu Weberling 1989). Thyrses vary between open and spreading (e.g. *Ailanthus*, *Eurycoma*, *Picrolemma*), and narrow, elongate and sparsely branched (e.g. *Brucea*, *Soulamea*, *Amaroria*). In *Simarouba* and *Picrolemma* the staminate thyrses are larger and have more flowers than the carpellate thyrses (Cronquist 1944b). *Picrasma* has a short, broad, rounded thyrse with a long peduncle (often described as a cyme), and in *Samadera* the inflorescence axis is condensed to form a pseudo-umbel (Nair and Joseph 1957). *Quassia amara* has a distinctive long raceme-like thyrse, occasionally branched at the base, and in *Castela* the flowers are occasionally solitary or clustered in the leaf axils (Cronquist 1944d), as in some *Samadera*. In *Leitneria* the inflorescence is an erect or occasionally pendulous catkin-like thyrse: in the staminate inflorescence the flowers cluster in cymules of three in the axils of large, spirally arranged bracts (Fig. 90E, F); in the carpellate inflorescence the flowers are solitary in the bract axils (Fig. 90I, J; Abbe and Earle 1940).

Flower Structure. Flowers in Simaroubaceae are small, actinomorphic, open and 4- or 5-merous (3-merous in *Soulamea*), with an intrastaminal nectary disk. Petals are usually red, pink, yellow, pale green or white. Unlike the majority of the family, *Quassia* has flowers with elongate, glabrous petals (sometimes with hairs at the base) that are coherent, forming a tube, and the stamens and style are exserted. *Leitneria* is unusual in having asepalous and apetalous flowers, although Abbe and Earle (1940) observed vestigial perianth structures in carpellate flowers. *Leitneria* also has a unicarpellate gynoecium, with vascular bundles suggesting reduction from a bicarpellate gynoecium (Abbe and Earle 1940). The androecium in the family is most commonly obdiplostemonous, although it is reduced to haplostemony in *Picrasma*, *Brucea*, *Picrolemma* and *Eurycoma*. In the latter two genera the stamens alternate with staminodes in the staminate flowers. In *Pierreodendron* the outer whorl of stamens is doubled. Adaxial scale-like appendages on the filaments occur in eleven genera, and vary in shape, length, pubescence and bifurcation. In unisexual flowers, vestigial staminodes and pistillodes are common. Filaments are inserted at the base of the nectary disk, which can vary between strongly lobed, cushion-like, tall and cylindrical, conical to inconspicuous. The disk usually enlarges in fruit. The gynoecium of *Soulamea* is reduced to two or three connate carpels (Fig. 91B), and is a single carpel in *Amaroria*.

Embryology. Embryology for the family was reviewed by Mauritzon (1935). Detailed studies of embryo anatomy are available for *Ailanthus* (Narayana 1957), *Samadera* (Nair and Joseph 1957), *Brucea* (Nair and Sukumaran 1960) and *Leitneria* (Pfeiffer 1912), and the following characteristics should be considered typical for the family: the anther wall consists of an epidermis, a fibrous endothecium, two to three middle layers and a multinucleate secretory tapetum (binucleate in *Ailanthus excelsa*); microsporogenesis is simultaneous; pollen tetrads are tetrahedral and decussate, shed at the two-celled stage; ovules are anatrapous or hemi-anatropous, crassinucellate and bitegmic, the inner integument forming the zig-zag micropyle; the nucellus is multinucleate, and the nucellar epidermis divides to form a cap; the archesporium can be multicellular or unicellular (*Ailanthus*), only one archesporial cell developing further; megaspores are arranged linearly (a solitary T-shaped tetrad is reported for *Ailanthus integrifolia*); the chalazal megaspore develops into a Polygonum type embryo sac; fertilisation may be chalazogamous, mesogamous or porogamous (Wiger 1935), but only porogamy is confirmed in *Samadera* and *Brucea*; endosperm development precedes embryo development, and is of the Nuclear type.

Pollen Morphology. Basak (1963, 1967) and Moncada and Machado (1987) used light microscopy to survey pollen morphology in *Quassia*, *Samadera*, *Simarouba*, *Simaba*, *Eurycoma*, *Soulamea*, *Ailanthus*, *Brucea*, *Castela* and *Picrasma*, and Zavada and Dilcher (1986) examined *Leitneria* with SEM and TEM. Pollen grains are 3-zonocolporate, typically 20–35 μm long by 13–30 μm wide, prolate in equatorial view (sometimes subspheroidal in *Castela* and spheroidal in *Samadera* and *Leitneria*) and planaperturate, with distinctly lalongate endoapertures; however, *Quassia* pollen grains are suboblate and angulaperturate, with a square type of endoaperture. Exine is 2–3 μm thick, and the surface pattern finely to coarsely reticulate, sometimes verrucate, in most genera. The exine is striate in *Soulamea* and *Brucea*, and striato-reticulate in *Quassia*.

Karyology. Simaroubaceae have base chromosome numbers of 8–13 (Stevens 2006). Bennett and Leitch (2005) record $2n = 64$ in *Ailanthus integrifolia*, which would suggest the plant is octoploid. Raven (1975) reports $x = 16$ for *Leitneria*, and *Castela coccinea* has $2n = 26$ (Bernardello et al. 1990).

Reproductive Biology. Simaroubaceae can be hermaphroditic, monoecious or dioecious. The extent of self-compatibility is unknown; however, flowers of *Quassia amara* have been shown to self-fertilise (Roubik et al. 1985). Insect-pollination predominates in the family, the flowers typically being small, actinomorphic, open, fragrant and borne in thyrses, attracting generalist small insects such as bees and moths (e.g. Aubréville 1962; Hardesty et al. 2005). *Quassia amara* is hummingbird-pollinated, as suggested by the raceme-like inflorescences bearing deep pink or red tubular flowers. Roubik et al. (1985) observed the role of nectar robbers in reproductive fitness of *Q. amara*, revealing that flowers were visited by nectar-robbing bees (*Trigona*) and hummingbirds, as well as the primary hummingbird pollinator. *Leitneria* shows strong morphological divergence towards wind-pollination in that the flowers lack a perianth and nectary disk, and are borne in catkin-like inflorescences that develop before the leaves emerge.

Fruits and Seeds. Fruits in the family are predominantly schizocarpous with drupaceous mericarps, and typically only 1–3 carpels reach maturity (Figs. 91, 92). The drupes have a thin pericarp, in which the exocarp can be fleshy (e.g. *Hannoa*, *Quassia*, *Simaba*), woody and fibrous (*Samadera*) or thin and dry (*Eurycoma*, *Leitneria*, some *Brucea*). The fleshy fruits can be pale yellow to red to deep purple-black, with a bitter taste, globose, obovoid, ovoid or ellipsoid, and between 0.3 and 10 cm long. The drupes are often carinate or bicarinate and flattened and, in *Samadera indica*, are strongly laterally dorso-ventrally compressed with a narrow, unilateral thinner edge in the apical half. In *Ailanthus* each carpel develops into a samaroid mericarp, elliptic in shape and tapering at each end (Fig. 89D). Variation in samara morphology is discussed in some detail by Nooteboom (1962) and Corbett and Manchester (2004). In *Soulamea* the carpels remain connate in fruit, forming a dry, narrowly to broadly winged, obcordate fruit. Fernando and Quinn (1992) discuss variation in pericarp anatomy in the family in detail. The exocarp varies in thickness and lignification and, in *Ailanthus*, is lacking except for the epidermal layer. Fernando and Quinn (1992) describe the endocarp as consisting of "a broad homogeneous zone of irregularly arranged isodiametric sclereids" with a strongly lignified inner epidermis. *Castela* and *Picrasma* lack the typical lightly lignified mesocarp and parenchymatous outer mesocarp. *Nothospondias* has an unusual *Spondias*-type endocarp, similar to that found in Anacardiaceae (Fernando and Quinn 1992).

The embryo is straight or curved, and consists of two large planoconvex cotyledons and a short plumule. Most Simaroubaceae have little or no endosperm, except for *Brucea* (Nair and Sukumaran 1960) and some *Soulamea* (Nooteboom 1962). Fatty oil and aleuron bodies are the most common seed storage products in the family, but starch is also reported from seeds of *Simaba* and *Perriera* (Netolitzki 1926) and *Leitneria* (Pfeiffer 1912), and reserve celluloses also occur (Czaja 1978; Stevens 2006). The seed coat is thin and hard, undistinguished or with scattered lignified cells (Stevens 2006), and is described as membranaceous in some genera.

Dispersal. Fleshy drupaceous fruits of Simaroubaceae are dispersed by fruit-eating birds and mammals, often primates (e.g. Hardesty et al. 2005). The samaroid mericarps of *Ailanthus* disperse over small distances by wind. Fruits of *Samadera indica*, a species that frequents alluvial and swamp forest, and *Soulamea amara*, a littoral species, are dispersed by water (Nooteboom 1962), which may account for their broad geographical distributions. *Leitneria* is also suspected to be water-dispersed, typically growing in freshwater and brackish swamps. In all cases, buoyancy is provided by an air cavity between seed and endocarp.

Phytochemistry. Simaroubaceae are characterised by their quassinoid chemistry. Quassinioids are triterpenoid derivatives, biosynthetically related to the limonoids of Rutaceae and Meliaceae (da Silva and Gottlieb 1987), and are considered (Dreyer 1983; Waterman 1983) to be further steps down the oxidative pathway of limonoids. Quassinoid structural and chemical characteristics are summarised in Waterman and Grundon (1983) and da Silva and Gottlieb (1987), who report 35 different structural types in *Picrasma* alone. Pentacyclic triterpenes are also common (Hegnauer 1983). Alkaloids have been reported in nine Simaroubaceae genera (Mester 1983), most commonly tryptophan derived, but also a quinolone alkaloid is reported in *Ailanthus*. Only a single simple coumarin has been detected in the family, in *Picrasma* and *Ailanthus* (Gray 1983). Of the flavonoid groups, flavonol glycosides and glycoflavones are reported in *Ailanthus* (Harborne 1983) and flavonols and flavones in *Leitneria* (Giannasi 1986). Essential oils contained within secretory cells and resin canals contain a low proportion of volatile compounds compared to Rutaceae and Meliaceae, and are in smaller amounts (Hegnauer 1983). Tannin content is low to considerable, and with relatively high levels of gallic and ellagic acid (Hegnauer 1983), although *Leitneria* lacks ellagic acid (Giannisi 1986).

Distribution and Habitats. Simaroubaceae have a primarily pantropical distribution; however, some species of *Brucea*, *Castela*, *Holacantha*, *Ailanthus* and *Picrasma* are subtropical, and *Ailanthus altissima*, *Picrasma quassioides* and *Leitneria floridana* grow in temperate climates. Generic diversity is split evenly among the New World, Africa, and Asia and Australasia; however, half of the species in the family occur in the New World. *Picrasma* is disjunct among Asia, SE Asia and Central and South America, *Brucea* is disjunct between Africa and SE Asia, and *Soulamea* has one species in the Seychelles, one widespread in Malesia and Polynesia, and the remainder endemic to New Caledonia. *Samadera* is primarily Australian and SE Asian, but *S. indica* occurs as far west as India and Madagascar. Several genera in the Simaroubaceae consist of one or two species with restricted geographic ranges, the majority of these genera being in Africa. *Simaba* is the most species-rich genus and is restricted to Central and South America.

Simaroubaceae are found in moist lowland tropical forest (although *Brucea mollis* is recorded as a high as 1,800 m in the Philippines, and *Odyendea gabonensis* at 2,500 m in Gabon), dry deciduous forest, and open sandy or savannah-type vegetation. *Soulamea amara* is a littoral species, *Castela* and *Holacantha* are found in desert and dry scrub environments, and *Leitneria*, *Samadera indica* and occasionally *Pierreodendron* inhabit swamp forest. *Eurycoma* is classified as silicicolous, showing a preference for acidic, leached sandy soils (Nooteboom 1962).

Dating and biogeographic analyses (Clayton et al. 2009) suggest the family originated in North America in the early Tertiary. However, ancient vicariant and dispersal patterns in the family are obscured by a multitude of more recent migration events, within and between the continents, post-Oligocene.

Fossil History. Fossils of the distinctive samaroid fruits of *Ailanthus* are found across the entire Northern Hemisphere, dating from the early Eocene up to the Pleistocene (Corbett and Manchester 2004). Three extinct species have been recognised, with the earliest occurrence a samara of *A. confucii* from the Green River Formation, Wyoming. Leaf fossils are also known with reasonable certainty from the Oligocene of Germany and Miocene of China, exhibiting distinctive basal teeth with enlarged glands on the leaflets, characteristic of extant *A. altissima* (Corbett and Manchester 2004). *Leitneria* has no fossil record from North America, but fossil fruits of it have been found in western Siberia from the Oligocene and in Europe from the Miocene to the Pliocene (Dorofeev 1994; Nikitin 2006); in transverse section, these have identical endocarp anatomy to extant *Leitneria floridana* (Dorofeev 1994). Less well understood are fossil fruits of *Chaneya*, an extinct genus from the Tertiary of North America, Europe and Eastern Asia (Wang and Manchester 2000; Teodoridis and Kvacek 2005). Teodoridis and Kvacek (2005) suggest an affiliation with the extant genus *Picrasma*, based on gynoecial morphology and persistent wing-like petals; however, the fossil has distinctive oil cells typical of Rutaceae. Fossil leaves formerly reported as *Leitneria* from the Eocene of Tennessee (Berry 1916) were subsequently reassigned to Rubiaceae, based on stipule configuration, epidermal anatomy and leaf architecture (Roth and Dilcher 1979). Fossil pollen of Simaroubaceae has been reported for *Ailanthus* (Song et al. 2004) and *Leitneria* (Machen 1971), but given the lack of distinctive morphological characteristics in extant Simaroubaceae pollen (Basa, 1963, 1967; Moncada and Machado 1987), these are considered unreliable.

Affinities. In the traditional circumscription, Simaroubaceae s.l. comprised six subfamilies (Engler 1931). However, molecular work by Fernando et al. (1995; also see Gadek et al. 1996) showed the family to be polyphyletic, with subfamilies originating in several places within eurosids I and II (sensu APG II 2003). Members of subfamily Simarouboideae, however, form a well-supported monophyletic group (excluding *Harrisonia*) within Sapindales. *Leitneria*, a genus traditionally segregated into the monotypic family Leitneriaceae on account of its wind-pollinated flowers (Cronquist 1981; Takhtajan 1997), was also found to be part of the Simarouboideae clade. Hence, the subfamily was recircumscribed as Simaroubaceae s.s., a clade of 20 genera and approximately 95 species by Fernando and Quinn (1995). *Nothospondias*, a monotypic genus sometimes placed in Anacardiaceae (Engler 1905), is a member of the family (Van der Veken 1960; Clayton et al. 2007). Also included is *Laumoniera* (Nooteboom 1987), a monotypic genus from Sumatra that was omitted from the family recircumscription of Fernando and Quinn (1995).

Simaroubaceae are well supported as a member of a Simaroubaceae+Rutaceae+Meliaceae clade in Sapindales (Gadek et al. 1996; Källersjö et al. 1998; Savolainen et al. 2000; Soltis et al. 2000), but the family's sister group is still undetermined, with data supporting three alternative topologies: Rutaceae sister to Simaroubaceae (Gadek et al. 1996); Meliaceae sister to Simaroubaceae (Chase et al. 1999; Muellner et al. 2006); Rutaceae sister to Meliaceae (Fernando et al. 1995; Stevens 2006). Traditional morphological and phytochemical classifications typically suggest an affiliation with Rutaceae and Meliaceae (e.g. Cronquist 1981; Takhtajan 1997).

Relationships Within the Family. Engler's (1931) classification of Simaroubaceae s.l. divided subfamily Simarouboideae (Simaroubaceae s.str.) into three tribes: Simaroubeae, Picrasmeae and Soulameae. Tribes were delimited by the presence or absence of filament appendages and the degree of connation of carpels. Of Engler's tribes, molecular data (Clayton et al. 2007) show only Soulameae, composed of *Soulamea* and *Amaroria*, to be monophyletic. Relationships based on DNA sequence data from the chloroplast genome (*rbc*L, *atp*B, *mat*K) and nuclear genome (1 kb of *phy*C) produced a well-resolved and well-supported phylogeny, excluding *Laumoniera* and *Iridosma* (Fig. 87; Clayton et al. 2007). Simaroubaceae consist of three major clades: the *Picrasma* clade (22 spp.), comprising *Picrasma*, *Castela* and *Holacantha*, sister to the rest of the family; the *Soulamea* clade (22–23 spp.), comprising *Leitneria*, *Brucea*, *Laumoniera* and *Soulamea* (with *Amaroria* nested within *Soulamea*); the *Simarouba* clade (45–48 spp.), composed of *Simarouba*, *Simaba*, *Pierreodendron*,

Hannoa, *Perriera*, *Gymnostemon*, *Odyendea*, *Iridosma* and *Eurycoma*. *Ailanthus* (5 spp.) diverges after the *Picrasma* clade, and a grade of *Nothospondias*, *Picrolemma*, *Quassia* and *Samadera* (10–11 spp.) follows the *Soulamea* clade (see Fig. 87). The relationship of *Laumoniera* to *Brucea* is unknown. The position of *Iridosma* is also unknown but likely to be within the *Simarouba* clade. Relationships among the four subclades of the *Simarouba* clade (*Hannoa* +*Perriera*+*Gymnostemon*, *Eurycoma*, *Odyendea* and *Pierreodendron*+*Simarouba*+*Simaba*) are poorly resolved; this lack of resolution is attributed to a rapid radiation that occurred in the Miocene (Clayton et al. 2009).

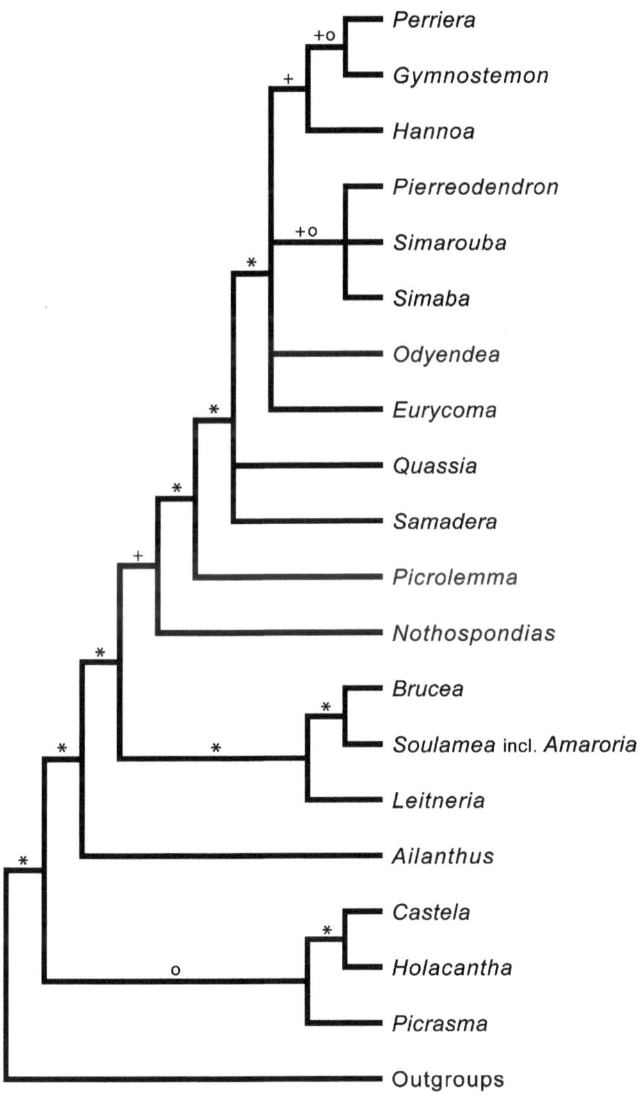

Fig. 87. Simaroubaceae. Phylogenetic hypothesis of relationships of Simaroubaceae based on three chloroplast genes and one nuclear gene, modified from Clayton et al. (2007). All genera except for *Simaba* have posterior probability (PP) = 1.0 and bootstrap support (BS) = 100%. *Laumoniera* and *Iridosma* were not sampled for this study. (orig.). *: PP = 1.0 and BS = 100%; +: PP > 0.95; o: BS > 70%

Economic Importance. A range of biological properties has been demonstrated by the quassinoids of Simaroubaceae, including antimalarial, antileukemic, antiviral, insecticidal and amoebicidal properties (Polonsky 1983; Klocke et al. 1985), and correspondingly, many genera are used locally as medicinal plants. *Quassia amara* and *Picrasma quassioides* have been used to aid digestion, and treat chronic dyspepsia. Fruits of *Brucea javanica* were imported into Europe as a drug (Nooteboom 1962), and the plant is used locally in Malaysia to treat malaria and dysentry. *Eurycoma* is used to treat malaria, diabetes, hypertension and stomach ache, typically by boiling the roots for drinking. *Ailanthus* is known in traditional Chinese and Korean medicine as a treatment for digestive complaints, haemorrhoids and mastitis. Simaroubaceae are not commercially harvested for timber but are used locally in building in some areas of both the Old and New World. *Leitneria* (corkwood) is one of the lightest known woods, and has been used traditionally by fisherman for net floats. *Ailanthus* (Tree of Heaven), *Simarouba* (paradise tree) and *Quassia amara* are cultivated and planted as ornamentals.

Key to the New World Genera

1. Perianth vestigial or absent; flowers surrounded by large, hirsute bracts **5. *Leitneria***
- Petals +, sepals +; bracts not large, not surrounding flowers 2
2. Stamens with appendaged filaments 3
- Filaments lacking appendage 5
3. Leaf rachis distinctly winged and jointed **12. *Quassia***
- Leaf rachis not winged, not jointed 4
4. Flowers unisexual; stigmas as long as style or longer, stellately spreading; leaflets alternate **21. *Simarouba***
- Flowers bisexual; stigmas capitate or lobed; leaflets typically opposite or subopposite **22. *Simaba***
5. Leaves unifoliolate or absent; plant often armed with thorns; stamens twice as many as petals 6
- Leaves pinnately compound; plant without thorns; stamens equal in number to petals 7
6. Plant with leaves; petals 4–5; stamens 8–10 **2. *Castela***

- Plant leafless, or leaves reduced to scales; petals 6–8; stamens 12–16 **3. *Holacantha***
7. Staminodes present in staminate flowers; inflorescence elongate, narrowing above; fruit ellipsoid, elongate, 20–30 mm long **11. *Picrolemma***
- Staminodes absent or in carpellate flowers only; inflorescence short, broad and rounded; fruit globose, less than 15 mm long **1. *Picrasma***

Key to the Old World Genera

1. Stamens with appendaged filaments 2
- Filaments lacking appendage 8
2. Leaves unifoliolate; inflorescence a pseudo-umbel **13. *Samadera***
- Leaves pinnately compound; inflorescence not umbellate 3
3. Leaf rachis jointed and often narrowly winged **12. *Quassia***
- Leaf rachis not jointed, not winged 4
4. Stamens alternating with outer whorl of staminodes or staminodial scales in staminate flowers; induplicate- valvate aestivation; Indomalesia **14. *Eurycoma***
- Staminodes absent in staminate flowers; contorted, imbricate, occasionally valvate aestivation; tropical Africa 5
5. Petals 7–8, valvate in bud; stamens 12–13 **19. *Iridosma***
- Petals 4–5, imbricate or contorted in bud; stamens 8–10(–15) 6
6. Stamens (10–)15(–18); leaves with 11–31 leaflets, up to 1 m long; leaflets apex with hard, pointed gland **20. *Pierreodendron***
- Stamens 8–10; leaves with 3–15 leaflets, less than 60 cm long; leaflets without hard pointed gland at apex 7
7. Calyx in bud irregularly rupturing into 2–3 lobes; 5 petals; 10 stamens; 5 carpels; fruits 15–35 mm long **17. *Hannoa***
- Calyx connate with 4(–5) very short obtuse lobes; 4(5) petals; 8(–10) stamens; 4 carpels; fruits 50–70 mm long **18. *Odyendea***
8. Gynoecium of 1 or 2(3) connate carpels 9
- Gynoecium of (2)3–5 carpels (if 2 then carpels distinct) 10
9. Gynoecium a single carpel; fruit ovoid, not winged; flowers 4- or 5-merous **7. *Amaroria***
- Gynoecium of two or three carpels; fruit obcordate, winged; flowers predominantly 3-merous **6. *Soulamea***
10. Fruit samaroid **4. *Ailanthus***
- Fruit drupaceous, fleshy or dry and nut-like 11
11. Stamens equal in number to petals 12
- Stamens twice as many as petals 14
12. Inflorescence short, broad and rounded; sepals and petals persistent in fruit, accrescent; fruit globose **1. *Picrasma***
- Inflorescence mostly unbranched, elongate; petals caducous in fruit; fruit ovoid or ellipsoid, or nut-like with 2 ribs when mature 13
13. Leaves imparipinnate; stigmas distinct, recurving; fruit 7–18 mm long **8. *Brucea***
- Leaves paripinnate; stigmas connate, discoid; fruit 45–60 mm long **9. *Laumoniera***
14. Carpels 2; inflorescence axillary; Madagascar **16. *Perriera***
- Carpels 4–5; inflorescence typically terminal; tropical Africa 15
15. Leaves with 19–43 leaflets; flowers 4-merous; fruits up to 45 mm in length; tropical west Africa **10. *Nothospondias***
- Leaves with 13–25 leaflets; flowers 5-merous; fruits about 100 mm in length; Côte d'Ivoire endemic **15. *Gymnostemon***

Genera of Simaroubaceae

1. *Picrasma* Blume Fig. 88

Picrasma Blume, Bijdr. Fl. Ned. Ind.: 247 (1825); Cronquist, Brittonia 5: 128–147 (1944), rev.
Aeschrion Vell. (1827).

Small trees, sometimes to 20 m, or shrubs, monoecious or dioecious. Leaves imparipinnate, pseudostipules present, early caducous; leaflets opposite to subopposite, petiolulate, entire or serrate-crenate, glabrous or nearly so, without glands. Flowers in axillary, short and broad, rounded determinate thyrses with puberulent axes; sepals 4(5), distinct or basally connate; petals 4(5), valvate, mostly glabrous; stamens 4(5), filaments lacking appendage; anthers dorsifixed; staminodes absent in staminate flowers; disk fleshy, sometimes conical, glabrous or hairy; carpels (2–)4(5), distinct; stylodia connate above, sometimes distinct; stigmatic branches filiform, recurved. Fruit 1–3(–5) drupaceous mericarps, globose, not carinate, 5–12 mm long, exocarp red to blue-black at maturity, pericarp fleshy.

Eight species, two in Asia and SE Asia, six in Mexico to Argentina, and Caribbean islands.

2. *Castela* Turpin

Castela Turpin, Ann. Mus. Natl. Hist. Nat. 7: 78 (1806), nom. cons.; Cronquist, J. Arnold Arb. 25: 122–128 (1944), rev.

Shrubs, erect or trailing, or small trees to 5 m, dioecious, armed with (occasionally branching) thorns. Leaves unifoliolate, petiolate, entire, glabrous to tomentose-pubescent, without glands. Flowers solitary, clustered in leaf axils, or in axillary, sparsely flowered determinate thyrses, with typically a single, sparsely to densely hairy axis; sepals

Fig. 88. Simaroubaceae. *Picrasma javanica*. **A** Flowering twig. **B** Full-grown pseudostipules. **C** Male flower. **D** Female flower. **E** Fruits. (Nooteboom 1962; artwork: C. van Crevel)

4(5), basally connate; petals 4(5), imbricate, glabrous to occasionally pubescent; stamens 8 or 10; filaments lacking appendage; anthers dorsifixed; staminodes absent in staminate flowers; disk fleshy, ring-like, glabrous; carpels 4(5), weakly united or distinct, stylodia connate at base, stigmatic branches linear, divergent or recurved, occasionally circinately rolled. Fruit 1–2(–4) drupaceous mericarps, lenticular, bicarinate, 6–12 mm long, exocarp red at maturity, pericarp fleshy.

Twelve species from southern United States to Argentina, the Caribbean islands and the Galápagos.

3. *Holacantha* A.Gray

Holacantha A.Gray, Mem. Am. Acad. Arts II, 5 (Pl. Nov. Thurb.): 310, t. 8 (post May 1855); Cronquist, Brittonia 5: 128–147 (1944), rev.

Depressed, ascending or erect shrubs or small trees to 5 m, dioecious, armed with thorns at branch tips, essentially leafless or leaves scale-like. Flowers in axillary, short, densely flowered determinate thyrses with one or two strongly hirsute axes, or appearing fasciculate in leaf axils; sepals 5–8, basally connate; petals 6–8, imbricate, strigose on abaxial surface; stamens 12–16, filaments lacking appendage, anthers dorsifixed, staminodes absent in staminate flowers; disk narrow and ring-like, densely hairy to glabrous; carpels 6–8, weakly united; style short and broad, stellately spreading into 5 stigmatic branches. Fruit 1–4 drupaceous mericarps, ovoid and slightly compressed, sometimes carinate on abaxial side, 5–9 mm long; exocarp red or greenish at maturity; pericarp fleshy.

Two species from southern California, southern and western Arizona to Mexico.

4. *Ailanthus* Desf. Fig. 89

Ailanthus Desf., Mém. Acad. Sci. Paris 1786: 270, t. 8 (1789), nom. cons.; Nooteboom, Fl. Males. I, 6: 215–220 (1962).

Large trees to 60 m, dioecious or monoecious. Leaves imparipinnate or paripinnate; leaflets opposite, subopposite or alternate below, petiolulate, entire to coarsely toothed, glabrous to densely pubescent, with sometimes large abaxial glands, occasionally domatia present as hair tufts at leaf base. Flowers in axillary or terminal determinate thyrses with glabrous to sparsely hairy axes; sepals 5(6), connate basally or calyx cupular with very short lobes; petals 5(6), induplicate-valvate, glabrous to pubescent; stamens 10, filaments lacking appendage; anthers ± ventrifixed; staminodes absent in staminate flowers; disk fleshy, glabrous; carpels 2–5; stylodia distinct to connate; stigmatic branches peltate, stellately spreading, sometimes recurved. Fruit 1–5 samaroid mericarps with elongate, membranous wings tapering towards the ends, 25–220 mm long; exocarp brown at maturity; pericarp dry.

Five species from Turkestan, India, China, SE Asia and northern Australia.

5. *Leitneria* Chapm. Fig. 90

Leitneria Chapm., Fl. S. U. St.: 427 (1860).

Small tree to 6 m; typically dioecious. Leaves unifoliolate, petiolate, entire, villous, without

Fig. 89. Simaroubaceae. *Ailanthus altissima*. **A** Flowering twig. **B** Male flower. **C** Female flower. **D** Fruit. (Takhtajan 1981; artwork: A. Schilitschkin)

Fig. 90. Simaroubaceae. *Leitneria floridana*. **A** Branch. **B** Young female catkins. **C** Young male catkins. **D** Elongating male catkins. **E** Male catkin with anthetic flowers. **F** Partial inflorescence thereof. **G** Stamen. **H** Twig with anthetic female catkins. **I** Female catkin with exposed stylodia. **J** Female flower. **K** Gynoecium in vertical section. **L** Seed. **M** Fruit. (Takhtajan 1980)

glands. Staminate flowers in axillary, catkin-like, highly reduced thyrses with 1–3-flowered cymules; carpellate flowers solitary in carpellate inflorescences, surrounded by densely hirsute bracts and arranged on an single glabrous axis; perianth 0 in male flowers, vestigial in carpellate flowers; stamens (1–)4 per flower in bract axil; filaments lacking appendage; anthers basifixed to dorsifixed; staminodes absent in staminate flowers; disk absent or rudimentary; carpel 1, stigmatic branch distally expanded, recurved. Fruit a drupe, narrowly ellipsoid, conspicuously flattened, bicarinate, 12–30 mm long; exocarp brown at maturity; pericarp dry to occasionally fleshy.

One species, *L. floridana* Chapm., in SE United States.

6. *Soulamea* Lam. Fig. 91

Soulamea Lam., Encyc. 1: 449 (1783); Jaffré & Fambart, Adansonia III, 24: 159–168 (2002).

Shrubs or small trees to 5(–15) m, dioecious or flowers bisexual (*S. amara*). Leaves unifoliolate or imparipinnate; leaflets opposite, petiolulate, leaves petiolate, entire and often revolute, densely pubescent or glabrous on adaxial surface, sometimes with glands. Flowers in axillary, elongate determinate thyrses, typically with a single, often ferruginous-tomentose major axis; sepals 3(–5), basally connate; petals 3(–5), glabrous to pubescent towards the base; stamens 6(–10); filaments lacking appendage; anthers basifixed to dorsifixed; staminodes absent in staminate flowers; disk fleshy, glabrous; carpels 2(3), connate; stylodia distinct, flattened, horizontally appressed to carpel, stigma fleshy, rarely reniform. Fruit samaroid, 2-celled, obcordate, flattened, with a distinct wing, 10–20 mm long, exocarp brown at maturity, pericarp dry.

Thirteen species, one widespread in SE Asia and Polynesia (*S. amara* Lam.), one endemic to the Seychelles (*S. terminalioides* Baker), and eleven species endemic to New Caledonia.

Fig. 91. Simaroubaceae. *Soulamea amara*. **A** Fruiting twig. **B** Flower. **C** Flower sectioned, pistil removed. **D** Fruit. (Nooteboom 1962; artwork: C. van Crevel)

7. *Amaroria* A.Gray

Amaroria A.Gray, Bot. U. St. Expl. Exped. 1: 356, t. 40 (1854); Smith, Fl. Vit. Nova 3: 479–487 (1985).

Small tree to 15(–20) m, dioecious. Leaves unifoliolate, petiolate, entire; glands unknown. Flowers in axillary, elongate determinate thyrses with a single major axis; sepals 4–5, basally connate; petals 4–5, glabrous or sometimes short strigillose along adaxial midline; stamens 8 or 10; filaments lacking appendage; anthers dorsifixed; staminodes absent in staminate flowers; disk fleshy, globose; carpel 1; stigma sessile. Fruit a drupe, ovoid to subglobose, slightly flattened, sometimes inconspicuously carinate, 17–30 mm long; exocarp greenish yellow, becoming white at maturity; pericarp fleshy.

One species, *A. soulameoides* A. Gray, endemic to Fiji.

8. *Brucea* J.F. Mill.

Brucea J.F. Mill., Icon.: t. 25 (1779), nom. cons.

Shrubs or small trees to 12 m, dioecious or polygamous. Leaves imparipinnate; leaflets opposite, petiolulate to subsessile, entire or crenate-serrate, ferruginous-pubescent to glabrous, with dotted glands associated with peripheral secondary venation underneath. Flowers in axillary, elongate determinate thyrses, typically with a single glabrous to densely pubescent major axis; sepals (3)4(5), basally connate; petals (3)4(5), imbricate, glabrous to densely pubescent; stamens (3)4(5), protruding between disk lobes, filaments lacking appendage; anthers basifixed to dorsifixed, staminodes absent in staminate flowers; disk fleshy, glabrous; carpels (3)4(5), distinct or united at the base; stylodia connate at base, stigmatic branches linear, recurved or bending inwards. Fruit 1–2(–4) drupaceous mericarps, ovoid, bicarinate, 4–18 mm long, exocarp red to black at maturity, pericarp dry to thinly fleshy.

Six to seven species, tropical Africa to tropical and subtropical Asia and northern Australia.

9. *Laumoniera* Noot.

Laumoniera Noot., Blumea 32: 383 (1987).

Small tree to 16 m, dioecious. Leaves paripinnate; leaflets petiolulate, entire, glands unknown. Flowers in axillary determinate thyrses, typically with a single pubescent axis; sepals 4, basally connate; petals 4, sparsely pubescent; stamens 4, filaments lacking appendage; staminodes absent in staminate flowers; disk fleshy, slightly hairy; carpels 4, distinct; stigmas sessile, connate, discoid, covering top of ovaries. Fruit 1–4 drupaceous mericarps, ellipsoid, 45–60 mm long; exocarp yellow at maturity, pericarp fleshy.

One species, *L. bruceadelpha* Noot., Indonesia.

10. *Nothospondias* Engl.

Nothospondias Engl., Bot. Jahrb. Syst. 36: 216 (1905); Van der Veken, Bull. Jard. Bot. Etat Brux. 30: 105–109 (1960).

Tree to 25 m, dioecious. Leaves imparipinnate; leaflets opposite to alternate, petiolulate, entire, glabrous, without glands. Flowers in axillary or terminal determinate thyrses, with multiple densely pubescent axes; sepals 4, basally connate; petals 4, slightly imbricate, glabrous to puberulent; stamens 8; filaments lacking appendage; anthers basifixed; staminodes absent in staminate flowers; disk fleshy, glabrous; carpels 4, distinct, style simple. Fruit 1–4 drupaceous mericarps, ovoid-ellipsoid, 20–45 mm long; exocarp yellow to orange at maturity; pericarp fleshy.

One species, *N. staudtii* Engl., in tropical west Africa.

11. *Picrolemma* Hook.f.

Picrolemma Hook.f. in Benth. & Hook., Gen. Pl. 1: 312 (1862); Cronquist, Brittonia 5: 128–147 (1944).

Small shrubs, up to 6 m, dioecious. Leaves imparipinnate; leaflets opposite to sometimes alternate below, petiolulate, entire, glabrous, punctate glands associated with secondary venation underneath. Flowers in terminal determinate thyrses, with multiple glabrous axes; sepals (4)5, basally connate; petals (4)5, imbricate, glabrous; stamens 5, filaments lacking appendage; anthers dorsifixed; staminodes alternating with petals in staminate flowers; disk fleshy, glabrous; carpels (4)5, distinct; stylodia distinct but cohering, stigmatic branches fleshy, club-like. Fruit 1–2 drupaceous mericarps, ellipsoid and slightly elongate, not carinate, 20–30 mm long; exocarp brown to red at maturity; pericarp fleshy.

Two species from Peru and Brazil.

12. *Quassia* L.

Quassia L., Sp. Pl. ed. 2: 553 (1762); Engl. in Engl. & Prantl, Nat. Pflanzenfam., 2nd edn, 19a: 377–379 (1931).

Shrub or small tree to 8 m. Leaves imparipinnate, rachis and petiole conspicuously winged in *Q. amara* L., narrowly winged or wingless in *Q. africana* Baill., articulated; leaflets opposite, sessile, entire, glabrous, with punctate glands towards leaf apex adaxially. Flowers bisexual, in axillary or terminal determinate thyrses, appearing raceme-like in *Q. amara*, with puberulent axes; sepals 5, distinct, overlapping at base; petals 5, contorted, glabrous or basally pubescent inside, cohering into a tube in *Q. amara*; stamens 10; filaments with basal appendage; anthers dorsifixed, staminodes absent in staminate flowers; disk fleshy, narrowing towards base, glabrous; carpels 5, distinct; style simple; stigma capitate or slightly lobed. Fruit 1–2 drupaceous mericarps, obovoid to ellipsoid, bicarinate, 10–25 mm long; exocarp dark red at maturity; pericarp fleshy.

Two species, 1 neotropical, 1 in tropical west Africa.

13. *Samadera* Gaertn.

Samadera Gaertn., Fruct. 2: 352, t. 156, 'f. 3' (1791), nom. cons. *Hyptiandra* Hook.f. (1862).

Small tree, occasionally up to 20 m. Leaves unifoliolate, entire, glabrous, with scattered punctate glands. Flowers bisexual, in axillary or terminal pseudo-umbels, axes glabrous to puberulent, or clustered in leaf axils; sepals (3)4(5), distinct or mostly connate with short lobes, occasionally with a concave gland; petals (3)4(5), imbricate or contorted, glabrous to pubescent abaxially; stamens 8 or 10; filaments with basal appendage; anthers dorsifixed; staminodes absent in staminate flowers; disk fleshy, conical or cylindrical, glabrous; carpels 4–5; stylodia conglutinate into a common style; stigma capitate or slightly lobed. Fruit 1(–5) drupaceous mericarps, ovoid, ellipsoid or semicircular and flattened, slightly to strongly carinate, 5–50 mm long; exocarp orange to red or brown at maturity; pericarp fleshy or dry and woody.

Five to six species from Madagascar, Indo-China, SE Asia and Australia.

14. *Eurycoma* Jack Fig. 92

Eurycoma Jack, Malay. Misc. 2, 7: 44 (1822); Nooteboom, Fl. Males. I, 6: 193–226 (1962).

Small trees to 10 m, or rarely shrubs, monoecious or dioecious. Leaves imparipinnate; leaflets opposite to subopposite, sessile or nearly so, sometimes appearing articulated, entire, glabrous, without glands. Flowers in axillary determinate thyrses, multiple axes with thick, capitate-glandular hairs; sepals 5(6), basally connate, with capitate-glandular hairs; petals 5(6), induplicate-valvate, pubescent, with capitate-glandular hairs; stamens 5(6); filaments with very small appendage near base; anthers dorsifixed; staminodes 5(6), alternating with stamens in

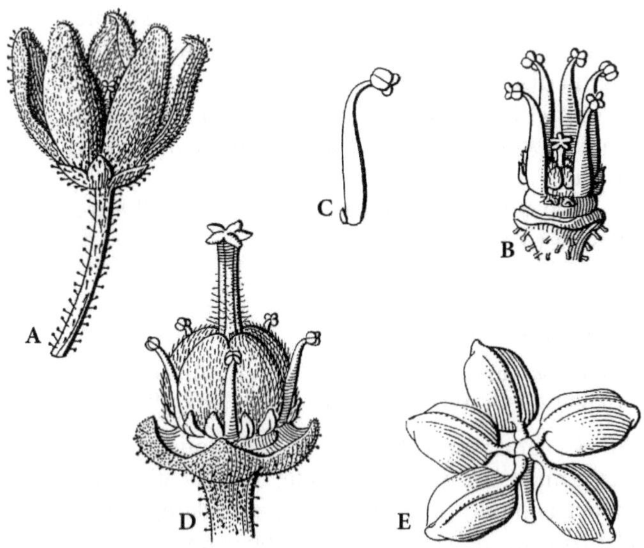

Fig. 92. Simaroubaceae. *Eurycoma longifolia*. **A** Male flower. **B** Same, perianth removed. **C** Stamen. **D** Female flower, petals removed. **E** Fruit. (Nooteboom 1962; artwork: C. van Crevel)

staminate flowers; disk inconspicuous; carpels 5 (6), distinct, stylodia connate or cohering; stigma lobed, peltate. Fruits 1–5 nut-like mericarps, ovoid, bicarinate, 10–20 mm long; exocarp brown at maturity; pericarp dry.

Three species, tropical SE Asia, Sumatra, Malay peninsula, Borneo, S. Philippines.

15. *Gymnostemon* Aubrév. & Pellegr.

Gymnostemon Aubrév. & Pellegr., Bull. Soc. Bot. France 84: 183 (1937).

Large tree, with bisexual flowers or polygamous. Leaves imparipinnate; leaflets opposite to subopposite, subsessile, entire, glabrous, with punctate glands regularly spaced towards apex adaxially. Flowers in axillary or terminal determinate thyrses, with multiple densely short-hairy axes; sepals 5, connate, with short lobes; petals 5, slightly imbricate, villous; stamens 10; filaments lacking appendage; anthers dorsifixed; staminodes absent in staminate flowers; disk fleshy, pubescent; carpels 5, distinct, style simple; stigma simple or slightly lobed. Fruit a single drupaceous mericarp, ovoid, up to 100 mm long; pericarp fleshy, fibrous.

One species, *G. zaizou* Aubrév. & Pellegr., endemic to Côte d'Ivoire.

16. *Perriera* Courchet

Perriera Courchet, Bull. Soc. Bot. France 52: 284 (1905); Capuron, Adansonia II, 1: 87 (1961).

Tree to 30 m. Leaves imparipinnate; leaflets opposite to subopposite, subsessile, entire, pubescent when young, becoming glabrous, with punctate glands regularly spaced towards apex adaxially. Flowers typically bisexual, in axillary determinate thyrses, with multiple pubescent axes; sepals 5, basally connate; petals 5, induplicate-valvate, slightly villous; stamens 10; filaments lacking appendage; anthers dorsifixed; staminodes absent in staminate flowers; disk fleshy; carpels 2, slightly united at base, style simple; stigmatic branches divergent. Fruit typically a single drupaceous mericarp, ovoid, up to 50 mm long; exocarp pale yellow at maturity; pericarp fleshy.

One or two species endemic to Madagascar.

17. *Hannoa* Planch.

Hannoa Planch., London J. Bot. 5: 566 (1846).

Trees to 50 m or shrubs, sometimes suffructescent. Leaves imparipinnate; leaflets opposite to alternate, subsessile to petiolulate, entire, glabrous, with punctate glands on upper surface, more so towards margins. Flowers typically bisexual, in terminal or occasionally axillary determinate thyrses, with multiple glabrous to sparsely pubescent axes; sepals 5, or often calyx rupturing into 2–3 irregular lobes; petals 5, imbricate, puberulent to densely tomentose; stamens 10; filaments with appendage; anthers dorsifixed; staminodes absent in staminate flowers; disk fleshy, sometimes with gynoecium sunken within, glabrous; carpels 5, distinct; style simple; stigmatic branches short, spindly lobes. Fruit 1–3 drupaceous mericarps, ellipsoid or ovoid, slightly bicarinate, 15–35 mm long; exocarp red to purplish brown at maturity; pericarp fleshy.

Five to seven species in tropical Africa.

18. *Odyendea* (Pierre) Engl.

Odyendea (Pierre) Engl. in Engl. & Prantl, Nat. Pflanzenfam. III, 4: 215 (1896); Aubrév. & Pellegr., Fl. Gabon 3: 33–52 (1962).

Tree to 30 m. Leaves imparipinnate; leaflets opposite to subopposite, petiolulate, entire, glabrous, with punctate glands on upper surface, more so towards margins. Flowers bisexual, in terminal or axillary determinate thyrses, with multiple glabrous axes; sepals 4(5), calyx cupular with short or absent lobes; petals 4(5), imbricate, puberulent adaxially; stamens 8(–10); filaments with densely hairy appendage; anthers dorsifixed; staminodes absent in staminate flowers; disk fleshy, subcylindrical, with gynoecium slightly immersed within, glabrous; carpels 4, distinct or united at base; style simple, with very short, divergent stigmatic branches. Fruit a single drupaceous mericarp, obovoid to ellipsoid, strongly carinate, up to 70 mm long; exocarp red at maturity; pericarp fleshy.

One species, *O. gabonensis* (Pierre) Engl., endemic to Gabon and Cameroon.

19. ***Iridosma*** Aubrév. & Pellegr.

Iridosma Aubrév. & Pellegr., Fl. Gabon 3: 47 (1962).

Trees. Leaves imparipinnate; leaflets opposite to subopposite, subsessile, entire, glabrous, glands unknown. Flowers bisexual, in determinate thyrses, with multiple pubescent axes; calyx cupular, irregularly undulating; petals (7)8, valvate, villous; stamens 12–13; filaments with appendage; anthers dorsifixed; staminodes absent in staminate flowers; disk fleshy, pubescent; carpels 4, distinct; stylodia spirally twisted to form single column; stigma peltate, stellate. Fruit unknown.

One species, *I. le-testui* (Pellegr.) Aubrév. & Pellegr., endemic to Gabon and Cameroon.

20. ***Pierreodendron*** Engl. (1906) non A. Chev. (1917).

Mannia Hook.f. (1862) non Opiz (1829) non Trevis (1857).

Tree to 15 m. Leaves imparipinnate; leaflets subopposite to alternate, petiolulate, entire, glabrous to sparsely pubescent below, without glands. Flowers bisexual, in axillary or terminal determinate thyrses, with one or two major axes; sepals 5; calyx cupular with short lobes; petals 5, imbricate or contorted; stamens (10–)15(–18); filament appendage short with small free tip; anthers basifixed; staminodes absent in staminate flowers; disk fleshy, sometimes with gynoecium sunken within, glabrous; carpels 5, distinct, style simple, with short, divergent stigmatic branches, or stigma discoid. Fruits 1–5 drupaceous mericarps, oblong-ellipsoid, laterally compressed, 70–80 mm long; exocarp yellow at maturity; pericarp fleshy, fibrous.

Two species, tropical Africa (Côte d'Ivoire, Nigeria to Angola and Congo).

21. ***Simarouba*** Aubl. Fig. 93

Simarouba Aubl., Hist. Pl. Guiane: 859 (1775), nom. cons.; Cronquist, Bull. Torrey Bot. Club 71: 226–234 (1944), rev.

Shrubs and trees to 35 m, dioecious. Leaves paripinnate or imparipinnate; leaflets alternate to occasionally subopposite, petiolulate, entire, glabrous or densely tomentose below, with punctate glands scattered on upper surface, more so towards apex. Flowers in terminal determinate thyrses, with multiple glabrous axes; sepals 5, basally connate; petals 5, imbricate or contorted, glabrous; stamens 10; filaments with glabrous to pubescent appendage; anthers dorsifixed; staminodes absent in staminate flowers; disk

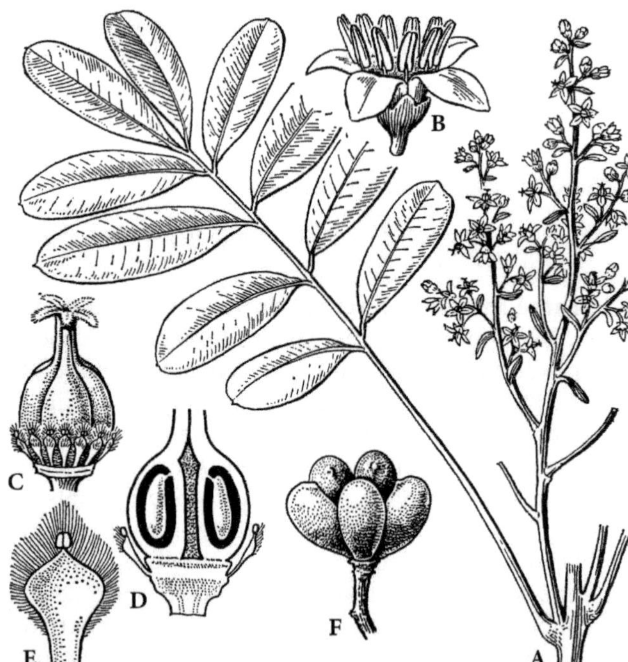

Fig. 93. Simaroubaceae. *Simarouba glauca*. **A** Flowering twig. **B** Male flower. **C** Female flower, perianth removed. **D** Longisection of ovary. **E** Staminode. **F** Fruits. (Takhtajan 1981; artwork: A. Schilitschkin)

fleshy, short, glabrous to pubescent; carpels 5, distinct or weakly united, stylodia connate below, stigmatic branches stellately spreading, recurved. Fruit 1–3 drupaceous mericarps, ovoid or ellipsoid, slightly flattened, bicarinate, 10–25 mm long; exocarp orange-red to black at maturity; pericarp fleshy.

Six species in Central and South America, the Caribbean islands and south Florida.

22. *Simaba* Aubl.

Simaba Aubl., Pl. Gui.: 409, t. 153 (1775); Cronquist, Lloydia 7: 81–92 (1944); Cavalcante, Rev. tax. Gen. *Simaba*. Publ. Avuls. Mus. Goeldi 37, 87 pp. (1983), rev.

Trees to 30 m, shrubs, rarely suffructescent with all leaves basal. Leaves paripinnate or imparipinnate, trifoliolate or rarely unifoliolate; leaflets usually opposite, petiolulate to sessile, entire, glabrous to occasionally pubescent, with punctate glands on upper surface, and occasionally with conspicuous apical gland. Flowers bisexual, in terminal or axillary determinate thyrses, with multiple glabrous to densely pubescent axes, or occasionally reduced to axillary clusters; sepals (4)5, basally connate; petals (4)5, imbricate, puberulent to densely pubescent; stamens (8–) 10; filaments with appendage; degree of fusion between filament and appendage variable; anthers dorsifixed; staminodes absent in staminate flowers; disk fleshy, cylindrical, glabrous to densely pubescent; carpels (4)5, distinct or weakly connate; style simple, stigma capitate or slightly lobed. Fruit 1(–5) drupaceous mericarps, ellipsoid to obovoid, lenticular, slightly carinate or occasionally strongly winged, 10–80(–100) mm long; exocarp orange, red, brown, black or yellow at maturity; pericarp fleshy.

Twenty five species in tropical South America, *S. cedron* Planch. extending into Central America.

Selected Bibliography

Abbe, E.C., Earle, T.T. 1940. Inflorescence, floral anatomy and morphology of *Leitneria floridana*. Bull. Torrey Bot. Club 67: 173–193.

APG II. 2003. An update of the angiosperm phylogeny group classification for the orders and families of flowering plants: APG II. Bot. J. Linn. Soc. 141: 399–436.

Aubréville, A. 1962. Simaroubacées. In: Aubréville, A., Hallé, N., Descoings, B., Koechlin, J., Cavaco, A. (eds.) Flore du Gabon. Vol. 3. Paris: Muséum National d'Histoire Naturelle, pp 33–52.

Basak, R.K. 1963. Pollen morphology of Indian Simaroubaceae. Bull. Bot. Surv. India 5: 381–397.

Basak, R.K. 1967. Studies on the pollen morphology of the Simaroubaceae. Bull. Bot. Surv. India 9: 63–67.

Bennett, M.D., Leitch, I.J. 2005. Nuclear DNA amounts in angiosperms: progress, problems and prospects. Ann. Bot. 95: 45–90.

Bernadello, L.M., Stiefkens, L.B., Piovano, M.A. 1990. Números cromosómicos en dicotiledóneas Argentinas. Bol. Soc. Argent. Bot. 26: 149–157.

Berry, E.W. 1916. The physical conditions and age indicated by the flora of the Alum Bluff Formation. Profess. Pap. U.S. Geol. Surv. 98E: 41–59.

Chase, M.W., Morton, C.M., Kallunki, J.A. 1999. Phylogenetic relationships of Rutaceae: A cladistic analysis of the subfamilies using evidence from *rbc*L and *atp*B sequence variation. Am. J. Bot. 86: 1191–1199.

Clayton, J.W., Fernando, E.S., Soltis, P.M., Soltis, D.E. 2007. Molecular phylogeny of the Tree-of-Heaven family (Simaroubaceae) based on chloroplast and nuclear markers. Int. J. Plant Sci. 168: 1325–1339.

Clayton, J. W., Soltis, P.S., Soltis, D.E. 2009. Recent long-distance dispersal overshadows ancient biogeographical patterns in a pantropical angiosperm family (Simaroubaceae, Sapindales). Syst. Biol. 58: 395–410.

Corbett, S.L., Manchester, S.R. 2004. Phytogeography and fossil history of *Ailanthus* (Simaroubaceae). Int. J. Plant Sci. 165: 671–690.

Cronquist, A. 1944b. Studies in the Simaroubaceae IV: Resume of American genera. Brittonia 5: 128–147.

Cronquist, A. 1981. An integrated system of classification of flowering plants. New York: Columbia University Press.

Czaja, A.T. 1978. Stärke und Stärkespeicherung bei Gefässpflanzen. Stuttgart: Gustav Fischer.

da Silva, M.F.D.G.F., Gottlieb, O.R. 1987. Evolution of quassinoids and limonoids in the Rutales. Biochem. Syst. Ecol. 15: 85–103.

Dorofeev, P.I. 1994. *Leitneria*. In: Budantsev, L. (ed.) Fossil flowering plants of Russia and adjacent states. Vol. 3. Leitneriaceae-Juglandaceae. St. Petersburg, Russia: Kamarov Botanical Institute, Russian Academy of Sciences, pp. 8–12, pls. 47–55.

Dreyer, D.L. 1983. Limonoids of the Rutaceae. In: Waterman, P.G., Grundon, M.F. (eds.) Chemistry and chemical taxonomy of the Rutales. London: Academic Press, pp. 215–246.

Engler, A. 1905. Anacardiaceae africanae III. Bot. Jahrb. Syst. 36: 216–217.

Engler, A. 1931. Simarubaceae. In: Engler, A., Prantl, K. (eds.) Die natürlichen Planzenfamilien, 2nd edn, 19a. Leipzig: Engelmann, pp. 359–405.

Fernando, E.S., Quinn, C.J. 1992. Pericarp anatomy and systematics of the Simaroubaceae s.l. Aust. J. Bot. 40: 263–289.

Fernando, E.S., Quinn, C.J. 1995. Picramniaceae, a new family, and recircumscription of Simaroubaceae. Taxon 44: 177–181.

Fernando, E.S., Gadek, P.A., Quinn, C.J. 1995. Simaroubaceae, an artificial construct: evidence from *rbc*L sequence variation. Am. J. Bot. 82: 92–103.
Gadek, P.A., Fernando, E.S., Quinn, C.J. Hoot, S.B., Terrazas, T., Sheahan, M.C., Chase, M.W. 1996. Sapindales: molecular delimitation and infraordinal groups. Am. J. Bot. 83: 802–811.
Giannasi, D.E. 1986. Phytochemical aspects of phylogeny in Hamamelidae. Ann. Missouri Bot. Gard. 73: 417–437.
Gray, A.I. 1983. Structural diversity and distribution of coumarins and chromones in the Rutales. In: Waterman, P.G., Grundon, M.F. (eds.) Chemistry and chemical taxonomy of the Rutales. London: Academic Press, pp. 97–146.
Harborne, J.B. 1983. The flavonoids of the Rutales. In: Waterman, P.G., Grundon, M.F. (eds.) Chemistry and chemical taxonomy of the Rutales. London: Academic Press, pp. 147–174.
Hardesty, B.D., Dick, C.W., Kremer, A., Hubbell, S., Bermingham, E. 2005. Spatial genetic structure of *Simarouba amara* Aubl. (Simaroubaceae), a dioecious, animal-dispersed Neotropical tree, on Barro Colorado Island, Panama. Heredity 95: 290–297.
Hegnauer, R. 1983. Chemical characters and the classification of the Rutales. In: Waterman, P.G., Grundon, M.F. (eds.) Chemistry and chemical taxonomy of the Rutales. London: Academic Press, pp. 401–440.
Jaffré, T., Fambart, J. 2002. Quatre nouvelles espèces de *Soulamea* (Simaroubaceae) de Nouvelle-Calédonie. Adansonia III, 24: 159–168.
Källersjö, M., Farris, J.S., Chase, M.W., Bremer, B., Fay, M. F., Humphries, C.J., Petersen, G., Seberg, O., Bremer, K. 1998. Simultaneous parsimony jackknife analysis of 2538 *rbc*L DNA sequences reveals support for major clades of green plants, land plants, and flowering plants. Plant Syst. Evol. 213: 259–287.
Klocke, J.A., Arisawa, M., Handa, S.S., Kinghorn, A.D., Cordell, G.A., Farnsworth, N.R. 1985. Growth inhibitory, insecticidal and antifeedant effects of some antileukemic and cytotoxic quassinoids on two species of agricultural pests. Cell. Mol. Life Sci. 4: 379–382.
Machen, J. 1971. Plant microfossils from Tertiary deposits of the Isle of Wight. New Phytol. 70: 851–872.
Mauritzon, J. 1935. Kritic von J. Wiger's Abhandlung "Embryological studies on the families Buxaceae, Meliaceae, Simarubaceae and Burseraceae. Bot. Not. 1935: 490–502.
Mester, I. 1983. Structural diversity and distribution of alkaloids in the Rutales. In: Waterman, P.G., Grundon, M.F. (eds.) Chemistry and chemical taxonomy of the Rutales. London: Academic Press, pp. 31–96.
Moncada, M., Machado, S. 1987. Los granos de polen de Simarubaceae. Acta Bot. Cub. 45: 1–7.
Muellner, A.N., Savolainen, V., Samuel, R., Chase, M.W. 2006. The mahogany family "out-of-Africa": divergence time estimation, global biogeographic patterns inferred from plastid *rbc*L DNA sequences, extant, and fossil distribution of diversity. Mol. Phylogenet. Evol. 40: 236–250.
Nair, N.C., Joseph, T.C. 1957. Floral morphology and embryology of *Samadera indica*. Bot. Gaz. 119: 104–115.
Nair, N.C., Joshi, R.K. 1958. Floral morphology of some members of the Simaroubaceae. Bot. Gaz. 120: 88–99.
Nair, N.C., Sukumaran, N.P. 1960. Floral morphology and embryology of *Brucea amarissima*. Bot. Gaz. 121: 175–185.
Narayana, L.L. 1957. Embryology of the Simaroubaceae. Curr. Sci. 26: 323–324.
Netolitzki, F. 1926. Anatomie der Angiospermen-Samen. In: Linsbauer, K. (ed.) Handbuch der Pflanzenanatomie, vol 2(X). Berlin: Borntraeger.
Nikitin, V.P. 2006. Paleocarpology and stratigraphy of the Paleogene and Neogene strata in Asian Russia. Novosibirsk: Academic Publishing House "Geo".
Nooteboom, H.P. 1962. Simaroubaceae. Flora Malesiana I, 6: 193–226.
Nooteboom, H.P. 1987. *Laumoniera*, a new genus of Simaroubaceae from Sumatra. Blumea 32: 383–384.
Pfeiffer, W.M. 1912. The morphology of *Leitneria floridana*. Bot. Gaz. 53: 189–203.
Polonsky, J. 1983. Chemistry and biological activity of the quassinoids. In: Waterman, P.G., Grundon, M.F. (eds.) Chemistry and chemical taxonomy of the Rutales. London: Academic Press, pp. 247–266.
Raven, P.H. 1975. The bases of angiosperm phylogeny: cytology. Ann. Missouri Bot. Gard. 68: 724–764.
Record, S. J., Hess, R.W. 1943. Timbers of the New World. New Haven: Yale University Press.
Roth, J. L., Dilcher, D.L. 1979. Investigations of angiosperms from the Eocene of North America: stipulate leaves of the Rubiaceae including a probable polyploid population. Am. J. Bot. 66: 1194–1207.
Roubik, D.W., Holbrook, N.M., Parra, G.V. 1985. Roles of nectar robbers in reproduction of the tropical treelet *Quassia amara* (Simaroubaceae). Oecologia 66: 161–167.
Savolainen, V., Chase, M.W., Hoot, S.B., Morton, C.M., Soltis, D.E., Bayer, C., Fay, M.F., de Bruijn, A.Y., Sullivan, S., Qiu, Y.-L. 2000. Phylogenetics of flowering plants based on a combined analysis of plastid *atp*B and *rbc*L gene sequences. Syst. Biol. 49: 306–362.
Soltis, D.E., Soltis, P.S., Chase, M.W., Mort, M.E., Albach, D.C., Zanis, M., Savolainen, V., Hahn, W.H., Hoot, S. B., Fay, M.F., Axtell, M., Swensen, S.M., Prince, L.M., Kress, W.J., Nixon, K.C., Farris, J.S. 2000. Angiosperm phylogeny inferred from 18S rDNA, *rbc*L, and *atp*B sequences. Bot. J. Linn. Soc. 133: 381–461.
Song, Z. -C., Wang, W. -M., Huang, F. 2004. Fossil pollen records of extant angiosperms in China. Bot. Rev. 70: 425–458.
Spiekerkoetter, H. 1924. Untersuchungen zur Anatomie und Systematik ostafrikanischer Meliaceen, Burseraceen und Simarubaceen. Bot. Arch. 7: 274–320.
Stevens, P.F. 2006. Angiosperm Phylogeny Website v. 7. http://www.mobot.org/MOBOT/research/APweb/
Takhtajan, A.L. (ed.) 1980. Flowering Plants, vol. V (1). Leningrad: Proswjeschtschenye.
Takhtajan, A.L. (ed.) 1981. Flowering Plants, vol. V (2). Leningrad: Proswjeschtschenye.
Takhtajan, A. 1997. Diversity and classification of flowering plants. New York: Columbia University Press.

Teodoridis, V., Kvacek, Z. 2005. The extinct genus *Chaneya* Wang et Manchester in the Tertiary of Europe – a revision of *Porana*-like fruit remains from Öhningen and Bohemia. Rev. Palaeobot. Palynol. 134: 85–103.

Van der Veken, P. 1960. *Nothospondias* Engl., Simaroubaceae africaine méconnue. Bull. Jard. Bot. Etat Brux. 30: 105–109.

Wang, Y., Manchester, S.R. 2000. *Chaneya*, a new genus of winged fruit from the Tertiary of North America and Eastern Asia. Int. J. Plant Sci. 161: 167–178.

Waterman, P.G. 1983. Phylogenetic implications of the distribution of secondary metabolites within the Rutales. In: Waterman, P.G., Grundon, M.F. (eds.) Chemistry and chemical taxonomy of the Rutales. London: Academic Press, pp. 377–400.

Waterman, P.G., Grundon, M.F. 1983. Chemistry and chemical taxonomy of the Rutales. London: Academic Press.

Webber, I.E. 1936. Systematic anatomy of the woods of the Simarubaceae. Am. J. Bot. 23: 577–587.

Weberling, F. 1989. Morphology of flowers and inflorescences. Cambridge: Cambridge University Press.

Weberling, F., Leenhouts, P.W. 1966. Systematisch-morphologische Studien an den Terebinthales-Familien. Akad. Wiss. Lit., Abh. math.-Nat. Kl. Jahrg. 1965, 10. 90 pp.

Wiger, J. 1935. Embryological studies in Buxaceae, Meliaceae, Simarubaceae and Burseraceae. Dissertation, University of Lund.

Zavada, M. S., Dilcher, D.L. 1986. Comparative pollen morphology and its relationship to phylogeny of pollen in the Hamamelidae. Ann. Missouri Bot. Gard. 73: 348–381.

Tetradiclidaceae

Tetradiclidaceae (Engl.) Takht. (1986).
Peganoideae Engl. (1896).
Peganaceae (Engl.) Tiegh. ex Takht. (1987).

M.C. Sheahan

Annual and perennial herbs or suffrutescent shrubs to 100 cm tall, prostrate to ascending, with alternate branching. Leaves alternate, rarely (*Tetradiclis*) opposite, simple, deeply divided into linear segments or rarely entire; pseudostipules minute. Flowers bisexual, terminal or leaf-opposed, solitary or in spike-like cymes; sepals 4–5, persistent; petals 4–5, white or yellow; stamens 4 or 15 with filaments dilated below, attached to the intrastaminal annular or angular disk; ovary sessile, 3–4-locular; style simple, clavate, with 3(4) decurrent stigmatic lobes; ovules several to many per loculus, with axile placentation. Fruit a 3–4-locular capsule or a berry. Seeds albuminous.

Three genera and 7–8 species, growing in dry sandy and saline environments of S and E Europe, the Middle East and Asia, with one species in Mexico and one (introduced) in the Americas.

Vegetative Morphology. *Peganum* is a genus of perennial herbs and low-growing, often chamaephytic shrubs. *Malacocarpus* is low growing with thin lax branches which are woody almost to the apex, and can form spreading mats up to 2 m across; it sometimes climbs up through taller plants. *Tetradiclis* is a small delicate annual which can be erect or spreading. Leaves are simple, sessile, fleshy and alternate (in *Tetradiclis*, opposite at the base). The leaf shape is deeply pinnatisect with narrow, more or less linear segments. In *Tetradiclis*, the leaves are much smaller and with fewer lobes, and sometimes ternate or entire. Pseudostipules (rudimentary basal leaf segments) are known from *Tetradiclis* and *Peganum* (Weberling and Leenhouts 1965); they are inconspicuous, setaceous, and apparently caducous. In dry conditions leaves and stems may fall, leaving a woody base from which new growth appears when conditions are more favourable.

Vegetative Anatomy. There is little information about *Tetradiclis*, except for a brief description in Chermezon (1910). In *Peganum* and *Malacocarpus*, the linear leaf segments are narrow, fleshy and nearly centric with chlorenchymatous tissue all round the leaf; in *Tetradiclis*, this tissue covers only the adaxial and two lateral sides. The interior of the leaf is of large, colourless water storage cells. There is a straight primary vein in each segment, and the secondary veins run more or less parallel to it; in transverse section, the veins form an arc and, in *Tetradiclis*, they are very slender. Marginal ultimate venation is looped, with sparse elongated veinlet termination cells at the vein endings and associated with the higher-order veins. Areoles are elongated and irregular. Leaf epidermal cells are in an unusual arrangement with bands 1–2 cells wide of large cells, elongated in the direction of the leaf axis, alternating with wider bands of shorter, more irregular cells. The stomata, which are large and anomocytic, are found only in these latter bands. *Peganum* and *Malacocarpus* are often described as glabrous, but there are sparse trichomes on the younger leaves and at the base of older leaves; they are glandular, multicellular and caducous. There are raphide crystals in *Peganum* and *Malacocarpus*, also short acicular crystals and solitary crystals throughout the leaves and stem.

The young stem in *Peganum* and *Malacocarpus* has a glabrous surface and thin cuticle; the epidermal cells are axially elongated and there is a 1–2-seriate hypodermis. There is a pericyclic ring of fibres encircling the phloem, and periderm differentiation is in the inner cortex. In the older stem, the phloem is well developed but sieve elements are usually narrow (diameter 5–8 μm), with compound sieve plates; sieve element plastids are reported to be S-type, with a diameter of 1.2 μm (Behnke 1988). The vessels are solitary or in short radial chains; vessel elements have simple perforations and alternate bordered pits, and there is some tertiary spiral thickening. Non-tracheary elements are libriform fibres with few simple pits and tracheids with bordered pits.

Flower Structure. The flowers are bisexual, actinomorphic and hypogynous. In *Peganum* and *Malacocarpus* they are terminal and solitary; in *Tetradiclis* they form terminal spike-like cymes. Sepals are distinct and persistent, narrow, sometimes pinnatisect. The petals are imbricate, white or pale yellow. *Tetradiclis* is tetramerous with a 4-locular ovary; *Peganum* and *Malacocarpus* are basically pentamerous but with a 3-locular ovary in *Peganum*, 2-locular in *Malacocarpus*. The disk is annular or angular, inconspicuous in *Tetradiclis*. *Peganum* and *Malacocarpus* have 15 stamens, with the five inner stamens antesepalous and the ten outer antepetalous; in *Peganum*, Ronse Decraene et al. (1996) interpreted this as secondary polyandry coupled to an obdiplostemenous androecium. There is a similar arrangement in *Malacocarpus*, but *Tetradiclis* is normally haplostemonous. Filaments in *Peganum* and *Malacocarpus* are slender and somewhat dilated at the base; in *Tetradiclis* they are subulate. The ovary is sessile and globose; the style is simple, divided above into decurrent vertical stigmatic lobes. The ovules are anatropous with axile placentation; in *Peganum* and *Malacocarpus* they are numerous, and in *Tetradiclis* there are up to six in each loculus.

Embryology. Information on *Peganum* is from Davis (1966) and Singh and Kaur (1998). *Peganum* has two fused layers of tapetum, which is of the glandular type; pollen tetrads are tetrahedral, and two-celled when shed; there is a thick cuticle on the anther epidermis. Ovules are anatropous, bitegmic, crassinucellar, with the micropyle formed by both integuments; endothelium is absent in *Peganum*; embryo sac development is of the Polygonum type; synergids are persistent, and synergid polyembryony has been reported in *Peganum*. Endosperm formation is nuclear, and wall formation starts from the micropylar end. The embryo is of the solanad type. For *Tetradiclis*, which has also a Polygonum type embryo sac and nuclear endosperm formation, see Kamelina (1994).

Pollen Morphology. Pollen grains are 3-colporate (*Tetradiclis* occasionally 6-rugorate), spheroidal or subspheroidal, prolate or subprolate; sexine thinner than nexine (as thick as nexine in *Peganum harmala*), exine ornamentation reticulate, finely striate in *Tetradiclis*; ora lalongate. Erdtman (1952) points out that the pollen shows some similarities with Rutaceae, although there are also similarities with Zygophyllaceae (Lahham and Al-Eisawi 1986; Xi and Zhou 1989). According to Singh and Kaur (1998), the pollen of *P. harmala* is exactly similar to that of *Zygophyllum simplex* and species of *Fagonia*.

Karyology. Ma et al. (1984) reported $2n = 22$ in *P. harmala*; however, in Ma et al. (1990) this was amended to $2n = 24$. Other references in Goldblatt and Johnson (2003) for *Peganum* agree with $2n = 24$ (no data available for *Malacocarpus*). *Tetradiclis* is reported as $2n = 14$ (Astanova 1993).

Fruit, Seed and Dispersal. The fruit of *Peganum* is a 2–3-valvate, loculicidally dehiscent capsule, slightly flattened above. *Malacocarpus* is soft-fruited, as the name suggests, with a brownish-red, 3-locular globose berry. *Tetradiclis* is somewhat different: it has a globose, loculicidally dehiscent capsule, 3 mm in diameter, flattened above, but it is 4-lobed and 4-locular, and has a complicated mechanism of seed dispersal (see below).

The seeds of both *Peganum* and *Malacocarpus* have a fleshy endosperm and slightly curved embryo. *Peganum* has numerous dark brown angular seeds with a rough mucilaginous surface. The structure of *P. harmala* seeds was described in detail by Corner (1976): they are brown, 3.5–4 mm long, angled; in the testa, the outer epidermis is formed of a palisade of large radially elongated cells with slightly thickened and lignified walls, the testa mesophyll has small thin-walled, slightly suberised or lignified cells which become crushed, and the inner epidermis is a palisade of thin-walled cells. The tegmen layer soon becomes crushed, except for the inner tegmen epidermis which is formed of lignified cells. Corner (1976) also reports that in *P. harmala* the endosperm is oily and rather thick-walled. The seeds of *Malacocarpus* are about 2 mm in length, dark brown and subovoid. *Tetradiclis* seeds are small, 0.6 mm in diameter, oblong, with a thin, partly mucous seed coat (see below) and scanty endosperm; the endotegmen is not lignified. The embryo is straight or slightly curved, the cotyledons fleshy and plano-convex.

Porter (1974) hypothesised that the numerous small seeds of *P. harmala* are probably fed upon by birds and animals. *Tetradiclis* has dimorphic seeds which are released sequentially. In its capsules, each locule is divided by false septa into three intercommunicating secondary compartments, the middle one containing up to four seeds and the other two one seed each. At maturity, first the four seeds of the middle locellus are released while the lateral seeds are still in their locelli. As a result of mechanical rupture of the epicarp tissues around the lateral seeds, these are discharged together with part of the surrounding epidermal tissue, which provides a membranous edging around the seed facilitating dispersal by wind. Thus, the seeds are dimorphic; those from the middle locellus, which have a tuberculate and mucous surface, are adapted for more local dispersal, whereas those from the lateral locelli, which are almost glabrous and not mucous, are prone to be dispersed over greater distances (Bunge 1840; Yanishevski 1940).

Phytochemistry. Information from Hegnauer (1973) and Umadevi et al. (1990). *Peganum* contains many alkaloids in the roots, leaves and seeds: the main ones are harmaline, harmine, harmalol, peganol and peganine; *Malacocarpus* has anabasin D in place of the harmaloids of *Peganum*, and is not foetid when bruised. The flavonoid profile of *P. harmala* is similar to that of Rutaceae but lacks the essential oils, limonoids and coumarins characteristic of Rutaceae and, unlike Zygophyllaceae, it does not contain steroidal saponins. Principal flavonoids are apigenin and kaempferol. The seeds contain oils but no starch, and have mucilage in the epidermis.

Subdivision and Relationships Within the Family. *Peganum* and *Malacocarpus* have always been recognised as very close; indeed, *M. crithmifolius* was originally known as *P. crithmifolium* (Engler 1931). The chief difference is in their fruit: a loculicidal capsule in *Peganum*, and a fleshy berry in *Malacocarpus*. *Tetradiclis* differs from these two genera in being basically tetramerous rather than pentamerous, in having 4 stamens, not 15, and in its unusual ovary with each locule divided by false septa into three compartments.

Affinities. *Peganum* and *Malacocarpus* were long considered to be either members of Zygophyllaceae or closely allied to them, on the grounds of similarities in pollen, flower and fruit; for example, Engler (1931) put *Peganum* and *Malacocarpus* (as *P. crithmifolium*) in their own subfamily Peganoideae, and *Tetradiclis* in Tetradiclidoideae, both within Zygophyllaceae. Hutchinson (1967) also included these three genera in the family. Singh and Kaur (1998) commented on similarities between *P. harmala* and some *Fagonia* and *Zygophyllum* species in pollen morphology, floral anatomy, embryology and chemotaxonomy.

More recently, however, there has been a general lack of agreement about their taxonomic position. They were excluded from Zygophyllaceae by Dahlgren (1980). El Hadidi (1975) reported the presence of certain compounds in *Peganum* and *Tetradiclis* which are not common in Zygophyllaceae, and later (El Hadidi 1977) he suggested they should be excluded from Zygophyllaceae on account of morphological and chemical differences. Souèges (1953), unlike Singh and Kaur (1998), considered that the embryology showed no relationship between *P. harmala* and Zygophyllaceae, and favoured placing *Peganum* in a separate family Peganaceae. Takhtajan (1987), among others, considered for *Peganum* and *Tetradiclis* a position close to Rutaceae but later changed his view (see Takhtajan 2009).

The little-known *Tetradiclis* has also proved difficult to place; it has been assigned not only to Zygophyllaceae, Rutaceae and Elatinaceae but also to its own family Tetradiclidaceae (Takhtajan 1987), on the grounds of many differences in leaf morphology and in the structure of the embryo and seed.

More recent molecular studies, in which these genera were analysed in an *rbc*L matrix including other members of Zygophyllaceae as well as a number of representatives from other eurosid families (Gadek et al. 1996; Sheahan and Chase 1996), have indicated that *Peganum* and *Malacocarpus* (together with *Nitraria*) form a well-supported clade within Sapindales, in the eurosid II group sensu APG (1998), close to Rutaceae and Meliaceae. Sheahan and Chase (1996) and Savolainen et al. (2000) also confirm the position of *Tetradiclis* as sister to *Peganum* and *Malacocarpus* within Tetradiclidaceae.

In APG II (2003), Tetradiclidaceae (and Peganaceae) are listed as possible synonyms of Nitrariaceae, though they are also recognised as acceptable alternatives. There are many dissimilarities between Tetradiclidaceae and Nitrariaceae. For example, leaves in *Nitraria* are simple and fasciculate but, in *Peganum* and *Malacocarpus*, they are pinnatifid with an unusual arrangement of epidermal cells; moreover, the fruit of *Nitraria* is drupaceous, unlike the loculicidal capsules of *Peganum* and *Tetradiclis* and the berry of *Malacocarpus*. Ronse Decraene et al. (1996) saw similarities between *Nitraria* and *Peganum* 'at first glance' but qualified this to find many more dissimilarities. Sheahan and Chase (1996) pointed out that the *rbc*L sequences of these two genera are as divergent as those of other families in Sapindales, and concluded that they should constitute separate families within Sapindales.

Distribution and Habitats. *Peganum* is mainly known from the southern temperate regions of the Mediterranean, SE Europe and Asia but there is one species, *P. mexicanum*, indigenous to the Chihuahuan Desert in Mexico. *P. harmala* has been introduced into S France, S Australia and the Americas (where it is known as African rue). *Peganum* grows in arid habitats, steppes and dry waste places, on sandy or saline soils. On saline soils it has high sodium and potassium ion concentrations—up to seven times higher than in surrounding soils (Moore et al. 1972). It has an acrid smell which may deter herbivores; according to Pantanelli (1900), even camels will not eat it but will, however, remove its competitors, allowing monospecific stands to develop.

The distribution of *Malacocarpus* is more limited; it grows in the area between the Caspian and the Aral Sea, in dry sandy environments on rocks, cliffs, stony slopes and gravel-silt deposits. *Tetradiclis* is found in the Middle East: Egypt, Palestine (but rare in both), Iran, Afghanistan, Jordan, Iraq and Syria; also W Pakistan and SE and S of Russia and in Central Asia (Turkmenistan). It grows in silty and rocky deposits, depressions and flood plains, and is also found in salt deserts, salt marshes and on sandy sea shores; according to Yanishevski (1940), it is bound to solonchac soils.

Economic Importance. *P. harmala* has a number of uses in traditional folk medicine. The seeds are a source of the drug harmal which is used as an anthelmintic and narcotic in small doses, also as a febrifuge, an aphrodisiac, an antispasmodic and a treatment for asthma. Ambasta (1986) lists several disorders which can be treated by the seeds or leaves, including hiccups, hysteria and rheumatism. According to Saleh and El-Hadidi (1975), vasicinone from *P. harmala* has a bronchodilator effect. The seeds are also a source of a fatty oil and a dye (Turkey red).

Key to the Genera

1. Fruit a loculidical capsule 2
– Fruit a fleshy 3-locular berry **2. *Malacocarpus***
2. Large perennial herbs or shrubs, flowers solitary, leaf-opposed **1. *Peganum***
– Small annual herb; flowers in terminal leafy spike-like racemes **3. *Tetradiclis***

Genera of Tetradiclidaceae

1. ***Peganum*** L.

Peganum L., Sp. pl. 1: 444 (1753).

Perennial herbs or low-growing shrubs 30–100 cm high, prostrate to ascending. Leaves alternate, sessile, fleshy, irregularly pinnatisect, with narrow branched linear-lanceolate acute lobes; pseudostipules variable: minute, setaceous, sometimes undetectable, caducous. Flowers solitary on long pedicel, bisexual, white, leaf-opposed; sepals (4)5, linear, entire or sometimes pinnatisect, persistent in fruit; petals (4)5, imbricate; disk annular or angular; stamens (12)15, with slender filaments dilated at the base; ovary ± sessile, globose, 3(4)-locular; style divided above into 3 vertical stigmatic lobes appressed to style; ovules numerous, axile, crassinucellate, with long funicle. Fruit a 3(4)-locular capsule, slightly depressed above, with numerous angular seeds with rough mucilaginous surface; endosperm fleshy; embryo slightly curved. $2n = 24$.

Five or six spp., desert regions of Old World from Mediterranean to Mongolia and China, also southern North America (Mexico and Texas).

2. ***Malacocarpus*** Fischer & C.A. Meyer

Malacocarpus Fischer & C.A. Meyer, Ind. Sem. Hort. Petrop. 9: 78 (1843).
Peganum sect. *Malacocarpus* Engl.

A low-growing (to 100 cm), profusely branching small shrub with thin, lax, spreading branches which are woody almost to the tip, and whitish-grey bark darker at the base. It may form extensive mats up to 2 m wide, and sometimes climbs up and through taller plants like a liana. Leaves small, alternate, fleshy, sessile, deeply divided into narrow tapering segments to 6 cm long; pseudostipules minute, oblong, fleshy, inconspicuous, sometimes branched. Flowers terminal, solitary, leaf-opposed; sepals 5, narrow, persistent, becoming reflexed; petals 5, white or pale yellow, oblong, imbricate; disk annular; stamens 15, with filaments dilated below; ovary 3-locular; style clavate, keeled in upper part, persistent in fruit; ovules numerous, axile. Fruit a reddish globose 3-locular berry 6–10 mm in diameter, with numerous seeds. Endosperm fleshy; embryo slightly curved. $2n = 24$.

Only one sp., *M. crithmifolius* Fischer & C.A. Meyer, found in arid rocky and gravelly areas in Russia near the Caspian Sea, Northern Iran and Central Asia around the Aral Sea.

3. ***Tetradiclis*** Stev. ex M. Bieb.

Tetradiclis Stev. ex M. Bieb., Fl. Taur. Cauc. 3: 648, cf. 277 (1819).
Anatropa Ehrenb. (1829).

A delicate low-growing annual with many slender, erect or trailing branches, to 25 cm. Leaves succulent, sessile, glabrous, simple and opposite on the main axis, pinnatisect and alternate on its branches; pseudostipules minute. Flowers many, small, bisexual, subsessile in terminal spike-like cymes; sepals (3)4, persistent; petals (3)4, short-clawed, twice as long as calyx, white; disk annular, inconspicuous; stamens (3)4 with persistent subulate filaments; anthers subglobose; ovary sessile, (3)4-locular, (3)4-angled, depressed in the centre; style with decurrent 4-lobed stigmas; ovules up to 6 in each loculus, axile. Fruit a (3)4-lobed, globose capsule c. 3 mm in diameter, flattened above, loculicidally dehiscent; each locule divided by false septa into 3 secondary compartments, the centre one with up to 4 seeds, free, the others 1-seeded. Seeds very small, c. 0.6 mm diameter, oblong; endosperm scanty; embryo straight or slightly curved; cotyledons fleshy, plano-convex. $2n = 14$.

Two spp., calcareous and saline desert and semi-desert areas in N Africa, Algeria, and Egypt to Turkestan.

Selected Bibliography

Ambasta, S.P. (ed.) 1986. The useful plants of India. New Delhi: CSIR.
APG II. 2003. An update of the Angiosperm Phylogeny Group classification for the orders and families of flowering plants: APG II. Bot. J. Linn. Soc. 141: 399–436.
Astanova, S.B. 1993. In: Takhtajan, A. (ed.) Numeri chromosomatum Magnoliophytorum Florae URSS, Moraceae–Zygophyllaceae. Petropoli: Nauka.
Behnke, H.-D. 1988. Sieve-element plastids and systematic relationships of Rhizophoraceae, Anisophylleaceae and allied groups. Ann. Missouri Bot. Gard. 75: 1387–1409.
Bobrov, E.G. 1974. Zygophyllaceae. In: Shishkin, B.K., Bobrov, E.G. (eds.) Flora of the U.S.S.R., vol. 14 (translation of the 1949 Russian edition). Jerusalem: Israel Program for Scientific Translations.
Bunge, A. 1840. Über die Gattung *Tetradiclis* Stev. Linnaea 14: 161–178, t. I.
Chermezon, H. 1910. Recherches anatomiques sur les plantes littorales. Ann. Sci. Nat. bot. 12: 117–313.
Corner, E.J.H. (1976). The seeds of dicotyledons. Cambridge: C.U.P.
Dahlgren, R. 1980. A revised system of classification of the angiosperms. Bot. J. Linn. Soc. 80: 91–124.
Davis, G.L. 1966. Systematic embryology of the angiosperms. New York: Wiley.
Elgin, N. 1950. Les caractères morphologiques et anatomiques du *Peganum harmala* L. (Zygophyllaceae). Revue Fac. Sci. Univ. Istanbul. B, 15: 333–361.
El Hadidi, M.M. 1975. Zygophyllaceae in Africa. Boissieria 24: 317–323.

El Hadidi, M.M. 1977. Tribulaceae as a distinct family. Publ. Cairo Univ. Herb. 7-8: 103–108.
Engler, A. 1931. Zygophyllaceae. In: Engler, A., Prantl, K., Die natürlichen Pflanzenfamilien, ed. 2, 19a. Leipzig: Engelmann.
Erdtman, G. 1952. Pollen morphology and plant taxonomy. Angiosperms. Stockholm: Almqvist and Wiksell.
Gadek, P., Fernando, E.S., Quinn, C.J., Hoot, S.B., Terrazas, T., Sheahan, M.C., Chase, M.W. 1996. Sapindales: molecular delimitation and infraordinal groups. Am. J. Bot. 83: 802–811.
Goldblatt, P., Johnson, D.E. 2003. Index to Plant Chromosome Numbers 1998-2000. Monogr. Syst. Bot. Missouri Bot. Gard. 94.
Hegnauer, R. 1973. Chemotaxonomie der Pflanzen, vol. 6. Basel: Birkhäuser.
Hutchinson, J. 1967. The genera of flowering plants, vol. II: Dicotyledons. Oxford: Clarendon Press.
Kamelina, O.P. 1994. Embryology and systematic position of *Tetradiclis* (Tatradiclidaceae) (in Russian with Engl. summary). Bot. Zhurn. 79: 11–27.
Lahham, J.N., Al-Eisawi, D. 1986. Pollen morphology of Jordanian Zygophyllaceae. Candollea 41: 325–328.
Lia, V.V., Confalonieri, V.A., Comas, C.I., Hunziger, J.H. 2001. Molecular phylogeny of *Larrea* and its allies (Zygophyllaceae): reticulate evolution and the probable time of creosote bush arrival to North America. Mol. Phylog. Evol. 21: 309–320.
Ma, X.H., Qin, R.L., Xing, W.B. 1984. Chromosome observations of some medical plants in Xinjiang. Acta Phytotax. Sinica 22: 243–249.
Ma, X.H., Ma, X.Q., Li, N. 1990. Chromosome observation of some drug plants in Xinjiang. Acta Bot. Boreali-Occid. Sinica 10: 203–210.
Moore, R.T., Breckle, S.-W., Caldwell, M.M. 1972. Mineral ion composition and osmotic relations of *Atriplex confertifolia* and *Eurotia lanata*. Oecologia 11: 67–78.
Pantanelli, E. 1900. Anatomia fisiologica delle Zygophyllaceae. Atti società nat. mat. Modena IV, 2: 93–181.
Porter, D.M. 1974. Disjunct distributions in the New World: Zygophyllaceae. Taxon 23: 339–346.
Ronse Decraene, L.P., De Laet, J., Smets, E.F. 1996. Morphological studies in Zygophyllaceae II. The floral development and vascular anatomy of *Peganum harmala*. Am. J. Bot. 83: 201–215.
Saleh, N.A.M., El-Hadidi, M.N. 1975. An approach to the chemosystematics of the Zygophyllaceae. Biochem. Syst. Ecol. 5: 121–128.
Savolainen, V., Fay, M.F., Albach, D.C., Backlund, A., Van der Bank, M., Cameron, K.J., Johnson, S.A., Lledó, M.D., Pintaud, J.-C., Powell, M., Sheahan, M.C., Soltis, D.E., Soltis, P.S., Weston, P., Whitten, W.M., Wurdack, K.J., Chase, M.W. 2000. Phylogeny of the eudicots: a nearly complete familial analysis based on *rbc*L gene sequences. Kew Bull. 55: 257–309.
Sheahan, M.C., Chase, M.W. 1996. A phylogenetic analysis of Zygophyllaceae R.Br. based on morphological, anatomical and *rbc*L DNA sequence data. Bot. J. Linn. Soc. 122: 279–300.
Sheahan, M.C., Cutler, D.C. 1993. Contribution of vegetative anatomy to the systematics of the Zygophyllaceae R.Br. Bot. J. Linn. Soc. 113: 227–262.
Singh, B.P., Kaur, I. 1998. Systematic position of the genus *Peganum*. J. Econ. Tax. Bot. 22: 705–708.
Souèges, R. 1953. Embryogénie des Péganacées: développement de l'embryon chez le *Peganum harmala* L. C. R. Acad. Sci. 236: 2185–2188.
Takhtajan, A. 1987. Systema Magnoliophytorum. Leningrad: Nauka.
Takhtajan, A. 2009. Flowering plants, 2nd edn. Dordrecht: Springer.
Umadevi, I., Daniel, M., Sabnis, S.D. 1990. Chemotaxonomy of some Rutaceae. Indian J. Bot. 13: 23–28.
Weberling, F., Leenhouts, P.W. 1965. Systematisch-morphologische Studien an Terebinthales-Familien. Akad. Wiss. Lit. Mainz, Abh. Math.-nat. Kl., Jg. 1965, 10: 495–584.
Xi, Y., Zhou, S. 1989. Pollen morphology and its exine ultrastructure of the Zygophyllaceae in China. Bot. Res. (China) 4: 75–86.
Yanishevski, D.E. 1940. *Tetradiclis tenella*, (Ehrenb.) Litv. as the example of an ephemeral on the solonchaks of the Mediterranean desert regions (in Russian, with Engl. summary). Trudy Bot. Inst. Akad. Nauk SSSR IV, 4: 236–248.

Index

References to accepted names in **bold-faced** print, to synonyms in upright print, to illustrations in *italics*.

MIX
Papier aus verantwortungsvollen Quellen
Paper from responsible sources
FSC® C105338

If you have any concerns about our products,
you can contact us on
ProductSafety@springernature.com

In case Publisher is established outside the EU,
the EU authorized representative is:
Springer Nature Customer Service Center GmbH
Europaplatz 3, 69115 Heidelberg, Germany

Printed by Libri Plureos GmbH
in Hamburg, Germany